T0339921

Mechanical Design Engineering Handbook

Mechanical Design Engineering Handbook

Peter RN Childs

Second edition

Butterworth-Heinemann is an imprint of Elsevier
The Boulevard, Langford Lane, Kidlington, Oxford OX5 1GB, United Kingdom
50 Hampshire Street, 5th Floor, Cambridge, MA 02139, United States

Notices

Knowledge and best practice in this field are constantly changing. As new research and experience broaden our understanding, changes in research methods, professional practices, or medical treatment may become necessary.

Practitioners and researchers must always rely on their own experience and knowledge in evaluating and using any information, methods, compounds, or experiments described herein. In using such information or methods they should be mindful of their own safety and the safety of others, including parties for whom they have a professional responsibility.

To the fullest extent of the law, neither the Publisher nor the authors, contributors, or editors, assume any liability for any injury and/or damage to persons or property as a matter of products liability, negligence or otherwise, or from any use or operation of any methods, products, instructions, or ideas contained in the material herein.

Library of Congress Cataloging-in-Publication Data
A catalog record for this book is available from the Library of Congress

British Library Cataloguing-in-Publication Data
A catalogue record for this book is available from the British Library

ISBN: 978-0-08-102367-9

Publisher: Matthew Deans
Acquisition Editor: Brian Guerin
Editorial Project Manager: John Leonard
Production Project Manager: R.Vijay Bharath
Cover Designer: Miles Hitchen

Typeset by SPi Global, India

Contents

Preface to the second edition

This edition of the Mechanical Design Engineering Handbook has been extensively updated. Each chapter has been reviewed and developed. Chapters 1 and 2 dealing with the design process and specification have been updated with a development of the total design process and an introduction to project management. Chapter 3 has been revised substantially incorporating developments in creativity and ideation processes relevant to engineering. Chapter 4 has been further developed to illustrate the scope and context of machine elements.

Chapters 5 and 6 introducing the first of the machine elements to be considered in detail, bearings, have been expanded to include flow charts illustrating the design of boundary lubricated, hydrodynamic and ball bearings. The introductory and extended worked examples have been retained throughout the chapters on machine elements in the book to enable the reader to follow the detailed analysis and associated design decisions. Chapter 7 addressing shaft design has been expanded to include consideration of a factor of safety according to the DE Goodman, DE Gerber, DE ASME elliptic and DE Soderberg criteria. Chapters 8–11 introducing gears have been expanded to include flow charts for the selection of spur gears, and the calculation of bending and contact stresses, and the design of gear sets using the AGMA equations. Similarly Chapter 13 introducing belt and chain drives has been extended with flow charts for the selection and design of wedge and synchronous belts, and roller chain drives.

Chapter 14 on seals has been updated to include additional examples. Chapter 15 has been extended to include selection and design flow charts for helical compression spring and helical extension spring design. Chapters 16–18 have been extended with additional examples of various fastener, wire rope, and pneumatic and hydraulic technologies, respectively. Three short case studies have been included in Chapter 19 illustrating the importance of a detailed consideration of tolerancing in precision engineering. Chapter 20 is a new chapter providing an overview of a diverse range of machine elements as building blocks for mechanism design.

Throughout the text an additional 100 or so images have been included in order to aid the reader in becoming familiar with the technology being considered.

Mechanical engineering design is an engaging subject area with many applications and this major revision has been a pleasurable undertaking enabling implementation of many updates from my engineering and design practice and associated interactions. I hope this text is able to aid the reader in the development of understanding of the principles and associated technology and in its implementation in worthwhile innovations and applications.

Acknowledgements

I have the delight of having developed and practised my design and engineering skills and their application with so many impressive individuals and organisations. I would like to thank my colleagues at Imperial College London and Q-Bot Ltd, and former colleagues from the University of Sussex for their extensive support and patience through the years. Without the chance to practise engineering and design on exciting and ambitious commercial applications and research projects, the opportunity to develop knowledge is limited. I have had the privilege of working with diverse companies and organisations including Rolls-Royce plc, Alstom, Snecma, DaimlerChrysler, BMW, MTU, Volvo, Johnson Matthey, Siemens, Industriales Turbinas Propulsores, Fiat Avio, Airbus, Ricardo Consulting Engineers, Ford, Rio Tinto, McLaren, Dyson, Naked Energy and Q-Bot Ltd, Innovate UK, the EPSRC and Horizon 2020. I would like to thank the engineers, designers and managers from these companies and organisations for the opportunity to engage in such exciting technologies.

Several colleagues in particular have been very helpful in the implementation of this edition including Shayan Sharifi and Andy Brand who assisted with proof reading, Ben Cobley and Ruth Carter who assisted with several of the images. Many collaborators and companies have kindly given permission to include key images to aid in the effective presentation of this book and this is gratefully acknowledged. Finally, I would like to thank my wife Caroline for her patience over the last year when I have expended a significant number of hours working on this project.

Peter Childs
Professor and Head of School, Dyson School of Design Engineering,
Imperial College London, United Kingdom

Design

1

Chapter Outline

Abbreviations

BS British Standard

CDIO conceive, design, implement, operate

CDR critical design review

CTP critical to process

Mechanical Design Engineering Handbook. https://doi.org/10.1016/B978-0-08-102367-9.00001-9

CTQ	critical to quality
DFA	design for assembly
DFM	design for manufacture
DFSS	design for six sigma
DMADV	design, measure, analyse, design, verify
DoE	design of experiments
ERP	Enterprise resource planning
FMEA	failure mode and effects analysis
ID	Identifier
IDOV	identify, design, optimise, verify
ISO	International Organisation for Standardisation
KPI	key point indicator
MDO	multiobjective design optimisation
PDS	product design specification
PID	project initiation document
QFD	quality function deployment
PDR	preliminary design review
PLR	Postlaunch review
PRINCE	Projects IN Controlled Environments
R&D	research and development
SDR	system design review
SMART	specific, measurable, achievable, relevant, time-bound
XP	extreme programming

1.1 Introduction

The aims of this book are to present an overview of the design process and to introduce the technology and selection of a number of specific machine elements that are fundamental to a wide range of mechanical engineering design applications. This chapter introduces the design process from an inventor's perspective and double diamond to more formal models such as 'total design' and systematic approaches to design. The chapter introduces a series of approaches to project management and concludes with an overview of the technology base serving as building blocks for machinery and mechanical design.

The term design is popularly used to refer to an object's aesthetic appearance with specific reference to its form or outward appearance as well as its function. For example we often refer to designer clothes, design icons and beautiful cars, and examples of some classically acclaimed vehicles are given in Figs 1.1 and 1.2. In these examples it is both visual impact, appealing to our visual perception, and the concept of function, that the product will fulfil a range of requirements, which are important in defining so-called good design.

The word 'design' is used as both a noun and a verb and carries a wide range of context-sensitive meanings and associations. George Cox (2005) stated "Design is what links creativity and innovation. It shapes ideas to become practical and attractive propositions for users or customers. Design may be described as creativity deployed to a specific end." The word design has its roots in the Latin word 'designare', which means to designate or mark out. Design can be taken to mean all the processes of conception, invention, visualisation, calculation, refinement and specification of details

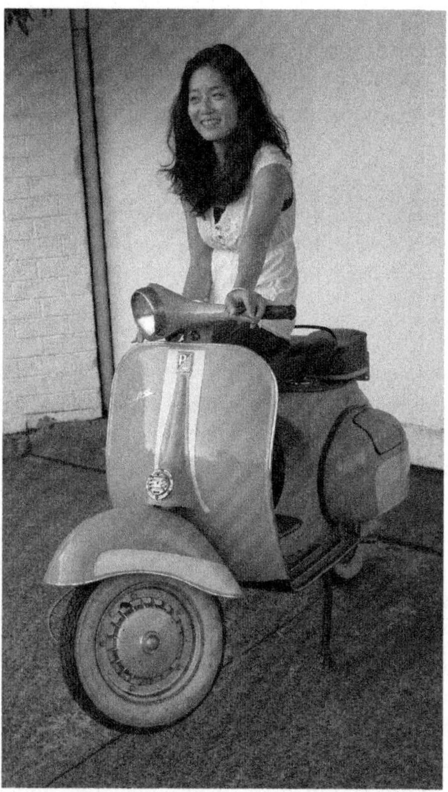

Fig. 1.1 Piaggio's Vespa launched in 1946. The Vespa was an early example of monocoque construction where the skin and frame are combined as a single construction to provide appropriate rigidity and mounting for the vehicle's components and riders.

that determine the form of a product. Design generally begins with either a need or requirement or, alternatively, an idea. It can end with a set of drawings or computer representations and other information that enables a product to be manufactured, a service or system realised and utilised. While recognising that there are no widely accepted single definitions, to clarify what the term design means the following statement can provide a basis.

> ***Design*** *is the process of conceiving, developing and realising products, artefacts, processes, systems, services, platforms and experiences with the aim of fulfilling identified or perceived needs or desires typically working within defined or negotiated constraints.*

This process may draw upon and synthesise principles, knowledge, methods skills and tools from a broad spectrum of disciplines depending on the nature of the design initiative and activity. Design can also be regarded as 'the total activity necessary to provide a product or process to meet a market need'. This definition comes from the SEED (Sharing Experience in Engineering Design, now DESIG the Design Education Special Interest Group of the Design Society) model, see Pugh (1990).

According to a Royal Academy of Engineering document, engineering can be defined as

> *The discipline, art and profession of acquiring and applying scientific, mathematical, economic, social and practical knowledge to design and build structures, machines, devices, systems, materials and processes that safely realise solutions to the needs of society.*

This definition is not attributed to a single individual and ABET (2011), the Institution of Mechanical Engineers and the National Academy of Engineering (2004) all have similar definitions for engineering involving the application of scientific and mathematic principles to design. The following statement provides an indication of the scope of engineering.

> ***Engineering*** *is the application of scientific and mathematic principles in combination with professional and domain knowledge, to design, develop and deliver artefacts, products and systems to realise a societal, commercial or organisation requirement or opportunity.*

The terms 'engineering design' and 'design engineering' are often used interchangeably. The inclusion of the word engineering in both suggests that they involve the application of scientific and mathematical knowledge and principles. It may be useful to think of 'engineering design' sitting alongside 'engineering science' as the strand of engineering that is concerned with application, designing, manufacture and building. Design engineering suggests a process in which engineering (scientific and mathematical) approaches are applied in the realisation of activities that began with a design concept or proposal (Childs and Pennington, 2015). However such distinctions remain subtle and subject to context.

1.2 The design process

Design processes abound and have been widely documented, with many design schools, consultancies and engineering corporations developing their own brand of approaches (see, e.g. Clarkson and Eckert 2005). Commonly cited methods include the educational approach CDIO (conceive, develop, implement, operate), total design, double diamond, concurrent engineering, six sigma, MDO (multiobjective design optimisation) and gated reviews. Design processes can be broadly categorised as activity-based, involving generation, analysis and evaluation, and stage-based, involving distinct phases of, for example, task clarification and conceptual design. It is also widely recognised that experienced practitioners approach design in a different manner to novice designers (see, e.g. Björklund 2013).

Fig. 1.2 The Audi TT, originally launched in 1998, and a contender for the most attractive sports car of the 20th century.
Courtesy of Audi.

Probably from your own experience you will know that design can consist of examining a need or opportunity and working on the problem by means of sketches, models, brain storming, calculations as necessary, development of styling as appropriate, making sure the product fits together and can be manufactured, and calculation of the costs. The process of design can be represented schematically to levels of increasing formality and complexity. Fig. 1.3 represents the traditional approach associated with lone inventors. This model comprises the generation of the 'bright idea', drawings and calculations giving form or shape to the idea, judgement of the design and reevaluation if necessary, resulting in the generation of the end product. The process of evaluation and reworking an idea is common in design and is represented in the model by the iteration arrow taking the design activity back a step so that the design can be improved. Fig. 1.4 illustrates the possible results from this process for a helmet providing peripheral and reverse vision.

Fig. 1.5 shows a more prescribed description of a design process that might be associated with engineers operating within a formal company management structure. The various terms used in Fig. 1.5 are described in Table 1.1.

Although Figs 1.3 and 1.5 at first sight show design occurring in a sequential fashion, with one task following another, the design process may actually occur in a step forward, step back fashion. For instance you may propose a solution to the design need and then perform some calculations or judgements, which indicate that the proposal is inappropriate. A new solution will need to be put forward and further assessments made. This is known as the iterative process of design and forms an essential part

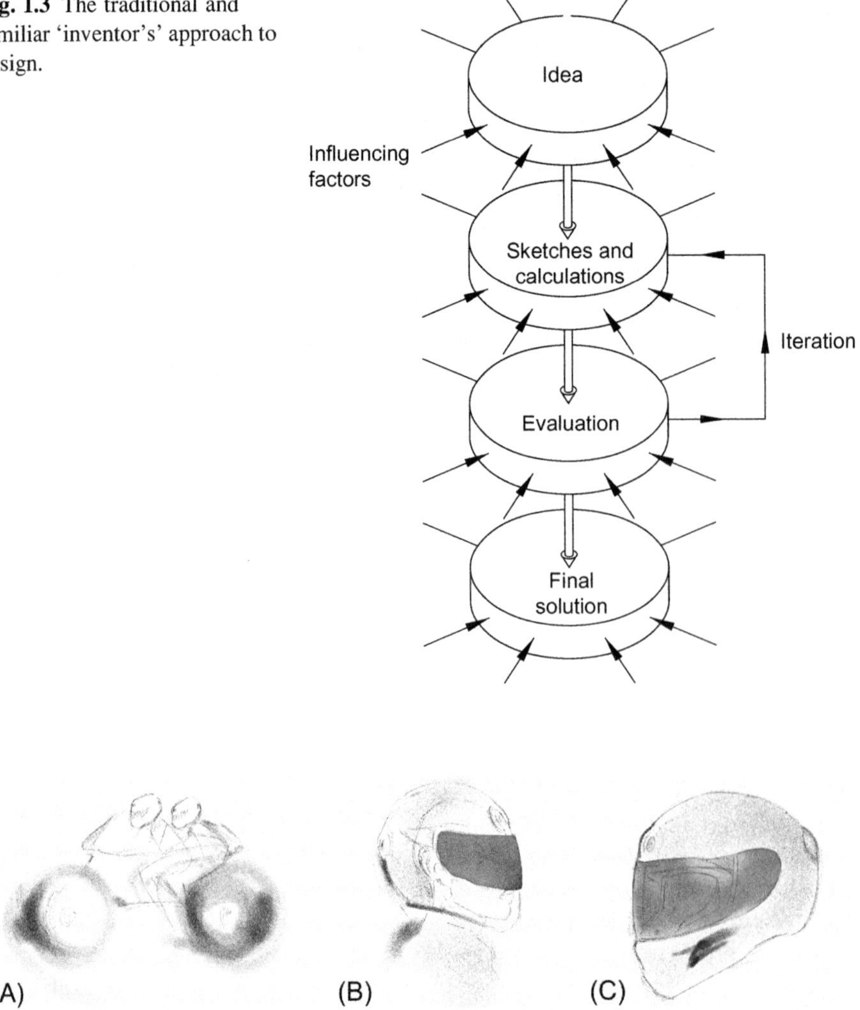

Fig. 1.3 The traditional and familiar 'inventor's' approach to design.

Fig. 1.4 Panoramic helmet. (A) The need: to be able to view around you. (B) The idea: Use of a VR visor and camera feeds to enable forward as well as peripheral vision. (C) Practical sketches showing the concept.
Sketches courtesy of Ruth Carter.

of refining and improving the product proposal. The nonlinear nature of design is considered by Hall and Childs (2009).

Note that the flow charts shown in Figs 1.3 and 1.5 do not represent a method of design, but rather a description of what actually occurs within the process of design. The method of design used is often unique to the engineering or design team. Design methodology is not an exact science and there are indeed no guaranteed methods of design. Some designers work in a progressive fashion, while others work on several

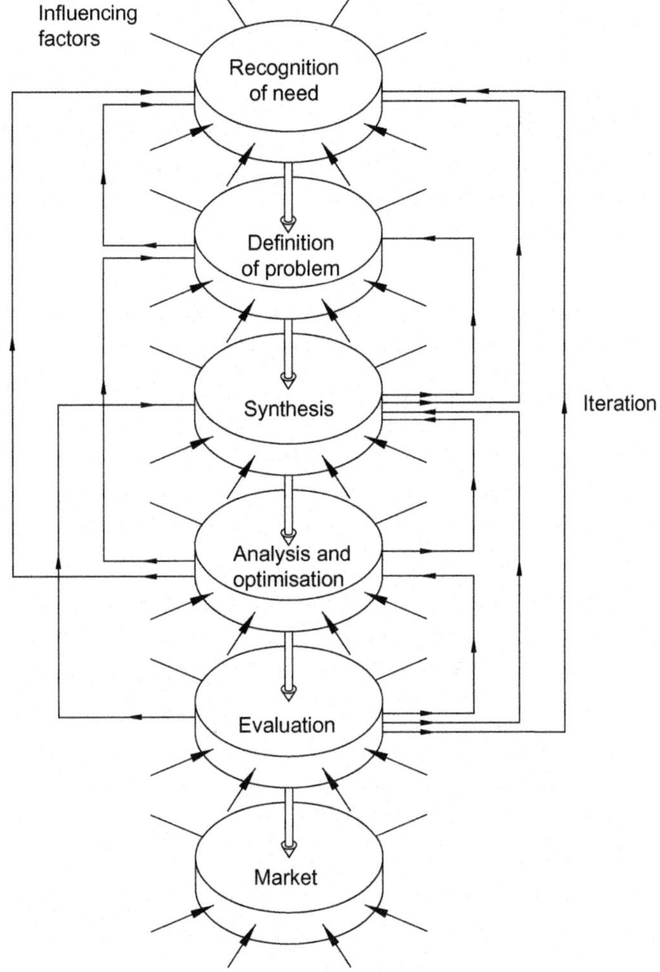

Fig. 1.5 The design process illustrating some of the iterative steps associated with the process.

aspects simultaneously. An example of design following the process identified in Fig. 1.5 is given in the following example (Section 1.2.1).

1.2.1 Case study

Following some initial market assessments, the Board of a plant machinery company has decided to proceed with the design of a new product for transporting pallets around factories. The Board have in mind a forklift truck but do not wish to constrain the design team to this concept alone. The process of the design can be viewed in terms of the labels used in Fig. 1.5.

Table 1.1 Design phases

Phase	Description
Recognition of need	Often design begins when an individual or company recognises a need, or identifies a potential market, for a product, device or process. Alternatively 'need' can be defined as when a company decides to reengineer one of its existing products (e.g. producing a new car model). The statement of need is sometimes referred to as the brief or market brief.
Definition of problem	This involves all the specification of the product or process to be designed. For example this could include inputs and outputs, characteristics, dimensions and limitations on quantities.
Synthesis	This is the process of combining the ideas developed into a form or concept, which offers a potential solution to the design requirement. The term synthesis may be familiar from its use in chemistry where it is used to describe the process of producing a compound by a series of reactions of other substances.
Analysis	This involves the application of engineering science; subjects explored extensively in traditional engineering courses such as statics and dynamics, mechanics of materials, fluid flow and heat transfer. These engineering 'tools' and techniques can be used to examine the design to give quantitative information such as whether it is strong enough or will operate at an acceptable temperature. Analysis and synthesis invariably go together. Synthesis means putting something together and analysis means resolving something into its constituent parts or taking it to pieces. Designers have to synthesise something before it can be analysed. The famous chicken and the egg scenario! When a product is analysed some kind of deficiency or inadequacy may be identified requiring the synthesis of a new solution prior to reanalysis and repetition of the process until an adequate solution is obtained.
Optimisation	This is the process of repetitively refining a set of often-conflicting criteria to achieve the best compromise.
Evaluation	This is the process of identifying whether the design satisfies the original requirements. It may involve assessment of the analysis, prototype testing and market research.

Recognition of need (or market brief)

The company has identified a potential market for a new pallet-moving device.

Definition of problem

A full specification of the product desired by the company should be written. This allows the design team to identify whether their design proposals meet the original request. Here a long list of information needs to be developed and clarified before design can proceed. For example for the pallet-moving device being explored here this would likely include aspects for consideration such as:

What sizes of pallet are to be moved?
What is the maximum mass on the pallet?
What is the maximum size of the load on the pallet?
What range of materials are to be moved and are they packaged?
What maximum height must the pallet be lifted?
What terrain must the pallet-moving device operate on?
What range is required for the pallet-moving device?
Is a particular energy source/fuel to be used?
What lifetime is required?
Are there manufacturing constraints to be considered?
What is the target sales price?
How many units can the market sustain?
Is the device to be automatic or manned?
What legal constraints need to be considered?

This list is not exhaustive and would require further consideration. The next step is to quantify each of the criteria. For instance the specification may yield information such as that standard size pallets (see Fig. 1.6) are to be used, the maximum load to be moved is 1000 kg, the maximum volume of load is 2 m^3, the reach must be up to 3 m, and use is principally on factory floor and asphalt surfaces. The pallet-moving device must be capable of moving a single pallet 100 m and must be able to repeat this task at least 300 times before refuelling if necessary, electricity, gas or diesel fuel, 7 year lifetime, production in an European country, target selling price 20,000 Euros, 12,000 units per year, manned use, design to ISO (International Organisation for Standardisation) and target country national standards (see, e.g. BS ISO 509, BS ISO 6780, BS EN ISO 445, BS EN 1726-1, BS EN 13545, 99/705213 DC, ISO 18334, 99/712554 DC, BS 3726, BS 5639-1 and BS ISO 2330). The task of specification is an involved activity and is considered more fully in Chapter 2.

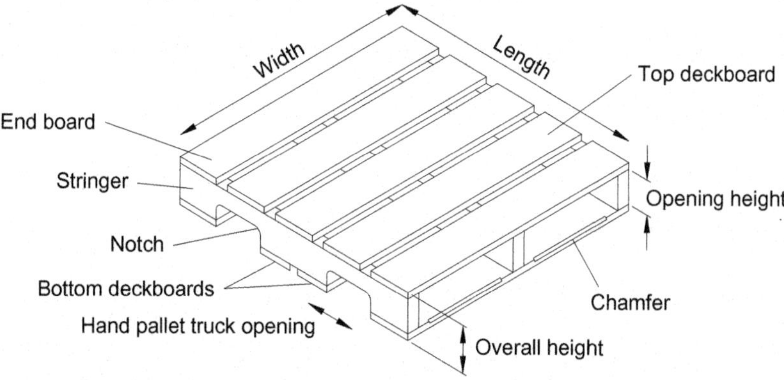

Fig. 1.6 Pallet dimensions and terminology (see BS ISO 509, 99/712554 DC and 99/712555 DC).

Synthesis

This is often identified as the formative and creative stage of design. Some initial ideas must be proposed or generated for them to be assessed and improved. Concepts can be generated by imagination, experience or by the use of design techniques such as morphological charts. Some evaluation should be made at this stage to reduce the number of concepts requiring further work. Various techniques are available for this including merit and adequacy assessments.

Analysis

Once a concept has been proposed it can then be analysed to determine whether constituent components can meet the demands placed on them in terms of performance, manufacture, cost and any other specified criteria. Alternatively analysis techniques can be used to determine the size of the components to meet the required functions.

Optimisation

Inevitably there are conflicts between requirements. In the case of the forklift truck, size, manoeuvrability, cost, aesthetic appeal, ease of use, stability and speed are not necessarily all in accordance with each other. Cost minimisation may call for compromises on material usage and manufacturing methods. These considerations form part of the optimisation of the product producing the best or most acceptable compromise between the desired criteria. Optimisation is considered further in Section 1.8.

Evaluation

Once a concept has been proposed and selected and the details of component sizes, materials, manufacture, costs and performance worked out, it is then necessary to evaluate it. Does the proposed design fulfil the specification? If it appears to, then further evaluation by potential customers and use of prototype demonstrators may be appropriate to confirm the functionality of the design, judge customer reaction and provide information of whether any aspects of the design need to be reworked or refined.

This case study is developed further in Chapter 3, Section 3.7.1.

1.3 Total design

The process of design has been the focus of research for many years and a number of design models and methodologies are available. Design methodology is a framework within which the designer can practise with thoroughness. One such approach called 'total design' has been proposed by the SEED programme (1985) and Pugh (1990) comprising core activities of design: marketing, specification, conceptual design, detailed design and marketing/selling. This model was developed from extensive

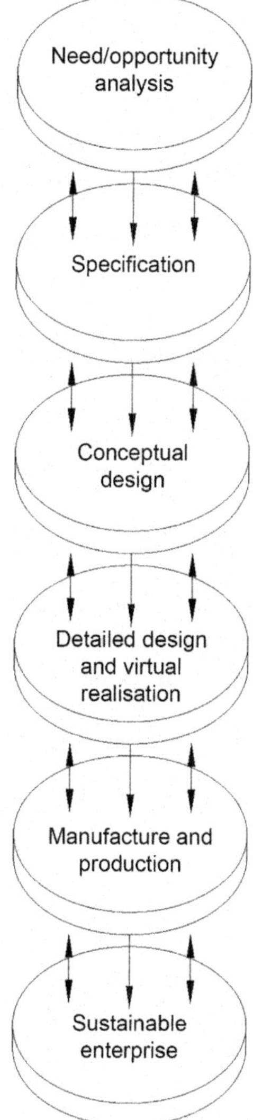

Fig. 1.7 The total design core. Originally developed by Pugh (1990) and updated here accounting for consideration of addressing both needs and opportunities, the ability to virtually model many attributes of a design at the detailed design phase, the applicability of the model beyond traditional manufacture and consideration of the business model and sustainability.

industrial consultation and experience and is shown in an updated form in Fig. 1.7, accounting for consideration of addressing both needs and opportunities, the ability to virtually model many attributes of a design at the detailed design phase, the applicability of the model beyond traditional manufacture and consideration of the business model and sustainability. As in Figs 1.3 and 1.5, the iterative nature of design is accounted for where work on design results in the need to go back and redo previous work to produce a better overall design to meet the requirements. Indeed it is sometimes necessary to go back a few or several levels. An example might be the discovery at manufacture that an item cannot be made as envisaged and a new concept altogether is required. Ideally such a discovery should not occur, as every other level of the design process illustrated in Fig. 1.7 should be considered at each stage. Each of the design activities illustrated in Fig. 1.7 is described in more detail in Sections 1.3.1–1.3.6. As it is the same fundamental process being described, these descriptions are similar to those which are dealt with Fig. 1.5.

1.3.1 Need/opportunity analysis

The need/opportunity analysis or marketing phase refers to the assessment of sales opportunities or perceived need to update an existing product, service, system or platform resulting in a statement sometimes called the market brief, design brief, brief or statement of need.

1.3.2 Specification

Specification involves the formal statement of the required functions, features and performance of the product or process to be designed. Recommended practice from the outset of design work is to produce a product design specification (PDS) that should be formulated from the statement of need. The PDS is the formal specification of the product to be designed. It acts as the control for the total design activity because it sets the boundaries for the subsequent design. Further details of the PDS are described in Chapter 2.

1.3.3 Conceptual design

The early stages of design where the major decisions are to be made is sometimes called conceptual design. During this phase a rough idea is developed as to how a product will function and what it will look like. The process of conceptual design can also be described as the definition of the product's morphology, how it is made up and its layout. Conceptual design is the generation of solutions to meet specified requirements. It can represent the sum of all subsystems and component parts that go on to make up the whole system. Ion and Smith (1996) describe conceptual design as an iterative process comprising of a series of generative and evaluative stages that converge to the preferred solution. At each stage of iteration the concepts are defined in greater detail allowing more thorough evaluation. It is important to generate as many concepts and ideas as possible or economically expedient. There is a temptation to accept the first promising concept and proceed towards detailed design and the final product. This should be resisted as such results can invariably be bettered. It is worth noting that sooner or later your design will have to compete against those from other manufacturers, so the generation of developed concepts is prudent. Some methods such as brainstorming, morphological analysis and SCAMPER used to aid the generation of concepts are described in Chapter 3.

1.3.4 Detailed design

The detailed design and virtual realisation phase consists of the determination of the specific shape and size of individual components, what materials should be used, how are materials and subsystems going to be recycled, how they fit together and the method of manufacture. Detailed design makes use of the many skills acquired and developed by engineers in the areas of analysis. It is the detailed design phase that can take up the bulk of the time spent on a design. However as implied earlier it is wise to only spend time on details once a sensible concept has been selected.

1.3.5 Manufacturing and production

The manufacture and production phase, although identified as distinct within the structure, is typical of other phases in that it influences all the others. The design of any item must be such that it is feasible to produce or manufacture the product, service, system or platform concerned. In the case of a physical product, the materials selected must be compatible with the manufacturing facilities and skills available, and at acceptable costs to match marketing requirements. Manufacturing is so important that design strategies to reinforce its importance have been developed such as design for assembly (DFA) and design for manufacture (DFM) (see for instance Boothroyd 1997). The concurrent engineering model provides a systematic approach that encourages the developer from the outset to consider all the elements of a product lifecycle or process from concept through disposal including quality control, scheduling and user requirements. This is illustrated in Fig. 1.8.

Market drivers such as a need for shorter innovation cycles, more complicated products and greater data volumes combined with requirements for individualisation, high productivity and energy and resource efficiency have led to the emergence of Industry 4.0. Industry 4.0 involves organisation and control of all design, engineering and business activities across the entire product lifecycle and value chain, leveraging insights from data flows to provide, in theory, seamless delivery of value to all stakeholders. Central to Industry are data flows and a digital model, which needs to be up to date and complete across all activities associated with product design, production planning, engineering and execution, and services. A schematic showing the

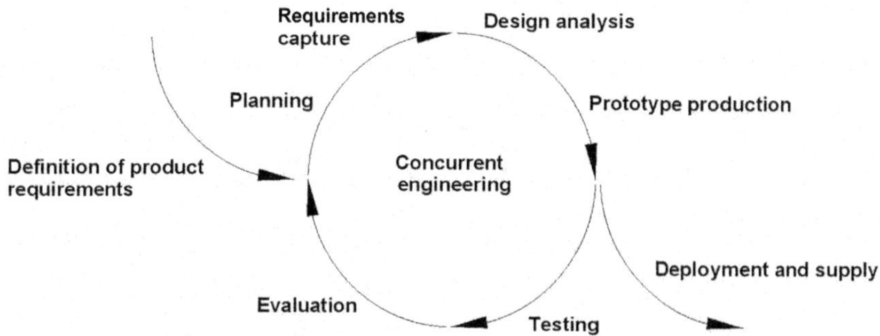

Fig. 1.8 Concurrent engineering.

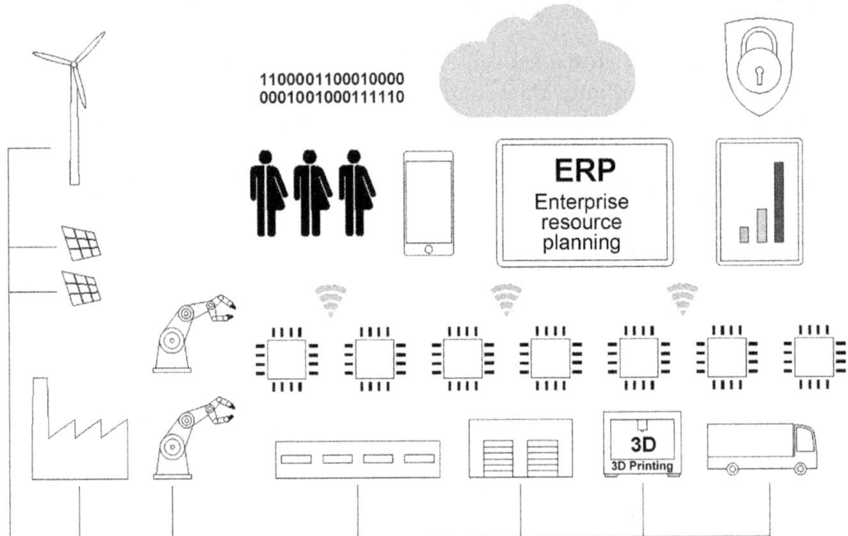

Fig. 1.9 Schematic indicating the Industry 4.0 opportunity for manufacturing.

opportunity Industry 4.0 offers to manufacturing is given in Fig. 1.9. Sensors, computing power, data analytics and networking are all key enablers for the Industry 4.0 vision.

Industry 4.0 first emerged as a concept at the Hannover fair in 2011 with aims including assisting in the long-term sustainability of national industry enabled by leverage of information for quality and bespoke provision. Since then the confluence of increasing data connectivity, digitisation and opportunities associated with this (see Fig. 1.10) have led to widespread consideration and adoption of Industry 4.0 principles. Many reports are available concerning the opportunity and challenges for industry aiming at engaging (see, e.g. McKinsey, 2015; IMechE 2016; PWC, 2016). Industry 4.0 enables many new paradigms ranging from seamless product design to realisation and product design for mass individualisation (see Sikhwal and Childs, 2017).

1.3.6 Sustainable enterprise

The last phase, sustainable enterprise, is of course essential for the success of any business. Sales should match the expectations of the initial marketing analysis, otherwise there may be a need to reconsider production runs. Business factors have a fundamental impact on other phases within the design core. Information such as customer reaction to the product, any component failures or wear, damage during packaging and transportation should be fed back to influence the design of product revisions and future designs. If sales revenues exceed design and production costs then it will be possible to invest in design updates for the product, service, system or platform concerned, as well as the development of brand new designs. The concept of sustainability

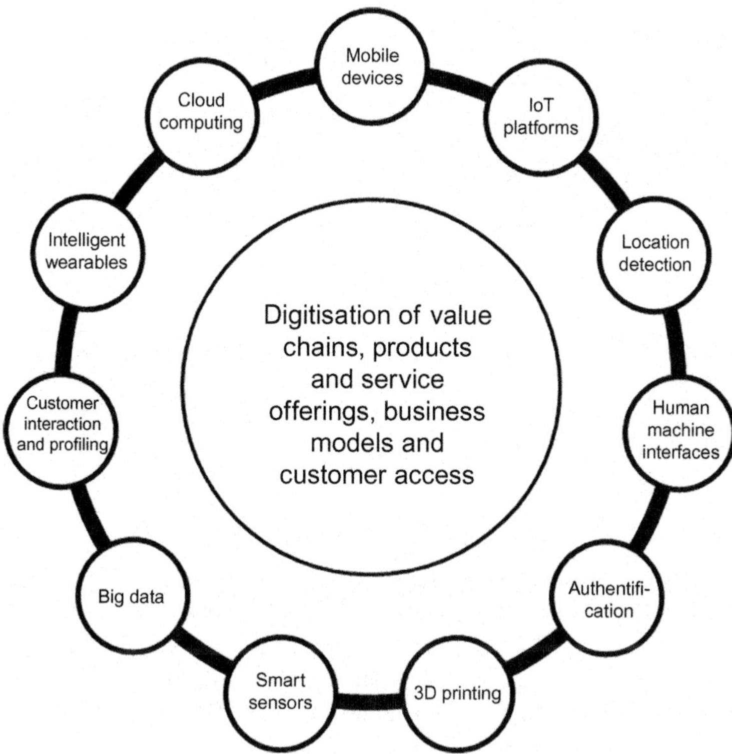

Fig. 1.10 Industry 4.0 principles.

and a sustainable enterprise can be intertwined with service models or incentives, for example by encouraging customers to return devices for recycling in return for an updated model, thereby enabling control of the end-of-life dismantling and recycling of components and retention of customers.

1.3.7 Total design information flows and activities

The double arrows shown in Fig. 1.7 represent the flow of information and control from one activity to another as well as the iterative nature of the process. For instance detailed design activity may indicate that an aspect of the conceptual design is not feasible and must be reconsidered. Alternatively conceptual work may yield features, which have potential for additional marketing opportunities. In other words, the activity on one level can and does interact dynamically with activities on other levels. Fig. 1.11 illustrates the possible flow of this process during the development of a product.

Almost any product such as a vacuum cleaner, kettle, automobile or cordless hand-tool requires input from people of many disciplines including engineering, legal and marketing and this requires considerable coordination. In industrial terms, the integration comes about as a result of the partial design inputs from each discipline. In

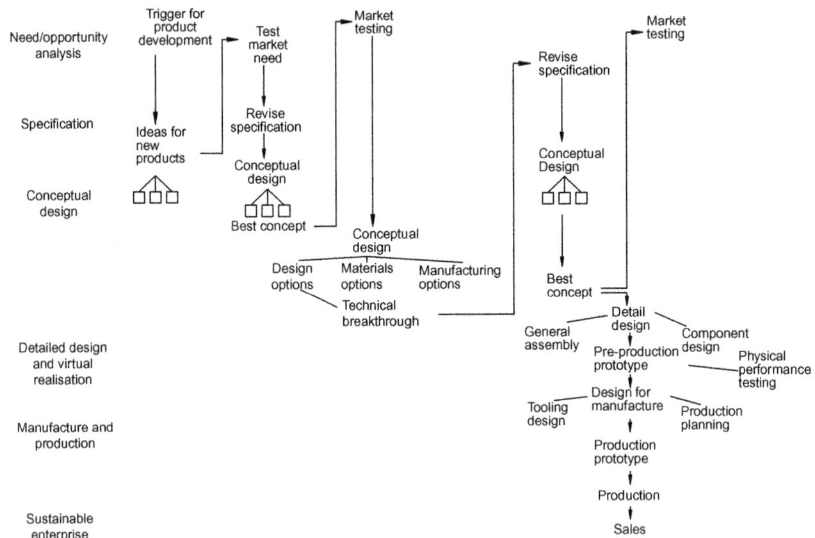

Fig. 1.11 Design activities at different stages in product development.
Adapted from Baxter, M., 1995. *Product Design*. Chapman and Hall, New York.

Fig. 1.12 additional activities, such as market analysis, stressing and optimisation, have been added to the design core as inputs. The effective and efficient design of any product invariably requires the use of different techniques and skills. The disciplines indicated are the designer's toolkit and indicate the multidisciplinary nature of design. The forklift truck example mentioned in Section 1.2.1 will require engine management and control systems as well as the design of mechanical components. Although this text concentrates on mechanical design, this is just one, albeit an important, interesting and necessary aspect of the holistic or total design activity.

A number of circumferential inputs have been shown as arrows in Figs 1.3, 1.5, 1.7 and 1.12. These represent elements of the specification listed in order of importance for each phase of design. The priority order of these specifications may alter for different phases of the design activity. The exact number will depend on the actual case under consideration.

Industry is usually concerned with total design. Total design is the systematic activity necessary from the identification of a market need to the commercialisation of the product to satisfy the market need. Total design can be regarded as having a central core of activities consisting of the market potential, product specification, conceptual design, detailed design, manufacture and marketing.

1.4 Systematic design

A systematic approach to design has been developed and proposed by Pahl and Beitz (1996) who divide their model (see Fig. 1.13) into four phases:

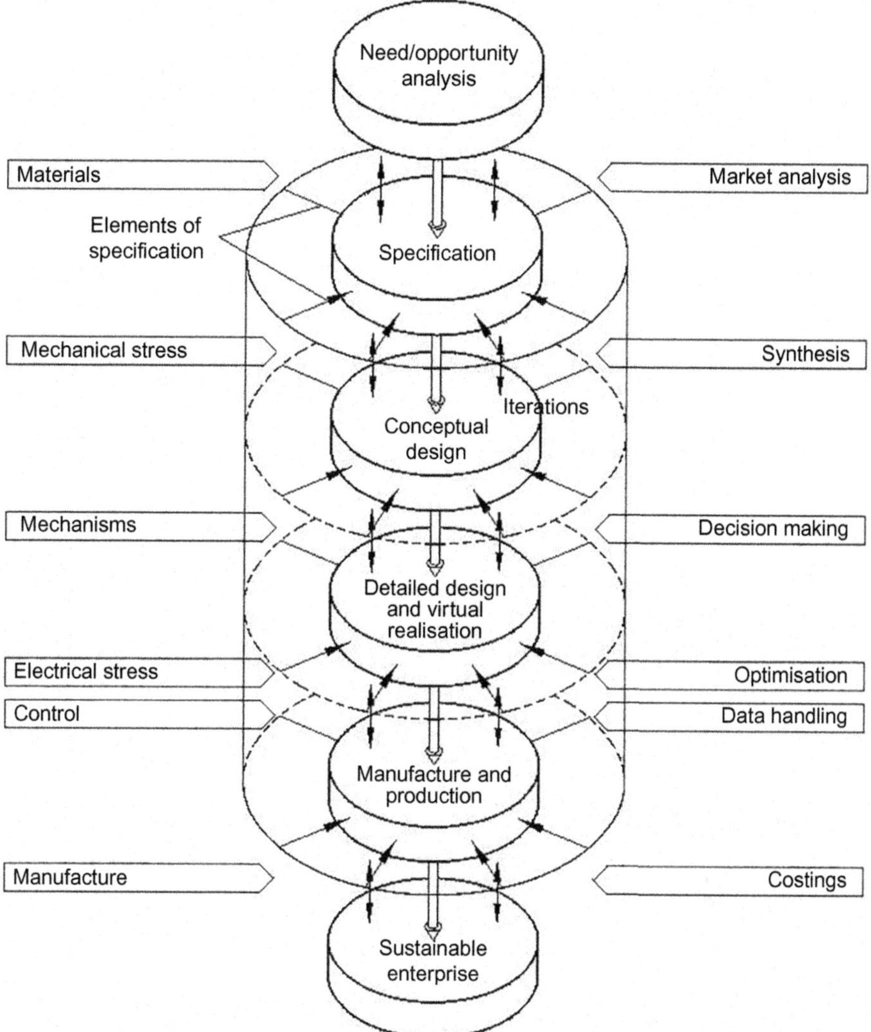

Fig. 1.12 The total design process. Originally developed by Pugh (1990) and updated here accounting for consideration of addressing both needs and opportunities, the ability to virtually model many attributes of a design at the detailed design phase, the applicability of the model beyond traditional manufacture and consideration of the business model and sustainability.

1. product planning and clarifying the task;
2. conceptual design;
3. embodiment design;
4. detail design.

The approach acknowledges that because of the complex nature of modern technology it is now rarely possible for a single person to undertake the design and development of

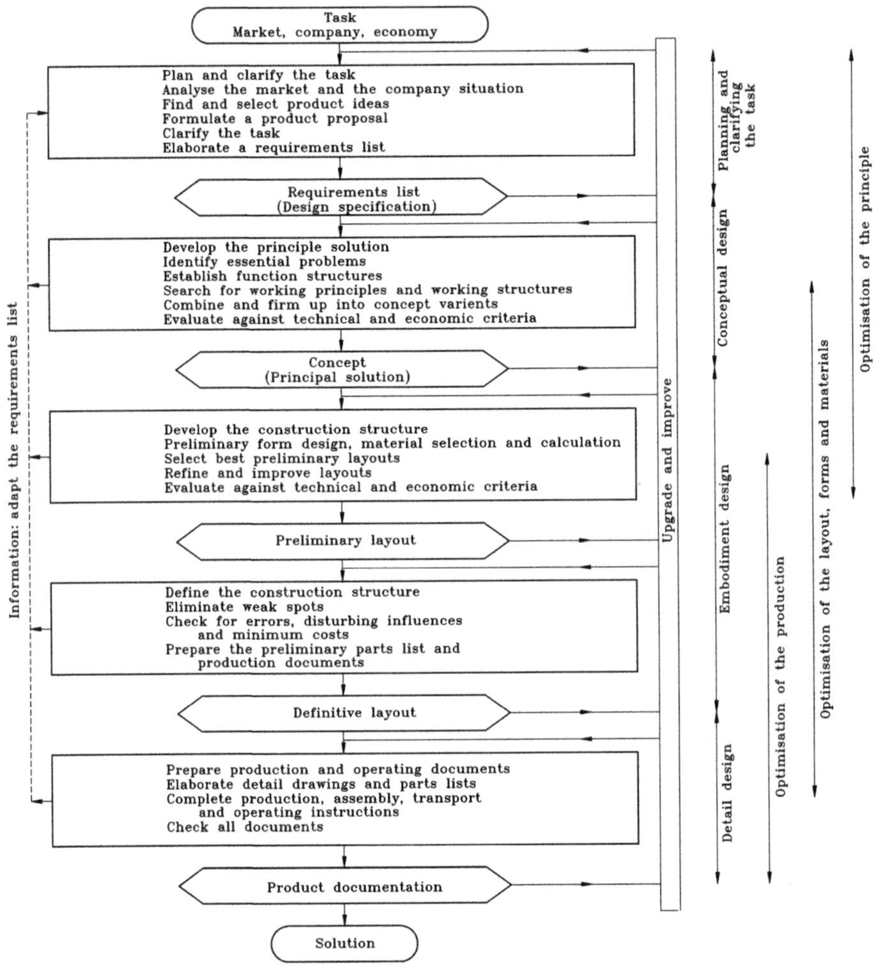

Fig. 1.13 The design process proposed by Pahl & Beitz.
Adapted from Pahl, G., Beitz, W., 1996. *Engineering Design: A Systematic Approach*, second ed. Springer, London.

a major project on their own. Instead a large team is involved and this introduces the problems of organisation and communication within a larger network. The aim is to provide a comprehensive, consistent and clear approach to systematic design.

Design models and methodologies encourage us to undertake careful marketing and specification. Because of their sequential presentation, 'design starts with a need' or 'design starts with an idea', they inherently encourage us to undertake tasks sequentially. This is not necessarily the intention of the models and indeed this approach is countered within the descriptions and instructions given by the proponents of the model who instead encourage an iterative feedback working methodology.

A criticism of the Pahl and Beitz and Pugh design models is that they tend to be encyclopaedic with consideration of everything possible. As such, their use can be viewed as a checklist against which a personal model can be verified. A further criticism of design models is that they are too serialistic as opposed to holistic and that because of the serious manner in which the models are portrayed and documented they can have a tendency to put the intuitive and impulsive designer off!

1.5 Double diamond

The Design Council (2007) reported a study of the design process in 11 leading companies and identified a four-step design process called the 'double diamond' design process model, involving the following phases: discover, define, develop and deliver. In Fig. 1.14, the divergent and convergent phases of the design process are indicated, showing the different modes of thinking that designers use.

1.6 Conceive, design, implement, operate

The CDIO (conceive, design, implement, operate) framework is widely used in design and engineering education and was developed in recognition of a divergence between academic culture and practical engineering requirements. The framework explicitly recognises the importance of holistic considerations for effective design outcomes with application of both engineering practice skills such as design, manufacture, personal, professional, interpersonnel and business in combination with disciplinary knowledge from the sciences and mathematics as well as the humanities (Crawley 2001).

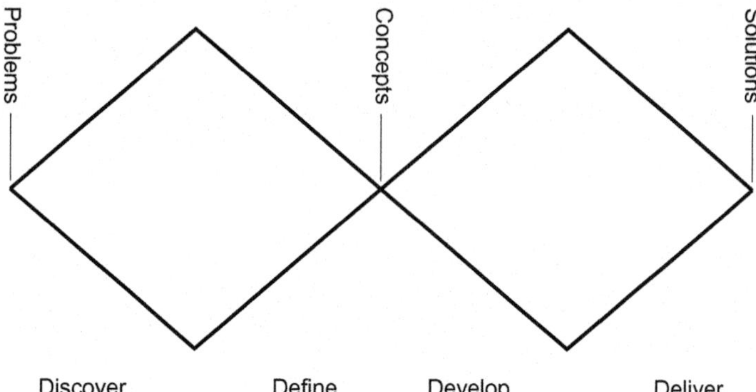

Fig. 1.14 Schematic describing the design process.
Adapted from Design Council. 2007. The 'double diamond' design process Model, 2007. www.designcouncil.org.uk/en/About-Design/managingdesign/The-Study-of-the-Design-Process/. Accessed December 2007.

1.7 Design for six sigma

Design for six sigma (DFSS) is an approach for designing a new product or service with a measurable high performance. This requires the development of an understanding of customer needs prior to launch rather than afterwards.

There are a number of methodologies applying six sigma principles including DMADV (design, measure, analyse, design, verify), IDOV (identify, design, optimise, verify) as well as DFSS. IDOV tends to focus on the final stages of engineering optimisation and may not address the selection of product features and attributes that actually address customer requirements (Tennant 2002). DFSS comprises a number of defined activities as outlined in Table 1.2.

Table 1.2 Design for six sigma

Phase	Characteristic activities
New product introduction	Selection of the concept or service to fulfil a perceived need. Benchmarking, customer surveys, R&D, sales and marketing input, risk analysis
Define	Benchmarking, customer surveys and analysis. Development of a team charter to provide a solid foundation for the project.
Customer (measure)	Identification of the full characteristics and needs of the customer. Use of Quality function deployment (QFD) (see Chapter 2) to identify the set of critical to quality (CTQ) metrics. This arises from the set of customer needs along with a list of potential parameters that can be measured and quantified defining the targets for each CTQ metric.
Concept (analyse—conceptual design)	A design is explored and developed for the new product and service. This requires a further round of QFD to identify the best features that have the potential to deliver the critical to quality metrics. During this phase there is a move from CTQ to critical to process (CTP) metrics. At the end of this stage a concept or concepts together with a set of CTP metrics that constrain the formal and technical design will have been produced.
Design (technical design)	The team handovers the design brief for the design team to complete using the CTP metrics. Typical tools applied in this phase include Design of Experiments (DoE) and statistical optimisation.
Implement	Use of Failure Mode Effect Analysis (FMEA) and other prototyping and product or system testing approaches to enable fine-tuning of the design.
Handover	Full-scale commercial roll-out

1.8 Design optimisation

Inevitably there are conflicts between the diversity of requirements driven by the stakeholders. Optimisation can be viewed as the process of repetitively refining a set of often-conflicting criteria to achieve the best compromise. In the case of a transportation system, size, manoeuvrability, cost, aesthetic appeal, ease of use, stability, safety and speed are not necessarily all in accordance with each other, for example (see Hall et al., 2012). Priorities can change within the lifetime of a product and vary within markets and cultures. Cost minimisation may call for compromises on material usage and manufacturing methods. These considerations form part of the optimisation of the product producing the best or most acceptable compromise between the desired criteria.

A traditional engineering design process often comprises a series of sequential steps, beginning with defined requirements or an opportunity and proceeding through ideation, synthesis, analysis and optimisation, to production. This process can be controlled by a series of gate reviews (see Section 1.9) in coordination with the stakeholders and the process can be iterative with phases being revisited when rework is recognised as necessary. This type of process can lead to bottlenecks in activity and a tendency to stick to a particular suboptimal solution as so much time and effort has already been allocated to it.

The range of optimisation tools used in design are reviewed by Roy et al. (2008). MDO, for example, combines tools and approaches from a number of disciplines to tackle the refinement of a set of parameters for a given problem area to deliver the best compromise between those parameters and has been widely applied in aerospace applications. A key characteristic of MDO is that the solution is better than that obtained by optimising each of the parameters sequentially. The approach is resource intensive in terms of computational power, however the Moore's law enhancement of processing means that this is not a hindrance to application of the approach. Optimisation approaches and numerical strategies typically employed have included decomposition, approximation, evolutionary and mimetic algorithms, response surface methodologies, reliability and multiobjective. The METUS methodology (METUS 2012), for example, has been used in Airbus development programmes to help provide a holistic approach to product development, covering the phases of conception and optimisation of product architecture, visualisation and integration of partners in the supply chain. A topological optimised cantilever is illustrated in Fig. 1.15.

MDO can be considered to be a methodology for the design of an engineering system that exploits the synergies between interacting parameters. The principle of MDO is that it provides the collection of tools and methods that enables and permits the trade-off between different disciplines inherent in design. Proponents of MDO suggest that this provides the justification for its application at an early stage in a product development programme (see, e.g. Sobieszczanski-Sobieski et al., 1984).

Ideally an MDO environment should permit the definition of the brief and specification constraints for all the various stakeholders (see, e.g. Kroll et al., 2009). This is typically achieved using a single parametric model for the whole system facilitating

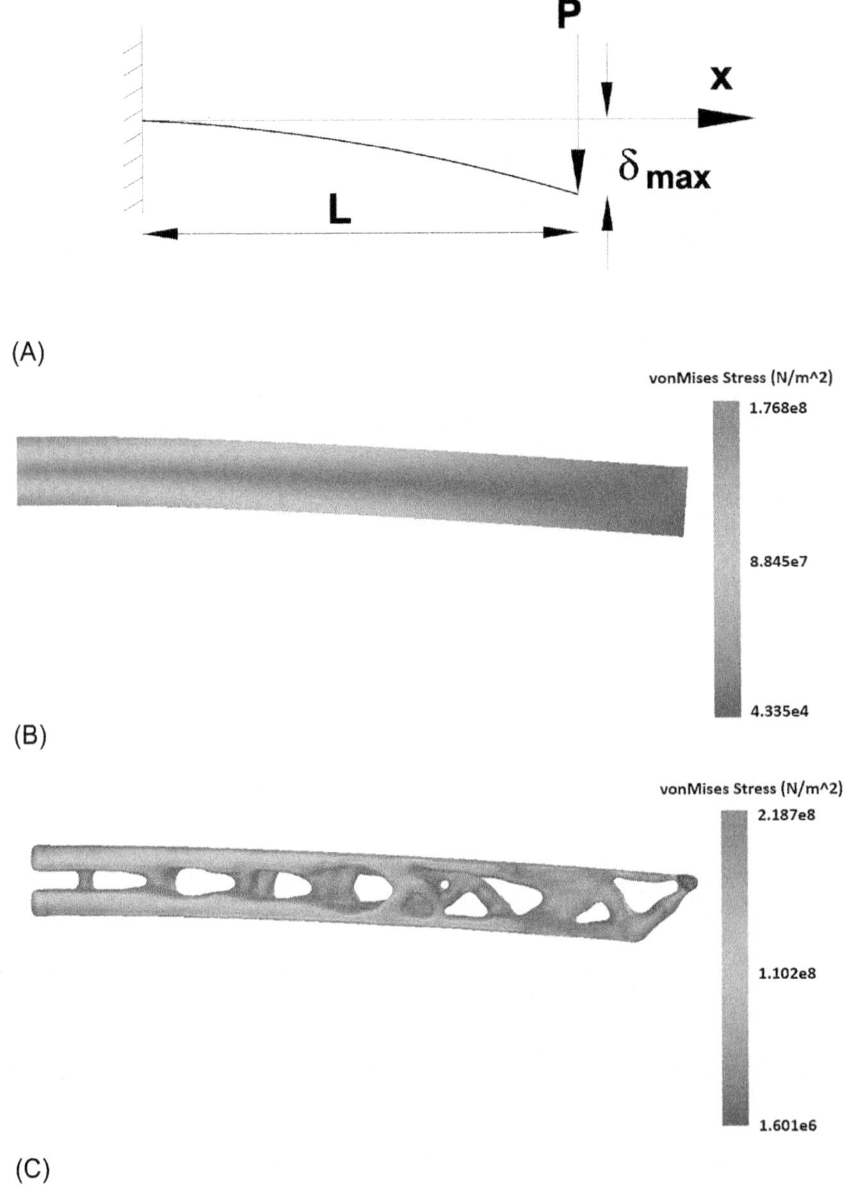

Fig. 1.15 Cantilever beam optimisation case study: (A) Loading and boundary conditions, (B) von Mises Stress field on initial model and (C) Topological optimised model (Zhu et al., 2018).

effective communication between the different stakeholders. MDO offers the potential for the interactions between subsystems and systems to be explored from an early stage in the design process by a number of stakeholders. The purpose being to find the minima for the cost functions and reach an optimal solution for the holistic system.

1.9 Project management

Project management is fundamentally concerned with getting significant activities achieved. More formally, project management involves the application of processes, methods, knowledge, skills and experience to achieve predefined objectives. A project is generally a unique endeavour undertaken over a specific time period to achieve the planned objective. A project is usually deemed successful if it achieves the objectives according to defined acceptance criteria relevant to the project, such as outputs, outcomes and benefits within an agreed timescale and budget. Fig. 1.16 illustrates the scope of a generic project at the core, and its links to time, quality and cost. Invariably these criteria are interlinked. For example, it may be possible to deliver a project sooner, but to the detriment of quality and cost.

The core components of project management can include

- defining reasons why a project is necessary;
- capturing project requirements;
- specifying deliverables;
- estimating the resources required;
- defining timescales for each aspect of the project;
- preparing a business case to justify the investment for the project;
- securing organisational agreement and funding to commence;
- developing and implementing a management plan for the project;
- assembling the team and resources to implement the project;
- leading and motivating the project delivery team;
- managing the risks, issues and changes on the project;
- monitoring progress against plan and communicating progress to all relevant stakeholders;
- managing the project budget;
- maintaining communications with internal and external stakeholders;
- provider management (ensuring that contractors and suppliers deliver according to requirements and managing these relationships);
- closing the project in a controlled fashion when the project has been completed or run its course.

Projects tend to be distinct from business-as-usual activities, requiring a team to work together temporarily, to focus on a specific project or subtask within a project and its

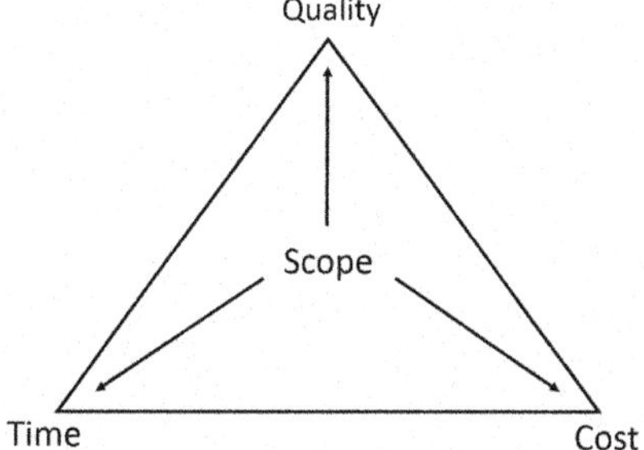

Fig. 1.16 The interrelationship between the scope of a project and time, quality and cost.

objectives (Association for Project Management 2017). Teamwork is often an essential component of project management with the need to leverage diverse resources to achieve the intended outcomes. A common approach to tackling a significant challenge is to subdivide it into a number or subtasks. In project management these sub tasks are often referred to as a work-package. Each work-package may itself be divided up into a number of discrete chunks of work. The way an overall challenge is broken down will depend on factors such as scale, complexity, function and significance of the activity. Designing a passenger aircraft, jet engine or new vehicle represents a difference in scale of activity to designing a bearing housing, disc brake or door lock. Both classes of examples may involve the application of similar principles, but represent different levels of risk and require significantly different levels of resource.

It can be helpful to express objectives in terms of outputs, outcomes, benefits or strategic aims. For example, an output could be that a new clutch will be produced, an outcome could be less staff required to manufacture a batch of clutches, a benefit could be enhanced reliability and a strategic objective could be entrance into a new market sector.

A wide range of approaches for project management have been codified including

- The traditional approach;
- PRINCE;
- Waterfall;
- V Model;
- Stage-gate;
- Agile.

In addition some of the design models such as total design, systematic design and concurrent engineering can be used as a basis for defined phases of activities. Each of the above methods are described in Sections 1.9.1–1.9.6.

Project management typically occurs under the auspices of a project manager. The position, responsibilities and contribution of a project manager can vary widely from organisation to organisation. It can range from monitoring the performance of others, to controlling their performance, to having the overall technical and operational responsibility for a complete activity, or project, or business. Project managers may be extra to the present teams, or an existing member may be given this additional task. In general the project manager will be the reporting link to the person in charge of the complete project, but being just the reporter is a minimal responsibility, as this carries no authority for the information being passed. To make project management a worthwhile position, there must be some element of control, even if this is the result of planning of the various activities as part of a critical path analysis (CPA). Just planning the work must not be mistaken for control.

A project manager can be the prime contact with the client, with suppliers and subcontractors. The project manager is in a position of considerable responsibility, with the ability to commit significant amounts of expenditure on behalf of the company. A full understanding of the specification is essential. Any doubts must be resolved at a very early stage of the project. Even if the project manager does not have direct control of the people on the project, the expectation is still to deliver on time. The project manager is the project champion and will be expected to promote the project with the company management in order to optimise the project performance. Fig. 1.17 shows the links that are essential for any project manager.

In a real world scenario, a project manager may be faced with the following situations:

- the specification is incomplete and ambiguous;
- the required resources are not available;
- the design is not correct first time;
- the production unit has made scrap;
- the supplies are late;
- the customer has changed his mind;
- the changes were not negotiated correctly due to a lack of understanding.

The project manager has to maintain his reputation throughout such scenarios and situations, and still convince all concerned that his or her requests are credible and important.

1.9.1 The traditional approach

The traditional approach to project management uses a set of well-established techniques and tools to deliver a product, service or outcome. These techniques and tools arise from substantial experience and involve establishing the objectives of a project, gathering information, leveraging resources, focussing on outcomes, establishing a schedule, consideration of risks and their mitigation, reviews and good

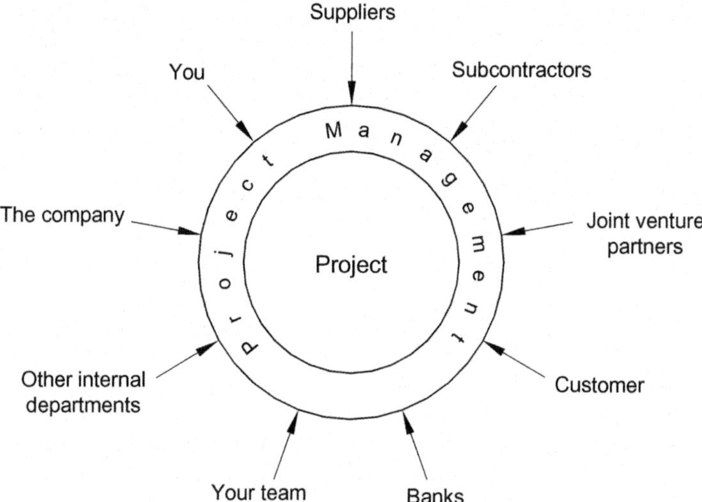

Fig. 1.17 Some essential links between a project manager and stakeholders for the project.

communication, to ensure progress and delivery within time and budget constraints and to the required level of quality.

It is important to ensure the project objectives and desired outcomes are defined at the outset. Knowledge of these can help motivate team members and guide activity towards these goals. The following steps represent well-established principles that can aid in establishing a project's rationale and buy-in to it from the various stakeholders involved.

- Identify the key stakeholders that you are performing the project for. The key stakeholders will invariably be represented by individuals in the organisations involved and are sometimes referred to as project drivers. The project drivers have authority to define the project objectives and desired outcomes.
- Collect information and documentation that supports the project's justification. This can include the strategic plan for the organisation, annual reviews, aspects of a company mission and vision statements, any key point indicators (KPIs) for the organisation that relate to the project.
- Pay attention to meeting etiquette. Planning and preparing for a meeting is important to get value from the time and effort that everyone will be putting in. Ensure a meeting has clear purpose and focus such as to address a problem, explore ideas, motivate personnel or to commence a new initiative. An agenda is important. This can be a simple list of items to consider with an approximate time allocation for each item. Attention should be given to the order of items to help the meeting make sense to the participants and aid the flow of ideas and decisions. Ensure that the relevant people are present as this will help leverage the help needed in achieving goals and carrying out tasks. A number of people can aid the dynamics of the meeting—by including relevant officers this can help in achieving goals, sharing duties and motivation. Experience suggests that if 5–10 people are in a meeting taking 30–60min then all parties are likely to contribute in some way. If the number is higher, then it is unlikely that everyone will be able to speak, but it may still be necessary to engage the various stakeholders and help with the flow of information. In meetings ensure notes are taken and circulated afterwards to the relevant parties and filed. In one-to-one meetings confirm any salient points and decisions in writing to provide a written record that can be scrutinised later if required. For informal meetings when speaking about important matters, do so with at least three people present so that any decisions can be corroborated.
- Focus on outcomes, not activities. SMART objectives can be helpful in focusing attention on defined outcomes. SMART is an acronym for specific or strategic, measurable, achievable, relevant, time-bound and trackable (Doran 1981).
- Use clear language, so that objectives are not ambiguous and that information and instructions can be understood.
- Ensure each objective has at least one measure that relates to a specific performance target.

As indicated previously the classic success criteria for a project are delivery on time, to cost and to the expected quality level. To deliver a project on time the traditional approach involves significant attention to scheduling and definition of resources. In general a schedule needs to be achievable, responsive to changes and understandable by the team involved. The following steps can be helpful in establishing a schedule.

- Identify the activities required for the project.
- Break down each of the activities into subtasks with sufficient detail so that if each of the tasks is understandable and completing the tasks looks plausible.

- Consider both the duration of time required to complete an activity as well as the order in which the activities need to be performed (these are known as the interdependencies).
- Consider what strategies might be appropriate to perform an activity and the resources required. This can assist in helping to define how long an activity will take.
- Consider the availability of each resource required. For example if a machining operation must be done on a particular piece of equipment and the organisation only has one of these, then the limited capacity needs to be accounted for in defining the duration required for the machining operations concerned. Some alternative strategies could be considered such as acquiring additional machines, running multiple shifts or outsourcing the manufacture operations, subject to budget and quality considerations.
- Consider what budget needs to be allocated to achieve a task. Document the assumptions associated with this.
- Consider the risks associated with each task. For each risk, consider and develop a plan that mitigates against the risk (see Tables1.3 and 1.4).
- Set out a draft schedule. Review this, considering whether each task has sufficient resource to achieve the task within the time defined. Estimate the costs for the resource allocated. Redefine the schedule, iterating the overall plan to see if a more plausible and realisable plan is possible. One of the most popular methods of laying out the activities associated with a schedule is to use a Gantt chart. Examples of information associated with these are given in Figs 1.18 and 1.19. Typically a Gantt chart will comprise a list of activities on the left and a

Table 1.3 Generic risk assessment table

Likelihood	Consequences				
	Insignificant Risk is easily mitigated by normal day-to-day processes	Minor Delays up to 10% of schedule Additional cost up to 10% of budget	Moderate Delays up to 30% of schedule Additional cost up to 30% of budget	Major Delays up to 50% of schedule Additional cost up to 50% of budget	Catastrophic Project abandoned
Certain >90% chance	High	High	Extreme	Extreme	Extreme
Likely 50%–90% chance	Moderate	High	High	Extreme	Extreme
Moderate 10%–50% chance	Low	Moderate	High	Extreme	Extreme
Unlikely 3%–10% chance	Low	Low	Moderate	High	Extreme
Rare <3% chance	Low	Low	Moderate	High	High

Table 1.4 Risk assessment example for a wheeled robot

ID	Risk item	Effect	Cause	Likelihood (*L*)	Severity (*S*)	Importance	Action	Owner
	Describe the risk briefly	**What is the effect on any or all of the project deliverables if the cause happens?**	**What is(are) the possible cause(s) of this risk?**			**L×S**	**What action(s) will you take, and by when, to prevent, reduce the impact or transfer the risk of this occurring?**	**Who is responsible for following through on mitigation?**
1	Speed failure	If the robot is too fast, we will not be able to brake adequately. Potential danger	Possible injury to people. Damage to objects	3	3	9	Reduce speed of robot; optimise robot speed	Hardware specialist
2	Failure of sensors	Collision with a person or object	Robot malfunction or operator error causes damage to sensors	2	3	6	Make sure code is operating properly. Specify back-up sensors	Software specialist

3	Failure to find location of robot	Possibly lost robot. Possible that final destination not reached	RFID positioning failure. RFID signal not strong enough. Wheel encoder failure.	2	2	4	Ensure encoders are functioning. Test signal strength for RFID	Hardware specialist
4	Battery failure	Robot does not move. Need to recharge battery	Charge cycle	2	2	4	Plan for battery recharging prior to 30% battery charge remaining	Project lead
5	Budget	Not able to complete project	Poor planning	1	2	2	Detailed resource requirement planning	Project lead

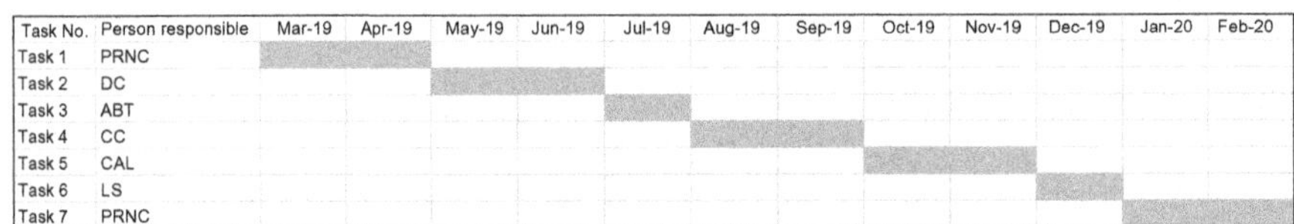

Fig. 1.18 Generic project management Gantt chart.

timescale along the top row. Each activity can be represented by a bar indicating when it should start and end, along with additional information such as who is involved and responsible for the task and an indication of what proportion of the task has been completed. There are many templates readily available for Gantt charts along with software platforms to aid project management.

- Once you are confident with the schedule, engage some of the project drivers in reviewing the plan and aim towards getting project buy-in.

Most projects require effort from many different people and facilitating effective professional relationships can aid in realising successful outcomes. The following general principles can help foster high levels of effective professional relationships between team members and stakeholders: clarify project objectives with the team; ensure the benefits for the organisation and team members are understood; engage team members in helping to form the project plan; ensure you address any issues, concerns and questions promptly; communicate regularly within the team on issues and progress; acknowledge the contributions being made by members of the team. A key facet of project management is responsibility. If an individual accepts responsibility then in doing so they will also be taking on a level of accountability for their performance. This link between responsibility and accountability is helpful in transcending some of the traditional line management and hierarchical relationships in an organisation. For example someone more senior may accept responsibility for a task and are then by implication responsible for that task and can be held to account for it.

Good communication with all concerned is essential, including, for example, manufacturing and commercial departments, to ensure that the specification is still being met in all respects, including the costs of design and manufacture. Good communication at the specification stage ensures 'ownership' of decisions. A desire or need to change the design from the specification must be agreed, particularly by the marketing department as it may affect the viability of the product in the marketplace. It cannot be assumed that the design team knows the capability of the manufacturing unit, or that the factory will understand, as intended, the information, which they receive from the designers. A balance must be found between performance, availability and cost. This decision must be taken at an early stage since the whole philosophy of the design is affected by these choices, which again depend upon the needs of the actual or potential customers.

The following 12 point list provides a set of steps that can aid the planning process for management of a project (Leach 1999).

1. Requirements capture and flow-down.
2. Define the deliverables.

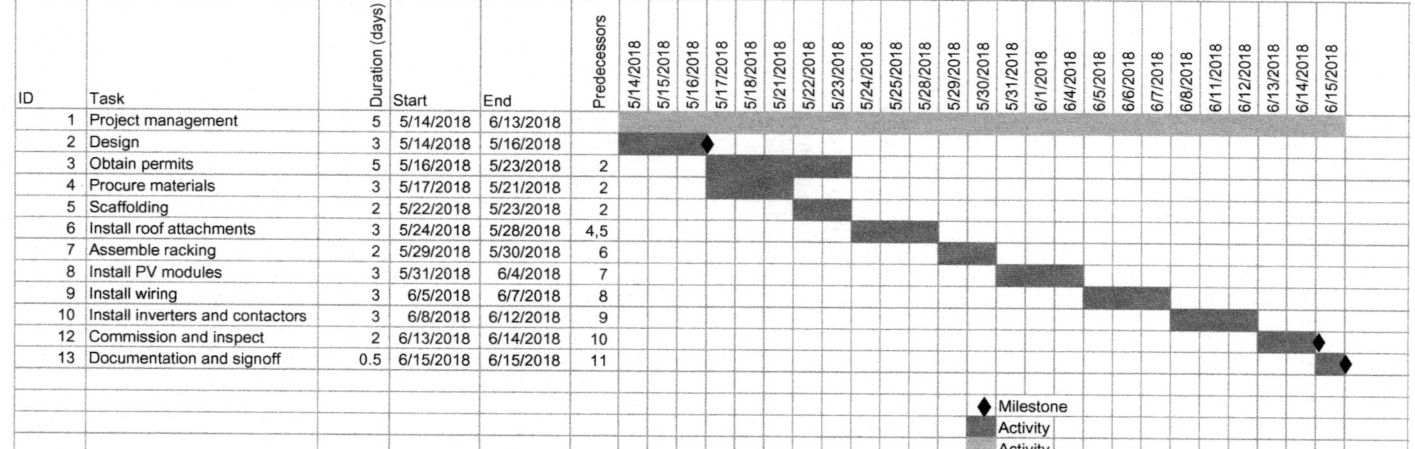

ID	Task	Duration (days)	Start	End	Predecessors
1	Project management	5	5/14/2018	6/13/2018	
2	Design	3	5/14/2018	5/16/2018	
3	Obtain permits	5	5/16/2018	5/23/2018	2
4	Procure materials	3	5/17/2018	5/21/2018	2
5	Scaffolding	2	5/22/2018	5/23/2018	2
6	Install roof attachments	3	5/24/2018	5/28/2018	4,5
7	Assemble racking	2	5/29/2018	5/30/2018	6
8	Install PV modules	3	5/31/2018	6/4/2018	7
9	Install wiring	3	6/5/2018	6/7/2018	8
10	Install inverters and contactors	3	6/8/2018	6/12/2018	9
12	Commission and inspect	2	6/13/2018	6/14/2018	10
13	Documentation and signoff	0.5	6/15/2018	6/15/2018	11

Fig. 1.19 Example of a Gantt chart for a 100kW commercial PV installation.

3. Create the work breakdown structure.
4. Define the performance measures.
5. Identify and assess the risks.
6. Define the activities.
7. Identify the key dependencies.
8. Define the milestones.
9. Produce the schedule.
10. Estimate resource and cost requirements.
11. Assemble the plan, review this and revise it to help ensure it meets the requirements, and is achievable within the resources and time frame available.
12. Obtain commitment acceptance.

1.9.2 PRINCE and PRINCE2

A wide range of commercial project management tools and supporting software has been developed. PRINCE is an acronym for Projects In Controlled Environments. PRINCE2 was released in 1996 as a generic project management method, aimed at helping users to organise, manage and direct projects on time and to budget. The methodology can be tailored according to the scope and scale of a project.

A key principle of PRINCE is to gather information quickly at the start, to establish whether a project is worth the investment of time and effort in planning in detail, and to provide source data for subsequent planning if the decision is made to take it forward. Some of the key elements of this initial phase include the following.

- Gather information quickly.
- The aim at the end of this phase is whether you are to progress to the next stage of full planning (known as initiation).
- Assign key roles such as project executive and project manager.
- Produce a project brief (see Chapter 2).
- Check whether the outline business case appears viable.
- Check whether the risks appear acceptable based on the information available at this stage.
- Set out a plan for the initiation stage that allows for more detailed consideration of risks and the business plan.

An important consideration in project management is the balance between planning and need for control. As mentioned previously there can be a trade-off between cost, timeliness and quality. Deciding the level of quality that is appropriate is an important criteria. Following initiation, key items of PRINCE are to

- Produce product and activity plans for the project.
- Produce product and activity plans for the first phases after initiation.
- Undertake a full risk analysis.
- Produce a full business case for the project.
- Implement simple controls and reporting procedures for the project.
- Assemble the Project Initiation Document (PID). This provides a foundation for the project and aims to help all the stakeholders understand why they are doing the project. Common items to cover in a PID include information on why the project is being undertaken, what is to be delivered, who is responsible, how and when the project will be delivered, what are the risks, whether there are any issues and constraints, how much it will cost.

- Set up a project board. The membership for this will vary depending on the nature of the project but it will typically comprise high-level stakeholders and advisors.

Project management concerns effective delivery rather than being seen to be busy. A key focus in PRINCE is checking progress against the Product Checklist, the list of tasks, and the quality of the outcome. Decisions need to be taken regularly about whether tasks are on track or whether an exception needs to be made. An exception is a piece of project management terminology for addressing a project plan issue. An exception plan can be made and then implemented to recover from a deviation caused by a task taking longer, costing more or quality being below that required. Exceptions need rapid and careful investigation. It will be necessary to try and find out the underlying reason for the exception, and to take a decision on whether to carry on, to develop a new plan or to stop the project. Key to progress of any plan is to leverage the resources available so it is important to allocate work within the team and to monitor progress. Progress can be reported regularly and at set intervals to the Project Board using a highlight report function. If there are any exceptions that cannot be met then these need to be reported to the Project Board at the earliest possible moment, to alert the Board members and to see what measures need to be taken to address the situation.

1.9.3 Waterfall

The central philosophy for the waterfall model is that time and effort is spent ensuring the requirements and design are suitable at the start of the project, saving time, effort and resources later. The waterfall model (Fig. 1.20) has a significant heritage with about 60 years of practice, particularly in software applications, systems integration, large engineering and government contracts.

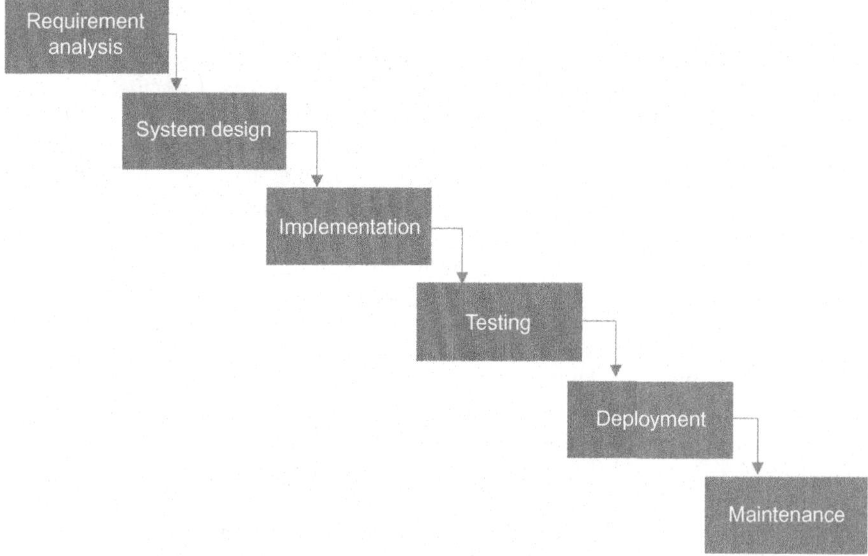

Fig. 1.20 The waterfall method.

The waterfall model is a sequential design process. In the waterfall model, each phase must be completed before the next phase can begin and there is no overlapping of the phases. Progress flows steadily downwards, like a waterfall, through phases of conception, initiation, analysis, design, construction, testing, production/implementation and maintenance.

Various versions of the waterfall model have been developed. Royce (1970) gives the following phases which are followed in order:

1. System and software requirements: captured in a product requirements document.
2. Analysis: resulting in models, schema and business rules.
3. Design: resulting in the software architecture.
4. Coding: the development, proving and integration of software.
5. Testing: the systematic discovery and debugging of defects.
6. Operations: the installation, migration, support, and maintenance of complete systems.

An alternative set of phases for waterfall project model includes the following stages:

- Definition and requirements analysis stage;
- Design stage;
- Development, implementation and production stage;
- Integration and test stage;
- Factory Acceptance Test.

The waterfall model maintains that a team should move to the next phase only when the preceding phase has been reviewed and verified. The sequential phases in a Waterfall model are outlined in Table 1.5.

Table 1.5 Outline description of the phases associated with a typical waterfall project.

Phase	Brief Description
Requirement gathering and analysis	All possible requirements of the system to be developed are captured and documented in a requirement specifications document.
System design	The requirement specifications from the first phase are studied and the system design prepared. System design helps in specifying hardware and system requirements and also in defining overall system architecture.
Implementation	With inputs from the system design, the system is developed in small programmes called units, which are integrated in the next phase. Each unit is developed and tested for its functionality which is referred to as unit testing.
Integration and testing	All the units developed in the implementation phase are integrated into a system after testing of each unit. Post integration the entire system is tested for any faults and failures.
Maintenance	There may be issues which arise in the client environment. To fix these issues patches can be released. To enhance the product better versions can be released. Maintenance is undertaken to deliver these changes in the customer environment.

The waterfall model is appropriate for tasks where work can be defined tightly and is part of a sequential process. The model is appropriate for activities where the team has significant experience (in the specific task application) and is therefore able to define and undertake the tasks with confidence, enabling work to be completed in each phase and 'flow' down the waterfall without revisiting of previous phases and without associated rework.

Some situations where the use of waterfall model is most appropriate include:

- Requirements are very well documented, clear and fixed.
- Product definition is stable.
- Technology is understood and is not dynamic.
- There are no ambiguous requirements.
- Ample resources with required expertise are available to support the product.
- The project is short.

The advantage of waterfall development is that it allows for departmentalisation and control. A schedule can be set with deadlines for each stage of development and a product can proceed through the development process model phases one by one. Development moves from concept, through design, implementation, testing, installation, troubleshooting, and ends up at operation and maintenance. Each phase of development proceeds in strict order. Benefits associated with waterfall model include:

- It is simple and easy to understand and use.
- It is easy to manage due to the rigidity of the model. Each phase has specific deliverables and a review process.
- Phases are processed and completed one at a time.
- It works well for smaller projects where requirements are very well understood.
- It has clearly defined stages.
- It provides well-understood milestones.
- It provides easy to arrange tasks.
- The process and results can be documented readily and professionally.

The disadvantage of waterfall development is that it does not allow for much reflection or revision. Once an application is in the testing stage, it is very difficult to go back and change something that was not well-documented or thought upon in the concept stage. Disadvantages of waterfall model include:

- No working software or functioning technology is produced until late during the lifecycle.
- High amounts of risk and uncertainty.
- Not suitable for complex and object-oriented projects.
- Poor for long and ongoing projects.
- Not suitable for the projects where requirements are at a moderate to high risk of changing. Risk and uncertainty is high with this process model.
- It is difficult to measure progress within stages.
- Cannot accommodate changing requirements.
- Adjusting scope during the lifecycle can end a project.
- Integration is done as a 'big-bang' in the final phases, which does not allow identification of any technological challenges or business bottlenecks at an early stage.

Various modified waterfall models (e.g. Royce 1970; Bell and Thayer 1976) include slight or major variations on the process. These variations include returning to the previous cycle after flaws have been found downstream, or returning all the way to the design phase if downstream phases are deemed insufficient.

1.9.4 Das V Modell

The term V Model has been applied widely in different domains to assist in tackling complex systems development, lifecycle models and project management. The 'Das V Modell' is the official project management method used by the German Government and provides guidance for planning and executing projects. A key feature of the V Model is definition of who has to do what and when in a project, and the use of decision gates to indicate a milestone in the progress of a project (see Fig. 1.21). In the V Model, an emphasis is placed on verification on the left hand side of the V and validation on the right hand side with the use of test cases to ensure adherence between equivalent activities on either side of the V. The key elements of the V Model have been widely translated and adapted resulting in many variations (see, e.g. Fig. 1.22). The Das V Modell has some equivalence with PRINCE2.

1.9.5 Stage-Gate

The Stage-Gate product innovation process was originally developed (Cooper 1986, 2011) for new product development taking a team or organisation from an initial idea to launch. The process is characterised by breaking down the effort of project management into stages separated by management decision gates, and is used in many diverse applications. The gates control the process and serve as quality control and checkpoints for 'Go', 'Kill', 'Hold' and 'Modify' decisions. The process has been emulated by processes such as the phase gate process and gated reviews.

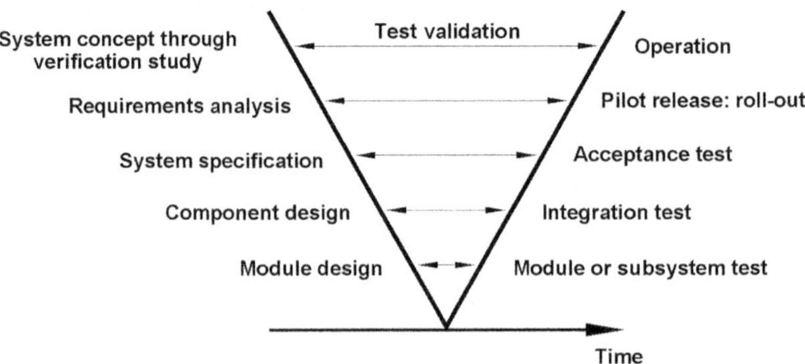

Fig. 1.21 Das V Model.

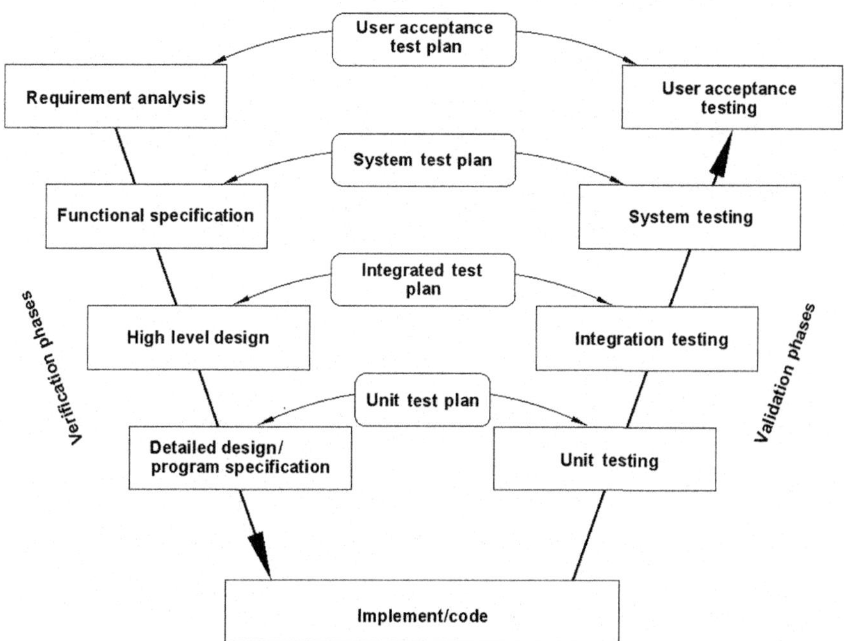

Fig. 1.22 An example of the V Model.

The term 'stage' is used to define where the detailed activity, progress and action associated with a project occurs. A stage enables information to be gathered to reduce or mitigate project uncertainties and risks. At the end of a stage, a management decision can be made about whether to move forward with a project. In a project each subsequent stage represents an increase in funds spent because the process is sequential. However with each successful phase project risks, unknowns and uncertainties decrease.

A gate represents a point in the project where the team converges and the information associated with that activity is brought together to produce a decision on go, kill, hold or modify the project plan. A gate provides a funnel where mediocre projects can be culled or stopped, and successful ones prioritised. A gateway meeting is usually attended by senior staff, the gatekeepers, who have the authority to commit further resources to the project. At a gate meeting it is important that the project manager presents the deliverables achieved by the project to date against the success criteria for the project so that progress can be judged effectively, allowing the project to move on to the next stage. An overview of a generic stage-gate project is illustrated in Fig. 1.23. The key stages and gates in a typical Stage-Gate model are outlined in Table 1.6.

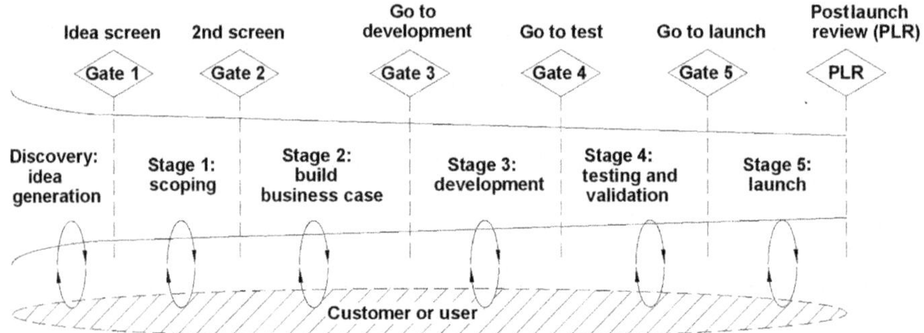

Fig. 1.23 A generic 5 stage idea to launch state gate—the loops represent a series of build, test, feedback and revision iterations with the customer or end-user.

Table 1.6 Typical stage-gate process activities

Phase	Activity
Discovery	Prework to reveal new opportunities and produce ideas.
Gate 1. Idea screen	Decision to commit tentative or initial resources to a particular project.
Stage 1. Scoping	A preliminary investigation of the project to enable the field of the project to be focussed.
Gate 2. Second screen	A more rigorous screen where the project is evaluated again in the light of the information found in Stage 1. If the project is deemed Go at this stage, more significant resources are provided for Stage 2 to go ahead.
Stage 2. Build the business case	A more detailed investigation involving primary research to produce a business case.
Gate 3. Go to development	This is the last point at which a project can be halted before committing to very significant resources for the organisation concerned. The Gate 3 evaluation will involve review and detailed scrutiny of the activities and material produced in Stage 2.
Stage 3. Development	Detailed design and development of the new product or system along with some product testing, preparation of production and launch plans.
Gate 4. Go to testing	Review of development work to check that the outcome is attractive and consistent with the agreed requirements. Financial analysis undertaken based on more accurate data.
Stage 4. Testing and validation	In house and marketplace tests and trials to validate and verify the new product and determine customer acceptance and the economics of the enterprise. Stage 4 can result in requirements for rework of some aspects and the need to go back to Stage 3.
Gate 5. Go to launch	Decision to go to full commercialisation and market launch, production and operations set-up.
Stage 5. Launch	Commercialisation, full production or full operation of the enterprise. Execution of the production, operations, distribution, quality assurance and postlaunch monitoring. Sales and revenue streams commence.
Postlaunch reviews	Typically at 3–5 months after launch, with initial data, and then again at 12–19 months, once the project is stable.

1.9.6 Agile

Agile project management involves continuous improvement, and enables scope flexibility, team input, and delivery of quality products. Agile project management emerged from a review of best practice approaches across a wide range of project management and methods to overcome some of the limitations associated with these. Methodologies associated with agile project management include scrum, extreme programming (XP) and lean. These methodologies all adhere to the Agile Manifesto and the 12 agile principles. The agile principles (Beck et al., 2001) are a set of guiding concepts that support project teams in implementing agile projects.

1. The highest priority is to satisfy the customer through early and continuous delivery of valuable product.
2. Welcome changing requirements, even late in development. Agile processes harness change for the customer's competitive advantage.
3. Deliver working product frequently (within weeks rather than months, with a preference to the shorter timescale).
4. Daily close cooperation between business people and developers throughout a project.
5. Projects should be built around motivated individuals. Provide the team with the environment, support and trust they need, to get the job done.
6. Face-to-face conversation and colocation provide the most efficient and effective method of conveying information to and within a development team.
7. Working product is the primary measure of progress.
8. Agile processes promote sustainable development. Sponsors, product developers, and users should be able to maintain a constant pace of activity.
9. Continuous attention to technical excellence and good design.
10. Simplicity—the art of maximising the amount of work not done—is essential.
11. The best architectures, requirements and designs emerge from self-organising teams.
12. At regular intervals throughout the project, the team reflects on how to become more effective, then adjusts its behaviour accordingly.

In general a cooperative team of employees and stakeholders is required to successfully complete a project. Agile project teams include the following five roles (Layton and Ostermiller 2017):

- *Development team*: The group of people who do the work of creating a product, such as programmers, testers, designers, writers and anyone who has a hands-on role in product development should be members of the development team.
- *Product owner*: This is the person responsible for bridging the gap between the customer, business stakeholders and the development team. They are an expert on the product and the customer's requirements and priorities. The product owner works with the development team daily to help clarify requirements. They are sometimes referred to as the customer representative.
- *Scrum master*: The person responsible for supporting the development team, clearing organisational roadblocks, and keeping the agile process consistent. A scrum master is sometimes referred to as the project facilitator.
- *Stakeholders*: These are anybody with an interest in the project. Stakeholders are not ultimately responsible for the product, but they provide input and will be affected by the project's outcome. The group of stakeholders will be diverse and can include people from different departments, or even different organisations.

- *Agile mentor*: This is a person who has experience implementing agile projects and can share their experience with the project team. The agile mentor can provide invaluable feedback and advice to new project teams and to project teams that want to perform at an ever higher level.

Project progress needs to be measurable. Agile project teams typically use six principal elements to help develop products and track progress (Layton and Ostermiller 2017), as listed here.

1. *Product vision statement*: This is an elevator pitch, or a quick summary, to communicate how the product supports the organisation's strategies. The product vision statement must clearly articulate the goals for the product.
2. *Product backlog*: The full list of what is within scope for the project, ordered by priority. Once you have the first requirement, a product backlog is generated.
3. *Product roadmap*: The product roadmap is a high-level view of the product requirements, with a loose time frame for when these requirements will be developed.
4. *Release plan*: This is a high-level timetable for the release of working software or product.
5. *Sprint backlog*: The goal, user stories and tasks associated with a sprint that is in progress.
6. *Increment*: The working product functionality at the end of each sprint.

The Agile approach uses a Roadmap to Value to provide a high-level view of a project. The stages of the Roadmap to Value are as follows (Layton and Ostermiller 2017).

- *Stage 1*: The product owner identifies the product vision. The product vision is a definition of what the product is, how it will support the company or organisation's strategy and who will use the product. On longer projects, recommended practice is to revisit the product vision at least once a year.
- *Stage 2*: The product owner creates a product roadmap. The product roadmap is a high-level view of the product requirements, with a loose estimate for the time frame when these requirements will be developed. This stage can be achieved quickly, say within a day.
- *Stage 3*: The product owner creates a release plan. The release plan identifies a high-level timetable for the release of working software or product. An agile project will have many releases, with the highest-priority features being launched first. A typical release includes three-to-five sprints. A release plan should be produced at the beginning of each release.
- *Stage 4*: The product owner, the master and the development team plan sprints. These are also known as iterations. A sprint represents a short cycle of development, where the team produces potentially shippable product functionality. Sprints typically last between 1 and 4 weeks, but can take as little time as a day. Sprint best practice indicates that they should remain the same length throughout the whole project. Sprint planning sessions take place at the start of each sprint, where the scrum team determines what requirements will be in the upcoming iteration and commits to a specific sprint goal.
- *Stage 5*: During each sprint, the development team has daily meetings known as a scrum. This is an opportunity for each member of the development team to state what they have completed during the previous day, what they will work on during that day and whether there are any roadblocks.
- *Stage 6*: The team holds a sprint review at the end of each sprint. In a sprint review, the development team demonstrates the working product or product functionality created during the sprint to the product stakeholders.
- *Stage 7*: The team holds a sprint retrospective at the end of each sprint, where the scrum team discusses what went well, what could change and how to make any changes.

1.10 Design reviews

Design reviews can be an important method of auditing a design and may be a formal part of a quality management system (e.g. ISO 9000 and associated standards). A design review brings together representatives of all the people who have an interest in the new product. It should be a critical examination of the design. It will, because it takes time, increase the cost of design, but will also increase the cost effectiveness if the recommendations are put into practice. Participants should therefore include, in addition to the design team, representatives of production, quality, safety, purchasing and of course marketing. The number of reviews will depend on the complexity of the product, but an initial review at the proposal stage and another as the design approaches maturity represents the minimum needed. A design review is an opportunity for chief engineers, project managers, supervisors and associate supervisors, quality managers, and engineers for interfacing components and systems to challenge and scrutinise your designs and the hard data that proves that requirements have been properly met.

The use of design reviews is standard practice in industry and an important stage particularly where complex systems are being developed to establish detailed requirements and solutions that are fit for purpose and practically realisable within the emergent constraints and limitations that are exposed during the detailed exploration of a design challenge. The management of design is a complex activity, as it needs to encourage creativity and ensure motivated staff whilst at the same time allowing work to be controlled and completed, especially when defined timescales and fixed costs are involved. Design reviews are a fundamental part of project management. The following three types of design review are typical within a waterfall project.

- SDR—System Design Reviews
- PDR—Preliminary Design Reviews
- CDR—Critical Design Reviews

1.11 The technology base

The range of existing and well-tried technology is extensive. Many items are available as standard components and can be purchased from specialist suppliers. In addition standard practice and design methodology has been specified for many others. The designer is able to exploit available technology as it is or by combination and adaptation. The scope of mechanical machine elements available to the designer is outlined in Table 1.7. This list is not exhaustive but does give an idea of some of the 'building blocks' available. Many of these machine elements are considered within this book and are identified by shading in the table.

Developing an overview of the available technology base is useful to the designer as it avoids time spent repeating the design process on a technology that has already been developed. In addition knowledge of science and engineering enables a designer to judge what is technically and scientifically feasible so that 'far-fetched' ideas can be objectively ruled out.

Table 1.7 An overview of the scope of machine elements

Energy conversion	Energy transmission	Energy storage	Locating	Friction reduction	Switching	Sealing	Sensors	Miscellaneous mechanisms
Turbomachinery	Gears	Flywheel	Threaded	Rolling element	Clutches	Dynamic seals	Motion	Hinges, pivots
Gas turbine	Spur	Springs	Fasteners	bearings	Square jaw	Mechanical	Dimensional	Linkages
engines	Helical	Helical	Bolts	Deep groove	Multiple	face	Mass	Levers
Rotodynamic	Bevel	Leaf	Nuts and	Cylindrical	serration	Lip ring	Force	Tools
pumps and	Worm	Belleville	lock nuts	roller	Sprag	Bush	Torque	Cutters
compressors	Conformal	Rubber	Grub	Needle roller	Roller	Labyrinth	Power	Shears
Fans	Belts	Spiral	screws	Taper roller	Disk	Brush	Pressure	Drills
Propellers	Flat	Garter	Studs	Angular	Drum	Ferrofluidic	Pitot tubes	Formers
Turbines	Bee	Fluid	Screws	contact	Cone	Rim seals	Static tappings	Grips
Internal	Wedge	accumulator	Washers	Self-aligning	Magnetic	O rings	Manometers	Guides
combustion engines	Round	Gas spring	Nails	Thrust	Synchromesh	Packings	Piezoelectric	Followers
Rotary	Synchronous	Reservoir	Pins	Recirculating	Ratchet and pawl	Piston rings	Sound	Housings
Reciprocating	Chains	Pressure	Cylindrical	ball	Geneva	Static seals	Flow	Frames
Boilers and	Roller	vessel	Taper	Sliding bearings	Mechanisms	Gaskets	Laser doppler	Casings
combustors	Leaf	Solid mass	Spring	Plain rubbing	Valves	O rings	Hot wire	Enclosures
Electric motors	Conveyor	Chemical	Rivets	Hydrodynamic	Latches	Gaskets	Ultrasonic	Sprayers
Alternators and	Silent	Charge	Tolerance rings	Hydrostatic	Triggers	Sealants	Level	Shutters
generators	Cables and ropes	Torsion bar	Expanding	Hydrodynamic	Bimetallic strips		Humidity	Hooks
Solenoids	Couplings		bolts	slideway			Temperature	Pulleys
Pneumatic and	Rigid		Keys	Wheels and rollers			Thermocouples	Handles
hydraulic	Flexible		Flat, round	Brushes			Resistance	Rollers and
actuators	Universal		Profiled				thermometers	drums
							Pyrometers	

Brakes Disk Drum and band Pumps Rockets Heat exchangers Guns Dampers and shock absorbers Fuel cells	Cranks Cams Power screws and Threads Levers and linkages Pipes, hoses and ducts Ball screws		Gib head Woodruff Splines Circlips Snap rings Clamps Retaining rings Shoulders Spacers Grooves Fits Clearance Transition Interference Adhesives Welds				Heat Flux Thermopile Gardon gauges Strain and stress Time Chemical composition	Centrifuges Filters

After private communication: C. McMahon, Bristol University, 1997

1.12 Conclusions

Design can be considered to be the process of conceiving, developing and realising products, artefacts, processes, systems, services and experiences with the aim of fulfilling identified or perceived needs or desires typically working within defined or negotiated constraints. Engineering involves the application of scientific and mathematic principles in combination with professional and domain knowledge, to design, develop and deliver artefacts, products and systems to realise a societal, commercial or organisation requirement or opportunity. This chapter has briefly reviewed a range of design and management processes relevant to both engineering and design domains. The models presented provide an indication of process, although a given individual or organisation is likely to develop their own bespoke approach relevant to their specific requirements or context.

References

ABET. Criteria for accrediting engineering programs. Effective evaluations during the 2010–2011 accreditation cycle. www.abet.org, Accessed 24 November 2012, 2011.

Association for Project Management, 2017. http://www.apm.org.uk/resources/what-is-project-management/. Accessed 22 September 2017.

Beck, K., Grenning, J., Martin, R.C., Beedle, M., Highsmith, J., Mellor, S., van Bennekum, A., Hunt, A., Schwaber, K., Cockburn, A., Jeffries, R., Sutherland, J., Cunningham, W., Kern, J., Thomas, D., Fowler, M., Marick, B. 2001. Principles behind the Agile Manifesto. Agile Alliance, http://agilemanifesto.org/principles.html. Accessed 2 April 2018.

Bell, T.E., and Thayer, T.A. Software requirements: are they really a problem? Proceedings of the 2nd International Conference on Software Engineering. IEEE Computer Society Press, 1976.

Björklund, T.A., 2013. Initial mental representations of design problems: differences between experts and novices. Des. Stud. 34, 135–160.

Boothroyd, G., 1997. Design for manufacture and assembly. In: ASM Handbook. vol. 20. ASM International, Metals Park, OH, pp. 676–686. Material Selection and Design.

Childs, P.R.N., Pennington, M., 2015. Industrial, and innovation design engineering. In: Chakrabarti, A., Lindemann, U. (Eds.), Impact of Design Research on Industrial Practice. Springer, Cham, Switzerland, pp. 133–149.

Clarkson, J., Eckert, C., 2005. Design Process Improvement. A Review of Current Practice. Springer, London.

Cooper, R.G., 1986. Winning at New Products. Wadsworth.

Cooper, R.G., 2011. Winning at New Products, fourth ed. Basic Books.

Cox, G. Cox Review of Creativity in Business: Building on the UK's Strengths. HM Treasury, UK, 2005.

Crawley, E.F. MIT CDIO Report #1. The CDIO Syllabus. A statement of goals for undergraduate engineering education. Department of Aeronautics and Astronautics, Massachusetts Institute of Technology. 2001.

Design Council. The 'double diamond' design process Model, 2007. www.designcouncil.org.uk/en/About-Design/managingdesign/The-Study-of-the-Design-Process/, Accessed December 2007.

Doran, G.T., 1981. There's a S.M.A.R.T. way to write management's goals and objectives. Manag. Rev. 70, 35–36.

Hall, A., Childs, P.R.N., 2009. Innovation design engineering: non-linear progressive education for diverse intakes. In: Clarke, A., Ion, B., McMahon, C., Hogarth, P. (Eds.), Proceedings of the 11th EPDE International Conference.pp. 312–317.

Hall, A., Wuggetzer, I., Mayer, T., and Childs, P.R.N. Future aircraft cabins and design thinking: Optimisation vs. win-win scenarios. Paper ISJPPE-2012-F0077, 4th ISJPPE Conference, Xi'an, China, 2012.

IMechE. Industry 4.0 Report. 2016, https://www.imeche.org/docs/default-source/1-oscar/reports-policy-statements-and-documents/bdo-industry-4-0-report5bdaad8d54216d0c8310ff0100d05193.pdf?sfvrsn=0, Accessed 2 April 2018.

Ion, B., Smith, D., 1996. The Conceptual Design Phase. SEED.

Kroll, N., Schwamborn, D., Becker, K., Rieger, H., Thiele, F. (Eds.), 2009. MEGADESIGN and MegaOpt—German Initiatives for Aerodynamic Simulation and Optimization in Aircraft Design. Results of the closing symposium of the MEGADESIGN and MegaOpt Projects. Springer.

Layton, M.C., Ostermiller, S.J., 2017. Agile Project Management for Dummies, second ed. Wiley.

Leach, S., 1999. Planning Guidelines: A Guide to the Rolls-Royce Planning Process. Rolls-Royce.

McKinsey. Industry 4.0 How to navigate digitization of the manufacturing sector, 2015, https://www.mckinsey.de/files/mck_industry_40_report.pdf, Accessed 2 April 2018.

McMahon, C., 1997. Private Communication. University of Bristol.

METUS, 2012. http://www.id-consult.com/en/metus-software/ [(Accessed 23 June 2012)].

National Academy of Engineering, 2004. The Engineer of 2020: Visions of Engineering in the New Century. National Academies Press, Washington, DC.

Pahl, G., Beitz, W., 1996. Engineering Design: A Systematic Approach, second ed. Springer, London.

Pugh, S., 1990. Total Design. Addison Wesley.

PWC. Industry 4.0: Building the digital enterprise 2016, https://www.pwc.com/gx/en/industries/industries-4.0/landing-page/industry-4.0-building-your-digital-enterprise-april-2016.pdf, Accessed 2 April 2018.

Roy, R., Hinduja, S., Teti, R., 2008. Recent advances in engineering design optimisation: Challenges and future trends. CIRP Ann. Manuf. Technol. 57, 697–715.

Royce, W., 1970. Managing the development of large software systems. Proc. IEEE WESCON 26 (August), 1–9.

Sharing Experience in Engineering Design, 1985. Curriculum for Design—Engineering Undergraduate Courses. SEED.

Sikhwal R.K., and Childs, P.R.N. Product design for mass individualisation for industrial application. International Conference on Industrial Engineering and Engineering Management (IEEM2017), 2017.

Sobieszczanski-Sobieski, J., Barthelemy, J.-F.M. and Giles, G.L. Aerospace engineering design by systematic decomposition and multilevel optimization. 14th Congress of the International Council of the Aeronautical Sciences (ICAS), 1984.

Tennant, G., 2002. Design for Six Sigma. Gower, Hampshire.

Zhu, L., Li, N., Childs, P.R.N., 2018. Light-weighting in aerospace component and system design. Propuls. Power Res. 7, 103–119.

Standards

99/705213 DC. ISO 18334. Timber pallets for materials handling. Quality of new wood pallet assembly.

99/712554 DC. PrEN 13698-1. Pallet product specification. Part 1. Construction specification for 800 mm x 1200 mm wooden pallet.

99/712555 DC. PrEN 13698-2. Pallet product specification. Part 2. Construction specification for 1000 mm x 1200 mm wooden pallet.

British Standards Institution. BS 3726:1978, ISO 1074-1975. Specification for counterbalanced lift trucks—stability—basic tests. 1978.

British Standards Institution. BS 5639-1:1978, ISO 2331-1974. Fork arms for fork lift trucks. Vocabulary for hook-on type form arms. 1978.

British Standards Institution. BS ISO 6780:1988. General purpose flat pallets for through transit of goods. Principal dimensions and tolerances. 1988.

British Standards Institution. BS ISO 509:1996. Pallet trucks. Principal dimensions. 1996.

British Standards Institution. BS EN 1726-1:1999. Safety of industrial trucks. Self-propelled trucks up to and including 10000 kg capacity and industrial tractors with a drawbar pull up to and including 20000 N. General requirements. 1999.

British Standards Institution. BS EN ISO 445:1999. Pallets for materials handling. Vocabulary. 1999.

British Standards Institution. BS EN 13545:2002. Pallet superstructures. Pallet collars. Test methods and performance requirements. 2002.

British Standards Institution. BS ISO 2330:2002. Fork-lift trucks. Fork arms. Technical characteristics and testing. 2002.

Websites

At the time of going to press, the world-wide-web contained useful information relating to this chapter at the following sites.

http://www.bsi.org.uk.

http://www.iso.org.

http://www.axelos.com/best-practice-solutions/prince2.

http://www.creativeproblemsolving.com.

http://www.dummies.com/how-to/content/agile-project-management-for-dummies-cheat-sheet.html.

http://www.dummies.com/how-to/content/project-management-for-dummies-cheat-sheet.html.

http://www.excel-easy.com/examples/gantt-chart.html.

http://www.gantt.com/creating-gantt-charts.htm.

http://www.pmi.org/pmbok-guide-and-standards/pmbok-guide.aspx.

http://www.resourceforyoursource.com/Get%20productive-.pdf.

http://www.smartsheet.com/blog/where-do-you-find-best-gantt-chart-spreadsheet-templates.

http://www.stage-gate.com.

http://www.tutorialspoint.com//management_concepts/traditional_project_management.htm.

http://ww.imperial.ac.uk/designengineering/tools/designvue

http://icreate-project.eu/index.php?t=98.

Further reading

Bak-Maier, M., 2012. Get Productive. Capstone, Oxford.
Baxter, M., 1995. Product Design. Chapman and Hall, New York.
German Directive 250, Software Development Standard for the German Federal Armed Forces, V-Model, Software Lifecycle Process Model, August 1992.
Manzini, E., 1989. The Material of Invention. Design Council, London.
March, L.J., 1976. The Architecture of Form. Cambridge University Press.

Specification

2

Chapter Outline

Abbreviations

PDS product design specification
QFD quality function deployment
VOC voice of the customer

2.1 Introduction

Specification represents an important part of the design process to ensure that a product, service or system is addressing the task required. Specification may be undertaken prior to design or as part of a gated process where some design work is necessary in order to define the scope of the product, artefact or system being considered. This chapter provides an overview of the principles of specification and describes a standard product design specification pro-forma and also quality function deployment, a technique widely used in industry that has been developed in order to ensure that the voice of the customer is addressed.

Design usually starts with a need or requirement which, when satisfied, will allow a product process or system to be offered to an existing or new market. The market need may have arisen to address increased competition, a response to social changes, opportunities to reduce hazards, inconvenience or costs, or exploitation of a new market opportunity. This is sometimes known as the 'front end' of design. People with front-end skills, such as market researchers and information scientists, are fundamental to the design process. The starting point for the majority of design activities in industry is the establishment of the market need situation. Common practise is to produce a document or statement of need, called 'the brief' or 'market brief', at this stage. The brief varies from a simple statement of the requirements to a comprehensive document that describes the criteria that are understood to meet the users' true needs.

Mechanical Design Engineering Handbook. https://doi.org/10.1016/B978-0-08-102367-9.00002-0

Specification is usually critical to effective design. Thorough specification provides limits defining the attributes and functions that a design should have. Common practice is to develop a product design specification (PDS) where each attribute and function required of the product, service or system is described and quantified. An alternative is the use of quality function deployment (QFD), where a series of matrices are used to capture the attributes, functions and features associated with a product or system, with a particular focus on ensuring the voice of the customer is captured in the specification with a view to ensuring that the product development cycle results in an outcome that matches the customer requirement.

There are many variants and forms of specification. Specification can represent a significant undertaking. The specification documentation associated with a new turbofan engine for an aviation application, for example may occupy over 500 pages. The precise form used to capture the specification required will likely depend on precedence within a particular organisation. The automotive industry, for example has tended to use QFD over the last few decades (see, e.g. Miller et al., 2005; Sugumaran et al., 2014), and such approaches may demand more resource than available and simpler pro-forma may enable a more timely and accessible approach to specification.

Specification is usually incorporated as a fundamental part of new product development project management. It is not always possible to produce a complete specification at the outset of a project. It may be necessary to take an iterative approach to specification with elaboration of details as a result of initial phases of design activity and concept ideas, with these providing insights to what attributes and functions for a product or system are possible and desirable. In essence, the early concepts and designs serve to inform the specification. Such an approach fits well with stage-gated project management processes, perhaps with phases of system design review, performance design review and technical design review with each referring back to an original specification and enabling the development of an agreed, technically feasible specification that provides a design that is fit for purpose. A common occurrence in industry is the use of an agreed specification to provide the basis for contractual negotiation enabling the associated parties to be able define completion of tasks. In general, attention to specification helps reduce the need for late changes in a project (Simpson 2015). Any variation being made to a specification needs to be managed by a change management process where changes to the specification need to be agreed, or where agreement cannot readily be obtained, these must be managed by an arbitration process.

On completion of the specification, conceptual design can proceed. This tends to be the most creative part of the design process where solutions to the specification are given form. Having generated the conceptual solutions, the next step is to express these so that they are communicable to all involved in the design process, perhaps following the total design process illustrated in Fig. 1.7. In practice, this may take the form of a drawn scheme or three-dimensional model, which may be physical or computer-generated. By the time the general scheme has been completed and calculations and analysis undertaken to determine the solution's compatibility to the PDS, the basis will have been established for the detailed design phase to commence.

Specification using a pro-forma table is considered in Section 2.2. QFD is described in Section 2.3.

2.2 Product design specification

Specification involves the formal statement of the required functions, features and performance of the product or process to be designed. One approach from the outset of design work is to produce a PDS that should be formulated from the statement of need. The PDS is the formal specification of the product to be designed. It acts as the control for the total design activity because it sets the boundaries for the subsequent design.

Specification has different meanings depending on the context and the individuals and organisation concerned. Here, specification refers to an activity that assists in providing a framework within which design can take place. Work on a design may commence with an identification of a need, which is sometimes summarised in the form of a statement commonly known as the 'design brief' or 'design intent'. A design brief may provide some direction to initial design activity but rarely sets sufficient bounds on it to allow a design team to know when their efforts are adequate, or not, to fulfil the requirements. In order to overcome this shortcoming and to compel the individuals concerned to explore what is actually required of the product or process concerned, a 'product design specification' (PDS) should be developed.

In developing a PDS, bounds should be set on a wide range of relevant parameters for the product or process concerned. The spider chart shown in Fig. 2.1 provides a helpful checklist of aspects for consideration using an updated and revised list originally developed by Pugh (1990). Each of these items can be considered in turn, or alternatively those relevant pulled out and focused on. A useful rule associated with specification, which will be stated frequently, is the mantra:

quantify wherever possible.

Having identified a general need or requirement and documented this in the form of a brief, the specification of the product to be developed should be generated. This is called the product design specification (PDS). The PDS acts as the controlling mechanism, mantle or envelope for the total design activity. Whatever you are concerned with, the PDS gives you the terms of reference. The starting point for design activity is market research, analysis of the competition, literature and patent searching. Once this has been undertaken, the PDS can be developed. Fig. 2.2 shows some areas of information and research required to produce a PDS and Tables 2.1 and 2.2 show a couple of alternatives for the generic information content of a typical PDS.

In a new design, it is preferable to consider each aspect identified in Fig. 2.1, however for the re-engineering of a current product, engineering expediency dictates that some aspects will be passed over. The PDS thus consists of a document covering a wide range of considerations. The PDS is usually dynamic. If during the design of a product there is a good reason for changing the basic PDS, this can be done. It should be noted, however, that the PDS may actually form part of contractual obligations and the legal implications must be addressed by an agreed change management process.

Undertaking a PDS will typically be an iterative process where initial parameters are identified and limits set but more information may be required in order to confirm these. The process of developing a PDS will require input from people with a variety of skills and knowledge. It will very rarely be simply 1 day's work and may extend to

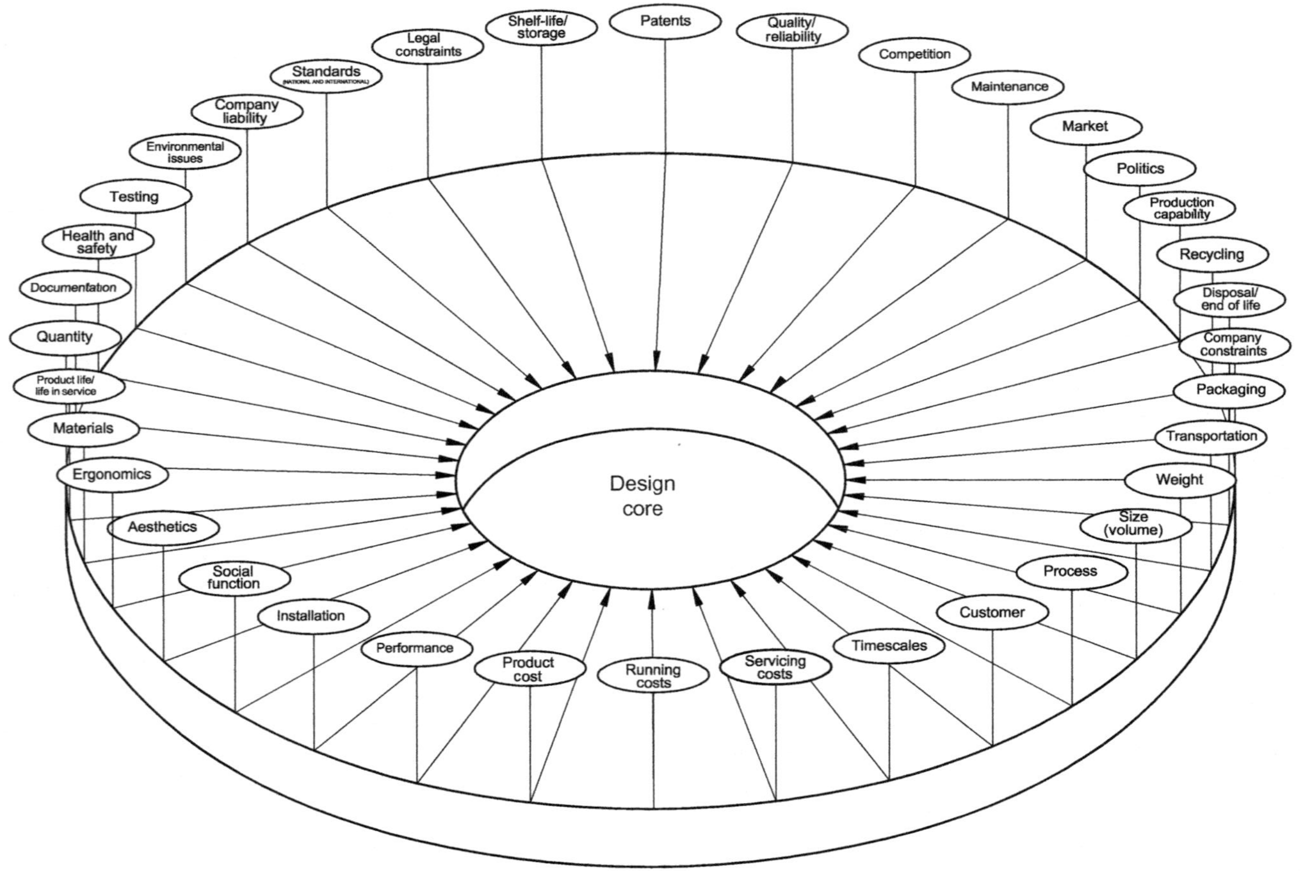

Fig. 2.1 Information content of a typical product design specification. List originally developed by Pugh (1990).

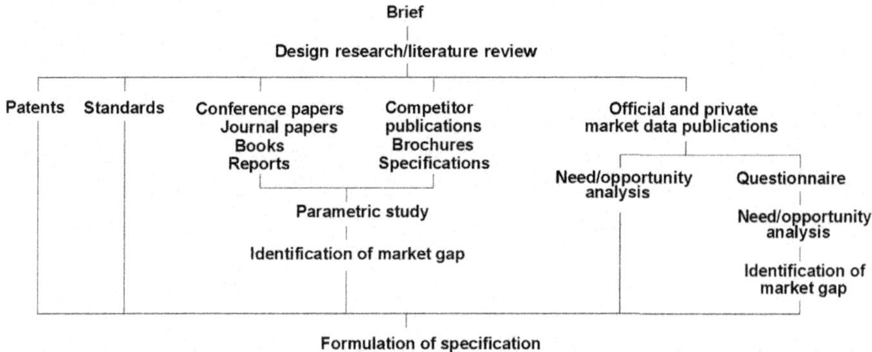

Fig. 2.2 Information required for the production of the product design specification.

Table 2.1 Format for the documentation of a product design specification (all relevant aspects identified in Fig. 2.1 should be considered)

Date:....................Product:...Issue:..............			
Aspect	**Objective**	**Criteria**	**Test conditions**
Aspect 1 Aspect 2			

After Pugh, S., 1990. Total Design. *Addison Wesley, New York.*

Table 2.2 Alternative format for the documentation of a product design specification

Date:....................Product:...Issue:..............				
	Parameters			
	Competition best	**Current model (ours)**	**This design (intent)**	**World class (target)**
Performance Description Safety Description				

After Pugh, S., 1990. Total Design. *Addison Wesley, New York.*

occupying many person-months of effort. Recall that conceptual design, let alone detailed design, cannot sensibly proceed towards any firm status until the specification has been completed.

Example 2.1. After a demonstration by competitive companies at a national ski exposition, the managing director of a recreational equipment company has outlined a requirement for a snow sports activity vehicle. The director has marched into your office and asked you to develop an initial PDS for this, by the end of the day. Although you know that doing this job fully will take orders of magnitude more time and resource than that available, outline the overall requirement, in a short paragraph and using a standard pro-forma for the PDS, develop the PDS.

Solution

A vehicle for snow sports is required to address market opportunities associated with the ski-market and winter sports activities. The vehicle should be associated with high-quality performance and manufacture, be reliable and be capable of operation across a wide range of winter conditions. The vehicle should be capable of a series of sports-associated manoeuvres such as turns, accelerations and leaps. The vehicle should have reasonable range and been capable of carrying up to two 95 percentile adults. An outline specification is given in Table 2.3.

Example 2.2. A small and medium enterprise (SME) company has investigated the potential market for a small electric motor driven air compressor. The market analysis has confirmed the requirement for an 80 m^3/h, 7 bar air supply driven by a 5 kW electric motor. Preliminary design has already resulted in the selection of an electric motor, with a synchronous speed of 2850 rpm, and the piston and cylinder and associated components for a reciprocating compressor. The compressor design speed is 1500 rpm. A schematic for the compressor concept is given in Fig. 2.3. The choice of transmission and its detailed design is yet to be undertaken. The task required here is to develop the specification for the transmission for this compressor.

Solution

The brief given outlines the general requirement for a transmission for a reciprocating compressor. Not all the information necessary to develop the design is given in the brief and must be developed by means of liaison with the company and application of design skills and experience.

A specification for the transmission is developed as follows:

1. The transmission must transmit at least 5 kW.
2. The input is from an electric motor with a synchronous speed of approximately 2850 rpm. Details of the electric motor are not given in the brief. Contact with the company has identified the critical interface motor dimensions.
3. The output from the transmission is to a reciprocating compressor running at 1500 rpm. The load is reciprocating and therefore constitutes medium-to-light shock.
4. Following contact with the preliminary design team, dimensions for the compressor shaft have been provided.
5. The preliminary design team has confirmed that the electric motor torque-speed characteristic is sufficient to start the compressor, provided the transmission gives the required ratio with reasonable efficiency.
6. A mechanical efficiency for the transmission has not been specified. However, a mechanical efficiency of at least 95% should be achievable and is proposed for the specification.
7. The means of connecting the motor and compressor to the transmission have not been detailed in the brief. Contact with the SME has confirmed that the connection of the transmission to the compressor and electric motors forms part of the brief. Flexible couplings permit some degree of misalignment and isolation of vibration and are therefore specified for connecting the input and output shafts of the transmission to the motor and compressor.

Table 2.3 Initial product design specification for the snow sports activity vehicle

Aspect	Objective	Criteria	Test conditions
Performance	Operation at low temperature	Assume arctic conditions. Operation at temperatures between −40°C and 25°C	Test components at extreme temperatures
	High speed	60 mph maximum	Rolling road test or dynamometer test
	High acceleration	Limit to the ability of an average person to hold on to and control vehicle	Test existing products and identify maximum acceleration suitable. Test prototype on rolling road or dynamometer
	High manoeuvrability	<5 m turning circle	Test model
	Easy steering	Steering must be responsive to driver intent and provide adequate feedback of driving conditions	Use a focus group to evaluate steering on a prototype model
	Easy starting	Push button start	Expose computer-aided design (CAD) model to a focus group for assessment. Expose a prototype model of starting to a focus group
	High range	2 h operation or 120 miles	Dynamometer or rolling road test
	Good braking	Must be capable of braking from 30 kph in 30 m on fresh snow	Rolling road test
Accommodation	Two people	Driver plus one passenger	Test accommodation of two people in the product using digital mannequins in the CAD model. Use a focus group to evaluate the layout and accommodation using a prototype mock-up of the design

Continued

Table 2.3 Continued

Aspect	Objective	Criteria	Test conditions
Cost	Acceptable cost for market sector	The works cost price must not exceed £3000	Obtain quotations for constituent components from Original Equipment Manufacturers (OEMs) and estimates for in-house manufactured components and assembly costs
Size	Small	Must be able to transport device on a small trailer	Determine size of concept from CAD model. Test concept on a CAD trailer model
Noise	Low	Noise levels must not exceed national standard limits	Test to national standards
Materials	Appropriate to design	Compatible with low temperatures and fuel (fire retardant, see standards)	Test to national standards
Weight	Low	One person must be able to move the device. Two people must be able to lift the device. Mass not to exceed 200 kg	Determine weight during design from CAD model. Weigh prototype components
Safety	High	Power plant must cut out when driver loses control	Bench test
		Materials must be fire retardant to national standards	Bench test
Fuel	Readily available	Any readily available fuel such as petrol, diesel, kerosene or electricity	Must comply with BS/ISO standards. Check fuel readily available from retailers and check no legal changes to fuel legislation is imminent

Table 2.3 Continued

Aspect	Objective	Criteria	Test conditions
Propulsion	Available technology	Any available technology power plant	Dynamometer test. Check ready availability of power plant from suppliers and check no legal changes to power plants are imminent
Aesthetics	Attractive styling	Current trend/modern sports style	Expose concept design ideas to focus groups for appraisal. Expose prototypes to focus groups for appraisal
Life	Maximise	The warrantee life of the machine should exceed 500 h of use	Test design under accelerated stress conditions
Manufacture	Ensure ready supply of components	Use stock machine elements where possible Design using manufacturing techniques that are readily available from a number of local plants	Identify components in stock supply inventories Use established suppliers
Design time	Minimise	Detailed design must take less than a maximum of 6 elapsed months	Use standard project management methods to track progress and rectify excursions from the plan

8. Following discussion with the company and examination of the market analysis, compact size for the compressor and therefore the transmission is desirable. Limits have not been identified in the market analysis.
9. The life of the compressor is not defined in the brief. Following discussion with the SME business team, the following information has been identified. The compressor is expected to operate for a maximum of 2 h a day on 200 days of the year. An 18-month trouble-free warrantee life is desirable.
10. The compressor will be enclosed to prevent casual contact with hot or moving components. However, the transmission should include means to support its functional components and retain all lubricants if present.
11. The transmission will operate in an internal environment at temperatures between −10°C and 50°C. The design should be tolerant to dust and high humidity.
12. Up to 1000 units, a year will be manufactured.

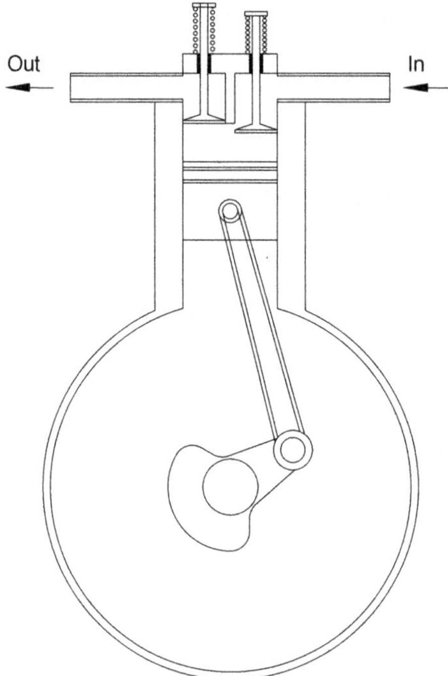

Fig. 2.3 Schematic for a 7 bar, 80 m^3/h reciprocating compressor (Childs, 2004).

13. The cost of the transmission should be minimised. A maximum limit for the works cost price has been set as 600 Euros.
14. The weight is not critical but very high weights are deemed undesirable. A maximum mass for the transmission has been set by the SME at 30 kg.
15. Ideally the transmission should use readily available components.
16. Unique transmission components should be manufactured locally to the SME in order to ensure reliability of parts supply.
17. The design should take a maximum of 6 man-months and be undertaken over a 2-month period.
18. Although aesthetics are not a primary factor in the design, the design should augment the overall aesthetics of the overall product.
19. The transmission should comply with all relevant safety criteria and legislation.
20. The design should not violate any existing patents in order to protect against royalty payments.
21. The design should provide recommendations for future recycling of components in order to comply with expected legislation on sustainability.
22. The transmission must be robust enough for transportation on standard pallets by road.

The aspects of the specification listed do not yet provide sufficient bounds in order to enable conceptual or detailed design to proceed sensibly at this stage. Use of a pro-forma such as those given in Tables 2.1 and 2.2 might be helpful in order to further develop definite limits on the design parameters in order to ensure that the work effort is being sensibly directed towards the product need defined by the marketing analysis referred to in the brief. The PDS is developed in Table 2.4.

Table 2.4 Product design specification for the compressor transmission

Aspect	Objective	Criteria	Test conditions
Performance	Transmit torque	Transmission must transmit 5 kW at 2850 rpm	Test transmission on a dynamometer
	Speed ratio	The transmission ratio between driving and driven shafts should be 2850/1500 = 1.9	If positive drive is employed, then assume specified ratio; or if a positive drive is not used, measure speed of output shaft at design conditions
	Efficiency	At least 95%	Dynamometer test
	Drive	Must connect to specified 1.5″ driving shaft and 1.5″ driven shaft using flexible couplings	Use standard coupling methods that are rated for the maximum torque and speed with a factor of safety suitable for the application
	Temperature	Operation in an internal environment at temperatures between −10°C and 50°C	Bench test prototype
	Humidity	Operate in an environment with a humidity of 100% at 20°C	Bench test prototype
Size	Minimise	Must be smaller or similar to competition. Competition best transmission fits within an envelope space of $120 \times 250 \times 250$ mm	Measure
Cost	Minimise cost	The cost for the transmission per unit should be <600 Euros assuming 1000 units per year	Calculate works cost price
Life	Maximise	The warrantee life of the machine should exceed $2850 \times 60 \times 2 \times 200 \times 1.5 = 1.02 \times 10^8$ cycles	Test design under accelerated stress conditions
Weight	Minimise	Must be <30 kg	Weigh it
Manufacture	Ensure ready supply of components	Use stock machine elements where possible	Identify components in stock supply inventories
		Design using manufacturing techniques that are readily available from a number of local plants	Use established suppliers
Design time	Minimise	Detailed design of the transmission must take less than a maximum of 6 man-months and occur over a 2-month period	Use standard project management methods to track progress and rectify excursions from the plan
Patents	Avoid	Do not infringe existing patents	Patent search

Continued

Table 2.4 Continued

Aspect	Objective	Criteria	Test conditions
Transport	Robust design	The transmission must be robust enough for transportation on standard pallets by road	Vibration test
Aesthetics	Attractive design	Design should augment the overall aesthetics of the overall product	Use focus group on concept graphics model
Safety	High	Must comply with existing and envisaged legislation	Check national and international standards and relevant legislation
	Noise	Must be less than legal limits	Measure
Recycling	Compliant	The design should provide recommendations for future recycling of components in order to comply with expected legislation on sustainability	Check national and international standards and legislation
Maintenance	Easy	Provide access to components than might fail to enable their replacement	Use graphics model of design to verify maintenance access

As part of the specification, it is also important to identify the evaluation criteria before the design process proceeds. These may be altered as the design proceeds if aspects of the evaluation are deemed inappropriate. Evaluation criteria are necessary because it is possible that a number of competing designs may be identified, all of which meet the specification. The evaluation criteria can be used to determine which of the design solutions should be adopted. Here, the evaluation criteria that seem appropriate at this stage include the following:

- cost,
- safety,
- size,
- reliability,
- maintenance,
- noise and vibration.

2.3 Quality function deployment

QFD, also known as the voice of the customer, is an approach widely used in industry to help a company achieve the goal of high-quality products, services or systems. By focussing on customer satisfaction, QFD provides a means of translating customer requirements into appropriate technical requirements for each stage of the product development process. The process arose originally in Japan and has been widely applied in the automotive sector as well as is industries with an attention on continuous process improvement (see Akao, 2004; Akao and Mazur, 2003; Cohen, 1995; Ju and Sohn, 2015).

Quality in this context is (Aurisicchio 2013)

- defined by the customer;
- problem avoidance and
- aspects that are value for money to the customer.

Quality is not

- inspection alone (i.e. problem detection);
- delivered by the manufacturing stage alone and
- achieving high standards of manufacture without direct link to a proven customer requirement.

QFD is a method to transform customer requirements into design quality, to deploy the functions forming quality and to deploy methods for achieving the design quality into subsystems and component parts, and ultimately to specify elements of the manufacturing process.

There are four phases associated with QFD:

- Product Planning (QFD1 or Phase I);
- Design Deployment (QFD2 or Phase II);
- Process Planning (QFD3 or Phase III);
- Production Planning (QFD4 or Phase IV).

The relationships between the phases are illustrated schematically in Fig. 2.4.

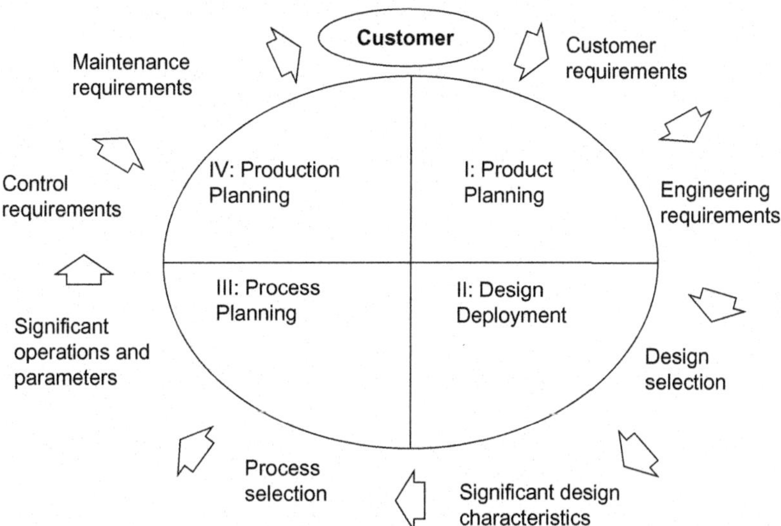

Fig. 2.4 QFD phases of the product delivery process.

Each of the phases of QFD relates to ensuring development of the specification and its subsequent implementation with a product development cycle. The characteristic activities with each of the phases are outlined in Table 2.5.

The QFD cycle of phases starts and ends at the customer. It is the customer who defines the quality standard in terms of their requirements, and it is the customer who ultimately determines if the product meets these requirements.

Note that the four phases of QFD are sequential in completion, but can be conducted in parallel. Process planning (QFD3) and design deployment (QFD2) must normally be conducted in parallel. Product planning (QFD1) should in principle be completed before design deployment (QFD2) commences, see Fig. 2.5.

Table 2.5 Characteristic activities associated with each of the QFD phases

QFD phase	Activity
QFD1/Phase I: Product Planning	Listen to and capture the *Voice of the Customer*. Interpret the customer's requirements. Translate the spoken and unspoken customer requirements into engineering requirements. Document these in a product design specification using a QFD template.
QFD2/Phase II: Design Deployment	Develop and select designs. Identify significant design characteristics (the specific assemblies, sub-assemblies, parts, dimensions, properties and attributes of the designs considered that directly relate to achievement of customer requirements). Demonstrate that the design fulfils the requirements of the specification (by applying the appropriate verification plan—calculations, simulations, tests, etc.).
QFD3/Phase III: Process Planning	Develop and select manufacturing processes. Identify significant process operations and parameters (the specific operations, e.g. critical assembly steps, and parameters, e.g. feed rates, temperature of operations and processes that directly relate to achievement of significant design characteristics).
QFD4/Phase IV: Production Planning	Implement the manufacturing processes of Phase III in the production environment (i.e. design, build, install and prove out the required tooling, machinery, fixtures, jigs, inspection equipment for the selected manufacturing processes). Develop the control requirements (i.e. the factors that should be inspected and monitored, in order to adequately control the significant process operations and parameters). Define maintenance requirements (i.e. prepare the maintenance regime for tooling, machinery, etc., to deliver appropriate availability and function).

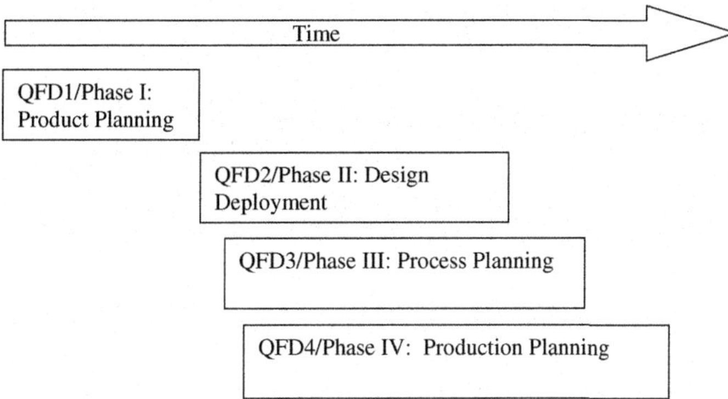

Fig. 2.5 QFD phase timeline.

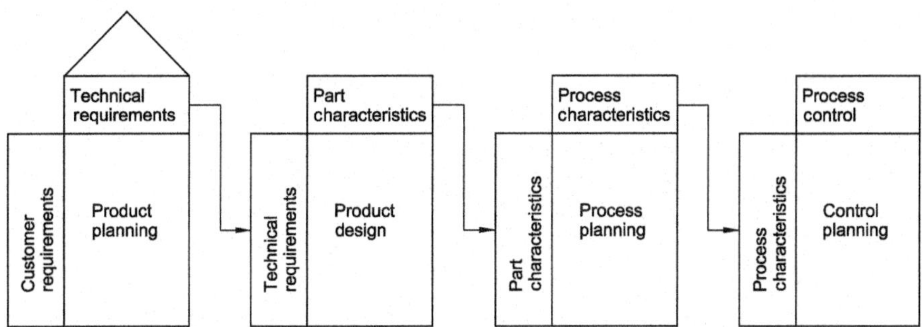

Fig. 2.6 Generic QFD matrices.

QFD uses matrix diagrams to assess the relationship between two sets of elements, for example two sets of requirements or a set of requirements and a set of solution concepts. The method is implemented by filling in four matrices as shown in Fig. 2.6 (see, e.g., Chan and Wu, 2002). In forming each of the matrices, the engineering design team needs to consider and define key requirements and the characteristics associated with these (see Bouchereau. and Rowlands, 2000; Franceschini et al., 2015). In this chapter, the focus is on QFD1 and QFD2 matrices, product planning and product design respectively.

2.3.1 QFD1 and QFD2 matrices

A QFD1 matrix helps capture and structure requirements during engineering projects. It is a systematic approach to translate customer requirements (or product attributes) into engineering requirements. The QFD1 method is especially useful when the opportunity for commercial success is significant and there is a need for rigour in mapping the project requirements. QFD1, compared to use of a standard PDS pro-forma table, represents a more structured approach to requirement management. The QFD1 matrix allows an individual or a team to identify customer and engineering requirements and

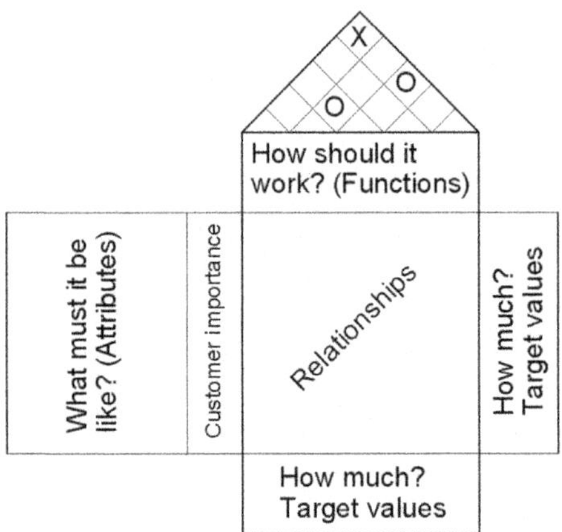

Fig. 2.7 Generic example of a QFD1 matrix. o, synergy; x, conflict.

set the relationships between these two groups of requirements. A generic QFD1 matrix is illustrated in Fig. 2.7. This provides an understanding of the correlation between customer requirements and functions.

The QFD matrices follow a prescribed format and process:

1. Fill in the rows;
2. Fill in the columns.

- Customer and engineering requirements are captured in the rows and columns of the matrix, respectively.
- The importance of the customer requirements (product attributes) is specified in the column after the customer requirements.
- Quantitative values for the requirements are captured in the 'How Much' cells at the end of rows and columns, see Fig. 2.7.
- The roof of a QFD matrix provides an understanding of the interactions and conflicts between the functions.
- It is noteworthy that functions have to be captured using the verb–noun format.

Example entries for a partially developed QFD1 table for a cordless lawn and garden hand tool with a few row and column entries are given in Fig. 2.8. Details of how to complete the matrix are considered in Example 2.3.

A QFD2 matrix represents a systematic approach to translate engineering requirements (or product functions) into product parts. It provides an understanding of the correlation between functions and product parts. The QFD2 method is especially useful when there is need for a rigorous mapping between product functions and parts. A generic QFD2 matrix is illustrated in Fig. 2.9. There are three principal steps for completing a QFD2 matrix.

1. Fill in the rows—Start QFD2 by asking what the product must do and how much.
1. Fill in the columns—Continue QFD2 by asking 'how does it do' what 'it must do'.
2. Establish the relationships—Using a numeric notation, identify the relationships between what is must do and how it does this.

Attributes	Customer importance	Engineering requirements: Loosening earth	Removing slug of earth	Holding bulb	Inserting bulb	Target values
Sufficient torque for producing hole	40%	9	1			> 4 N m
Price associated with a quality product	20%	4		4	4	<60 Euros
Long battery use period	30%	4	4			Battery must last > 1 h in normal use
Short battery charge period	10%	1	1			<1 h
How much		>750 cm^3; diameter 75 mm; depth 100 mm	> 400 cm^3 earth	Max diameter 60 mm; Max length 90 mm	1 bulb; upright; max depth 90 mm	

Fig. 2.8 Example entries for a partially developed QFD1 matrix.

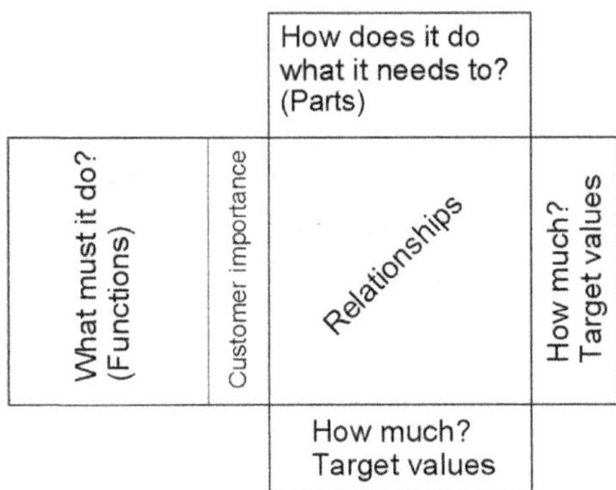

Fig. 2.9 Generic QFD2 matrix.

Details of how to complete the matrix are considered in Examples 2.3 and 2.4.

QFD with its emphasis on rigorous requirement management before solution development represents a step change from traditional way of approaching product development processes, which tended to develop solutions and commit to designs early in the process. Designers and engineers would not have ignored the voice of the customer completely; but by choosing designs too early, there was an increased risk that the solution offered did not fully meet requirements. A product specification may have been written, but it may have been a retrospective record of the finished design. In this way, products would go into production and be delivered to the customer with relatively high probability of failing to actually meet customer requirements, that is be of low quality, not necessarily because they were manufactured badly but because the design did not provide the functions that the customer required. This 'traditional' process versus QFD is illustrated in Fig. 2.10.

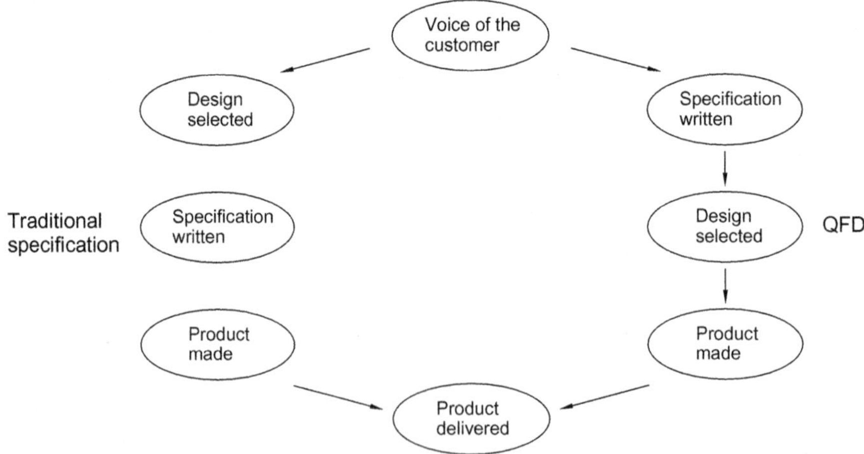

Fig. 2.10 QFD versus a traditional industrial approach.

Example 2.3. In the 1970s, corrosion of car bodies and frames was widespread around the world. In many markets, you could have expected many makes of car to have serious body and chassis rust, within a few years after manufacture. Toyota Auto Body used QFD to resolve this problem in their new vehicle design programmes (Eureka 1987).

The voice of the customer was already loud and clear, and car companies were aware of the problem, but were consistently failing to improve their designs. The reason for this was that they were following a 'traditional' engineering approach. 'Engineering' units within each company knew the problem existed, and no doubt had a strong desire to fix it. The traditional approach, however, led to early selection of a preferred design, which usually meant using similar body construction designs and techniques as used on previous models. Attempts to address corrosion then became a 'sticky plaster' approach, as an afterthought late in the development cycle. Perhaps more paint was specified or more sealer, but the rust problem remained when the product was delivered to the customer. By using QFD, the problem was firmly recognised at all levels in the company, including high-level management. As a result, the focus, discipline and resources needed to solve the problem were generated and applied. The insights from the QFD phases relevant to this particular case are considered in each of the following subheadings.

QFD1/Phase I: Product Planning

- The problem was clearly understood in customer terms (e.g. what level of corrosion, if any, was acceptable within what time period);
- The issue was written into the PDS clearly, with both its importance to the customer, and specific targets identified (e.g. types and locations of corrosion that could be accepted, after what period of use);
- The requirement was embraced by senior management, who then had the commitment to follow through the rest of the process.

QFD2/Phase II: Design Deployment

- Design concepts had to consider the requirement to address corrosion from the very beginning of design activity;
- Design deployment refers to all aspects of design activity in the product development process. It may be that the only way corrosion could be eliminated from designs was by iterations of hardware testing (e.g. both corrosion chamber testing of body components, and full vehicle testing). This would likely be extremely expensive, but the senior management understanding the importance of the requirement from Phase I means that resources could be justified and
- Significant design characteristics need to identified, for example material selection, body section/panel designs, paint type and thickness, and metal preparation parameters.

QFD3/Phase III: Process Planning

- In parallel, the design and process engineers developed the body designs and manufacturing processes (a two way activity with designs affecting process selection, and vice versa), for example material selection, metal forming, metal cleaning, coatings, joining, sealing and painting;
- The significant process operations and parameters were defined, for example cleaning process durations, cleaning solution concentrations/temperatures.

QFD4/Phase IV: Production Planning

- The control requirements need to be defined, for example paint thickness inspection, sealer bead dimensions, sealant curing hardness and
- Maintenance requirements need to be defined, for example cleaning solution changeover intervals, spray gun cleaning schedules.

Toyota may not have fixed the problem of rust completely in one new model; but over a series of model releases, the corrosion performance improved, and other automotive manufactures took note and amended their processes in response—today as customers we expect minimal, if any, body corrosion in the first 6 or more years of use.

One of the fundamental principles of QFD is that every decision and selection made in the whole product development cycle can be referred back to the customer requirements. From design concepts right through to what things are inspected and controlled in manufacture all relate back to customer requirements.

Example 2.4. An application of the QFD method to the design of a hand-held hairdryer product is explored here. As there are many hair dryers in the market, the aim is to provide a quality solution that has added attributes in comparison to the many 'me-too' competing products.

The planning and design of a hairdryer are documented in the QFD1 and QFD2 matrices in Figs. 2.11 and 2.12.

The QFD 1 matrix enables the relationship between the following two sets of elements, customer requirements (or product attributes) and engineering requirements (or product functions), to be explored. Customer requirements are captured in the rows of the matrix. In this example, the customer requirements are broken down in a tree structure including three levels. The root concepts are safety, performance, market competitiveness and ergonomics, see Fig. 2.11. Safety, for example is subdivided into low risk to user, and fire protection, overheating protection and protection from burning. The last column of the matrix captures the 'How Much' values for the customer requirements. Protection from burning, for example has a corresponding 'How much' cell where it is noted that the maximum allowable temperature for the scalp is 60°C. Another important feature of QFD1 is the column marked in grey where the importance of the requirements is set based on discussions with customers. In this example, the importance is defined distributing 100 points between the 21 customer requirements.

The engineering requirements are captured in the columns of the matrix. Removing moisture from hair, for example is one of the main functions of a hairdryer. The last row of the matrix captures the 'How much' values for the functions. The function removing moisture from hair, for example has a corresponding 'How much' cell where it is noted that the air flow requirement must meet or exceed a defined quantity.

The elements in the rows and columns of a QFD matrix can have relationships of different strength. In QFD, three types of relationships are generally used:

- low (1);
- medium (4) and
- high (9).

			Importance	Draws air in	Contains rear air flow	Removes debris from air flow	Locates debris filter	Supplies front air flow	Heats front air flow	Guides front air flow	Accelerates front air flow	Lowers static p of front air flow to atmospheric	Insulates front housing	Dries hair	Positions front housing	Positions insulator	Offers holdability	Locates AC neutral supply	Acquires AC voltage	Allows power control	Allows cooling control or hot shot	Guides switch assembly	Secures most handle and body parts	Allows the device to be stored	Allows the cord to be stowed	Positions spring	Positions cool shot switch bracket	Positions cord	Protects cord	How much
Safety	Low risk to user	Fire protection	3.8			9			9									4	4				1						4	Compatible BS/ISO standards numbers ...
		Overheating protection	3.8			9			9									4	4											Compatible BS/ISO standards numbers ...
		Burning protection	3.8						9				9	9		4														Max T (scalp): 60°C
Performance	Quality	Positive switch positions	8.9																	4	4									Switch must engage positively for 99.99% of attempts
		Gentle and even heating	7.9						9																					$h < 120$ W/m^2 K
	Efficiency	Dries hair quickly	1.5						9																					Power consumption: 1875 W
	Reliability	Seldom breaks down (reliable)	7.8					9	9									4	4											Last for 2000 h of drying time
	Robustness	Withstands rough treatment	10.5		9			9	9							9		4	4				9			4	4	4		Resists a drop test from 1.5 m
Market competitiveness	Appearance	Pleasing colour-way	7.7																											>80% of focus group must acclaim colour-way
		Pleasing shape/form/aesthetic	7		9		9								4	9	9													>80% of focus group must acclaim form
	Price	Justifiable price	6		4		4	4	9						4	4														<£400GBP, >£50GBP
Ergonomics design	Usability	Easy storage	3.2																					9	9					Must fit in 0.3 m by 0.2 m by 0.15 m storage volume
		Easy use at distance from the mains	3															9	9											Cord > 3 m
		Low noise	4					9																						Max noise: 72 dB
	Easy controls	Easy one hand use	2.1														9			9	9	4								80% of focus groups to acclaim single handed use
		Choice of air temperature setting	1.5						4												9	1								2 settings; max T (air): 55°C
		Choice of air speed setting	0.8																	9		1								Min 2 speeds; max speed: 16.44 m/s
	Comfortable to hold	Comfortable hand grip	3.8														9			4	4									80% of focus groups to acclaim hand grip
		Well balanced	3.5					9	4								9													Hand moment < 0.3 N m
	Size and weight	Light weight	7.8	9	4	4		9	9	4	4	4	4				9	4	4	4	4									<0.5 kg
		Appropriate size	1.6		9										9	9	9													< 220 mm L × 200 mm H × 80 mm W
		How much	100	MFR (kg/s): 0.05	Leakage minimal (<0.2)	Screen apertures: 1 mm	+/− 0.2 mm on axial	VFR (L/min): 2685	Max T (air): 55°C	Within +/−7 mm of axis	Max air speed > 16.44	101325 Pa +/− 1000 Pa	Casing T < 25°C	Mositure content < 0.1 g	+/− 0.2 mm on axial	+/− 0.2 mm on axial	80% of focus group to	+/− 0.2 mm on axial	Voltage: 220 V	3 positions	2 positions	+/− 0.2 mm on axial	+/− 0.2 mm on axial	15 to 25 mm hook	1.5 m< cord length < 2.5	+/− 0.2 mm on axial	+/− 0.2 mm on axial	+/− 0.5 mm	To BS/ISO/IEC standards	

Fig. 2.11 QFD1—product planning matrix for a hairdryer example. *Note*: All aspects should be quantified if possible in the How Much column and row.

		Hair dryer												
		Body					Handle assembly							
	Importance	Debris filter	Front housing	Insulator	Rear housing	Motor casing assembly	Cap	Cord support	Handle casing	Hanger	Spring	Spring bracket	Switch assembly	How much
Draws air in	7					9								MFR (kg/s): 0.05
Contains rear air flow	4				9									Leakage minimal (<0.2 g/s)
Removes debris from air flow	7	9												Screen apertures: 1 mm
Locates debris filter	2				9									+/−0.2 mm on axial datum
Supplies front air flow	7					9								VFR (L/min): 2685
Heats front air flow	8					9								Max T (air): 55°C
Guides front air flow	2		1	9										Within +/−7 mm of axis
Accelerates front air flow	4			9										Max air speed > 16.44 m/s
Lowers static p of front air flow to atmospheric	4			9										101325 Pa +/−1000 Pa
Insulates front housing	7			9										Casing T <25°C
Dries hair	7					9								Moisture content < 0.1 g per 10 g hair
Positions front housing	2				9									+/−0.2 mm on axial datum
Positions insulator	2		9											+/−0.2 mm on axial datum
Offers holdability	5								9					80% of focus group to acclaim holdability
Locates AC neutral supply	4												9	+/−0.2 mm on axial datum
Acquires AC voltage	4												9	Voltage: 220 +/−20 V AC
Allows power control	4												9	Voltage: 50 to 240 V
Allows cooling control or hot shot	4												9	3 positions
Guides switch assembly	2								9					2 positions
Secures most handle and body parts	2								9					+/−0.2 mm on axial datum
Allows the device to be hung up	2									9				15 to 25 mm hook diameter
Allows the cord to be stowed	2							9						1.5 m < cord length < 2.5 m
Positions spring	2											4		+/−0.2 mm on axial datum
Positions cool shot switch bracket	2										4			+/−0.2 mm on axial datum
Positions cord	2							9						+/−0.5 mm
Protects cord	2							9						To BS/ISO/IEC standards Nos.
How much	100	One plastic filter	One plastic front housing	One insulator	One plastic rear housing	One motor, etc.	One cap	One cord support	One handle casing	One hanger	One spring	One spring bracket	One switch, etc.	

Fig. 2.12 QFD2—design deployment matrix for a hairdryer. *Note*: All aspects should be quantified if possible in the how much column and row.

A QFD 2 matrix explores the relationship between engineering requirements (or product functions) and product parts. The functions are inherited from the QFD 1 matrix, so the first column in Fig. 2.12 matches the top row in Fig. 2.11. The process to complete this matrix is similar to that illustrated for QFD1. However, the meaning of the matrix is significantly different, as the QFD2 matrix focuses on translating functions into physical components.

Fig. 2.13 illustrates the Dyson supersonic hairdryer that came onto the market in 2016 and had a series of innovative features including low noise and high air flow enabled by its handle mounted high-speed electric motor driven fan. This is an example of how a product can enter a crowded market place and yet still succeed by addressing the needs of multiple stakeholders including the customer.

Fig. 2.13 The Dyson supersonic hairdryer, along with an impeller and motor–impeller assembly that came onto the market in 2016.

Some of the benefits claimed for QFD are as follows:

- The rigorous approach to specification helps ensure a systematic approach to the design process that is expected to decrease design and manufacturing costs and consequently time to market;
- The approach shifts the focus of an organisation from engineering capabilities to customer needs;
- The method makes requirements more evident and visible allowing an organisation to prioritise them and to deliver them based on customer needs;
- The approach helps ensure that an organisation captures a complete and accurate set of requirements;
- It allows key requirements to be considered throughout the process;
- The approach captures design data in a centralised format that can be accessed by different teams;
- It improves communication and sharing of information within a product development team and
- It allows identification of 'holes' in the current knowledge of a design team.

Some of the pitfalls or disadvantages of QFD include the following:

- Generating the information needed for each matrix is a complex and detailed process which demands commitment and resource;
- The process of establishing the relationships between the rows and columns of a matrix is generally subjective.

2.4 Conclusions

There is no single proven optimum approach to innovation to date. Differing innovation strategies have their proponents and detractors. The approach adopted for managing innovation and the development of a commercial enterprise is often dictated by historical precedence within a given industry or organisation. The identification of the requirements that a design needs to fulfil is a fundamental aspect of the majority of product innovation and management processes. This process of identification may involve a number of cycles of detailing with increasing veracity with each iteration as a result of more information becoming available. For some applications, it is not possible to clearly identify the requirements prior to a design concept becoming available for scrutiny. Once some concepts have been produced it is then possible to explore the innovation space and, for example determine potentially new functionalities that a focus group or management team would not necessarily have envisaged prior to the insights enabled by the new emergent design.

For the majority of industry applications, common practice is to use detailed specification to define the scope for a new product development activity. This specification serves to both inform and limit the activity for the multiple stakeholders involved. A specification enables a design team to know the limit of functionality associated with a particular feature, subsystem or system. It enables a design team to know when they have delivered sufficiently to fulfil the requirement. It enables management of an innovation process. Critically, if the specification has accurately identified and embodied user requirements and the commercial opportunity, then if a design and development programme has met these requirements, a specification provides confidence that the commercial opportunity can actually be realised.

There are a wide range of approaches to specification. Some organisations start with a brief statement of need, which is then subsequently worked up into a formal detailed specification. A principle associated with detailed specification is to ensure that each aspect is quantified addressing the range of functionalities, for example technical, aesthetic, economic and social function. Detailed specifications can be very large documents occupying hundreds of pages. A specification for a new gas turbine engine for an aviation application might typically be a document consisting of several hundred pages. An alternative to a discursive or tabular-based specification is to use QFD. QFD uses a matric approach to capture and quantify customer requirements and enable transfer of information through the typical phases associated with product development and delivery.

This chapter has introduced the principles of specification and provided examples of typical pro-forma tables used for detailed PDS along with an introduction to QFD. Specification is typically used within a project management process to define given tasks to enable an assessment of whether the task has been achieved or completed.

References

Akao, Y., 2004. Quality Function Deployment. Productivity Press, New York.

Akao, Y., Mazur, G.H., 2003. The leading edge in QFD: past present and future. Int. J. Qual. Reliab. Manag. 20, 20–35.

Aurisicchio, M. Quality function deployment. Private Communication, 2013.

Bouchereau, V., Rowlands, H., 2000. Methods and techniques to help quality function deployment (QFD). Benchmark. Int. J. 7, 8–20.

Chan, L., Wu, M.L., 2002. Quality function deployment: a comprehensive review of its concepts and methods. Qual. Eng. 15, 23–35.

Childs, P.R.N., 2004. Mechanical Design, second ed. Elsevier Butterworth Heinemann, Oxford.

Cohen, L., 1995. Quality Function Deployment: How to Make QFD Work for You. Prentice Hall, Upper Saddle River, NJ.
Eureka, W.E., 1987. Introduction to QFD: Collection of Presentations and Case Studies. American Supplier Institute, Detroit, MI.
Franceschini, F., Galetto, M., Maisano, D., Mastrogiacomo, L., 2015. Prioritisation of engineering characteristics in QFD in the case of customer requirements orderings. Int. J. Prod. Res. 53, 3975–3988.
Ju, Y., Sohn, S.Y., 2015. Patent-based QFD framework development for identification of emerging technologies and related business models: a case of robot technology in Korea. Technol. Forecast. Soc. Chang. 94, 44–64.
Miller, K., Brand, C., Heathcote, N., Rutter, B., 2005. Quality function deployment and its application to automotive door design. Proc. IMechE D: J. Automob. Eng. 219, 1481–1493.
Pugh, S., 1990. Total Design. Addison-Wesley, New York.
Simpson, P.A., 2015. Design specification. In: FPGA Design. Springer, New York, pp. 9–13.
Sugumaran, C., Muthu, S., Devadasan, S.R., Srinivasan, K., Sivaram, N.M., Rupavathi, N., 2014. Integration of QFD and AHP with TPM: an implementation study in an automotive accessories manufacturing company. Int. J. Prod. Qual. Manag. 14, 263–295.

Websites

At the time of going to press, the world-wide-web contained useful information relating to this chapter at the following sites.
http://www.qfdi.org.
http://www.qfdonline.com/templates.
http://www.npd-solutions.com.

Further reading

Ficalora, J., Cohen, L., 2010. Quality Function Deployment and Six Sigma. A QFD Handbook, second ed. Prentice Hall, Upper Saddle River, NJ.
Sharing Experience in Engineering Design, 1985. Curriculum for Design: Engineering Undergraduate Courses. SEED.

Ideation

Chapter Outline

3.1 Introduction

Engineering design relies upon the generation of alternative ideas and their effective evaluation in order to ensure that attention is directed in a worthwhile manner. This chapter explores a number of creativity tools that are widely used for the generation of ideas including flip chart; post-it notes; alphabet and grid brainstorming; the creative problem solving process, also known as CPS; morphological analysis, a matrix

Mechanical Design Engineering Handbook. https://doi.org/10.1016/B978-0-08-102367-9.00003-2

combinational creativity tool; and TRIZ the theory of inventive problem solving. The chapter includes consideration of group dynamics and the evaluation of ideas as well as approaches for encouraging the realisation of value from creativity through innovation.

Ideas mean money. Money means business. These notions are familiar to any entrepreneur, innovator and manager. To come up with ideas, you either need to embrace creativity directly or surround yourself with people who do. On its own, creativity is not enough to sustain a business. We are familiar with bright, able and creative individuals who are endlessly generating new ideas but are seemingly unable to turn these ideas into reality or make a living out of them. The effective exploitation of an idea is the realm of business, and the skill sets for creativity and commerce are often at odds with each other. If we endlessly generate ideas but do not put in the effort to develop them sufficiently so that an idea is robust, combine it with an appropriate business strategy to exploit it and turn it into a revenue stream that is sufficient to pay for the development of the product, service or system concerned together with the ongoing business costs for its marketing, production, distribution and future developments, then the ideas will fail to benefit the originators, employers or sponsors.

The need for robust ideas that have been appropriately developed is important. If there is a flaw in a product, then it is possible that a customer will accept it for a while. The product martyr is the dream of the warranty department and we have all put up with inadequate devices, products and processes in our lives. In a competitive world however, it is a matter of time until we are wooed by a competitor whose products appear more reliable and fulfil the function to a higher standard. Remember that for every unit of currency spent on marketing to attract a customer for the first time, it can cost 10 times that quantity to win back a customer who has become dissatisfied with the quality of a brand (Kochan, 1996).

It is the blend of methods, work practices, culture and infrastructure that makes an organisation succeed. IDEO, the design and innovation consultancy, expound the importance of understanding, observing, visualising, evaluating, refining and implementing. We are all familiar with the phrase that a picture tells a thousand words. In idea generation, it is important not just to use pictures but also to get physical and explore ideas in three dimensions where appropriate. In idea generation, it is helpful to leave our performance anxieties behind, and using words and symbols, expand into imagery and three-dimensional prototyping. This is embodied in the IDEO process of ideation: generating, developing and communicating an idea. As part of this process the production of physical prototypes is emphasised. These may be crude representations of the idea, sometimes made with card, wires and found objects, but they provoke discussion and engagement in design and this can lead to significant iterative and step improvements in its implementation. In the case of IDEO this is combined with capacity, expertise and experience of innovation enabling the fast development and trialling of ideas.

Observation is important. Many people in response to a question on a meal might say that it was '*very good*'. Similarly, if you ask someone what they think of a product or service they may say it is '*fine*' or '*very good*'. Part of the reason for the somewhat uninformative response is that the person concerned may just be being polite, lack the vocabulary to express their opinion, be hesitant to offer criticism, constrained by cultural norms and possibly a lack of willingness to provide competitive advantage. If we are enthusiasts or experts in the area concerned, then we may be able to express our opinion more specifically. It is for this reason that surveys and market-based information needs to be carefully assessed, as significant information can be lost or never accumulated in such processes. Observation provides key insights into the way that customers actually use a product, service or system and can reveal particular issues that surveys miss. Observation is a key activity in user-based design, and astute observers tend to produce ideas as a result of their observations because they identify needs and are able to suggest methods to meet these.

Our understanding of creativity stems not just from the insights from successful companies but from the endeavours of individuals through history, from research specifically into creative processes and people, and from the conditions that have enabled or stymied activity. This chapter aims to provide a series of tools and approaches that can aid engineering and design processes. To this end, an introduction to creative process is introduced in Section 3.2 and a variety of approaches to brainstorming in Section 3.3. A process based on implementation of the principles of brainstorming called Creative Problem Solving is described in Section 3.4. A useful and simple ideation tool called SCAMPER is given in Section 3.5. A process developed in industry showing the integration of ideation techniques called the Create Process is introduced in Section 3.6. A combinational creativity tool called morphological analysis is described in Section 3.7. The theory of inventive problem solving (TRIZ), a systematic inventive tool that exploits prior experience, is introduced in Section 3.8. A technique that can be useful in challenging conventional approaches to working with a specification, called boundary shifting, is described in Section 3.9.

3.2 Creative process

Goldenberg and Mazursky (2002) in their review of research on creativity identified that many investigators have used one of three approaches to explain how creativity happens: the creative person; the creative process; the creative idea.

- A creative person is able to choose high-quality ideas from a wider set of ideas. To solve problems, a creative person often focuses on finding methods to generate a number of ideas. Some methods are suggested such as brainstorming (see Section 3.3), synectics, random stimulation and lateral thinking.
- Research at the end of the 1980s showed that restricted processes of thinking are more reliable for creativity than unrestricted scopes.
- Some research showed that through the use of the previous creative ideas it is possible to formulate new, successful, creative ideas. For example, Genrich Altschuller's TRIZ (Altshuller (1984) and described in Section 3.8) uses a systematic method, based on previous successful inventions, to guide engineers and designers towards creative solutions. After years of research and reviewing more than 200,000 successful patents and invention disclosures, Altshuller and his colleagues defined 40 patterns of inventions that could inspire engineers and designers to generate creative solutions. In some of the modern TRIZ platforms this analysis has been extended to over two million patents and has confirmed the original selection of just 40 patterns.

Wallas (1926) in his pioneering work suggested that the creative process can be modelled in four phases.

1. Preparation—becoming immersed in a problematic and interesting set of issues.
2. Incubation—churning ideas around consciously and unconsciously.
3. Illumination or insight.
4. Verification or evaluation.

Such phases need to be combined and followed by elaboration, or detailing, to work the idea up into something that can be realised and effectively used. It is one thing having a good idea; it is quite another turning this into reality. As Thomas Edison famously noted "*Two percent is genius and ninety-eight percent is hard work*" (The *Ladies' Home Journal*, 1898). Of course, such a model of the creative process is a generalisation and simplification of the complex processes involved which will differ from person to person, or group concerned, as well as topic.

To solve a problem, it is necessary not only to understand the issues but also to use creative thinking. A problem can be a conflict or a tension that needs to be resolved. Some problems are well formulated with everybody aware of what the issue is but without a solution. Problems can arise from experience, requirements of the domain and social pressure. In some cases, stumbling across a new, unexpected problem can lead to a creative discovery without preparation. Examples of this sometimes occur in scientific research where investigation of unusual results provides new insight.

The Design Council (2007) reported a study of the design process in 11 leading companies and identified a four-step process referred to as the 'double diamond' design process model, involving phases of: discover, define, develop and deliver. In Fig. 3.1, the divergent and convergent stages of the design process are indicated, showing the different modes of thinking that engineers and designers use.

Coming up with new ideas is difficult. This is why creativity is acclaimed and held in high esteem in our societies. Part of the reason that creativity is difficult is that so much has been done before with a population of billions thinking so much so fast. Jacob Rabinow (Csikszentmihalyi, 1996) notes that you need three things to be an

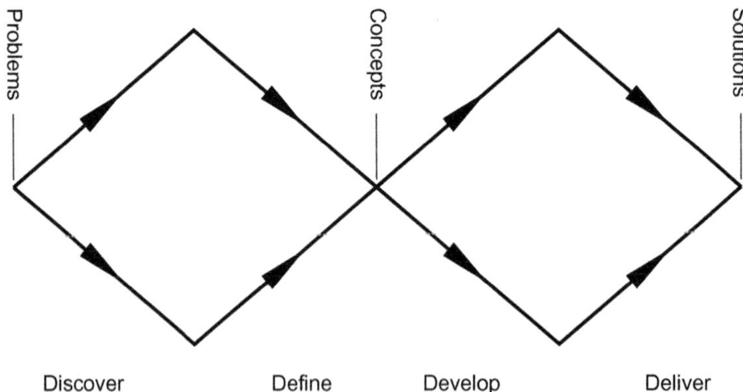

Fig. 3.1 Double-diamond schematic describing the design process (Design Council, 2007).

original thinker: a tremendous amount of information, willingness to come up with ideas and put the effort into this, and the ability to discard the weaker ideas. Creativity tools provide a means to augment innate generative activity. The scope of creativity tools available is extensive (see, e.g. Lidwell et al., 2003 and The Create Project, 2005) with hundreds having been identified and a selection of these are presented in Table 3.1. Different creativity tools have been found to be suitable for different personality traits and different applications, so care needs to be exercised in the selection of a particular tool (see Yan et al., 2013).

Table 3.1 Some example creativity tools

Tool	Brief description
'Post-it' and 'Grid' brainstorming (Michalko, 2001; Osborn, 1963)	In 'Post-it' participants record ideas on Post-it notes and these are collected, reviewed and analysed. In 'Grid' a participant records some solutions to a problem on a grid that is completed by other members of the group. These tools encourage simultaneous generative activity but some participants may remain reluctant to share an idea for fear of ridicule or loss of ownership. This latter concern may be addressed by implementation of IPR protocol.
Checklists (Osborn, 1963)	A series of brief questions and statements is used to stimulate creativity when it is proving difficult to think in original ways. Can be used for idea generation *and* evaluation but is a systematic method that may not appeal to all personality traits.
Lateral thinking (de Bono, 1970)	Solutions are proposed by looking at a problem using random associations, provocation, challenging current solutions and divergence. Requires both curiosity and confidence, and encourages a wide range of alternative solutions. Robust evaluation of ideas is necessary to identify worthwhile concepts to develop.
Mind mapping (Buzan and Buzan, 2006)	Connections between associated pieces of information are emphasised by clustering the information on a visual map: this can stimulate creativity. Some people are hesitant to reveal a perceived weakness in sketching but this can be overcome with practice.
Six hats (de Bono, 1999)	Parallel thinking process in which team members wear coloured hats representing data, creativity, positivity, feelings, criticism and control. The approach can minimise conflict, and encourage participation and consideration of a problem from a wide range of perspectives.

Continued

Table 3.1 Continued

Tool	Brief description
Morphological analysis (Childs, 2004; Zwicky, 1969)	A matrix-based tool in which a problem is broken down into component parts and a range of approaches suggested for each of these elements. Encourages combinations of features that otherwise might not have been considered but the large quantity of combinations generated means that good ideas can be overlooked. Weighting criteria should therefore be used to guide selection of solutions.
Synectics (Gordon, 1961)	Exploits our capacity to connect apparently irrelevant elements to spark new ideas and solutions. The approach helps participants break existing mindsets and internalise abstract concepts but is time-consuming, requiring practice and expert facilitation.
TRIZ (Altshuller, 1984)	Russian acronym for the Theory of Inventive Problem Solving. Provides a framework and toolbox for systematic, inventive problem solving but is sometimes viewed as complicated and difficult to use; this can be overcome by sustained use and practice.

Most creativity tools try to enhance:

- Fluency—the quantity of responses
- Flexibility—ideas that are distinct from each other
- Originality—the level of uniqueness of the ideas generated

A creativity tool will not produce ideas. Instead it can be used to assist in the generative process. Most creativity tools can actually be used at any stage in a problem-solving process but tend to be mainly focused on problem exploration, idea generation and concept evaluation.

- Problem exploration: tools such as Fishbone Diagram, Question asking, QFD.
- Idea generation: tools such as Brainstorming, Checklists, Gallery Method, Lateral Thinking, Morphological Analysis, TRIZ and SCAMPER.
- Concept evaluation: tools such as SWOT Analysis, Taguchi Methods, Value Engineering.

To be effective, a creativity tool needs to influence the thinker at the time of idea generation. Human memory and the process of information retrieval are crucial in this process. Memory can be defined as the ability to store, retain and subsequently access information. Models of memory include the simple multistore memory model of Atkinson and Shriffin (1968). The capacity of short-term memory is limited to approximately seven items plus or minus two (see Miller, 1956; Poirrier and Saint-Aubin, 1995). Cowan (2001) also showed that the duration of short-term memory retention is between 2 and 30 s. This limited capacity presents a significant challenge to problem solving. If, for example, trial and error is used to tackle an issue then the available short-term memory is likely to be dominated by information relating to the problem definition, which is rarely useful for solving it. Most solutions can be characterised by the connection of diverse thoughts from previous experience. Most creativity tools implicitly acknowledge the limited capacity of short-term memory and function by ensuring that a problem can be understood in relatively simple terms, thereby occupying only a fraction of short-term memory, supplying cues to make the search of long-term memory more efficient and providing cues to ensure refreshing of short-term memory and thereby retention of key information.

3.3 Brainstorming

Brainstorming typically involves a multidisciplinary group meeting together to propose and generate ideas to solve a stated problem. Certain forms of brainstorming can be performed individually, with other people at the same time or with a time lapse between communications. The emphasis within brainstorming is on quantity rather than quality of ideas, and criticism of another person's idea is strictly forbidden during phases of the process. Following a brainstorming session, a review should be undertaken to identify ideas or aspects of the concepts generated that have merit for further consideration. A typical outcome is that aspects of a few of the concepts generated can be combined to form a solution to the problem that gives a better overall solution than any one of the individual concepts from the original session. Brainstorming is a familiar concept within industry and widely used within stage-gate, waterfall and agile management processes.

Osborn (1963) observed that results of generative activity improved if there was deferment of judgement and also that quantity breeds quality. These principles have been embedded in a series of rules for brainstorming activities. During a brainstorming session, criticism, evaluation or judging an idea is forbidden. This encourages people to come up with and offer ideas without fear of criticism, ridicule or judgement. When sharing an idea, it needs to be recognised that a person is sharing something intimate—a set of thoughts. Respect for this is important. In a critical environment people are less likely to offer ideas, or may limit the ideas they offer to ones that they have evaluated and think reasonable. There have been many ideas that at first airing may seem unviable but with the benefit of attention from the collective mind of a group can be built on, improving the viability of the thought. If the original idea is not shared this is not possible. It should be noted that individuals will develop coping strategies within a critical environment. Many offices and organisations stumble forward in the midst of a cut-and-thrust, competitive and often highly critical working environment. In such cases it is common to hear comments along the lines of

'Why should I share an idea when all that will happen is …'
'This has been done before'; 'Been there, done that'
'This just won't work'

'Someone else will take or claim the idea and all the credit for it;'
'The idea will be ignored because it was not suggested by the people in charge (or who matter);' (Not-invented-here syndrome)
'Managers won't like it'

These scenarios are common-place, and strategies to avoid them include:

1. Explaining the rationale and rules of brainstorming prior to the session. This can be achieved by a premeeting to define these for the participants.
2. Using a system to document the originator of an idea such as associating an idea with an individual by including their initials alongside it.
3. Using a quasiquantitative technique to evaluate the ideas generated, such as an evaluation matrix (see Section 3.3.11).
4. Explaining to managers the benefits of brainstorming and that ideas will occur under the manager's direction and therefore will reflect positively on the manager.

The principal rules commonly associated with brainstorming are:

- brainstorming should be a group activity undertaken by an interdisciplinary team;
- there should be no more than 10 people present;
- no criticism or mockery of any idea, statement or individual is allowed;
- flip charts, post-its, cards, marker pens, pencils and paper should be provided;
- the design brief should be clearly stated;
- a set period of time should be allocated;
- an individual should be identified to introduce the brief.

Tom Kelley of IDEO (2001) suggests that effective brainstorming sessions should not take longer than 60–90 min. The IDEO organisation uses brainstorming regularly with each person involved several times a month. Brainstorming is the idea engine for many industries and over 70% of business leaders use it in their organisations (Anderson et al., 2004). There is a substantial body of evidence providing convincing data showing the effectiveness of brainstorming and creative tools based on divergence and convergence when appropriately implemented. Evidence for the effectiveness of brainstorming and associated tools for problem solving, techniques for using them and strategies for their selection and implementation are reviewed by Treffinger et al. (1994), Isaksen and DeSchryver (2000), Torrence (1972) and Rose and Lin (1984).

3.3.1 Intellectual property rights and brainstorming

During a brainstorming session, many ideas can be generated and shared. Some of these will be shared during the session as a result of prior consideration over many years. Others may be sparked by interactions within the session itself. Just as in a contractual relationship, it is important to establish what mechanisms will be implemented to deal with any potential intellectual property rights (IPR). Options include: Chatham house rules; direct attribution; equal attribution, company retention of IPR. Regardless of the type of attribution applied within a brainstorming session, care needs to be taken regarding the assertion or claim of a company or organisation to ideas developed.

- Chatham house rules. Chatham house rules state that there is no attribution to a particular individual for anything that is said or shared within a discussion.
- Direct attribution with each participant's initials or moniker associated with each idea. This can be readily implemented on post-its, for example, where each participant writes their ideas on a single post-it along with, say, their initials. These can then be scanned following the sessions and embedded within a database.
- Equal attribution—any IPR reward is divided equally between all participants.
- IPR is retained solely by the organisation or client. Such a system may be compatible with the terms of employment.

At the start of a brainstorming session the rules for IPR need to be stated.

3.3.2 Brainstorming methods

Osborn (1963) suggested that it was important that "especially in approaching a creative problem we should give imagination priority over judgement and let it roam around our objective. We might even make a conscious effort to think up the wildest ideas that could possibly apply." The emphasis in brainstorming on quantity increases the number of alternative solutions and hence the potential for a viable and worthwhile end-result.

The description of brainstorming given here suggests that there is much more to brainstorming than a simple free-wheeling event where people turn up and share ideas. It is necessary to prepare for the session. Work needs to be undertaken to understand the task or problem to be solved. This is compatible with the need to ensure that the task can be understood in simple terms ensuring that our limited short-term memory is not occupied by multiple facets of the problem. By freeing up our short-term memory, we enhance our ability to access potential ideas from our long-term memory. In addition, participants need to be forewarned of the task to enable them to develop and rehearse sharing of these ideas prior to a session. Necessary preparation by the organiser prior to the session also includes booking facilities, determining appropriate participants and distributing the brief.

Brainstorming can be used for both individual and group efforts (Isaksen, 2000). A potentially fruitful approach is to generate the core idea by yourself and, subsequently, to leave the development phase to the others.

Rule-breaking and wacky ideas are welcome in brainstorming. It is the rule-breaker who might discover the short cut in a journey by doing something normally forbidden. Although such behaviour might rightly be frowned upon in some contexts, the possibilities revealed may provide useful information that had not previously been considered. In idea generation, the concepts of rule-breaking and boundary shifting are useful and to be encouraged. Similarly, wacky ideas should therefore be welcome. Some suggestions can make a team crack up with laughter and still be amusing several days later. Whether these ideas have merit in terms of the task will be determined at the evaluation phase.

A range of creativity tools have been developed embedding the principles of 'deferment of judgement' and 'quantity breeds quality'. Some of these are group-based involving direct verbal interactions between those present, while others are silent tools known as brainwriting methods and hybrids. Flip chart, post-it and alphabet brainstorming techniques are described in Sections 3.3.3–3.3.5. Brainwriting and grid are described in Sections 3.3.6 and 3.3.7. Although each method is presented in isolation, a typical idea generation session may comprise a sequence of techniques with, for example, a period of flip chart brainstorming, followed by post-it and then a warm-up prior to using grid. A warm-up is a mental exercise designed to encourage a particular behaviour such as suppressing mental editing.

3.3.3 Flip chart

The flip chart technique is the poor relation of brainstorming as it does not encourage simultaneous contributions and therefore does not fully embrace the power of the collective mind. Once the brief has been shared with participants, the technique involves the facilitator inviting suggestions for its solution. Traditionally, the facilitator may use a flip chart, whiteboard or large sheet of paper to record the ideas (Fig. 3.2). Alternatively, these can be recorded on a computer system, as in Fig. 3.3 showing the case of holding a gear hub onto a shaft. A typical procedure for flip chart brainstorming follows.

1. Prepare for the session by ensuring that participants are familiar with the principles of brainstorming and have undertaken relevant training as appropriate. Ensure that the brief defines the problem in an accessible manner. Distribute the brief to the group prior to the brainstorming session. Ensure that the room has been booked and the relevant participants have been invited. Make sure that a flip chart, whiteboard or equivalent and relevant markers, or means to record the ideas, are available.
2. Organise the group so that they are comfortably able to see the display medium.
3. Remind participants of the brief.
4. The facilitator invites solutions to the brief for a period of time. During this period, no criticism of ideas is allowed. If criticism does occur, this needs to be dealt with by the facilitator (see Section 3.3.9).
5. Either the facilitator or a scribe records the suggestions from the group on a flip chart or another recording medium. If it has been agreed that attribution of ideas is to occur, then it is important that the facilitator records the moniker of the relevant participant against the idea. If the participants are unknown to the scribe, then badges or name tags can be worn

Fig. 3.2 'Flip chart' brainstorming.

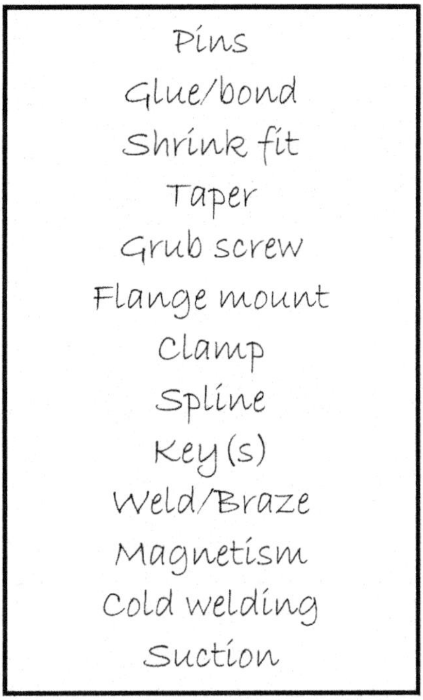

Pins
Glue/bond
Shrink fit
Taper
Grub screw
Flange mount
Clamp
Spline
Key(s)
Weld/Braze
Magnetism
Cold welding
Suction

Fig. 3.3 Some ideas for holding a hub for a gear onto a shaft.

to assist identification. In addition, it is useful if a sequential number is associated with each idea. This provides a means to refer back to a specific idea in a long list.

6. The time for a flip chart session can be until the frequency of ideas being suggested begins to reduce or until it drops to an uncomfortable level for the group. At this point, the facilitator can either suggest that it is time to review the material suggested or alternatively use a warm-up, group exercise or series of props to reinvigorate the group prior to another session using the flip chart to record ideas.
7. Once ideas begin to become less frequent, the facilitator can either suggest that it is time to review the material suggested or use a warm-up, group exercise or props to reinvigorate the group prior to another session using the flip chart to record ideas. Groups tend to be capable of determining when it is time to move on.
8. Once a sufficient quantity of ideas has been generated, by say two, 10–20-min sessions, then participants can be invited to review the material to see if additional ideas are inspired by, for instance, a combination of two of the suggestions.

Flip chart is the commonly known brainstorming technique. A number of factors, however, combine to reduce its effectiveness as a creativity tool. These include:

- participants finding it difficult to get a word in;
- shyer participants not feeling or being able to make suggestions out loud;
- people not being willing to submit ideas within such a session for either selfish reasons, such as, 'Why should I bother to help this organisation', or because they are not convinced that they will receive due recognition for the idea;

- the facilitator not hearing/taking notice of an idea;
- the facilitator misinterpreting an idea;
- the facilitator failing to pursue the direction of an idea;
- interruption of each other's thought processes by the constant flood of input to the session.

Such factors have been cited as being responsible for the output of brainstorming, in terms of number and quality of ideas, not being as effective as the collective output of individuals working on the task concerned on their own (Furnham, 2000). However, this criticism of brainstorming needs to be limited to flip chart brainstorming and should not be applied to all types of brainstorming. In addition, some facilitators chose to implement it in such a way that further reduces its effectiveness such as dominating the session, suggesting too many of their own ideas, dismissing a particular direction of thought and suggestions too early with implied criticism of the ideas, favouring comments from certain individuals and failing to encourage wider participation.

The flip chart approach can serve as a good entry activity to a brainstorming session, but should be used sparingly. It can, for example, be used to determine the important areas of a problem. The group can 'flip chart' brainstorm these for several minutes. Following this, a more structured brainstorming technique such as post-its or grid can be employed, which encourage simultaneous mental activity and contribution to the task for all the participants.

3.3.4 Post-it brainstorming

Post-its and their equivalents are widely used in brainstorming. They provide a readily accessible means to record an idea and then order or collect similar ideas together (see, e.g. Fig. 3.4). The asynchronous nature of flip chart brainstorming can be overcome by asking all participants in a brainstorming session to record their ideas on post-its, with one idea per post-it. After a set period of time, say 10–20min, participants can be invited to consider a stimulus, such as the use of found objects to inspire ideas or standard mental warm-up exercises. A further session of recording ideas on post-its with

Fig. 3.4 Random post-its.

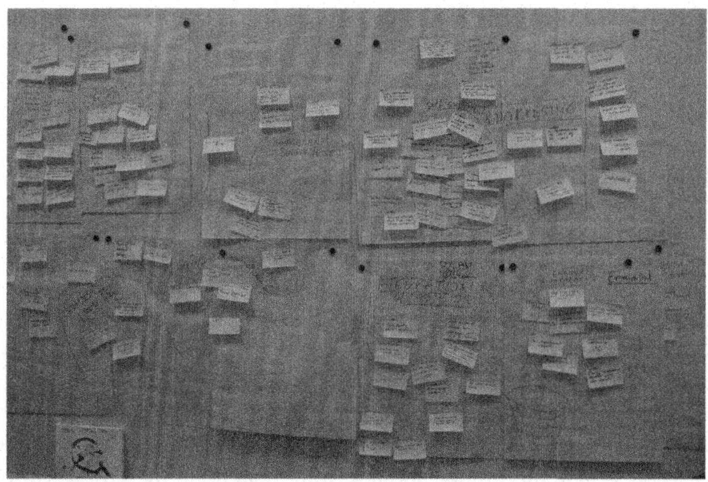

Fig. 3.5 Organised post-its.

one idea per post-it can then take place. Following this all the ideas generated can be collected together and an attempt made to review these and if possible organise them (Fig. 3.5). As post-its can be peeled off a surface and reapplied several times, different classifications can be explored. It should be noted that an idea generation session using post-its may not be quick and can take 1 or 2 h.

A typical session is outlined below but many variants on this are possible.

1. Ensure that the participants are familiar with the principles of brainstorming and have had relevant training.
2. If the task to be tackled is not well understood, hold a focus group to define the brief.
3. Distribute the brief to the participants in advance of the session.
4. Prepare for the session by ensuring sufficient post-it pads are available (different colours or shapes can be used for each participant).
5. Organise the group so that they are comfortably able to record their ideas.
6. Remind participants of the brief.
7. The facilitator invites solutions to the brief for a period of time, say 10–20 min, with participants recording one idea per post-it.
8. After about 20 min, the facilitator can halt recording of ideas and either undertake a review of ideas generated to date or prompt the session with a warm-up or structured thought process to inspire additional ideas.
9. A further idea-generation session using post-its can then take place for say another 20 min.
10. The ideas generated can be reviewed collectively. This can be done by means of asking each member to share a few (say two or three) of their ideas with the group. This allows the facilitator to identify some potential categories into which the different ideas can potentially be placed. The group can then be invited to cluster their different ideas into the categories selected. This can be done by placing the post-its on white boards, flip charts or appropriate wall or table surfaces. A camera can be used to record the grouping of ideas.
11. During the review, ideas suggested by the group should be recorded on a flip chart.
12. The review may provoke discussion and the facilitator can decide whether a further round of post-it or other technique is appropriate.

A range of software systems are available to emulate and facilitate group brainstorming sessions. Electronic notice-boards and forums can be used for a group to post ideas, organise and collate them. Such systems allow groups that cannot physically be present to participate in idea generation, and can also enable individuals to feed ideas into a session at disparate times.

3.3.5 Alphabet brainstorming

In this brainstorming technique, participants are invited to work through the alphabet sequentially letter by letter, suggesting a few ideas for each letter that are relevant to its potential suggestion. Once a few ideas beginning with the letter A have been suggested and recorded, the group then moves on to consider a few ideas beginning with the letter B and so on to Z. Alphabet brainstorming serves to encourage consideration of ideas across a wide range of possibilities. At the end of an alphabet brainstorming session, which might take about 45 min to an hour, the group will typically have between 75 and 150 ideas. Alphabet brainstorming is a good method to help a group to widen their horizons and develop a large range of options to explore. A typical sequence, in this case for possible solutions to climate change, has been listed in Table 3.2.

Care needs to be taken in introducing alphabet brainstorming to a group. For most of us we gave up doing alphabet sequencing in primary school and it is possible that one or more members of a group may ridicule or resent an approach based on such a simple concept. Nevertheless the approach can be productive and the technique can be introduced by explaining not only the procedure for its implementation but also its rationale.

As with any brainstorming technique, the session should be followed up with a review of the ideas generated. If necessary this can be immediately following the session, exploring opportunities to combine ideas and seeing whether any of the suggestions provoke additional avenues for exploration. Following this an evaluation of the ideas can be taken.

3.3.6 Brainwriting

The weakness of flip chart brainstorming with one person's thoughts interrupting another's in a serial manner, can be overcome by one of a number of techniques based on parallel activity. Brainwriting is the phrase commonly associated with a series of techniques developed by Horst Geschka and colleagues at the Batelle Institute in Frankfurt (Geschka, 1993). A typical brainwriting session comprises the following phases.

1. The facilitator introduces the session and if necessary the principle of the technique.
2. A discussion of the problem to clarify the issues to be addressed follows.
3. The participants are grouped in a circle say around a large table.
4. Each participant in the brainwriting session is given blank cards, such as index cards, and is invited to record their ideas on these with one idea per card.

Table 3.2 Example of ideas generated in an alphabet brainstorming session

Letter	Suggestions
A	Abstinence; air travel reduction; agricultural reform
B	Bio-fuel; bio-energy; better use of electricity
C	Car pooling; carbon budgets; carbon taxes; combined heat and power; consumption charges
D	Distributed micro generation; diesel substitutes (bio-diesel)
E	Energy efficiency; education; emissions policy; ethics and equity
F	Fuel cells; fuel farming
G	Geo-thermal energy; global public good
H	Hydro-power; hybrid vehicles; hydrogen fuel cells; hydrogen economy
I	Insulation; incentives; ice caps
J	Joined up thinking; justice
K	Knowledge
L	Light emitting diodes; low carbon growth; low energy life styles; localisation; lichens; lifestyle
M	Methane; many approaches; multidisciplinary; motivation
N	Nuclear power; natural resources; not for profit/pay back period
O	Oil; oil substitutes; organic; ocean power
P	Public transport; petrol; packaging; population; power; penalties
Q	Quotas; quality of life; quadruple wind power; quantify product life
R	Resilience; robustness; rate; regeneration; reflectors
S	Solar; sewage gas
T	Technological quantum leap
U	Uncertainty; usage
V	Volcano emissions; vernal pools
W	Waste; wave power
X	Xanthium and xyridales (grow lots); Xerox solar panels
Y	Yew trees; Yield enhancement; Yachts for transport
Z	Zero-emissions; Zambezi hydro-electric power; Zonal policy

InQbate, University of Sussex 2008, Design Solutions for Climate Change.

5. As each member in the group generates ideas, silently, they then pass them in one direction, say to the right, in the group.
6. The intent of the cards is to stimulate ideas. On receipt of a card a participant reads it to see if it inspires an idea. Each participant can then write down the idea inspired by the stimulus and then pass this new card on to the right.
7. Within a few minutes, several idea cards will be being passed around the group.
8. After about 20–30 min all the cards generated can be collected.
9. The cards can then be sorted by the facilitator with guidance from the group into different categories of ideas with a generic title given to each category.
10. Each participant in the session can be invited to evaluate the ideas with a traffic light system. Each participant can be given a sheet of dots and can allocate a set number of these to ideas that they think warrant further consideration. For example, if each person is allowed to use five dots they can choose to place all five on one idea or divide them between cards.

3.3.7 Grid brainstorming

The grid method of brainstorming enables all participants to contribute simultaneously. In grid brainstorming, each person in the group is given a grid, say a piece of A4 paper divided into nine rectangles. The facilitator then explains the brief and invites the participants to develop three potential partial or full solutions to the problem, one in each sector of the first row. The solutions developed need not fully resolve the problem and can be simply ideas that seem relevant. The solutions can be portrayed using a mixture of sketches and words as illustrated as per the example in Fig. 3.6. Once a row with three part-solutions has been completed, the participant places their sheet of paper in the middle of the group. Another group member, if they have also finished their row, can then pick up a set of solutions that they have not been working on and commence developing solutions in the next row of the grid, based on the ideas developed by the previous person. The method has many similarities with the game Pictionary. In essence, once the participants are working with a set of ideas partially developed by someone else, each person is able to blend at least two sets of thoughts, their own and those of the previous contributor. The grid method allows the participants time to think at their own pace and is good at enabling those who are more reticent about participating to contribute.

The different pace at which participants work can present some challenges in running a grid session. Some people may finish their row of ideas very quickly while others can take longer. A number of strategies can be employed to deal with this. The facilitator can prepare some sheets in advance with the first row completed with their own ideas. If someone has finished very quickly, then such sheets can be handed out to keep all participants occupied on the task.

The grid technique can be summarised in the following steps.

- Each member of the group is given a grid divided into, say, nine rectangles.
- The brief is communicated by the facilitator.
- Participants start by sketching three ideas in the top row.
- When finished, they place their sheet in the middle for someone else to work on the next row developing the ideas further.

Fig. 3.6 Grid brainstorming.

- Once the sheets have been completed, a review can be undertaken to look at the ideas generated to see if additional avenues to explore in the session arise.
- Following the session, the sheets can be collected and the ideas evaluated.

Many variations on the grid technique are possible. For example, 3 by 4 grid with the top row seeded with suggestions; 1 by 4 grid with the top row seeded with a suggestion. It should be noted that the number of ideas to be developed, three suggested here, is arbitrary, but three provides some diversity and also involves a socially acceptable length of time before the exchange.

The grid technique is a particularly powerful brainstorming method. It allows for the collective power of a group's minds to be employed. All participants are able to contribute in simultaneous mental activity applied to the task.

3.3.8 Computational aids for brainstorming

Machine learning and artificial intelligence (AI) algorithms have led to the emergence of a wide range of tools for computational approaches to the generation of ideas. Brainstorming can be performed by an individual or group. The activity can be augmented through the use of software tools such as web crawlers and search engines providing lists of responses to a search input, which can provide inspiration. More sophisticated approaches for brainstorming have been developed including the B-Link tool (Shi et al., 2016, 2017a,b; Chen et al., 2017) which mines a scientific publication-derived database to provide a visual map relating to the inputs provided and gives a visual map as a basis for further inspiration and insight. An example for the case of bearings and fault diagnosis is given in Fig. 3.7. A further example of a

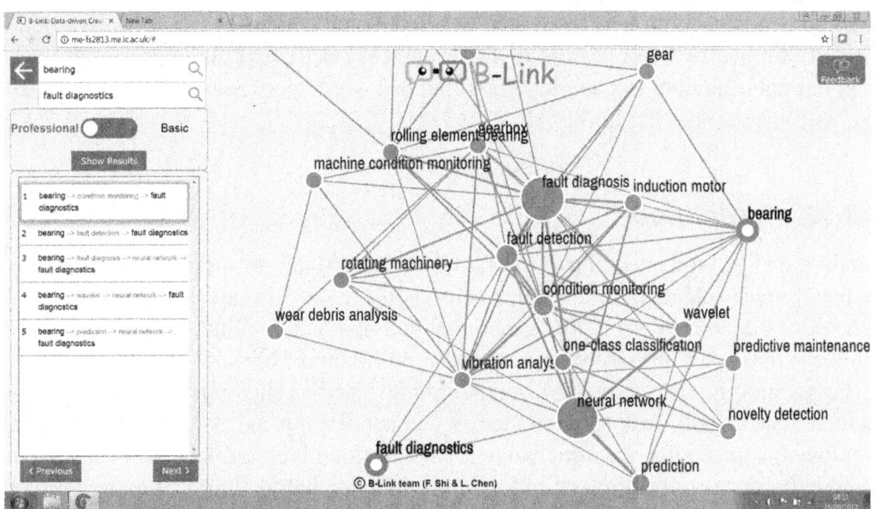

Fig. 3.7 : Output from the B-Link software which can be used to aid brainstorming activities. From http://www.imperial.ac.uk/design-engineering/research/engineering-design/creativity/b-link/.

tool that can aid in brainstorming is the combinator visual aid developed by Han et al. (2016, 2017), which provides blends of images to prompt ideas.

3.3.9 Group dynamics

The ideal brainstorming group should be derived from diverse disciplines and backgrounds and include experts and nonexperts. For example, a brainstorming group looking at new marketing ideas could comprise a customer, someone from manufacturing, an engineer, a receptionist, as well as staff from the marketing department. A brainstorming group for the design of a new product could comprise a chemist, a materials specialist, a manufacturing and a mechanical engineer, a personnel manager and a legal representative as well as a product designer. Participants in a brainstorming session need to regard each other as equals. If a participant feels or is made to feel that their opinion is not as valid as someone else's within the group then they are less likely to be able to contribute effectively. Alternatively, if someone feels their status means that their ideas are more significant than another's then it is possible that this person will draw attention to themselves and become the focus, thereby inhibiting the creative potential of the group. Thus, for the purpose of a brainstorming session, if a person is normally used to having their view prevail in the workplace, then this privilege needs to be temporarily surrendered. Similarly, if a person is used to withholding a view or idea or keeping silent then again this privilege needs to be abandoned for the brainstorming session.

Criticism is a fundamental part of many of our developed and learnt thought processes. Some of us have great difficulty in suspending our critical natures during idea generation sessions. In brainstorming, suspension of judgement is very important to enable the quantity of ideas to be generated taking advantage of the collective mind virtually formed by those present in the group. One tactic that can help people adapt their normal behaviour is to avoid saying 'but that won't work because …' and instead say 'yes and another possibility is …'.

3.3.10 Facilitation

Facilitation is a fundamental part of brainstorming and many other creative methods. In the absence of facilitation, groups can readily revert to critical and judgemental behaviour that normally inhibits generative activity. The facilitator's role is to help the ideas inherently present in a group come to life and flourish.

Preparation is a key aspect of brainstorming. Encouraging participants to devote limited time to any task is a challenge within all projects. A tactic that can be employed is to require all participants to submit three ideas at the start of a session as an entry ticket to be allowed in to the session (Michalko, 2001). If experience of a particular group suggests that production of ideas may be limited then an idea quota can be stated to encourage the quantity of ideas. For example, the facilitator may state that the idea quota required is 100 or 200 ideas (see Litchfield, 2008). While some of these may be less valuable than others, interesting suggestions may arise that might not otherwise be offered.

The facilitator may often take on the role of data recording. For flip chart sessions the facilitator can undertake both facilitation and scribe, or the role of scribe can be assigned to someone else. A key function is to capture not only the group activity but also the ideas generated by individuals. The facilitator therefore needs to ensure that post-its and other materials generated within the session are collected, organised and evaluated following the session. Grids, and post-its, for example, can be scanned and circulated by email, or embedded in a database for subsequent retrieval and evaluation.

Facilitation serves the purpose of giving a structure to a session, guiding attention to a particular focus and reassurance to the team as appropriate that the process is safe. If brainstorming becomes a natural part of your organisation's process, it is possible that a high level of structured facilitation, as described here, is not necessary. Kelley (2001) notes six methods to kill off the productivity of brainstorming.

- Letting the boss speak first setting an agenda that every idea needs to count or be patentable. This will immediately limit the spontaneity of the session.
- Letting everybody have a turn in sequence. If there are 10 of you in the room with each taking a minute or two to share an idea it can take 15 min to get round to you, by which time you may be begging for the world to end and thinking that brainstorming is less effective than the worst committee meeting.
- Stating that experts only should suggest an idea. Even if a domain is particularly complex, valuable ideas can still be offered by nonexperts. Discipline specific experts may be inspired by the suggestions made.
- Arranging the brainstorming session off-site. If idea generation is important to your organisation then brainstorming should become a regular activity in the workplace. If the session is in the workplace, then the buzz of news from the session will rapidly spread through the organisation with a net positive benefit.
- Stating that no 'silly stuff' is to be suggested. Again, this will limit the ideas offered.
- Note taking. If individuals are focussed on writing down everything that is said then their attention to idea generation is reduced. Encourage everyone to focus on idea generation rather than note recording, which in any case, is captured by the scribe.

3.3.11 Evaluation

Evaluation is an essential part of the decision process associated with idea generation. If a brainstorming session has generated over a hundred possible ideas then filtering these to a manageable number for more detailed consideration is necessary. It is important to consider the ideas while people still remember what they are about and while you have the attention of the group. Product development, detailed design and idea implementation teams cannot get going until decisions have been taken about which idea or ideas to follow up. A quasiobjective technique is recommended for evaluation. These range from the use of a voting system, say using sticky dots to signify a vote against ideas, to an evaluation matrix.

Voting systems can be sophisticated or simple. In a simple system, each person may have a limited number of votes, say 5 or 10, which they are allowed to issue with one or more votes per idea they wish to see considered further. Such voting systems can result in the loss of ideas unless some prior ordering and filtering of ideas has already occurred during the facilitated session. If, for example, a large number of minor variations on a particular idea have been recorded as separate concepts, then it is possible that votes will be shared between these and when votes are summed, a rarer, but possibly weaker, concept may receive more votes because all the votes were concentrated. To overcome this potential problem, it is necessary for the facilitator and group to review the ideas generated and to group similar ideas together. In this case fewer ideas are put forward to the evaluation phase, making the task easier to handle and avoiding dilution of votes.

Votes can be implemented by issuing each person with a set number of dots. The participants are asked to assign their dots to any idea they consider to be among the best ideas for the resolution of the task. Dots can be assigned with one per idea or in multiple numbers with a participant giving all of their dots or vote to an individual idea. Alternatively, participants can be given a red, amber and green dots representing traffic lights. Green indicates an idea is viable and a 'go', amber signifies caution and consideration, a 'wait', and red means 'stop' and is a clear 'no'. The dots provide a quick visual indication of where the team think attention should be focussed and can be quickly counted or assessed to see which ideas are favoured.

Another method of evaluation is the use of tabular evaluation matrices as illustrated in Table 3.3. These consist of a series of criteria against which the concepts must be marked. The importance of the criteria can be weighted, if appropriate, and the most suitable concept identified as the one with the highest overall mark. This method provides a structured technique of evaluation and makes it more difficult for individuals within a team to push their own ideas for irrational reasons. The outcome is a quasiquantitative measure of the merits of a specific idea and ranking of concepts.

An example of the use of an evaluation matrix is given in Table 3.4 considering four different concepts for a pallet-moving device. The criteria considered important here are safety, positional accuracy, ease of control, ease of manufacture and durability. These criteria have been weighted with positional accuracy deemed the most important. The scores are given with a maximum of 10. The marks in the columns for the weighted scores have been summed and the totals are given at the bottom. The highest weighted score indicates the concept that should be selected and developed.

3.4 Creative problem solving

In addition to developing and formulating effective rules for brainstorming, Osborn and Parnes also considered the process of solving a problem (Parnes et al., 1977). They developed an approach for this, comprising 12 steps and 6 stages, called the Creative Problem-Solving process (CPS). The process is based on phases of divergent and

Table 3.3 Generic evaluation matrix

Selection criteria	Weight	Concepts					
		Concept A		Concept B		Concept N	
		Rating	Weighted score	Rating	Weighted score	Rating	Weighted score
Aspect 1	Percentage importance of Aspect 1	Mark out of 10 for Aspect 1 of Concept A	Rating × weight for aspect 1 of Concept A	Mark out of 10 for Aspect 1 of Concept B	Rating × weight for Aspect 1 of Concept B	Mark out of 10 for Aspect 1 of Concept N	Rating × weight for Aspect 1 of Concept N
Aspect 2	Percentage importance of Aspect 2	Mark out of 10 for Aspect 2 of Concept A	Rating × weight for Aspect 2 of Concept A	Mark out of 10 for Aspect 2 of Concept B	Rating × weight for Aspect 2 of Concept B	Mark out of 10 for Aspect 2 of Concept N	Rating × weight for Aspect 2 of Concept N
…							
Aspect n	Percentage importance of Aspect n	Mark out of 10 for Aspect n of Concept A	Rating × weight for Aspect n of Concept A	Mark out of 10 for Aspect n of Concept B	Rating × weight for Aspect n of Concept B	Mark out of 10 for Aspect n of Concept N	Rating × weight for Aspect n of Concept N
Total weighted score			Sum of weighted scores		Sum of weighted scores		Sum of weighted scores
Ranking							

Table 3.4 Evaluation tabular matrix for a pallet-moving device

		Concepts							
		Hydraulic		Cable hoist		Linkage mechanism		Rack and pinion	
Selection criteria	**Weight**	**Rating**	**Weighted score**	**Rating**	**Weighted score**	**Rating**	**Weighted score**	**Rating**	**Weighted score**
Safety	20%	2	0.4	3	0.6	4	0.8	2	0.4
Positional accuracy	35%	6	2.1	5	1.75	7	2.45	6	2.1
Ease of control	10%	6	0.6	5	0.5	7	0.7	6	0.6
Ease of manufacture	20%	7	1.4	8	1.6	5	1.0	6	1.2
Durability	15%	7	1.05	6	0.9	5	0.75	6	0.9
Total Score		5.55		5.35		5.7		5.2	
Rank		2		3		1		4	
Continue?		No		No		Develop		No	

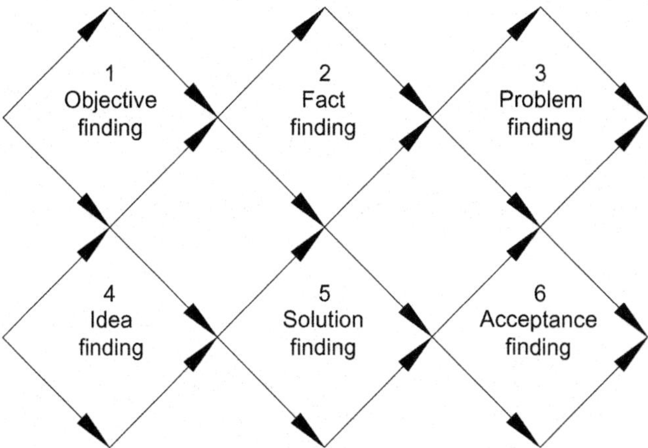

Fig. 3.8 The creative problem-solving process.

convergent thinking, illustrated schematically in Fig. 3.8, and is one of the oldest structured frameworks for problem solving.

The six stages of the creative problem-solving process which feed sequentially into each other are:

- Objective finding—An objective can arise from experience and day-to-day life, requirements of the domain and social pressure.
- Fact finding—Information relevant to the issue that needs to be defined. For example, limits on allowable size or mass, how many users there are, and are there any legal requirements.
- Problem finding—For any task, there are associated challenges that need to be resolved. In the context of creative problem solving, problem finding relates to exploring what the deficiencies of existing solutions really are or which issues or challenges arise in tackling the problem.
- Idea finding—The generation of multiple concepts that may be useful and relevant to the challenge.
- Solution finding—Elaboration is necessary to turn any idea into a viable solution. Elaboration may involve exploring design issues in some further detail providing embodiment to the idea so that there is confidence that it will work.
- Acceptance finding—Clients and peers need to be persuaded that the idea developed is going to work. A bright idea on its own is not enough. Instead any issues relating to it need to be explored in sufficient and convincing enough a manner that all relevant parties, clients, colleagues, end users and customers can be persuaded that the solution is the right one.

Osborn and Parnes recognised that CPS requires convergence as well as divergence, and each stage of CPS involves this. The divergent phase requires creative thinking where you encounter paradoxes, opportunities, challenges and concerns, and search and generate possible solutions. This may involve generating many new and varied potential solutions and details that may contribute towards a solution. The convergent phase requires critical thinking with examination of the possibilities and focusing of thoughts. This may involve organising and analysing the potential solutions and refining them to improve their viability. Potential solutions can be ranked and prioritised so that attention can be concentrated on developing particular ideas. Each stage can involve use of brainstorming with deferment of judgement, production of many ideas and looking for combinations of ideas, prior to sifting and refining these ideas.

Effective tools are required to help evaluate competing ideas to overcome the temptation to slam an idea as unworkable at too early a stage in its development. Tools for focusing options, recommended by Treffinger et al. (2006), include:

- Hits and hotspots—Identifying promising possibilities and clustering and organising them to see if meaningful insights can be identified.
- ALoU—Advantages, Limitations and Unique features—Consider the advantages, limitations and possible methods to overcome them and the unique features of an idea in a constructive manner with a view to strengthening the idea.
- PCA—Paired Comparison Analysis.
- Sequencing—Short, Medium and Long-term actions.
- Evaluation matrices.

In application of CPS or any creative method, considering the magnitude of the challenge is as important as assessing whether there is a high level of tension or conflict in the group of personnel involved. If an initial exploration of the data available reveals that the task to be undertaken is large, then subdivision of the activity may be appropriate with delegation of duties to specific subteams. If there is already a high level of tension or conflict in a team, then conflict resolution or team-building techniques will need to be employed.

CPS can be used when working individually or in a team (Isaksen, 2000). If working in a team, then this will require more planning and resourcing than if working alone as it will require coordination of several people and provision of resources.

The six stages of CPS can be grouped in phases.

- Identify and clarify the challenge: Stages 1–3
- Finding solutions and evaluation: Stages 4 and 5
- Putting the solutions into action: Stage 6

A statement of opportunity does not represent a statement of a problem. The purpose of identifying and clarifying the challenge, stages 1–3, is to enable define the problem you need to resolve and to give a direction to the effort. Asking questions *who*, *what*, *where*, *when* and *how* can be useful in identifying key data relevant to the problem. In addition, considering a future desired state in the form of an ideal final result can inspire formulation of the problem statement to enable you to get from the current status to fulfilment of the objective.

Problems sometimes arise as a result of dire prospects or from aspirations and wishes. Two facilitating constructs used in CPS to explore these situations are WIBAI and WIBNI. These represent '*Wouldn't it be awful if …*?' and '*Wouldn't it be nice if…?*' These constructs can be used to explore and develop an opportunity. An issue may commence with '*Wouldn't it be awful if…?*' and be turned into a constructive opportunity by '*Wouldn't it be nice if…*?'

Experience in using the CPS technique has resulted in the use of a series of facilitating open-ended problem statements that invite ideas, including:

'How might we...?' or *'How might I...?'* HMW, HMI.
'In what ways might we...?' or *'In what ways might I...?'* IWWMW, IWWMI.
'How to ...?' H2.

These constructs can be combined with a verb and objective and used to explore and construct an opportunity. Examples of this are 'How might we extend the useful life of this product', and 'In what ways might we increase the chance of winning this bid?'

Constructs for encouraging divergent thinking include consideration of what options and alternatives there are or might be. Convergent thinking involves focusing attention on the hits or most appealing alternatives in an extensive list of ideas.

Once an idea has been explored and selected as the best option, it will be necessary to persuade other stakeholders, such as managers and colleagues, that it is the right idea. It is helpful to identify people and factors that can assist this. The 5Ws (*who, what, where, when, with whom*) can be used to consider, for example, who can help, what resources are necessary for successful implementation of the idea, when is the best time to implement the plans, where the ideas can be developed, and why the idea should be taken forward. In a similar fashion, detractors or resistance to an idea can be explored to mitigate against things that could go wrong. Again, the 5Ws questions can be used to identify who might criticise the ideas and potential responses for them, what resources might be missing or unavailable, when might the worst possible time be to implement or reveal the idea, where might the worst place or factory be and what are the least persuasive aspects of the proposition.

3.5 SCAMPER

Osborn developed sets of words and phrases called idea-spurring questions. Bob Eberle subsequently developed these ideas into a readily memorable acronym called SCAMPER (Eberle, 1997). SCAMPER stands for

- Substitute
- Combine
- Adapt
- Magnify/Minify/Modify
- Put to other uses
- Eliminate
- Reverse/Rearrange

Each letter can be used to remind us of a word or phrase and this can be used to inspire and provoke related ideas. SCAMPER can be particularly useful when we feel stuck or uninspired, and it can be employed to get ideas flowing or to provoke thinking from another perspective. Some possible provocations for SCAMPER are given in Table 3.5.

SCAMPER can be applied to almost any domain. We do this almost instinctively in addressing an issue.

Table 3.5 SCAMPER provocations

Substitute	What might you use instead?
	What can be done instead?
	Who else?
	Can the rules be altered?
	Alternative materials?
	Another place?
	Other processes or procedures?
	Another approach?
Combine	What ideas can be combined?
	What materials can be combined?
	What methods can be combined?
	What procedures or tasks can be combined?
	What functions can be combined?
	What can be blended?
	What can be synthesised?
	What can be mixed?
	Can an assortment fulfil the requirements?
Adapt	What might be changed?
	What might be used in a different way?
	How can this be adapted for different customers?
	What other ideas does this suggest?
	What else is like this?
	What can be copied to produce this?
	Who could be emulated?
	What different contexts can this concept be applied to?
	What ideas from other disciplines can be applied?
Magnify/modify	Can the idea or object be scaled?
	What can be made larger?
	What can be made smaller?
	Can it be multiplied?
	Can the colour be changed?
	Can the shape or/and form be changed?
	Can the sound be altered?
	Can the scent be changed?
	What changes can be made elsewhere in the system?
Put to other uses	What else can it be used for?
Eliminate	What can be omitted?
	Can it be divided?
	Can it be separated?
	What can be trimmed?
	What can we do without?
	Can it be streamlined to reduce losses?
	Can it be miniaturised?
	Can something be subtracted?
	Can something be deleted or removed?
	What is unnecessary?
Reverse/rearrange	What can be made opposite?
	What can be mirrored?
	Can it be turned inside out?
	Can roles be reversed?
	What are the negatives?
	What can be restructured?

3.6 Create process

The Create Project (2005) offered a framework, called the Create Process, for generative activity developed in association with industry and academia. The Create Process comprises phases on predisposition, external mapping, internal mapping, idea generation and evaluation. This process is summarised in Table 3.6 and provides a validated framework within which to undertake the design process. The Create Process can be implemented using a series of practices readily recognisable and familiar to the practicing engineer.

Possible techniques suggested for implementation of each phase of the Create Process are listed below (Childs et al., 2006).

- Predisposition—Providing a work environment that allows expression of individual creativity, promotes a creative culture and identifies leaders and facilitators. This can be achieved by training in creativity and innovation, and change management processes.
- External mapping—Analysing the environment outside the organisation to identify, for example, new needs and opportunities, talented individuals and the economic implications

Table 3.6 The Create Process phases (The Create Project, 2005)

Phase	Summary
Predisposition	This concerns the creation of those essential circumstances in an organisation that are the preliminary step to the development of creativity: • company atmosphere; • leadership style; • culture; • organisation structure and systems.
External mapping	In this phase, a series of activities are included, such as: • the search for entrepreneurial ideas linked to the environment outside a company; • identification of new and unexpressed needs; • utilisation of talents; • new opportunities and abilities; • evaluation of the economic implications of possible changes in competition.
Internal mapping	This phase concerns the identification and evaluation of all the company resources to promote the development of new opportunities/business concepts by making the most of its potential. It also includes an internal study to identify malfunctions, processes and products that could be improved.
Idea generation	This phase concerns the production of ideas that can occur at an individual and joint level.
Evaluation	This involves identification and selection of the best ideas according to the organisation's, group's or company's own internal assessment criteria.

of entering a market. A questionnaire survey and analysis, known as Attribute Value Chain, can be used to produce a map to stimulate the next phase of idea generation.

- Internal mapping—Analyse the organisation concerned using, for example, SWOT analysis to identify the Strengths, Weaknesses, Opportunities and Threats.
- Idea generation—Augment with, for example, brainstorming, morphological analysis and TRIZ.
- Evaluation—Select the best idea(s) using, for example, the criteria matrix and six thinking hats (see De Bono, 1999; Childs, 2012).

3.7 Morphological analysis

Morphological analysis is a technique that can be used to generate additional ideas for product concepts that would not normally spring to mind. The technique involves considering the function of a generic solution to a problem and breaking it down into a number of systems or subfunctions. The next step is to generate a variety of means to fulfil each of these systems or subfunctions. The subfunctions and potential means of fulfilling each of these subfunctions are arranged in a grid. An overall solution is then formulated by selecting one means for each subfunction and the combination of these forms the overall solution. A generic morphological analysis table is illustrated in Table 3.7. The grid can be populated by text or sketches depicting the potential means of fulfilling the subsystem requirement (Childs, 2004). The technique relies upon the user's selection of subfunction options to synthesise an overall solution or set of possibilities. The user can explore the design space systematically, using their experience to guide the outcome, or experimentally. The matrix given in Table 3.7 is abstract and

Table 3.7 Generic morphological analysis chart

Subsystem	Means			
Subsystem 1	Method 1 of fulfilling subsystem 1	Method 2 of fulfilling subsystem 1	Method 3 of fulfilling subsystem 1	Method n of fulfilling subsystem 1
Subsystem 2	Method 1 of fulfilling subsystem 2	Method 2 of fulfilling subsystem 2	Method 3 of fulfilling subsystem 2	Method n of fulfilling subsystem 2
Subsystem 3	Method 1 of fulfilling subsystem 3	Method 2 of fulfilling subsystem 3	Method 3 of fulfilling subsystem 3	Method n of fulfilling subsystem 3
Subsystem 4	Method 1 of fulfilling subsystem 4	Method 2 of fulfilling subsystem 4	Method 3 of fulfilling subsystem 4	Method n of fulfilling subsystem 4
Subsystem 5	Method 1 of fulfilling subsystem 5	Method 2 of fulfilling subsystem 5	Method 3 of fulfilling subsystem 5	Method n of fulfilling subsystem 5

generic in its formulation. Care needs to be applied in defining the functional requirements to avoid the specification of functions and subfunctions with a preconceived solution. Specific examples arising from practical implementation are given in Sections 3.7.1 and 3.7.2.

Morphological analysis is a phrase common to a number of subject disciplines. In this chapter, however, the term morphological analysis is used to refer to a particular creativity tool for the generation of alternative concepts by means of development of generic subfunctions for a product or process, and consideration of alternative means for implementation of each subfunction.

Morphological analysis was originally developed and applied to jet engine technology following the Second World War by F Zwicky (see, e.g. Zwicky (1969) and the Swedish Morphological Society), although Allen calls his 1962 book the Allen morphologizer. Zwicky identified the important parameters of aviation engine technology, including the thrust mechanism, oxidiser and fuel type. He then broke each of these functions down into constituent parts under each parameter or subsystem heading. He then assembled all possible permutations, such as a ramjet that used, for example, atmospheric oxygen and a solid fuel. For some of the permutations produced, an existing product and solution using this approach already existed. The new or original permutations were viewed by Zwicky as stimuli for creativity and worthy of the question 'why not?'. Morphological analysis is a proven ideation method that results in organised invention and has been demonstrated to be appropriate for new product development, services, patent definition and value management. The technique comprises two key elements:

- a systematic analysis of the current and future structure of a domain as well as identification of key gaps in that structure;
- a strong stimulus for the invention of new alternatives that fill these gaps and meet any imposed requirements.

A typical morphological field can contain thousands to millions of formal configurations. This number of permutations is far too many to inspect by hand and to overcome this, the internal relationships between the field parameters can be examined and the field reduced by weeding out all mutually contradictory conditions. In addition, increased levels of sophistication can be applied to morphological analysis, such as weighting of ideas and the addition of costing information to guide users in the impact of the individual selection of each subfunctional option. In the absence of coherent strategies to identify preferred solutions, it has been observed that once a few sets of plausible alternative selections have been tried, searching for a better solution gets abandoned (see Jones, 1970). Whiting (1958) notes that in practice too many possibilities can be developed leading to problems in evaluation of the alternative concepts effectively. Recently, however, computational approaches have enabled nonconsistent combinations to be excluded (see Childs and Garvey, 2015; Garvey, 2016).

To illustrate the technique, a case study for a pallet-moving device is considered in Section 3.7.1, prior to the introduction of case studies across diverse applications in Section 3.7.2.

3.7.1 Pallet case study

Following some initial market assessments, it is assumed that the board of a plant machinery company has decided to proceed with the design of a new product for transporting pallets around factories.

The use of morphological charts requires the designer to consider the function of components rather than their specific details. A pallet-moving device would probably comprise of support, propulsion, power, transmission, steering, stopping and lifting systems. Within each of these criteria all plausible options should be listed as indicated in Table 3.8. In this table only a selection of possible means for fulfilling each subfunction have been included to illustrate the technique. In practice, such tables can occupy large sheets of paper or many columns in a spreadsheet. The task of the designer is to look at this chart and select a justifiable option for each of the criteria producing the make-up of the pallet-moving device. In this case there would not necessarily be any surprises in the final make-up of the product and selection of wheels, driven wheels, petrol, gears and shafts, turning wheels, brakes, hydraulic ram, and seated at the front would result in the traditional forklift truck epitomised by the E12 to E20 Linde trucks. However, the use of this method can assist in the production of alternatives.

In a regular exercise, engineering and design students are challenged to use this table and produce three radically different alternatives for consideration, and then sketch them. For example, air cushion for support, air thrust for propulsion, bottled gas for power, gears and shafts for transmission, air thrust for steering, reverse thrust for stopping, hydraulics for lifting, and seated for the operation could be selected for each of the subsystems respectively. Images from this exercise are illustrated in Fig. 3.9. While some aspects of the images given in Fig. 3.9 may promote some alarm

Table 3.8 Morphological chart for a pallet-moving device

Feature	Means				
Support	Track	Wheels	Air cushion	Slides	Pedipulators
Propulsion	Driven wheels	Air thrust	Moving cable	Linear induction	Toothed wheel and rack
Power	Electric	Diesel	Petrol	Bottled gas	Steam
Transmission	Belts	Chains	Gears and shafts	Hydraulics	Flexible cable
Steering	Turning wheels	Air thrust	Rails	Magnetism	Drag
Stopping	Brakes	Reverse thrust	Ratchet	Magnetism	Anchor
Lifting	Hydraulic ram	Rack and pinion	Screw	Chain or rope hoist	Telescopic linkage
Operator	Standing	Walking	Seated at front	Seated at rear	Remote control

Fig. 3.9 Sketches following selection of alternative concepts (University of Sussex, 2004; Beihang University, 2008; RCA, 2008).

among experienced plant designers and health and safety experts, the intent of the technique is to provide alternatives for consideration. Morphological analysis serves to augment the range of possibilities to be considered and does not negate the responsibilities of a professional in carrying out their task responsibly and utilising their detailed analytical skills. The morphological analysis chart repeated in Table 3.9 shows the selection of a telescopic based lifting method and an implementation of this is illustrated in Fig. 3.10.

Table 3.9 Morphological chart for a pallet-moving device with choices identified

Feature	Means				
Support	Track	Wheels	Air cushion	Slides	Pedipulators
Propulsion	Driven wheels	Air thrust	Moving cable	Linear induction	Toothed wheel and rack
Power	Electric	Diesel	Petrol	Bottled gas	Steam
Transmission	Belts	Chains	Gears and shafts	Hydraulics	Flexible cable
Steering	Turning wheels	Air thrust	Rails	Magnetism	Drag
Stopping	brakes	Reverse thrust	Ratchet	Magnetism	Anchor
Lifting	Hydraulic ram	Rack and pinion	Screw	Chain or rope hoist	Telescopic linkage
Operator	Standing	Walking	Seated at front	Seated at rear	Remote control

Fig. 3.10 The Teletruk concept by JCB.
Courtesy of J.C. Bamford Excavators Ltd.

3.7.2 Applications of morphological analysis

Although originally developed to tackle technological challenges in the US aerospace industry and military sector, the formulation of morphological analysis does not limit it to a particular discipline. As examples, Tables 3.10 and 3.11 show the development of morphological analysis charts to an image recording device, and vehicle design. The method has also been applied to applications such as protein identification, interior design and creative writing. The diverse nature of these examples indicates the potential for this technique.

Morphological analysis has a long history of application in industry and in education. The technique involves formulation of the design space in terms of subfunctions and potential options for each of the subfunctions. The technique relies upon the user's selection of subfunction options to synthesise an overall solution or set of possibilities. The user can explore the design space systematically, using their experience to guide the outcome, or experimentally. This section has illustrated the use of morphological analysis across a wide range of disciplines focusing on technology, mobility and healthcare. The case studies highlighted demonstrate that the technique is accessible across a range of disciplines and assists in provoking consideration of alternatives. The process of selection and generation of a quick sketch or outline provides a prompt for dealing with the integration of the subfunctions into a holistic proposal. Some concerns remain that too many possibilities can be developed leading to problems in evaluating the alternative concepts effectively. This can be offset by means of weighted functions relating to, for example, cost of implementation or marketing advantage. A structured approach using morphological analysis supported by software to perform a cross consistency analysis and refine the number of plausible options is reported by Garvey and Childs (2013), Childs and Garvey (2015) and Garvey (2016). An example of the cross-consistency analysis used to reduce the number of combinations that are to be considered is given in Fig. 3.11 and a solution space in Fig. 3.12.

The diverse case studies presented reveal that the technique can be accepted and readily used by diverse groupings of individuals. It is proposed that morphological analysis is valuable in producing additional conceptual solutions and ideas for consideration, but it is not a substitute for the application of experience, use of skills and innate talent.

3.8 Standard solutions

Many organisations rely on new products and services to sustain their activity and grow. Examples include Proctor and Gamble, Unilever, Sony, Motorola, Hewlett-Packard and 3M where more than 30% of their revenue is due to products introduced in the previous 2 years. Sony introduces about 50 new products every year. These examples show the importance of the realisation of new ideas in a form that can generate revenue. In the United Kingdom, the Cox review (Cox, 2005) confirmed this by stating that 75% of company turnover stems from products developed within the previous 5 years.

Table 3.10 Morphological chart for an image recording device

Subfunction	Means							
Image capture	Lens	Detector array	Static charge	Photosensitive bacteria	Chemical			
Image storage	Chemical film	Chemical bath	Digital memory	Static charge	Photosensitive bacteria	Instant processing		
Actuation	Tilt button	Joystick	Push button	Programmable	Timer	Tilt	Light activated	
Image display	Project image	Screen	Lens	Printed image	None	Liquid crystal display	Use computer or phone screen	Hologram projection
Storage	Clamshell mould	Compliant material case	None	Metal case	Contained in another product			
Image development	Chemical processing	Memory manipulation	None	Printing	Electromagnetic waves	Ultrasonic		
Communications	Blue tooth	Infrared	Cable digital	Cable analogue	Physical transfer			
Protection of subsystems	Moulded casing	Bag	Frame	C-shell	Compliant material case	Shell		
Power	Manual	None	Biological	Battery	Ac mains	Clockwork	Solar	Chemical
Target illumination	Flash	None	Infrared	Spot lamp	Reflector	LASER	Natural light	
Added functionality	Video	Zoom	Sound recording	Image manipulation	Lock	Time reference	Manual focus	Automatic focus

Table 3.11 Morphological chart for vehicle design

Subsystem	Means									
Stopping	Disc brake	External drum brake	Internal drum brake	Parachute	Air jet	Momentum exchange	Magnetic induction	Anchor		
Steering	Steering wheel	Joy stick	Handle bars	Automatic	Reins	Remote control	Air jet	Levers	Prosthetic implant on wrists and static sensors	Magnetic
Styling	Futuristic	Space age	Old school	Retro 1960s	Retro 1970s	Sporty	Aerodynamic	Functional	Chunky	Off road
Prime mover	4 stroke	2 stroke	GTE	Human powered	Electric motor	Electromagnetic induction	Hybrid	Clockwork	Rocket	Jet
Transmission	Gears	Belts	Chains	Hydraulics	Pneumatics					
Propulsion	Wheels	Balls	Tracks	Air cushion	Glide on smooth surface	Electromagnetic induction	Rails	Rollers		
Accessory pack	GPS, e windows, air con	MP3 stereo, back seat speaker	Sun roof, alloys, blaser hook	Two bar, caravan mirror	DVD players	Underpan neon lighting				
Entry	Teleport	Electric ramp	Door	Hole	Gull-wing	Cockpit entry	Curtain			
Storage	Boot	Bonnet	Top box	Under floor	Overhead storage	Panniers	Roof rack			
Purpose	Family	Commercial	Sport	Professional female	City	Off road	Racing	Business	Wideboy	Manual worker
Seating	Standing	2+2	2+3	Bench seat	2 front +bed	Single seater	None – commercial only	Kneeling pads	Double decker	
Fuel	Cow manure	Diesel	Petrol	Cooking oil	H_2	Compost	Steam	Electricity	Nuclear	Bio-fuel
Performance	Sports	Efficient	High octane	All round	High load capacity	Aid	Mobility +	MPV	Camping	

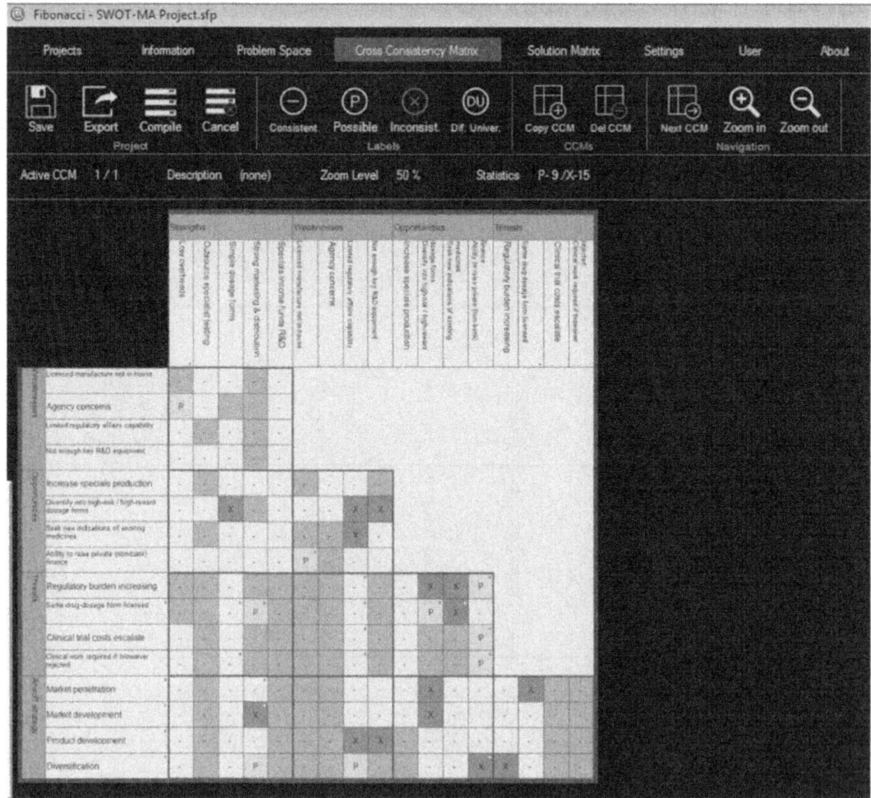

Fig. 3.11 Use of cross-consistency analysis to reduce the problem space in morphological analysis.

Fibonacci - Chemical Spillage.sfp

Projects | Information | Problem Space | Cross Consistency Matrix | Solution Matrix | Settings | User | About

Save | Export | Freeze | Defreeze | Backtrack

Showing Single | Include Possible | Exclude Possible | CCM 1 | Total solutions including/excluding possible scenarios: 2122 / 2122 | Selected solution size 2122 | Solution #23888

TYPE OF PLAN	TRAINING & EDUCATION	PERSONNEL	EQUIPMENT AVAILABILITY	WHO AUTHORISES?	STOPPING RELEASE	PUBLIC WARNING	TREATMENT POSSIBILITIES
Full preparedeness plan	Broad co-op training	11 or more	Specialised	Minister	Reduce by 80% in 15 min	Warn in 5 minutes	Help many within 30 minutes
Advanced for specific cases	Training for specific cases	8 to 10	Basic for all cases	Regional director	Reduce by 80% in 30 min	Warn in 30 minutes	Help some within 15 minutes
Standard for specific cases	Basic education & training	5 to 7	Less than basic for all cases	Manager	Reduce by 50% in 15 min	No warning possible	Help some within 30 minutes
Standard plan for all cases	Basic education only	4 or less		Employee	Reduce by 50% in 30 min		No help within 30 minutes
Only an alert plan					No measures in 30 min		

Fig. 3.12 A solution space enabled by the use of general morphological analysis software.

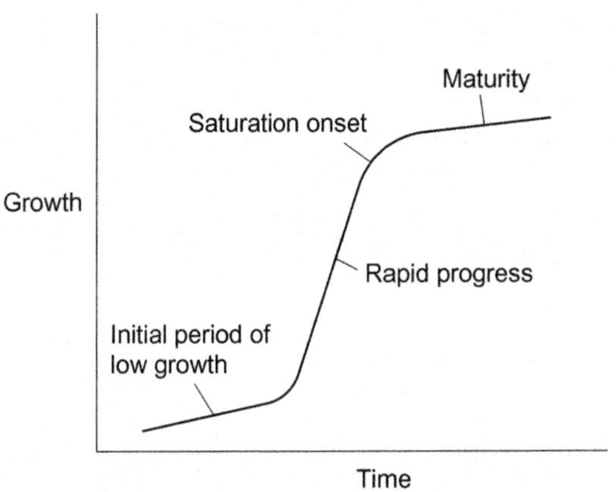

Fig. 3.13 A Sigmoid S curve showing infancy in growth or progress, followed by rapid growth, onset of saturation and maturity.

The requirement for new products and services is embodied in the Sigmoid S curve (Fig. 3.13). Many processes that involve growth, such as scientific progress and sales, tend to follow this S curve. The process is characterised by an initial period of low growth where progress is slow. This is followed typically by a period of rapid progress and growth until a point where the process starts to saturate bringing a reduction in the rate of growth and the progress beginning to level off. These periods can be classed as infancy, rapid progress, saturation onset and maturity.

However, this is not the end of the story. The S curve nature of many products and services is subject to intervention and deviations as a result of competitive and collaborative factors. Companies within a market sector during a period of high sales and rapid growth will be aware that at some stage the sales successes are likely to level off, with most consumers likely to want to purchase the item having considered it or bought it. In the case of scientific development, the 'easy' hits of knowledge and capability may have been obtained, and advances level off as a result of the law of diminishing returns where more and more effort is required for less and less progress. As a result of intervention or simultaneous action, development may start on a new related area within the field. Initially the performance in terms of output in this new area may be below that of the existing approach. However, as effort is applied, progress begins to take off and the new approach goes into a period of rapid growth with performance exceeding that of the previous idea. This process can then be repeated as illustrated in Fig. 3.14 in the form of a series of S curves.

The repeated experience of trends following a family of S curves for a wide range of products and processes demonstrates the need for realisation of new ideas to sustain an organisation. Changing from making proven products that a company is familiar and feels safe with, to innovative new products is not easy. Things we might have cherished in the past and found security in, such as specialist skills, habits, paradigms, or working environment, can nourish mental blocks that constrain our innovative thinking. Such constraints can act as psychological inertia, preventing a domain from moving forward.

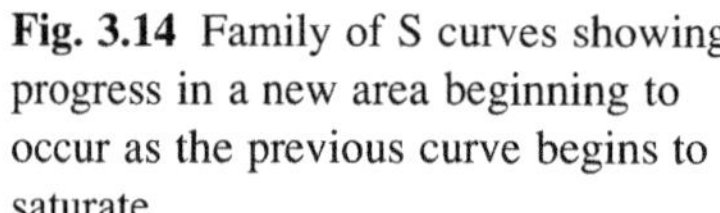
Fig. 3.14 Family of S curves showing progress in a new area beginning to occur as the previous curve begins to saturate.

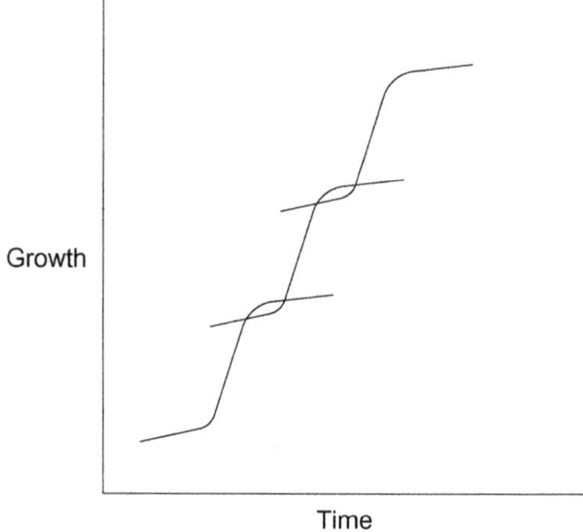

TRIZ is a technology-based systematic methodology that aims to overcome 'psychological inertia' and generate a large range of solution concepts. The emphasis is on finding innovative solutions and concepts, from other fields that use available resources. TRIZ is a Russian acronym for *Teoria Resheneya Isobretatelskih Zadach*. This can be translated as the 'Theory of Inventive Problem Solving' which would give the acronym TIPS. Nearly all practitioners, however, use its original acronym, TRIZ.

TRIZ was originally developed by Genrich Altshuller. In 1946 Genrich Altshuller, aged 20, was a reviewer for the registry of inventions/patents at the Russian naval patent office. In reviewing inventions being submitted he became convinced that there were definite patterns to the way innovation was occurring in technical systems. Altshuller and his collaborators based their initial findings on an analysis of 46,000 registrations of inventions/patents. This was then expanded to a study of 200,000 patents. Since then analysis of the world-wide patent base has continued and some of the modern TRIZ databases rely on analysis of over 2 million patents. Altshuller and his collaborators found that the same fundamental solutions were used over and over again. A series of tools was developed (Altshuller, 1984, 1994, 2005) which pulled this experience together and could be applied to solving new problems.

TRIZ represents a short-cut to experience. You could spend your life studying, reading, going around museums looking at iconic designs, taking note of the latest scientific breakthroughs and generally expanding your knowledge base. This will, combined with your own activities in a domain, provide a valuable set of experiences with which you can tackle problems. Each time you see how someone else has solved a problem you are likely to internalise the experience and hopefully be able to apply the principle when needed. This gathering of experience takes time. The principles embodied in TRIZ provide a faster route to the tool-box of life-times of experience.

TRIZ involves the transformation of a specific problem into that of a general problem, the identification of solutions that have previously been successful for the general problem identified and then the application of these principles to the specific problem.

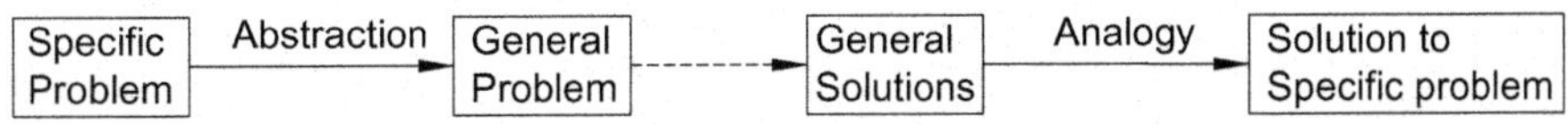

Fig. 3.15 The TRIZ process.

If the problem is made general, then it is possible to take advantage of the knowledge available from the solution of similar problems. The general solutions can then be transformed into specific ideas to resolve the specific problem (Fig. 3.15).

3.8.1 Contradiction and trade-off

A common characteristic of problem solving is that in trying to improve one desirable parameter or property, another desirable property degrades. This characteristic of problem solving can be viewed as a trade-off or contradiction. Examples of contradictions in tackling a problem can be observed in the following examples.

- Bin design: A large bin is desirable so that it does need to be emptied often but a small is desirable so that it does not occupy too much space.
- The forming of a horseshoe: A horseshoe needs to be hot so that it is easier to alter the shape of the material but it is also desirable for the shoe to be cold so that it is easy to handle.
- Axe design: If you make the blade heavier it can strike a more effective blow with better splitting power, but it becomes more awkward to handle.

Conventional problem solving generally leads to a 'compromise' solution. Altshuller noted that the most 'inventive' solution is obtained when a technical problem containing a 'contradiction' is solved by completely eliminating the contradiction.

From the initial research, Altshuller found that there were only 39 generic parameters which either improve or degrade. Examples of these parameters include items such as the weight of a moving object, speed, force, temperature, accuracy of measurement and ease of repair. In the development of the list of generic parameters, specific meaning was given to a moving and stationary object. Objects which can easily change position in space, either on their own or as a result of external forces, are classed as moving objects. Objects or vehicles that are designed to be portable are the basic members of this class. Objects which do not change position in space, either on their own or as a result of external forces, are classed as stationary objects. The 39 generic parameters and a brief explanation are given in Table 3.12.

If the concept of contradiction in problem solving is accepted, then every problem can be described as a conflict between a pair of parameters. Altshuller and his collaborators observed that many patents had, in the past, resolved these individual conflicts in several different fields. He also found that the principles used to resolve these contradictions fully, not as a trade-off or compromise solution, could be described in a list of just 40 principles. These 40 principles are listed in Table 3.13. TRIZ has developed over several decades and the specific names and numbering for the principles vary according to author and translation. Along with the names for the principles used in this text, a list of other commonly used names are given in the table.

Table 3.12 The 39 parameters of TRIZ

No	Title	Explanation
1	Weight of moving object	The mass of the object in a gravitational field. The force that the body exerts on its support or suspension.
2	Weight of stationary object	The mass of the object in a gravitational field. The force that the body exerts on its support or suspension, or on the surface on which it rests.
3	Length of moving object	Any one linear dimension, not necessarily the longest, is considered a length.
4	Length of stationary object	Any one linear dimension, not necessarily the longest, is considered a length.
5	Area of moving object	A geometrical characteristic described by the part of a plane enclosed by a line. The part of a surface occupied by the object, OR the square measure of the surface, either internal or external, of an object.
6	Area of stationary object	A geometrical characteristic described by the part of a plane enclosed by a line. The part of a surface occupied by the object, OR the square measure of the surface, either internal or external, of an object.
7	Volume of moving object	The cubic measure of space occupied by the object. Length x width x height for a rectangular object, height x area for a cylinder and so on.
8	Volume of stationary object	The cubic measure of space occupied by the object. Length x width x height for a rectangular object, height x area for a cylinder, and so on.
9	Speed	The velocity of an object; the rate of a process or action in time.
10	Force	Force measures the interaction between systems. In Newtonian physics, force = mass x acceleration. In TRIZ, force is any interaction that is intended to change an object's condition.
11	Stress or pressure	Force per unit area. Also, tension.
12	Shape	The external contours; appearance of a system.
13	Stability of the object's composition	The wholeness or integrity of the system; the relationship of the system's constituent elements. Wear, chemical decomposition and disassembly are all decreases in stability. Increasing entropy is decreasing stability.
14	Strength	The extent to which the object is able to resist changing in response to force. Resistance to breaking.
15	Duration of action by a moving object	The time that the object can perform the action. Service life. Mean time between failures is a measure of the duration of action. Also, durability.
16	Duration of action by a stationary object	The time that the object can perform the action. Service life. Mean time between failures is a measure of the duration of action. Also, durability.

Table 3.12 Continued

No	Title	Explanation
17	Temperature	The thermal condition of the object or system. Loosely includes other thermal parameters, such as heat capacity, that affect the rate of change of temperature.
18	Illumination intensity	Light flux per unit area, also any other illumination characteristics of the system such as brightness, light quality and so on.
19	Use of energy by moving object	The measure of the object's capacity for doing work. In classical mechanics, energy is the product of force x distance. This includes the use of energy provided by the super-system (such as electrical energy or heat). Energy required doing a particular job.
20	Use of energy by stationary object	The measure of the object's capacity for doing work. In classical mechanics, energy is the product of force x distance. This includes the use of energy provided by the super-system (such as electrical energy or heat.) Energy required doing a particular job.
21	Power	The time rate at which work is performed. The rate of use of energy.
22	Loss of energy	Use of energy that does not contribute to the job being done. See 19. Reducing the loss of energy sometimes requires different techniques from improving the use of energy, which is why this is a separate category.
23	Loss of substance	Partial or complete, permanent or temporary, loss of some of a system's materials, substances, parts or subsystems.
24	Loss of information	Partial or complete, permanent or temporary, loss of data or access to data in or by a system. Frequently includes sensory data such as aroma, texture, sound.
25	Loss of time	Time represents the duration of an activity. Improving the loss of time means reducing the time taken for the activity. 'Cycle time reduction' is a common requirement.
26	Quantity of substance/ the matter	The number or amount of a system's materials, substances, parts or subsystems which might be changed fully or partially, permanently or temporarily.
27	Reliability	A system's ability to perform its intended functions in predictable ways and conditions.
28	Measurement accuracy	The closeness of the measured value to the actual value of a property of a system. Reducing the error in a measurement increases the accuracy of the measurement.
29	Manufacturing precision	The extent to which the actual characteristics of the system or object match the specified or required characteristics.

Continued

Table 3.12 Continued

No	Title	Explanation
30	External harm affects the object	Susceptibility of a system to externally generated (harmful) effects.
31	Object-generated harmful factors	A harmful effect is one that reduces the efficiency or quality of the functioning of the object or system. These harmful effects are generated by the object or system, as part of its operation.
32	Ease of manufacture	The degree of facility, comfort or effortlessness in manufacturing or fabricating the object/system.
33	Ease of operation	Simplicity: A process is not easy if it requires a large number of people, large number of steps in the operation, needs special tools and so on. "Hard' processes have low yield and 'easy' processes have high yield; they are easy to do right.
34	Ease of repair	Quality characteristics such as convenience, comfort, simplicity and time to repair faults, failures or defects in a system.
35	Adaptability or versatility	The extent to which a system/object positively responds to external changes. Also, a system that can be used in multiple ways in a variety of circumstances.
36	Device complexity	The number and diversity of elements and element interrelationships within a system. The user may be an element of the system that increases the complexity. The difficulty of mastering the system is a measure of its complexity.
37	Difficulty of detecting and measuring	Measuring or monitoring systems that are complex, costly, require much time and labour to set up and use, or that have complex relationships between components or components that interfere with each other all demonstrate 'difficulty of detecting and measuring'. Increasing cost of measuring to a satisfactory error is also a sign of increased difficulty of measuring.
38	Extent of automation	The extent to which a system or object performs its functions without human interface. The lowest level of automation is the use of a manually operated tool. For intermediate levels, humans programme the tool, observe its operation, and interrupt or reprogramme as needed. For the highest level, the machine senses the operation needed, programmes itself, and monitors its own operations.
39	Productivity	The number of functions or operations performed by a system per unit time. The time for a unit operation. The output per unit time, or the cost per unit output.

Table 3.12 Continued

No	Title	Explanation
	Title	*Explanation*
	Moving objects	Objects which can easily change position in space, either on their own, or as a result of external forces. Vehicles and objects designed to be portable are the basic members of this class.
	Stationary objects	Objects which do not change position in space, either on their own, or as a result of external forces. Consider the conditions under which the object is being used.

Adapted from Domb, E., 1998. The 39 features of Altshuller's contradiction matrix. *TRIZ J.*

Table 3.13 The 40 principles of TRIZ

Number	Principle	Alternative names
1	Segmentation	Fragmentation; Segmentation
2	Extraction	Separation; Take out
3	Local quality	Local property
4	Asymmetry	Symmetry change
5	Combining	Merging; Combination; Consolidation; Unite
6	Universality	Multifunctionality
7	Nesting	Nested doll; Matryoshka
8	Counterweight	Antiweight; Weight compensation
9	Prior counteraction	Preliminary antiaction; Preliminary counteraction
10	Prior action	Preliminary action; Do it in advance
11	Cushion in advance	Beforehand cushioning; Beforehand compensation; Previously installed cushions
12	Equipotentiality	
13	Inversion	The other way round; Inverse action
14	Spheroidality	Curvature
15	Dynamics	Dynamicity; Dynamism
16	Partial or excessive action	
17	Transition into another dimension	Dimensionality; Another dimension
18	Mechanical vibration	Use of mechanical oscillations
19	Periodic action	
20	Continuity of useful action	Uninterrupted useful function
21	Rushing through	Skipping; Quick jump
22	Convert harm into benefit	Blessing in disguise; Transform damage into use; Lemons to lemonade
23	Feedback	
24	Mediator	Intermediary

Continued

Table 3.13 Continued

Number	Principle	Alternative names
25	Self-service	
26	Copying	
27	Inexpensive short life	Cheap disposables; Cheap short-living
28	Mechanical substitution	Another sense; Replacement of a mechanical system; Use of fields; Replacement of mechanical matter
29	Pneumatics and hydraulic construction	Pneumatics and hydraulics
30	Flexible membranes and thin films	Flexible shells and thin films
31	Porous materials	
32	Colour change	Optical property changes
33	Homogeneity	
34	Discard and renewal	Rejecting and regenerating parts; Discarding and recovering
35	Transforming the physical or chemical state of an object	Parameter change; Transforming physical or chemical states; Transformation of properties; Change in the aggregate state of an object
36	Phase change	Phase transition
37	Thermal expansion	
38	Strong oxidants	Accelerated oxidisation
39	Inert environment	Inert atmosphere
40	Composite materials	

Principles for resolving design contradictions.

3.8.2 The 40 principles

The 40 principles are powerful tools on their own. Each one represents an approach to solving a challenge. Some principles are found to be generally more effective than others and this section concentrates on describing these along with relevant examples.

Principle 1: Segmentation—The Segmentation principle encourages consideration of the division of an object or system into smaller independent parts, making it sectional, making it easy to assemble or disassemble, and increasing the degree of its divisibility or fragmentation. Examples of segmentation include, distributed loading, multistorey housing, division of an organisation into different product or process centres, franchise outlets, modular furniture, modular computer components, provision of separate bins for recycling different materials, use of office cubicles to enable the layout to be easily changed, use of temporary workers on short-term projects, a folding wooden ruler, garden hoses sections that can be joined together to give the total length required, Venetian blinds, and multipane windows. An advantage of separation is that each part can be treated differently. Parts can be made of different materials or have different material treatments or different shapes.

Principle 2: Extraction—Extraction involves removing or separating an undesirable or disturbing part or property from an object or system. Alternatively, it may involve extracting the desirable part or property of a body or isolating the critical part of a system. An example is the location of noisy equipment such as an air conditioning plant away from the room being cooled so that the occupants are not disturbed by noise. Another example illustrating this principle is embodied in water mist fire suppression systems where pumps are located away from the potential fire zone, thereby protecting the pumps, and only small diameter pipes are necessary to convey the water to the spray nozzles.

Principle 3: Local quality—It need not be assumed that the current use or location of parts in a system cannot be changed. Local quality involves identifying specific parts and changing or moving them so that they can operate at optimal conditions. This may involve changing an object's structure from uniform to nonuniform, or making each part of a system fulfil a different or complementary function. An example of local quality can be found in coal mining to combat dust where a fine mist of water in a conical form is sprayed on parts of the drilling and loading machinery. The smaller the droplets, the greater the effect in suppressing the dust, but fine mist can hinder work. A solution to this is to develop a layer of coarse mist around the cone of fine mist. Another example of local quality is the combination pencil and eraser in the same unit whereby additional local functionality is provided.

Principle 4: Asymmetry—Symmetry is a powerful organising factor. We tend to find symmetric products more aesthetically pleasing. Symmetric objects also tend to be easier to form and manufacture. However, a symmetric object may not be the easiest to handle or organise. Asymmetry involves replacing a symmetrical form with an asymmetrical form. Alternatively, if an object is already asymmetrical, the degree of asymmetry can be increased. Asymmetric stirrers and paddles mix more effectively than symmetric ones. Asymmetric scissors handles provide increased ease of use. Asymmetry is widely used in fashion to produce interesting and engaging styles. As the asymmetry principle involves changing the geometric form it is often an easy method for exploiting the available resource inherent in the geometric form of an object.

Principle 5: Combining—Combining involves bringing together or uniting things that happen at the same time or in the same place. The combination may be permanent or temporary. An example is a combination of washing machine and tumble drier that provides added functionality within the same spatial footprint. The principle of combining is embodied in networked computers and parallel computing providing added functionality and increased processing power. Other examples include double glazing and mixer taps. Fragile glass can be combined with a frame to provide increased strength. An example of temporary union is the use of booster rockets for space payloads where the boosters are fixed to the craft for the first stage of a launch and then discarded.

Principle 6: Universality—The principle of universality refers to making an object perform multiple functions, thereby eliminating the need for some other object or objects. An example is a sofa that can convert into a bed or a vehicle seat that can be adjusted to provide sleeping or cargo carrying capacity. The principle is embodied in the classic Swiss army knife and other multifunction tools. The principle can also be applied to the workplace where an individual in a manufacturing cell can be multi-skilled so that he/she can attend to several tasks. Roof-lights can be designed to provide the multiple functions of illumination, weather proofing, ventilation, thermal and sound insulation.

Principle 7: Nesting—Nesting involves putting one object inside another which, in turn, is placed inside a third. The principle in Russian is called matryoshka after the well-known Russian dolls. An object that is inside another is protected, and reduces the overall volume occupied by the components. Alternatively, nesting refers to passing an object through a cavity of another object. Examples of nesting include the telescope where the casing provides protection for the optical components and the telescopic assembly reduces the total volume for storage and transport, telescopic antennae, nested measuring cups, chairs that can be stacked on each other to facilitate storage, stuffing a chicken or turkey so that two foodstuffs are cooked at once mutually enhancing the overall flavours, and the mechanical pencil with a series of leads stored within its barrel.

Principle 8: Counter-weight—The counterweight principle involves compensating for an undesirable force in one direction by introducing a force in the opposite direction. Counterweights are used, for example, in sash windows with a heavy mass slung over a pulley wheel connected to the window frame so that force required to open or close the window is reduced. Alternatively, the counterweight principle involves compensating for the weight of an object by interaction with an environment providing aerodynamic or hydrodynamic lift or drag or buoyancy force. An example of this principle is the inverted rear wing in racing cars which increases the downward force on the car. An example of weight-compensation is the use of helium or hot-air blimps for advertising. A business example of the counterweight principle is to increase falling sales by associating a product with a rising symbol such as a new movie or trendy star.

Principle 9: Prior counteraction—Prior counteraction involves performing a mitigating task in advance. If you know that something is going to happen then preparations for this can be made or action taken in advance to either prevent it from happening or reduce the impact of the event occurring. Examples of prior counteraction are the use of reinforced concrete for a column or floor slab, prestressed bolts and dust removal systems for power tools so that a work-piece can be seen and so that harmful dust does not block cooling vents. The principle of prior counteraction is addressed in sustainable design where the potentially harmful impacts of a product are considered and either minimised or eliminated.

Principle 10: Prior action—Prior action involves carrying out all or part of a required action in advance or arranging objects so they can go into action in a timely manner and from a convenient location. An example is embodied in modular assembly where modules are manufactured in advance and then joined together, such as prefabricated windows for buildings, prefabricated building sections and gearboxes for installation in a transmission. Another example is the notching of a craft knife enabling a used blunt section to be snapped off to reveal a new sharp edge. Blanks for machining can be produced to resemble the final component thereby requiring only partial or finish machining. The Kanban principles for just-in-time supply in factory production involves prior action with an order being made for resupply of components as soon as the level in a parts bin drops below a critical value.

Principle 11: Cushion in advance—Cushion in advance involves preparing measures in advance to compensate for the relatively low reliability of an object or system. Examples include: uninterruptible power supplies to protect key equipment in case of electrical supply failure; airbags; life boats and jackets; emergency exits with inflatable ramps on aircrafts; run-off strips on hills in case of brake failure; fire extinguishers which are conveniently located and easy to use. Emergency instructions and contingency planning are also examples of cushioning in advance of an event. A domestic example is the use of an aerator in a tap to make water feel 'softer'.

Principle 12: Equipotentiality—Many tasks involve lifting and lowering objects. Equipotentiality involves altering working conditions so that lifting or lowering of an object is avoided. Chests of drawers for example negate the need to lift everything out of a chest to access objects at a lower level. Another example is the use of a pit to facilitate access to an automobile engine. The principle can be applied to a gravitational field or any other potential field. Grounding straps can be used to reduce the risk of static shocks on key equipment in factories.

Principle 13: Inversion—Inversion involves deliberately implementing a solution that is opposite to that initially specified or envisaged. Examples include heating an object instead of cooling it. Inversion can be achieved in a number of ways.

a. Instead of an action dictated by the specifications of the problem, deliberately implement an opposite action.
b. Make a moving part of the object or the outside environment fixed, and the fixed part movable. An example of this is embodied in a fitness running machine where the runner remains stationary in the room but the treadmill belt moves.
c. Turn the object upside down or around. For example, a chiselling tool in a process can be applied from below the work-piece so that chips of waste material fall away from the item being machined.

Other examples of inversion include: slow food instead of fast food, such as a slowly cooked casserole; working at home to avoid travelling; problem-based learning where students explore the material to be learned rather than being given lectures; broadcasting events such as concerts directly into homes so that people can participate in the experience without having to travel; using a small screen or display in front of an eye as part of a head up display system instead of using a large screen. Another example of inversion includes expanding instead of contracting a business during a recession or downturn and preparing products for an assumed or engineered up-turn.

Principle 14: Spheroidality—Many objects are designed to use or be used on flat surfaces. The principle of spheroidality involves considering using curvature or rotation and can be implemented by a number of means.

a. Changing linear parts or flat surfaces to curved ones; changing cube and parallel-piped features to spherical shapes. An historic example of this principle is the use of arches and domes in place of lintels for increased structural strength. To reduce stress concentrations in formed products, a fillet radius can be introduced between features at different angles. Curved furniture can be used for play rooms to mitigate against injuries.

b. Using rollers, balls, spirals or springs. Rolling element bearings reduce friction by providing reduced surface-to-surface contact and extra freedom of movement. A spiral can be used for the filament of a classic filament light bulb to enable a longer length of wire to be located in a given space and reduce stresses.
c. Replacing a linear motion with rotating movement using centrifugal effects. A rotating cylinder with brushes can be used to disturb the surface of a carpet so that it is easier for dirt particles to become airborne and get carried into a vacuum cleaner. Centripetal acceleration is exploited in cyclone separators so that heavy particles are separated from lighter ones. A further example is use of multiaxis rotation in rotational moulding.

Principle 15: Dynamics—Objects or systems that are rigidly connected together are less able to respond adaptively to changes in conditions than those with some degree of freedom. Dynamics involves designing a system so that it can adjust readily to an external influence. The principle of dynamism can be implemented in a number of ways.

a. Make an object or its environment adjust automatically so that performance is optimised at each stage of operation. An example is the use of an adjustable swept wing as on the Tornado, F14 Tomcat, Su 17 and Tu 160 to enable efficient aerodynamic performance of an aircraft for both subsonic and supersonic flights. Other examples include photo and thermochromic glass.
b. Divide an object into subelements that can change position relative to each other. An example to prevent illegal parking is to use posts that are hinged at the base with a lock to set the position of the post in its vertical or horizontal position. Other examples embodying this principle include roll-on and roll-off ferries with a hinged bow to enable access, and the Beluga Airbus transportation aircraft with its hinged cabin section that provides efficient aerodynamics in its closed position but facilitates the transportation of aircraft wings between factories.
c. Make a fixed object moveable or interchangeable. For example, PV solar panels that track the sun.

Repeated use of the dynamism principle in combination with segmentation is characterised by an increase in interactions and transition to a micro level. This pattern can be illustrated by the sequence: rigid system; single hinge; multiple hinges; elastic system; a field in place of a physical system. An example illustrating such development can be found in plumbing where fixed hard-soldered pipework has been substituted by a system of pipes and compression joints, which in turn has been replaced by flexible pipes and hoses. A similar trend can be found in desk lamps where the functionality of a standard lamp is improved by an anglepoise lamp and, in turn, by articulated joints.

Principle 16: Partial or excessive action—If it is difficult to obtain 100% of a desired effect, consider whether achieving a bit more or a bit less makes the problem easier. This is the principle embodied in partial or excessive action. An example is to apply excess paint to a component by dipping it in a vat of paint and then removing the excess paint by spinning the component to throw off the excess by centrifugal effects. Another example is continuously filling a hopper, or constant head device, so that the force of the material above causes the material to flow out of a discharge point at an even rate. By removing some material in advance, perforations in packaging make it easier to open.

Principle 17: Transition to a new dimension- Transition to a new dimension involves a shift in approach to another plane. This can involve consideration of another physical dimension, say changing linear to planar motion, or going upwards or around a corner. The principle of transition to a new dimension can be formalised as follows.

a. Move an object or system in two-dimensional or three-dimensional space.
b. Change the motion of an object from linear to movement in a plane (two dimensions). Change two-dimensional, planar motion, to three-dimensional motion.
c. Use a multilayered assembly of objects or components instead of a single layer. Examples include multistorey car parks, underground rail systems running in tunnels on different levels, and apartments in multistorey buildings.
d. Incline an object or rotate it. A square mirror tile can be hung or supported using a corner rather than its edge. A drawing or painting surface can be inclined to the user to facilitate use. The side or inside of packaging can be used to list the instructions for a game. Inclined reflectors are used in some greenhouses and bio-spheres to improve insolation and hours of illumination by daylight in housing.

Principle 18: Mechanical vibration—Mechanical vibration can be used to put energy into a system. This energy can then be used to excite an object or system so that, for example, it can more easily be broken, moved or mixed. Mechanical vibration can be achieved by a number of means including shaking, sound waves and ultrasonics. Means of implementing the principle are as follows.

a. Cause an object to vibrate. Vibration instead of noise can be used as an alert for a message. To remove a cast from a limb, a vibrating knife can be used in place of a hand saw. Vibration can be used to improve the flow and structural properties of a casting in a mould.
b. If an object is already vibrating or oscillating, raise the frequency of oscillation. Consider ultrahigh frequency vibration. An ultrasonic bath can be used to clean bearings by exciting dirt and grime so that it detaches from the bearing's surfaces.
c. Use the resonant frequency to excite an object. To break up stones in gallbladders and kidneys, the stones can be excited at their resonant frequency using ultrasound negating the need for invasive surgery.
d. Instead of mechanical vibrations, use piezo-vibrators.
e. Use ultrasonic vibrations in conjunction with an electromagnetic field.

Principle 19: Periodic action—The principle of periodic action involves introducing gaps into an action. This principle can be implemented in a number of ways.

a. Replace a continuous with a periodic or pulsed action. Warning lamps, such as bicycle lights, can be made to flash so they are more noticeable and require less power than if powered continuously. In a hammer action drill, the continuous rotation associated with a twist drill is replaced by a periodic reciprocating motion, superimposed on the rotating motion so that the drill imparts an impulse of energy to the work area.
b. If an action is already periodic, alter its frequency. A pulsed siren can be replaced by a siren with variations in both amplitude and frequency.
c. Use the period between impulses to provide an additional action. If the periods of demand from several devices are synchronised, then a single limited energy storage device such as a capacitor, battery or flywheel can be used to provide energy to them all. If two propellers are used on a boat or ship, their blades can be synchronised so that the impulse of one blade

provides an impulse while its equivalent has moved on. The period between demands for energy can be used to store energy in, for example, a capacitor so that energy can be quickly supplied if required giving greater power.

Principle 20: Continuity of a useful action—Some actions, components or processes, do not provide a useful function throughout their entire period of operation. The principle of continuity of a useful action involves making all parts of an action provide a useful function throughout the operation. The principle can be subdivided as follows.

a. Carry out an action continuously, without interruptions, where all parts of an object operate at full capacity. Lean and just-in-time manufacturing, continuous casting of metals and continuously variable transmission are examples of this principle.
b. Remove idle motions and interruptions. This principle is used in self-emptying filters that dump contaminants when the load is low.

It should be noted that the principles of continuity of useful action and periodic action contradict each other. If all intermittent actions are eliminated then there will not be any gaps to exploit. As with any guiding principle, application of common sense and judgement is necessary in their use.

Principle 21: Rushing through—Rushing through a task can reduce the time during which problems can occur. This principle involves performing harmful or hazardous operations at high speed. When a soft material is cut slowly, it can deform making the task difficult and inaccurate. By speeding up the cutting process it is sometimes possible to complete the task before the material has time to distort significantly. For example, when plastic pipes are cut quickly, heat from the cutting process does not propagate significantly within the surrounding material to cause distortion. Another example of the rushing through principle is harvesting, as a crop remains ripe for a very short period of time and leaving it for too long increases the risk of over-ripening and damage by weather and other environmental factors.

Principle 22: Convert harm into benefit—Sometimes undesirable effects such as the generation of waste result from a process. The convert-harm-into-benefit principle encourages conversion of an undesirable effect into a benefit. Many industries and commercial ventures have developed out of developing inventive means to treat or convert waste products into something harmless or useful. The following examples illustrate this principle.

a. Use damaging factors or harmful environmental effects to obtain a positive result. An example is the conversion of methane emitted from decomposing rubbish into useful power.
b. Remove a harmful factor by combining it with other harmful factors. For example, toxic chemicals can be used to treat timber to prevent or limit rot and infestation.
c. Increase the amount of harmful action or tolerate it until it ceases to be harmful. Each time a computer system is exposed to a virus attack, the systems developed to counteract this increase the protection available for the system.

Principle 23: Feedback—Feedback involves taking a signal from a system and using this to control the system itself. Feedback can be used to exaggerate an effect or to cancel out something undesirable. Feedback is used in many heating systems. A thermostat measures the temperature of the system and when the temperature of the thermostat exceeds a preset value, the thermostat causes a heating element to turn off or reduce its output. Feedback can take various forms. If the feedback signal causes the system to respond such that the feedback then increases, it is known as positive feedback. If the feedback signal causes the system to respond in such a way that the feedback decreases, it is known as negative feedback. If the feedback signal does not cause the system to change, then the system is in equilibrium. If feedback is not present in a system, it can be introduced to improve control. If feedback already exists in a system, then options for changing it can be considered. Other examples of the use of feedback include statistical process control, motion sensitive flush, and tap systems. An example of adjusting the feedback in a system is the use of variable sensitivity for a thermostat so that it responds to rising temperature in a different way to falling temperature to improve comfort by reducing temperature fluctuations or reduce energy consumption.

Principle 24: Mediator—A task may need to be undertaken that cannot readily be performed by an existing system. The mediator principle involves introducing a new object to transfer or transmit an action, or temporarily introducing an object to perform a function that can subsequently be removed. Examples of the mediator principle include: coating of weights in fitness machines with a sound-deadening material such as rubber; using fixtures and jigs that make positioning and processing of parts easier; using moulds for forming shapes which are removed after the shape has been formed; using coatings to protect a surface from corrosion; and introducing abrasive particles to a water jet, to enhance water-jet cutting.

Principle 25: Self-service—The self-service principle involves configuring the object or system so that it services itself and carries out supplementary and repair operations, and also considering making use of waste energy and material. Examples include a tyre that automatically repairs itself by releasing fluid from an integrated container filled with repair compound; if a surface is subject to wear or corrosion, the prevention of wear or deterioration by use of a layer of the abrasive or corrosive material itself; heating of a vehicle using fluid that transfers harmful heat away from the engine; designating customers as the waiters so that they can serve themselves and clear away their mess in a self-service, carvery or fast-food restaurant; advancing the rod in a welding gun using a solenoid. Halogen lamps invoke the self-service principle, regenerating the filament by redepositing material on the filament that has evaporated off. Parasitic power generation systems exploit energy available from the host. Examples include quartz watches powered by the wearer's movements, a bicycle hub, heel strike and light generators.

Principle 26: Copying—The copying principle suggests:

a. using a simple and inexpensive copy in place of an object which is complex, expensive, fragile or inconvenient to operate;
b. replacing an object or system of objects by an optical copy or image. Scaling can be used to reduce or enlarge the image;
c. if visible copies are used already, replace them with infrared or ultraviolet copies.

Examples include using: 2D cut-outs for shop window dummies to display clothes; holographic projections to display valuable sculptures, manuscripts or paintings; video conferencing instead of travelling; a virtual reality model to explore or test an environment, or train on it in advance of the environment being available. Measurements can be made from images rather than directly from the original item. Examples of this approach include measurements for maps and surveys based on satellite images and ultrasonic images of foetuses in the womb. The height of tall objects can be measured by means of their shadow.

Principle 27: Inexpensive short-life—The inexpensive short-life principle involves replacement of an expensive object by a series of inexpensive ones, possibly forgoing some, or all, of the original object's attributes such as longevity. Examples include the use of recyclable paper cups in place of ceramic crockery in cafes to avoid the cost of cleaning and storing durable objects; the use of disposable nappies in place of reusable ones; the use of fuses to protect expensive electrical hardware; medical swabs, poultices and sample containers. Numerical simulation, such as computational fluid dynamics can be used to model the effect of a change in conditions on a system without having to go to the expense of testing the actual system.

Principle 28: Mechanical substitution—Many mechanical systems, due to their reliance on physical parameters, are inherently bulky and expensive. The mechanical substitution principle involves replacement of the mechanical system by a sensory system or field. The principle can be invoked in a number of ways.

a. Replace a mechanical system by a sensory (optical, acoustic or olfactory) system. An example of this is the inclusion of an additional gas with a distinct odour in natural gas supplies to alert customers of leaks. Require customers to enter data by means of a touch screen, instead of filling out a form that must be keyed in by employees. Replace a physical fence for confining animals such as dogs with an acoustic 'fence' (where signal is audible to the animal). Using motion or infrared sensors to avoid the user having to locate or touch a control, such as hands-free taps and dryers.

b. Use electrical, magnetic or electromagnetic fields to interact with objects. An example embodying this principle are electronic swipe keys and 'Oyster' cards, GPS for monitoring fleet vehicle movement, and the use of electromagnet stays to hold open fire doors until released by the alarm.

c. Replace static fields with dynamic fields, replace temporarily fixed fields with flexible ones, change unstructured fields to structured (e.g. load-based traffic and zonal heating).

d. Use a field in conjunction with field-activated (e.g. ferromagnetic) particles. Examples include intelligent tidal traffic flow control (using for example road cameras), integrating a ferromagnetic material in a sample to allow it to be manipulated or sorted.

Principle 29: Pneumatic or hydraulic construction—The pneumatic or hydraulic construction principle involves replacing solid parts of an object by a gas or a liquid which can be more easily channelled and have different properties that may be useful such as compressibility. This may enable parts to be inflated, filled, supported or separated using hydrodynamic effects and hydrostatics as in bearings. Examples also include air tents for exhibitions, inflatable boats, play-castles, pneumatic and hydraulic clamps with an elastic cushion that holds a part with evenly distributed pressure, and the use of bubble wrap to protect fragile goods. Fluid pressure can be used to produce mechanical advantage by altering the bore of pipes in a system, as in pneumatic and hydraulic rams and jacks with a small force but large motion resulting in a large force but small motion (see Chapter 18).

Principle 30: Flexible membranes or thin films—Thin films and membranes offer potential advantages in terms of cost, flexibility, space usage and separation. The flexible-membranes or thin-films principle involves using flexible membranes or thin film to replace traditional construction, or to isolate an object from its environment. Examples include the use of sprayed or painted coatings to protect a surface, and the thin walled structures of green houses. Other examples include webbed structures, bubble wrap and egg boxes. To enable a drug to release over a period of time, a pill can be formed with layers of material that dissolve at different rates, encapsulating the active drug.

The thin-film principle can be combined with hydraulics or pneumatics to produce strength in the structure as in aluminium drinks cans where a very thin light-weight vessel is pressurised, giving it increased structural properties.

Principle 31: Porous materials—A porous material allows some substances to pass through it while filtering out or blocking others. The porous-materials principle can be useful for separating or filtering out undesirable items or substances and involves making an object porous or adding porous elements (inserts, covers, etc.), or, if an object is already porous and this is undesirable, filling the pores in advance with some kind of substance. An example of this principle include drilling or forming holes in a structure to make it lighter without impacting significantly on the structure properties. Cinder and breeze blocks formed by coal and coke ash bonded with cement provide a light weight construction material as much of the volume of the block is air. Microfibres are widely used in sports clothing with the small pores of the fibres preventing moisture from passing through it as liquid, but still allowing moisture to evaporate. Semipermeable films are used for some adhesive skin plasters which allow moisture to evaporate but exclude water, dirt and bacteria. To avoid installing equipment for pumping coolant around a system, for a short working life, the system can be prefilled with just enough coolant to operate for the period concerned.

Principle 32: Colour change—Colour influences our aesthetic judgements and can also be used for providing information. The colour change principle can be implemented in several ways.

a. Changing the colour of an object or its surroundings. Many warning lights involve a change of a colour to signify a status. The six hats thinking technique provokes changes in viewpoint by adoption of different coloured hats.
b. Changing the level of transparency of an object or its environment. A transparent bandage enables a wound to be inspected without removing the dressing.
c. Using colour additives to aid observation of objects or processes which are difficult to see. Fluorescent strips can be used on cycle jackets to aid visibility to other road users. Coloured threads and strips are used to aid the security of bank notes and bonds. Another example is the addition of particles to a flow of fluid to aid observation of its detailed flow paths.
d. If such additives are already used, adding lighting or employing luminescent traces or tracer elements.

Principle 33: Homeogeneity—A homogeneous material has the same physical properties throughout its structure. The homogeneity principle involves making objects which interact with a primary object out of the same material or out of one with similar properties. If all the components are made from the same material then this may result in cheaper material sourcing because of economies of scale, and less corrosion due to chemical or thermo-chemical reactions between components. For example, the contacts in an electric circuit are usually made of the same material as the circuit to avoid corrosion problems. If the circuit is made from copper, gold can be used for the small surfaces associated with the contacts as it offers similar electrical properties to copper, has a low resistance and good surface corrosion resistance. The homogeneity principle can also be applied to components. All connectors in a design can be chosen so that they are of the same kind. In building design, the same type of motor could be specified for a wide range of different applications. To avoid distortion of assemblies, components can be made from materials with similar coefficients of linear expansion.

Principle 34: Discard and renewal—Once a component has served a useful purpose it may not be required again, or it may need to be replaced if its functionality has been impaired. The discard and renewal principle can be considered in two different ways.

a. After a component has fulfilled its function, or become useless, discard it. This can be achieved by, for example ejecting, dissolving or causing it to evaporate. Biodegradable components used in surgery such as polylactide pins and screws dissolve slowly in body fluids providing a temporary fastening while the body heals. An example of discard after use is the separation and release of rocket boosters after firing.

b. As soon as a component has become worn or depleted, replace it immediately. An example of this is the ejection of a bullet from a gun after firing, then its replacement from a magazine. Similarly, many machine tools have the facility to replace a broken or worn cutting tool by a new one from a magazine. Some business organisations follow this principle in their selection of which projects to support. If, say, a new type of snack has been launched with a shelf life of 12 weeks, then this period can be used to ascertain whether it should be supported into the future based on the sales figures over the 12-week period in a test stall or supermarket. If sales are disappointing the snack line can be discarded and replaced by a new idea. Other examples of this principle include grey water recycling and heat exchangers.

Principle 35: Transforming the physical or chemical state of an object—Altering an object in some way such as changing its temperature, the concentration of compounds or its density can sometimes provide overall benefits. Transforming the physical or chemical state of an object involves such considerations and can be implemented in several ways.

a. Change an object's physical state. Examples include use of adhesives in place of mechanical fastening and virtual prototyping in place of physical prototyping.

b. Change concentration or consistency. Washing powders can be concentrated to reduce packaging and transportation costs. Powdered paint that can be made up with water offers reduced packaging costs and longer storage life.

c. Change the degree of flexibility. Software can be offered with a range of options or levels, such as novice to expert user, with increasing levels of sophistication.

d. Change the temperature or pressure. Examples include climate control and laminar flow curtains used for some shop doorways, and open refrigerators.

e. Change other parameters.

f. Use the transition to or through a pseudo state, such as a pseudo liquid state, or use a transitionary phase, such as plastic deformation.

Principle 36: Phase change—It may be advantageous to exploit the phenomena that alter during the phase change of a substance to produce a desired effect. Physical properties that change during phase transitions include volume, heat release and heat absorption. When exploiting phase change, it is important to consider how the process is started and how it will stop. The most common phase changes are melting from a solid to a liquid phase, boiling and evaporation, freezing and condensation for solids, liquids, gases and plasmas. Other transitional phenomena include paramagnetic to ferromagnetic, conductor to superconductor, crystallographic changes and superfluidity. An example is the use of sodium acetate for heat pads, which contain a supersaturated solution. Crystallisation of the fluid can be triggered by releasing crystals into the solution which act as nucleation sites and cause the solution to crystallise suddenly, thereby releasing crystal lattice energy which is manifested by a temperature rise.

Principle 37: Thermal expansion—When an object is heated it tends to expand. Thermal expansion of materials can be exploited in tackling a range of problems. It should, however, be noted that there are some exceptions to this tendency, with water in its solid phase being less dense than as a liquid (this is why lumps of ice tend to float with only a small proportion of their surface exposed at the free surface). Another substance where expansion is minimal with increased temperature is Invar, which has a near zero coefficient of linear expansion. Thermal expansion is used in bimetallic thermometers and thermostats. In these devices two strips of dissimilar materials such as brass and steel are welded together (Childs, 2001). As the materials have different coefficients of linear expansion, with brass having the higher value, if the temperature rises then the brass will expand more than the steel causing the assembly to bend. The magnitude of the distortion of the strip can be calibrated to give an indication of temperature and the strip either connected to a dial indicator or the strip's used to close an electric circuit for control purposes. Another example of the use of thermal expansion is in interference fits (see Chapter 19) where the diameter of two components overlap and assembly is achieved by either forcing the components together or by heating up the outer component and cooling the inner component, so that the two components can slide readily together. With subsequent heat transfer, the temperature of both components will tend to equalise with the outer component cooling and the inner component heating up causing the outer component's inner dimension to reduce, thereby clamping onto the inner.

Principle 38: Strong oxidants—Oxygen reacts readily with many elements and compounds. The strong oxidants principle involves enhancing this effect by using materials that combine with oxygen readily, or by adding more oxygen to the system. Examples of the principle include:

a. replacing normal air with enriched air with a higher concentration of oxygen;
b. replacing enriched air with oxygen;
c. treating an object in air or in oxygen with ionising radiation;
d. using ozone in combination with oxygen;
e. using ionised oxygen.

For example, stainless steel can be cut with plasma arcs in a stream of oxygen.

Principle 39: Inert environment—If chemical reactions in an environment are a problem, then an option is to remove the chemicals. The inert environment principle involves replacing the normal medium or environment with an inert one or carrying out the process in a vacuum. For example, to reduce the risk of cotton from catching fire in a warehouse, inert gas can be used to purge it of air while it is being transported from one location to another. Some welding processes are undertaken with the workpiece held in a purge flow of inert gas. Some light bulbs are partially evacuated and then filled with an inert gas to reduce oxidation of the light element.

Principle 40: Use composite materials—By using materials that have complementary or synergistic properties, it is sometimes possible to construct an assembly which has the desirable properties of the constituent elements. The use of composite-materials principle involves replacing a homogeneous material with a composite or combination. Examples include rebar reenforced concrete, the reenforcement of wedge belts with hoop wires (see Chapter 12), and Kevlar. Some military and civil aircraft have wings made of composites of plastics and carbon fibres for high strength and low weight.

The 40 principles represent a tool-kit for problem solving in their own right. The list can be browsed to provoke ideas. Another strategy for using the principles is to select them based on trade-offs. Principles that negate trade-offs can be selected and explored. Alternatively use can be made of the contradiction matrix (see Section 3.8.3) to select the principles that are most likely to be useful based on experience to resolve a contradiction between two features.

The examples presented for each of the principles have been sourced from a wide range of applications and their inclusion here does not necessarily mean that the originator used or was even aware of TRIZ. The examples nevertheless show that the principles are wide ranging in their possible application to technology and business. TRIZ comprises various tools that can be used in isolation to assist in creative problem solving. Tools such as the 40 principles can be applied within a brainstorming session to provoke ideas. The TRIZ tools, with their built-in knowledge of generic solutions, represent a short-cut to experience, giving an extremely powerful tool. It is widely used in industry and embedded in various inventive problem-solving software packages. Further information on TRIZ can be found in Gadd (2011), the *TRIZ journal* and in Vaneker and Lutters (2008).

3.8.3 Contradiction matrix

As noted earlier, while trying to improve one desirable property, another desirable property may deteriorate. Conventional problem solving generally leads to a 'compromise' solution. The most 'inventive' solution is obtained when a technical problem containing a 'contradiction' is solved by completely eliminating the contradiction where one property improves and another degrades. Altshuller, from his research on over 40,000 inventive patents, found that there are only 39 features, defined in Table 3.12, which either improve or degrade. As a result, every problem can be described as a conflict between a pair of parameters (2 out of 39 parameters).

Altshuller found that many patents and invention disclosures had, in the past, resolved these individual conflicts in several different fields. He concluded that only 40 inventive principles, as defined in Table 3.13, were used to resolve these contradictions fully without a trade-off or compromise. He argued that, if these results had been known earlier, the problems being tackled in the original patents would have been solved with more ease.

From knowledge of the 40 principles, combined with a method to identify the similarity of a new problem to previous problems, it is possible to tackle an issue with greater ease. One of the processes Altshuller developed was based on analysis of contradiction that was seen to have been resolved in a patent. Alsthuller categorised these, along with the principles used to resolve the contradiction, in a tool called the contradiction matrix. This gives an indication of which of the 40 inventive principles are most likely to be useful in resolving the contradiction. This tool has been found to be very powerful and has since been embodied in a wide range of TRIZ software packages.

The matrix is organised in the form of 39-improving parameters and 39-worsening parameters (a 39 by 39 matrix) with each cell entry giving the most often used inventive principles that may be used to eliminate the contradiction. This matrix is known as the contradiction matrix and is one of the main TRIZ tools. The contradiction matrix consists of a list of parameters so that the parameter that is getting better can be identified. In addition, the corresponding parameter that is getting worse can also be identified. The improving parameters are listed in the first column and the worsening parameter in the first row. In the corresponding intersection, between the improving and worsening parameter, the table gives a list of principles that can be explored to resolve the contradiction. A full version of the matrix is shown in Table 3.14.

A four-step process can be followed for using the contradiction matrix.

1. Use the 39 parameters to identify critical features in the problem.
2. Identify the contradictions between the parameters where one causes problems with another.
3. Use the contradiction matrix to identify principles that can be used to resolve the contradictions.
4. Use the numbers from the contradiction matrix to look up resolution principles and use these principles to find solutions to the problem.

It should be noted that the principles in the table for a given entry are listed in order of priority based on analysis of previous solutions. The principle most likely from experience to yield a helpful solution is listed first. If exploring this principle proves unhelpful, then the next principle in the list can be tried and so on until a solution is apparent or options have been exhausted.

Example 3.1. The example of contradiction in the design of conventional axes for splitting wood was introduced earlier. If the blade is made heavier, it can strike a more effective blow with better splitting power, but the quantity of material used increases making the device more awkward to handle.

Solution

The improving feature in this case is force and the worsening feature is the quantity of material. The corresponding feature numbers are 10 and 26 respectively.

Table 3.14 TRIZ Contradiction Matrix

Improving parameter		1	2	3	4	5	6	7	8	9	10	11	12	13	14	15	16	17	18	19	20
		Worsening parameter																			
1	Weight of moving object		All	15 8 29 34	All	29 17 38 34	All	29 2 40 28	All	2 8 15 38	8 10 18 37	10 36 37 40	10 14 35 40	1 35 19 39	28 27 18 40	5 34 31 35	All	6 29 4 38	19 1 32	35 12 34 31	All
2	Weight of stationary object	All		All	10 1 29 35	All	35 30 13 2	All	5 35 14 2	All	8 10 19 35	13 29 10 18	13 10 29 14	26 39 1 40	28 2 10 27	All	2 27 19 6	28 19 32 22	19 32 35	All	18 19 28 1
3	Length of moving object	8 15 29 34	All		All	15 17 4	All	7 17 4 35	All	13 4 8	17 10 4	1 8 35	1 8 10 29	1 8 15 34	8 35 29 34	19	All	10 15 19	32	8 35 24	All
4	Length of stationary object	All	35 28 40 29	All		All	17 7 10 40	All	35 8 2 14	All	28 10	1 14 35	13 14 15 7	39 37 35	15 14 28 26	All	1 10 35	3 35 38 18	3 25	All	All
5	Area of moving object	2 17 29 4	All	14 15 18 4	All		All	7 14 17 4	All	29 30 4 34	19 30 35 2	10 15 36 28	5 34 29 4	11 2 13 39	3 15 40 14	6 3	All	2 15 16	15 32 19 13	19 32	All
6	Area of stationary object	All	30 2 14 18	All	26 7 9 39	All		All	All	All	1 18 35 36	10 15 36 37	All	2 38	40	All	2 10 19 30	35 39 38	All	All	All
7	Volume of moving object	2 26 29 40	All	1 7 4 35	All	1 7 4 17	All		All	29 4 38 34	15 35 36 37	6 35 36 37	1 15 29 4	28 10 1 39	9 14 15 7	6 35 4	All	34 39 10 18	2 13 10	35	All-
8	Volume of stationary object	All	35 10 19 14	19 14	35 8 2 14	All	All	All		All	2 18 37	24 35	7 2 35	34 28 35 40	9 14 17 15	All	35 34 38	35 6 4	All	ALL	ALL
9	Speed	2 28 13 38	All	13 14 8	All	29 30 34	All	7 29 34	All		13 28 15 19	6 18 38 40	35 15 18 34	28 33 1 18	8 3 26 14	3 19 35 5	All	28 30 36 2	10 13 19	8 15 35 38	All

Improving parameter		21	22	23	24	25	26	27	28	29	30	31	32	33	34	35	36	37	38	39
1	Weight of moving object	12 36 18 31	6 2 34 19	5 35 3 31	10 24 35	10 35 20 28	3 26 18 31	1 3 11 27	28 27 35 26	28 35 26 18	22 21 18 27	22 35 31 39	27 28 1 36	35 3 2 24	2 27 28 11	29 5 15 8	26 30 36 34	28 29 26 32	26 35 18 19	35 3 24 37
2	Weight of stationary object	15 19 18 22	18 19 28 15	5 8 13 30	10 15 35	10 20 35 26	19 6 18 26	10 28 8 3	18 26 28	10 1 35 17	2 19 22 37	35 22 1 39	28 1 9	6 13 1 32	2 27 28 11	19 15 29	1 10 26 39	25 28 17 15	2 26 35	1 28 15 35
3	Length of moving object	1 35	7 2 35 39	4 29 23 10	1 24	15 2 29	29 35	10 14 29 40	28 32 4	10 28 29 37	1 15 17 24	17 15	1 29 17	15 29 35 4	1 28 10	14 15 1 16	1 19 26 24	35 1 26 24	17 24 26 16	14 4 28 29
4	Length of stationary object	12 8	6 28	10 28 24 35	24 26	30 29 14	All	15 29 28	32 28 3	2 32 10	1 18	All	15 17 27	2 25	3	1 35	1 26	26	All	30 14 7 26
5	Area of moving object	19 10 32 18	15 17 30 26	10 35 2 39	30 26	26 4	29 30 6 13	29 9	26 28 32 3	2 32	22 33 28 1	17 2 18 39	13 1 26 24	15 17 13 16	15 13 10 1	15 30	14 1 13	2 36 26 18	14 30 28 23	10 26 34 2
6	Area of stationary object	17 32	17 7 30	10 14 18 39	30 16	10 35 4 18	2 18 40 4	32 35 40 4	26 28 32 3	2 29 18 36	27 2 39 35	22 1 40	40 16	16 4	16	15 16	1 18 36	2 35 30 18	23	10 15 17 7
7	Volume of moving object	35 6 13 18	7 15 13 16	36 39 34 10	2 22	2 6 34 10	29 30 7	14 1 40 11	25 26 28	25 28 2 16	22 21 27 35	17 2 40 1	29 1 40	15 13 30 12	10	15 29	26 1	29 26 4	35 34 16 24	10 6 2 34
8	Volume of stationary object	30 6	All	10 39 35 34	All	35 16 32 18	35 3	2 35 16	All	35 10 25	34 39 19 27	30 18 35 4	35	All	1	All	1 31	2 17 26		35 37 10 2
9	Speed	19 35 38 2	14 20 19 35	10 13 28 38	13 26	All	10 19 29 38	11 35 27 28	28 32 1 24	10 28 32 25	1 28 35 23	2 24 35 21	35 13 8 1	32 28 13 12	34 2 28 27	15 10 26	10 28 4 34	3 34 27 16	10 18	All

10	Force (intensity)	8 1 37 18	18 13 1 28	17 19 9 36	28 10	19 10 15	1 18 36 37	15 9 12 37	2 36 18 37	13 28 15 12		18 21 11	10 35 40 34	35 10 21	35 10 14 27	19 2	All	35 10 21	All	19 17 10	1 16 36 37	19 35 18 37	14 15	8 35 40 5	All	10 37 36	14 29 18 36	3 35 13 21	35 10 23 24	28 29 37 36	1 35 40 18	13 3 36 24	15 37 18 1	1 28 3 25	15 1 11	15 17 18 20	26 35 10 18	36 37 10 19	2 35	3 28 35 37
11	Stress or pressure	10 36 37 40	13 29 10 18	35 10 36	35 1 14 16	10 15 36 28	10 15 36 37	6 35 10	35 24	6 35 36	36 35 21		35 4 15 10	35 33 2 40	9 18 3 40	19 3 27	All	35 39 19 2	All	14 24 10 37	All	10 35 14	2 36 25	10 36 3 37	All	37 36 4	10 14 36	10 13 19 35	6 28 25	3 35	22 2 37	2 33 27 18	1 35 16	11	2	35	19 1 35	2 36 37	35 24	10 14 35 37
12	Shape	8 10 29 40	15 10 26 3	29 34 5 4	13 14 10 7	5 34 4 10	All	14 4 15 22	7 2 35	35 15 34 18	35 10 37 40	34 15 10 14		33 1 18 4	30 14 10 40	14 26 9 25	All	22 14 19 32	13 15 32	2 6 34 14	All	4 6 2	14	35 29 3 5	All	14 10 34 17	36 22	10 40 16	28 32 1	32 30 40	22 1 2 35	35 1	1 32 17 28	32 15 26	2 13 1	1 15 29	16 29 1 28	15 13 39	15 1 32	17 26 34 10
13	Stability of object composition	21 35 2 39	26 39 1 40	13 15 1 28	37	2 11 13	39	28 10 19 39	34 28 35 40	33 15 28 18	10 35 21 16	2 35 40	22 1 18 4		17 9 15	13 27 10 35	39 3 35 23	35 1 32	32 3 27 16	13 19	27 4 29 18	32 35 27 31	14 2 39 6	2 14 30 40	All	35 27	15 32 35	All	13	18	35 24 30 18	35 40 27 39	35 19	32 35 30	2 35 10 16	35 30 34 2	2 35 22 26	35 22 39 23	1 8 35	23 35 40 3
14	Strength	1 8 40 15	40 26 27 1	1 15 8 35	15 14 28 26	3 34 40 29	9 40 28	10 15 14 7	9 14 17 15	8 13 26 14	10 18 3 14	10 3 18 40	10 30 35 40	13 17 35		27 3 26	All	30 10 40	35 19	19 35 10	35	10 26 35 28	35	35 28 31 40	All	29 3 28 10	29 10 27	11 3	3 27 16	3 27	18 35 37 1	15 35 22 2	11 3 10 32	32 40 25 2	27 11 3	15 3 32	2 13 25 28	27 3 15 40	15	29 35 10 14
15	Duration of action by moving object	19 5 34 31	All	2 19 9	All	3 17 19	All	10 2 19 30	All	3 35 5	19 2 16	19 3 27	14 26 28 25	13 3 35	27 3 10		All	19 35 39	2 19 4 35	28 6 35 18	All	19 10 35 38	All	28 27 3 18	10	20 10 28 18	3 35 10 40	11 2 13	3	3 27 16 40	22 15 33 28	21 39 16 22	27 1 4	12 27	29 10 27	1 35 13	10 4 29 15	19 29 39 35	6 10	35 17 14 19
16	Duration of action by stationary object	All	6 27 19 16	All	1 40 35	All	All	All	35 34 38	All	All	All	All	39 3 35 23	All	All		19 18 36 40	All	All	All	16	All	27 16 18 38	10	28 20 10 16	3 35 31	34 27 6 40	10 26 24	All	17 1 40 33	22	35 10	1	1	2	All	25 34 6 35	1	20 10 16 38
17	Temperature	36 22 6 38	22 35 32	15 19 9	15 19 9	3 35 39 18	35 38	34 39 40 18	35 6 4	2 28 36 30	35 10 3 21	35 39 19 2	14 22 19 32	1 35 32	10 30 22 40	19 13 39	19 18 36 40		32 30 21 16	19 15 3 17	All	2 14 17 25	21 17 35 38	21 36 29 31	All	35 28 21 18	3 17 30 39	19 35 3 10	32 19 24	24	22 33 35 2	22 35 2 24	26 27	26 27	4 10 16	2 18 27	2 17 16	3 27 35 31	26 2 19 16	15 28 35
18	Illumination intensity	19 1 32	2 35 32	19 32 16	All	19 32 26	All	2 13 10	All	10 13 19	26 19 6	All	32 30	32 3 27	35 19	2 19 6	All	32 35 19		32 1 19	32 35 1 15	32	13 16 1 6	13 1	1 6	19 1 26 17	1 19	All	11 15 32	3 32	15 19	35 19 32 39	19 35 28 26	28 26 19	15 17 13 16	15 1 19	6 32 13	32 15	2 26 10	2 25 16
19	Use of energy by moving object	12 18 28 31	All	12 28	All	15 19 25	All	35 13 18	All	8 15 35	16 26 21 2	23 14 25	12 2 29	19 13 17 24	5 19 9 35	28 35 6 18	All	19 24 3 14	2 15 19		All	6 19 37 18	12 22 15 24	35 24 18 5	All	35 38 19 18	34 23 16 18	19 21 11 27	3 1 32	All	1 35 6 27	2 35 6	28 26 30	19 35	1 15 17 28	15 17 13 16	2 29 27 28	35 38	32 2	12 28 35
20	Use of energy by stationary object	All	19 9 6 27	All	All	All	All	All	All	All	36 37	All	All	27 4 29 18	35	All	All	All	19 2 35 32	All		All	All	28 27 18 31	All	All	3 35 31	10 36 23	All	All	10 2 22 37	19 22 18	1 4	All	All	All	All	19 35 16 25	All	1 6

Continued

Table 3.14 Continued

Improving parameter		1	2	3	4	5	6	7	8	9	10	11	12	13	14	15	16	17	18	19	20	21	22	23	24	25	26	27	28	29	30	31	32	33	34	35	36	37	38	39
		Worsening parameter																																						
21	Power	8 36 38 31	19 26 17 27	1 10 35 37		19 38	17 32 13 38	35 6 38	30 6 25	15 35 2	26 2 36 35	22 10 35	29 14 2 40	35 32 15 31	26 10 28	19 35 10 38	16	2 14 17 25	16 6 19	16 6 19 37	all		10 35 38	28 27 18 38	10 19	35 20 10 6	4 34 19	19 24 26 31	32 15 2	32 2	19 22 31 2	2 35 18	26 10 34	26 35 10	35 2 10 34	19 17 34	20 19 30 34	19 35 16	28 2 17	28 35 34
22	Loss of energy	15 6 19 28	19 6 18 9	7 2 6 13	6 38 7	15 26 17 30	17 7 30 18	7 18 23	7	16 35 38	36 38	All	All	14 2 39 6	26	All	All	19 38 7	1 13 32 15	All	All	3 38		35 27 2 37	19 10	10 18 32 7	7 18 25	11 10 35	32	All	21 22 35 2	21 35 2 22	All	35 32 1	2 19	All	7 23	35 3 15 23	2	28 10 29 35
23	Loss of substance	35 6 23 40	35 6 22 32	14 29 10 39	10 28 24	35 2 10 31	10 18 39 31	1 29 30 36	3 39 18 31	10 13 28 38	14 15 18 40	3 36 37 10	29 35 3 5	2 14 30 40	35 28 31 40	28 27 3 18	27 16 18 38	21 36 39 31	1 6 13	35 18 24 5	28 27 12 31	28 27 18 38	35 27 2 31		All	15 18 35 10	6 3 10 24	10 29 39 35	16 34 31 28	35 10 24 31	33 22 30 40	10 1 34 29	15 34 33	32 28 2 24	2 35 34 27	15 10 2	35 10 28 24	35 18 10 13	35 10 18	28 35 10 23
24	Loss of information	10 24 35	10 35 5	1 26	26	30 26	30 16		2 22	26 32		All	All	All	All	10	10	All	19	All	All	10 19	19 10	All		24 26 28 32	24 28 35	10 28 23	All	All	22 10 1	10 21 22	32	27 22	All	All	All	35 33	35	13 23 15
25	Loss of time	10 20 37 35	10 20 26 5	15 2 29	30 24 14 5	26 4 5 16	10 35 17 4	2 5 34 10	35 16 32 18	All	10 37 36 5	37 36 4	4 10 34 17	35 3 22 5	29 3 28 18	20 10 28 18	28 20 10 16	35 29 21 18	1 19 26 17	35 38 19 18	1	35 20 10 6	10 5 18 32	35 18 10 39	24 26 28 32		35 38 18 16	10 30 4	24 34 28 32	24 26 28 18	35 18 34	35 22 18 39	35 28 34 4	4 28 10 34	32 1 10	35 28	6 29	18 28 32 10	24 28 35 30	All
26	Quantity of substance	35 6 18 31	27 26 18 35	29 14 35 18	All	15 14 29	2 18 40 4	15 20 29	All	35 29 34 28	35 14 3	10 36 14 3	35 14	15 2 17 40	14 35 34 10	3 35 10 40	3 35 31	3 17 39	All	34 29 16 18	3 35 31	35	7 18 25	6 3 10 24	24 28 35	35 38 18 16		18 3 28 40	13 2 28	33 30	35 33 29 31	3 35 40 39	29 1 35 27	35 29 25 10	2 32 10 25	15 3 29	3 13 27 10	3 27 29 18	8 35	13 29 3 27
27	Reliability	3 8 10 40	3 10 8 28	15 9 14 4	15 29 28 11	17 10 14 16	32 35 40 4	3 10 14 24	2 35 24	21 35 11 28	8 28 10 3	10 24 35 19	35 1 16 11	all	11 28	2 35 3 25	34 27 6 40	3 35 10	11 32 13	21 11 27 19	36 23	21 11 26 31	10 11 35	10 35 29 39	10 28	10 30 4	21 28 40 3		32 3 11 23	11 32 1	27 35 2 40	35 2 40 26	All	27 17 40	1 11	13 35 8 24	13 35 1	27 40 28	11 13 27	1 35 29 38
28	Measurement accuracy	32 35 26 28	28 35 25 26	28 26 5 16	32 28 3 16	26 28 32 3	26 28 32 3	32 13 6	All	28 13 32 24	32 2	6 28 32	6 28 32	32 35 13	28 6 32	28 6 32	10 26 24	6 19 28 24	6 1 32	3 6 32	All	3 6 32	26 32 27	10 16 31 28	All	24 34 28 32	2 6 32	5 11 1 23		All	28 24 22 26	3 33 39 10	6 35 25 18	1 13 17 34	1 32 13 11	13 35 2	27 35 10 34	26 24 32 28	28 2 10 34	10 34 28 32
29	Manufacturing precision	28 32 13 18	28 35 27 9	10 28 29 37	2 32 10	28 33 29 32	2 29 18 36	32 23 2	25 10 35	10 28 32	28 19 34 36	3 35	32 30 40	30 18	3 27	3 27 40	All	19 26	3 32	32 2	All	32 2	13 32 2	35 31 10 24	All	32 26 28 18	32 30	11 32 1	All		26 28 10 36	4 17 34 26	All	1 32 35 23	25 10	All	26 2 18	All	26 28 18 23	10 18 32 39

30	Object affected harmful factors	22 21 27 39	2 22 13 24	17 1 39 4	1 18	22 1 33 28	27 2 39 35	22 23 37 35	34 39 19 27	21 22 35 28	13 35 39 18	22 2 37	22 1 3 35	35 24 30 18	18 35 37 1	22 15 33 28	17 1 40 33	22 33 35 2	1 19 32 13	1 24 6 27	10 2 22 37	19 22 31 2	21 22 35 2	33 22 19 40	22 10 2	35 18 34	35 33 29 31	27 24 2 40	28 33 23 26	26 28 10 18		All	24 35 2	2 25 28 39	35 10 2	35 11 22 31	22 19 29 40	22 19 29 40	33 3 34	22 35 13 24
31	Object generated harmful factors	19 22 15 39	35 22 1 39	17 15 16 22	All	17 2 18 39	22 1 40	17 2 40	30 18 35 4	35 28 3 23	35 28 1 40	2 33 27 18	35 1	35 40 27 39	15 35 22 2	15 22 33 31	21 39 16 22	22 35 2 24	19 24 39 32	2 35 6	19 22 18	2 35 18	21 35 2 22	10 1 34	10 21 29	1 22	3 24 39 1	24 2 40 39	3 33 26	4 17 34 26	All		All	All	All	All	19 1 31	2 21 27 1	2	22 35 18 39
32	Ease of manufacture	28 29 15 16	1 27 36 13	1 29 13 17	15 17 27	13 1 26 12	16 40	13 29 1 40	35	35 13 8 1	35 12	35 19 1 37	1 28 13 27	11 13 1	1 3 10 32	27 1 4	35 16	27 26 18	28 24 27 1	28 26 27 1	1 4	27 1 12 24	19 35	15 34 33	32 24 18 16	35 28 34 4	35 23 1 24		1 35 12 18	All	24 2	All		2 5 13 16	35 1 11 9	2 13 15	27 26 1	6 28 11 1	8 28 1	35 1 10 28
33	Convenience of use	25 2 13 15	6 13 1 25	1 17 13 12	All	1 17 13 16	18 16 15 39	1 16 35 15	4 18 39 31	18 13 34	28 13 35	2 32 12	15 34 29 28	32 35 30	32 40 3 28	29 3 8 25	1 16 25	26 27 13	13 17 1 24	1 13 24	All	35 34 2 10	2 19 13	28 32 2 24	4 10 27 22	4 28 10 34	12 35	17 27 8 40	25 13 2 34	1 32 35 23	2 25 28 39	All	2 5 12		12 26 1 32	15 34 1 16	32 26 12 17	All	1 34 12 3	15 1 28
34	Ease of repair	2 27 35 11	2 27 35 11	1 28 10 25	3 18 31	15 13 32	16 25	25 2 35 11	1	34 9	1 11 10	13	1 13 2 4	2 35	11 1 2 9	11 29 28 27	1	4 10	15 1 13	15 1 28 16	All	15 10 32 2	15 1 32 19	2 35 34 27	All	32 1 10 25	2 28 10 25	11 10 1 16	10 2 13	25 10	35 10 2 16	All	1 35 11 10	1 12 26 15		7 1 4 16	35 1 13 11	All	34 35 7 13	1 32 10
35	Adaptability or versatility	1 6 15 8	19 15 29 16	35 1 29 2	1 35 16	35 30 29 7	15 16	15 35 29	All	35 10 14	15 17 20	35 16	15 37 1 8	35 30 14	35 3 32 6	13 1 35	2 16	27 2 3 35	6 22 26 1	19 35 29 13	All	19 1 29	18 15 1	15 10 2 13	All	35 28	3 35 15	35 13 8 24	35 5 1 10	All	35 11 32 31	All	1 13 31	15 34 1 16	1 16 7 4		15 29 37 28	1	27 34 35	35 28 6 37
36	Device complexity	26 30 34 36	2 26 35 39	1 19 26 24	26	14 1 13 16	6 36	34 26 6	1 16	34 10 28	26 16	19 1 35	29 13 28 15	2 22 17 19	2 13 28	10 4 28 15	All	2 17 13	24 17 13	27 2 29 28	All	20 19 30 34	10 35 13 2	35 10 28 29	All	6 29	13 3 27 10	13 35 1	2 26 10 34	26 24 32	22 19 29 40	19 1	27 26 1 13	27 9 26 24	1 13	29 15 28 37		15 10 37 28	15 1 24	12 17 28
37	Difficulties of detecting and measuring	27 26 28 13	6 13 28 1	16 17 26 24	26	2 13 18 17	2 39 30 16	29 1 4 16	2 18 26 31	3 4 16 35	30 28 40 19	35 36 37 32	27 13 1 39	11 22 39 30	27 3 15 28	19 29 39 25	25 34 6 35	3 27 35 16	2 24 26	35 38	19 35 16	18 1 16 10	35 3 15 19	1 18 10 24	35 33 27 22	18 28 32 9	3 27 29 18	27 40 28 8	26 24 32 28	All	22 19 29 28	2 21	5 28 11 29	2 5	12 26	1 15	15 10 37 28		34 21	35 18
38	Extent of automation	28 26 18 35	28 26 35 10	14 13 17 28	23	17 14 13	All	35 13 16	All	28 10	2 35	13 35	15 32 1 13	18 1	25 13	6 9	All	26 2 19	8 32 19	2 32 13	All	28 2 27	23 28	35 10 18 5	35 33	24 28 35 30	35 13	11 27 32	28 26 10 34	28 26 18 23	2 33	2	1 26 13	1 12 34 3	1 35 13	27 4 1 35	15 24 10	34 27 25		5 12 35 26
39	Productivity	35 26 24 37	28 27 15 3	18 4 28 38	30 7 14 26	10 26 34 31	10 35 17 7	2 6 34 10	35 37 10 2	All	28 15 10 36	10 37 14	14 10 34 40	35 3 22 39	29 28 10 18	35 10 2 18	20 10 16 38	35 21 28 10	26 17 19 1	35 10 38 19	1	35 20 10	28 10 29 35	28 10 35 23	13 15 23	All	35 38	1 35 10 38	1 10 34 28	18 10 32 1	22 35 13 24	35 22 18 39	35 28 2 24	1 28 7 10	1 32 10 25	1 35 28 37	12 17 28 24	35 18 27 2	5 12 35 26	

A blank means no principle suggested; try exploring a different contradiction. – Means no sensitive principles; try exploring a different contradiction.
Adapted from Altshuller, G., 2005. 40 Principles Extended Edition: TRIZ Keys to Technical Innovation. Technical Innovation Center.

Table 3.15 Extract from the contradiction matrix showing the principles recommended for resolving a contradiction between force and weight of the moving object

	Worsening feature Improving feature	**25**	**26**	**27**
9	Speed	all	10 19 29 38	11 35 27 28
10	Force	10 37 36	14 29 18 36	3 35 13 21

Examination can be made of Table 3.14 to identify which of the 40 principles are recommended. By identifying row 10 for force and column 26 for quantity of substance, the recommended principles are seen to be 14, 29, 18 and 36 corresponding to curvature, pneumatics and hydraulics, mechanical vibration, and phase transition, respectively. The relevant part of the contradiction matrix is illustrated in Table 3.15. The axe developed for Fiskars (Fig. 3.16) makes use of a hollow handle to resolve the issue using the pneumatics and hydraulics principle. By using a hollow handle (i.e. replacing solid parts of an object with) in this case, air, the centre of gravity is moved forward and the effective blow from the axe becomes more powerful from a lighter tool.

Example 3.2. In competitive high diving, it is important for the diver to engage in much practice but repetitive diving involves multiple impacts between the diver and the water potentially leading to injuries. Here the improving feature is speed (the rate of action in time), and the worsening feature is reliability (the ability of the body to perform healthily as intended).

Solution

Examination can be made of Table 3.14 to identify which of the 40 principles are recommended. By identifying row 9 for speed and column 27 for reliability, the recommended principles are 11, 35, 27 and 28 corresponding to cushion in advance, parameter changes, cheap short living and mechanics substitution respectively. The cushion-in-advance principle has been exploited in one solution by using aeration of water to make it softer during diving practice enabling more dives to be undertaken.

Fig. 3.16 : Fiskars axe using a hollow handle to make the tool lighter for handling.

Additional examples of contradiction and identification of possible parameters to explore with the contradiction matrix shown in the square brackets include:

- Horseshoe (hot to work but cold to handle) [12; 17];
- Icebreaker (fast but low power) [9; 21];
- Flag (big but quiet) [6; 31];
- Pipe connection (fast but low force) [9; 10];
- Training (lots but short) [16; 25];
- Bin (big but small) [8; 26].

3.9 Boundary shifting

Boundary shifting involves challenging the constraints defined in the problem design specification (PDS) to identify whether they are necessary. For example, the PDS may define that steel should be used for a component. Boundary shifting challenges this specification to see whether it is appropriate and, if not, what other materials could be considered.

Example 3.3. The mayor of Stockholm has outlined funding for a sculpture, based upon an architect's sketches, to span the entrance of Stockholm harbour and has presented you as a leading engineering company with the contract to design and construct this (see Matthews, 1998).

Solution

There are often problems with designing something new. This sculpture is not an exception. Consideration of fatigue due to wind loading is critical for a structure of this sort. In this case, the designer challenged the original specification for a sculpture that spanned the harbour and in place proposed a half arch structure with the remainder of the geometry completed by a stream of water (Fig. 3.17).

3.10 The creativity and innovation engine

In essence, creativity represents a significant risk, and yet in a fast-moving business climate, 'no creativity' is even riskier. The definition and embodiment of details that enable an idea to be realised in practice is essential for realising financial or societal value from an idea. The tensions involved with creativity, the value and use of creativity tools, and the importance of design in realising commercial potential, can be embodied in the moniker 'commercivity', the commercial exploitation of creativity through the implementation of game-changing and sustainable ideas (see Childs and Fountain, 2011).

Creativity tools provide a means to augment innate generative activity. Most creativity tools can be used at any stage in a problem-solving process and tend to be focused on problem exploration, idea generation and concept evaluation. Creativity tools in expert hands are as important to the engineer or designer and their challenge as the surgeon's tools to the saving of life in a medical procedure. Used recklessly,

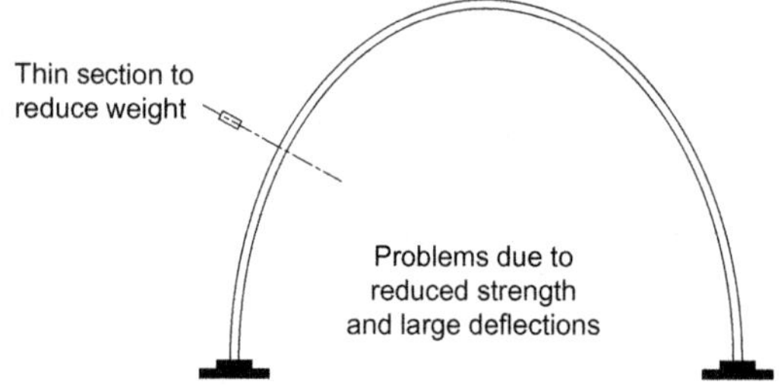

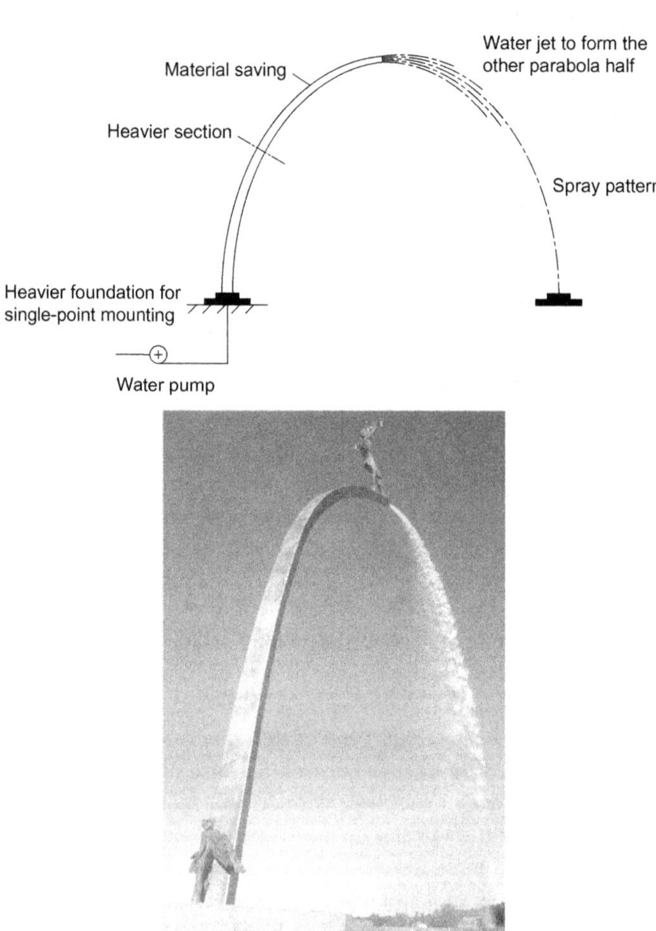

Fig. 3.17 Conceptual and final solutions (Matthews, 1998).

they can be dangerous and counterproductive, yet without them there is scant chance of success. Deciding on the appropriate tool and then managing its use with great care is one of the defining skills of creativity. The tools presented within this chapter such as brainstorming, SCAMPER, morphological analysis, six hats and TRIZ can be readily applied in engineering and design activities. Some of these approaches can be aided or augmented by use of software (see, e.g. Shi et al., 2016, 2017a, b; Han et al., 2016, 2017; Chen et al., 2017).

Creativity tools can certainly be of assistance in developing the quantity of ideas, and if a process of convergence and divergence is followed, as in the creative solving process (CPS), it will involve evaluation and refinement of the ideas considered as a means to improve quality. It is questionable whether the tools will provoke game-changing concepts. More likely game-changing ideas that lead to new commercial opportunities will arise from sustained effort and an environment that encourages and permits risk taking. Creativity tools may, however, be of significant value in enabling the design space to be explored and for developing the details for each step of the idea to overcome challenges and issues inherent in idea realisation.

The conditions for creativity to occur include the combination of expertise, sponsorship, motivation and good communication skills. Creativity can be enhanced by ensuring that our short-term memory is free enough to be able to consider memories. This can be achieved in a number of ways including: the use of creative tools by ensuring that a problem can be understood in relatively simple terms; supplying cues to make the search of long-term memory more efficient; providing cues to refresh short-term memory and thereby retain key information as appropriate. The combination of these factors for encouraging creativity is illustrated in the concept of an idea engine as shown in Fig. 3.18 which uses the analogy of a jet engine with the multiple inputs of motivation, expertise, sponsorship and communication skills used for air, fuel, oil and control, the short-term memory modelled by a compressor connected to the power-house of the long-term memory or turbine. In an engine, the compressor is usually directly connected to the turbine, and this connectivity is preserved in the analogy with the short-term memory fed with ideas by the long-term memory. The 'refresh' for ensuring retention of an idea in working memory is taken as analogous to a glow or spark-plug.

To take an idea forward, the developed knowledge in the field of innovation and experience in commercial activities can be applied. Creativity and innovation are integrally linked by design where embodiment is given to the idea with refinement and

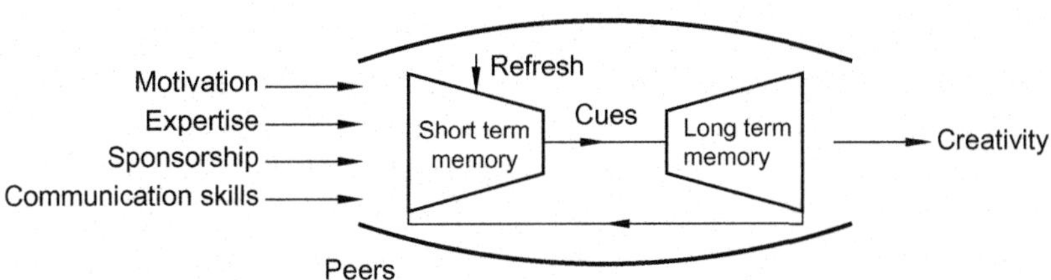

Fig. 3.18 The idea engine © Peter Childs.

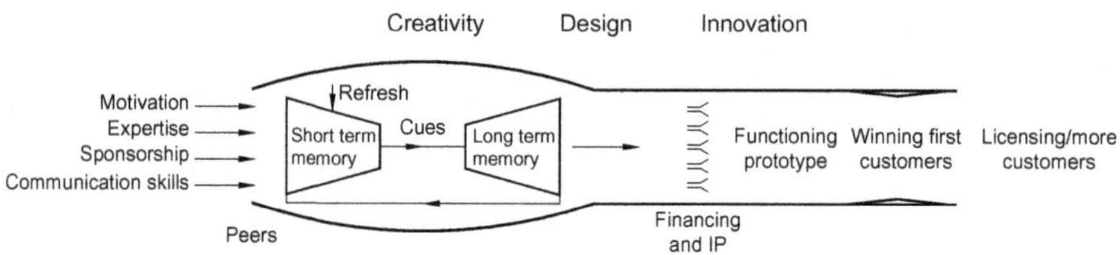

Fig. 3.19 The creativity and innovation engine © Peter Childs.

elaboration of all of the details required to enable the idea to be detailed. Innovation, the realisation of value from creativity, has been extensively studied and can be represented by a simple model in an extension to the idea engine shown in Fig. 3.19 comprising the steps of financing and intellectual property considerations, development of a functional prototype or system, winning initial customers and thereby securing the finance for either licensing or winning more customers. The extension to the idea engine uses the analogy of an afterburner. The innovation engine encompasses the notions of testing whether the venture is concerned with a physical product or artefact or a process or system. Just as with a physical prototype an essential part of the development process is to explore customer reaction and to explore its functions, so too with a process or system. The creativity and innovation engine also promotes a systems approach where there needs to be an incentive for engagement in the commercial activity.

There are many ways to enable the processes indicated in Fig. 3.19. The Create Project (2005) could be adopted to provide an overall framework for the activity, with creativity tools such as morphological analysis, post-it and grid brainstorming used to facilitate the generation of ideas. A standard design process could be applied and the ideas taken to market through the processes of securing finance, prototyping and marketing to initial customers prior to the generation of sufficient finance to take the idea forward to wider markets. The innovation project process of building a star team, agitating to generate key insights and breakthroughs, embodying and detailing ideas, piloting and roll-out (Neill, 2010) has been demonstrated as a highly effective, albeit linear, approach to turning ideas to reality.

Turning ideas into reality generally requires small, balanced, empowered and well-resourced teams of people or a single person fulfilling multiple roles. A single project champion for the duration of the project provides many advantages including ensuring that the vision is preserved. A challenging but uncomplicated project plan ensures focus without diverting attention to process. Early conceptualisation of ideas and prototyping provides clear advantages in building confidence in ideas and demonstrating the potential viability of a concept. Iteration and product or system optimisation enables cycles of improvements to be made to an idea. Facilitation of problem solving at decision gates assist in ensuring that ideas are addressed and rationale used in the judgement of issues.

Design thinking offers valuable insight into enhancing many activities. Designing can be used as a verb or noun. If you are designing, a verb, then you are designing an experience, whereas if you are designing, a noun, then you are designing 'stuff'.

With about 80% of the Western economy currently based on services and an increasing proportion of China's and India's economies (currently about 40 and 50% respectively), much greater opportunities for commercial activity reside with services and the associated user experience. By addressing the design of a socio-cultural system rather than just concentrating on a technical system, you are more likely to be able to develop game-changing and disruptive innovations that serve to raise the market capitalisation and value of the business or organisation concerned.

3.11 Conclusions

There are a wide number of creativity tools that can be used to aid the generation of ideas. For example, some of the brainstorming tools such as post-it, alphabet and grid, can be useful in the early phases of a project and whenever progress is blocked. Some organisations are serial users of brainstorming, such as IDEO (Kelley, 2001), and over 70% of business people use them in their organisations (Anderson et al., 2004). The Creativity Problem-Solving process has been validated extensively and provides a valuable framework for tackling a challenge and developing ideas. Other tools such as morphological analysis can be used to aid the development of ideas based on functional analysis of the subsystems. The theory of inventive problem solving provides a short-cut to experience, and various software tools are emerging that can aid and augment idea generation. It is suggested that regular use of these tools within engineering design activity can aid significantly in the generation of ideas.

References

Altshuller, G., 1994. (Translation by L. Shulyak). And Suddenly the Inventor Appeared: TRIZ, the Theory of Inventive Problem Solving. Technical Innovation Center, Inc..

Altshuller, G., 2005. 40 Principles Extended Edition: TRIZ Keys to Technical Innovation. Technical Innovation Center.

Altshuller, G.S., 1984. Creativity as an Exact Science—The Theory of the Solution of Inventive Problems. Gordon and Breach Science Publishers, New York.

Anderson, A., Heibeler, R., Kelley, T., Ketteman, C., 2004. Best Practices Building your Business with Customer Focussed Solutions. (Kindle Edition). Touchstone.

Atkinson, R.C., Shriffin, R.M., 1968. Human memory: a proposed system and its control processes. In: Spence, W.K., Spence, J.T. (Eds.), The Psychology of Learning and Innovation: Advances in Research and Theory, Vol. 2. Academic Press, pp. 89–195.

de Bono, E., 1970. Lateral Thinking. Penguin.

de Bono, E., 1999. Six Thinking Hats. Penguin.

Buzan, T., Buzan, B., 2006. The Mind Map Book. BBC Active.

Chen, L., Shi, F., Han, J, and Childs, P.R.N. A Network-Based computational model for creative knowledge discovery bridging human-computer interaction and data mining. ASME/IDETC Conference, 2017.

Childs, P.R.N., 2001. Practical Temperature Measurement. Butterworth Heinemann.

Childs, P.R.N., 2004. Mechanical Design, second ed. Butterworth Heinemann.

Childs, P.R.N. Use of six hats in STEM subjects. STEM Annual Conference 2012, HEA Academy, 2012.

Childs, P.R.N. and Fountain, R. Commercivity. E&PDE11, 2011. Design education for creativity and business innovation. 13th International Conference on Engineering and Product Design Education. Kovacevic, A., Ion, W., McMahon, C., Buck, L., and Hogarth, P. (Editors), Design Society, DS69–1, 2011, pp. 3–8.

Childs, P.R.N., Garvey, B., 2015. Using morphological analysis to tackle uncertainty at the design phase for a safety critical application. Propulsion Power Res. 4, 1–8.

Childs, P.R.N., Hamilton, T., Morris, R.D. and Johnston, G. Centre for technology enabled creativity. In: Rothbucher, B., Kolar, M., Ion, B. and Clarke, A. (Eds.), Proceedings of the 4th Engineering and Product Design Education International Conference, Salzburg, Austria, 7th–8th September 2006, 2006, Hadleys, pp. 367–372.

Cowan, N., 2001. The magical number 4 in short term memory: a reconsideration of mental storage capacity. Behav Brain Sci 24, 1–185.

Cox, G., 2005. Cox Review of Creativity in Business: Building on the UK's Strengths. HM Treasury.

CREATE Project, 2005. http://www.diegm.uniud.it/create/. (Accessed 10 October 2008).

Csikszentmihalyi, M., 1996. Creativity- Flow and the Psychology of Discovery and Invention. Harper-Perennial.

Design Council. The 'double diamond' design process Model, 2007. http://www.designcouncil.org.uk/en/About-Design/managingdesign/The-Study-of-the-Design-Process/, Accessed December 2007 2007.

Furnham, A., 2000. The brainstorming myth. Bus. Strateg. Rev. 11, 21–28.

Gadd, K., 2011. TRIZ for Engineers. Wiley.

Garvey, B. Combining quantitative and qualitative aspects of problem structuring in computational morphological analysis. Phd thesis, Imperial College London. 2016.

Garvey, B., and Childs, P.R.N. Applying problem structuring methods to the design process for safety helmets. Paper HPD-2013-12, Helmet Performance and Design, 2013.

Geschka, H., 1993. The development and assessment of creative thinking techniques: a German perspective. In: Isaksen, S.G., Murdock, M.C., Firestein, R.L., Treffinger, D.J. (Eds.), Nurturing and Developing Creativity: The Emergence of a Discipline. Norwood, pp. 215–236.

Goldenberg, J., Mazursky, D., 2002. Creativity in Product Innovation. Cambridge University Press.

Gordon, W.J.J.S., 1961. The Development of Creative Capacity. Harper & Row. pp. 3–7 and 33–56.

Han, J., Shi, F., and Childs, P.R.N. The combinator: a computer-based tool for idea generation. Proceedings of the Design 2016 2016.

Han, J, Shi, F., Chen, L., and Childs, P.R.N. The analogy retriever—an idea generation tool. ICED17, 2017.

Isaksen, S.G. (Ed.), 2000. Facilitative Leadership: Making a Difference with CPS. Kendall-Hunt.

Isaksen, S.G., DrShryver, L., 2000. Making a difference with CPS: a summary of the evidence. In: Isaksen, S.G. (Ed.), Facilitative Leadership: Making a Difference with CPS. Kendall-Hunt, pp. 187–248.

Jones, J.C., 1970. Design Methods: Seeds for Human Futures. John Wiley.

Kelley, T., 2001. The Art of Innovation. Profile books.

Kochan, N. (Ed.), 1996. The World's Greatest Brands. Palgrave Macmillan.

Lidwell, W., Holden, K., Butler, J., 2003. Universal Principles of Design. Rockport.

Litchfield, R.C., 2008. Brainstorming reconsidered: a goal based view. Acad. Manag. Rev. 33, 649–668.

Matthews, C., 1998. Case Studies in Engineering Design. Arnold.

Michalko, M., 2001. Cracking Creativity. Ten Speed Press.

Miller, G., 1956. The magical number seven, plus or minus two: some limits on our capacity for processing information. Psychol. Rev. 63, 81–97.

Neill, R. Per Diem. Private communication, 2010.

Osborn, A.F., 1963. Applied Imagination: Principles and Procedures of Creative Problem-Solving, third ed. Scribner's.

Parnes, S.J., Noller, R.B., Biondi, A.M., 1977. Guide to Creative Action. Scribners.

Poirier, M., Saint-Aubin, J., 1995. Memory for related and unrelated words: further evidence on the influence of semantic factors in immediate serial recall. Q. J. Exp. Psychol. 48A, 384–404.

Rose, L.H., Lin, H.T., 1984. A meta-analysis of long-term creativity training programs. J. Creat. Behav. 18, 11–22.

Shi, F., Han, J., Childs, P.R.N., 2016. A data mining approach to assist design knowledge retrieval based on keyword associations. Design. .

Shi, F., Chen, L., Han, J., Childs, P.R.N., 2017a. A data-driven text mining and semantic network analysis for design information retrieval. J. Mech. Des.

Shi, F., Chen, L., Han, J, and Childs, P.R.N. Implicit knowledge discovery in design semantic network by applying pythagorean means on shortest path searching. ASME/IDETC, 2017b.

The Ladies' Home Journal, 1898. The Anecdotal Side of Edison, Subsection: His Estimate of Genius. Curtis Publishing Company, Philadelphia, PA, pp. 7–8.

Torrence, E.P., 1972. Can we teach children to think creatively? J. Creat. Behav. 6, 114–143.

Treffinger, D.J., Isaksen, S.G., Dorval, K.B., 1994. Creative learning and problem solving: an overview. In: Runco, M. (Ed.), Problem finding, problem solving, and creativity. Ablex, pp. 223–236.

Treffinger, D.J., Isaksen, S.G., Stead-Dorval, K.B., 2006. Creative problem solving: an introduction, fourth ed. Prufrock Press.

Vaneker, T., Lutters, E., 2008. TRIZ-Future Conference '08 Synthesis of Innovation. Ipskamp.

Wallas, G., 1926. The Art of Thought. Jonathan Cape.

Whiting, C.S., 1958. Creative Thinking. Van Nostrand.

Yan, Y., Childs, P.R.N., and Hall, A. An assessment of personality traits and their implication for creativity amongst Innovation Design Engineering masters students using the MBTI and KTS instruments. 19th International Conference on Engineering Design, 2013.

Zwicky, F., 1969. Discovery, Invention, Research through the Morphological Analysis. The Macmillan Company.

Further reading

Allen, M.S., 1962. The Allen Morphologizer. Prentice-Hall.

B Link, 2017. http://www.imperial.ac.uk/design-engineering/research/engineering-design/creativity/b-link/. (Accessed 15 September 2017).

Bauer, L., 2003. Introducing Linguistic Morphology, second ed. Georgetown University Press.

Boden, M.A., 1990. The Creative Mind: Myths and Mechanisms. Weidenfeld, London.

Childs, P.R.N., 2006a. CETL in Creativity. Annual Report 2005–2006. InQbate, University of Sussex.

Childs, P.R.N., 2006b. The Mercenary Use of Creative Methods. HEEG, University of Sussex, UK.

Domb, E., 1998. The 39 features of Altshuller's contradiction matrix. TRIZ J. .

Eberle, B., 1997. Scamper: Creative Games and Activities for Imagination Development. Prufrock Press.

Guildford, J.P., 1967. In: Moonly, R.L., Razik, T.A. (Eds.), Intellectual factors in productive thinking, Explorations in creativity. Harper & Row, New York, Evanston, and London, pp. 95–106.

Harris, R. Introduction to creative thinking, 1998. http://www.virtualsalt.com/crebook1.htm, Accessed July 2006.

Isaksen, S.G., Treffinger, D.J., 1985. Creative Problem Solving: The Basic Course. Bearly Limited, Buffalo, New York.

Isaksen, S.G., Dorval, K.B., Treffinger, D.J., 1993. Creative Approaches to Problem Solving, third test ed. Buffalo, New York.

Rantanen, K., Domb, E., 2002. Simplified TRIZ. St Lucie Press.

Swedish Morphological Society, 2008. http://www.swemorph.com. (Accessed 18 October 2018).

The TRIZ Journal. http://www.triz-journal.com/.

Thompson, G., Lordan, M., 1999. A review of creative principles applied to engineering design. Proc. Inst. Mech. Eng. 213 (Part E), 17–31.

Vidal, R.V.V., 2005. Creativity for operational researchers. Investigacao Operational 25 (1), 1–24.

Zusman, A., Zlotin, B., 1999. Overview of creative methods. TRIZ J 7.

Machine elements

4

Chapter Outline

Nomenclature

Generally, preferred SI units have been stated.

a	acceleration (m/s^2)
F	force (N)
L	length (m)
m	mass (kg)
OEM	original equipment manufacturer

4.1 Introduction

The design activity involves endowing form and ensuring functionality of a given product, service or system. The function of many products depends on devices that modify force or motion and consist of a number of interrelated units. An example is a gearbox that modifies the torque-speed characteristic of a prime mover to match that of the load and comprises gears, shafts, bearings and seals. Such devices that modify force or motion are called machines and the interrelated units are sometimes known as machine elements. The technology base has developed to a mature status in this area providing a number of such devices that can in some cases be used as they are or designed fit-for-purpose.

In the design of a machine and its machine elements, it is the engineer's or designer's task to determine the motion, forces and changes in energy involved so that sizes, shapes

Mechanical Design Engineering Handbook. https://doi.org/10.1016/B978-0-08-102367-9.00004-4

and materials for each machine element making up the whole machine can be determined. Although design work may involve concentrating on one component at a given time, it is essential to consider its interrelationship with the whole product following the total design philosophy, taking into account the market requirement, detailed specifications, the conceptual design, design of other components and manufacture requirements.

Engineers and designers are familiar with the concept of delivering products, services and systems with functionalities. The term function is generally used to refer to the purpose, objective, intent or aim of the subject or object concerned (see Aurisicchio et al., 2011). Traditionally in engineering, attention has been focused on technical function, but a wide scope of types of function exist including technical, aesthetic, social, economic, psychological, latent and emergent (see Childs, 2013). Establishing the function required or analysing what the features of a design actually perform can be a key step in machine design. An effective tool to this end is the use of Functional Analysis Diagrams (FADs) as illustrated in Fig. 4.1. An FAD provides a snapshot of a design with features and their purpose identified (see Aurisicchio et al., 2011; Aurisicchio and Bracewell, 2013; Michalakoudis et al., 2016).

The goal in machine design is to determine the size, shape and selection of materials and manufacturing processes for each of the parts of the machine so that the machine will perform its intended function without failure. This requires the engineer or designer to model and predict the performance and operation of each component and overall assemblies studying the mode of failure for each and ensuring that conditions expected in service for the product are met. This may involve determining stresses, deflections, temperatures and material degradation for each and every component and overall assembly as well as its influence within the total design. As stresses are a function of forces, moments and torques, consideration of the dynamics of the system will often be necessary before detailed stress analysis can be undertaken. Furthermore, as the strength of materials is a function of temperature, a thermal analysis may also be necessary in order to determine operating temperatures throughout the performance cycle of the machine.

If a machine has no moving parts, then the task for the designer is simpler since only static force analysis and thermal analysis is necessary. However, even seemingly static structures are subject to loading from external loads such as weather conditions of wind and rain, earth tremors and traffic. As such a consideration of the environment into which the machine is to be installed should be made and appropriate failure modes mitigated against. Mitigation involves considering what may go wrong and determining what action to take to either prevent it from going wrong in the first case or if it does go wrong what is going to be done about it.

If the motions involved in a machine are slow, then it is likely that a static force analysis will be sufficient. In a static structure such as the floor of a casing designed to support a particular weight, the payload that the structure can support can be increased by adding appropriately distributed weight (dead weight) to its structure. If however the components of a machine have significant accelerations, a dynamic force analysis will be necessary. Adding additional mass to a moving component increases, the forces involved and it is possible for such components to become victims of their own mass. This is because some of the loading causing stress in a moving

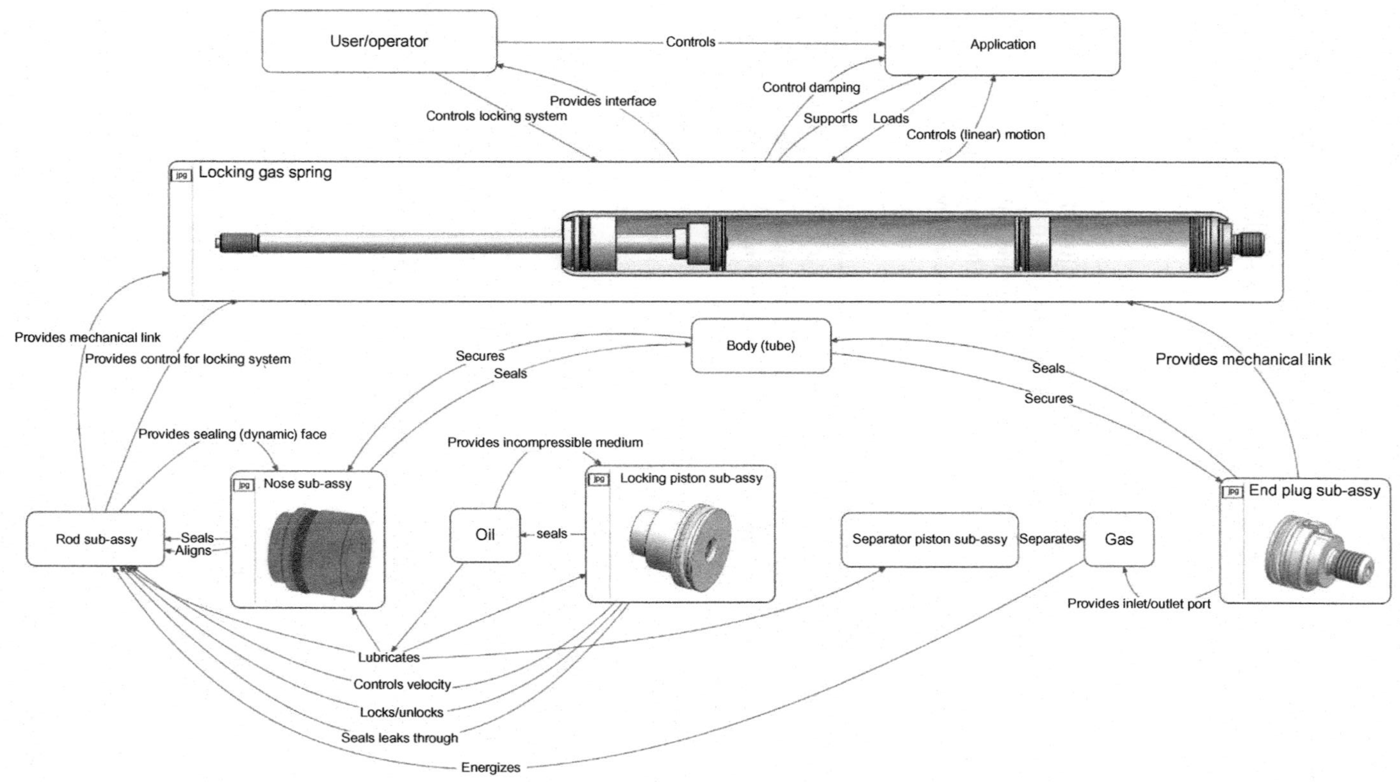

Fig. 4.1 Top and subassembly level FAD for a lockable gas spring (Michalakoudis et al., 2016).

component is due to inertial forces modelled by Newton's second law of motion, embodied in the equation

$$F = ma. \tag{4.1}$$

Although addition of mass may increase the strength of a component, the benefit may be reduced by the increases in inertial forces.

To meet a design requirement, the designer needs to conceive a form and express the connectivity of its parts in order to meet the desired goal. A proportion of the mental sculpting involved is geometric in nature because shape often permits the achievement of a given function. The design procedures presented here follow this approach with an initial proposal for the geometric configuration followed by analysis to justify the proposal or provide guidance for its modification. Boden (1990) discusses creativity in terms of the connection of conceptual spaces in the mind. In order to create a design, it is often necessary to combine ideas from different fields or disciplines. This may or may not take place as part of a structured design procedure or in a prescribed time period and overreliance on procedure should in some cases actually be guarded against as this may stymie the activity.

The variety of machine elements and technologies already available to the designer as a result of years of development is extensive as indicated in Table 1.3 in Chapter 1. Machine elements considered in this book are those more commonly found in machines transmitting power by means of moving components such as motors and engines. The diagram given in Fig. 4.2 illustrates the range of machine elements involved in the design of a gear box for a single cylinder air compressor and comprises the following: shafts, bearings, gears, seals and enclosure. The machine elements considered in this book provide skills in the design of such applications and cover bearings, shafts, gears, belts and chains, seals, clutches and brakes, springs, fasteners and enclosures. These subjects are introduced in Sections 4.2–4.9 prior to more thorough consideration in the relevant chapters later in this book.

A fundamental element for many machines is the lever and the principle of leverage is used in many machine elements as well as machines. The fixed point of the lever about which it moves is known as the fulcrum. A lever can be used with a fulcrum (also known as a pivot) to allow a small force moving over a large distance to create a large force moving over a short distance. The position of the force and the load are interchangeable and by moving them to different points on the lever, different effects can be produced. There are three generic orders or classes of lever: first, second and third. A first-order lever is like a see-saw or balance, with the load and the force separated by the fulcrum. As one side moves up, the other moves down. The amount and the strength of the movement are proportional to the distance from the fulcrum. If L_2 is 10 times of L_1, then F_1 will be 10 times of F_2.

$$F_1 = F_2 \frac{L_2}{L_1}. \tag{4.2}$$

A wheel barrow is an example of a second-order lever. Here, the load is between the force and the fulcrum. This uses mechanical advantage to ease lifting of a large weight. In a third-order lever, the force is between the fulcrum and the load.

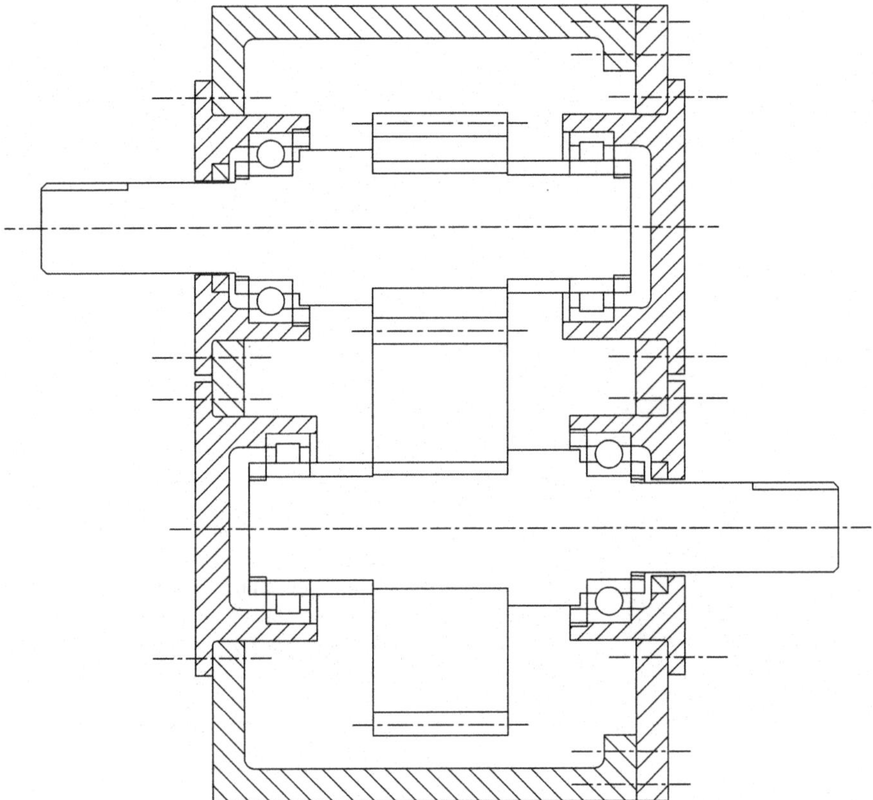

Fig. 4.2 Transmission for an engine driven compressor.

Mechanical advantage is reduced but the movement at the load point is increased. Tweezers are an example of a third-order lever. The characteristic of the three types of lever are given in Table 4.1.

A further fundamental building block in machine design and many mechanisms is the linkage. A linkage is a system of links, rods or spars connected at joints with rotary or linear bearings. Linkages can be used to change direction, alter speed and change the timing of moving parts. An example of the use of linkages is shown in Fig. 4.3 where the articulation of the grab bucket is enabled by a four-bar linkage giving prescribed rotation in response to the hydraulic ram actuation. A four-bar linkage is also used to provide mechanical advantage for grasping a work-piece in mole grips.

In the example shown in Fig. 4.4, two linked linkages are used to convert the small linear movement of the drive shaft (bottom left) into first a rotational body movement and secondly a fast hammer movement. In this case, the speed of the input drive shaft is amplified by the use of the linkages providing high-speed hammer action.

Many machine elements involve relative motion between components and associated friction and wear. Tribology is the study of friction, lubrication and wear of surfaces in relative motion. Because of the desire to produce efficient machines where the

Table 4.1 The three generic orders or classes of lever, first, second and third

Type of lever	Example	Characteristics	Useful equations
First-order lever		The load and the force are separated by the fulcrum. As one moves, the other moves in the opposite direction. The magnitude and movement is proportional to the distance from the fulcrum	$F_1 = F_2 \frac{L_2}{L_1}$
Second-order lever		The load is between the force and the fulcrum. This type of lever uses mechanical advantage to ease the lifting of a large weight or the application of a large load	$F_e = F_l L_l / L_e$
Third-order lever		In a third-order lever, the force is between the fulcrum and the load. Mechanical advantage is reduced but the movement at the load point is increased	$F_e = F_l L_l / L_e$

Fig. 4.3 Multiple use of linkages is made in construction equipment.

Fig. 4.4 Two linked linkages.

loss of energy due to friction is minimal, extensive research and effort have been put into studying tribology. As a subject, it has considerable interrelationships and cross-over with materials and fluid mechanics. The study on reduction of friction has been of concern to humans for as long as machines have been in use and the subject of tribology includes consideration of machine elements such as bearings, gears, belts, chains, clutches, brakes and seals. Tribology as an engineering science is now a well-established subject and has resulted in significant breakthroughs, for example, in bearing technology where some companies are now able to offer high-efficiency bearings with significantly reduced friction in comparison with previous products.

4.2 Bearings

The term bearing, in its general sense, is used to refer to the assembly formed by two surfaces that have the capacity for relative motion. A wide range of bearings have been developed some of which involve rolling or sliding motion and some both. Lubrication of a bearing is frequently used to reduce friction and therefore the wear and power absorbed by the bearing, and to remove heat to ensure operation of the bearing assembly at temperatures compatible with the materials and lubricant used.

Typical examples of bearings are shown in Figs. 4.5 and 4.6. Fig. 4.5 illustrates rolling bearings. These come in many forms each designed for a given set of performance characteristics such as speed and load capability. Rolling element bearings usually need to be lubricated by grease or oil with the choice depending on the speed and

Fig. 4.5 Examples of rolling element bearings available from existing suppliers.

Fig. 4.6 Multi-lobe journal bearings.

temperature of operation. Rolling element bearings are typically available as stock items. This is of significant potential advantage to a machine designer, as the task becomes that of bearing selection rather than bearing design. Bearing design is a significant undertaking involving consideration of very high-point contact stresses and use of special materials. The technology involved relies heavily on the science of tribology and there are a number of worldwide companies already operating in this market. A designer needs to carefully consider the issue of whether time is better spent producing a new design for a bearing that is already available from an existing supplier or whether it is better spent on designing and adding value to the product in hand.

Plain bearings rely on either rubbing contact between the bearing surfaces or pressure to separate the two surfaces. The pressure required can be induced in a lubricant by movement of one component relative to another or by the external supply of the lubricant under the required pressure. Plain bearings are also available from specialist bearing suppliers. In principle, they can be significantly cheaper than rolling element bearings as they involve less moving components. The selection of an appropriate bearing for a given task, however, is an involved activity, which needs to take into account, amongst other factors:

- load,
- speed,
- location,
- size,
- cost,
- starting torque,
- noise, and
- lubrication supply.

The design of plain bearings and the selection of rolling element bearings as stock items from bearing supplies are considered more fully in Chapters 5 and 6. The related topic of shafts is introduced in Chapter 7. A consequence of bearings being supplied by a select number of specialist manufactures is that they have been subject to standardisation and are usually only available in a limited number of standard sizes. Standardisation has the advantage that parts from one original equipment manufacturer (OEM) will be interchangeable in terms of geometry with another. This does not, however, mean that the performance of two similarly sized components from different manufacturers will be the same.

4.3 Gears, belts and chains

Gears, belts and chains can be used to transmit power from one shaft to another. Typical examples are illustrated in Figs. 4.7–4.9.

Many power-producing machines, or prime movers, such as internal combustion engines, gas turbines and electric motors produce power in the form of rotary motion. The operating torque versus speed characteristics of prime movers vary according to

Fig. 4.7 Spur gears incorporated within an epicyclic gear box for a cordless hand-tool transmission.

Fig. 4.8 Flat belt and synchronous belt drives driving multiple devices from a single belt.

Fig. 4.9 Chain drive for powering multiple elements.

their type and size as do the driven loads. It is common for the torque versus speed characteristics to be mismatched requiring the need for gearing and perhaps also a clutch to enable the prime mover to attain sufficient momentum to engage a load.

When transmitting power from a source to the required point of application, a series of devices are available including belts, pulleys, chains, hydraulic and electrical systems, and gears. Generally, if the distances of power transmission are large, gears are not suitable and chains and belts can be considered. However, when a compact, efficient or high-speed drive is required, gear trains offer a competitive and suitable solution.

For example, for a small electric motor driven air compressor options that might be possible for the transmission include the following: belt drive, chain drive, gear drive, belt and chain, belt and gear, and gear and chain. Three different types of machine element drive along with direct drive are all potential solutions that could be viable. Schematics for the different concepts are illustrated in Fig. 4.10.

Gears, belts and chains are all available from specialist suppliers. The manufacture of gears requires specialist machinery and the stresses in the region of contact between gears in mesh can be exceptionally high requiring the use of specialist materials. Similarly, the design of roller chain and associated sprockets involves high contact

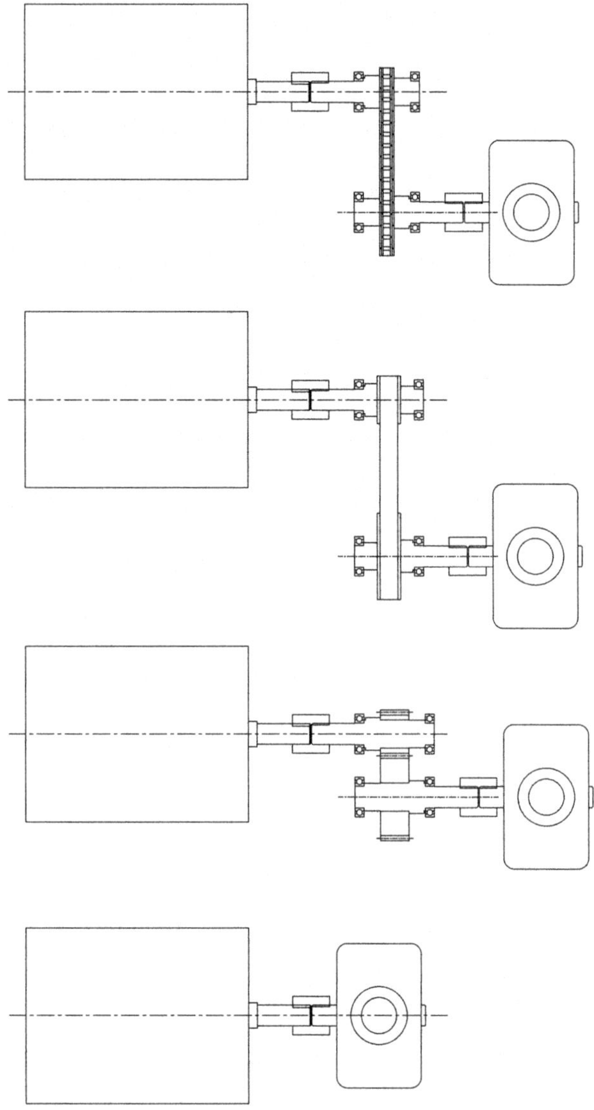

Fig. 4.10 Conceptual arrangements for a transmission.

stresses; and in addition, the design of an articulated joint and associated bearings. For these reasons, combined with financial expediency, leaving the cost of specialist manufacture infrastructure to other companies, gear, chain and belt manufacture is usually best left to the specialist and the task of the machine designer becomes that of selecting and specifying appropriate components from the specialist stock suppliers and OEMs. Gears are considered in Chapters 8–11 and belts and chains in Chapter 12.

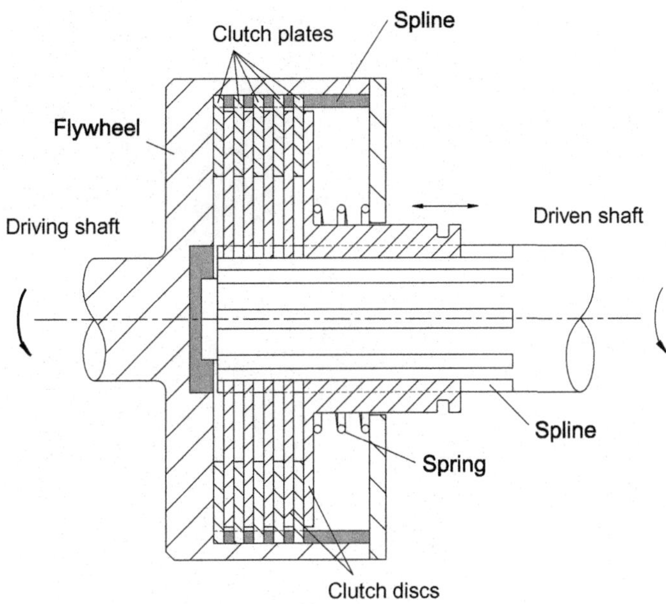

Fig. 4.11 A multidisc clutch.

4.4 Clutches and brakes

Clutches and brakes provide frictional, magnetic, hydraulic or mechanical connection between two machine elements, usually shafts. There are significant similarities between clutches and brakes. If both shafts rotate, then the machine element will be classed as a clutch and the usual function is to connect or disconnect a driven load from a driving shaft. An example of a multidisc clutch is given in Fig. 4.11. If one shaft rotates and the other is stationary, then the machine element is classed as a brake and the likely function is to decelerate a shaft, e.g., see Fig. 4.12. In reality, however, the same device can function as a brake or clutch by fixing its output element to a shaft that can rotate or to ground, respectively.

Brakes and clutches are familiar devices from their use in automotive applications. They are, however, also used extensively in production machinery. Clutches allow a high-inertia load to be rotated with a smaller electric motor than would otherwise be required if it was directly connected. Clutches are also used to maintain a constant torque on a shaft for tensioning of webs and filaments. They can be used, in emergencies, to disconnect a driven machine from a motor in the event of jamming of the machinery. In such cases, a brake may also be installed in order to bring the machinery to a rapid stop.

Complete clutches and brakes are available from specialist manufacturers. In addition, key components such as discs and hydraulic and pneumatic actuators are also

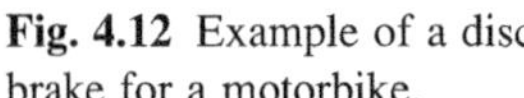

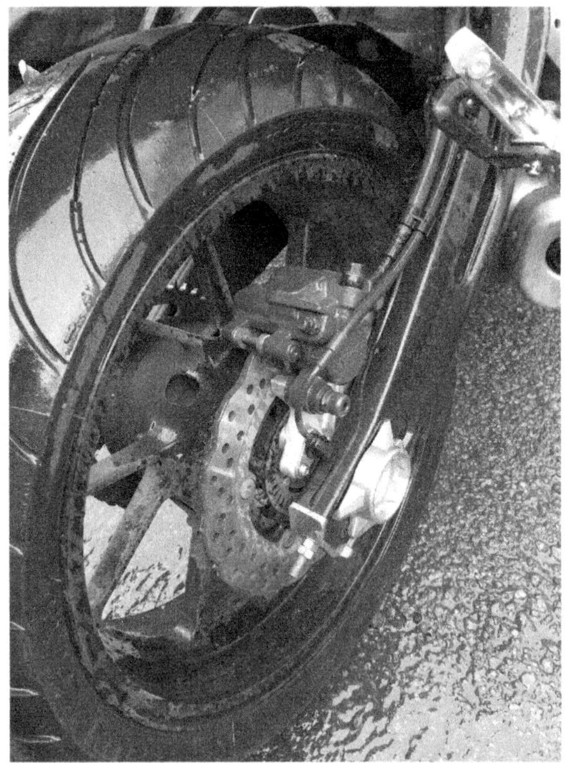

Fig. 4.12 Example of a disc brake for a motorbike.

available enabling the designer to opt, if appropriate, to design the overall configuration of the clutch or brake and buy in say the frictional lining surface. Clutches and brakes are considered more fully in Chapter 13.

4.5 Seals

Seals are devices that are used to reduce or eliminate the flow of fluid or particulates between two locations. Seals are an important part of a machine design where fluids under relatively high differential pressures must be contained within a particular region. An example is the fluid within the cylinder of an internal combustion engine. Here, the working fluid needs to be contained in the cylinder for part of the cycle and excluded, in as much as possible, from contaminating the lubricating oil for the crankshaft bearings. This is conventionally achieved by means of piston rings, Fig. 4.13.

There are many types of seal, but they can generally be classified as either static or dynamic depending on whether relative movement is involved. Many static seals involve use of a gasket or sealing material such as an elastomer. These are widely available from specialist suppliers and the task facing the designer is the selection of appropriate materials for the application. Some dynamic seals are also

Fig. 4.13 Piston ring seals.

commercially available such as radial lip seals and mechanical face seals. However, there is not the level of standardisation in shaft design and the range of possible applications is seemingly infinite with differing pressure ratios and fluids being sealed. As a result, the range of seal geometries required is too extensive to warrant an OEM designing or stocking a wide number of seal sizes and types. Instead, many dynamic seals are designed specifically for the purpose. Typical considerations in the selection and design of seals include the nature of the fluid or particulates to be contained, the differential pressure, the nature of any relative motion between components, the level of sealing required, operating conditions, life expectancy and total costs. Seals are introduced more fully in Chapter 14.

4.6 Springs

Springs are flexible elements used to exert force or torque and store energy. The force exerted by a spring can be a linear push or pull or a radial force similar to that of a rubber band around a roll of paper. Alternatively, a spring can be configured to produce a torque with applications including door closers. Springs store energy when they are deformed and return energy when the force that is causing the deflection is removed. A further type of spring is the power springs or spring motor, which once wound up can dissipate energy at a steady pace.

Springs can be made from round or rectangular wire bent into a suitable form such as a coil or made from flat stock and loaded as a cantilevered beam. Many standard size spring configurations are available as stock items from specialist manufacturers and suppliers, e.g., see Fig. 4.14, and it is usually expedient to use these where possible. As well as paying attention to the performance characteristics of the spring, attention should also be focussed on the spring ends or terminations. Many options are available

Fig. 4.14 A coil spring applied to part of a suspension system.

such as twist loops or hooks, side loops and extended hooks. Some applications require a custom spring design. Spring design involves consideration of many variables such as length, wire diameter, forces, spring rate, spring index, number of coils, pitch, pitch angle, installation considerations, coil clearance, materials, types of loading and allowable stresses. The selection and design of springs is considered in Chapter 15.

4.7 Fasteners

Fasteners connect or join two or more components. It has been estimated that there were over 2,500,000 fasteners in the classic Boeing jumbo 747 jet. For more modern airliners such as the Airbus Industries A340, this figure has been dramatically reduced but the total is still significant. This type of example illustrates the importance of fasteners to the designer.

There are thousands of types of fasteners commercially available from specialist suppliers. The most common includes threaded fasteners such as bolts, screws, nuts, studs and set screws, rivets, fast-operating fasteners requiring say a quarter turn to connect or disconnect and snap joints commonly used in plastic component assembly. In addition, there is the range of permanent fastening techniques by means of welding, brazing, soldering and use of adhesive. Some of the principal types of fastener are featured in Table 4.2.

Given the wide range of fastening techniques available and the costs involved in sourcing, stocking and assembly, the correct choice of fastener is very important. Chapter 16 serves to introduce a variety of types of fastener and fastening and their selection based on appropriate analysis and consideration of the intended function.

Table 4.2 Some examples of types of fastener

Type of fastener	Example	Characteristics
Bolts	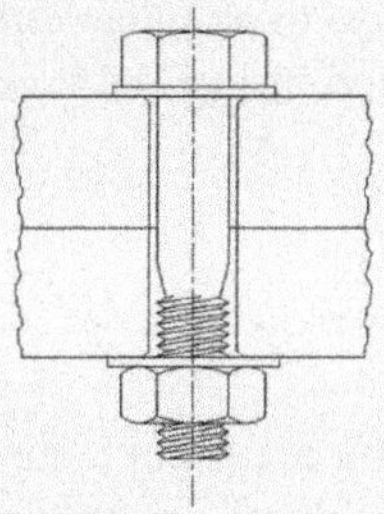	Screw fasteners are highly versatile and can be removed readily. The common element of a screw fastener is the helical thread that causes a screw to advance into a component or nut when rotated. The range of threaded fasteners available is extensive including nuts and bolts, machine screws, set screws and sheet metal screws
Rivets	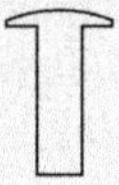	Rivets are permanent nonthreaded fasteners that are usually manufactured from steel or aluminium. They consist of a preformed head and shank, which is inserted into the material to be joined and the second head that enables the rivet to function as a fastener is formed on the free end by a variety of means known as setting
Adhesives	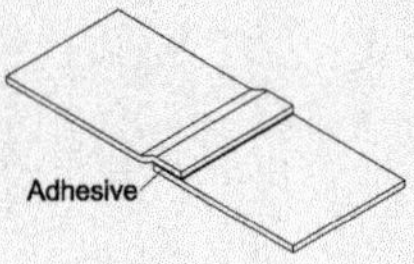	Adhesive bonding involves the joining of similar or dissimilar materials using natural or synthetic substances, which form a rigid or semirigid interface. Advantages include distribution of stresses over the bonded area, stiffness, smooth surface finishes, fatigue resistance, low-temperature fabrication, connection of dissimilar materials and sealing
Welding	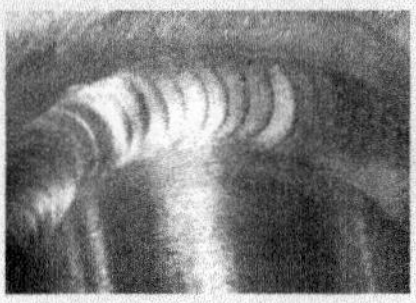	Welding is the process of joining material together by raising the temperature of the surfaces to be joined so that they become plastic or molten. Welding provides a permanent join that can have good strength, high temperature and pressure characteristics.
Snap fits	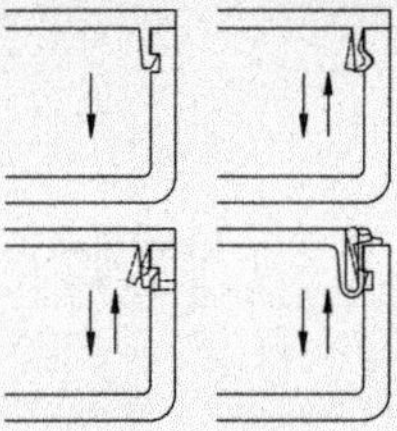	Snap fasteners comprise a component that deforms during assembly and then positively locates against a corresponding feature. Common deforming components include hooks, knobs, protrusions or bulges and corresponding depressions, undercuts, detents and openings on the other part to be joined. Snap fasteners can be moulded directly into a component reducing part count and the need to purchase separate fasteners

4.8 Wire rope

Wire rope is sometimes used in hoisting, haulage and conveyer applications. By using lots of small diameter wires, twisted around a central core, some flexibility in the wire rope can be achieved and it is possible for the wire rope to articulate drums and other radial segments. An example of wire rope is illustrated in Fig. 4.15, and the topic is introduced in Chapter 17.

4.9 Pneumatics and hydraulics

Significant mechanical advantage can be achieved by exposing a surface area to a pressurised fluid. This principle is exploited in many hydraulic and pneumatic devices which have been used extensively in machine design and are available as standard stock items. A hydraulic actuator is illustrated in Fig. 4.16 and the topics of pneumatics and hydraulics are considered in Chapter 18.

Fig. 4.15 Wire rope.

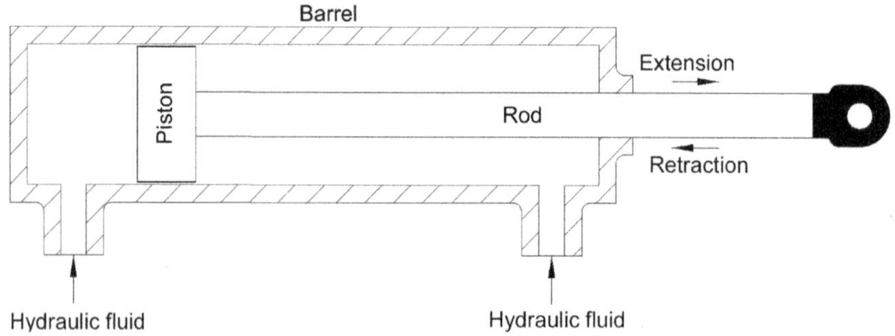

Fig. 4.16 Hydraulic actuator.

4.10 Enclosures

Individual machine elements such as shafts on bearings must themselves be supported and this is the function of an enclosure or frame. The total design model requires attention to be applied to all other phases of the design process whilst undertaking any one task. As such the design of any given machine element should not occur in isolation but instead give due consideration to how it will fit in and interact with the overall assembly. The enclosure for a machine provides not only the support for the machine but also provides form to a product. Attention to visual attractiveness is not always given the attention it warrants by designers of machine elements. However, we produce our components to be purchased and current thinking indicates that visual perception is the dominant amongst the senses with over half the cerebral cortex dedicated to visual processing. The shape of our products is therefore important and we should design accordingly.

Enclosures are usually unique to a given class of machine, as each machine tends to be different in terms of the number, size and function of its constituent components. It is not practical, therefore, to have a completely general approach to enclosure and frame design. Common examples of the variation of frames and enclosures are those for vehicles and toys. Toys must be safe, functional and fun whilst minimising the material and personnel effort in producing the toy in order to minimise the production cost and maximise profit.

The design of an enclosure or frame is to some extent an art, in that, the constituent machine components must be accommodated. There may be constraints on where supports can be placed in order not to interfere with the function and operation of the machine or in order to ensure access to particular components. The frame for the bus shown in Fig. 4.17 illustrates this point, in that, access for doors, windows and occupants must be provided.

Some of the more important design parameters to consider for an enclosure or frame include the following: strength, size, assembly, appearance, corrosion resistance, stiffness, vibration, weight, noise, cost to produce and maintain, and life. A list such as this may be too all-embracing to be of much use at the start of a design process and a more helpful list of factors to consider as the starting point for a frame or enclosure include the following:

- forces exerted by the machine components through mounting points such as bearings, pivots, brackets and feet;
- how the frame itself will be supported;
- allowable deflections of constituent machine elements;
- cooling requirements;
- the environments into which the machine will be transported and installed and the relationship to other machines and infrastructure;
- the quantity required and production facilities available and
- the expertise available for the design task.

The complexity involved when considering more than one machine element is evident and because of the need to keep the bigger picture in mind, models for the design process such as total design are useful.

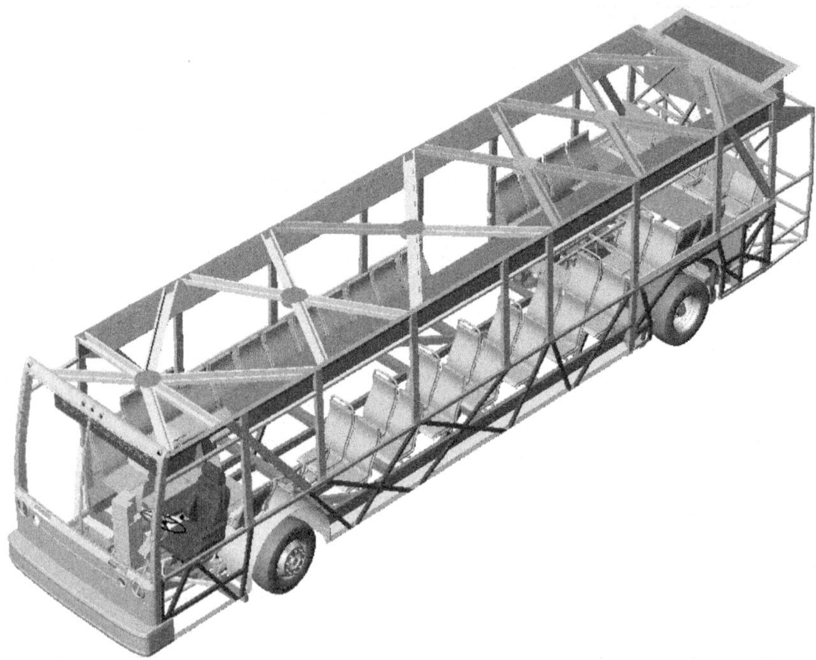

Fig. 4.17 Bus frame design (Allin and Simon, 2002).

4.11 Conclusions

The technology base available to us includes a wide range of machine elements such as bearings, shafts, gears, chains, belts, seals, clutches, brakes and fasteners. This chapter has served to introduce these items, many of which are available as stock items from specialist manufacturers. This means that the designer's task becomes that of selecting and specifying the appropriate item and relieves him or her of the task of the detailed design of for instance a roller chain so the design of the overall machine can be concentrated on.

It is appropriate to take time in overviewing the technologies available to the designer in order to be able and consider and take into account the interrelationships between individual machine elements. This is in keeping with the total design model where each phase takes into account all other phases and aspects of the design process.

References

Allin, K., Simon, D., 2002. A new approach to vehicle architecture design. In: Stobart, R.K., Childs, P.R.N. (Eds.), Total Vehicle Technology, How Do We Get the Innovation Back into Vehicle Design. Professional Engineering Publishing, pp. 49–56.

Aurisicchio, M., Bracewell, R.H., 2013. The Function analysis diagram: intended benefits and co-existence with other functional models. Artif. Intell. Eng. Des. Anal. Manuf.

Aurisicchio, M., Eng, N., Ortiz Nicolas, J.C., Childs, P.R.N., Bracewell, R., 2011. On the functions of products. In: ICED11. Paper 367.
Boden, M.A., 1990. The Creative Mind: Myths and Mechanisms. Weidenfeld and Nicolson.
Childs, P.R.N., 2013. Engineering freakout. Int. J. Mech. Eng. Educ. 41, 297–305.
Michalakoudis, I., Childs, P., Aurisicchio, M., Harding, J., 2016. Using functional analysis diagrams to improve product reliability and cost. Adv. Mech. Eng. 8, 1–11.

Websites

jolisfukyu.tokai-sc.jaea.go.jp/fukyu/tayu/ACT01E/02/0202.htm.
ocw.mit.edu/OcwWeb/Mechanical-Engineering/2-007Spring-2005/LectureNotes/index.htm.
science.howstuffworks.com/gear7.htm.
video_demos.colostate.edu/mechanisms/index.html.
www.acmcf.org.tw/model/page/simulation/ncku/D01.htm.
www.coinopgamemuseum.com/mechanisms/photoboothcamera/photoboothcamera.html.
www.cs.cmu.edu/~rapidproto/mechanisms/tablecontents.html.
www.flying-pig.co.uk/mechanisms/.
www.howstuffworks.com/transmission.htm.
www.ijn.dreamhost.com/Optics/Optics%20-%2012%20cm%20binoc%20model%205.htm.
www.jcb.com/products/MachineProduct.aspx?PID=7&RID=2.
www.mechanisms101.com/theo_jansen.shtml.
www.mystery-productions.com/hyper/Hypermedia_2003/Muirhead/website/index.htm.
www.precisionmotion.co.uk/page3.htm.
www.sandsmuseum.com.
www.stanleyworks.co.uk/Clamps+and+Pliers/Mole+Grips.ctlg.
www.tc.gc.ca/roadsafety/safevehicles/mvstm_tsd/tsd/2060rev0_e.htm.
www.technologystudent.com/cams/cam1.htm.
www.ul.ie/~kirwanp/linkstoanimations.htm.

Journal bearings

Chapter Outline

Nomenclature

Generally, preferred SI units have been stated.

a	radius (m)
b	outer radius (m)
b_o	characteristic length (m)
c	radial clearance (m)
D	journal diameter (m)
e	eccentricity
Eu	Euler number (dimensionless)
f	coefficient of friction
F_g	geometrical factor
Fr	Froude number (dimensionless)
F_x	body force component in the x direction (N/m^3)
F_y	body force component in the y direction (N/m^3)
F_z	body force component in the z direction (N/m^3)
g	acceleration due to gravity (m/s^2)
h	film thickness (m)
h_o	minimum film thickness (m)
I_c	cosine load integral
I_s	sine load integral
L	characteristic length, length (m), bearing length (m)
l_o	characteristic length (m)
N	speed (normally in rpm)

Mechanical Design Engineering Handbook. https://doi.org/10.1016/B978-0-08-102367-9.00005-6

N_s	journal speed (revolutions per second)
p	static pressure (Pa)
p_{max}	maximum annulus static pressure (N/m^2)
P	load capacity (N/m^2), load (N)
q_x	volumetric flow rate per unit width in the x direction (m^2/s)
q_y	volumetric flow rate per unit width in the y direction (m^2/s)
Q	lubricant flow rate (m^3/s)
Q_π	flow through the minimum film thickness (m^3/s)
Q_s	side flow rate (m^3/s)
r	radius (m)
r_m	mean radius (m)
R	specific gas constant (J/kg K)
Re	Reynolds number (dimensionless)
Re_x	Reynolds number based on x component of velocity (dimensionless)
Re_y	Reynolds number based on y component of velocity (dimensionless)
Re_z	Reynolds number based on z component of velocity (dimensionless)
S	geometrical factor, bearing characteristic number
SAE	Society of Automotive Engineers
t	time (s)
t_o	characteristic time scale (s)
t_*	dimensionless time ratio (dimensionless)
T	temperature (°C, K)
T_{av}	average temperature (°C)
T_1	temperature of the lubricant supply (°C)
T_2	temperature of the lubricant leaving the bearing (°C)
Ta	Taylor number (dimensionless)
Ta_m	Taylor number based on mean radius (dimensionless)
$Ta_{m,cr}$	critical Taylor number based on mean annulus radius (dimensionless)
u_r	velocity component in the r direction (m/s)
u_x	velocity component in the x direction (m/s)
$u_{x,o}$	characteristic velocity component in the x direction (m/s)
u_y	velocity component in the y direction (m/s)
$u_{y,o}$	characteristic velocity component in the y direction (m/s)
u_z	velocity component in the z direction (m/s)
$u_{z,o}$	characteristic velocity component in the z direction (m/s)
u_ϕ	velocity component in the tangential direction (m/s)
$\overline{u}$	average velocity (m/s)
U_o	characteristic velocity (m/s), reference velocity (m/s)
V	journal velocity (m/s); velocity (m/s)
W	applied load (N)
x	distance along the x axis (m),
x_*	dimensionless location (dimensionless)
y_*	dimensionless location (dimensionless)
z	distance along the z axis (m)
z_*	dimensionless axial location (dimensionless)
α	angle (degree or rad)
ε	eccentricity ratio
$\phi_{p_{max}}$	angular position of maximum pressure (degree)
ϕ_{h_o}	angular position of minimum film thickness (degree)

ϕ_{p_o}	film termination angle (degree)
ϕ	position of the minimum film thickness (degree), attitude angle (degree), azimuth angle (rad)
ϕ'	relative azimuth angle (rad)
$\phi*$	nondimensional circumferential location (dimensionless)
μ	viscosity (Pa s)
μ_o	reference viscosity (Pa s)
$\mu*$	dimensionless viscosity (dimensionless)
ν	kinematic viscosity (m^2/s)
ρ	density (kg/m^3)
$\rho*$	dimensionless density (dimensionless)
ρ_o	reference density (kg/m^3)
σ_s	squeeze number (dimensionless)
τ_{zx}	viscous shear stress acting in the x direction, on a plane normal to the z direction (N/m^2)
τ_{zy}	viscous shear stress acting in the y direction, on a plane normal to the z direction (N/m^2)
ω	angular velocity (rad/s)
Ω	angular velocity magnitude (rad/s)

5.1 Introduction

The purpose of a bearing is to support a load, typically applied to a shaft, whilst allowing relative motion between two elements of a machine. The two general classes of bearings are journal bearings, also known as sliding or plain surface bearings, and rolling element bearings, sometimes also called ball-bearings. The aims of this chapter are to describe the range of bearing technology, to outline the identification of which type of bearing to use for a given application, and to introduce journal bearing design with specific attention to boundary lubricated bearings and full film hydrodynamic bearings. The selection and use of rolling element bearings is considered in Chapter 6.

The term 'bearing' typically refers to contacting surfaces through which a load is transmitted. Bearings may roll or slide or do both simultaneously. The range of bearing types available is extensive, although they can be broadly split into two categories: sliding bearings also known as journal or plain surface bearings, where the motion is facilitated by a thin layer or film of lubricant, and rolling element bearings, where the motion is aided by a combination of rolling motion and lubrication. Lubrication is often required in a bearing to reduce friction between surfaces and to remove heat. Fig. 5.1 illustrates two of the more commonly known bearings: a journal bearing and a deep groove ball bearing. A general classification scheme for the distinction of bearing types is given in Fig. 5.2.

As can be seen from Fig. 5.2 the scope of choice for a bearing is extensive. For a given application it may be possible to use different bearing types. For example in a small gas turbine engine rotating as say 50,000 rpm, either rolling bearings or journal bearings could typically be used although the optimal choice will depend on a number of factors such as life, cost and size. Fig. 5.3 can be used to give guidance for which kind of bearing has the maximum load capacity at a given speed and shaft size and

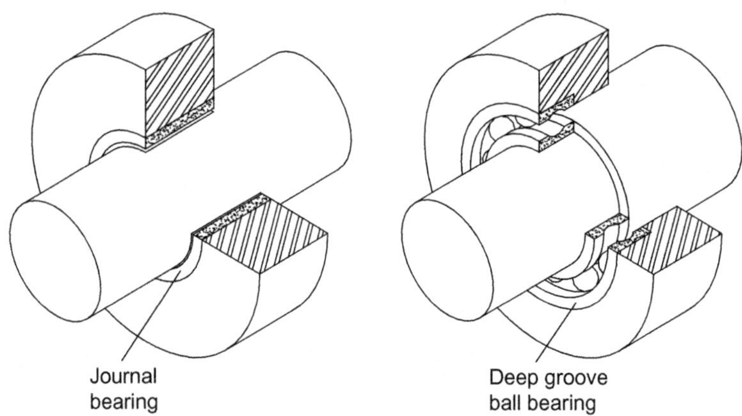

Fig. 5.1 A journal bearing and a deep grove ball bearing.

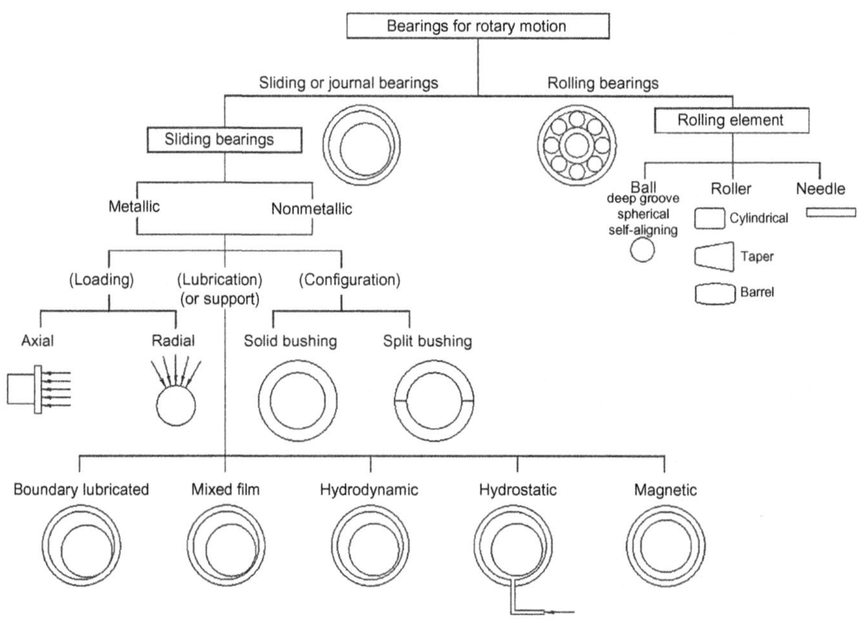

Fig. 5.2 Bearing classification.
Based on a taxonomy originally developed by Hindhede, U., Zimmerman, J.R., Hopkins, R.B., Erisman, R.J, Hull, W.C., Lang, J.D. Machine Design Fundamentals. A Practical Approach. Wiley, 1983.

Table 5.1 gives an indication of the performance of the various bearing types for some criteria other than load capacity.

This and the following chapter provide an introduction to bearings. Lubricant film sliding bearings are introduced in Section 5.2, the design of boundary lubricated

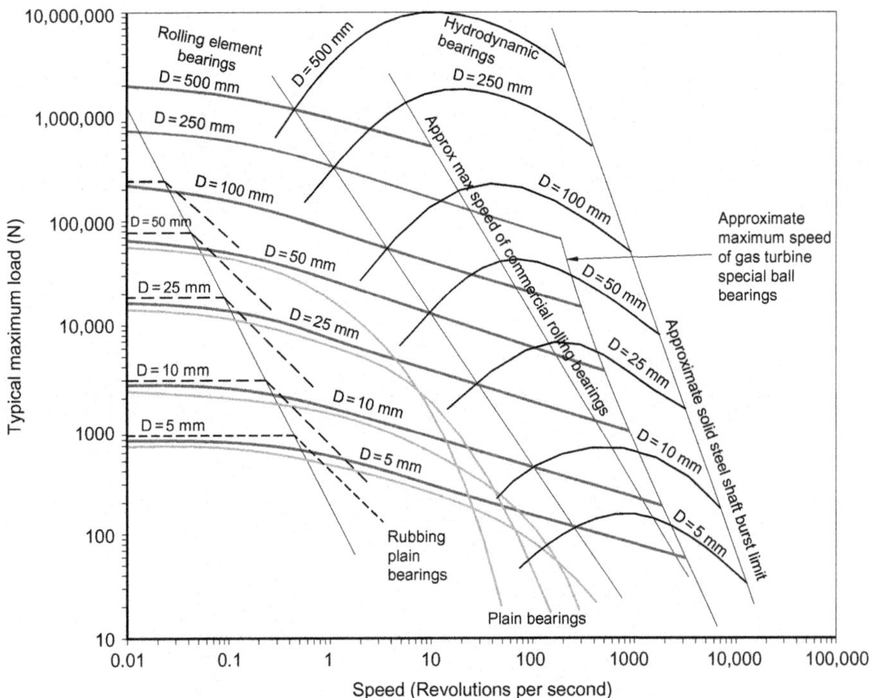

Fig. 5.3 Bearing type selection by load capacity and speed.
Reproduced with permission from Neale, M.J. (ed.), 1995. The Tribology Handbook. Butterworth Heinemann.

bearings which are typically used for low speed applications are considered in Section 5.3, the design of full-film hydrodynamically lubricated bearings is described in Section 5.4 and Chapter 6 considers rolling element bearings. For further reading the texts by Khonsari and Booser (2017), Harris (2001) and Brandlein et al. (1999) provide an extensive overview of bearing technologies.

5.2 Sliding bearings

The term sliding bearing refers to bearings where two surfaces move relative to each other without the benefit of rolling contact. The two surfaces slide over each other and this motion can be facilitated by means of a lubricant which gets squeezed by the motion of the components and can generate sufficient pressure to separate them, thereby reducing frictional contact and wear.

A typical application of sliding bearings is to allow rotation of a load-carrying shaft. The portion of the shaft at the bearing is referred to as the journal and the stationary part, which supports the load, is called the bearing (Fig. 5.4). For this reason,

Table 5.1 Comparison of bearing performance for continuous rotation

Bearing type	Accurate radial location	Combined axial and radial load capability	Low starting torque capability	Silent running	Standard parts available	Lubrication simplicity
Rubbing plain bearings (nonmetallic)	Poor	Some in most cases	Poor	Fair	Some	Excellent
Porous metal plain bearings oil impregnated	Good	Some	Good	Excellent	Yes	Excellent
Fluid film hydrodynamic bearings	Fair	No; separate thrust bearing needed	Good	Excellent	Some	Usually requires a recirculation system
Hydrostatic bearings	Excellent	No; separate thrust bearing needed	Excellent	Excellent	No	Poor special system needed
Rolling bearings	Good	Yes in most cases	Very good	Usually satisfactory	Yes	Good when grease lubricated

Reproduced with permission from Neale, M.J. (ed.), 1995. The Tribology Handbook. Butterworth Heinemann.

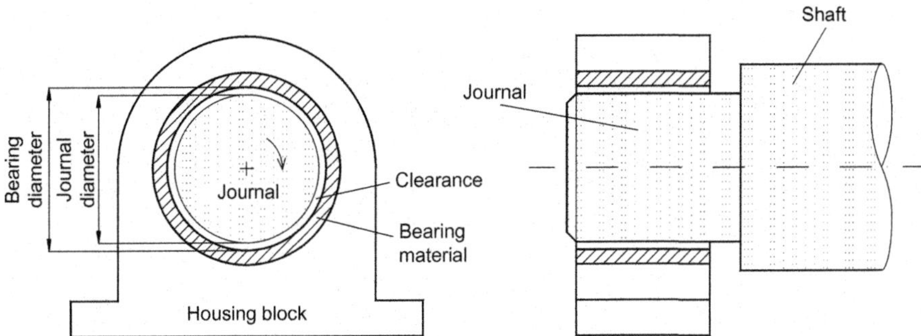

Fig. 5.4 A plain surface, sliding or journal bearing.

sliding bearings are often collectively referred to as journal bearings, although this term ignores the existence of sliding bearings that support linear translation of components. Another common term frequently used in practice is plain surface bearings. This section is principally concerned with bearings for rotary motion and the terms journal and sliding bearing are used interchangeably.

There are three regimes of lubrication for sliding bearings:

(i) boundary lubrication,
(ii) mixed film lubrication,
(iii) full film lubrication.

Boundary lubrication typically occurs at low relative velocities between the journal and the bearing surfaces and is characterised by actual physical contact. The surfaces, even if ground to a low value of surface roughness, will still consist of a series of peaks and troughs as illustrated schematically in Fig. 5.5. Although some lubricant may be present, the pressures generated within it will not be significant and the relative motion of the surfaces will bring the corresponding peaks periodically into contact. Mixed film lubrication occurs when the relative motion between the surfaces is

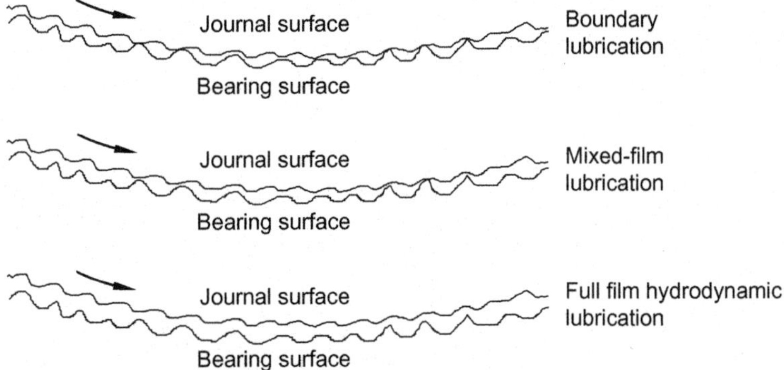

Fig. 5.5 A schematic representation of the surface roughness for sliding bearings and the relative position depending on the type of lubrication occurring.

sufficient to generate high enough pressures in the lubricant film, which can partially separate the surfaces for periods of time. There is still contact in places around the circumference between the two components. Full film lubrication occurs at higher relative velocities. Here the motion of the surface generates high pressures in the lubricant, which separate the two components and journal can 'ride' on a wedge of fluid. All of these types of lubrication can be encountered in a bearing without external pressuring of the bearing. If lubricant under high enough pressure is supplied to the bearing to separate the two surfaces it is called a hydrostatic bearing (Rowe, 2012).

The performance of a sliding bearing differs markedly depending on which type of lubrication is physically occurring. This is illustrated in Fig. 5.6, which shows the variation of the coefficient of friction with a group of variables called the bearing parameter which is defined by:

$$\frac{\mu N}{P} \tag{5.1}$$

where:

μ = viscosity of lubricant (Pa s),
N = speed (for this definition normally in rpm),
P = load capacity (N/m^2) given by

$$P = \frac{W}{LD} \tag{5.2}$$

W = applied load (N),
L = bearing length (m),
D = journal diameter (m).

The bearing parameter, $\mu N/P$, groups several of the bearing design variables into one number. Normally, of course, a low coefficient of friction is desirable to ensure that only a small amount of power is necessary to rotate or drive the component concerned.

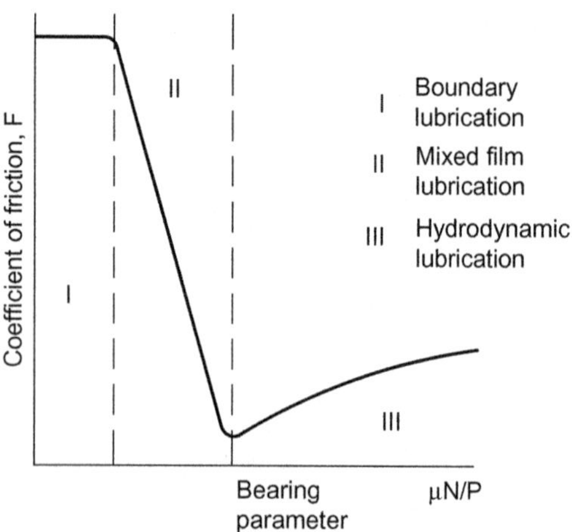

Fig. 5.6 Schematic representation of the variation of bearing performance with lubrication.

Typical coefficients of friction for a boundary lubricated bearing are between approximately 0.05 and 0.1. By contrast the coefficient of friction for a rolling element bearing is typically of the order of 0.005.

In general, boundary lubrication is used for slow speed applications where the surface speed is less than approximately 1.5 m/s. Mixed-film lubrication is rarely used because it is difficult to quantify the actual value of the coefficient of friction and small changes in, for instance, the viscosity will result in large changes in friction (note the steep gradient in Fig. 5.6 for this zone). The design of boundary-lubricated bearings is outlined in Section 5.3 and full-film hydrodynamic bearings in Section 5.4.

5.2.1 Lubricants

As can be seen from Fig. 5.6 bearing performance is dependent on the type of lubrication occurring and the viscosity of the lubricant. The viscosity is a measure of a fluid's resistance to shear. Lubricants can be solid, liquid or gaseous although the most commonly known are oils and greases. The principal classes of liquid lubricants are mineral oils and synthetic oils. Their viscosity is highly dependent on temperature as illustrated in Fig. 5.7. They are typically solid at −35°C, thin as paraffin at 100°C and burn above 240°C. Many additives are used to affect their performance. For example EP (extreme pressure) additives add fatty acids and other compounds to the oil, which attack the metal surfaces to form 'contaminant' layers, which protect the surfaces and

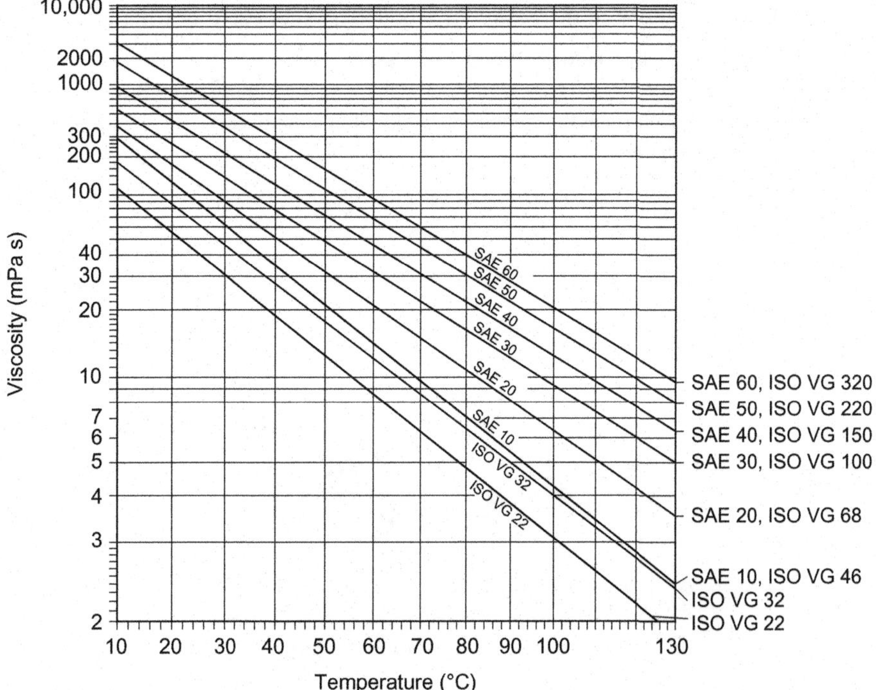

Fig. 5.7 Variation of absolute viscosity with temperature for various lubricants.

reduce friction even when the oil film is squeezed out by high contact loads. Greases are oils mixed with soaps to form a thicker lubricant that can be retained on surfaces.

The viscosity variation with temperature of oils has been standardised and oils are available with a class number, for example SAE 10, SAE 20, SAE 30, SAE 40, SAE 5 W and SAE 10 W. The origin of this identification system developed by the Society of Automotive Engineers was to class oils for general-purpose use and winter use, the 'W' signifying the latter. The lower the numerical value, the thinner or less viscous the oil. Multigrade oil (e.g. SAE 10 W/40) is formulated to meet the viscosity requirements of two oils giving some of the benefits of the constituent parts. An equivalent identification system is also available from the International Organisation for Standardisation (ISO 3448).

5.3 Design of boundary-lubricated bearings

As described in Section 5.2, the journal and bearing surfaces in a boundary lubricated bearing are in direct contact in places. These bearings are typically used for very low speed applications such as bushes and linkages where their simplicity and compact nature are advantageous. Examples include household lawnmower wheels, garden hand tools such as shears, ratchet wrenches and domestic, automotive door hinges incorporating the journal as a solid pin riding inside a cylindrical outer member and even for cardboard wheels (Fig. 5.8). A further example, involving linear motion is the slider in an ink jet printer.

General considerations in the design of a boundary lubricated bearing are:

- the coefficient of friction (both static and dynamic),
- the load capacity,
- the relative velocity between the stationary and moving components,
- the operating temperature,
- wear limitations and
- the production capability.

Fig. 5.8 The Move-it wheel for box transportation—Courtesy of Move-it.

A useful measure in the design of boundary-lubricated bearings is the *PV* factor (load capacity × peripheral speed), which indicates the ability of the bearing material to accommodate the frictional energy generated in the bearing. At the limiting *PV* value the temperature will be unstable and failure will occur rapidly. A practical value for *PV* is half the limiting *PV* value. Values for *PV* for various bearing materials are given in Table 5.2, and Lancaster (1971), Yamaguchi and Kashiwagi (1982), and Koring (2012). The preliminary design of a boundary-lubricated bearing essentially consists of setting the bearing proportions, its length and its diameter, and selecting the bearing material such that an acceptable *PV* value is obtained. This approach is set out step-by-step in Table 5.2.

Example 5.1. A bearing is to be designed to carry a radial load of 700N for a shaft of diameter 25mm running at a speed of 75rpm (Fig. 5.9). Calculate the *PV* value and by comparison with the available materials listed in Table 5.3 determine a suitable bearing material.

Solution

The primary data are $W=700\,\text{N}$, $D=25\,\text{mm}$ and $N=75\,\text{rpm}$.

Use $L/D=1$ as an initial suggestion for the length to diameter ratio for the bearing. $L=25\,\text{mm}$.

Calculating the load capacity, P

$$P=\frac{W}{LD}=\frac{700}{0.025\times0.025}=1.12\ \text{MN/m}^2$$

$$V=\omega r=\left(\frac{2\pi}{60}\right)N\frac{D}{2}=0.1047\times75\frac{0.025}{2}=0.09816\ \text{m/s}$$

$$PV=0.11\left(\text{MN/m}^2\right)(\text{m/s})$$

Multiplying this by a factor of 2 gives $PV=0.22$ (MN/m s).

A material with *PV* value greater than this, such as filled PTFE (limiting *PV* value up to 0.35), or PTFE with filler bonded to a steel backing (limiting *PV* value up to 1.75), would give acceptable performance.

Example 5.2. The paper feed for a photocopier is controlled by two rollers which are sprung together with a force of approximately 20N. The rollers each consist of a 20mm outer diameter plastic cylinder pressed onto a 10mm diameter steel shaft (Fig. 5.10). The maximum feed rate for the copier is 30 pages per minute. Select bearings to support the rollers (After IMechE, 1994).

Solution

The light load and low speed, together with the requirement that the bearings should be cheap, clean and maintenance free suggest that dry rubbing bearings be considered.

The length of A4 paper is 297mm.

The velocity of the roller surface is given by:

$$V_{\text{roller}}=\frac{30\times0.297}{60}=0.1485\,\text{ms}^{-1}.$$

Table 5.2 Step-by-step guideline for boundary lubricated bearing design

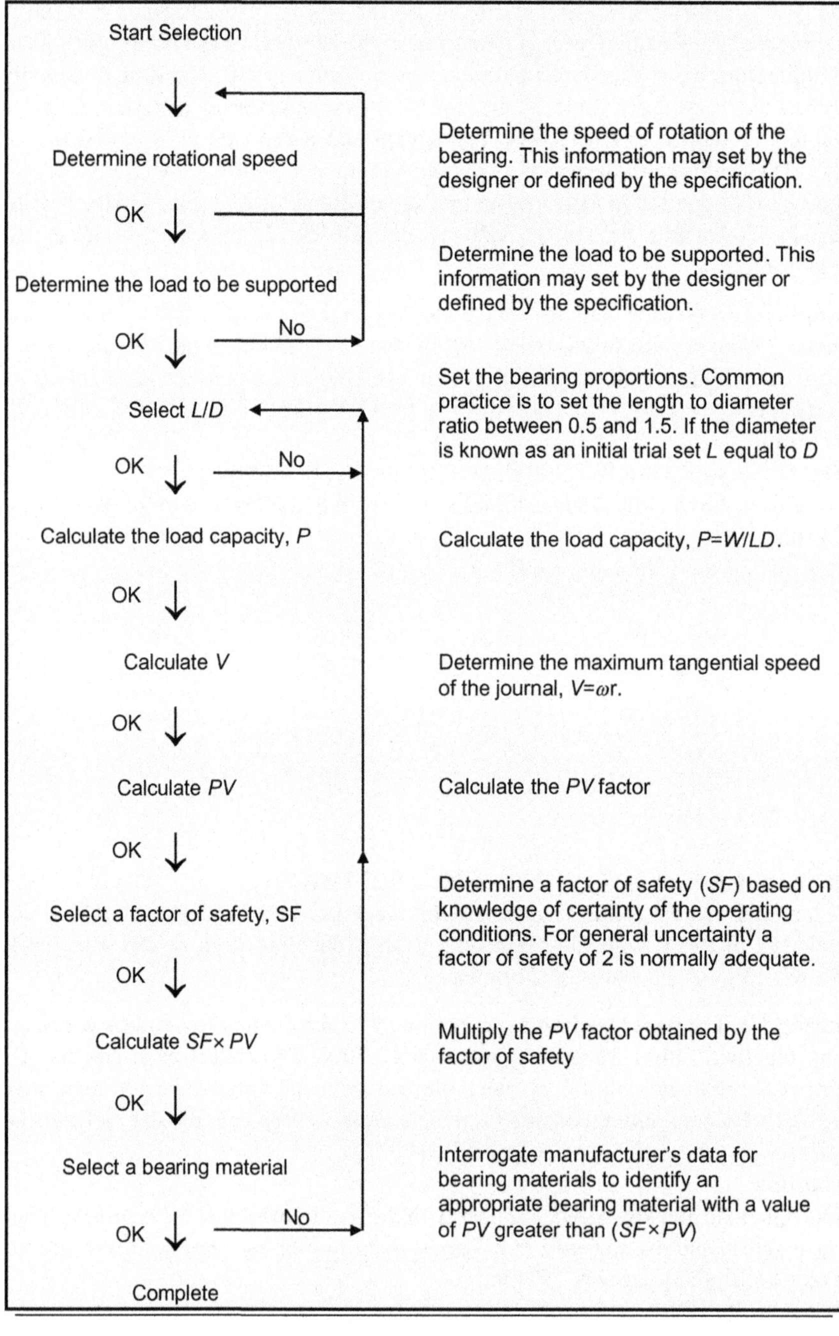

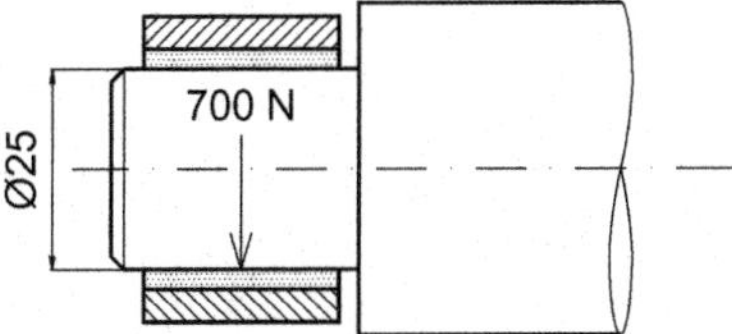

Fig. 5.9 Boundary lubricated bearing design example (dimensions in mm).

The angular velocity of the rollers is 0.1485/0.01 = 14.85 rad/s.
The velocity of the bearing surface is therefore.

$$V = 14.85 \times 0.005 = 0.07425\,\text{m/s}.$$

An *L*/*D* ratio of 0.5 could be suitable giving a bearing length of 5 mm. The load capacity is:

$$P = \frac{W}{LD} = \frac{20}{0.005 \times 0.01} = 0.4\text{ MPa}..$$

$$PV = 0.4 \times 0.07425 = 0.0297\left(\text{MNm}^{-2}\right)\left(\text{ms}^{-1}\right).$$

For this low value of *PV* the cheapest bearing materials, thermoplastics are adequate.

5.4 Design of full film hydrodynamic bearings

In a full film hydrodynamic bearing, the load on the bearing is supported on a continuous film of lubricant so that no contact between the bearing and the rotating journal occurs. The motion of the journal inside the bearing creates the necessary pressure to support the load, as indicated in Fig. 5.11. Hydrodynamic bearings are commonly found in internal combustion engines for supporting the crankshaft, and in turbocharger applications to support the rotor assembly (Fig. 5.12). Hydrodynamic bearings can consist of a full circumferential surface or a partial surface around the journal (Fig. 5.13).

Bearing design can involve so many parameters that it is a challenge to develop an optimal solution. One route towards this goal would be to assign attribute points to each aspect of the design and undertake an optimisation exercise using multiobjective design optimisation or a similar scheme (Hashimoto and Matsumoto, 2001; Hirania and Suh, 2005). This can however be time consuming if the software was not already in a developed state and would not necessarily produce an optimum result due to inadequacies in modelling and incorrect assignment of attribute weightings. An alternative

Table 5.3 Characteristics of some rubbing materials

Material	Maximum load capacity P(MN/m^2)	Limiting PV value (MN/m S)	Maximum operating temperature (°C)	Coefficient of friction	Coefficient of expansion ($\times 10^{-6}$/°C)	Comments	Typical application
Carbon/ graphite	1.4–2	0.11	350–500	0.1–0.25 dry	2.5–5.0	For continuous dry operation	Food and textile machinery
Carbon/ graphite with metal	3.4	0.145	130–350	0.1–0.35 dry	4.2–5		
Graphite impregnated metal	70	0.28–0.35	350–600	0.1–0.15 dry	12–13		
Graphite/ thermo-setting resin	2	0.35	250	0.13–0.5 dry	3.5–5	Suitable for sea water operation	
Reinforced thermo-setting plastics	35	0.35	200	0.1–0.4 dry	25–80		Roll-neck bearings
Thermo-plastic material without filler	10	0.035	100	0.1–0.45 dry	100		Bushes and thrust washers
Thermo-plastic with filler or metal backed	10–14	0.035–0.11	100	0.15–0.4 dry	80–100		Bushes and thrust washers

Thermo-plastic material with filler bonded to metal back	140	0.35	105	0.2–0.35 dry	27	With initial lubrication only	For conditions of intermittent operation or boundary lubrication (e.g. ball joints, suspension, steering)
Filled PTFE	7	Up to 0.35	250	0.05–0.35 dry	60–80	Glass, mica, bronze, graphite	For dry operations where low friction and wear required
PTFE with filler, bonded to steel backing	140	Up to 1.75	280	0.05–0.3 dry	20	Sintered bronze bonded to steel backing impregnated with PTFE/lead	Aircraft controls, linkages, gearbox, clutch, conveyors, bridges
Woven PTFE reinforced and bonded to metal backing	420	Up to 1.6	250	0.03–0.3	–	Reinforcement may be interwoven glass fibre or rayon	Aircraft and engine controls, linkages, engine mountings, bridge bearings

Reproduced with permission from Neale, M.J. (ed.), 1995. The Tribology Handbook. Butterworth Heinemann.

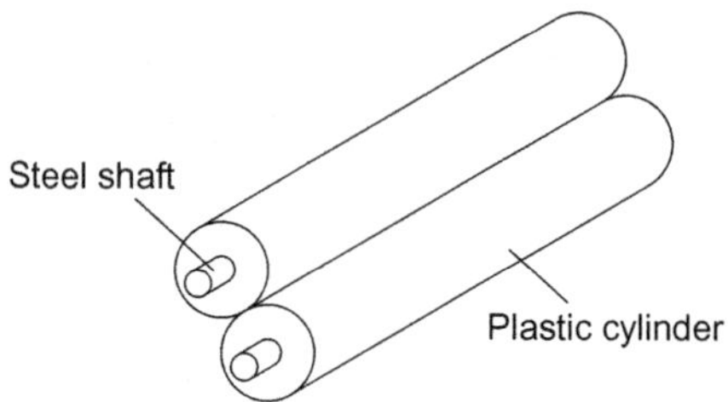

Fig. 5.10 Copier rollers.
Reproduced from Institution of Mechanical Engineers, 1994. Tribological Design Data. Part 1: Bearings. I. Mech. E.

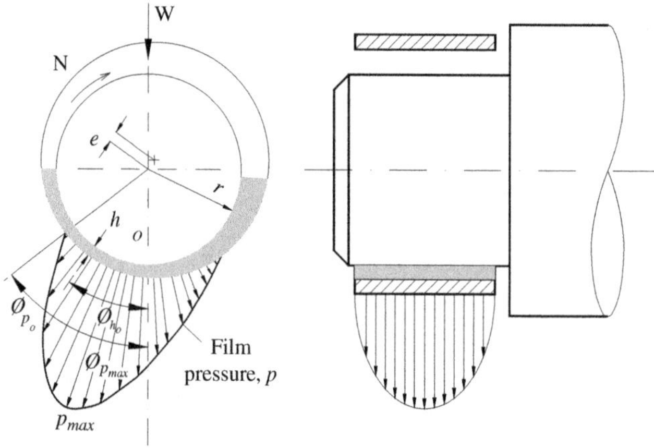

Fig. 5.11 Motion of the journal generates pressure in the lubricant separating the two surfaces. Beyond h_o, the minimum film thickness, the pressure terms go negative and the film is ruptured.

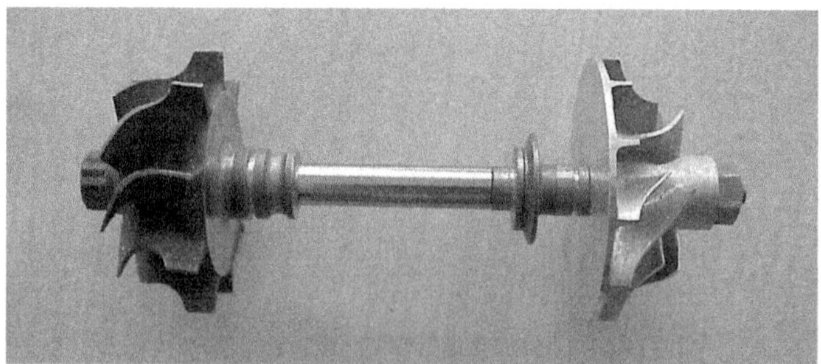

Fig. 5.12 Turbocharger journal.

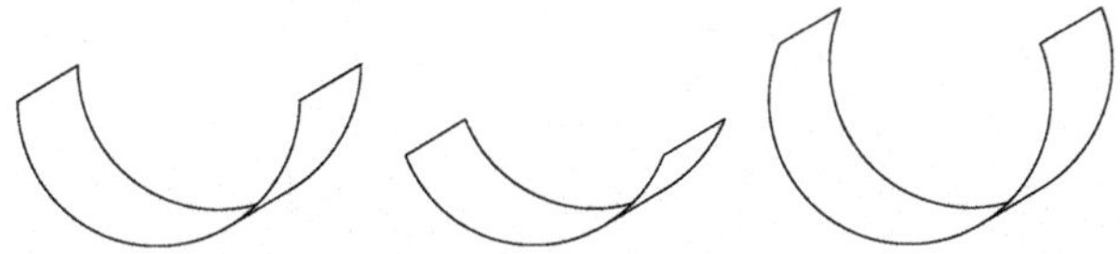

Fig. 5.13 Partial surface journal bearings.

approach which is sensible as a starting point and outlined here is to develop one or a number of feasible designs and use judgement to select the best, or combine the best features of the proposed designs.

The design procedure for a journal bearing, recommended here as a starting point, includes specification of the:

- journal radius, r;
- radial clearance, c;
- axial length of the bearing surface, L;
- type of lubricant and its viscosity, μ;
- journal speed, N;
- load, W.

Values for the speed, the load and possibly the journal radius are usually specified by the machine requirements, and stress and deflection considerations. As such, journal bearing design consists of the determination of the radial clearance, the bearing length and the lubricant viscosity. The design process for a journal bearing is usually iterative. Trial values for the clearance, the length and the viscosity are chosen, various performance criteria calculated and the process repeated until a satisfactory or optimised design is achieved. Criteria for optimisation may be minimising of the frictional loss, minimising the lubricant temperature rise, minimising the lubricant supply, maximising the load capability, and minimising production costs.

The clearance between the journal and the bearing depends on the nominal diameter of the journal, the precision of the machine, surface roughness and thermal expansion considerations. An overall guideline is for the radial clearance, c, to be in the range $0.001r < c < 0.002r$, where r is the nominal bearing radius ($0.001\text{D} < 2c < 0.002\text{D}$). Fig. 5.14 shows values for the recommended diametral clearance ($2 \times c$) as a function of the journal diameter and rotational speed for steadily loaded bearings.

For a given combination of a, c, L, μ, N and W, the performance of a journal bearing can be calculated. This requires determining the pressure distribution in the bearing, the minimum film thickness h_o, the location of the minimum film thickness $\theta_{p_{\max}}$, the coefficient of friction f, the lubricant flow Q, the maximum film pressure p_{max} and the temperature rise ΔT of the lubricant.

The pressure distribution in a journal bearing (Fig. 5.11) can be determined by solving the relevant form of the Navier–Stokes fluid flow equations, which in the reduced form for journal bearings is called the Reynolds equation, Eq. (5.3).

$$\frac{\partial}{\partial x}\left(\frac{h^3}{\mu}\frac{\partial p}{\partial x}\right) + \frac{\partial}{\partial z}\left(\frac{h^3}{\mu}\frac{\partial p}{\partial z}\right) = 6V\frac{\partial h}{\partial x} + 6h\frac{\partial V}{\partial x} \tag{5.3}$$

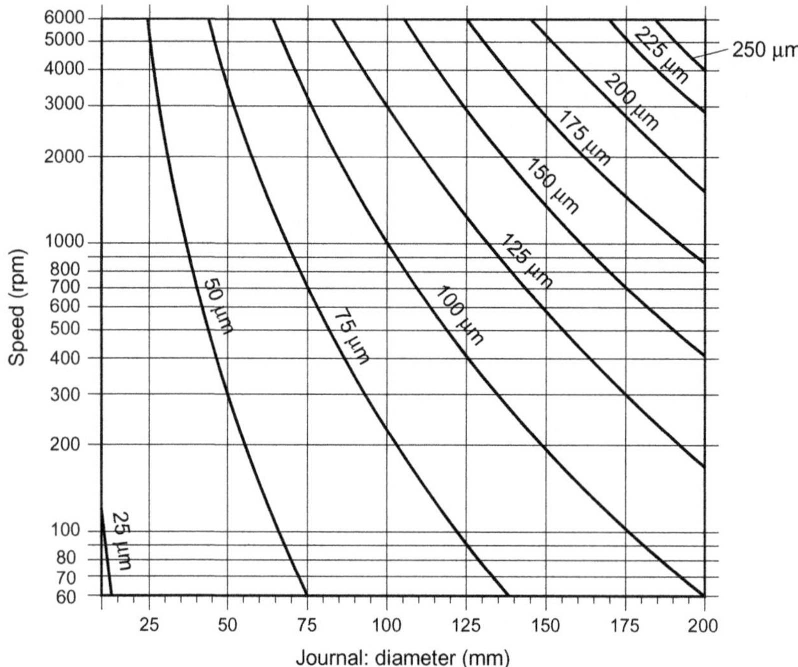

Fig. 5.14 Minimum recommended values for the diametral clearance ($2 \times c$) for steadily loaded journal bearings.
Reproduced with permission from Welsh, R.J., 1983. Plain Bearing Design Handbook. Butterworth.

where:

h = film thickness (m);
μ = dynamic viscosity (Pa s);
p = fluid pressure (Pa);
V = journal velocity (m/s);
x, y, z = rectangular coordinates (m).

The Reynolds equation was first derived by Reynolds (1886) and further developed by Harrison (1913) to include the effects of compressibility. Here the derivation of the general Reynolds equation is given in Section 5.4.1 and is developed following the general outline given by Hamrock (1994) and Childs (2011). Many readers may prefer to omit detailed consideration of the derivation as there is no need to go through this for practical design applications and instead proceed directly to Section 5.4.2. Nevertheless the derivation is included here for reference.

5.4.1 Reynolds equation derivation

The Navier–Stokes equations in a Cartesian coordinate system, for compressible flow assuming constant viscosity are given by Eqs (5.4)–(5.6). Here x is taken as the coordinate in the direction of sliding, y is the coordinate in the direction of side leakage and z is the coordinate across the lubricant film.

$$\rho\left(\frac{\partial u_x}{\partial t}+u_x\frac{\partial u_x}{\partial x}+u_y\frac{\partial u_x}{\partial y}+u_z\frac{\partial u_x}{\partial z}\right)=-\frac{\partial p}{\partial x}+\mu\left(\frac{\partial^2 u_x}{\partial x^2}+\frac{\partial^2 u_x}{\partial y^2}+\frac{\partial^2 u_x}{\partial z^2}\right)+ \\ \frac{\mu}{3}\frac{\partial}{\partial x}\left(\frac{\partial u_x}{\partial x}+\frac{\partial u_y}{\partial y}+\frac{\partial u_z}{\partial z}\right)+F_x \tag{5.4}$$

$$\rho\left(\frac{\partial u_y}{\partial t}+u_x\frac{\partial u_y}{\partial x}+u_y\frac{\partial u_y}{\partial y}+u_z\frac{\partial u_y}{\partial z}\right)=-\frac{\partial p}{\partial y}+\mu\left(\frac{\partial^2 u_y}{\partial x^2}+\frac{\partial^2 u_y}{\partial y^2}+\frac{\partial^2 u_y}{\partial z^2}\right)+ \\ \frac{\mu}{3}\frac{\partial}{\partial y}\left(\frac{\partial u_x}{\partial x}+\frac{\partial u_y}{\partial y}+\frac{\partial u_z}{\partial z}\right)+F_y \tag{5.5}$$

$$\rho\left(\frac{\partial u_z}{\partial t}+u_x\frac{\partial u_z}{\partial x}+u_y\frac{\partial u_z}{\partial y}+u_z\frac{\partial u_z}{\partial z}\right)=-\frac{\partial p}{\partial z}+\mu\left(\frac{\partial^2 u_z}{\partial x^2}+\frac{\partial^2 u_z}{\partial y^2}+\frac{\partial^2 u_z}{\partial z^2}\right)+ \\ \frac{\mu}{3}\frac{\partial}{\partial z}\left(\frac{\partial u_x}{\partial x}+\frac{\partial u_y}{\partial y}+\frac{\partial u_z}{\partial z}\right)+F_z \tag{5.6}$$

If the density is assumed constant then the dilation (Eq. 5.7) is zero.

$$\frac{\partial u_x}{\partial x}+\frac{\partial u_y}{\partial y}+\frac{\partial u_z}{\partial z}=0 \tag{5.7}$$

The Navier–Stokes equations for constant density and constant viscosity reduce to

$$\rho\left(\frac{\partial u_x}{\partial t}+u_x\frac{\partial u_x}{\partial x}+u_y\frac{\partial u_x}{\partial y}+u_z\frac{\partial u_x}{\partial z}\right)=-\frac{\partial p}{\partial x}+\mu\left(\frac{\partial^2 u_x}{\partial x^2}+\frac{\partial^2 u_x}{\partial y^2}+\frac{\partial^2 u_x}{\partial z^2}\right)+F_x \tag{5.8}$$

$$\rho\left(\frac{\partial u_y}{\partial t}+u_x\frac{\partial u_y}{\partial x}+u_y\frac{\partial u_y}{\partial y}+u_z\frac{\partial u_y}{\partial z}\right)=-\frac{\partial p}{\partial y}+\mu\left(\frac{\partial^2 u_y}{\partial x^2}+\frac{\partial^2 u_y}{\partial y^2}+\frac{\partial^2 u_y}{\partial z^2}\right)+F_y \tag{5.9}$$

$$\rho\left(\frac{\partial u_z}{\partial t}+u_x\frac{\partial u_z}{\partial x}+u_y\frac{\partial u_z}{\partial y}+u_z\frac{\partial u_z}{\partial z}\right)=-\frac{\partial p}{\partial z}+\mu\left(\frac{\partial^2 u_z}{\partial x^2}+\frac{\partial^2 u_z}{\partial y^2}+\frac{\partial^2 u_z}{\partial z^2}\right)+F_z \tag{5.10}$$

For conditions known as slow viscous motion, where pressure and viscous terms predominate simplifications are possible for the Navier–Stokes equations making their solution more amenable to analytical and numerical techniques.

The Navier–Stokes equations can be nondimensionalised to enable a generalised solution (Childs, 2011). Here the process is repeated to give dimensionless groups of specific relevance to journal bearings using the characteristic parameters given in Eqs (5.11)–(5.20).

$$x_* = \frac{x}{l_o} \tag{5.11}$$

where l_o is a characteristic length in the x direction.

$$y_* = \frac{y}{b_o} \tag{5.12}$$

where b_o is a characteristic length in the y direction.

$$z_* = \frac{z}{h_o} \tag{5.13}$$

where h_o is a characteristic length in the z direction.

$$t_* = \frac{t}{t_o} \tag{5.14}$$

where t_o is a characteristic time.

$$u_{x^*} = \frac{u_x}{u_{x,o}} \tag{5.15}$$

where $u_{x,o}$ is a characteristic velocity in the x direction.

$$u_{y^*} = \frac{u_y}{u_{y,o}} \tag{5.16}$$

where $u_{y,o}$ is a characteristic velocity in the y direction.

$$u_{z^*} = \frac{u_z}{u_{z,o}} \tag{5.17}$$

where $u_{z,o}$ is a characteristic velocity in the z direction.

$$\rho_* = \frac{\rho}{\rho_o} \tag{5.18}$$

where ρ_o is a characteristic density.

$$\mu_* = \frac{\mu}{\mu_o} \tag{5.19}$$

where μ_o is a characteristic viscosity.

$$p_* = \frac{h_o^2 p}{\mu_o u_{x,o} l_o} \tag{5.20}$$

Substitution of the above dimensionless parameters, Eqs (5.11)–(5.20) in Eq. (5.4), gives

$$\frac{l_o}{u_{x,o}t_o}\frac{\partial u_{x^*}}{\partial t_*}+u_{x^*}\frac{\partial u_{x^*}}{\partial x_*}+\frac{l_o u_{y,o}}{b_o u_{x,o}}u_{y^*}\frac{\partial u_{x^*}}{\partial y_*}+\frac{l_o u_{z,o}}{h_o u_{x,o}}u_{z^*}\frac{\partial u_{x^*}}{\partial z_*}=\frac{l_o g}{u_{x,o}^2}-\frac{\mu_o}{\rho_o u_{x,o} l_o}\left(\frac{l_o}{h_o}\right)^2\frac{1}{\rho_*}\frac{\partial p_*}{\partial x_*}-$$
$$\frac{2}{3}\frac{\mu_o}{\rho_o u_{x,o} l_o}\frac{1}{\rho_*}\frac{\partial}{\partial x_*}\left[\mu_*\left(\frac{\partial u_{x^*}}{\partial x_*}+\frac{u_{y,o}l_o}{u_{x,o}b_o}\frac{\partial u_{y^*}}{\partial y_*}+\frac{u_{z,o}l_o}{u_{x,o}h_o}\frac{\partial u_{z^*}}{\partial z_*}\right)\right]+\frac{2\mu_o}{\rho_o u_{x,o} l_o}\frac{1}{\rho_*}\frac{\partial}{\partial x_*}\left(\mu_*\frac{\partial u_{x^*}}{\partial x_*}\right)+$$
$$\frac{\mu_o}{\rho_o u_{x,o} l_o}\left(\frac{l_o}{b_o}\right)^2\frac{1}{\rho_*}\frac{\partial}{\partial y_*}\left[\mu_*\left(\frac{\partial u_{x^*}}{\partial y_*}+\frac{u_{y,o}b_o}{u_{x,o}l_o}\frac{\partial u_{y^*}}{\partial x_*}\right)\right]+\frac{\mu_o}{\rho_o u_{x,o} l_o}\left(\frac{l_o}{h_o}\right)^2\frac{1}{\rho_*}\frac{\partial}{\partial z_*}\left[\mu_*\left(\frac{\partial u_{x^*}}{\partial z_*}+\frac{u_{z,o}h_o}{u_{x,o}l_o}\frac{\partial u_{z^*}}{\partial x_*}\right)\right] \tag{5.21}$$

The relative importance of inertia and viscous forces can be determined by examination of the value of the Reynolds number

$$Re=\frac{\rho_o u_{x,o} l_o}{\mu_o} \tag{5.22}$$

The inverse of the Reynolds number occurs throughout Eq. (5.21).

In fluid film lubrication, because of the dominance of the viscous term $\partial^2 u_{x^*}/\partial^2 z_*$, a modified form of the Reynolds number is used, defined for the x component of velocity by

$$Re_x=\frac{\rho_o u_{x,o} h_o^2}{\mu_o l_o} \tag{5.23}$$

and similarly for the y and z directions

$$Re_y=\frac{\rho_o u_{y,o} h_o^2}{\mu_o b_o} \tag{5.24}$$

$$Re_z=\frac{\rho_o u_{z,o} h_o}{\mu_o} \tag{5.25}$$

The squeeze number is defined by

$$\sigma_s=\frac{\rho_o h_o^2}{\mu_o t_o} \tag{5.26}$$

Typically in hydrodynamically lubricated journal bearings, viscous forces are much greater than inertia forces and a typical Reynolds number, using Eq. (5.23), might be of the order of 1×10^{-4}.

Substitution of the Reynolds and squeeze numbers into Eq. (5.21), gives

$$\sigma_s \frac{\partial u_{x*}}{\partial t_*} + Re_x u_{x*} \frac{\partial u_{x*}}{\partial x_*} + Re_y u_{y*} \frac{\partial u_{x*}}{\partial y_*} + Re_z u_{z*} \frac{\partial u_{x*}}{\partial z_*} = \frac{l_o g}{u_{x,o}^2} Re_x - \frac{1}{\rho_*} \frac{\partial p_*}{\partial x_*} + \frac{1}{\rho_*} \frac{\partial}{\partial z_*} \left(\mu_* \frac{\partial u_{x*}}{\partial z_*} \right) -$$
$$\frac{2}{3} \left(\frac{h_o}{l_o} \right)^2 \frac{1}{\rho_*} \frac{\partial}{\partial x_*} \left[\mu_* \left(\frac{\partial u_{x*}}{\partial x_*} + \frac{u_{y,o} l_o}{u_{x,o} b_o} \frac{\partial u_{y*}}{\partial y_*} + \frac{u_{z,o} l_o}{u_{x,o} h_o} \frac{\partial u_{z*}}{\partial z_*} \right) \right] +$$
$$\left(\frac{h_o}{b_o} \right)^2 \frac{1}{\rho_*} \frac{\partial}{\partial y_*} \left[\mu_* \left(\frac{\partial u_{x*}}{\partial y_*} + \frac{u_{y,o} b_o}{u_{x,o} l_o} \frac{\partial u_{y*}}{\partial x_*} \right) \right] + 2 \left(\frac{h_o}{l_o} \right)^2 \frac{1}{\rho_*} \frac{\partial}{\partial x_*} \left(\mu_* \frac{\partial u_{x*}}{\partial x_*} \right) + \frac{1}{\rho_*} \frac{\partial}{\partial z_*} \left(\mu_* \frac{u_{z,o} h_o}{u_{x,o} l_o} \frac{\partial u_{z*}}{\partial x_*} \right) \tag{5.27}$$

Examination of the order of magnitude of the terms in Eq. (5.27) provides an indication of the relative significance of the terms and which need to considered. The inertia terms, gravity term and $u_{z,o}/u_{x,o}$ are of order h_o/l_o. The pressure gradient term and the first viscous term are of order unity. The remaining viscous terms are of order $(h_o/l_o)^2$ or $(h_o/b_o)^2$ and therefore very small in comparison to the other terms. Neglecting terms of the order of $(h_o/l_o)^2$ and $(h_o/b_o)^2$ gives

$$\sigma_s \frac{\partial u_{x*}}{\partial t_*} + Re_x u_{x*} \frac{\partial u_{x*}}{\partial x_*} + Re_y u_{y*} \frac{\partial u_{x*}}{\partial y_*} + Re_z u_{z*} \frac{\partial u_{x*}}{\partial z_*} = \frac{l_o g}{u_{x,o}^2} Re_x - \frac{1}{\rho_*} \frac{\partial p_*}{\partial x_*} + \frac{1}{\rho_*} \frac{\partial}{\partial z_*} \left(\mu_* \frac{\partial u_{x*}}{\partial z_*} \right) \tag{5.28}$$

Similarly for the y and z components

$$\sigma_s \frac{\partial u_{y*}}{\partial t_*} + Re_x u_{x*} \frac{\partial u_{y*}}{\partial x_*} + Re_y u_{y*} \frac{\partial u_{y*}}{\partial y_*} + Re_z u_{z*} \frac{\partial u_{y*}}{\partial z_*} = \frac{b_o g}{u_{y,o}^2} Re_y - \frac{1}{\rho_*} \frac{\partial p_*}{\partial y_*} + \frac{1}{\rho_*} \frac{\partial}{\partial z_*} \left(\mu_* \frac{\partial u_{y*}}{\partial z_*} \right) \tag{5.29}$$

$$\frac{\partial p}{\partial z_*} = 0 \tag{5.30}$$

Examination of Eqs (5.28)–(5.30) shows that the pressure is a function of x_*, y_* and t_*,

$$p = f(x_*, y_*, t_*) \tag{5.31}$$

The continuity equation can be expressed as

$$\sigma_s \frac{\partial p_*}{\partial t_*} + Re_x \frac{\partial}{\partial x_*} \left(\rho_* u_{x*} \right) + Re_y \frac{\partial}{\partial y_*} \left(\rho_* u_{y*} \right) + Re_z \frac{\partial}{\partial z_*} \left(\rho_* u_{z*} \right) = 0 \tag{5.32}$$

Taylor vortices can be formed in a fluid filled annulus with inner cylinder rotation as illustrated in Figs 5.15 and 5.16. Vortices involve significant acceleration and deceleration of the flow and therefore dissipation of energy and are normally undesirable in

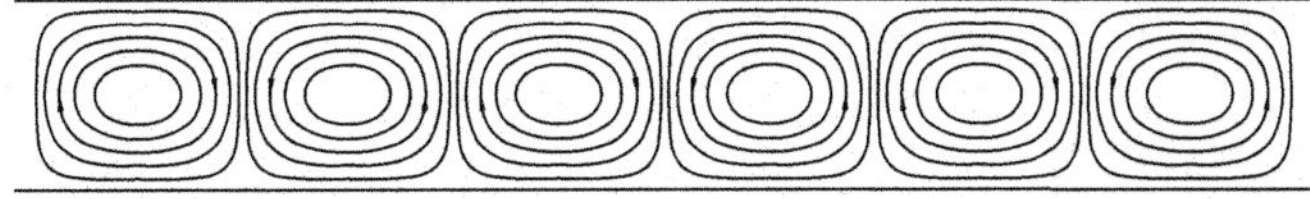

Fig. 5.15 Taylor vortices.

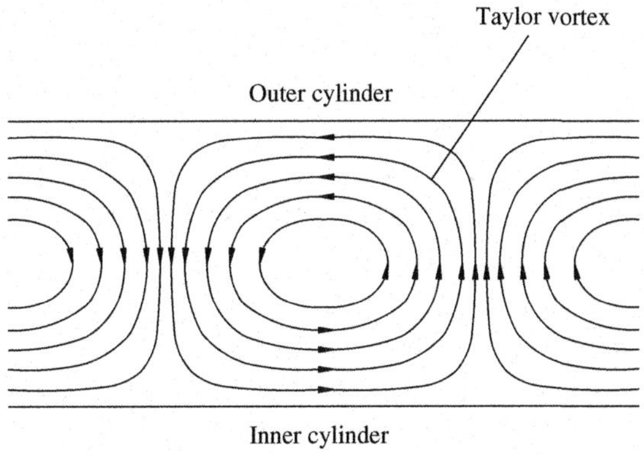

Fig. 5.16 Taylor vortex cells.

any bearing lubrication Scheme. A check can readily be made to identify if Taylor vortices are likely to occur. If the Taylor number, Ta_m, based on the mean annulus radius (Eq. 5.33) is greater than approximately 41.2, then Taylor vortices may form and laminar flow conditions no longer hold invalidating the use of the equations developed in this section.

$$Ta_m = \frac{\Omega r_m^{0.5}(b-a)^{1.5}}{\nu} \tag{5.33}$$

$$r_m = 0.5(b+a) \tag{5.34}$$

$$Ta_{m,cr} = \sqrt{1697} = 41.19 \tag{5.35}$$

The critical angular velocity for this case is given by

$$\Omega_{cr} = \frac{41.19\nu}{r_m^{0.5}(b-a)^{1.5}} \tag{5.36}$$

For an annulus with a finite gap,

$$\Omega_{cr} = \frac{41.19\nu F_g}{r_m^{0.5}(b-a)^{1.5}} \tag{5.37}$$

where F_g is a geometrical factor defined by

$$F_g = \frac{\pi^2}{41.19\sqrt{S}}\left(1 - \frac{b-a}{2r_m}\right)^{-1} \tag{5.38}$$

and S is given in an alternative form by

$$S = 0.0571\left(1 - 0.652\frac{(b-a)/r_m}{1-(b-a)/2r_m}\right) + 0.00056\left(1 - 0.652\frac{(b-a)/r_m}{1-(b-a)/2r_m}\right)^{-1} \tag{5.39}$$

The Froude number (Eq. 5.40) provides an indication of the relative importance of inertia and gravity forces. A typical Froude number for a journal bearing might be of the order of 10, providing an indication that gravity forces can be neglected in comparison with viscous forces.

$$Fr = \frac{U_o}{\sqrt{gL}} \tag{5.40}$$

The importance of pressure relative to inertia can be judged by examination of the Euler number (Eq. 5.41). For a typical journal bearing the Euler number may be of the order of 100, giving an indication that the pressure term is much larger than the inertia term.

$$Eu = \frac{p - p_o}{0.5\rho U_o^2} \tag{5.41}$$

If in addition to neglecting terms of order $(h_o/l_o)^2$ or $(h_o/b_o)^2$, terms of the order of h_o/l_o and h_o/b_o are neglected and only terms of the order of unity are considered, the Navier–Stokes equations reduce to

$$\frac{\partial p}{\partial x} = \frac{\partial}{\partial z}\left(\mu\frac{\partial u_x}{\partial z}\right) \tag{5.42}$$

$$\frac{\partial p}{\partial y} = \frac{\partial}{\partial z}\left(\mu\frac{\partial u_y}{\partial z}\right) \tag{5.43}$$

Eq. (5.31) shows that the pressure is only a function of x and y for steady state conditions. Eqs (5.42), (5.43) can therefore be integrated directly to give general expressions for the velocity gradients as follows.

$$\frac{\partial u_x}{\partial z}=\frac{z}{\mu}\frac{\partial p}{\partial x}+\frac{A}{\mu} \tag{5.44}$$

$$\frac{\partial u_y}{\partial z}=\frac{z}{\mu}\frac{\partial p}{\partial y}+\frac{C}{\mu} \tag{5.45}$$

where A and C are constants of integration.

The temperature across the thin layer of layer lubricant in a journal bearing may vary significantly. As viscosity is highly dependent on temperature, this leads to increased complexity in obtaining a solution to Eqs 5.44 and 5.45. It has been found acceptable, however, in many fluid film applications to model the viscosity of a fluid film using an average value for the viscosity across the film. With μ taken as the average value of viscosity across the fluid film, then integration of Eqs 5.44 and 5.45 gives the velocity components as follows.

$$u_x=\frac{z^2}{2\mu}\frac{\partial p}{\partial x}+A\frac{z}{\mu}+B \tag{5.46}$$

$$u_y=\frac{z^2}{2\mu}\frac{\partial p}{\partial y}+C\frac{z}{\mu}+D \tag{5.47}$$

where B and D are constants of integration.

If the no-slip condition is assumed at the fluid solid interface then the boundary conditions are as follows and illustrated in Fig. 5.17.

$$u_x=u_{x,a}, u_y=u_{y,a} \text{ at } z=0$$

$$u_x=u_{x,b}, u_y=u_{y,b} \text{ at } z=h$$

Application of the boundary conditions to the equations for the velocity gradients and velocity components (Eqs 5.46 and 5.47) gives

$$\frac{\partial u_x}{\partial z}=\frac{2z-h}{2\mu}\frac{\partial p}{\partial x}-\frac{u_{x,a}-u_{x,b}}{h} \tag{5.48}$$

$$\frac{\partial u_y}{\partial z}=\frac{2z-h}{2\mu}\frac{\partial p}{\partial y}-\frac{u_{y,a}-u_{y,b}}{h} \tag{5.49}$$

$$u_x=\frac{z(z-h)}{2\mu}\frac{\partial p}{\partial x}+\frac{h-z}{h}u_{x,a}+\frac{z}{h}u_{x,b} \tag{5.50}$$

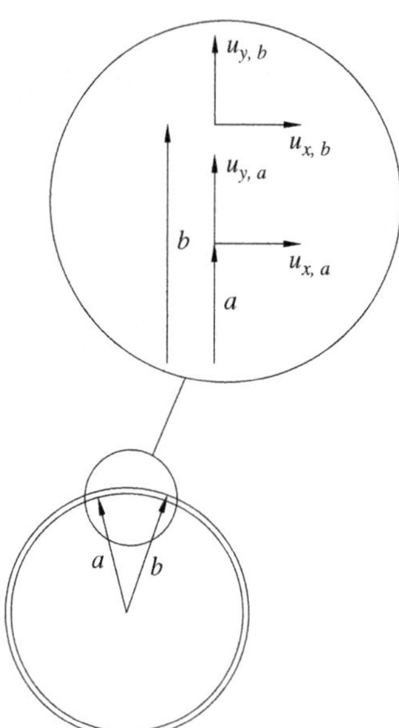

Fig. 5.17 Journal bearing boundary conditions.

$$u_y = \frac{z(z-h)}{2\mu}\frac{\partial p}{\partial y} + \frac{h-z}{h}u_{y,a} + \frac{z}{h}u_{y,b} \tag{5.51}$$

The viscous shear stresses are defined by

$$\tau_{zx} = \mu\left(\frac{\partial u_z}{\partial x} + \frac{\partial u_x}{\partial z}\right) \tag{5.52}$$

$$\tau_{zy} = \mu\left(\frac{\partial u_z}{\partial y} + \frac{\partial u_y}{\partial z}\right) \tag{5.53}$$

The order of magnitude of $\partial u_z/\partial x$ and $\partial u_z/\partial y$ are much smaller than $\partial u_x/\partial z$ and $\partial u_y/\partial z$ so the viscous shear stresses can be approximated by

$$\tau_{zx} = \mu\frac{\partial u_x}{\partial z} \tag{5.54}$$

$$\tau_{zy} = \mu\frac{\partial u_y}{\partial z} \tag{5.55}$$

From Eqs (5.48), (5.49), the viscous shear stresses acting on the solid surfaces can be expressed by

$$(\tau_{zx})_{z=0} = \left(\mu\frac{\partial u_x}{\partial z}\right)_{z=0} = -\frac{h}{2}\frac{\partial p}{\partial x} - \frac{\mu(u_{x,a} - u_{x,b})}{h} \tag{5.56}$$

$$(-\tau_{zx})_{z=h} = -\left(\mu\frac{\partial u_x}{\partial z}\right)_{z=h} = -\frac{h}{2}\frac{\partial p}{\partial x} + \frac{\mu(u_{x,a} - u_{x,b})}{h} \quad (5.57)$$

$$(\tau_{zy})_{z=0} = \left(\mu\frac{\partial u_y}{\partial z}\right)_{z=0} = -\frac{h}{2}\frac{\partial p}{\partial y} - \frac{\mu(u_{y,a} - u_{y,b})}{h} \quad (5.58)$$

$$(-\tau_{zy})_{z=h} = -\left(\mu\frac{\partial u_y}{\partial z}\right)_{z=h} = -\frac{h}{2}\frac{\partial p}{\partial y} + \frac{\mu(u_{y,a} - u_{y,b})}{h} \quad (5.59)$$

The negative signs for the viscous shear stresses in Eqs (5.56)–(5.59) indicate that the stress acts in a direction opposite to the motion.

The volumetric flow rates per unit width in the x and y directions are defined by

$$q_x = \int_0^h u_x dz \quad (5.60)$$

$$q_y = \int_0^h u_y dz \quad (5.61)$$

Substituting for the velocity components using Eqs (5.50), (5.51) in Eqs (5.60), (5.61) gives

$$q_x = -\frac{h^3}{12\mu}\frac{\partial p}{\partial x} + \frac{u_{x,a} + u_{x,b}}{2}h \quad (5.62)$$

$$q_y = -\frac{h^3}{12\mu}\frac{\partial p}{\partial y} + \frac{u_{y,a} + u_{y,b}}{2}h \quad (5.63)$$

The Reynolds equation is formed by substituting the expressions for the volumetric flow rate into the continuity equation.

Integrating the continuity equation gives

$$\int_0^h \left[\frac{\partial \rho}{\partial t} + \frac{\partial}{\partial x}(\rho u_x) + \frac{\partial}{\partial y}(\rho u_y) + \frac{\partial}{\partial z}(\rho u_z)\right] dz = 0 \quad (5.64)$$

The integral

$$\int_0^h \frac{\partial}{\partial x}[f(x,y,z)]dz = -f(x,y,h)\frac{\partial h}{\partial x} + \frac{\partial}{\partial x}\left[\int_0^h f(x,y,z)dz\right] \quad (5.65)$$

If the density is assumed to be the mean density of the fluid across the film, then the u_x term in Eq. (5.64) is

$$\int_0^h \frac{\partial}{\partial x}(\rho u_x)dz = -(\rho u_x)_{z=h}\frac{\partial h}{\partial x} + \frac{\partial}{\partial x}\left(\int_0^h \rho u_x dz\right)$$
$$= -\rho u_{x,b}\frac{\partial h}{\partial x} + \frac{\partial}{\partial x}\left(\int_0^h \rho u_x dz\right) \quad (5.66)$$

Similarly for the u_y term in Eq. (5.64).

$$\int_0^h \frac{\partial}{\partial y}(\rho u_y)dz = -\rho u_{y,b}\frac{\partial h}{\partial y} + \frac{\partial}{\partial y}\left(\int_0^h \rho u_y dz\right) \quad (5.67)$$

The u_z term can be integrated directly giving

$$\int_0^h \frac{\partial}{\partial z}(\rho u_z)dz = \rho(u_{z,b} - u_{z,a}) \quad (5.68)$$

The integral form of the continuity equation (Eq. 5.64) on substitution of Eqs (5.66)–(5.68), can be stated as

$$h\frac{\partial \rho}{\partial t} - \rho u_{x,b}\frac{\partial h}{\partial x} + \frac{\partial}{\partial x}\left(\rho\int_0^h u_x dz\right)$$
$$-\rho u_{y,b}\frac{\partial h}{\partial y} + \frac{\partial}{\partial y}\left(\rho\int_0^h u_y dz\right) + \rho(u_{z,b} - u_{z,a})$$
$$= 0 \quad (5.69)$$

The integrals in Eq. (5.69) represent the volumetric flow rates per unit width. Substitution of the values for these from Eqs (5.62), (5.63) gives the general Reynolds equation.

$$\frac{\partial}{\partial x}\left(-\frac{\rho h^3}{12\mu}\frac{\partial p}{\partial x}\right) + \frac{\partial}{\partial y}\left(-\frac{\rho h^3}{12\mu}\frac{\partial p}{\partial y}\right) + \frac{\partial}{\partial x}\left[\frac{\rho h(u_{x,a} + u_{x,b})}{2}\right] + \frac{\partial}{\partial y}\left[\frac{\rho h(u_{y,a} + u_{y,b})}{2}\right] +$$
$$\rho(u_{z,b} - u_{z,a}) - \rho u_{x,b}\frac{\partial h}{\partial x} - \rho u_{y,b}\frac{\partial h}{\partial y} + h\frac{\partial p}{\partial t} = 0$$
$$(5.70)$$

The first two terms of Eq. (5.70) are the Poiseuille terms and describe the net flow rates due to pressure gradients within the lubricated area. The third and fourth terms are the Couette terms and describe the net entrained flow rates due to surface velocities. The fifth to the seventh terms are due to a squeezing motion. The eighth term describes the net flow rate due to local expansion as a result of density changes with time. Eq. (5.70) is repeated below with the identification of the various terms emphasised.

$$\underbrace{\frac{\partial}{\partial x}\left(-\frac{\rho h^3}{12\mu}\frac{\partial p}{\partial x}\right)+\frac{\partial}{\partial y}\left(-\frac{\rho h^3}{12\mu}\frac{\partial p}{\partial y}\right)}_{\text{Poiseuille}}+\underbrace{\frac{\partial}{\partial x}\left[\frac{\rho h\left(u_{x,a}+u_{x,b}\right)}{2}\right]+\frac{\partial}{\partial y}\left[\frac{\rho h\left(u_{y,a}+u_{y,b}\right)}{2}\right]}_{\text{Couette}}$$

$$\underbrace{+\rho(u_{z,b}-u_{z,a})-\rho u_{x,b}\frac{\partial h}{\partial x}-\rho u_{y,b}\frac{\partial h}{\partial y}}_{\text{Squeeze}}+\underbrace{h\frac{\partial p}{\partial t}}_{\text{Net flow due to local expansion}}=0$$

For tangential only motion, where

$$u_{z,b}=u_{x,b}\frac{\partial h}{\partial x}+u_{y,b}\frac{\partial h}{\partial y} \tag{5.71}$$

and $u_{z,a}=0$, Eq. (5.70) reduces to

$$\frac{\partial}{\partial x}\left(\frac{\rho h^3}{\mu}\frac{\partial p}{\partial x}\right)+\frac{\partial}{\partial y}\left(\frac{\rho h^3}{\mu}\frac{\partial p}{\partial y}\right)=12\overline{u}_x\frac{\partial(\rho h)}{\partial x}+12\overline{u}_y\frac{\partial(\rho h)}{\partial y} \tag{5.72}$$

where

$$\overline{u}_x=\frac{u_{x,a}+u_{x,b}}{2}=\text{constant} \tag{5.73}$$

$$\overline{u}_y=\frac{u_{y,a}+u_{y,b}}{2}=\text{constant} \tag{5.74}$$

For hydrodynamic lubrication, the fluid properties do not vary significantly through the bearing and can be considered constant. In addition for hydrodynamic lubrication, the motion is pure sliding and $\overline{u}_y=0$.

The Reynolds equation can therefore be simplified to

$$\frac{\partial}{\partial x}\left(h^3\frac{\partial p}{\partial x}\right)+\frac{\partial}{\partial y}\left(h^3\frac{\partial p}{\partial y}\right)=12\overline{u}_x\mu_o\frac{\partial h}{\partial x} \tag{5.75}$$

In some lubrication applications, side leakage can be neglected and Eq. (5.72) can be restated as

$$\frac{\partial}{\partial x}\left(\frac{\rho h^3}{\mu}\frac{\partial p}{\partial x}\right)=12\overline{u}_x\frac{\partial(\rho h)}{\partial x} \tag{5.76}$$

Eq. (5.76) can be integrated giving

$$\frac{1}{\mu}\frac{dp}{dx}=\frac{12\overline{u}_x}{h^2}+\frac{A}{\rho h^3} \tag{5.77}$$

With boundary conditions $dp/dx=0$ when $x=x_m$, $\rho=\rho_m$, $h=h_m$ gives $A=-12\overline{u}_x\rho_m h_m$. The subscript m refers to the condition for which $dp/dx=0$ such as the point of maximum pressure. Substituting for A in Eq. (5.77) gives

$$\frac{dp}{dx}=12\overline{u}_x\mu\frac{\rho h-\rho_m h_m}{\rho h^3} \tag{5.78}$$

If the density can be considered constant then Eq. (5.78) becomes

$$\frac{dp}{dx}=12\overline{u}_x\mu\frac{h-h_m}{h^3} \tag{5.79}$$

For the case of a gas lubricated bearing the density, using the ideal gas law, is given by

$$\rho=\frac{p}{RT} \tag{5.80}$$

From Eq. (5.72), and taking the viscosity as constant

$$\frac{\partial}{\partial x}\left(\rho h^3\frac{\partial p}{\partial x}\right)+\frac{\partial}{\partial y}\left(\rho h^3\frac{\partial p}{\partial y}\right)=12\overline{u}_x\mu_o\frac{\partial(\rho h)}{\partial x} \tag{5.81}$$

Expressing Eq. (5.72) in cylindrical coordinates

$$\frac{\partial}{\partial r}\left(\frac{r\rho h^3}{\mu}\frac{\partial p}{\partial r}\right)+\frac{1}{r}\frac{\partial}{\partial\phi}\left(\frac{\rho h^3}{\mu}\frac{\partial p}{\partial\phi}\right)=12\left[\overline{u}_r\frac{\partial(\rho rh)}{\partial r}+\overline{u}_\phi\frac{\partial(\rho h)}{\partial\phi}\right] \tag{5.82}$$

where

$$\overline{u}_r=\frac{u_{r,a}+u_{r,b}}{2} \tag{5.83}$$

$$\overline{u}_\phi=\frac{u_{\phi,a}+u_{\phi,b}}{2} \tag{5.84}$$

If the viscosity and density can be assumed to be constant then Eq. (5.82) becomes

$$\frac{\partial}{\partial r}\left(rh^3\frac{\partial p}{\partial r}\right)+\frac{1}{r}\frac{\partial}{\partial\phi}\left(h^3\frac{\partial p}{\partial\phi}\right)=12\mu_o\left[\overline{u}_r\frac{\partial(rh)}{\partial r}+\overline{u}_\phi\frac{\partial h}{\partial\phi}\right] \tag{5.85}$$

Eq. (5.85) applies to a thrust bearing where the fluid film is in the z direction and the bearing dimensions are in the r, ϕ directions.

For turbulent flow, the Reynolds equation is

$$\frac{\partial}{\partial x}\left(\frac{h^3}{\mu k_x}\frac{\partial\overline{p}}{\partial x}\right)+\frac{\partial}{\partial y}\left(\frac{h^3}{\mu k_y}\frac{\partial\overline{p}}{\partial y}\right)=\frac{\overline{u}_x}{2}\frac{\partial h}{\partial x} \tag{5.86}$$

where (from Constantinescu (1962))

$$k_x = 12 + 0.026\left(\frac{\rho\Omega rh}{\mu}\right)^{0.102} \tag{5.87}$$

$$k_y = 12 + 0.0198\left(\frac{\rho\Omega rh}{\mu}\right)^{0.091} \tag{5.88}$$

For an infinitely wide journal bearing the pressure in the axial direction can be assumed to be constant. This approach is valid for a length to diameter ratio of $L/D > 2$. The integrated form of the Reynolds Equation (Eq. 5.79) on substitution for $\overline{u}_x = (u_a + u_b)/2 = \Omega a/2$, as $u_b = 0$ for a stationary bearing surface, gives

$$\frac{dp}{dx} = \frac{6\mu_o a\Omega(h - h_m)}{h^3} \tag{5.89}$$

$$dx = ad\phi \tag{5.90}$$

$$\frac{dp}{d\phi} = \frac{6\mu_o a^2\Omega(h - h_m)}{h^3} \tag{5.91}$$

$$\cos\alpha = \frac{1}{b}[h + a + e\cos(\pi - \phi)] \tag{5.92}$$

$$h = b\cos\alpha - a + e\cos\phi \tag{5.93}$$

Hence, using trigonometric relationships,

$$h = b\left[1 - \left(\frac{e}{b}\right)^2 \sin^2\phi\right]^{0.5} - a + e\cos\phi \tag{5.94}$$

Expanding the series

$$\left[1 - \left(\frac{e}{b}\right)^2 \sin^2\phi\right]^{0.5} = 1 - \frac{1}{2}\left(\frac{e}{b}\right)^2 \sin^2\phi - \frac{1}{8}\left(\frac{e}{b}\right)^4 \sin^4\phi - \ldots \tag{5.95}$$

Hence

$$h = b\left[1 - \frac{1}{2}\left(\frac{e}{b}\right)^2 \sin^2\phi - \frac{1}{8}\left(\frac{e}{b}\right)^4 \sin^4\phi - \ldots\right] - a + e\cos\phi \tag{5.96}$$

$$b - a = \mathrm{c} \tag{5.97}$$

$$h = c + e\left[\cos\phi - \frac{1}{2}\left(\frac{e}{b}\right) \sin^2\phi - \frac{1}{8}\left(\frac{e}{b}\right)^3 \sin^4\phi - \ldots\right] \tag{5.98}$$

The ratio e/b is typically of the order of 1×10^{-3}. As a result terms in Eq. (5.98) with this ratio can readily be neglected, giving

$$h = c + e\cos\phi = c\left(1 + \frac{e}{c}\cos\phi\right) = c(1 + \varepsilon\cos\phi) \tag{5.99}$$

where ε is the eccentricity ratio defined by

$$\varepsilon = \frac{e}{c} \tag{5.100}$$

Substitution of Eq. (5.99) in Eq. (5.91) gives

$$\frac{dp}{d\phi} = 6\mu\Omega\left(\frac{a}{c}\right)^2\left[\frac{1}{(1+\varepsilon\cos\phi)^2} - \frac{h_m}{c(1+\varepsilon\cos\phi)^3}\right] \tag{5.101}$$

Integration of Eq. (5.101) gives the following expression for the pressure distribution.

$$p = 6\mu\Omega\left(\frac{a}{c}\right)^2 \int \left[\frac{1}{(1+\varepsilon\cos\phi)^2} - \frac{h_m}{c(1+\varepsilon\cos\phi)^3}\right] d\phi + A \tag{5.102}$$

where A is a constant of integration.

For a journal bearing,

$$x = a\phi \tag{5.103}$$

and

$$\bar{u} = \frac{u_{x,a}}{2} = \frac{a\Omega}{2} \tag{5.104}$$

From Eq. (5.75),

$$\frac{\partial}{\partial\phi}\left(h^3\frac{\partial p}{\partial\phi}\right) + a^2\frac{\partial}{\partial y}\left(h^3\frac{\partial p}{\partial y}\right) = 6\mu_o\Omega a^2\frac{\partial h}{\partial\phi} \tag{5.105}$$

Substituting for the film thickness gives

$$\frac{\partial}{\partial\phi}\left(h^3\frac{\partial p}{\partial\phi}\right) + a^2h^3\frac{\partial^2 p}{\partial y^2} = -6\mu_o\Omega a^2 e\sin\phi \tag{5.106}$$

The Reynolds equation can be solved by approximate mathematical methods or numerically. Once the pressure distribution has been established the journal performance can be determined in terms of the bearing load capacity, frictional losses, lubricant flow requirements and the lubricant temperature rise.

5.4.2 Design charts for full-film hydrodynamic bearings

If the designer wishes to avoid the direct solution of Eq. (5.3), use can be made of a series of design charts. These were originally produced by Raimondi and Boyd (1958a, b, c) who used an iterative technique to solve the Reynolds equations. These charts give the film thickness, coefficient of friction, lubricant flow, lubricant side flow ratio, minimum film thickness location, maximum pressure ratio, maximum pressure ratio position and film termination angle versus the Sommerfield number (Figs 5.18–5.25).

The Sommerfield number, which is also known as the bearing characteristic number, is defined in Eq. (5.107). It is used as it encapsulates all the parameters usually defined by the designer. Great care needs to be taken with units in the design of journal bearings. Many design charts have been produced using English units (psi, reyn, Btu etc.). As long as a consistent set of units is maintained, use of the charts will yield sensible results. In particular note the use of revolutions per second in the definition of speed in the Sommerfield number.

$$S=\left(\frac{r}{c}\right)^2\frac{\mu N_s}{P} \tag{5.107}$$

where:

S is the bearing characteristic number;
r is the journal radius, (m);
c is the radial clearance (m);
μ is the absolute viscosity (Pa s);
N_s is the journal speed (revolutions per second, rps);
$P=W/LD$ is the load per unit of projected bearing area (N/m^2);
W is the load on the bearing (N);
D is the journal diameter (m);
L is the journal bearing length (m).

Consider the journal shown in Fig. 5.26. As the journal rotates it will pump lubricant in a clockwise direction. The lubricant is pumped into a wedge-shaped space and the journal is forced over to the opposite side. The angular position where the lubricant film is at its minimum thickness h_o is called the attitude angle, ϕ. The centre of the journal is displaced from the centre of the bearing by a distance e called the eccentricity. The ratio of the eccentricity e to the radial clearance c is called the eccentricity ratio ($\varepsilon = e/c$). The relationship between the film thickness, the clearance, eccentricity and eccentricity ratio are defined in Eqs (5.108)–(5.110).

$$h_o=c-e \tag{5.108}$$

$$\varepsilon=\frac{e}{c} \tag{5.109}$$

$$\frac{h_0}{c}=1-\varepsilon \tag{5.110}$$

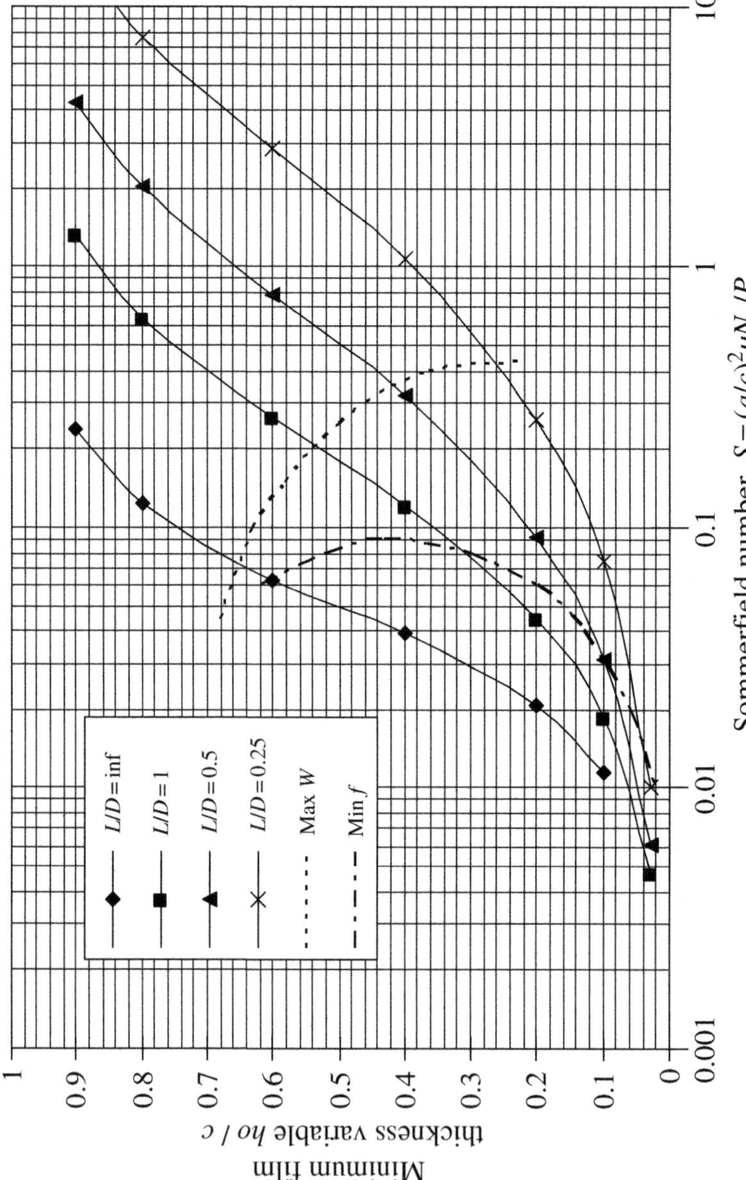

Fig. 5.18 Chart for the minimum film thickness variable, (h_o/c) versus the Sommerfield number. Data from Raimondi and Boyd (1958c).

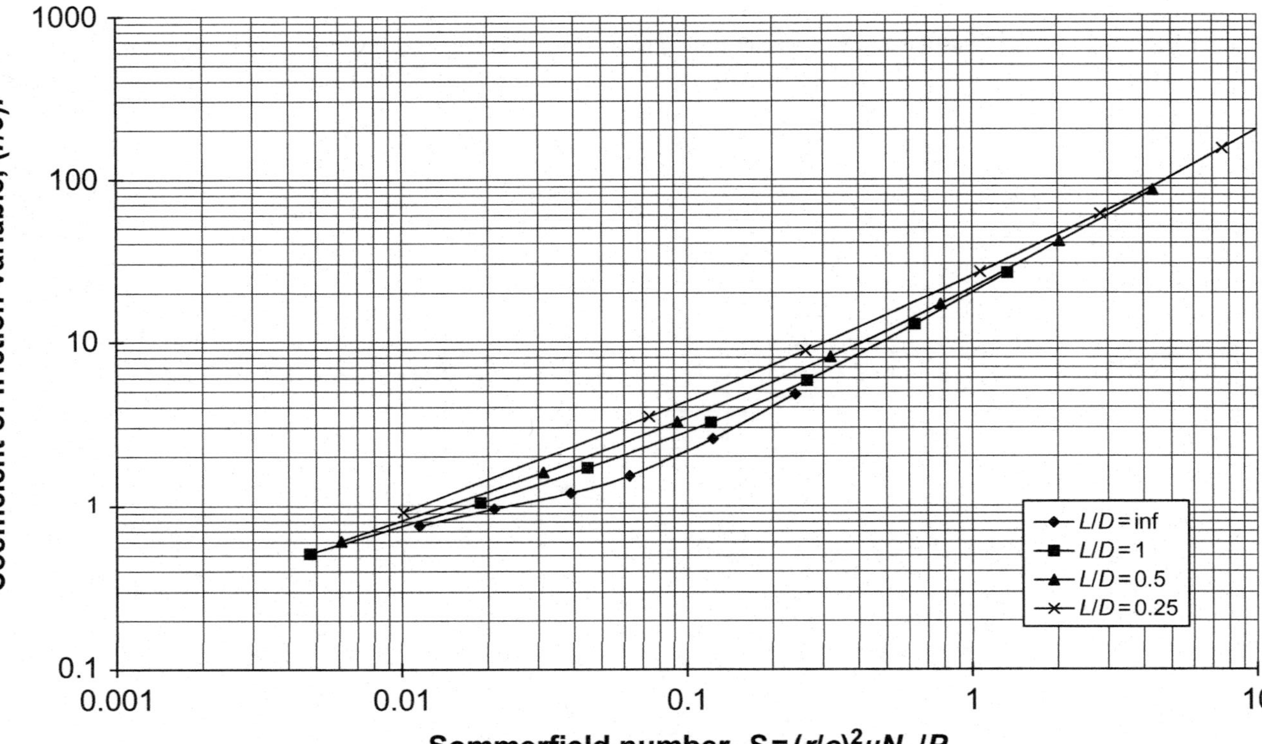

Fig. 5.19 Chart for determining the coefficient of friction variable, $(r/c)f$.
Data from Raimondi, A.A., Boyd, J., 1958c. A solution for the finite journal bearing and its application to analysis and design: III. ASLE Trans. 1, 194–209.

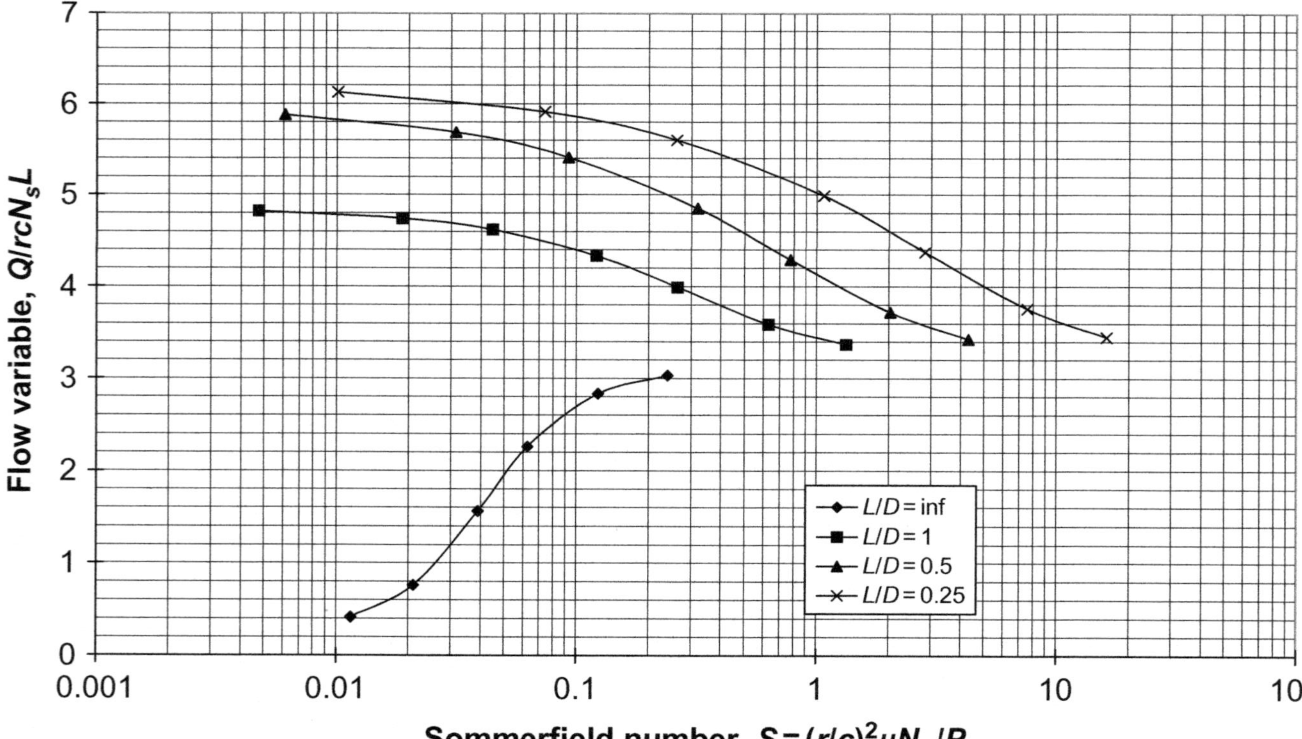

Fig. 5.20 Chart for determining the flow variable, $Q/(acN_sL)$.
Data from Raimondi, A.A., Boyd, J., 1958c. A solution for the finite journal bearing and its application to analysis and design: III. ASLE Trans. 1, 194–209.

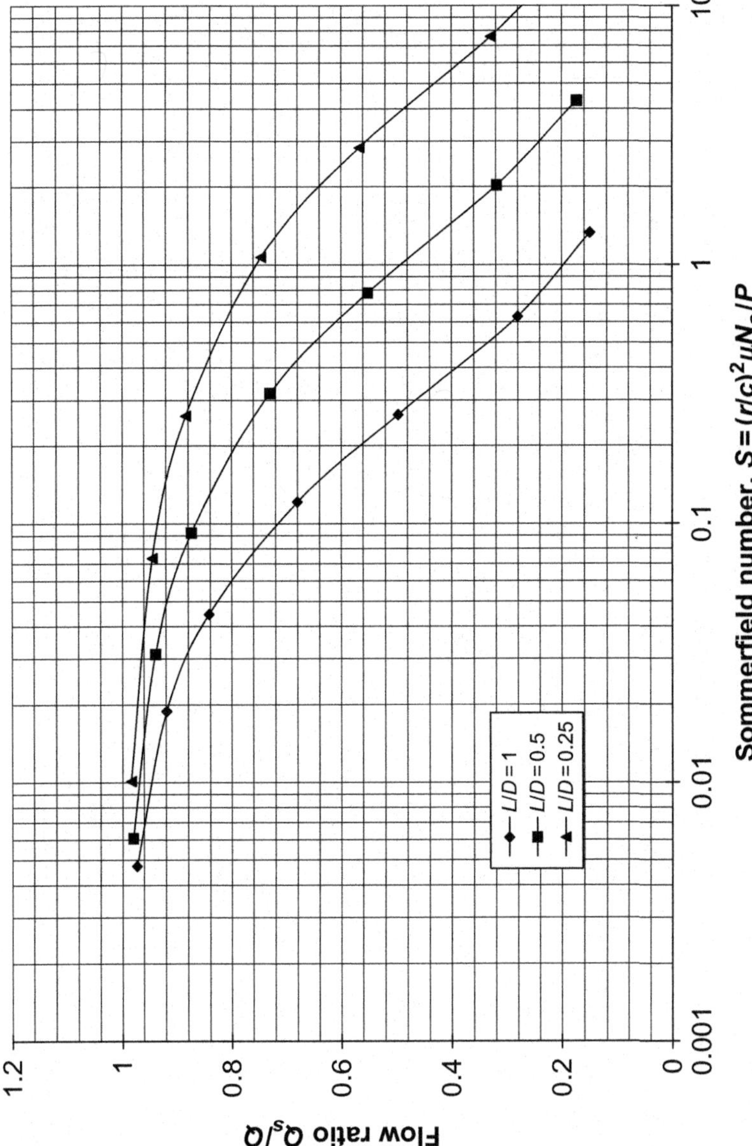

Fig. 5.21 Chart for determining the ratio of side flow, Q_s, to total flow, Q.
Data from Raimondi, A.A., Boyd, J., 1958c. A solution for the finite journal bearing and its application to analysis and design: III. ASLE Trans. 1, 194–209.

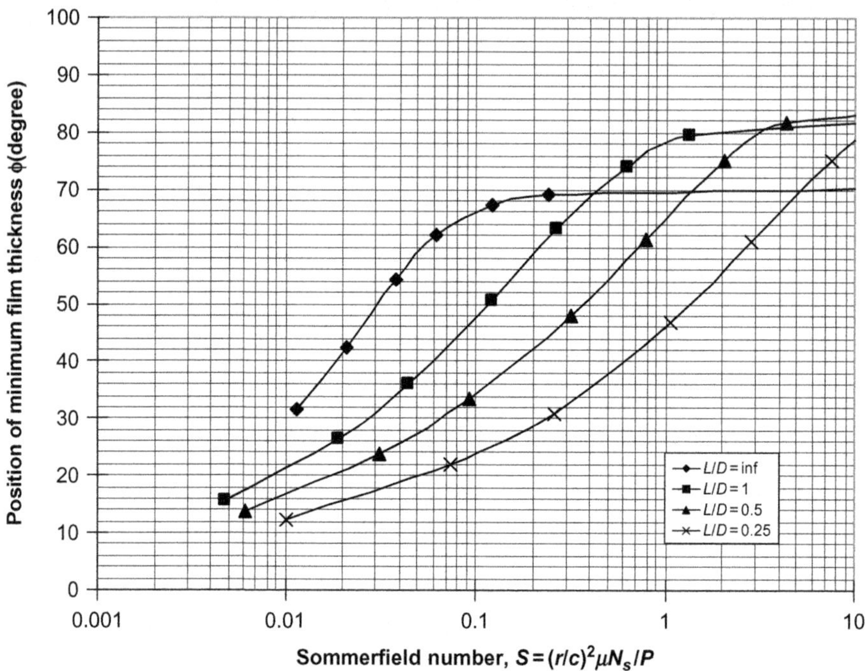

Fig. 5.22 Chart for determining the position of the minimum film thickness ϕ.
Data from Raimondi, A.A., Boyd, J., 1958c. A solution for the finite journal bearing and its application to analysis and design: III. ASLE Trans. 1, 194–209.

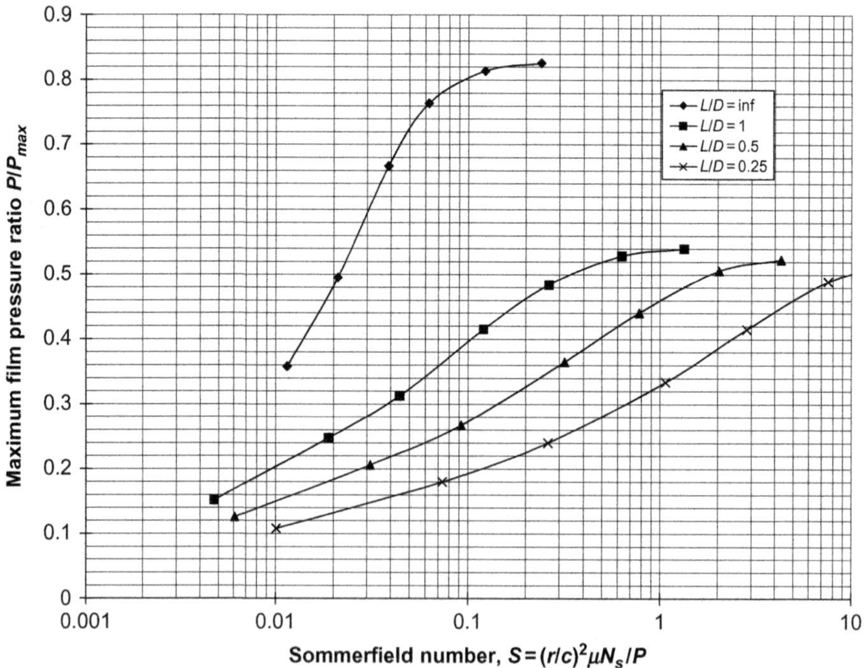

Fig. 5.23 Chart for determining the maximum film pressure ratio, p/p_{max}.
Data from Raimondi, A.A., Boyd, J., 1958c. A solution for the finite journal bearing and its application to analysis and design: III. ASLE Trans. 1, 194–209.

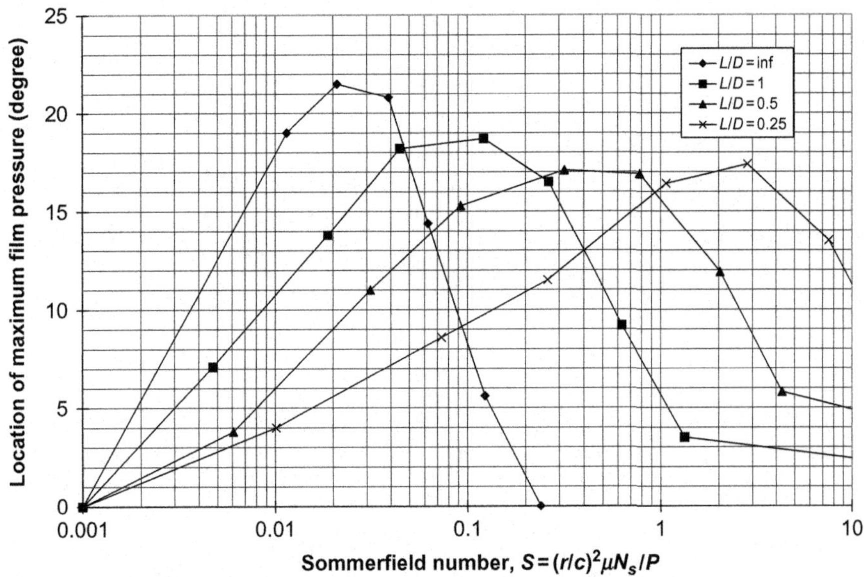

Fig. 5.24 Chart for determining the position of maximum pressure, $\varphi_{p_{max}}$.
Data from Raimondi, A.A., Boyd, J., 1958c. A solution for the finite journal bearing and its application to analysis and design: III. ASLE Trans. 1, 194–209.

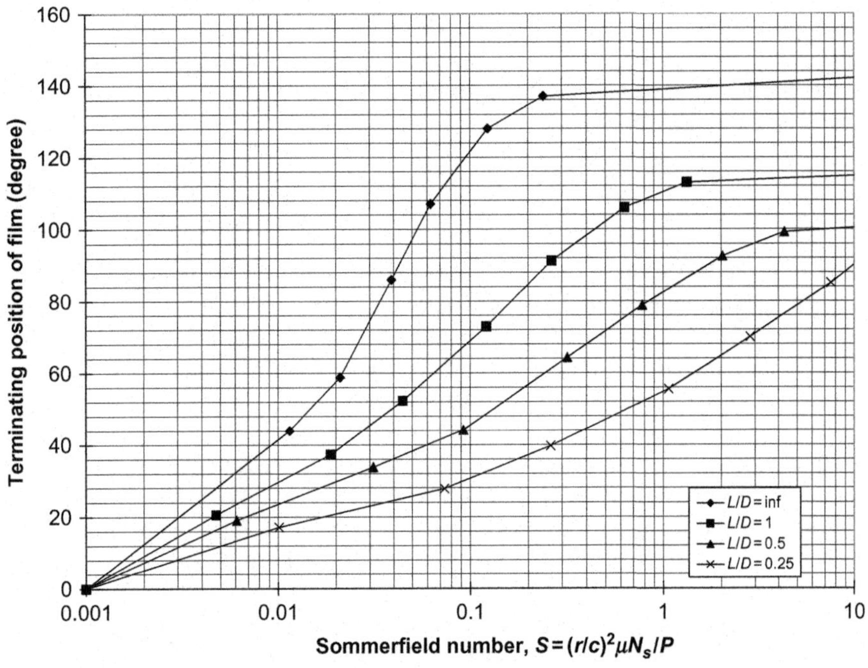

Fig. 5.25 Chart for determining the film termination angle, φ_{p_o}.
Data from Raimondi, A.A., Boyd, J., 1958c. A solution for the finite journal bearing and its application to analysis and design: III. ASLE Trans. 1, 194–209.

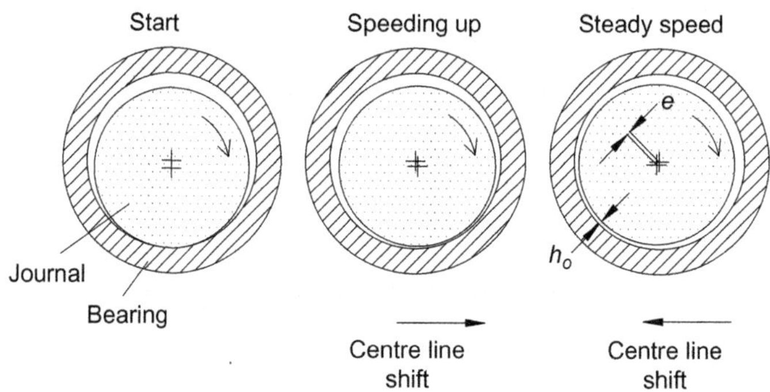

Fig. 5.26 Full film hydrodynamic bearing motion from start up.

One of the assumptions made in the analysis of Raimondi and Boyd is that the viscosity of the lubricant is constant as it passes through the bearing. However, work is done on the lubricant in the bearing and the temperature of the lubricant leaving the bearing zone will be higher than the entrance value. Fig. 5.7 shows the variation of viscosity with temperature for some of the SAE and ISO defined lubricants and it can be seen the value of the viscosity for a particular lubricant is highly dependent on the temperature. Some of the lubricant entrained into the bearing film emerges as side flow which carries away some of the heat. The remainder flows through the load-bearing zone and carries away the remainder of the heat generated. The temperature used for determining the viscosity can be taken as the average of the inlet and exit lubricant temperatures as given by Eq. (5.111).

$$T_{av} = T_1 + \frac{T_2 - T_1}{2} = T_1 + \frac{\Delta T}{2} \tag{5.111}$$

where T_1 is the temperature of the lubricant supply and T_2 is the temperature of the lubricant leaving the bearing. Generally petroleum lubrication oils should be limited to a maximum temperature of approximately 70°C to prevent excessive oxidation.

One of the parameters that needs to be determined is the bearing lubricant exit temperature T_2. This is a trial and error or iterative process. A value of the temperature rise, ΔT, is guessed, the viscosity for a standard oil, corresponding to this value, determined and the analysis performed. If the temperature rise calculated by the analysis does not correspond closely to the guessed value the process should be repeated using an updated value for the temperature rise until the two match. If the temperature rise is unacceptable it may be necessary to try a different lubricant or modify the bearing configuration.

Given the length, the journal radius, the radial clearance, the lubricant type and its supply temperature the steps for determining the various bearing operating parameters are given in Table 5.4.

The temperature rise of the lubricant through the bearing can be determined from

Table 5.4 Full-film hydrodynamic bearing design guideline

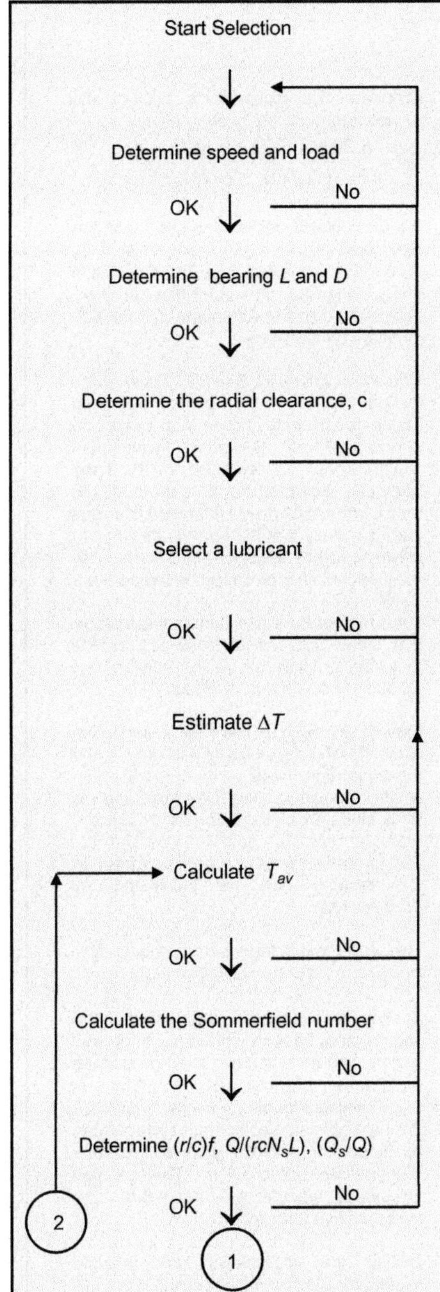

Determine the speed of the journal and the load to be supported.

If L and D not already determined, set the proportions of the bearing so that the load capacity, $P = W/LD$, is somewhere between 0.34 MN/m^2 for light machinery to 13.4 MN/m^2 for heavy machinery

Determine a value for the radial clearance, c, of the bearing using the data presented in Figure 5.14.

If not already specified select a lubricant. Lubricant oil selection is a function of speed or compatibility with other lubricant requirements. Generally as the design speed rises, oils with a lower viscosity should be selected.

Estimate a value for the temperature rise ΔT across the bearing. The value taken for the initial estimate is relatively unimportant. As a guide, a value of $\Delta T = 10$ °C is generally a good starting guess. This value can be increased for high speed bearings and for low bearing clearances

Determine the average lubricant temperature $T_{av} = T_1 + \Delta T/2$ and find the corresponding value for the viscosity for the chosen lubricant.

Calculate the Sommerfield number, $S = (r/c)^2 \mu N_s/P$ and the length to diameter ratio.

Use the charts (Figs. 5.19, 5.20 and 5.21) to determine values for the coefficient of friction variable, the total lubricant flow variable, and the ratio of the side flow to the total lubricant flow with the values for the Sommerfield number and the L/D ratio.

Continued

Table 5.4 Continued

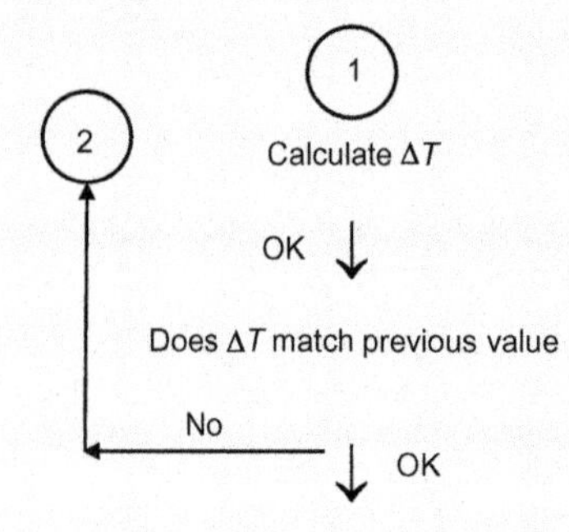

Calculate ΔT OK ↓	Calculate the temperature rise of the lubricant through the bearing using $$\Delta T = \frac{8.30\times10^{-6}P}{1-\frac{1}{2}(Q_s/Q)} \times \frac{(r/c)f}{Q/rcN_sL}$$
Does ΔT match previous value No → 2 ↓ OK	If the calculated value does not match the estimated value for ΔT to within say 1°C repeat the procedure from the calculation of T_{av} using the updated value of the temperature rise to determine the average lubricant temperature.
Check operation in optimal zone OK ↓	Check the values for S and *L/D* ratio give a design that is in the optimal operating region for minimal friction and maximum load capability on the chart for the minimum film thickness variable (Fig. 5.18). If the operating Sommerfield number and *L/D* ratio combination do not fall within this zone, then it is likely that the bearing design can be improved by altering values for *c*, *L*, *D*, the lubricant and the operating temperature.
Calculate Q, Q_s, f OK ↓	If the temperature rise across the bearing has converged, calculate values for the total lubricant flow rate, Q, the side flow rate Q_s and the coefficient of friction f.
Calculate h_o, p/p_{max}, ϕ_{pmax}, ϕ_{po} OK ↓	Figs. 5.18 to 5.25 can be used to determine values for the maximum film thickness, the maximum film pressure ratio, the location of the maximum film pressure and its terminating angular location as required.
Calculate the torque OK ↓	The torque required to overcome friction in the bearing can be calculated by Torque = fWr
Calculate the power lost OK ↓	The power lost in the bearing is given by Power = $\omega \times$ Torque = $2\pi N_s \times$ Torque
Specify surface roughness OK ↓	Specify the surface roughness for the journal and bearing surfaces. A ground journal with an arithmetic average surface roughness of 0.4 to 0.8 μm is recommended for bearings of good quality. For high precision equipment surfaces can be lapped or polished to give a surface roughness of 0.2 to 0.4 μm. The specified roughness should be less than the minimum film thickness.
Design lubricant system OK ↓ Complete	Design the recirculation and sealing system for the bearing lubricant.

$$\Delta T = \frac{8.30 \times 10^{-6} P}{1 - \frac{1}{2}(Q_s/Q)} \times \frac{(r/c)f}{Q/rcN_s L} \tag{5.112}$$

The torque required to overcome friction in the bearing can be calculated by

$$\text{Torque} = \text{fWr} \tag{5.113}$$

The power lost in the bearing is given by

$$\text{Power} = \omega \times \text{Torque} = 2\pi N_s \times \text{Torque} \tag{5.114}$$

where ω is the angular velocity (rad/s).

For values of the length to diameter (L/D) ratio other than those shown in the charts given in Figs 5.18–5.25, values for the various parameters can be found by interpolation using Eq. (5.115)(Raimondi and Boyd, 1958b).

$$y = \frac{1}{(L/D)^3}\left[\begin{array}{l} -\frac{1}{8}\left(1-\frac{L}{D}\right)\left(1-2\frac{L}{D}\right)\left(1-4\frac{L}{D}\right)y_\infty + \\ \frac{1}{3}\left(1-2\frac{L}{D}\right)\left(1-4\frac{L}{D}\right)y_1 - \frac{1}{4}\left(1-\frac{L}{D}\right)\left(1-4\frac{L}{D}\right)y_{1/2} + \frac{1}{24}\left(1-\frac{L}{D}\right)\left(1-2\frac{L}{D}\right)y_{1/4} \end{array}\right] \tag{5.115}$$

Here y is the desired variable and y_∞, y_1, $y_{1/2}$ and $y_{1/4}$ are the variable at $L/D = \infty$, 1, 1/2 and 1/4 respectively.

Example 5.3. A full journal bearing has a nominal diameter of 50.0 mm and a bearing length of 25.0 mm (Fig. 5.27). The bearing supports a load of 3000 N, and the journal design speed is 3000 rpm. The radial clearance has been specified as 0.04 mm. An SAE 10 oil has been chosen and the lubricant supply temperature is 50°C.

Find the temperature rise of the lubricant, the lubricant flow rate, the minimum film thickness, the torque required to overcome friction and the heat generated in the bearing.

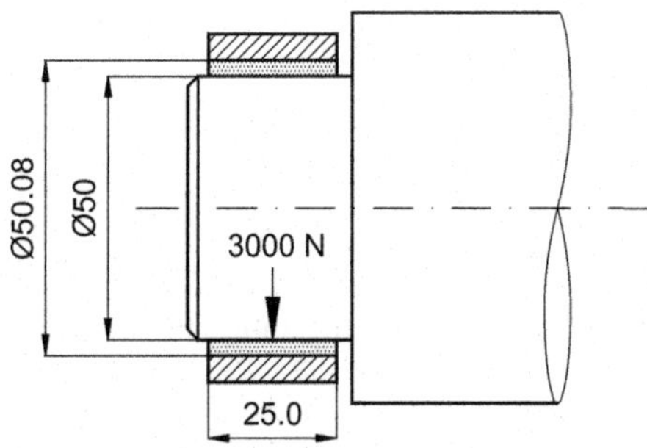

Fig. 5.27 Bearing design example.

Solution

The primary data are $D=50.0\,\text{mm}$, $L=25\,\text{mm}$, $W=3000\,\text{N}$, $N=3000\,\text{rpm}$, $c=0.04\,\text{mm}$, SAE 10, $T_1=50°\text{C}$.

Guess a value for the lubricant temperature rise ΔT across the bearing to be say $\Delta T=20°\text{C}$.

$$T_{av}=T_1+\frac{\Delta T}{2}=50+\frac{20}{2}=60^{\circ}\text{C}$$

From Fig. 5.7 for SAE 10 at 60°C, $\mu=0.014\,\text{Pa s}$.

The mean radius is given by

$$r_m=\frac{a+b}{2}=\frac{25+25.04}{2}=25.02\text{ mm}$$

$$b-a=0.04\,\text{mm}$$

Assuming a density of $900\,\text{kg/m}^3$,

$$\nu=\mu/\rho=0.014/900=1.55\times10^{-5}\,\text{m}^2/\text{s}$$

So the Taylor number using this approximate value for the kinematic viscosity is given by

$$Ta_m=\frac{\Omega r_m^{0.5}(b-a)^{1.5}}{\nu}=0.81$$

As this is much less than the critical Taylor number, 41.19, laminar flow can be assumed.

$$N_s=3000/60=50\,\text{rps}, L/D=25/50=0.5.$$

$$P=\frac{W}{LD}=\frac{3000}{0.025\times0.05}=2.4\times10^6\,\text{N/m}^2$$

$$S=\left(\frac{r}{c}\right)^2\frac{\mu N_s}{P}=\left(\frac{25\times10^{-3}}{0.04\times10^{-3}}\right)^2\frac{0.014\times50}{2.4\times10^6}=0.1139$$

From Fig. 5.19 with $S=0.1139$ and $L/D=0.5$, $(r/c)f=3.8$.

From Fig. 5.20 with $S=0.1139$ and $L/D=0.5$, $Q/(rcN_sL)=5.34$.

From Fig. 5.21 with $S=0.1139$ and $L/D=0.5$, $Q_s/Q=0.852$.

The value of the temperature rise of the lubricant can now be calculated using Eq. (5.112):

$$\Delta T=\frac{8.3\times10^{-6}P}{1-0.5\frac{Q_s}{Q}}\times\frac{\frac{r}{c}f}{\frac{Q}{rcN_sL}}=\frac{8.3\times10^{-6}\times2.4\times10^6}{1-(0.5\times0.852)}\times\frac{3.8}{5.34}=24.70^{\circ}\text{C}$$

As the value calculated for the lubricant temperature rise is significantly different to the estimated value, it is necessary to repeat the above calculation but using the new improved estimate for determining the average lubricant temperature.

Using $\Delta T = 24.70°C$ to calculate T_{av} gives: $T_{av} = 50 + \frac{24.70}{2} = 62.35^{\circ}C$.

Repeating the procedure using the new value for T_{av} gives:

$$\mu = 0.0136 \text{Pas},$$

$$S = 0.1107,$$

from Fig. 5.19 with $S = 0.1107$ and $L/D = 0.5$, $(r/c)f = 3.7$,

from Fig. 5.20 with $S = 0.1107$ and $L/D = 0.5$, $Q/(rcN_sL) = 5.35$,

from Fig. 5.21 with $S = 0.1107$ and $L/D = 0.5, Q_s/Q = 0.856$,

$$\Delta T = \frac{8.3 \times 10^{-6} \times 2.4 \times 10^{6}}{1 - (0.5 \times 0.856)} \times \frac{3.7}{5.35} = 24.08^{\circ}C,$$

$$T_{av} = 50 + \frac{24.08}{2} = 62.04^{\circ}C.$$

This value for T_{av} is close to the previous calculated value suggesting that the solution has converged. For $T_{av} = 62.04°C$, $\mu = 0.0136$ Pas and $S = 0.1107$.

The other parameters can now be found.

$$Q = rcN_sL \times 5.35 = 25 \times 0.04 \times 50 \times 25 \times 5.35 = 6688 \text{mm}^3/\text{s}.$$

From Fig. 5.18$h_o/c = 0.22$. $h_o = 0.0088$ mm.

$$f = 3.7 \times (c/r) = 3.7 \times (0.04/25) = 0.00592.$$

The torque is given by Torque $= fWr = 0.00592 \times 3000 \times 0.025 = 0.444$ N m.

The power dissipated in the bearing is given by Power $= 2\pi \times$ Torque $\times N_s = 139.5$ W.

Example 5.4. A full journal bearing is required for a 30 mm diameter shaft, rotating at 1200 rpm supporting a load of 1500 N (Fig. 5.28). From previous experience, the lubrication system available is capable of delivering the lubricant to the bearing at 40°C. Select an appropriate radial clearance, length and lubricant for the bearing and determine the temperature rise of the lubricant, the total lubricant flow rate required and the power absorbed in the bearing. Check that your chosen design operates within the optimum zone indicated by the dotted lines on the minimum film thickness ratio chart (Fig. 5.18).

Solution

The primary design information is: $D = 30.0$ mm, $W = 1500$ N, $N = 1200$ rpm, SAE 10, $T_1 = 40°C$.

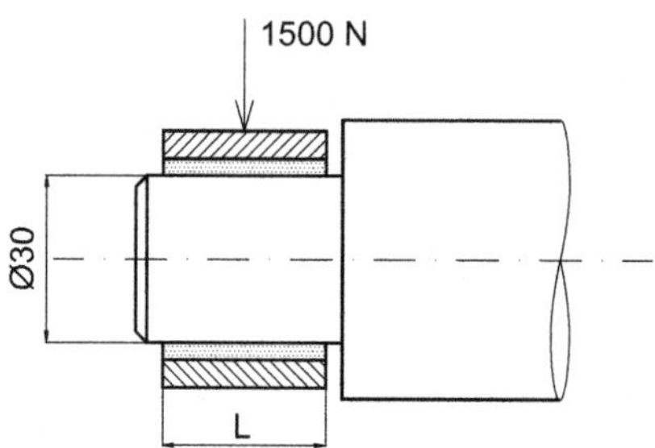

Fig. 5.28 Bearing design example.

From Fig. 5.14 for a speed of 1200 rpm and a nominal journal diameter of 30 mm, a suitable value for the diametral clearance is 0.05 mm. The radial clearance is therefore 0.025 mm. The next step is to set the length of the bearing. Typical values for L/D ratios are between 0.5 and 1.5. Here a value for the ratio is arbitrarily set as $L/D=1$ (i.e. $L=30$ mm).

Given the speed of the bearing, 1200 rpm, as an initial proposal an SAE 10 lubricant is selected.

The procedure for determining the average lubricant temperature is the same as for the previous example. As a first estimate the temperature rise of the lubricant is taken as 10°C and the results of the iterative procedure are given in Table 5.5.

$$P=\frac{W}{LD}=\frac{1500}{0.03\times 0.03}=1.667\times 10^6\ \mathrm{N/m^2}.$$

$$S=\left(\frac{r}{c}\right)^2\frac{\mu N_s}{P}=\left(\frac{15\times 10^{-3}}{0.025\times 10^{-3}}\right)^2\frac{\mu 20}{1.667\times 10^6}=4.319\mu.$$

$$T_{av}=T_1+\frac{\Delta T}{2}=40+\frac{10}{2}=45^{\circ}\mathrm{C}.$$

The mean radius is given by

$$r_m=\frac{a+b}{2}=\frac{15+15.025}{2}=15.01\,\mathrm{mm}$$

$$b-a=0.025\,\mathrm{mm}$$

Table 5.5 Tabular data for Example 5.4

Temperature rise, ΔT (°C)	10	15.90	13.76	13.76
Average lubricant temperature T_{av} (°C)	45	47.95	46.88	46.88
Average lubricant viscosity μ(Pa s)	0.028	0.023	0.024	Converged
Sommerfield number, S	0.1209	0.09934	0.1037	
Coefficient of friction variable $(r/c)f$	3.3	2.8	2.8	
Flow variable Q/rcN_sL	4.35	4.4	4.4	
Side flow to total flow ratio Q_s/Q	0.68	0.72	0.72	

An estimate for the kinematic viscosity can be found, taking the average temperature of the oil as 45°C and assuming a density of 900 kg/m^3. At 45°C, the viscosity for SAE 10 is approximately 0.028 Pa s

$$\nu = \mu/\rho = 0.028/900 = 3.11 \times 10^{-5}\,\text{m}^2/\text{s}$$

So the Taylor number using this approximate value for the kinematic viscosity is given by

$$Ta_m = \frac{\Omega r_m^{0.5}(b-a)^{1.5}}{\nu} = 0.062$$

As this is much less than the critical Taylor number, 41.19, laminar flow can be assumed.

The total lubricant flow rate required is given by $Q = rcN_sL \times 4.4 = 15 \times 0.025 \times 20 \times 30 \times 4.4 = 990\,\text{mm}^3/\text{s}$.

With $S = 0.1037$ and $L/D = 1$ the design selected is within the optimum operating zone for minimum friction and optimum load capacity indicated in Fig. 5.18.

The friction factor, $f = 2.8 \times (c/r) = 2.8 \times 0.025/15 = 0.004667$.

Torque $= fWr = 0.004667 \times 1500 \times 0.015 = 0.105\,\text{N m}$.

Power $= 2\pi N_s \times$ Torque $= 13.19\,\text{W}$.

Example 5.5. A full journal bearing has a nominal diameter of 25.0 mm and bearing length 25.0 mm. The bearing supports a load of 1450 N and the journal design speed is 2500 rpm. Two radial clearances are proposed (i) 0.025 mm and (ii) 0.015 mm. An SAE 10 oil has been chosen and the lubricant supply temperature is 50°C. Using the standard Raimondi and Boyd design charts, find the temperature rise of the lubricant, the lubricant flow rate, the torque required to overcome friction and the heat generated in the bearing for both of the proposed clearances. Determine which clearance, if any, should be selected for the bearing.

Solution

$D = 25\,\text{mm}$, $L = 25\,\text{mm}$, $W = 1450\,\text{N}$, $N = 2500\,\text{rpm}$, $N_s = 2500/60\,\text{rps} = 41.67\,\text{rps}$, SAE 10, $T_1 = 50°\text{C}$, $L/D = 1$.

$$P = \frac{W}{LD} = \frac{1450}{0.025 \times 0.025} = 2.32 \times 10^6\,\text{N m}^{-2}.$$

(i) $c = 0.025\,\text{mm}$.

$$S = \left(\frac{r}{c}\right)^2 \frac{\mu N_s}{P} = \left(\frac{12.5 \times 10^{-3}}{0.025 \times 10^{-3}}\right)^2 \frac{\mu 41.67}{2.32 \times 10^6} = 4.49\mu.$$

The mean radius is given by

$$r_m = \frac{a+b}{2} = \frac{12.5+12.525}{2} = 12.51\,\text{mm}$$

$$b-a = 0.025\,\text{mm}$$

An estimate for the kinematic viscosity can be found, taking the average temperature of the oil as 55°C and assuming a density of 900 kg/m^3. At 55°C, the viscosity for SAE 10 is approximately 0.017 Pa s

$$\nu = \mu/\rho = 0.017/900 = 1.89 \times 10^{-5}\,\text{m}^2/\text{s}$$

So the Taylor number using this approximate value for the kinematic viscosity is given by

$$Ta_m = \frac{\Omega r_m^{0.5}(b-a)^{1.5}}{\nu} = 0.19$$

As this is much less than the critical Taylor number, 41.19, laminar flow can be assumed. Reducing the clearance results in a smaller Taylor number so laminar flow can also be assumed for the smaller clearance.

The solution for the iterative procedure using the Raymondi and Boyd charts (Figs 5.19–5.21) to find the average temperature rise of the lubricant across the bearing is given in tabular form below. The first estimate for the temperature rise is taken as $\Delta T = 10°\text{C}$.

Temperature rise, ΔT (°C)	10	16.13	15.36
Average lubricant temperature T_{av} (°C)	55	58.07	57.68
Average lubricant viscosity μ(Pa s)	0.017	0.015	Converged
Sommerfield number, S	0.07633	0.06735	
Coefficient of friction variable $(r/c)f$	2.3	2.2	
Flow variable Q/rcN_sL	4.45	4.52	
Side flow to total flow ratio Q_s/Q	0.766	0.78	

$$\Delta T \approx 15.4°\text{C}.$$

$$Q = rcN_sL \times 4.52 = 12.5 \times 0.025 \times 41.67 \times 25 \times 4.52 = 1471\,\text{mm}^3/\text{s}.$$

$$f = 2.2 \times (c/r) = 2.2 \times 0.025/12.5 = 0.0044.$$

$$T = fWr = 0.0044 \times 1450 \times 0.0125 = 0.07975\,\text{Nm}.$$

$$\text{Heat generated} = 2\pi N_s \times \text{Torque} = 20.88\,\text{W}.$$

With $S = 0.06735$ and $L/D = 1$ the design selected is outside the optimum operating zone for minimum friction and optimum load capacity indicated in Fig. 5.18.

(ii) $c = 0.015$ mm.

$$S = \left(\frac{r}{c}\right)^2 \frac{\mu N_s}{P} = \left(\frac{12.5 \times 10^{-3}}{0.015 \times 10^{-3}}\right)^2 \frac{\mu 41.67}{2.32 \times 10^6} = 12.47\mu.$$

Assuming $\Delta T = 10°\text{C}$.

Temperature rise, ΔT (°C)	10	27.01	25.17	25.85
Average lubricant temperature T_{av} (°C)	55	63.51	62.58	62.93
Average lubricant viscosity μ(Pa s)	0.017	0.0125	0.013	Converged
Sommerfield number, S	0.2120	0.1559	0.1621	
Coefficient of friction variable $(r/c)f$	4.8	3.8	3.9	
Flow variable Q/rcN_sL	4.08	4.22	4.21	
Side flow to total flow ratio Q_s/Q	0.55	0.622	0.62	

$$\Delta T \approx 25.9^{\circ}\text{C}.$$

$$Q = rcN_sL \times 4.21 = 12.5 \times 0.015 \times 41.67 \times 25 \times 4.21 = 822.3\,\text{mm}^3/\text{s}.$$

$$f = 3.9 \times (c/r) = 3.9 \times 0.015/12.5 = 0.00468.$$

$$T = fWr = 0.00468 \times 1450 \times 0.0125 = 0.08483\,\text{Nm}.$$

$$\text{Heat generated} = 2\pi N_s \times \text{Torque} = 22.21\,\text{W}.$$

With $S = 0.1621$ and $L/D = 1$ the design selected is inside the optimum operating zone for minimum friction and optimum load capacity indicated in Fig. 5.18.

Therefore this clearance ($c = 0.015$ mm) should be selected.

Example 5.6. A full film hydrodynamic journal bearing is required for a shaft supporting a radial load of 7500 N whilst rotating at 800 rpm. The nominal diameter of the shaft is 80 mm. Select a suitable lubricant and appropriate values for the radial clearance and bearing length. Determine the overall temperature rise in the bearing and the flow rate of lubricant required. Establish that the selected design lies within the optimum zone for minimum friction and maximum load on the minimum film thickness ratio chart. The lubricant can be supplied at an inlet temperature of 50°C using an available cooling system.

Solution

From Fig. 5.14 for $N = 800$ rpm and $D = 80$ mm, $2c = 80$ μm.

Hence $c = 40$ μm.

$$13.4 > P > 0.34\ (\text{MPa})$$

$$P = \frac{W}{LD} = \frac{7500}{0.08L}$$

Try $L/D = 0.5$

$$P = \frac{W}{LD} = \frac{7500}{0.08 \times 0.04} = 2.34\,\text{MPa}$$

Try SAE 20. As $N_s \Downarrow$ SAE number $\Uparrow$

The mean radius is given by

$$r_m = \frac{a+b}{2} = \frac{40+40.04}{2} = 40.02\ \text{mm}$$

$$b - a = 0.04\,\text{mm}$$

An estimate for the kinematic viscosity can be found, taking the average temperature of the oil as 65°C and assuming a density of 900 kg/m^3. At 55°C, the viscosity for SAE 20 is approximately 0.017 Pa s

$$\nu = \mu/\rho = 0.017/900 = 1.89 \times 10^{-5}\ \text{m}^2/\text{s}$$

So the Taylor number using this approximate value for the kinematic viscosity is given by

$$Ta_m = \frac{\Omega r_m^{0.5}(b-a)^{1.5}}{\nu} = 0.22$$

As this is much less than the critical Taylor number, 41.19, laminar flow can be assumed.

$$S = \left(\frac{r}{c}\right)^2 \frac{\mu N_s}{P} = \left(\frac{40 \times 10^{-3}}{0.04 \times 10^{-3}}\right)^2 \frac{\mu 13.33}{2.34 \times 10^6} = 5.69\mu$$

Temperature rise, ΔT (°C)	30	21.6	23.2
Average lubricant temperature T_{av} (°C)	65	60.8	61.66
Average lubricant viscosity μ(Pa s)	0.017	0.02	0.02
Sommerfield number, S	0.0967	0.114	Converged
Coefficient of friction variable $(r/c)f$	3.4	3.7	
Flow variable Q/rcN_sL	5.4	5.35	
Side flow to total flow ratio Q_s/Q	0.866	0.84	

With $S = 0.114$ and $L/D = 0.5$, the design is well within the optimum zone for minimum friction and maximum load.

$$Q = rcN_sL \times 5.35 = 40 \times 0.04 \times 13.33 \times 40 \times 5.35 = 4565\,\text{mm}^3/\text{s}.$$

Example 5.7. A design proposal for a full film hydrodynamic bearing comprises a bearing of nominal diameter 30 mm, length 15 mm using ISO VG 22 lubricant supplied at 50°C. The bearing load is 1280 N and the rotational speed is 6000 rpm. The radial clearance for the bearing is 0.022 mm. Use the viscosity chart given in Fig. 5.7 and the Raimondi and Boyd design charts given in Figs 5.18–5.21 to determine the lubricant flow rate, lubricant temperature rise and minimum film thickness. The initial guess for the temperature rise across the bearing can be taken as 30°C.

If in place of ISO VG 22 oil, SAE 10 oil was used in the bearing determine the resulting temperature rise of the lubricant.

Comment on the suitability of the proposed design and justify the selection of either ISO VG 22 or SAE 10 as a lubricant.

Solution

Initial data: $L = 15$ mm, $D = 30$ mm. $L/D = 0.5$.

$$N = 6000\,\text{rpm}, N_s = 100\,\text{rps},$$

$$c = 0.022\,\text{mm}.$$

ISO VG 22.

$$T_1 = 60^{\circ}\text{C}.$$

$$\Delta T = 30^{\circ}\text{C}$$

$$\text{W} = 1280\,\text{N}$$

$$T_{av} = 50 + \frac{30}{2} = 65^{\circ}\text{C}$$

The mean radius is given by

$$r_m = \frac{a+b}{2} = \frac{15 + 15.022}{2} = 15.01\,\text{mm}$$

$$b - a = 0.022\,\text{mm}$$

An estimate for the kinematic viscosity can be found, taking the average temperature of the oil as 65°C and assuming a density of 900 kg/m^3. At 55°C, the viscosity for ISO VG 22 is approximately 0.0074 Pa s

$$\nu = \mu/\rho = 0.0074/900 = 8.22 \times 10^{-6}\,\text{m}^2/\text{s}$$

So the Taylor number using this approximate value for the kinematic viscosity is given by

$$Ta_m = \frac{\Omega r_m^{0.5}(b-a)^{1.5}}{\nu} = 0.97$$

As this is much less than the critical Taylor number, 41.19, laminar flow can be assumed.

$$P = \frac{W}{LD} = \frac{1280}{0.015 \times 0.03} = 2.844 \times 10^6 \text{ Pa}$$

Temperature rise, ΔT (°C)	30	30.8
Average lubricant temperature T_{av} (°C)	65	65.4
Average lubricant viscosity μ(Pa s)	0.0074	Converged
Sommerfield number, S	0.12	
Coefficient of friction variable $(r/c)f$	4	
Flow variable Q/rcN_sL	5.3	
Side flow to total flow ratio Q_s/Q	0.842	

$$\Delta T = 31^\circ\text{C}$$

$$Q = rcN_sL \times 5.3 = 15 \times 0.022 \times 100 \times 15 \times 5.3 = 2624 \text{mm}^3/\text{s}.$$

$$h_o/c = 0.24$$

$$h_o = 0.24 \times 0.022 \text{mm} = 5.28 \times 10^{-3} \text{mm}$$

$$h_o = 5.3\,\mu m$$

From the graph for h_o/c versus Sommerfield number, the design is in the optimum zone for maximum load and minimum friction.

For SAE 10:

Temperature rise, ΔT (°C)	30	43.2	36.9	38.6
Average lubricant temperature T_{av} (°C)	65	71.6	68.5	69.3
Average lubricant viscosity μ(Pa s)	0.0116	0.0092	0.0102	0.01
Sommerfield number, S	0.19	0.15	0.167	Converged
Coefficient of friction variable $(r/c)f$	5.6	4.8	5	
Flow variable Q/rcN_sL	5.1	5.2	5.15	
Side flow to total flow ratio Q_s/Q	0.8	0.82	0.812	

$$\Delta T \approx 38.6^\circ\text{C}$$

With $S = 0.167$ and $L/D = 0.5$, the design also falls within the optimum zone for maximum load and minimum friction.

Use ISO VG 22 as the lubricant temperature rise is lower unless other components or considerations dictate the use of SAE 10.

5.4.3 Alternative method for the design of full film hydrodynamic bearings

An alternative method to using the design charts of Raimondi and Boyd is to use an approximate method developed by Reason and Narang (1982). They defined an approximation for the pressure variation in a journal bearing given by:

$$p = \frac{\frac{3\mu V}{rc^2}\left[\frac{L^2}{4} - z^2\right]\left\{\frac{\varepsilon \sin\theta}{(1+\varepsilon\cos\theta)^3}\right\}}{1 + \frac{2+\varepsilon^2}{2r^2}\left[\frac{L^2}{4} - z^2\right]\left\{\frac{1}{(1+\varepsilon\cos\theta)(2+\varepsilon\cos\theta)}\right\}} \quad (4.116)$$

where:

μ = viscosity (Pa s),
V = surface velocity (m/s),
r = journal radius (m),
c = radial clearance (m),
L = bearing length (m),
z = axial coordinate (m),
ε = eccentricity variable,
θ = circumferential coordinate.

This equation was integrated using an approximate method and the solution given in terms of two integrals I_s and I_c, the sine and cosine load integrals respectively. Both I_s and I_c are functions of the L/D ratio and the eccentricity ratio ε. The values of these integrals are given in Table 5.6 along with the corresponding Sommerfield number calculated using Eq. (5.117).

$$S = \frac{1}{6\pi\sqrt{I_c^2 + I_s^2}} \quad (5.117)$$

Knowing values for the integrals I_s and I_c enables the bearing parameters to be calculated as shown below.

- Given: W, N_s, L, D, c, the type of lubricant and T_1.
- Guess a value for the temperature rise of the lubricant ΔT and calculate T_{av}. Determine the viscosity at T_{av} from Fig. 5.7.
- Calculate the bearing loading pressure $P = \frac{W}{LD}$.
- Calculate the Sommerfield number $S = \left(\frac{r}{c}\right)^2 \frac{\mu N_s}{P}$.
- Knowing the length to diameter ratio and the Sommerfield number, search the correct L/D column in Table 5.6 for a matching Sommerfield number. A matching Sommerfield number will give the eccentricity ratio reading across the table to the left. Interpolation may be necessary for values of S between the values listed in the table. Once the correct eccentricity has been found, determine the values for the sine and cosine integrals I_s and I_c to correspond with the value of the Sommerfield number.

Table 5.6 Values of the sine and cosine integrals, I_s and I_c, and the Sommerfield number S, as functions of the L/D ratio and the eccentricity ratio ε(Reason and Narang (1982)

ε	$L/D = 0.25$	$L/D = 0.5$	$L/D = 0.75$	$L/D = 1.0$	$L/D = 1.5$	$L/D = 2$	$L/D = \infty$
0.10	$I_s = 0.0032$	0.0120	0.0244	0.0380	0.0636	0.0839	0.1570
	$I_c = -0.0004$	−0.0014	−0.0028	−0.0041	−0.0063	−0.0076	−0.0100
	$S = 16.4506$	4.3912	2.1601	1.3880	0.8301	0.6297	0.3372
0.20	$I_s = 0.0067$	0.0251	0.0505	0.0783	0.1300	0.1705	0.3143
	$I_c = -0.0017$	−0.0062	−0.0118	−0.0174	−0.0259	−0.0312	−0.0408
	$S = 7.6750$	2.0519	1.0230	0.6614	0.4002	0.3061	0.1674
0.30	$I_s = 0.0109$	0.0404	0.0804	0.1236	0.2023	0.2628	0.4727
	$I_c = -0.0043$	−0.0153	−0.0289	−0.0419	−0.0615	−0.0733	−0.0946
	$S = 4.5276$	1.2280	0.6209	0.4065	0.2509	0.1944	0.1100
0.40	$I_s = 0.0164$	0.0597	0.1172	0.1776	0.2847	0.3649	0.6347
	$I_c = -0.0089$	−0.0312	−0.0579	−0.0825	−0.1183	−0.1391	−0.1763
	$S = 2.8432$	0.7876	0.4058	0.2709	0.1721	0.1359	0.0850
0.50	$I_s = 0.0241$	0.0862	0.1656	0.2462	0.3835	0.4831	0.8061
	$I_c = -0.0174$	−0.0591	−0.1065	−0.1484	−0.2065	−0.2391	−0.2962
	$S = 1.7848$	0.5076	0.2694	0.1845	0.1218	0.0984	0.0618
0.60	$I_s = 0.0363$	0.1259	0.2345	0.3396	0.5102	0.6291	0.9983
	$I_c = -0.0338$	−0.1105	−0.1917	−0.2590	−0.3474	−0.3949	−0.4766
	$S = 1.0696$	0.3167	0.1752	0.1242	0.0859	0.0714	0.0480
0.70	$I_s = 0.0582$	0.1927	0.3430	0.4793	0.6878	0.8266	1.2366
	$I_c = -0.0703$	−0.2161	−0.3549	−0.4612	−0.5916	−0.6586	−0.7717
	$S = 0.5813$	0.1832	0.1075	0.0798	0.0585	0.0502	0.0364
0.80	$I_s = 0.1071$	0.3264	0.5425	0.7220	0.9771	1.1380	1.5866
	$I_c = -0.1732$	−0.4797	−0.7283	−0.8997	−1.0941	−1.1891	−1.3467
	$S = 0.2605$	0.0914	0.0584	0.0460	0.0362	0.0322	0.0255

0.90	$I_s = 0.2761$	0.7079	1.0499	1.3002	1.6235	1.8137	2.3083
	$I_c = -0.6644$	−1.4990	−2.0172	−2.3269	−2.6461	−2.7932	−3.0339
	$S = 0.0737$	0.0320	0.0233	0.0199	0.0171	0.0159	0.0139
0.95	$I_s = 0.6429$	1.3712	1.8467	2.1632	2.5455	2.7600	3.2913
	$I_c = -2.1625$	−3.9787	−4.8773	−5.3621	−5.8315	−6.0396	−6.3776
	$S = 0.0235$	0.0126	0.0102	0.0092	0.0083	0.0080	0.0074
0.99	$I_s = 3.3140$	4.9224	5.6905	6.1373	6.6295	6.8881	8.7210
	$I_c = -22.0703$	−28.5960	−30.8608	−31.9219	−32.8642	−33.2602	−33.5520
	$S = 0.0024$	0.0018	0.0017	0.0016	0.0016	0.0016	0.0015

Recall that for a linear function, $y = f(x)$, the value for y between the bounding points (x_1, y_1) and (x_2, y_2) for a given value of x can be found by linear interpolation:

$$y = \left(\frac{x - x_1}{x_2 - x_1}\right)(y_2 - y_1) + y_1 \tag{5.118}$$

- Calculate the minimum film thickness $h_o = c(1 - \varepsilon)$.
- Calculate the attitude angle ϕ, (degree)

$$\phi = \tan^{-1}\left(-\frac{I_s}{I_c}\right) \tag{5.119}$$

- Calculate the flow variable $Q/(rcN_sL)$, where Q is the lubricant supply.

$$\frac{Q}{rcN_sL} = \pi(1+\varepsilon) - \pi\varepsilon\eta_1\left[1 - \eta_1\left(\frac{D}{L}\right)^2 \frac{1}{\sqrt{\left(\frac{D}{L}\right)^2 \frac{\eta_1}{2} + \frac{1}{4}}}\tanh^{-1}\left[\frac{1}{2\sqrt{\left(\frac{D}{L}\right)^2 \frac{\eta_1}{2} + \frac{1}{4}}}\right]\right] \tag{5.120}$$

where

$$\eta_1 = \frac{(1+\varepsilon)(2+\varepsilon)}{(2+\varepsilon^2)} \tag{5.121}$$

- Calculate the flow through the minimum film thickness Q_π.

$$\frac{Q_\pi}{rcN_sL} = \pi(1-\varepsilon) + \pi\varepsilon\eta_2\left[1 - \eta_2\left(\frac{D}{L}\right)^2 \frac{1}{\sqrt{\left(\frac{D}{L}\right)^2 \frac{\eta_2}{2} + \frac{1}{4}}}\tanh^{-1}\left[\frac{1}{2\sqrt{\left(\frac{D}{L}\right)^2 \frac{\eta_2}{2} + \frac{1}{4}}}\right]\right] \tag{5.122}$$

where

$$\eta_2 = \frac{(1-\varepsilon)(2-\varepsilon)}{(2+\varepsilon^2)} \tag{5.123}$$

Note that

$$\tanh^{-1}x = \frac{1}{2}\ \ln\left(\frac{1+x}{1-x}\right) \tag{5.124}$$

- The flow ratio can be calculated using Eq. (5.125).

$$\frac{Q_s}{Q} = 1 - \frac{Q_\pi}{Q} \tag{5.125}$$

- Calculate the friction variable $(r/c)f$

$$f\left(\frac{r}{c}\right) = 6\pi S\left\{\frac{\pi}{3\sqrt{(1-\varepsilon^2)}} + \frac{\varepsilon}{2}I_s\right\} \tag{5.126}$$

- Calculate the temperature rise:

$$\Delta T = \frac{8.30\times10^{-6}P}{1-\frac{1}{2}(Q_s/Q)} \times \frac{(r/c)f}{Q/rcN_sL}$$

Although these equations may appear complex the advantage of the method developed by Reason and Narang is that it is suitable for conventional programming or utilisation within a spreadsheet.

Example 5.8. A full journal bearing has a nominal diameter of 20.0 mm and a bearing length of 20.0 mm (Fig. 5.29). The bearing supports a load of 1000 N, and the journal design speed is 6000 rpm. The radial clearance has been specified as 0.02 mm. An SAE 20 oil has been chosen to be compatible with requirements elsewhere in the machine and the lubricant supply temperature is 60°C. Using the approximate method of Reason and Narang find the temperature rise of the lubricant, the total lubricant flow rate, the minimum film thickness, the torque required to overcome friction and the heat generated in the bearing.

Solution

The primary design data are $D = 20\,\text{mm}$, $L = 20\,\text{mm}$, $W = 1000\,\text{N}$, $N = 6000\,\text{rpm}$, $c = 0.02\,\text{mm}$, SAE 20, $T_1 = 60°\text{C}$.

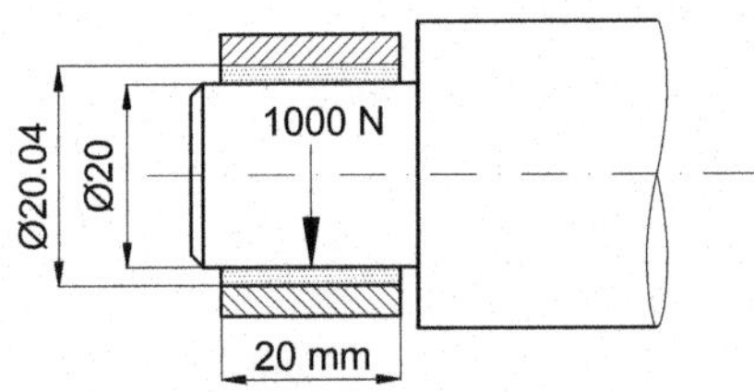

Fig. 5.29 Bearing design example.

Guess $\Delta T = 15°\text{C}$. Hence: $T_{av} = T_1 + \frac{\Delta T}{2} = 60 + \frac{15}{2} = 67.5^\circ\text{C}$.
For SAE 20, at 67.5°C, from Fig. 5.7, $\mu = 0.016\,\text{Pa s}$.
The mean radius is given by

$$r_m = \frac{a+b}{2} = \frac{20+20.02}{2} = 20.01\,\text{mm}$$

$$b - a = 0.02\,\text{mm}.$$

Assuming a density of $900\,\text{kg/m}^3$.

$$\nu = \mu/\rho = 0.016/900 = 1.78 \times 10^{-5}\,\text{m}^2/\text{s}$$

So the Taylor number using this approximate value for the kinematic viscosity is given by

$$Ta_m = \frac{\Omega r_m^{0.5}(b-a)^{1.5}}{\nu} = 0.45$$

As this is much less than the critical Taylor number, 41.19, laminar flow can be assumed.

$$P = \frac{W}{LD} = \frac{1000}{0.02 \times 0.02} = 2.5\,\text{MPa}.$$

$$S = \left(\frac{r}{c}\right)^2 \frac{\mu N_s}{P} = \left(\frac{0.01}{0.00002}\right)^2 \frac{0.016 \times 100}{2.5 \times 10^6} = 0.16.$$

From Table 5.6 for $L/D = 1$, the nearest bounding values of the Sommerfield number to $S = 0.16$, are for $\varepsilon = 0.5$ ($S = 0.1845$) and $\varepsilon = 0.6$ ($S = 0.1242$).
Interpolating to find the values of ε, I_s and I_c which correspond to $S = 0.16$:

$$\varepsilon = \frac{0.16 - 0.1845}{0.1242 - 0.1845}(0.6 - 0.5) + 0.5 = 0.5406,$$

$$I_s = \frac{0.16 - 0.1845}{0.1242 - 0.1845}(0.3396 - 0.2462) + 0.2462 = 0.2841,$$

$$I_c = \frac{0.16 - 0.1845}{0.1242 - 0.1845}(-0.2590 - -0.1484) + -0.1484 = -0.1933.$$

From Eq. (5.120) with $\eta_1 = 1.708$,

$$\frac{Q}{rcN_s L} = 4.84 - 2.901\left(1 - 1.708 \times 0.9517 \tanh^{-1}(0.4759)\right) = 4.379.$$

From Eq. (5.122) with $\eta_2 = 0.2925$,

$$\frac{Q_\pi}{rcN_sL} = 1.443 + 0.4968\left(1 - 0.2925 \times 1.589\tanh^{-1}(0.7943)\right) = 1.690.$$

From Eq. (5.125), $Q_s/Q = 0.6141$.

From Eq. (5.126), $f\left(\frac{r}{c}\right) = 3.016(1.245 + 0.07679) = 3.987$.

$$\Delta T = \left(\frac{8.3 \times 10^{-6}P}{1 - 0.5(Q_s/Q)}\right)\frac{(r/c)f}{Q/rcN_sL} = 27.26^\circ\text{C}.$$

Repeating the procedure with this value for ΔT gives:

$T_{av} = 73.6°\text{C}$, hence $\mu = 0.013$ Pa s,
$S = 0.13$, hence $\varepsilon = 0.5904$,
$Q/(rcN_sL) = 4.492$,
$Q_\pi/(rcN_sL) = 1.531$,
$(r/c)f = 3.418$.

Using these values gives $\Delta T = 23.56°\text{C}$.

Repeating the procedure with this value for ΔT gives:

$T_{av} = 71.78°\text{C}$, hence $\mu = 0.014$ Pa s,
$S = 0.14$, hence $\varepsilon = 0.5738$,
$Q/(rcN_sL) = 4.454$,
$Q_\pi/(rcN_sL) = 1.585$,
$(r/c)f = 3.613$.

Using these values gives $\Delta T = 24.83°\text{C}$. Hence $T_{av} = 72.4°\text{C}$. This is close enough to the previous value, indicating that the correct value for T_{av} has been determined.

The values for the total lubricant supply, minimum film thickness, coefficient of friction, torque and the heat generated in the bearing can now be determined:

$Q = 1782\,\text{mm}^3/\text{s}$,
$h_o = c(1 - \varepsilon) = 0.009\,\text{mm}$,
$f = 0.007226$,
Torque $= 0.072$ N m,
Power $= 45.4$ W.

Example 5.9. A full journal bearing has a nominal diameter of 80.0 mm, bearing length 60.0 mm. The bearing supports a load of 7500 N, and the journal design speed is 800 rpm. The radial clearance has been specified as 0.05 mm. An SAE 30 oil has been chosen and the lubricant supply temperature is 50°C. Using the approximate method of Reason and Narang for determining the design and performance of full film hydrodynamic bearings, find the temperature rise of the lubricant, the lubricant flow rate, the torque required to overcome friction, the heat generated in the bearing and the minimum film thickness.

Solution

$L/D = 0.75$, $P = 1.562\,\text{MN}\,\text{m}^{-2}$.

The mean radius is given by

$$r_m = \frac{a+b}{2} = \frac{40+40.05}{2} = 40.03\,\text{mm}$$

$$b - a = 0.05\,\text{mm}$$

An estimate for the kinematic viscosity can be found, taking the average temperature of the oil as 60°C and assuming a density of 900 kg/m^3. At 55°C, the viscosity for SAE 30 is approximately 0.032 Pa s

$$\nu = \mu/\rho = 0.032/900 = 3.56 \times 10^{-5}\,\text{m}^2/\text{s}$$

So the Taylor number using this approximate value for the kinematic viscosity is given by

$$Ta_m = \frac{\Omega r_m^{0.5}(b-a)^{1.5}}{\nu} = 0.17$$

As this is much less than the critical Taylor number, 41.19, laminar flow can be assumed.

Temperature rise, ΔT (°C)	20	19.16
Average lubricant temperature T_{av} (°C)	60	59.58
Average lubricant viscosity μ(Pa s)	0.032	Converged
Sommerfield number, S	0.1748	
ε	0.6006	
I_s	0.2352	
I_c	−0.1928	
Flow variable Q/rcN_sL	4.715	
Q_π/rcN_sL	1.440	
Side flow to total flow ratio Q_s/Q	0.6946	
Coefficient of friction variable $(r/c)f$	4.547	

$$\Delta T \approx 19.2^\circ\text{C}.$$

$$h_o = 0.05 \times 10^{-3}(1 - 0.6006) = 19.97\,\mu m \approx 20.0\,\mu m.$$

$$Q = 4.715 \times 40 \times 0.05 \times (800/60) \times 60 = 7544\,\text{mm}^3/\text{s}.$$

$$\text{Torque} = fWr = 4.547(0.05/40) \times 7500 \times 0.04 = 1.705\,\text{Nm}.$$

Heat generated $= 2\pi N_s \times \text{Torque} = 142.8\,\text{W}$.

5.5 Conclusions

Bearings are used to support a load whilst allowing relative motion between two elements of a machine. Typical applications are shafts where the shaft rotates relative to a housing but bearings can also be used to facilitate linear relative motion for, say, machine slides. This chapter has introduced the wide range of bearing types and their selection and the design of boundary lubricated and full-film hydrodynamic journal bearings. Boundary lubricated bearings represent an important class of bearings where the combination of low speed and load make it possible for surface-to-surface contact to occur without significant wear and this type of bearing is widely used in plant machinery and product design. In full film hydrodynamic bearings, rotation results in high pressures in the lubricant which act to separate the journal from the bearing enabling prolonged high speed operation.

References

Brandlein, J., Eschmann, P., Hasbargen, L., Weigand, K., 1999. Ball and Roller Bearings. Theory, Design and Application. John Wiley & Sons, Chichester, New York.

Childs, P.R.N., 2011. Rotating Flow. Elsevier, Amsterdam, Boston.

Constantinescu, V.N., 1962. Analysis of bearings operating in turbulent regime. Trans. ASME J. Basic Eng. 84, 139–151.

Hamrock, B.J., 1994. Fundamentals of Fluid Film Lubrication. McGraw Hill, New York.

Harris, T.A., 2001. Rolling Bearing Analysis, fourth ed. John Wiley & Sons, New York.

Harrison, W.J., 1913. The hydrodynamical theory of lubrication with special reference to air as a lubricant. Trans. Cambridge Philos. Soc. xxii (1912–1925), 6–54.

Hashimoto, H., Matsumoto, K., 2001. Improvement of operating characteristics of high-speed hydrodynamic journal bearings by optimum design: Part I Formulation of methodology and its application to elliptical bearing design. J. Tribol. 123, 305–312.

Hirania, H., Suh, N.P., 2005. Journal bearing design using multiobjective genetic algorithm and axiomatic design approaches. Tribol. Int. 38, 481–491.

Institution of Mechanical Engineers, 1994. Tribological Design Data. Part 1: Bearings. I. Mech. E.

Khonsari, M.M., Booser, E.R., 2017. Applied tribology: Bearing Design and Lubrication, third ed. Wiley, Hoboken, NJ.

Koring, R., 2012. Changes in Plain Bearing Technology. SAE International.

Lancaster, J.K., 1971. Estimation of the limiting PV relationships for thermoplastic bearing materials. Tribology 4, 82–86.

Raimondi, A.A., Boyd, J., 1958a. A solution for the finite journal bearing and its application to analysis and design: I. ASLE Trans. 1, 159–174.

Raimondi, A.A., Boyd, J., 1958b. A solution for the finite journal bearing and its application to analysis and design: II. ASLE Trans. 1, 175–193.

Raimondi, A.A., Boyd, J., 1958c. A solution for the finite journal bearing and its application to analysis and design: III. ASLE Trans. 1, 194–209.

Reason, B.R., Narang, I.P., 1982. Rapid design and performance evaluation of steady state journal bearings—a technique amenable to programmable hand calculators. ASLE Trans. 25 (4), 429–444.

Reynolds, O., 1886. On the theory of lubrication and its application to Mr Beauchamp Tower's experiments including an experimental determination of the viscosity of olive oil. Philos. Trans. Royal Soc. 177, 157–234.

Rowe, W.B., 2012. Hydrostatic, Aerostatic and Hybrid Bearing Design. Butterworth Heinemann, Oxford.

Yamaguchi, Y., Kashiwagi, K., 1982. The limiting pressure-velocity(PV) of plastics under unlubricated sliding. Polym. Eng. Sci. 22, 248–253.

Websites

At the time of going to press the world-wide-web contained useful information relating to this chapter at the following sites.

www.abc-bearing.com.
www.arb-bearings.com.
www.bocabearings.com.
www.buntingbearings.com/data.html.
www.championbearings.com.
www.cooperbearings.com.
www.fag.de.
www.ggbearings.com.
www.gtweed.com/docs/ebrochures/PlasticBearingDesignGuide-0901.pdf.
www.hb-bearings.com.
www.ina.com.
www.jskbearings.com.
www.mrcbearings.com.
www.nationalbearings.com.
www.newindo.com.
www.nsk-rhp.com.
www.ntn.ca.
qbcbearings.com/TechInfo/PDF/TechT32_34.pdf.
www.railko.co.uk.
www.rbcbearings.com.
www.revolve.com.
www.rosebearings.com.
www.skf.com.
www.snfa-bearings.co.uk.
www.spn.com.tw.
www.sun-sunshine.com.
www.thordonbearings.com.
www.timken.com.
www.torrington.com.
www.toyobearings.com.
hwww.tribology-abc.com/calculators/e3_11.htm.

Standards

American National Standards Institute. (1990). ANSI/AFBMA Standard 9-1990.*Load ratings and fatigue life for ball bearings*. ANSI.

American National Standards Institute. (1990). ANSI/AFBMA Standard 11-1990.*Load ratings and fatigue life for roller bearings*. ANSI.

American National Standards Institute. (1995). ANSI/ABMA Standard 7-1995.*Shaft and housing fits for metric radial ball and roller bearings (except tapered roller bearings) conforming to basic boundary plans.* ANSI.

British Standards Institution. (1979). BS 3134-1:1979.*Metric tapered roller bearings. Specification for dimensions and tolerances of single row bearings.*

British Standards Institution. (1982). BS 292: Part 1: 1982. *Rolling bearings: ball bearings, cylindrical and spherical roller bearings. Specification for dimensions of ball bearings, cylindrical and spherical roller bearings (metric series).*

British Standards Institution. (1984). BS 6560:1984. *Glossary of terms for rolling bearings.*

British Standards Institution. (1991). BS5512: 1991. *Method of calculating dynamic load rating and rating life of rolling bearings.*

British Standards Institution. (1992). BS 3134-4:1992, ISO 10317:1992. *Metric tapered roller bearings. Specification for designation system.*

British Standards Institution. (1992). BS 4480: Part 1: 1992. *Plain bearings metric series. Sintered bushes. Dimensions and tolerances.*

British Standards Institution. (1995). BS ISO 12128:1995. *Plain bearings. Lubrication holes, grooves and pockets. Dimensions, types, designation and their application to bearing bushes.*

British Standards Institution. (1998). BS ISO 15:1998. *Rolling bearings. Radial bearings. Boundary dimensions, general plan.*

British Standards Institution. (1999). BS ISO 113:1999. *Rolling bearings. Plummer block housings. Boundary dimensions.*

British Standards Institution. (1999). BS ISO 3547-1:1999.*Plain bearings. Wrapped bushes. Dimensions.*

British Standards Institution. (1999). BS ISO 3547-3:1999.*Plain bearings. Wrapped bushes. Lubrication holes, lubrication grooves and lubrication indentations.*

British Standards Institution. (1999). BS ISO/TS 16799:1999. *Rolling bearings. Dynamic load ratings and rating life. Discontinuities in the calculating of basic dynamic load ratings.*

British Standards Institution. (2001). BS ISO 12128:2001. *Plain bearings. Lubrication holes, grooves and pockets. Dimensions, types, designation and their application to bearing bushes.*

British Standards Institution. (2001). BS ISO 2982-2:2001.*Rolling bearings. Accessories. Locknuts and locking devices. Dimensions.*

British Standards Institution. (2001). BS ISO 3290:2001. *Rolling bearings. Balls. Dimensions and tolerances.*

British Standards Institution. (2002). BS ISO 104:2002. *Rolling bearings. Thrust bearings. Boundary dimensions, general plan.*

British Standards Institution. (2002). BS ISO 492:2002. *Rolling bearings. Radial bearings. Tolerances.*

International Organization for Standardization. (1992). ISO 3448 *Industrial liquid lubricants—ISO viscosity classification.*

International Organization for Standardization. (2001). ISO 14728-1.*Rolling bearings. Linear motion rolling bearings. Part 1: Dynamic load ratings and rating life.*

International Organization for Standardization. (2001). ISO 14728-2.*Rolling bearings. Linear motion rolling bearings. Part 2: Static load ratings.*

International Organization for Standardization. (2001). ISO 15243. *Rolling bearings. Damages and failures. Terms, characteristics and causes.*

Society of Automotive Engineers. (1989). SAE J300. *SAE standard engine oil viscosity classification.*

Further reading

Harnoy, A., 2002. Bearing Design in Machinery. Marcel Dekker, New York.

Hindhede, U., Zimmerman, J.R., Hopkins, R.B., Erisman, R.J., Hull, W.C., Lang, J.D., 1983. Machine Design Fundamentals. A Practical Approach. Wiley, New York.

Neale, M.J. (Ed.), 1995. The Tribology Handbook. Butterworth Heinemann, Oxford.

Welsh, R.J., 1983. Plain Bearing Design Handbook. Butterworth, London, Boston.

Rolling element bearings

6

Chapter Outline

Nomenclature

Generally, preferred SI units have been stated.

a_1	life adjustment factor for reliability
a_{SKF}	SKF life modification factor
B	width (m)
C	basic dynamic load rating (N)
C_o	basic static load rating (N)
d	bore diameter (m)
d_m	mean diameter (m)
D	outer diameter (m)
ISO	International organisation for standardisation
k	constant
L	life (number of revolutions)
L_1	life at load P_1 (number of revolutions)
L_{10}	rated life at the rated load (number of revolutions)
L_2	life at load P_2 (number of revolutions)
L_{cat}	catalogue rating (N)
L_d	design life (number of revolutions)
L_{nm}	rating life (at $100-n\%$ reliability) in millions of revolutions;
L_o	original length (m)
P	load (N)
P_1	load (N)
P_2	load (N)

Mechanical Design Engineering Handbook. https://doi.org/10.1016/B978-0-08-102367-9.00006-8

P_d	design load (N)
P_u	fatigue load limit (N)
R	applied radial load (N)
SAE	Society of Automotive Engineers
T	thrust load (N)
V	rotating factor
X	radial factor
Y	thrust factor
α	coefficient of expansion (K^{-1})
ΔL	change in length (m)
ΔT	temperature rise (°C)
η_c	contamination factor
κ	viscosity ratio
ν	operating kinematic viscosity of the lubricant (m^2/s)
ν_1	rated kinematic viscosity depending on the bearing mean diameter and rotational speed (m^2/s)

6.1 Introduction

The purpose of a bearing is to support a load, typically applied to a shaft, whilst allowing relative motion between two components of a machine. The selection of bearing type was introduced in Chapter 5 along with the design of boundary lubricated and hydrodynamically lubricated journal bearings. In this chapter rolling element bearings, sometimes also known as rolling contact or ball-bearings, are considered with a specific focus on the selection of stock items from original equipment manufacturers.

The term rolling element bearings, or rolling contact bearings encompasses the wide variety of bearings that use spherical balls or some type of roller between the stationary and moving elements as illustrated in Fig. 6.1. The most common type of bearing supports a rotating shaft resisting a combination of radial and axial (or thrust) loads. Some bearings are designed to carry only radial or only thrust loads. As can be seen in Fig. 6.1 the form of rolling element used varies widely from spherical balls, to cylinders, and conical sections. Owing to the very common use of spherical balls in rolling element bearings, the class of bearings is sometimes referred to as ball bearings, even though the rolling element for the specific bearing concerned may not actually be a spherical ball.

Selection of the type of bearing to be used for a given application can be aided by the comparison charts, an example of which is given in Table 5.1. Several bearing manufacturers produce excellent online catalogues (e.g. NSK, SKF, FAG, INA) including design guides. The reader is recommended to gain access to this information, which is readily available via internet sources.

The comparative ratings shown in Table 6.1 can be justified for say radial load carrying capacity; roller bearings are better than ball bearings because of their shape and the area over which the load is spread. Thrust load capacity varies dramatically with design. The grooves in the races of the deep groove ball bearing permit the transfer of

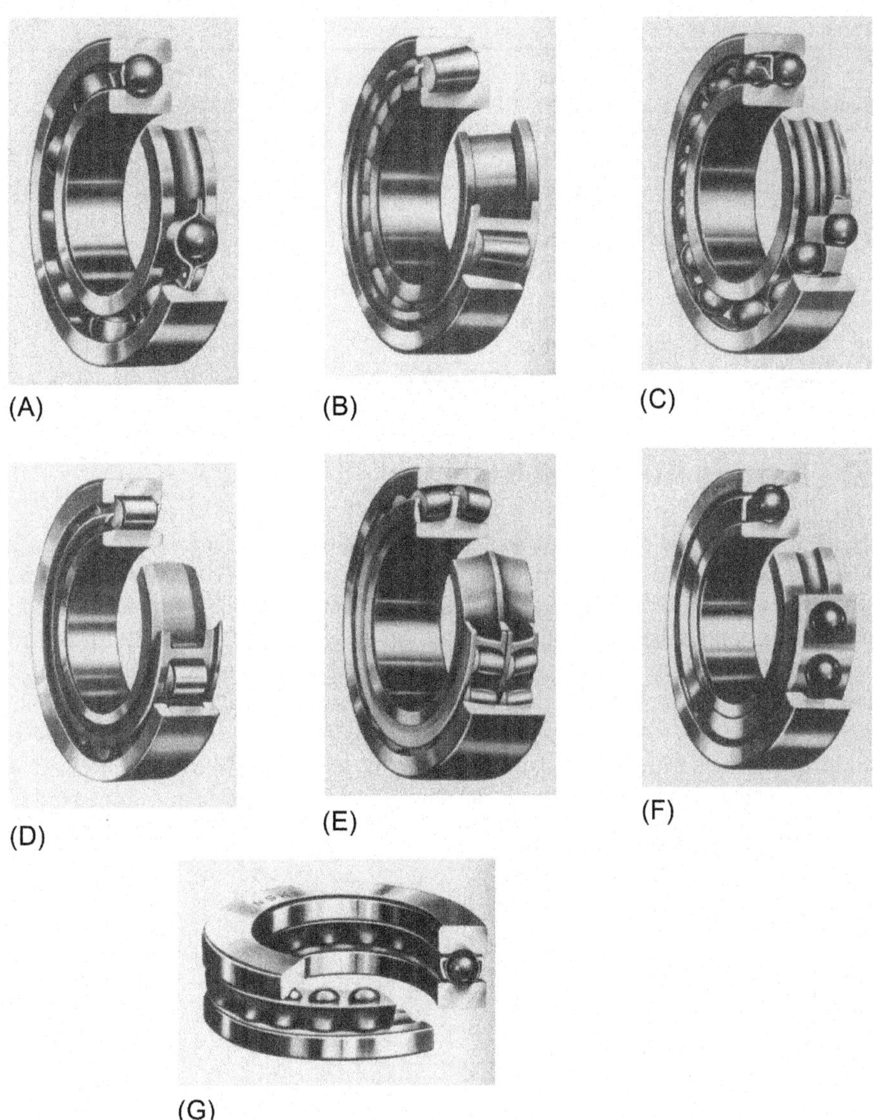

Fig. 6.1 A selection of the various types of rolling element bearings. (A) Deep groove ball bearing. (B) Taper roller bearing. (C) Self-aligning ball bearing. (D) Cylindrical roller bearing. (E) Spherical roller bearing. (F) Angular contact ball bearing. (G) Thrust ball bearing. Photographs courtesy of NSK RHP Bearings.

moderate thrust in combination with radial load. The angular contact bearing (Fig. 6.1F) is better than the single row ball bearing because the races are higher on one side providing a more favourable load path. Cylindrical and needle bearings should not generally be subjected to any thrust load.

Table 6.1 Merits of different rolling contact bearings

Bearing type	Radial load capacity	Axial or thrust load capacity	Misalignment capability
Single row	Good	Fair	Fair
Double row deep groove ball	Excellent	Good	Fair
Angular contact	Good	Excellent	Poor
Cylindrical roller	Excellent	Poor	Fair
Needle roller	Excellent	Poor	Poor
Spherical roller	Excellent	Fair/Good	Excellent
Tapered roller	Excellent	Excellent	Poor

6.2 Bearing life and selection

The theoretical region of contact between a spherical ball and a surface with a greater radius is a point. In the case of practical materials experiencing a force, there will be some deformation of the surfaces concerned and instead of point contact, the load on a rolling contact bearing will be exerted on a very small area, as illustrated in Fig. 6.2. The resulting contact stresses are very high and can be of the order of 2000 MPa. The arising stress of the load on the contact area will cause the materials to wear. Despite very strong steels (e.g. BS 970534 A99, AISI 52100), all bearings have a finite life and will eventually fail due to fatigue. Wear is a key issue in rolling element bearings and debris arising from wear can lead to rapid degradation of bearing performance and failure (El-Thalji and Jantunen, 2015; Sayles and Ioannides, 1988).

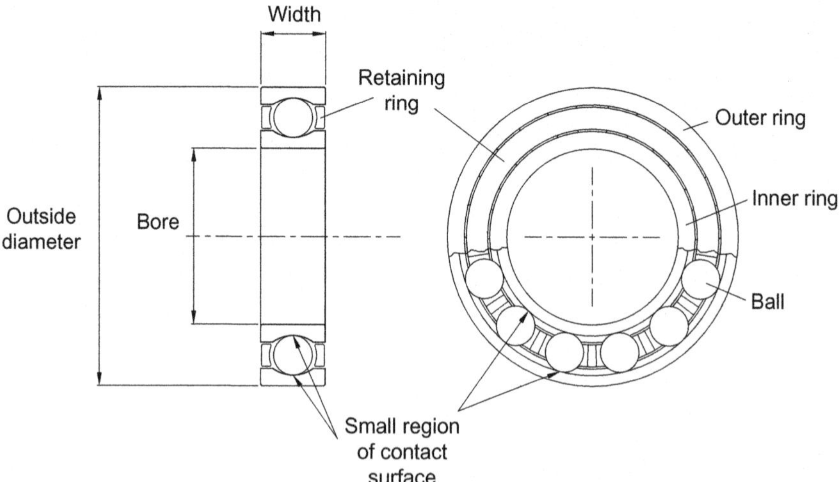

Fig. 6.2 Contact area for a ball bearing.

For two groups of apparently identical bearings tested under loads P_1 and P_2, the respective lives L_1 and L_2 are related by the life equation given in Eq. (6.1).

$$\frac{L_1}{L_2} = \left(\frac{P_2}{P_1}\right)^k \tag{6.1}$$

where:

L_1 = life at load P_1 (number of revolutions).
L_2 = life at load P_2 (number of revolutions).
P_1 = load (N).
P_2 = load (N).

and

$k = 3$ for ball bearings,
$k = 3.33$ for cylindrical roller bearings.

The dimensions for rolling element bearings have been standardised and can be purchased as stock items from specialist manufacturers and suppliers. The selection of a bearing from a manufacturer's catalogue involves consideration of the bearing load carrying capacity and the bearing geometry. For a given bearing the load carrying capacity is given in terms of the basic dynamic load rating and the basic static load rating. Various commonly used definitions for rolling element bearing life specification are outlined as follows.

The basic dynamic load rating, C, is the constant radial load which a rolling element bearing can endure for 1×10^6 revolutions without evidence of the development of fatigue in any of the bearing components.

The life of a ball bearing, L, is the number of revolutions (or hours at some constant speed), which the rolling element bearing runs before the development of fatigue in any of the bearing components.

Fatigue occurs over a large number of cycles of loading. For a bearing this would mean a large number of revolutions. Fatigue is a statistical phenomenon with considerable spread of the actual life of a group of bearings of a given design.

The rated life is the standard means of reporting the results of many tests of bearings. It represents the life that 90% of the bearings would achieve successfully at a rated load. The rated life is referred to as the L_{10} life at the rated load. The rated life, L_{10}, of a group of apparently identical bearings is defined as the number of revolutions (or hours at some constant speed) that 90% of the group of bearings will complete before the first evidence of fatigue develops.

If in Eq. (6.1), $P_2 = \text{C}$ and the corresponding life $L_2 = 1 \times 10^6$, then the life of a bearing L, with basic dynamic load rating C with a load P is given by:

$$L = \left(\frac{C}{P}\right)^k \text{ million revolutions} \tag{6.2}$$

where:

L = life (millions of revolutions).
C = basic dynamic load rating (N).
P = load (N).

When selecting a particular bearing from a manufacturer's catalogue, it is useful to know the required basic dynamic load rating C for a given load P and life L, which is given by:

$$C = P\left(\frac{L}{10^6}\right)^{(1/k)} \tag{6.3}$$

Example 6.1. A straight cylindrical roller bearing operates with a load of 7.5 kN. The required life is 8760 h at 1000 rpm. What load rating should be used for selection from the catalogue?

Solution

Using Eq. (6.3),

$$C = P_d\left(L_d/10^6\right)^{1/k} = 7500\left(\frac{8760 \times 1000 \times 60}{10^6}\right)^{1/3.33} = 49.2\text{ kN}$$

Example 6.2. A catalogue lists the basic dynamic load rating for a ball bearing to be 33,800 N for a rated life of 1 million revolutions. What would be the expected L_{10} life of the bearing if it were subjected to 15,000 N and determine the life in hours that this corresponds to if the speed of rotation is 2000 rpm. Comment on the value obtained and its suitability for a machine.

Solution

C_{cat} = 33,800 N, P_d = 15,000 N, $L_{cat} = 10^6$ (L_{10} life at load C), k = 3 (ball). Using Eq. (6.2), the basic rating life is given by:

$$L = 10^6\left(\frac{33,800}{15,000}\right)^3 = 11.44 \times 10^6 \text{ revolutions} (= L_{10} \text{ life at } 15,000\text{N}).$$

If the rotational speed is 2000 rpm, $L = 11.44 \times 10^6/(2000 \times 60) = 95$ h operation. This is not very long and illustrates the need to use a bearing with a high basic dynamic load rating.

An outline procedure for bearing selection is shown in Table 6.2.

Eqs (6.1)–(6.3) have been traditionally used to model the life of a bearing. However the life of practical rolling element bearings in service depends on a series of factors that include lubrication, contamination, misalignment, installation and environmental conditions as well as the quality of manufacture and design. ISO 281:1990/Amd 2:2000 provides a modified life equation to be used as a supplement to the basic rating

Table 6.2 Outline procedure for the selection of a rolling element bearing

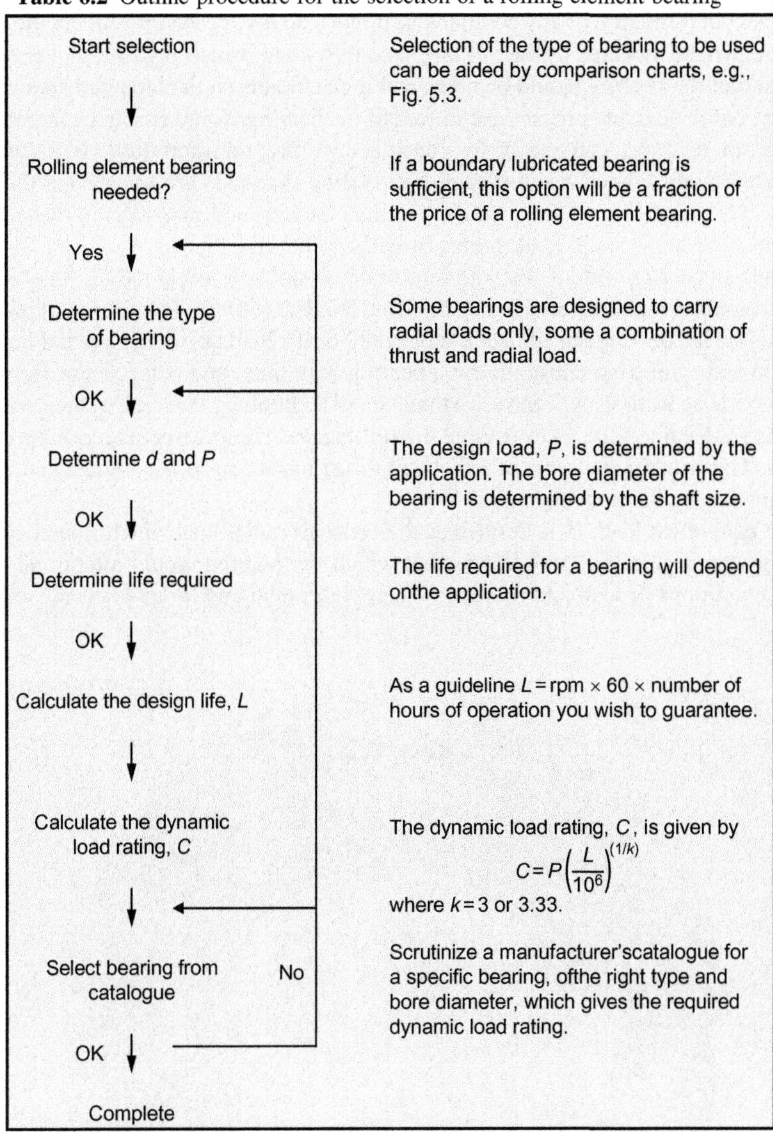

life. The modified life equation is presented in Section 6.2.1. The modified life equation can result in predicted lives that are an order of magnitude greater than the standard life equation (Eq. 6.1 or 6.3) depending of course on the various parameters for the application concerned.

Start selection
Selection of the type of bearing to be used can be aided by comparison charts, e.g., Fig. 5.3.
Rolling element bearing needed?
If a boundary lubricated bearing is sufficient, this option will be a fraction of the price of a rolling element bearing.
Yes
Determine the type of bearing
Some bearings are designed to carry radial loads only, some a combination of thrust and radial load.
OK
Determine d and P
The design load, P, is determined by the application. The bore diameter of the bearing is determined by the shaft size.
OK
Determine life required
The life required for a bearing will depend onthe application.
OK
Calculate the design life, L
As a guideline $L = \text{rpm} \times 60 \times$ number of hours of operation you wish to guarantee.
Calculate the dynamic load rating, C
The dynamic load rating, C, is given by
$C = P\left(\frac{L}{10^6}\right)^{(1/k)}$
where $k = 3$ or 3.33.
Select bearing from catalogue
No
Scrutinize a manufacturer'scatalogue for a specific bearing, ofthe right type and bore diameter, which gives the required dynamic load rating.
OK
Complete

The basic static load rating, C_o, is the load the bearing can withstand without any permanent deformation of any component. If this load is exceeded it is likely the bearing races will be indented by the rolling elements (called Brinelling). Subsequently the operation of the bearing would be noisy and impact loads on the indented area would produce rapid wear and progressive failure of the bearing would ensue. Unintentional damage to bearings can occur for machines during transportation. If a rotor is unsecured it can vibrate and although not rotating the loads are transferred through the small areas of contact between the bearing elements and raceways. If these loads are above the basic static load rating, Brinelling can take place.

Loads on bearings often vary with time and may not be entirely radial. An example is the thrust bearing illustrated in Fig. 6.3 from the Rolls-Royce Trent series of engines (Fig. 6.4). This bearing has to take a substantial thrust load as well as a radial load. In addition and somewhat unusually for a bearing both inner and outer raceways rotate. This is because Rolls-Royce plc use a three-spool technology for some of their engines (see the Jet Engine (2015)) to operate the turbine and compressor stages at optimum speeds. The combined effects of radial and thrust loads can be accommodated in the life equation by an equivalent load.

The equivalent load, P, is defined as the constant radial load which if applied to a bearing would give the same life as that which the bearing would attain under the actual conditions of load and rotation. When both radial and thrust loads are exerted

Fig. 6.3 Thrust bearing from the Rolls-Royce Trent series of engines.

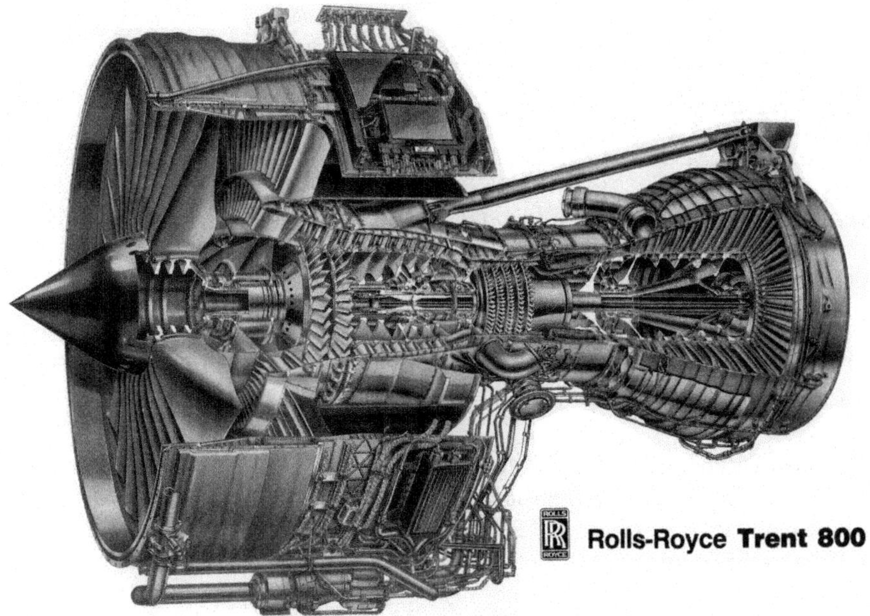

Fig. 6.4 The Trent 800 engine illustrating three spools, each rotating at a different speed. Figure courtesy of Rolls-Royce plc.

on a bearing the equivalent load is the constant radial load that would produce the same rated life for the bearing as the combined loading.

$$\text{Normally } P = VXR + YT \tag{6.4}$$

where:

P = equivalent load (N),
V = 1.2 if mounting rotates is recommended, V = 1.0 if shaft rotates,
X = radial factor (given in bearing catalogues, see Table 6.3 for example data),
R = applied radial load (N),
Y = thrust factor (given in bearing catalogues, see Table 6.3 for example data),
T = applied thrust load (N).

6.2.1 Modified life equation

The basic rating life of a rolling element bearing for a specific application can deviate significantly from the actual service life. Influencing factors include lubrication, contamination, misalignment, installation and environmental conditions. Careful attention to tolerancing and management of these factors in a rolling element bearing can result in significantly improved performance. ISO 281:1990/Amd 2:2000 gives a provision for a modification to the basic rating life accounting for lubrication,

Table 6.3 Values for the radial and thrust factors for determining the equivalent load for deep groove ball single bearings and bearing pairs arranged in tandem.

	Normal clearance			C3 clearance			C4 clearance		
F_a/C_0	e	X	Y	e	X	Y	e	X	Y
0.025	0.22	0.56	2	0.31	0.46	1.75	0.4	0.44	1.42
0.04	0.24	0.56	1.8	0.33	0.46	1.62	0.42	0.44	1.36
0.07	0.27	0.56	1.6	0.36	0.46	1.46	0.44	0.44	1.27
0.13	0.31	0.56	1.4	0.41	0.46	1.3	0.48	0.44	1.16
0.25	0.37	0.56	1.2	0.46	0.46	1.14	0.53	0.44	1.05
0.5	0.44	0.56	1	0.54	0.46	1	0.56	0.44	1

Data from SKF.

contamination conditions and the fatigue limit of the material. The standard also makes provision for a specific manufacturer to recommend how to calculate the life modification factor. In the case of SKF, for example, the factor a_{SKF} is used, and the equation for SKF rating life is given by

$$L_{nm} = a_1 a_{SKF} \left(\frac{C}{P}\right)^k \tag{6.5}$$

where:

L_{nm} = rating life (at $100 - n\%$ reliability) in millions of revolutions;
a_1 = the life adjustment factor for reliability;
a_{SKF} = SKF life modification factor.

Values of a_1 are given in Table 6.4. Values for the life modification factor can be found from tables provided by the specific manufacture in catalogues, on line or using an on line tool where the user inputs data for the fatigue load limit ratio, the lubrication condition and the contamination level. The approach requires calculation of the kinematic viscosity ratio (Eq. 6.6), of the actual operating kinematic viscosity of the lubricant to the rated kinematic viscosity depending on the mean diameter and the rotational speed.

Table 6.4 Modified life equation factors

Reliability, %	Failure probability, n%	SKF rating life, L_{nm}	a_1
90	10	L_{10m}	1
95	5	L_{5m}	0.62
96	4	L_{4m}	0.53
97	3	L_{3m}	0.44
98	2	L_{2m}	0.33
99	1	L_{1m}	0.21

$$\kappa = \frac{\nu}{\nu_1} \tag{6.6}$$

where:

κ = viscosity ratio;
ν = operating kinematic viscosity of the lubricant (m^2/s);
ν_1 = rated kinematic viscosity depending on the bearing mean diameter and rotational speed (m^2/s).

A guideline for the contamination factor is given in Table 6.5. Tables for a_{SKF} are provided as a function of the viscosity ratio, κ, and the product of the contamination factor and the fatigue load limit and load ratio, $\eta_c P_u / P$.

Tables 6.6–6.8 give an overview of the information typically available in bearing manufacturers' catalogues.

Example 6.3. A bearing is required to carry a radial load of 2.8 kN and provide axial location for a shaft of 30 mm diameter rotating at 1500 rpm. An L_{10} life of 10,000 h is required.

(i) Select and specify an appropriate bearing using the standard bearing life equation (Eq. 6.3).
(ii) Determine the rating life using the modified life equation assuming that the rated oil kinematic viscosity is 14 mm^2/s and the actual kinematic viscosity at operating conditions is 40 mm^2/s, normal cleanliness with a contamination factor of 0.6 and an SKF life modification factor of 14. The desired reliability is 90%.

Solution

(i) Axial shaft location is required, so a deep groove ball bearing, which provides axial location capability in both directions, would be suitable.

The total number of revolutions in life is.

$$10{,}000 \times 1500 \times 60 = 900\,\text{million}.$$

So $L = 900$.
The load is purely radial so $P = 2800$ N.

Table 6.5 Typical values for the contamination factor

Condition	$d_m < 100$ mm	$d_m \geq 100$ mm
Extreme cleanliness	1	1
High cleanliness	0.8–0.6	0.9–0.8
Normal cleanliness	0.6–0.5	0.8–0.6
Slight contamination	0.5–0.3	0.6–0.4
Typical contamination	0.3–0.1	0.4–0.2
Severe contamination	0.1–0	0.1–0
Very severe contamination	0	0

Table 6.6 Single row deep groove ball bearings.

Selected deep groove ball bearings.
Single row.

d (mm)	D (mm)	B (mm)	C (kN)	C_0 (kN)	Fatigue load limit, P_u (kN)	Reference speed (rpm)	Limiting speed (rpm)	Mass (kg)	Designation
3	10	4	0.54	0.18	0.007	130,000	80,000	0.002	623
4	9	3.5	0.54	0.18	0.007	140,000	70,000	0.001	628/4-2Z
4	9	4	0.54	0.18	0.007	140,000	70,000	0.001	638/4-2Z
4	9	2.5	0.54	0.18	0.007	140,000	85,000	0.001	618/4
4	11	4	0.715	0.232	0.01	130,000	80,000	0.002	619/4
4	12	4	0.806	0.28	0.012	120,000	75,000	0.002	604
4	13	5	0.936	0.29	0.012	110,000	67,000	0.003	624
4	16	5	1.11	0.38	0.016	95,000	60,000	0.005	634
5	11	5	0.637	0.255	0.011	120,000	60,000	0.002	638/5-2Z
5	11	4	0.637	0.255	0.011	120,000	60,000	0.001	628/5-2Z
5	11	3	0.637	0.255	0.011	120,000	75,000	0.001	618/5
5	16	5	1.14	0.38	0.016	95,000	60,000	0.005	625[a]
7	19	6	2.34	0.95	0.04	85,000	53,000	0.007	607-Z[a]
7	22	7	3.45	1.37	0.057	70,000	45,000	0.013	627[a]
8	16	5	1.33	0.57	0.024	–	26,000	0.004	628/8-2RS1

8	16	4	1.33	0.57	0.024	90,000	56,000	0.003	618/8
8	16	6	1.33	0.57	0.024	90,000	45,000	0.004	638/8-2Z
8	19	6	1.9	0.735	0.031	80,000	50,000	0.007	619/8
8	19	6	2.34	0.95	0.04	85,000	43,000	0.007	607/8-2Z[a]
8	22	7	3.45	1.37	0.057	75,000	48,000	0.012	608[a]
8	22	11	3.45	1.37	0.057	–	22,000	0.016	630/8-2RS1
8	24	8	3.9	1.66	0.071	63,000	40,000	0.017	628 [a]
8	28	9	4.62	1.96	0.083	60,000	30,000	0.03	638-2RZ
9	17	5	1.43	0.64	0.027	85,000	43,000	0.004	628/9-2Z
9	17	4	1.43	0.64	0.027	85,000	53,000	0.003	618/9
9	20	6	2.08	0.865	0.036	80,000	48,000	0.008	619/9
9	24	7	3.9	1.66	0.071	70,000	43,000	0.014	609[a]
9	26	8	4.75	1.96	0.083	60,000	38,000	0.02	629[a]
9.525	22.225	7.144	3.32	1.4	0.06	60,000	43,000	0.012	EEB 3-2Z
9.525	22.225	5.556	3.32	1.4	0.06	60,000	43,000	0.009	EE 3 TN9
10	19	5	1.38	0.585	0.025	80,000	48,000	0.005	61,800
10	22	6	2.08	0.85	0.036	75,000	45,000	0.01	61,900
10	26	8	4.75	1.96	0.083	67,000	40,000	0.019	6000[a]
10	26	12	4.62	1.96	0.083	–	19,000	0.025	63,000-2RS1
10	30	9	5.4	2.36	0.1	56,000	34,000	0.032	6200[a]
10	30	14	5.07	2.36	0.1	–	17,000	0.04	62,200-2RS1
10	35	17	8.06	3.4	0.143	–	15,000	0.06	62,300-2RS1
10	35	11	8.52	3.4	0.143	50,000	32,000	0.053	6300[a]
12	21	5	1.43	0.67	0.028	70,000	43,000	0.006	61,801
12	21	5	1.43	0.67	0.028	70,000	36,000	0.006	61,801-2Z
12	24	6	2.25	0.98	0.043	67,000	40,000	0.011	61,901
12	28	8	5.4	2.36	0.1	60,000	38,000	0.022	6001[a]
12	28	12	5.07	2.36	0.1	–	17,000	0.029	63,001-2RS1
12	30	8	5.07	2.36	0.1	56,000	28,000	0.023	16,101-2Z
12	32	10	7.28	3.1	0.132	50,000	32,000	0.037	6201[a]
12	32	14	6.89	3.1	0.132	–	15,000	0.045	62,201-2RS1

Continued

Table 6.6 Continued

Selected deep groove ball bearings. Single row.									
d (mm)	D (mm)	B (mm)	C (kN)	C_0 (kN)	Fatigue load limit, P_u (kN)	Reference speed (rpm)	Limiting speed (rpm)	Mass (kg)	Designation
12	37	17	9.75	4.15	0.176	–	14,000	0.07	62,301-2RS1
12.7	28.575	6.35	5.4	2.36	0.1	60,000	38,000	0.023	R 8
12.7	28.575	7.938	5.4	2.36	0.1	56,000	28,000	0.024	R 8-2Z
12.7	33.337	9.525	6.89	3.1	0.132	45,000	32,000	0.037	RLS 4
15	24	5	1.56	0.8	0.034	60,000	38,000	0.007	61,802
15	28	7	4.36	2.24	0.095	56,000	34,000	0.016	61,902
15	32	8	5.85	2.85	0.12	50,000	32,000	0.025	16,002[a]
15	32	9	5.85	2.85	0.12	–	14,000	0.03	6002-2RSH[a]
15	35	11	8.06	3.75	0.16	43,000	22,000	0.045	6202-2Z[a]
15	42	17	11.4	5.4	0.228	–	12,000	0.11	62,302-2RS1
15	42	13	11.9	5.4	0.228	38,000	24,000	0.082	6302[a]
15.875	39.688	11.113	9.56	4.75	0.2	34,000	24,000	0.065	RLS 5
15.875	46.038	15.875	13.5	6.55	0.275	30,000	22,000	0.12	RMS 5
17	26	5	1.68	0.93	0.039	56,000	34,000	0.008	61,803
17	35	10	6.37	3.25	0.137	45,000	28,000	0.039	6003[a]
17	35	8	6.37	3.25	0.137	45,000	28,000	0.032	16,003[a]
17	35	14	6.05	3.25	0.137	–	13,000	0.052	63,003-2RS1
17	40	16	9.56	4.75	0.2	–	12,000	0.083	62,203-2RS1
17	40	9	9.56	4.75	0.2	38,000	24,000	0.048	98,203
17	40	12	9.95	4.75	0.2	38,000	24,000	0.065	6203[a]
17	47	14	14.3	6.55	0.275	34,000	22,000	0.12	6303[a]
17	47	19	13.5	6.55	0.275	–	11,000	0.15	62,303-2RS1
17	62	17	22.9	10.8	0.455	28,000	18,000	0.27	6403
19.05	47.625	14.287	12.7	6.55	0.28	30,000	20,000	0.11	RLS 6

19.05	50.8	17.462	15.9	7.8	0.335	28,000	19,000	0.14	RMS 6
20	32	7	4.03	2.32	0.104	45,000	28,000	0.018	61,804
20	37	9	6.37	3.65	0.156	43,000	26,000	0.038	61,904
20	42	8	7.28	4.05	0.173	38,000	24,000	0.05	16,004[a]
20	42	12	9.95	5	0.212	38,000	24,000	0.069	6004[a]
20	42	9	7.93	4.5	0.19	38,000	24,000	0.051	98,204 Y
20	42	16	9.36	5	0.212	–	11,000	0.086	63,004-2RS1
20	47	18	12.7	6.55	0.28	–	10,000	0.13	62,204-2RS1
20	47	14	13.5	6.55	0.28	32,000	20,000	0.11	6204[a]
20	52	21	15.9	7.8	0.335	–	9500	0.2	62,304-2RS1
20	52	15	16.8	7.8	0.335	30,000	19,000	0.14	6304[a]
20	72	19	30.7	15	0.64	24,000	15,000	0.4	6404
22	50	14	14	7.65	0.325	30,000	19,000	0.12	62/22
22	56	16	18.6	9.3	0.39	28,000	18,000	0.18	63/22
22.225	47.625	9.525	11.7	6.3	0.265	30,000	20,000	0.073	EE 8 TN9
22.225	50.8	14.287	14	7.65	0.325	26,000	19,000	0.12	RLS 7
22.225	57.15	17.462	18.6	9.3	0.39	26,000	18,000	0.18	RMS 7
25	37	7	4.36	2.6	0.125	38,000	24,000	0.022	61,805
25	42	9	7.02	4.3	0.193	36,000	22,000	0.045	61,905
25	47	12	11.9	6.55	0.275	32,000	20,000	0.08	6005[a]
25	47	8	8.06	4.75	0.212	32,000	20,000	0.06	16,005[a]
25	52	15	14.8	7.8	0.335	28,000	14,000	0.13	6205-2RSL[a]
25	62	17	23.4	11.6	0.49	24,000	16,000	0.23	6305[a]
25	62	24	22.5	11.6	0.49	–	7500	0.32	62,305-2RS1
25	80	21	35.8	19.3	0.815	20,000	13,000	0.53	6405
25.4	57.15	15.875	17.8	9.65	0.405	24,000	17,000	0.17	RLS 8
25.4	57.15	15.875	17.8	9.65	0.405	24,000	12,000	0.17	RLS 8-2Z
25.4	63.5	19.05	22.5	11.6	0.49	22,000	16,000	0.23	RMS 8
28	58	16	16.8	9.5	0.405	26,000	16,000	0.18	62/28
28	68	18	25.1	13.7	0.585	22,000	14,000	0.29	63/28
28.575	63.5	15.875	19.5	11.2	0.475	22,000	15,000	0.2	RLS 9

Continued

Table 6.6 Continued

Selected deep groove ball bearings. Single row.									
d (mm)	D (mm)	B (mm)	C (kN)	C_0 (kN)	Fatigue load limit, P_u (kN)	Reference speed (rpm)	Limiting speed (rpm)	Mass (kg)	Designation
28.575	71.438	20.637	28.1	16	0.67	19,000	13,000	0.35	RMS 9
30	42	7	4.49	2.9	0.146	32,000	20,000	0.027	61,806
30	55	9	11.9	7.35	0.31	28,000	17,000	0.085	16,006[a]
30	55	13	13.8	8.3	0.355	28,000	17,000	0.12	6006[a]
30	55	13	13.8	8.3	0.355	–	8000	0.12	6006-2RS1[a]
30	55	19	13.3	8.3	0.355	–	8000	0.16	63,006-2RS1
30	62	20	19.5	11.2	0.475	–	7500	0.24	62,206-2RS1
30	62	16	20.3	11.2	0.475	24,000	15,000	0.2	6206[a]
30	72	27	28.1	16	0.67	–	6300	0.48	62,306-2RS1
30	72	19	29.6	16	0.67	20,000	13,000	0.35	6306[a]
30	90	23	43.6	23.6	1	18,000	11,000	0.74	6406
31.75	69.85	17.462	22.5	13.2	0.55	20,000	14,000	0.3	RLS 10
31.75	79.375	22.225	33.2	19	0.815	17,000	12,000	0.5	RMS 10
34.925	76.2	17.462	27	15.3	0.655	18,000	13,000	0.35	RLS 11
34.925	88.9	22.225	41	24	1.02	15,000	11,000	0.63	RMS 11
35	47	7	4.75	3.2	0.166	28,000	18,000	0.03	61,807
35	55	10	9.56	6.8	0.29	26,000	16,000	0.08	61,907
35	62	14	16.8	10.2	0.44	24,000	15,000	0.16	6007[a]
35	62	9	13	8.15	0.375	24,000	15,000	0.11	16,007[a]
35	62	20	15.9	10.2	0.44	–	7000	0.21	63,007-2RS1
35	72	23	25.5	15.3	0.655	–	6300	0.37	62,207-2RS1
35	72	17	27	15.3	0.655	20,000	13,000	0.29	6207[a]
35	80	31	33.2	19	0.815	–	6000	0.66	62,307-2RS1
35	80	21	35.1	19	0.815	19,000	12,000	0.46	6307[a]

35	100	25	55.3	31	1.29	16,000	10,000	0.95	6407
38.1	82.55	19.05	30.7	19	0.8	16,000	11,000	0.45	RLS 12
38.1	95.25	23.813	41	24	1.02	15,000	11,000	0.79	RMS 12
40	52	7	4.94	3.45	0.186	26,000	16,000	0.034	61,808
40	62	12	13.8	10	0.425	24,000	12,000	0.12	61,908-2RZ
40	68	9	13.8	9.15	0.44	22,000	14,000	0.13	16,008[a]
40	68	15	17.8	11.6	0.49	22,000	14,000	0.19	6008[a]
40	68	21	16.8	11.6	0.49	–	6300	0.26	63,008-2RS1
40	80	23	30.7	19	0.8	–	5600	0.44	62,208-2RS1
40	80	18	32.5	19	0.8	18,000	11,000	0.37	6208[a]
40	80	18	32.5	19	0.8	18,000	9000	0.37	6208-2Z[a]
40	90	33	41	24	1.02	–	5000	0.89	62,308-2RS1
40	90	23	42.3	24	1.02	17,000	11,000	0.63	6308[a]
40	90	23	42.3	24	1.02	17,000	11,000	0.63	6308-Z[a]
40	110	27	63.7	36.5	1.53	14,000	9000	1.25	6408
41.275	88.9	19.05	33.2	21.6	0.915	15,000	11,000	0.51	RLS 13
41.275	101.6	23.813	52.7	31.5	1.34	14,000	9500	0.91	RMS 13
44.45	95.25	20.637	35.1	23.2	0.98	14,000	10,000	0.66	RLS 14
44.45	107.95	26.988	61.8	38	1.6	12,000	8500	1.15	RMS 14
45	58	7	6.63	6.1	0.26	22,000	14,000	0.04	61,809
45	68	12	14	10.8	0.465	20,000	13,000	0.14	61,909
45	75	16	22.1	14.6	0.64	20,000	12,000	0.25	6009[a]
45	75	10	16.5	10.8	0.52	20,000	12,000	0.17	16,009[a]
45	75	23	20.8	14.6	0.64	–	5600	0.34	63,009-2RS1
45	85	19	35.1	21.6	0.915	17,000	11,000	0.41	6209[a]
45	100	36	52.7	31.5	1.34	–	4500	1.15	62,309-2RS1
45	100	25	55.3	31.5	1.34	15,000	9500	0.83	6309[a]
45	120	29	76.1	45	1.9	13,000	8500	1.55	6409
50	65	7	6.76	6.8	0.285	20,000	13,000	0.052	61,810
50	72	12	14.6	11.8	0.5	19,000	12,000	0.14	61,910
50	80	10	16.8	11.4	0.56	18,000	11,000	0.18	16,010[a]

Continued

Table 6.6 Continued

Selected deep groove ball bearings. Single row.									
d (mm)	D (mm)	B (mm)	C (kN)	C_0 (kN)	Fatigue load limit, P_u (kN)	Reference speed (rpm)	Limiting speed (rpm)	Mass (kg)	Designation
50	80	16	22.9	16	0.71	18,000	11,000	0.26	6010[a]
50	80	16	22.9	16	0.71	18,000	11,000	0.26	6010-Z[a]
50	90	20	37.1	23.2	0.98	15,000	10,000	0.46	6210[a]
50	90	20	37.1	23.2	0.98	15,000	8000	0.46	6210-2RZ[a]
50	110	27	65	38	1.6	13,000	8500	1.05	6310[a]
50	130	31	87.1	52	2.2	12,000	7500	1.9	6410
50.8	101.6	20.637	35.1	23.2	0.98	14,000	10,000	0.7	RLS 16
55	72	9	9.04	8.8	0.375	19,000	12,000	0.083	61,811
55	80	13	16.5	14	0.6	17,000	11,000	0.19	61,911
55	90	18	29.6	21.2	0.9	16,000	10,000	0.39	6011[a]
55	100	21	46.2	29	1.25	14,000	9000	0.61	6211[a]
55	120	29	74.1	45	1.9	12,000	8000	1.35	6311
55	140	33	99.5	62	2.6	11,000	70,00	2.3	6411
57.15	114.3	22.25	52.7	36	1.53	11,000	8000	0.96	RLS 18
57.15	127	31.75	81.9	52	2.2	10,000	7000	1.7	RMS 18
60	78	10	11.9	11.4	0.49	17,000	11,000	0.11	61,812
60	85	13	16.5	14.3	0.6	16,000	10,000	0.2	61,912
60	95	11	20.8	15	0.735	15,000	9500	0.28	16,012[a]
60	95	18	30.7	23.2	0.98	15,000	9500	0.42	6012[a]
60	110	22	55.3	36	1.53	13,000	8000	0.78	6212[a]
60	130	31	85.2	52	2.2	11,000	7000	1.7	6312[a]
63.5	127	23.813	60.5	45	1.9	10,000	7000	1.25	RLS 20
63.5	139.7	31.75	92.3	60	2.5	9500	6700	2.05	RMS 20
65	85	10	12.4	12.7	0.54	16,000	10,000	0.13	61,813

65	100	18	31.9	25	1.06	14,000	9000	0.44	6013[a]
65	100	11	22.5	16.6	0.83	14,000	9000	0.3	16,013[a]
65	120	23	58.5	40.5	1.73	12,000	7500	0.99	6213[a]
65	140	33	97.5	60	2.5	10,000	6700	2.1	6313[a]
65	160	37	119	78	3.15	9500	6000	3.3	6413
69.85	158.75	34.925	104	68	2.75	9000	6300	3	RMS 22
70	90	10	12.4	13.2	0.56	15,000	9000	0.14	61,814
70	100	16	23.8	21.2	0.9	14,000	7000	0.35	61,914-2RZ
70	110	13	29.1	25	1.06	13,000	8000	0.43	16,014[a]
70	110	20	39.7	31	1.32	13,000	8000	0.6	6014[a]
70	125	24	63.7	45	1.9	11,000	7000	1.05	6214[a]
70	150	35	111	68	2.75	9500	6300	2.5	6314[a]
70	180	42	143	104	3.9	8500	5300	4.85	6414
75	95	10	12.7	14.3	0.61	14,000	8500	0.15	61,815
75	105	16	24.2	19.3	0.965	13,000	8000	0.37	61,915
75	110	12	28.6	27	1.14	13,000	8000	0.38	16,115
75	115	20	41.6	33.5	1.43	12,000	7500	0.64	6015[a]
75	130	25	68.9	49	2.04	10,000	6700	1.2	6215[a]
75	160	37	119	76.5	3	9000	5600	3	6315[a]
75	190	45	153	114	4.15	8000	5000	6.8	6415
76.2	146.05	26.988	70.2	55	2.2	8500	6000	1.4	RLS 24
80	100	10	13	15	0.64	13,000	6300	0.15	61,816-2RZ
80	110	16	25.1	20.4	1.02	12,000	7500	0.4	61,916
80	125	22	49.4	40	1.66	11,000	7000	0.85	6016[a]
80	125	14	35.1	31.5	1.32	11,000	7000	0.6	16,016[a]
80	140	26	72.8	55	2.2	9500	6000	1.4	6216[a]
80	170	39	130	86.5	3.25	8500	5300	3.6	6316[a]
80	200	48	163	125	4.5	7500	4800	8	6416
85	110	13	19.5	20.8	0.88	12,000	7500	0.27	61,817
85	130	14	35.8	33.5	1.37	11,000	6700	0.63	16,017[a]
85	130	22	52	43	1.76	11,000	6700	0.89	6017[a]

Continued

Table 6.6 Continued

Selected deep groove ball bearings. Single row.									
d (mm)	D (mm)	B (mm)	C (kN)	C_0 (kN)	Fatigue load limit, P_u (kN)	Reference speed (rpm)	Limiting speed (rpm)	Mass (kg)	Designation
85	150	28	87.1	64	2.5	9000	5600	1.8	6217[a]
85	180	41	140	96.5	3.55	8000	5000	4.25	6317[a]
85	210	52	174	137	4.75	7000	4500	9.5	6417
88.9	165.1	28.575	95.6	73.5	2.8	7500	5300	2.85	RLS 28
88.9	206.375	44.45	153	117	4.15	6300	4500	6.5	RMS 28
90	115	13	19.5	22	0.915	11,000	7000	0.28	61,818
90	125	18	33.2	31.5	1.23	11,000	6700	0.59	61,918
90	140	24	60.5	50	1.96	10,000	6300	1.15	6018[a]
90	140	16	43.6	39	1.56	10,000	6300	0.85	16,018[a]
90	160	30	101	73.5	2.8	8500	5300	2.15	6218[a]
90	190	43	151	108	3.8	7500	4800	4.9	6318[a]
90	225	54	186	150	5	6700	4300	11.5	6418
95	120	13	19.9	22.8	0.93	11,000	6700	0.3	61,819
95	130	18	33.8	33.5	1.43	10,000	6300	0.61	61,919
95	145	16	44.8	41.5	1.63	9500	6000	0.89	16,019[a]
95	145	24	63.7	54	2.08	9500	6000	1.2	6019[a]
95	170	32	114	81.5	3	8000	5000	2.6	6219[a]
95	200	45	159	118	4.15	7000	4500	5.65	6319[a]
100	125	13	19.9	24	0.95	10,000	6300	0.31	61,820
100	140	20	42.3	41	1.63	9500	6000	0.83	61,920
100	150	24	63.7	54	2.04	9500	5600	1.25	6020[a]
100	150	16	46.2	44	1.73	9500	5600	0.91	16,020[a]
100	180	34	127	93	3.35	7500	4800	3.15	6220[a]
100	215	47	174	140	4.75	6700	4300	7	6320

101.6	215.9	44.45	174	140	4.75	6000	4300	6.55	RMS 32
105	130	13	20.8	19.6	1	10,000	6300	0.32	61,821
105	145	20	44.2	44	1.7	9500	5600	0.87	61,921
105	160	18	54	51	1.86	8500	5300	1.2	16,021[a]
105	160	26	76.1	65.5	2.4	8500	5300	1.6	6021[a]
105	190	36	140	104	3.65	7000	4500	3.7	6221[a]
105	225	49	182	153	5.1	6300	4000	8.25	6321
110	140	16	28.1	26	1.25	9500	5600	0.6	61,822
110	150	20	43.6	45	1.66	9000	5600	0.9	61,922
110	170	28	85.2	73.5	2.4	8000	5000	1.95	6022[a]
110	170	19	60.2	57	2.04	8000	5000	1.45	16,022[a]
110	200	38	151	118	4	6700	4300	4.35	6222[a]
110	240	50	203	180	5.7	6000	3800	9.55	6322
114.3	238.125	50.8	203	180	5.7	5300	3800	8.9	RMS 36
120	150	16	29.1	28	1.29	8500	5300	0.65	61,824
120	165	22	55.3	57	2.04	8000	5000	1.2	61,924
120	180	19	63.7	64	2.2	7500	4800	1.6	16,024[a]
120	180	28	88.4	80	2.75	7500	4800	2.05	6024[a]
120	215	40	146	118	3.9	6300	4000	5.15	6224
120	260	55	208	186	5.7	5600	3400	12.5	6324
130	165	18	37.7	43	1.6	8000	4800	0.93	61,826
130	180	24	65	67	2.28	7500	4500	1.85	61,926
130	200	33	112	100	3.35	7000	4300	3.15	6026[a]
130	200	22	83.2	81.5	2.7	7000	4300	2.35	16,026[a]
130	230	40	156	132	4.15	5600	3600	5.8	6226
130	280	58	229	216	6.3	5000	4500	17.5	6326 M
140	175	18	39	46.5	1.66	7500	4500	0.99	61,828
140	190	24	66.3	72	2.36	7000	5600	1.7	61,928 MA
140	210	22	80.6	86.5	2.8	6700	4000	2.5	16,028
141	210	33	111	108	3.45	6700	4000	3.35	6028
140	250	42	165	150	4.55	5300	3400	7.45	6228

Continued

Table 6.6 Continued

Selected deep groove ball bearings. Single row.									
d (mm)	D (mm)	B (mm)	C (kN)	C_0 (kN)	Fatigue load limit, P_u (kN)	Reference speed (rpm)	Limiting speed (rpm)	Mass (kg)	Designation
140	300	62	251	245	7.1	4800	4300	22	6328 M
150	190	20	48.8	61	1.96	6700	4300	1.4	61,830
150	210	28	88.4	93	2.9	6300	5300	3.05	61
140	300	62	251	245	7.1	4800	4300	22	6328 M
150	190	20	48.8	61	1.96	6700	4300	1.4	61,830
150	210	28	88.4	93	2.9	6300	5300	3.05	61,930 MA
150	225	35	125	125	3.9	6000	3800	4.8	6030
150	225	24	92.2	98	3.05	6000	3800	3.15	16,030
150	270	45	174	166	4.9	5000	3200	9.4	6230
150	320	65	276	285	7.8	4300	4000	26	6330 M
160	200	20	49.4	64	2	6300	4000	1.45	61,832
160	220	28	92.3	98	3.05	6000	5000	3.25	61,932 MA
160	240	25	99.5	108	3.25	5600	3600	3.7	16,032
160	240	38	143	143	4.3	5600	3600	5.9	6032
160	290	48	186	186	5.3	4500	3000	14.5	6232
160	340	68	276	285	7.65	4000	3800	29	6332 M
170	215	22	61.8	78	2.4	6000	3600	1.9	61,834
170	230	28	93.6	106	3.15	5600	4800	3.4	61,934 MA
170	260	42	168	173	5	5300	4300	7.9	6034 M
170	260	28	119	129	3.75	5300	3200	5	16,034
170	310	52	212	224	6.1	4300	3800	17.5	6234 M
170	360	72	312	340	8.8	3800	3400	34.5	6334 M
180	225	22	62.4	81.5	2.45	5600	3400	2	61,836
180	250	33	119	134	3.9	5300	4300	5.05	61,936 MA

180	280	31	138	146	4.15	4800	4000	6.6	16,036
180	280	46	190	200	5.6	4800	4000	10.5	6036 M
180	320	52	229	240	6.4	4000	3600	18.5	6236 M
180	380	75	351	405	10.4	3600	3200	42.5	6336 M
190	240	24	76.1	98	2.8	5300	3200	2.6	61,838
190	260	33	117	134	3.8	5000	4300	5.25	61,938 MA
190	290	46	195	216	5.85	4800	3800	11	6038 M
190	290	31	148	166	4.55	4800	3000	7.9	16,038
190	340	55	255	280	7.35	3800	3400	23	6238 M
190	400	78	371	430	10.8	3400	3000	49	6338 M
200	250	24	76.1	102	2.9	5000	3200	2.7	61,840
200	280	38	148	166	4.55	4800	3800	7.4	61,940 MA
200	310	34	168	190	5.1	4300	2800	8.85	16,040
200	310	51	216	245	6.4	4300	3600	14	6040 M
200	360	58	270	310	7.8	3600	3200	28	6240 M
220	270	24	78	110	3	4500	2800	3	61,844
220	300	38	151	180	4.75	4300	3600	8	61,944 MA
220	340	56	247	290	7.35	4000	3200	18.5	6044 M
220	340	37	174	204	5.2	4000	2400	11.5	16,044
220	400	65	296	365	8.8	3200	3000	37	6244 M
220	460	88	410	520	12	3000	2600	72.5	6344 M
240	300	28	108	150	3.8	4000	2600	4.5	61,848
240	320	38	159	200	5.1	4000	3200	8.6	61,948 MA
240	360	37	203	255	6.3	2800	2200	14	16,048 MA
240	360	56	255	315	7.8	3600	3000	19.5	6048 M
240	440	72	358	465	10.8	3000	2600	51	6248 M
240	500	95	442	585	12.9	2600	2400	92.5	6348 M
260	320	28	111	163	4	3800	2400	4.8	61,852
260	360	46	212	270	6.55	3600	3000	14.5	61,952 MA
260	400	65	291	375	8.8	3200	2800	29.5	6052 M
260	400	44	238	310	7.2	2600	2000	22.5	16,052 MA

Continued

Table 6.6 Continued

Selected deep groove ball bearings. Single row.									
d (mm)	**D (mm)**	**B (mm)**	**C (kN)**	**C_0 (kN)**	**Fatigue load limit, P_u (kN)**	**Reference speed (rpm)**	**Limiting speed (rpm)**	**Mass (kg)**	**Designation**
260	480	80	390	530	11.8	2600	2400	65.5	6252 M
260	540	102	507	710	15	2400	2200	115	6352 M
280	420	44	242	335	7.5	2400	1900	23.5	16,056 MA
280	420	65	302	405	9.3	3000	2600	31	6056 M
280	500	80	423	600	12.9	2600	2200	71	6256 M
280	580	108	572	850	17.3	2200	2000	140	6356 M
300	380	38	172	245	5.6	3200	2600	10.5	61,860 MA
300	420	56	270	375	8.3	3000	2400	24.5	61,960 MA
300	460	74	358	500	10.8	2800	2400	44	6060 M
300	460	50	286	405	8.8	2800	2400	32	16,060 MA
300	540	85	462	670	13.7	2400	2000	88.5	6260 M
320	400	38	172	255	5.7	3000	2400	11	61,864 MA
320	440	56	276	400	8.65	2800	2400	25.5	61,964 MA
320	440	37	216	310	6.7	2800	2400	16	60,964 MA
320	480	50	281	405	8.65	2600	2200	34	16,064 MA
320	480	74	371	540	11.4	2600	2200	46	6064 M
320	580	92	527	800	16	2200	1900	110	6264 MA
330	460	56	281	425	9	2600	2200	30	306,728
340	420	38	178	275	6	2800	2400	11.5	61,868 MA
340	460	56	281	425	9	2600	2200	26.5	61,968 MA
340	480	60	291	430	9	2600	2200	36	306,890
340	520	82	423	640	13.2	2400	2200	62	6068 M
340	520	57	345	520	10.6	2400	2000	45	16,068 MA
340	620	92	559	900	17.3	2000	1800	110	6268 MA

350	500	70	319	475	9.8	2400	2000	46	306,674
360	440	25	121	212	4.5	2600	2200	6.5	60,872 MA
360	440	38	182	285	6.1	2600	2200	12	61,872 MA
360	480	56	291	450	9.15	2600	2200	28	61,972 MA
360	540	82	462	735	15	2400	1900	64.5	6072 M
360	540	57	351	550	11	1800	1400	49	16,072 MA
380	480	46	242	390	8	2400	2000	20	61,876 MA
380	480	46	242	390	8	2400	1500	18.5	61,876 TN9
380	520	65	338	540	10.8	2400	1900	40	61,976 MA
380	560	57	377	620	12.2	2200	1800	51	16,076 MA
380	560	82	462	750	14.6	2200	1800	67.5	6076 M
400	500	31	163	280	5.6	2400	1900	15.5	60,880 MA
400	500	46	247	405	8.15	2400	1900	20.5	61,880 MA
400	540	65	345	570	11.2	2200	1800	41.5	61,980 MA
400	540	44	265	440	8.5	2200	1800	27.5	60,980 M
400	590	74	436	710	13.4	2000	1700	70	306,614
400	600	90	520	865	16.3	2000	1700	87.5	6080 M
400	720	130	663	1160	20.8	1800	500	235	BB1B 362,809
420	520	46	251	425	8.3	2200	1800	21.5	61,884 MA
420	560	65	351	600	11.4	2200	1800	43	61,984 MA
420	620	90	507	880	16.3	2000	1600	91.5	6084 M
440	540	31	159	290	5.6	2200	1800	16.5	60,888 MA
440	540	46	255	440	8.5	2200	1800	22.5	61,888 MA
440	599	80	410	720	13.2	2000	1600	66	BB1B 363,471
440	600	74	410	720	13.2	2000	1600	60.5	61,988 MA
440	600	50	312	550	10.2	2000	1600	40	60,988 MB
440	650	94	553	965	17.6	1900	1500	105	6088 M
460	580	56	319	570	10.6	2000	1600	35	61,892 MA
460	620	74	423	750	13.7	1900	1600	62.5	61,992 MA
460	680	100	582	1060	19	1800	1500	120	6092 MB
480	600	56	325	600	10.8	1900	1600	36.5	61,896 MA

Continued

Table 6.6 Continued

Selected deep groove ball bearings. Single row.									
d (mm)	D (mm)	B (mm)	C (kN)	C_0 (kN)	Fatigue load limit, P_u (kN)	Reference speed (rpm)	Limiting speed (rpm)	Mass (kg)	Designation
480	650	78	449	815	14.6	1800	1500	74	61,996 MA
480	700	100	618	1140	20	1700	1400	125	6096 MB
500	620	37	229	440	7.8	1800	1500	20	608/500 MA
500	670	78	462	865	15	1700	1400	77	619/500 MA
500	689	100	475	865	15	1700	480	110	BB1B 363,472
500	720	100	605	1140	19.6	1600	1300	135	60/500 N1MAS
530	650	56	332	655	11.2	1700	1400	39.5	618/530 MA
530	710	57	423	815	13.7	1600	1300	61	609/530 MA
530	710	82	488	930	15.6	1600	1300	90.5	619/530 MA
530	760	100	585	1120	18.6	1500	1300	150	360,476 [A]
530	780	112	650	1270	20.8	1500	1200	185	60/530 N1MAS
540	625	40	225	425	7.5	1700	1400	14	BB1B 362,692
560	680	56	345	695	11.8	1600	1300	42	618/560 MA
560	680	37	229	465	8	1600	1300	30.5	608/560 MA
560	750	85	494	980	16.3	1500	1200	105	619/560 MA
560	820	115	663	1470	22	1400	1200	210	60/560 N1MAS
600	730	42	265	550	9	1500	1200	40	608/600 MA
600	730	60	364	765	12.5	1500	1200	52	618/600 MA
600	800	90	585	1220	19.6	1400	1100	125	619/600 MA
600	870	118	728	1500	23.6	1300	1100	230	60/600 MA
630	780	48	364	765	12.2	1400	1100	41	608/630 MA
630	780	69	442	965	15.3	1400	1100	73	618/630 MA
630	850	100	624	1340	21.2	1300	1100	160	619/630 N1MA
630	850	71	488	1060	16.6	1300	1100	110	609/630 MB
630	920	128	819	1760	27	1200	1000	285	60/630 N1MBS
650	920	118	780	1730	26	1200	1000	250	306,708 D

670	820	69	442	1000	15.6	1300	1100	83.5	618/670 MA
670	820	69	442	1000	15.6	1300	800	77.5	618/670 TN
670	900	103	676	1500	22.4	1200	1000	185	619/670 MA
670	900	73	553	1220	18.6	1200	1000	145	609/670 MA
670	980	136	904	2040	30	1100	900	345	60/670 N1MAS
710	870	74	475	1100	16.6	1200	1000	93.5	618/710 MA
710	950	106	663	1500	22	1100	900	220	619/710 MA
710	950	78	559	1290	19	1100	900	150	609/710 MA
710	1000	140	832	1900	27.5	1100	900	335	306,704C
730	940	100	650	1500	22	1100	900	175	361,840
750	920	78	527	1250	18.3	1100	900	110	618/750 MA
750	1000	112	761	1800	25.5	1000	850	255	619/750 MA
750	1090	150	995	2360	33.5	950	800	485	60/750 MA
760	1080	150	923	2200	31	950	800	430	306,474 D
800	980	57	403	1000	14.3	1000	850	100	608/800 MA
800	1060	115	832	2040	28.5	950	800	275	619/800 MA
800	1080	115	819	2040	28.5	950	750	320	361,844
800	1150	155	1010	2550	34.5	900	750	535	60/800 N1MAS
850	1030	57	397	1020	14	950	750	75	608/850 MB
850	1030	82	559	1430	19.6	950	750	140	618/850 MA
850	1120	118	832	2160	29	850	750	310	619/850 MA
850	1220	165	1040	2700	35.5	800	670	630	306,493 AA
900	1090	85	618	1600	21.6	850	700	160	618/900 MA
900	1180	122	852	2280	30	800	670	350	619/900 MB
900	1280	170	1140	3100	39	750	630	720	60/900 MB
950	1150	90	637	1730	22.4	800	670	190	618/950 MA
950	1250	132	1010	2800	36	750	630	390	619/950 MB
950	1360	180	1170	3250	40.5	700	560	860	60/950 MB
970	1125	75	488	1340	17.6	800	670	125	BB1B 363,297
1000	1220	71	553	1560	20	750	600	175	608/1000 MB
1000	1220	100	637	1800	22.8	750	600	245	618/1000 MA
1000	1320	140	1010	2800	35.5	700	560	515	619/1000 MB

Continued

Table 6.6 Continued

Selected deep groove ball bearings. Single row.									
d (mm)	D (mm)	B (mm)	C (kN)	C_0 (kN)	Fatigue load limit, P_u (kN)	Reference speed (rpm)	Limiting speed (rpm)	Mass (kg)	Designation
1000	1420	185	1350	3900	47.5	630	530	930	60/1000 MB
1030	1250	100	605	1760	22	700	600	250	BB1B 630,533
1060	1280	100	728	2120	26.5	670	560	260	618/1060 MA
1060	1400	150	1010	3000	36	630	530	620	619/1060 MB
1060	1500	195	1350	3900	46.5	600	500	1080	60/1060 MB
1120	1360	106	741	2200	26.5	630	530	315	618/1120 MA
1120	1460	150	1040	3100	36.5	600	480	650	619/1120 MB
1120	1580	200	1460	4400	50	530	450	1250	60/1120 MB
1180	1420	106	761	2360	27.5	560	480	330	618/1180 MB
1180	1540	160	1140	3600	41.5	220	280	775	619/1180 MB
1250	1500	112	852	2750	31.5	530	430	385	618/1250 MB
1320	1600	122	956	3150	35.5	480	400	500	618/1320 MA
1320	1720	128	1210	4050	44	450	360	830	609/1320 MB
1400	1700	132	1120	4000	42.5	430	360	615	618/1400 MA
1400	1820	185	1590	5500	57	400	340	1250	619/1400 MB
1500	1820	140	1210	4400	46.5	380	240	690	618/1500 TN
1500	1950	195	1720	6100	62	380	320	1500	619/1500 MB
1600	1950	155	1270	4800	48	320	280	965	618/1600 MB
1600	2060	200	1860	6950	69.5	300	260	1650	619/1600 MB
1700	2060	160	1270	4900	48	300	240	1100	618/1700 MB
1700	2180	212	1990	7650	73.5	280	220	1950	619/1700 MB
2000	2200	75	936	4500	41.5	220	190	290	BB1B 363,270
2390	2690	120	1300	6200	52	120	100	975	BB1-8001

Note this table presents a selection only of the bearings available. Table after SKF.
[a]SKF explorer bearing.

Table 6.7 Single row cylindrical roller bearings

Selected cylindrical roller bearings.
Single row.

d (mm)	D (mm)	B (mm)	C (kN)	C_0 (kN)	Fatigue load limit (kN)	Reference speed (rpm)	Limiting speed (rpm)	Mass (kg)	Designation
15	35	11	12.5	10.2	1.22	22,000	26,000	0.047	NU 202 ECP
15	35	11	12.5	10.2	1.22	22,000	34,000	0.047	NU 202 ECPHA
15	35	11	12.5	10.2	1.22	22,000	26,000	0.049	NJ 202 ECP
17	40	12	17.2	14.3	1.73	19,000	22,000	0.07	NJ 203 ECP
17	40	16	23.8	21.6	2.65	19,000	22,000	0.095	NJ 2203 ECP
17	47	14	24.6	20.4	2.55	15,000	20,000	0.12	N 303 ECP
20	47	18	29.7	27.5	3.45	16,000	19,000	0.14	NJ 2204 ECP
20	47	14	25.1	22	2.75	16,000	19,000	0.11	NJ 204 ECP
20	47	14	25.1	22	2.75	16,000	19,000	0.11	NU 204 ECP
20	52	15	35.5	26	3.25	15,000	18,000	0.15	N 304 ECP[a]
20	52	21	47.5	38	4.8	15,000	18,000	0.21	NU 2304 ECP[a]
25	47	12	14.2	13.2	1.4	18,000	18,000	0.084	NU 1005
25	52	15	28.6	27	3.35	14,000	16,000	0.13	N 205 ECP
25	52	18	34.1	34	4.25	14,000	16,000	0.17	NUP 2205 ECP
25	52	15	28.6	27	3.35	14,000	16,000	0.14	NU 205 ECJ
25	52	18	34.1	34	4.25	14,000	16,000	0.18	NJ 2205 ECP

Continued

Table 6.7 Continued

Selected cylindrical roller bearings. Single row.									
d (mm)	D (mm)	B (mm)	C (kN)	C_0 (kN)	Fatigue load limit (kN)	Reference speed (rpm)	Limiting speed (rpm)	Mass (kg)	Designation
25	52	18	34.1	34	4.25	14,000	26,000	0.17	NU 2205 ECML
25	62	17	46.5	36.5	4.55	12,000	22,000	0.29	NJ 305 ECML[a]
25	62	24	64	55	6.95	12,000	15,000	0.42	NJ 2305 ECML[a]
25	62	17	46.5	36.5	4.55	12,000	15,000	0.28	NU 305 ECP[a]
25	62	24	64	55	6.95	12,000	15,000	0.39	NJ 2305 ECP[a]
30	55	13	17.9	17.3	1.86	15,000	15,000	0.12	NU 1006
30	62	16	44	36.5	4.55	13,000	22,000	0.22	NUP 206 ECML[a]
30	62	20	55	49	6.1	13,000	22,000	0.29	NU 2206 ECML[a]
30	62	16	44	36.5	4.55	13,000	14,000	0.22	NUP 206 ECP[a]
30	62	16	44	36.5	4.55	13,000	22,000	0.23	NU 206 ECML[a]
30	72	19	58.5	48	6.2	11,000	12,000	0.44	NU 306 ECM[a]
30	72	19	58.5	48	6.2	11,000	12,000	0.4	NU 306 ECP[a]
30	72	19	58.5	48	6.2	11,000	12,000	0.41	NJ 306 ECP[a]
30	72	27	83	75	9.65	11,000	19,000	0.58	NJ 2306 ECML[a]
35	62	14	35.8	38	4.55	12,000	20,000	0.16	NU 1007 ECMP
35	72	17	56	48	6.1	11,000	18,000	0.36	NUP 207 ECML[a]
35	72	23	69.5	63	8.15	11,000	12,000	0.4	N 2207 ECM[a]
35	80	21	75	63	8.15	9500	11,000	0.51	NUP 307 ECP[a]
35	80	31	106	98	12.7	9500	11,000	0.75	NUP 2307 ECP[a]
35	80	21	75	63	8.15	9500	11,000	0.55	N 307 ECM[a]
35	100	25	76.5	69.5	9	8000	9500	1.05	N 407
40	68	15	25.1	26	3	12,000	18,000	0.22	NU 1008 ML
40	80	23	81.5	75	9.65	9500	11,000	0.54	NU 2208 ECP[a]
40	80	18	62	53	6.7	9500	11,000	0.42	NU 208 ECJ[a]
40	80	18	62	53	6.7	9500	11,000	0.42	NU 208 ECP[a]

40	90	23	93	78	10.2	8000	15,000	0.81	NJ 308 ECML[a]
40	90	33	129	120	15.3	8000	9500	0.96	NJ 2308 ECJ[a]
40	110	27	96.8	90	11.8	7000	8500	1.35	NJ 408
40	110	27	96.8	90	11.8	7000	11,000	1.45	NJ 408 MA
45	75	16	44.6	52	6.3	9500	11,000	0.26	NU 1009 ECP
45	75	19	52.8	64	7.8	9000	11,000	0.34	NJ 2009 ECP
45	85	23	69.5	64	8.15	9000	9500	0.46	NUB 209 ECJ
45	85	19	69.5	64	8.15	9000	9500	0.48	NU 209 ECJ[a]
45	100	36	160	153	20	7500	13,000	1.45	NJ 2309 ECML[a]
45	100	25	112	100	12.9	7500	8500	0.95	NUP 309 ECP[a]
45	100	25	112	100	12.9	7500	8500	0.88	N 309 ECP[a]
45	100	25	112	100	12.9	7500	13,000	0.95	NUP 309 ECML[a]
45	100	25	112	100	12.9	7500	8500	1	NU 309 ECNP[a]
45	120	29	106	102	13.4	6700	7500	1.65	NU 409
50	80	16	30.8	34.5	4	10,000	15,000	0.31	NU 1010 ML
50	90	23	90	88	11.4	8500	9000	0.59	NUP 2210 ECNP[a]
50	90	20	73.5	69.5	8.8	8500	14,000	0.51	NUP 210 ECML[a]
50	90	20	73.5	69.5	8.8	8500	9000	0.49	NU 210 ECKP[a]
50	90	20	73.5	69.5	8.8	8500	14,000	0.57	NU 210 ECML[a]
50	90	20	73.5	69.5	8.8	8500	14,000	0.5	NJ 210 ECML[a]
50	110	27	127	112	15	6700	8000	1.15	NJ 310 ECP[a]
50	110	40	186	186	24.5	6700	12,000	2.05	NJ 2310 ECML[a]
50	130	31	130	127	16.6	6000	7000	2	NU 410
55	90	18	57.2	69.5	6.3	8000	8500	0.4	NU 1011 ECP
55	100	25	96.5	95	12.2	7500	8000	0.68	NUB 211 ECJ
55	100	21	96.5	95	12.2	7500	8000	0.72	NU 211 ECM[a]
55	120	43	232	232	15.3	6000	11,000	2.48	NJ 2311 ECML[a]
55	120	29	156	143	18.6	6000	11,000	1.58	NJ 311 ECML[a]
60	95	18	37.4	44	5.3	8000	13,000	0.48	NU 1012 ML
60	110	28	146	153	20	6700	7500	1.1	NU 2212 ECM[a]
60	110	22	108	102	13.4	6700	7500	0.81	NU 212 ECP[a]

Continued

Table 6.7 Continued

Selected cylindrical roller bearings. Single row.									
d (mm)	D (mm)	B (mm)	C (kN)	C_0 (kN)	Fatigue load limit (kN)	Reference speed (rpm)	Limiting speed (rpm)	Mass (kg)	Designation
60	110	22	108	102	13.4	6700	11,000	0.86	NUP 212 ECML[a]
60	130	46	260	265	34.5	5600	10,000	3.2	NJ 2312 ECML[a]
60	130	31	173	160	20.8	5600	10,000	2	NJ 312 ECML[a]
60	130	46	260	265	34.5	5600	6700	2.85	NUP 2312 ECP[a]
60	130	31	173	160	20.8	5600	6700	1.95	NU 312 ECM[a]
60	150	35	168	173	22	5000	6000	3.1	NJ 412
65	100	26	105	146	18.3	6300	7500	0.7	NJ 3013 ECP
65	100	18	38	46.5	5.5	7500	12,000	0.51	NU 1013 ML
65	120	23	122	118	15.6	6300	10,000	1.1	NUP 213 ECML[a]
65	120	31	170	180	24	6300	6700	1.5	NUP 2213 ECP[a]
65	140	48	285	290	38	5300	9500	3.6	NJ 2313 ECML[a]
65	160	37	183	190	24	4800	5600	3.65	NJ 413
70	110	20	56.1	67.2	8	7000	11,000	0.8	NU 1014 ML
70	125	31	180	193	25.5	6000	10,000	1.7	NJ 2214 ECML[a]
70	125	24	137	137	18	6000	6300	1.15	NU 214 ECP[a]
70	125	24	137	137	18	6000	6300	1.25	NJ 214 ECM[a]
70	150	35	236	228	29	4800	5600	2.85	NUP 314 ECM[a]
70	150	51	315	325	41.5	4800	5600	4.15	NUP 2314 ECP[a]
75	115	20	58.3	71	8.5	6700	10,000	0.74	NU 1015 ML
75	130	31	150	156	20.4	5600	6000	1.35	NUB 215 ECJ
75	130	25	150	156	20.4	5600	6000	1.35	NU 215 ECM[a]
75	160	37	280	265	33.5	4500	5300	3.45	NUP 315 ECP[a]
75	160	55	380	400	50	4500	5300	5.1	NUP 2315 ECP[a]
75	190	45	264	280	34	4000	4800	6.9	NJ 415
80	125	22	99	127	16.3	5600	9500	1.1	NJ 1016 ECML

80	140	33	212	245	31	5300	5600	2.1	NUP 2216 ECJ[a]
80	140	26	160	166	21.2	5300	8500	1.65	NUP 216 ECML[a]
80	170	58	415	440	55	4300	7500	6.6	NJ 2316 ECML[a]
80	170	39	300	290	36	4300	7500	4.7	NJ 316 ECML[a]
80	200	48	303	320	39	3800	4500	7.45	NJ 416
85	130	22	68.2	86.5	10.5	6000	6000	1.05	NU 1017 M
85	150	36	190	200	24.5	4800	5300	2.05	NUB 217 ECJ
85	150	28	190	200	24.5	4800	5300	2.05	NU 217 ECM[a]
85	180	41	340	335	41.5	4000	4800	4.9	NUP 317 ECP[a]
85	180	60	455	490	60	4000	4800	7	NUP 2317 ECP[a]
90	140	24	80.9	104	12.7	5600	8500	1.35	NU 1018 ML
90	160	40	280	315	39	4500	5000	3.15	NU 2218 ECP[a]
90	160	30	208	220	27	4500	5000	2.35	NU 218 ECJ[a]
90	190	64	500	540	65.5	3800	6700	8.8	NJ 2318 ECML[a]
90	190	43	365	360	43	3800	6700	6	NJ 318 ECML[a]
90	190	53	365	360	43	3800	4500	5.55	NJ 318 ECJ[a]
90	225	54	380	415	48	3400	4000	10.5	NU 418
95	145	24	84.2	110	13.2	5300	8000	1.4	NU 1019 ML
95	170	32	255	265	32.5	4300	4800	2.85	NU 219 ECJ[a]
95	170	43	325	375	45.5	4300	4800	3.85	NU 2219 ECM[a]
95	200	45	390	390	46.5	3600	4300	6.25	NUP 319 ECP[a]
95	200	67	530	585	69.5	3600	4300	9.75	NUP 2319 ECP[a]
100	150	24	85.8	114	13.7	5000	5000	1.45	NU 1020 M
100	180	46	285	305	36.5	4000	4500	3.75	NUB 220 ECJ
100	180	34	285	305	36.5	4000	4500	3.45	NU 220 ECJ[a]
100	215	73	670	735	85	3200	6000	13.2	NJ 2320 ECML[a]
100	215	47	450	440	51	3200	6000	9	NJ 320 ECML[a]
100	250	58	429	475	53	3000	3600	14	NU 420
105	160	26	101	137	16	4800	4800	1.85	NU 1021 M
105	190	36	300	315	36.5	3800	6300	4	NUP 221 ECML[a]
105	225	49	500	500	57	3200	5600	8.9	NU 321 ECML[a]

Continued

Table 6.7 Continued

Selected cylindrical roller bearings. Single row.									
d (mm)	D (mm)	B (mm)	C (kN)	C_0 (kN)	Fatigue load limit (kN)	Reference speed (rpm)	Limiting speed (rpm)	Mass (kg)	Designation
105	260	60	501	570	64	2800	3400	19	NU 421 M
110	170	28	128	166	19.3	4500	4500	2.3	NU 1022 M
110	200	53	335	365	42.5	3600	4000	5.2	NUB 222 ECJ
110	200	38	335	365	42.5	3600	4000	4.8	NU 222 ECP[a]
110	240	50	530	540	61	3000	3400	11.2	NUP 322 ECP[a]
110	240	80	780	900	102	3000	4500	18.5	NU 2322 ECKMA[a]
110	280	65	532	585	64	2600	3200	20	NU 422
120	180	28	134	183	20.8	4000	6300	2.5	NJ 1024 ML
120	215	58	520	630	72	3400	3600	9	NUP 2224 ECJ[a]
120	215	40	390	430	49	3400	5600	6	NUP 224 ECML[a]
120	260	55	610	620	69.5	2800	5000	15.1	NJ 324 ECML[a]
120	260	86	915	1040	116	2800	4300	24.3	NJ 2324 ECMA[a]
120	260	86	915	1040	116	2800	3200	24	NU 2324 ECM[a]
120	310	72	644	735	78	2400	2800	28	NU 424
127	254	50.8	484	585	65.5	2800	3200	12	CRM 40 AMB
130	200	33	165	224	25	3800	3800	3.8	NJ 1026 M
130	200	42	297	440	51	3200	4800	4.6	NU 2026 ECMA
130	230	40	415	455	51	3200	5300	7	NUP 226 ECML[a]
130	230	64	610	735	83	3200	5300	12.2	NUP 2226 ECML[a]
130	280	58	720	750	81.5	2400	4500	19.4	NJ 326 ECML[a]
130	280	93	1060	1250	137	2400	3800	30.5	NJ 2326 ECPA[a]
140	210	33	179	255	28	3600	5300	4.05	NU 1028 ML
140	250	42	450	510	57	2800	3200	8.6	NU 228 ECJ[a]
140	250	68	655	830	93	2800	4800	15.6	NUP 2228 ECML[a]

140	300	62	780	830	88	2400	4300	23	NJ 328 ECML[a]
140	300	102	1200	1430	150	2400	3600	37.5	NUP 2328 ECMA[a]
150	225	35	198	290	31.5	3200	5000	4.85	NU 1030 ML
150	270	45	510	600	64	2600	4500	11	NUP 230 ECML[a]
150	270	73	735	930	100	2600	2800	18.5	NU 2230 ECM[a]
150	320	65	900	965	100	2200	3400	27.7	NJ 330 ECM[a]
150	320	108	1370	1630	166	2200	3400	46	NJ 2330 ECMA[a]
160	240	48	418	670	72	2600	4000	7.9	NU 2032 ECMA
160	240	38	229	325	35.5	3000	3000	5.95	NU 1032 M
160	290	80	930	1200	129	2400	3400	25	NUP 2232 ECMA[a]
160	290	48	585	680	72	2400	4000	14.5	NUP 232 ECML[a]
160	340	68	1000	1080	112	2000	2400	32.2	NJ 332 ECM[a]
160	340	114	1250	1730	176	2000	3200	53.5	NJ 2332 ECMA
160	340	68	1000	1080	112	2000	3200	32.5	NUP 332 ECMA[a]
170	260	54	473	735	41.5	2400	3600	11	NU 2034 ECMA
170	260	42	275	400	41.5	2800	4300	8.15	NJ 1034 ML
170	310	52	695	815	85	2400	3200	20	NUP 234 ECMA[a]
170	310	86	1060	1430	140	2400	3200	30	NU 2234 ECMA[a]
170	360	120	1450	2040	204	1700	3000	63	NJ 2334 ECMA
170	360	72	952	1180	116	1700	2200	38.5	N 334 ECM
180	280	46	336	475	51	2600	2600	10.5	NU 1036 M
180	320	52	720	850	88	2200	3200	21	NUP 236 ECMA[a]
180	320	86	1100	1430	146	2200	2400	31.5	NU 2236 ECM[a]
180	380	75	1020	1290	125	1600	2200	45	NU 336 ECM
180	380	126	1610	2240	216	1600	2800	71.5	NU 2336 ECMA
180	440	95	1250	1600	150	1600	2400	80	NUP 436 MA
190	260	33	251	400	40.5	2600	4000	5.5	NJ 1938 ECML
190	290	46	347	500	53	2600	3800	11	NJ 1038 ML
190	340	55	800	965	98	2000	2200	24	NU 238 ECM[a]
190	340	92	1220	1600	160	2000	3000	39	NU 2238 ECM[a]
190	400	132	1830	2550	236	1500	2600	82.5	NU 2338 ECMA

Continued

Table 6.7 Continued

Selected cylindrical roller bearings. Single row.									
d (mm)	*D* (mm)	*B* (mm)	*C* (kN)	C_0 (kN)	Fatigue load limit (kN)	Reference speed (rpm)	Limiting speed (rpm)	Mass (kg)	Designation
190	400	78	1140	1500	143	1500	2000	50	NU 338 ECM
200	250	30	183	345	32.5	2800	3600	3.55	NJ 2840 ECMA
200	280	38	341	540	55	2600	3800	7.4	NJ 1940 ECMP
200	310	51	380	570	58.5	2400	2400	14.5	NU 1040M
200	360	58	850	1020	100	1900	2200	28.5	NU 240 ECM[a]
200	360	128	1540	2450	236	1600	2400	59	NU 3240 ECM
200	420	80	1230	1630	150	1400	2400	57.5	NU 340 ECMA
200	420	138	1980	2800	255	1400	2400	96.5	NU 2340 ECMA
220	300	38	336	560	55	2400	3000	8.3	NJ 1944 ECMA
220	300	48	457	830	83	2400	3600	10	NU 2944 M2P
220	340	56	495	735	73.5	2200	3200	19	NJ 1044 MP
220	400	65	1060	1290	125	1600	1900	38	NU 244 ECM[a]
220	400	108	1570	2280	212	1600	2400	62.5	NU 2244 ECMA
220	460	145	2380	3450	310	1300	2200	120	NU 2344 ECMA
220	460	88	1210	1630	150	1500	1700	72.5	NU 344 M
240	300	28	201	365	32	2400	3600	4.5	NUZ 1848 ECMA
240	360	56	523	800	78	2000	2600	20	NU 1048 MA
240	440	72	952	1370	129	1600	2200	52.5	NJ 248 MA
240	440	120	1450	2360	224	1500	2200	85	NJ 2248 MA
240	500	155	2600	3650	320	1200	2000	155	NU 2348 ECMA
240	500	95	1450	2000	180	1300	1600	94.5	NU 348 M
260	400	65	627	965	96.5	1800	1800	29.5	NU 1052 M
260	400	104	1420	2320	232	1500	2400	49.5	NU 3052 ECMA
260	440	144	2090	3450	310	1300	2200	98	NU 3152 ECMA

260	480	80	1170	1700	156	1400	2000	72	NUP 252 MA
260	480	130	1790	3000	275	1300	2000	112	NJ 2252 MA
260	540	165	3190	4500	380	1100	1800	190	NU 2352 ECMA
260	540	102	1940	2700	236	1100	1800	125	NU 352 ECMA
280	420	82	1230	2160	204	1400	2200	40	NU 2056 ECMA
280	420	65	660	1060	102	1700	2200	31.5	NU 1056 MA
280	460	146	2290	3900	335	1200	2000	105	NU 3156 EMA/HB1
280	500	80	1140	1700	153	1400	1900	71.5	NU 256 MA
280	500	165.1	2920	5000	430	1200	1500	160	BC1B 322,312
280	500	130	2200	3250	285	1200	1900	115	NU 2256 ECMA
280	580	175	2700	4300	365	1000	1700	230	NU 2356 MA
300	380	48	473	980	88	1700	1700	14	NU 2860 ECM
300	460	95	1510	2750	255	1300	2000	59.5	NU 2060 ECMA
300	460	74	858	1370	129	1500	2000	46.5	NU 1060 MA
300	540	140	2090	3450	300	1200	1800	145	NU 2260 MA
300	540	85	1420	2120	183	1300	1400	89.5	NU 260M
300	620	185	4020	5850	480	950	1600	270	NU 2360 ECMA
320	400	48	495	1060	93	1600	2400	14.5	NU 2864 ECMA
320	400	38	369	710	60	1600	2400	11	NU 1864 ECMP
320	440	72	765	1500	140	1400	1400	34.5	314,756
320	440	56	693	1200	110	1500	2000	27	NU 1964 ECMA
320	480	121	1830	3200	285	1200	1900	75	NU 3064 ECM
320	480	95	1450	265	245	1200	1900	62.5	NU 2064 ECMA
320	480	74	880	1430	132	1400	1400	49	NJ 1064 MA
320	540	176	3140	5400	450	1000	1700	175	NU 3164 ECMA
320	580	92	1610	2450	204	1200	1600	115	NU 264 MA
320	580	150	3190	5000	415	1000	1600	180	NU 2264 ECMA
340	420	48	512	1120	98	1500	2000	15	NJ 2868 ECMA
340	420	35	212	400	33.5	1500	2000	9.6	316,197
340	460	72	809	1660	146	1400	1400	36	NU 2968 M
340	460	56	682	1200	108	1400	1900	27.5	NU 1968 ECMA

Continued

Table 6.7 Continued

Selected cylindrical roller bearings. Single row.									
d (mm)	D (mm)	B (mm)	C (kN)	C_0 (kN)	Fatigue load limit (kN)	Reference speed (rpm)	Limiting speed (rpm)	Mass (kg)	Designation
340	520	82	1080	1760	156	1300	1700	68	NJ 1068 MA
340	580	190	3190	5700	475	950	1600	210	NU 3168 ECMA
340	620	165	2640	4500	365	1000	1500	220	NU 2268 MA
350	480	85	1060	2160	196	1200	1300	48	612,129 A
360	540	106	1940	3600	320	1100	1600	88.5	NU 2072 ECMA
360	540	82	1100	1830	163	1300	1600	67.5	NU 1072 MA
360	600	192	3410	5700	415	900	1500	225	NU 3172 ECMA
360	650	170	2920	4900	400	950	1400	250	NU 2272 MA
360	750	224	5010	8150	630	850	1300	510	NU 2372 ECMA
380	480	46	539	1060	86.5	1300	1300	23	NU 1876 ECM
380	480	40	270	520	37.5	1300	1300	15.5	316,010 A
380	560	135	2380	4750	415	1000	1800	115	NU 3076 ECMP
380	560	106	1980	3750	335	1000	1600	92.5	NU 2076 ECMA
380	560	82	1140	1930	170	1200	1600	71	NU 1076 MA
380	680	175	3960	6400	510	850	1300	275	NU 2276 ECMA
400	500	46	572	1180	100	1300	1900	21.8	NJ 1880 MP
400	540	82	1400	2850	255	1200	1600	57	NU 2980 ECMA
400	540	65	952	1730	153	1200	1600	41	NU 1980 ECMA
400	600	118	2200	4750	380	950	1400	120	NU 2080 ECMA
400	600	148	2810	5500	440	950	1400	175	NU 3080 MA6
400	600	90	1380	2320	204	1100	1500	92.5	NU 1080 MA
420	520	75	935	2240	196	1100	1600	33	NJ 3884 MA
420	520	46	550	1120	95	1200	1600	20.5	NJ 1884 MA6
420	520	46	550	1120	95	1200	1600	20	NU 1884 MA

420	560	82	1210	2550	228	1100	1500	59	NU 2984 MA
420	560	65	968	1800	156	1100	1500	48	NJ 1984 ECMA
420	620	150	2920	5400	430	900	1400	160	NU 3084 ECMA
420	620	118	2460	4750	380	900	1400	125	NU 2084 ECMA
420	620	90	1420	2450	212	1100	1400	96	NU 1084 MA
420	700	224	4950	9000	695	750	1300	380	NU 3184 ECMA/HB1
440	540	60	809	1900	153	1100	1500	34	NU 2888 ECMA
440	540	40	292	610	41.5	1100	1500	18.5	316,011
440	600	74	1060	2000	17	1100	1400	64	NU 1988 MA
440	600	95	1720	3600	310	1100	1400	84	NJ 2988 ECM/HB1
440	600	118	1980	4250	355	900	1100	105	NUP 3988 ECM
440	650	122	2550	4900	390	8500	1300	145	NU 2088 ECMA
440	650	94	1510	2650	212	1000	1300	105	NU 1088 MA
440	720	226	5120	9650	735	700	1200	395	NU 3188 ECMA/HB1
460	580	56	825	1700	140	1100	1400	37	N 1892 M2P/HB3
460	580	72	1080	2400	193	1100	1400	48	NJ 2892 ECMA
460	620	95	1720	3600	310	1000	1300	89	NJ 2992 ECMA
460	680	163	3470	6400	500	800	1200	210	NU 3092 ECMA
460	680	128	2810	5400	415	800	1200	165	NU 2092 ECMA
460	680	100	1650	2850	224	950	1200	115	NU 1092 MA
460	760	240	5280	9650	735	670	1100	455	NU 3192 ECMA/HB1
460	830	212	5120	8650	655	700	1100	530	NU 2292 MA
460	830	165	4180	6800	510	750	1100	415	NU 1292 MA
480	600	56	765	1630	120	1000	1300	39	NU 1896 MA
480	600	72	1100	2450	200	1000	1300	47.5	NJ 2896 ECMA
480	700	128	2860	5600	430	750	1200	170	NU 2096 ECMA
480	700	100	1680	3000	232	900	1200	130	NU 1096 MA
480	790	248	5940	10,800	800	630	1100	500	NU 3196 ECMA/HB1
500	620	72	1170	2700	204	950	1300	50	NU 28/500 ECMA
500	620	45	396	830	55	1000	1000	29	316,198
500	620	56	825	1730	129	950	950	39	NF 18/500 ECM

Continued

Table 6.7 Continued

Selected cylindrical roller bearings. Single row.									
d (mm)	*D* (mm)	*B* (mm)	*C* (kN)	C_0 (kN)	Fatigue load limit (kN)	Reference speed (rpm)	Limiting speed (rpm)	Mass (kg)	Designation
500	670	128	2330	5200	440	800	1200	130	N 39/500 ECMB/W33
500	670	78	1210	2360	193	900	1200	79	NU 19/500 MA
500	720	128	2920	5850	430	750	1100	180	NU 20/500 ECMA
500	720	167	4020	8000	610	750	1100	225	NU 30/500 ECFR
500	720	100	1720	3100	236	900	1100	135	NU 10/500 MA
500	830	264	6440	12,000	880	600	1000	595	NU 31/500 ECMA/HB1
500	920	185	5280	8500	620	670	950	585	NU 12/500 MA
530	650	45	418	900	62	2200	3400	30	315,835 A
530	650	72	1170	2750	220	900	1200	52	NJ 28/530 ECMA
530	710	82	1570	3050	240	850	1100	94	NJ 19/530 ECM/HB1
530	710	106	2380	5000	415	850	1100	120	NU 29/530 ECMA
530	780	145	3740	7350	550	670	1000	255	NU 20/530 ECMA
530	780	112	2290	4050	305	800	1000	190	NU 10/530 MA
530	870	272	7480	14,600	1040	560	950	660	NU 31/530 ECMA/HB1
560	680	56	809	1830	134	850	1100	44.5	NJ 18/560 ECMA6/343016
560	680	45	429	950	62	2000	3200	32	BC1B 320,938/HB3
560	750	85	1650	3250	260	800	1000	110	NU 19/560 ECMA
560	750	112	2460	5400	425	800	1000	145	NU 29/560 ECMA
560	820	150	3800	7650	560	630	1000	290	NU 20/560 ECMA
560	820	115	2330	4250	310	750	1000	210	NU 10/560 MA
560	1030	206	7210	11,200	780	560	800	805	NU 12/560 MA
600	730	60	897	2040	146	800	1000	50.5	NU 18/600 ECMA/HB1
600	730	52	468	1020	65.5	1800	2400	41	315,836
600	730	78	1320	3450	260	750	1000	70	NU 28/600 ECMA

600	800	118	2920	6550	475	750	950	165	NU 29/600 ECMA/HB1
600	800	90	1900	3800	275	750	950	130	NU 19/600 ECMA/HA1
600	870	118	2750	5100	365	700	900	245	NU 10/600 MA
600	870	200	5390	11,000	780	600	900	415	NU 30/600 ECMA/HA1
600	870	155	4180	8000	570	600	900	325	NU 20/600 ECMA
630	780	88	1570	3800	285	750	950	98	NU 28/630 MA
630	780	56	605	1290	85	1700	2600	55	315,933
630	780	88	1570	3900	305	750	950	100	N 28/630 MB
630	780	112	2200	5700	450	670	670	120	N 38/630 MB
630	780	69	1100	2500	183	800	1000	75	NJ 18/630 ECMA/HB1
630	850	128	3300	7200	510	700	900	230	NJ 29/630 ECMAS/HB1
630	850	100	2240	4400	315	700	900	165	NU 19/630 ECMA
630	920	128	3410	6200	430	630	850	285	NU 10/630 ECMA/HA2
630	920	212	6440	14,300	1000	530	850	490	NU 30/630/342771 A
630	920	170	4730	9500	670	560	850	400	NU 20/630 ECMA
630	1150	230	8580	13,700	915	450	700	1100	NU 12/630 ECMA
640	790	56	605	1290	83	1700	2200	56	315,837
670	820	69	1230	2800	204	670	900	84.5	NJ 18/670 MA6/343016
670	820	56	616	1370	83	1600	2200	60	316,012
670	900	103	2330	4750	325	630	850	195	NU 19/670 ECMA
670	980	136	3740	6800	465	600	800	350	NU 10/670 ECMA
670	980	180	5390	11,000	750	500	800	480	NU 20/670 ECMA
670	980	230	6600	14,000	950	500	800	600	NU 30/670 MA/342771
710	870	74	1450	3350	236	630	850	93.5	NU 18/710 ECMA/HB1
710	870	95	1940	5000	365	630	850	130	N 28/710 ECMB
710	950	140	3740	8300	570	600	800	295	NU 29/710 ECMA/HA1
710	1030	140	4680	8500	570	560	750	415	NU 10/710 ECN2MA
710	1030	185	5940	12,000	815	480	700	540	NU 20/710 ECMA
720	880	62	704	1560	91.5	1400	1900	74.5	315,799
750	920	78	1470	3450	228	600	800	110	NU 18/750 MA
750	1000	112	2700	5600	365	560	700	255	319,166 A

Continued

Table 6.7 Continued

Selected cylindrical roller bearings. Single row.									
d (mm)	*D* (mm)	*B* (mm)	*C* (kN)	C_0 (kN)	Fatigue load limit (kN)	Reference speed (rpm)	Limiting speed (rpm)	Mass (kg)	Designation
750	1090	150	4730	8800	585	430	670	490	NU 10/750 ECN2MA
750	1090	195	7040	14,600	980	430	670	635	NU 20/750 ECMA
800	980	82	1720	4150	280	530	700	145	NJ 18/800 ECMA
800	1150	200	7040	14,600	950	400	630	715	NU 20/800 ECMA
800	1150	155	5500	10,600	695	480	630	560	NU 10/800 ECMA/HB1
820	990	72	858	1960	112	1200	1600	100	315,800 B
820	990	72	858	1960	112	1200	1600	100	315,800
850	1030	106	2120	6000	405	500	670	195	N 28/850 MB
850	1030	74	935	2120	122	1200	1600	115	316,200
850	1030	106	2120	6000	405	500	670	190	NU 28/850 MA
850	1120	118	3190	6950	440	480	600	330	NU 19/850 ECMA/HA1
850	1120	155	4680	11,200	720	480	600	430	N 29/850 ECMB6
900	1090	112	2700	7200	490	450	600	240	N 28/900 MB
900	1090	85	1120	2550	156	1100	1500	145	319,161
900	1090	112	2700	7200	490	450	600	235	NU 28/900 MA/343017
900	1180	165	5830	14,000	880	430	560	560	NU 29/900 ECMA/HB1
900	1180	122	4130	8800	560	430	560	380	NU 19/900 ECMA
950	1150	90	1340	3100	190	1000	1300	170	315,869 A
950	1250	175	5830	14,000	880	400	530	745	NU 29/950 ECMA
1000	1220	100	2640	6550	400	400	530	265	NU 18/1000 MA/HB1
1000	1220	128	3690	10,000	620	400	530	350	NF 28/1000 ECMP/HA1
1000	1320	185	7040	17,300	1080	360	480	700	NU 29/1000 ECFR
1030	1250	100	1510	3450	216	320	380	225	BC1B 319,579
1060	1280	128	3580	10,400	640	300	360	360	N 28/1060 MB

1060	1400	195	7210	17,300	1060	280	340	870	NU 29/1060 ECMA/HB1
1060	1400	250	9130	24,000	1460	260	320	1070	NU 39/1060 ECKMA/HA1
1060	1500	325	13,000	32,500	1950	220	280	1900	N 30/1060/241663
1120	1360	104	1650	3800	224	220	280	285	316,201
1120	1580	345	15,700	39,000	2320	190	240	2150	NU 30/1120 K/VA901
1180	1420	106	3030	7800	455	260	320	350	NJ 18/1180 ECMA
1180	1540	206	8970	21,600	1290	190	240	1090	NUP 29/1180 ECMA/HA1
1180	1540	272	11,200	29,000	1730	200	260	1400	N 39/1180 MB
1180	1540	206	8970	21,600	1290	190	240	1050	NU 29/1180 ECMA/HA1
1250	1500	106	1720	4150	240	220	280	330	315,913
1250	1750	290	12,800	30,500	1760	170	200	2320	N 20/1250 MB
1320	1600	122	3800	10,000	570	260	340	530	NU 18/1320 MAS/343016
1320	1720	230	11,400	30,500	1760	240	320	1500	NU 29/1320 ECFR
1320	1720	300	13,200	34,000	1960	220	320	1900	N 39/1320 MB
1320	1850	400	21,600	55,000	3100	190	240	3550	N 30/1320/VE901
1400	1700	175	6600	18,300	1040	220	280	860	N 28/1400 ECMP
1500	1820	140	3300	8000	440	200	260	665	319,301
1700	2060	160	3690	9150	490	160	200	925	319,286
1700	2060	160	7210	19,300	1040	160	200	1150	NU 18/1700 ECMA/HB1

Note this table presents a selection only of the bearings available. Table after SKF.
[a]SKF explorer bearing.

Table 6.8 Single row angular contact ball bearings

Selected angular contact ball bearings.
Single row.

d (mm)	D (mm)	B (mm)	C (kN)	C_0 (kN)	Fatigue load limit (kN)	Reference speed (rpm)	Limiting speed (rpm)	Mass (kg)	Designation
10	30	9	7.02	3.35	0.14	30,000	30,000	0.03	7200 BECBP
10	30	9	7.02	3.35	0.14	30,000	30,000	0.03	7200 BEP
12	32	10	7.61	3.8	0.16	26,000	26,000	0.036	7201 BECBP
12	32	10	7.61	3.8	0.16	26,000	26,000	0.036	7201 BEGAP
12	32	10	7.61	3.8	0.16	26,000	26,000	0.036	7201 BEP
12	37	12	10.6	5	0.208	24,000	24,000	0.063	7301 BEP
15	35	11	9.5	5.1	0.216	26,000	26,000	0.045	7202 BECBP[a]
15	35	11	9.5	5.1	0.216	26,000	26,000	0.045	7202 BEGAP[a]
15	35	11	9.5	5.1	0.216	26,000	26,000	0.045	7202 BEGBP[a]
15	35	11	8.84	4.8	0.204	24,000	24,000	0.045	7202 BEP
15	42	13	13	6.7	0.28	20,000	20,000	0.081	7302 BECBP
15	42	13	13	6.7	0.28	20,000	20,000	0.081	7302 BEP
17	40	12	11	5.85	0.25	22,000	22,000	0.07	7203 BECBM[a]
17	40	12	10.4	5.5	0.236	20,000	20,000	0.064	7203 BEP
17	47	14	15.9	8.3	0.355	19,000	19,000	0.11	7303 BECBP
17	47	14	15.9	8.3	0.355	19,000	19,000	0.11	7303 BEGAP

17	47	14	15.9	8.3	0.355	19,000	19,000	0.11	7303 BEP
20	47	14	13.3	7.65	0.325	18,000	19,000	0.11	7204 BECBM
20	47	14	14.3	8.15	0.345	19,000	19,000	0.11	7204 BECBP[a]
20	52	15	19	10	0.425	18,000	18,000	0.15	7304 BECBM[a]
20	52	15	17.4	9.5	0.4	16,000	16,000	0.14	7304 BEP
25	52	15	15.6	10	0.43	17,000	17,000	0.14	7205 BECBM[a]
25	52	15	15.6	10.2	0.43	15,000	15,000	0.13	7205 BECBY
25	62	17	26.5	15.3	0.655	15,000	15,000	0.24	7305 BECBM[a]
25	62	17	24.2	14	0.6	14,000	14,000	0.23	7305 BEP
25	80	21	39.7	23.6	1	11,000	11,000	0.61	7405 BCBM
30	62	16	24	15.6	0.655	14,000	14,000	0.19	7206 BECAP[a]
30	72	19	35.5	21.2	0.9	13,000	13,000	0.37	7306 BECBM[a]
30	72	19	34.5	21.2	0.9	12,000	12,000	0.37	7306 BECBY
30	72	19	32.5	19.3	0.815	12,000	12,000	0.33	7306 BEP
30	90	23	47.5	29	1.22	10,000	10,000	0.85	7406 BM
35	72	17	31	20.8	0.88	12,000	12,000	0.3	7207 BECBM[a]
35	80	21	41.5	26.5	1.14	11,000	11,000	0.49	7307 BECBM[a]
35	80	21	39	24.5	1.04	10,000	10,000	0.45	7307 BEP
35	100	25	60.5	38	1.6	8500	9000	1.1	7407 BCBM
40	80	18	36.4	26	1.1	10,000	10,000	0.38	7208 BECBJ
40	80	18	34.5	24	1.02	10,000	10,000	0.39	7208 BEM
40	90	23	50	32.5	1.37	10,000	10,000	0.61	7308 BECAP[a]
40	110	27	70.2	45	1.9	8000	8000	1.4	7408 BCBM
45	85	19	37.7	28	1.2	9000	9000	0.43	7209 BECBJ
45	85	19	38	28.5	1.22	10,000	10,000	0.44	7209 BECBM[a]
45	100	25	61	40.5	1.73	9000	9000	0.82	7309 BECAP[a]
45	100	25	60.5	41.5	1.73	8000	8000	0.86	7309 BECBJ
45	120	29	85.2	55	2.36	7000	7500	1.8	7409 BCBM
47.625	114.3	26.988	83.2	58.5	2.28	8000	8000	1.25	AMS 15 ABP
50	90	20	39	30.5	1.29	8500	8500	0.47	7210 BECBJ
50	90	20	40	31	1.32	9000	9000	0.51	7210 BECBM[a]

Continued

Table 6.8 Continued

Selected angular contact ball bearings. Single row.									
d (mm)	D (mm)	B (mm)	C (kN)	C_0 (kN)	Fatigue load limit (kN)	Reference speed (rpm)	Limiting speed (rpm)	Mass (kg)	Designation
50	90	20	39	30.5	1.29	8500	8500	0.47	7210 BEY
50	110	27	74.1	51	2.2	7500	7500	1.13	7310 BECBJ
50	110	27	75	51	2.16	8000	8000	1.16	7310 BEGAM[a]
50	130	31	95.6	64	2.7	6300	6700	2.25	7410 BM
50.8	114.3	26.988	83.2	58.5	2.28	8000	8000	1.2	AMS 16 ABP
55	100	21	48.8	38	1.63	7500	7500	0.62	7211 BECBJ
55	100	21	49	40	1.66	8000	8000	0.66	7211 BECBM[a]
55	120	29	85.2	60	2.55	6700	6700	1.48	7311 BECBJ
55	120	29	85.2	60	2.55	6700	6700	1.48	7311 BECBY
55	140	33	111	76.5	3.25	6000	6300	2.75	7411 BCBM
55	140	33	111	76.5	3.25	6000	6300	2.75	7411 BM
57.15	127	31.75	111	80	3.35	6700	6700	1.6	AMS 18 ABP
60	110	22	57.2	45.5	1.93	7000	7000	0.83	7212 BECBJ
60	110	22	61	50	2.12	7500	7500	0.85	7212 BECBM[a]
60	130	31	104	76.5	3.2	6700	6700	1.71	7312 BECAP[a]
60	150	35	119	86.5	3.55	5600	5600	3.3	7412 BCBM
63.5	139.7	31.75	124	91.5	3.8	6300	6300	2	AMS 20 ABP
65	120	23	66.3	54	2.28	6300	6300	1	7213 BECBJ
65	120	23	69.5	57	2.45	6700	6700	1.1	7213 BECBM
65	140	33	108	80	3.35	5600	5600	2.35	7313 BECBF
65	160	37	130	96.5	3.8	5000	5300	3.85	7413 BGAM
70	125	24	71.5	60	2.5	6000	6000	1.1	7214 BECBJ
70	125	24	75	64	2.7	6300	6300	1.1	7214 BECBP[a]
70	150	35	127	98	3.9	5600	5600	2.55	7314 BECAP[a]

70	180	42	159	127	4.8	4500	4800	5	7414 BCBM
75	130	25	73.5	65.5	2.7	6300	6300	1.29	7215 BECBM
75	160	37	132	104	4.15	5300	5300	3.26	7315 BECBM[a]
75	190	45	168	140	5.1	4300	4500	6.85	7415 BCBM
80	140	26	85	75	3.05	5600	5600	1.59	7216 BECBM[a]
80	170	39	143	118	4.5	5000	5000	4.03	7316 BECBM[a]
80	170	39	143	118	4.5	4500	4500	3.7	7316 BEGAY
80	200	48	178	153	5.5	4000	4300	8	7416 CBM
80	200	48	178	153	5.5	4000	4300	8	7416 M
85	150	28	95.6	83	3.25	5000	5000	1.9	7217 BECBJ
85	150	28	102	90	3.55	5300	5300	1.83	7217 BECBP[a]
85	180	41	156	132	4.9	4800	4800	4.74	7317 BECBM[a]
85	180	41	153	132	4.9	4300	4300	4.59	7317 BECBY
85	210	52	190	166	5.85	3800	4000	10.3	7417 BACBMC
90	160	30	108	96.5	3.65	4500	4500	2.34	7218 BECBJ
90	160	30	108	96.5	3.65	4500	4800	2.41	7218 BECBM
90	190	43	165	146	5.2	4000	4000	5.22	7318 BECBJ
90	190	43	166	146	5.3	4500	4500	5.53	7318 BECBM[a]
90	225	54	216	200	6.7	3600	3800	11.5	7418 CBM
90	225	54	216	200	6.7	3600	3800	11.5	7418 GAM
95	170	32	129	118	4.4	4800	4800	2.95	7219 BECBM[a]
95	200	45	180	163	5.7	4300	4300	6.41	7319 BECBM[a]
95	200	45	168	150	5.2	3800	3800	6.41	7319 BEGAF
95	250	55	251	245	7.8	3400	3400	13.5	7419 M
100	180	34	135	122	4.4	4000	4300	3.61	7220 BECBM
100	215	47	216	208	6.95	4000	4000	8	7320 BECBM[a]
100	215	47	203	190	6.4	3600	3600	8	7320 BEGAF
100	265	60	276	275	8.65	3200	3200	15.5	7420 CBM
105	190	36	148	137	4.8	3800	4000	4.18	7221 BECBM
105	225	49	203	193	6.4	3400	3600	9.12	7321 BECBM

Continued

Table 6.8 Continued

Selected angular contact ball bearings. Single row.									
d (mm)	*D* (mm)	*B* (mm)	*C* (kN)	C_0 (kN)	Fatigue load limit (kN)	Reference speed (rpm)	Limiting speed (rpm)	Mass (kg)	Designation
110	200	38	153	143	4.9	3600	3800	4.95	7222 BECBM
110	240	50	225	224	7.2	3200	3400	10.7	7322 BECBM
120	180	28	87.1	93	3.2	3800	4000	2.4	7024 BGM
120	215	40	165	163	5.3	3400	3600	5.89	7224 BM
120	260	55	238	250	7.65	3000	3200	13.8	7324 BGBM
130	230	40	186	193	6.1	3200	3400	6.76	7226 BCBM
130	280	58	276	305	9	2800	2800	17.1	7326 BCBM
140	210	33	114	129	4.15	3200	3400	3.85	7028 BGM
140	250	42	199	212	6.4	2800	3000	8.83	7228 BCBM
140	300	62	302	345	9.8	2600	2600	21.3	7328 BCBM
150	225	35	133	146	4.55	3000	3200	4.7	7030 BGM
150	270	45	216	240	6.95	2600	2800	10.8	7230 BCBM
150	320	65	332	390	10.8	2400	2400	25	7330 BCBM
160	290	48	255	300	8.5	2400	2600	13.6	7232 BCBM
160	340	68	358	440	11.8	2200	2400	30.3	7332 BCBM
170	310	52	281	345	9.5	2400	2400	16.7	7234 BCBM
170	360	72	390	490	12.7	2000	2200	34.6	7334 BCBM
180	280	46	195	240	6.7	2400	2600	10	7036 BGM
180	320	52	291	375	10	2200	2400	17.6	7236 BCAM
180	320	52	291	375	10	2200	2400	17.6	7236 BCBM
180	380	75	410	540	13.7	2000	2000	40	7336 BCBM
190	340	55	307	405	10.4	2000	2200	21.9	7238 BCBM
190	400	78	442	600	14.6	1900	1900	48.3	7338 BCBM
200	360	58	325	430	11	1800	2000	25	7240 BCBM

200	420	80	462	655	15.6	1800	1800	52.8	7340 BCBM
220	340	56	255	355	9	2000	2000	18	7044 BGM
220	400	65	390	560	13.4	1800	1800	35.2	7244 BCBM
220	460	88	494	720	16.6	1600	1700	70	7344 BCBM
240	360	56	260	375	9.15	1800	1900	19	7048 BGM
240	440	72	364	540	12.5	1600	1700	49	7248 BCBM
240	500	95	559	865	19	1400	1500	88.5	7348 BCBM
260	480	80	507	780	17.3	1400	1500	66	7252 BM
280	420	65	338	540	12.2	1500	1600	30	7056 BGM
280	500	80	520	850	18.3	1400	1400	69.5	7256 BM
300	460	74	423	695	15.3	1500	1600	42.5	7060 AGM
320	440	56	351	585	12.9	1500	1600	25.5	71,964 AC
320	480	74	390	670	14	1300	1400	44.5	7064 BGM
335	450	56	281	480	10.4	1400	1400	25.5	466,952
340	460	56	338	585	12.5	1500	1500	24	71,968 ACMB
340	520	82	449	815	16.6	1200	1300	61.5	7068 BGM
340	620	92	702	1340	26	1100	1100	125	7268 BGM
360	440	38	234	425	9	1500	1500	12	71,872 ACM/ P5
360	480	56	351	630	13.2	1400	1500	28.5	71,972 ACMB
360	540	82	520	950	19	1300	1300	62.5	7072 AGM
380	480	31	190	345	8	1300	1400	13.5	70,876 AMB
380	560	82	507	950	18.6	1200	1300	65.5	7076 AM
400	540	65	423	780	15.6	1200	1300	42	71,980 AM
400	600	90	605	1180	22.4	950	1200	89.5	307,238
400	720	103	728	1500	26.5	900	950	190	7280 BM
410	560	70	423	830	16	1100	1100	49.5	468,431
420	560	65	423	800	16	1200	1200	44.5	71,984 AM
420	620	90	605	1180	22	1100	1100	95	7084 AM
440	650	94	650	1320	24	1000	1100	100	7088 AM
460	580	37	265	560	10.4	1100	1100	24.5	70,892 AM

Continued

Table 6.8 Continued

Selected angular contact ball bearings. Single row.									
d (mm)	*D* (mm)	*B* (mm)	*C* (kN)	C_0 (kN)	Fatigue load limit (kN)	Reference speed (rpm)	Limiting speed (rpm)	Mass (kg)	Designation
460	620	74	507	1040	19	100	1100	58	71,992 ACM
460	680	100	689	1460	25.5	950	1000	120	7092 AM
465	635	76	468	980	17.6	900	950	70.5	307,352
480	700	100	702	1530	26.5	950	1000	125	7096 AM
500	620	37	276	620	11	1000	1000	27	708/500 AMB
500	670	78	553	1220	21.2	950	1000	78	719/500 AGMB
500	720	100	715	1600	27.5	900	950	130	70/500 AM
530	650	56	390	900	15.6	950	1000	39.5	718/530 AMB
530	710	82	618	1340	23.6	850	900	92	719/530 ACM
530	760	100	702	1600	27.5	750	800	150	307,368 B
530	780	112	832	1900	31.5	800	850	175	70/530 AM
540	630	45	260	600	10.6	850	900	21.5	BA1B 311,585
560	750	85	592	1290	22.8	800	850	105	719/560 AMB
600	730	42	338	735	14	800	850	38.5	708/600 AGMB
600	800	90	715	1730	27.5	750	800	125	719/600 ACM
600	870	118	884	2160	34	700	750	230	70/600 AGMB
630	700	22	135	440	7.2	750	750	11.5	BA1B 311,712
630	920	128	956	2450	37.5	560	700	270	70/630 AMB
670	820	69	553	1290	22.4	700	750	77	718/670 ACMB
670	980	136	1170	3100	45.5	560	600	340	70/670 AMB
710	870	74	605	1630	24.5	630	670	93.5	718/710 ACMB

710	1030	140	1190	3250	46.5	560	600	370	70/710 AMB
750	920	78	650	1800	26.5	600	630	110	718/750 ACMB
750	1090	150	1300	3650	52	530	560	445	70/750 AMB
762	889	63.5	449	1270	18.6	560	600	58	BA1B 311,576
800	1130	120	1080	3200	44	450	480	395	BA1B 311,745
800	1150	155	1330	3800	52	480	500	500	70/800 AMB
850	1030	82	689	1860	29	500	530	140	718/850 AMB
850	1220	165	1530	4650	61	450	480	595	70/850 AMB
900	1030	63	416	1270	17.3	450	480	90	BA1B 307,788
900	1280	170	1560	4900	63	400	430	665	70/900 AMB
950	1360	180	1630	5200	65.5	380	400	805	70/950 AMB
1000	1220	100	923	2750	39	400	430	245	718/1000 AMB
1000	1320	103	832	2450	34.5	340	360	370	307,101C
1000	1420	185	1630	5400	67	360	380	890	70/1000 AMB
1060	1500	195	1680	5700	71	320	340	1050	70/1060 AMB
1120	1360	106	1060	3750	45	340	360	320	718/1120 AMB
1120	1580	200	1720	5850	71	300	320	1150	70/1120 AMB
1180	1660	212	1740	6200	73.5	280	280	1350	70/1180 AMB
1250	1500	80	806	2700	36	280	300	295	708/1250 AMB
1250	1750	218	1780	6550	76.5	240	260	1600	70/1250 AMB
1700	1900	80	975	4550	45.5	160	170	310	BA1B 307,756

[a]SKF explorer bearing.

Note this table presents a selection only of the bearings available. Table after SKF.

The required dynamic loading is given by.

$$C = PL^{1/3} = 2800 \times 900^{1/3} = 27{,}033\,\text{N}.$$

Reference to the deep groove bearing chart (Table 6.6) shows a suitable bearing could be:

ISO designation 6306 [a],
bore diameter 30 mm, outer diameter 72 mm,
width 19 mm,
$C = 29{,}600$ N, $C_o = 16{,}000$ N,
Reference speed 20,000 rpm,
Limiting speed 13,000 rpm,

(ii) Normally it would be necessary to determine the viscosity ratio, κ, and $\eta_c P_u/P$ to determine a_{SKF}. In this case the SKF life modification factor has been given. Nevertheless the steps are included below.

The viscosity ratio is given by $\kappa = 40/14 = 2.86$.
The contamination factor is that for high cleanliness, $\eta_c = 0.6$.

$$\eta_c \frac{P_u}{P} = 0.6 \frac{0.67}{2.8} = 0.6 \times 0.239 = 0.14$$

From manufacturer's tables, using κ and $\eta_c P_u/P$, $a_{SKF} = 14$.
Hence from Eq. (6.5), for a desired reliability of 90%,

$$L_{nm} = a_1 a_{SKF} \left(\frac{C}{P}\right)^p = 1 \times 14 \times \left(\frac{29600}{2800}\right)^3 = 16540 \text{ million revolutions}$$

This high life suggests either very extended operation or an opportunity for an alternative design that does not require this quality of bearing installation.

6.3 Bearing installation

The practical use of rolling element bearings requires consideration of their installation as well as correct selection. Bearing installation considerations include the bearing combination, the mounting of the bearings and the provision of lubrication.

A typical application of rolling element bearings is the support of a rotating shaft. If the operating temperature of the machine varies, the shaft length can grow relative to the casing or mounting arrangement. An idea of the magnitude of the axial shaft growth can be estimated by:

$$\Delta L = L_o \alpha \Delta T \tag{6.7}$$

where:

ΔL is the change in length (m),
L_o is the original length (m),
α is the coefficient of linear thermal expansion (°C^{-1}),
ΔT is the temperature rise (°C).

For a gas turbine engine the difference in temperature between the casing and the shaft can readily be of the order of 50°C. If the original length of the steel shaft was 1.0 m the growth of the shaft would be

$$\Delta L = L_o \alpha \Delta T = 1.0 \times 11 \times 10^{-6} \times 50 = 5.5 \times 10^{-4}\,\text{m} = 0.55\,\text{mm}.$$

This is a considerable axial movement within a machine and must be allowed for, if significant loadings and resultant stresses and possible contact between stationary and rotating components are to be avoided. A typical solution to this kind of situation is to allow for a limited axial movement on one bearing as illustrated in Fig. 6.5. Here one bearing has the location of its inner and outer races 'fixed' relative to the shaft and housing by means of shoulders and locking rings. The location of the other bearing is fixed only for the inner race. If the shaft expands, the axial movement can be accommodated by limited sliding motion of the outer race of the right hand bearing within the housing bore. For this kind of arrangement the bearings are referred to as fixed and floating bearings. Similar movement is also possible for the arrangement shown in Fig. 6.6. Here the right hand bearing is a cylindrical roller bearing and the axial location of the roller is not fixed and can move or float axially to the limited extents of the race to take up any axial movement or expansion of the shaft.

Correct lubrication is essential to ensure the calculated life of the bearing is achieved. Too much lubrication can result in increased levels of viscous dissipation of energy in the bearings and resultant overheating of the lubricant and the bearing. Too little lubrication can result in excessive wear and early failure. The form of lubrication depends on that required by the bearing application. Grease lubrication is

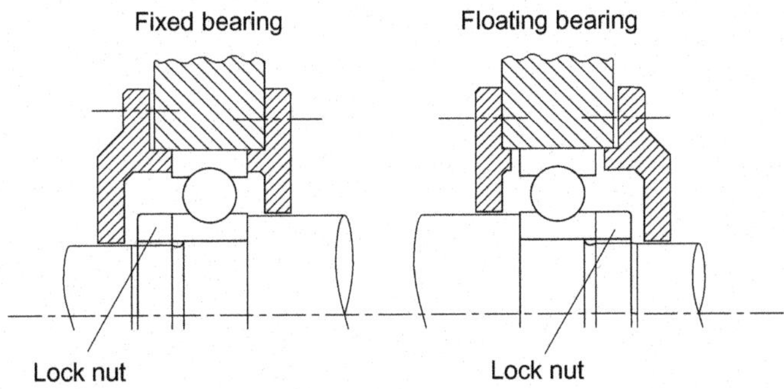

Fig. 6.5 Basic bearing mounting using two deep groove ball bearing for a rotating horizontal shaft for moderate radial and axial loading.

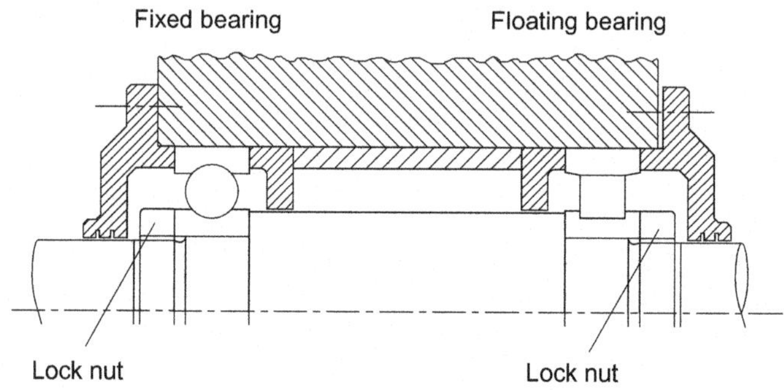

Fig. 6.6 Basic bearing mounting using a deep groove and a cylindrical roller bearing for moderate radial loads at the 'locating deep groove bearing' and high radial load capacity at the cylindrical roller bearing.

normally the easiest and requires injection of a specific quantity of grease into the bearing and some method of dirt and dust exclusion (Lugt, 2009, 2013; Morales-Espejel et al., 2014). Bearings can be purchased which are grease filled and sealed for life. Oil lubrication can be supplied by means of partial submersion as shown in Fig. 6.7, a recirculation circuit as shown in Fig. 6.8 or oil spot lubrication. A recirculation system typically consists of a cooler, an oil pump, a filter, oil jets, scavenge or collector holes and a reservoir. Oil spot lubrication involves the application of very small quantities of lubricant directly to the bearing by means of compressed air. Once again the bearing manufacturers' handbooks are a good source of information for lubricant supply rates and systems.

A key issue with bearings is ensuring their effective operation in service. If a bearing fails then the outage can be costly both in terms of loss of service and repair of any damage caused to the equipment as well as replacement of the bearings. A series of diagnostics technologies based on monitoring, for example, vibration and acoustics

Fig. 6.7 Partial submersion lubrication arrangement.

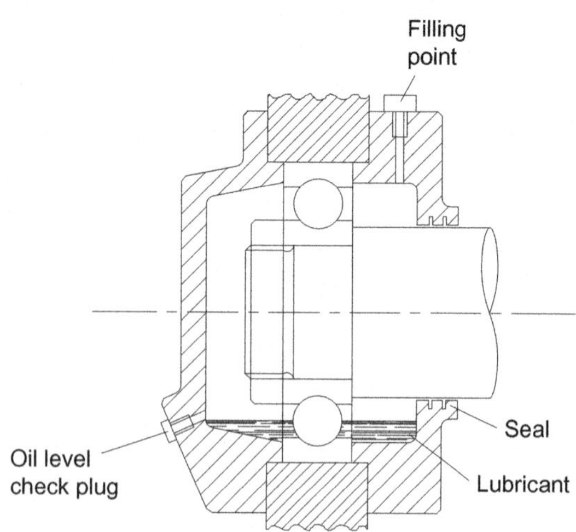

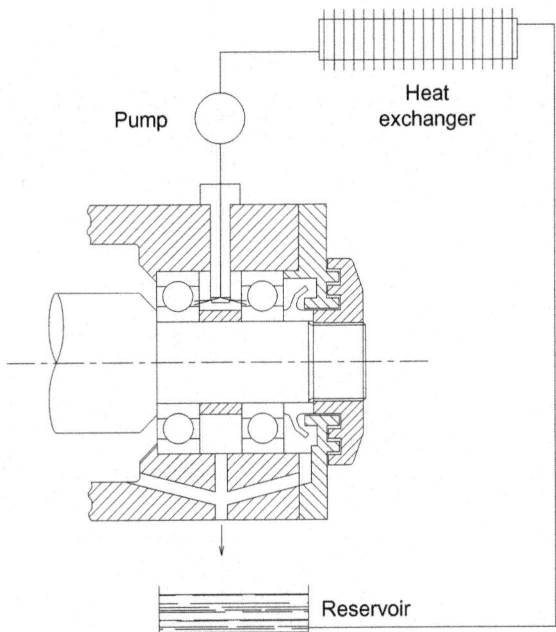

Fig. 6.8 Recirculation lubrication system.

(Chacon et al., 2015; Cong et al., 2013; McFadden and Smith, 1984; Tandon and Choudhury, 1999), as well as associated data analysis methods have been developed (Unal et al., 2014; Yang et al., 2005).

6.4 Radial location

The load capacity for a bearing, as defined by test procedures, within catalogues and by the basic dynamic rating, is based upon full support of the raceways at the inner and outer diameters, in the case of a radial bearing and the rings and washers in the case of, for instance, a thrust or taper roller bearing. The mating bearing seats for a bearing must be manufactured with adequate accuracy to provide the necessary support and prevent them from turning within their seats under load. The surfaces should be uninterrupted without grooves or other such features. Generally an interference fit is required for one ring and loose fit for the other. Table 6.9 provides an indication of the type of fit for a given application and Tables 6.10 and 6.11 provide recommended fits for the case of radial bearings with a solid steel shaft and housings respectively.

If a bearing is mounted on a hollow cylindrical shaft it is generally necessary to use a slightly higher interference fit than the equivalent sized solid shaft to provide a similar surface pressure between the shaft and inner raceway bearing seat. The fit is not normally appreciably affected until the ratio $c_i = d_i/d \geq 0.5$. For values above this a correction factor, normally available from bearing manufacturers, can be applied to the mean probable interference based on a solid shaft application to provide the requisite interference for the hollow shaft application.

Table 6.9 Recommend type of fit depending on operating conditions for radial bearings

Operating conditions	Load condition	Example application	Recommended fits
Rotating inner ring Stationary outer ring Constant load direction	Rotating load on inner ring Stationary load on outer ring	Belt drives	Interference fit for the inner ring Loose fit for the outer ring
Stationary inner ring Rotating outer ring Constant load direction	Stationary load on inner ring Rotating load on outer ring	Conveyer idlers Wheel hub bearings	Loose fit for the inner ring Interference fit for the outer ring
Rotating inner ring Stationary outer ring Load rotates with the inner ring	Stationary load on inner ring Rotating load on outer ring	Vibrating applications Motors	Interference fit for the outer ring Loose fit for the inner ring
Stationary inner ring Rotating outer ring Load rotates with the outer ring	Rotating load on inner ring Stationary load on outer ring	Gyratory crusher Merry go round	Interference fit for the inner ring Loose fit for the outer ring

6.5 Preload

An inherent characteristic associated with the fabrication and assembly of a rolling element bearing is clearance between the rolling elements and raceways. The total distance through which a bearing raceway can move relative to the other radially is called radial internal clearance or axially is called axial internal clearance. Radial internal clearance is sometimes referred to as radial play. It is important to distinguish between the internal clearance of a bearing prior to mounting versus that of a mounted bearing that has reached its operational conditions. The initial clearance of a bearing prior to mounting are likely to be greater than the operational clearance due to the fits used and differences in thermal expansion between the raceways and mountings.

Preload of a bearing involves managing these clearances to improve the running performance of the bearing. If a bearing in operation has radial play then one raceway can move radially and axially to another. This looseness results in wobble or runout. Such motion is unacceptable in high precision machinery such as prime movers, pinion bearings in automotive axles, machine tool spindles and high speed cordless and corded hand tools. Preload to the raceways forces the rolling elements in contact with the raceways, and the establishment of a stable and defined contact angle such that the set of balls will rotate in a uniform circumferential plane about the bearing axis. By preloading a bearing, the life, noise and vibration can all be improved. Preload is an important consideration in high precision and high speed application, particularly when high levels of positional accuracy are required.

Table 6.10 Recommended radial rolling element bearing fits for solid cylindrical steel shafts (after NSK)

Load conditions		Examples	Shaft diameter (mm)			Tolerance of shaft	Remarks
			Ball Brgs	Cylindrical roller Brgs, Tapered roller Brgs	Spherical roller Brgs		
Rotating outer ring load	Easy axial displacement of inner ring on shaft desirable	Wheels on stationary axles	All shaft diameters			g6	Use g5 and h5 where accuracy is required. In case of large bearings, f6 can be used to allow easy axial movement.
	Easy axial displacement of inner ring on shaft unnecessary	Tension Pulleys Rope Sheaves				h6	
Rotating Inner ring load or direction of load indeterminate	Light loads or variable loads (<0.06Cr(1))	Electrical home appliances pumps, blowers, transport vehicles, precision machinery, machine tools	<18	–	–	js5	
			18 to 100	<40	–	js6(j6)	
			100 to 200	40 to 140	–	k6	
			–	140 to 200	–	m6	
	Normal loads (0.06 to 0.13Cr(1))	General bearing applications, medium and large motors, turbines, pumps, engine main bearings, gears, woodworking machines	<18	–	–	js5 or js6 (j5 or j6)	k6 and m6 can be used for single-row tapered roller bearings and single row angular contact ball bearings instead of k5 and m5.
			18 to 100	<40	<40	k5 or k6	
			100 to 140	40 to 100	40 to 65	m5 or m6	
			140 to 200	100 to 140	65 to 100	m6	
			200 to 280	140 to 200	100 to 140	n6	
			–	200 to 400	140 to 280	p6	
			–	–	280 to 500	r6	
			–	–	Over 500	r7	

Continued

Table 6.10 Continued

Load conditions		Examples	Shaft diameter (mm)			Tolerance of shaft	Remarks
			Ball Brgs	**Cylindrical roller Brgs, Tapered roller Brgs**	**Spherical roller Brgs**		
	Heavy loads or shock loads (>0.13Cr(1))	Railway axleboxes, industrial vehicles, traction motors, construction equipment, crushers	– – – –	50 to 140 140 to 200 Over 200 –	50 to 100 100 to 140 140 to 200 200 to 500	n6 p6 r6 r7	More than CN bearing internal clearance is necessary.
Axial loads only			All shaft diameters			js6 (j6)	–

Table 6.11 Recommended radial rolling element bearing fits with housings. This table is applicable to cast iron and steel housings. For housings made of light alloys, the interference should be tighter than those in this table (after NSK)

Load conditions			Examples	Tolerances for housing bores	Axial displacement of outer ring	Remarks
Solid housings	Rotating outer ring load	Heavy loads on bearing in thin-walled housing or heavy shock loads	Automotive wheel hubs (roller bearings) crane travelling wheels	P7	Impossible	–
		Normal or heavy loads	Automotive wheel hubs (ball bearings) Vibrating screens	N7		
		Light or variable loads	Conveyor Rollers Rope sheaves Tension pulleys	M7		
	Direction of load indeterminate	Heavy shock loads	Traction motors	M7		
		Normal or heavy loads	Pumps Crankshaft main bearings Medium and large motors	K7	Generally impossible	If axial displacement of the outer ring is not required.
Solid or split housings		Normal or light loads		JS7 (J7)	Possible	Axial displacement of outer ring is necessary.
	Rotating inner ring load	Loads of all kinds	General bearing applications, railway axleboxes	H7	Easily possible	–
		Normal or light loads	Plummer blocks	H8		
		High temperature rise of inner ring through shaft	Paper dryers	G7		

Continued

Table 6.11 Continued

Load conditions			Examples	Tolerances for housing bores	Axial displacement of outer ring	Remarks
Solid housing		Accurate running desirable under normal or light loads	Grinding Spindle rear Ball bearings High speed Centrifugal Compressor Free Bearings	JS6 (J6)	Possible	–
	Direction of load indeterminate		Grinding Spindle Front Ball Bearings High Speed Centrifugal Compressor Fixed Bearings	K6	Generally impossible	For heavy loads, interference fit tighter than K is used. When high accuracy is required, very strict tolerances should be used for fitting.
	Rotating inner ring load	Accurate running and high rigidity desirable under variable loads	Cylindrical roller bearings for machine tool main spindle	M6 or N6	Impossible	
		Minimum noise is required.	Electrical home appliances	H6	Easily possible	–

The clearances associated with the manufacture of a bearing can result in bearing noise, vibration and compromised life. An axial preload to one of the raceways can be applied. This increases the stiffness of the bearing, has benefits including increased rotational positional accuracy and more precise shaft location, reduction or elimination of ball skidding, reduction of deflections, reduction of noise and improved load sharing between bearings.

The higher the preload, the higher the bearing stiffness. However if the preload is excessive this can result in reduced bearing life and failure. If inadequate preload is applied, vibration and fretting can result.

The calculation of the optimum preload for a bearing can be based on the theoretical consideration of the optimum surface stress that arises from elastic deformation of the contact areas between the rolling elements and raceways. In practical applications preload parameters can generally be obtained from the bearing manufactures for the specific bearing concerned.

There are two principal approaches to preloading: solid preload and spring preload, as illustrated in Fig. 6.9. Solid preload, also known as rigid or fixed-constraint preload, can be achieved by locking both inner and outer raceway in position, so that they cannot move axially relative to each other, thus fixing the 'as preloaded' positions of the raceways. Such a design, however, is likely to be subject to growth or shrinkage of components due to temperature changes which can result in changes to the expected preload. In addition components can wear and the level of preload reduces as a result. Spring preload, also known as constant pressure preload, can be applied to a bearing raceway using a coil spring, bevel spring or wave spring washer, with the spring pressing the raceways together or apart, depending on the requirement. The advantage of spring preload is stable loading with temperature changes. However this comes at the expense of increased machine complexity. In most applications the spring is configured to press against the nonrotating raceway.

Bearing manufacturers produce matched pairs of bearings, known as duplex bearings, with built-in preload, achieved by grinding the inner or outer ring faces with defined dimension known as the bearing offset. The bearing offset corresponds with

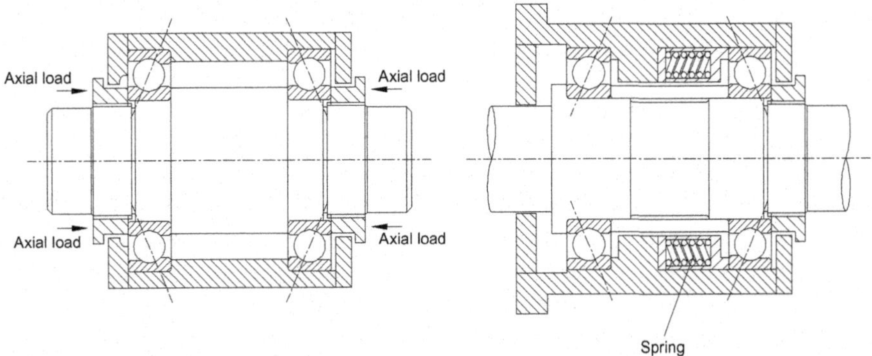

Fig. 6.9 Solid and spring preload.

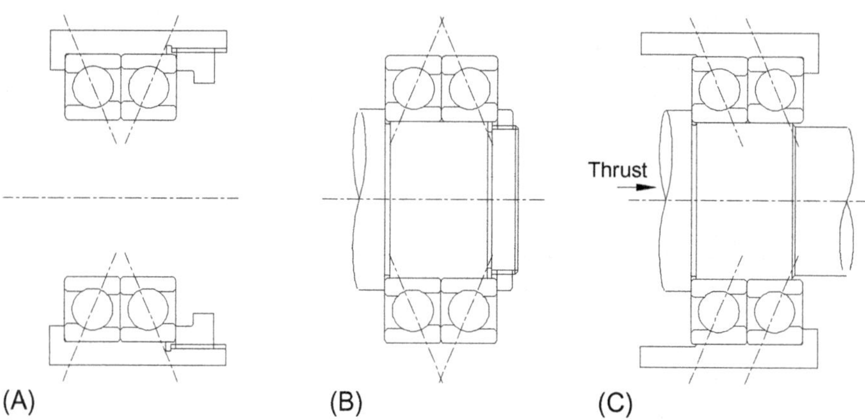

Fig. 6.10 (A) duplex face to face (DF), (B) duplex back to back (DB) and (C) tandem (DT) preload configurations.

the axial movement of the raceway if a specific preload is applied. When a duplex bearing pair is assembled the offset is removed as the bearings are clamped together establishing the preload in the bearing set. Preload can be applied in various configurations such as duplex face to face (DF), duplex back to back (DB) and tandem (DT) as illustrated in Fig. 6.10. Duplex bearings exhibit increased axial and radial rigidity. The stiffness of DB is higher than that associated with DF configurations. DB and DF configurations can handle bidirectional thrust loads and DT bearings can take heavy unidirectional thrust loads. As a rule of thumb a preloaded bearing pair can handle axial loads up to three times the preload. Beyond this limit one of the bearings in the set may become unloaded and the deflection will be that of a single bearing.

6.6 Conclusions

Bearings are used to support a load whilst allowing relative motion between two elements of a machine. Typical applications are shafts where the shaft rotates relative to a housing but bearings can also be used to facilitate linear relative motion for, say, machine slides. This chapter has introduced the selection and specification of rolling element bearings.

An important consideration in a design involving rolling element bearings is to ensure that the bearings have sufficient load capacity to accommodate the forces generated by thermal growth, and misalignment in addition to any functional loading. An alternative is to make sure the bearing mounting configuration allows for displacements to occur without imposing significant loads on the bearings. A typical configuration for bearing mounting is the use of a fixed and a floating bearing, with the all of the shoulders of the fixed bearing axially constrained, while a raceway or the rolling elements on the floating bearing are not axially constrained.

Rolling element bearings are produced by specialist manufacturers and many types and sizes are available as stock items. For further information on bearings the reader is recommended to view specific manufacturers' literature and the texts by Harris (2001), Harris and Kotzalas (2006a, b) and Brandlein et al. (1999).

References

Brandlein, J., Eschmann, P., Hasbargen, L., Weigand, K., 1999. Ball and Roller Bearings. Theory, Design and Application. John Wiley & Sons, Chichester.

Chacon, J.L.F., Kappatos, V., Balachandran, W., Gan, T.H., 2015. A novel approach for incipient defect detection in rolling bearings using acoustic emission technique. Appl. Acoust. 89, 88–100.

Cong, F., Chen, J., Dong, G., Pecht, M., 2013. Vibration model of rolling element bearings in a rotor-bearing system for fault diagnosis. J. Sound Vib. 332, 2081–2097.

El-Thalji, I., Jantunen, E., 2015. Dynamic modelling of wear evolution in rolling bearings. Tribol. Int. 84, 90–99.

Harris, T.A., 2001. Rolling Bearing Analysis, fourth ed. John Wiley & Sons, New York.

Harris, T.A., Kotzalas, M.N., 2006a. Advanced Concepts of Bearing Technology: Rolling Bearing Analysis, fifth ed CRC Press, Boca Raton.

Harris, T.A., Kotzalas, M.N., 2006b. Essential Concepts of Bearing Technology, fifth ed CRC Press, Boca Raton.

Lugt, P.M., 2009. A review on grease lubrication in rolling bearings. Tribol. Trans. 52, 470–480.

Lugt, P.M., 2013. Grease Lubrication in Rolling Bearings. Wiley-Blackwell.

McFadden, P.D., Smith, J.D., 1984. Vibration monitoring of rolling element bearings by the high-frequency resonance technique—a review. Tribol. Int. 17, 3–10.

Morales-Espejel, G.E., Lugt, P.M., Pasaribu, H.R., Cen, H., 2014. Film thickness in grease lubricated slow rotating rolling bearings. Tribol. Int. 74, 7–19.

Rolls-Royce, 2015. The Jet Engine, fifth ed Wiley-Blackwell.

Sayles, R.S., Ioannides, E., 1988. Debris damage in rolling bearings and its effects on fatigue life. ASME Trans. J. Tribol. 110, 26–31.

Tandon, N., Choudhury, A., 1999. A review of vibration and acoustic measurement methods for the detection of defects in rolling element bearings. Tribol. Int. 32, 469–480.

Unal, M., Onat, M., Demetgul, M., Kucuk, H., 2014. Fault diagnosis of rolling bearings using a genetic algorithm optimized neural network. Measurement 58, 187–196.

Yang, H., Mathew, J., Ma, L., 2005. Fault diagnosis of rolling element bearings using basis pursuit. Mech. Syst. Signal Process. 19, 341–356.

Standards

ANSI/ABMA Standard 11-1990(R1999). Load ratings and fatigue life for roller bearings.

ANSI/ABMA Standard 7-1995(R2001). Shaft and housing fits for metric radial ball and roller bearings (except tapered roller bearings) conforming to basic boundary plans.

ANSI/ABMA Standard 9-1990(R2000). Load ratings and fatigue life for ball bearings.

BS 4231:1992, ISO 3448:1992. Classification for viscosity grades of industrial liquid lubricants.

BS ISO 10317:2008. Rolling bearings. Tapered roller bearings. Designation system.

BS ISO 104:2002. Rolling bearings. Thrust bearings. Boundary dimensions, general plan.

BS ISO 113:2010. Rolling bearings. Plummer block housings. Boundary dimensions.
BS ISO 14728-1:2004. Rolling bearings. Linear motion rolling bearings. Dynamic load ratings and rating life.
BS ISO 14728-2:2004. Rolling bearings. Linear motion rolling bearings. Static load ratings.
BS ISO 15:2011. Rolling bearings. Radial bearings. Boundary dimensions, general plan.
BS ISO 15243:2004. Rolling bearings. Damage and failures. Terms, characteristics and causes.
BS ISO 281:2007. Rolling bearings. Dynamic load ratings and rating life.
BS ISO 2982-2:2001. Rolling bearings. Accessories. Locknuts and locking devices. Dimensions.
BS ISO 3290-1:2008. Rolling bearings. Balls. Steel balls.
BS ISO 355:2007 + A1:2012. Rolling bearings. Tapered roller bearings. Boundary dimensions and series designations.
BS ISO 492:2002. Rolling bearings. Radial bearings. Tolerances.
BS ISO 5593:1997. Rolling bearings. Vocabulary.
BS ISO 76:2006. Rolling bearings. Static load ratings.
PD ISO/TR 1281-2:2008. Rolling bearings. Explanatory notes on ISO 281. Modified rating life calculation, based on a systems approach to fatigue stresses.
Society of Automotive Engineers. SAE J300. Engine oil viscosity classification. 2013.

Websites

At the time of going to press the world-wide-web contained useful information relating to this chapter at the following sites.

www.abcbearings.com.
www.arb-bearings.com.
www.bocabearings.com.
www.championballbearings.com/.
www.cooperbearings.com.
www.fag.de.
www.hb-bearings.com.
www.ina.com.
www.nationalbearings.com.
www.newindo.com.
www.nsk.com/.
www.ntnamericas.com/en/.
www.peerbearing.com.
www.rbcbearings.com.
www.skf.com.
www.thordonbearings.com.
www.timken.com.
www.toyobearings.com.

Further reading

Harnoy, A., 2002. Bearing Design in Machinery. Marcel Dekker, New York.
Hindhede, U., Zimmerman, J.R., Hopkins, R.B., Erisman, R.J., Hull, W.C., Lang, J.D., 1983. Machine Design Fundamentals. A Practical Approach. Wiley.

Shafts

7

Chapter Outline

Nomenclature

Generally, preferred SI units have been stated.

a	influence coefficient (m)
c	distance from the neutral axis (m)
d	diameter (m)
E	Young's modulus (N/m^2)
F	force (N)
F_m	mid-range steady component of force (N)
F_{max}	maximum force (N)
F_{min}	minimum force (N)
F_a	amplitude of alternating force (N)
g	acceleration due to gravity (m^2/s)
I	second moment of area (m^4)
J	polar second moment of area (m^4)
k_a	surface factor
k_b	size factor
k_c	reliability factor
k_d	temperature factor
k_e	duty cycle factor
k_f	fatigue stress concentration factor

Mechanical Design Engineering Handbook. https://doi.org/10.1016/B978-0-08-102367-9.00007-X

K_f	component fatigue stress concentration factor
k_g	miscellaneous effects factor
k_t	geometric stress concentration factor
K_f	fatigue stress concentration factor for bending
K_{fs}	fatigue stress concentration factor for torsion
L	length (m)
m	mass (kg)
M	moment (N m)
M_a	alternating bending moment (N m)
M_m	mid-range bending moment (N m)
n_s	factor of safety
q	notch sensitivity
R	reaction (N), stress ratio
S_a	alternating strength (N/m^2)
S_e	endurance limit (N/m^2)
S_m	mid-range strength (N/m^2)
S_{ut}	ultimate strength (N/m^2)
S_y	yield stress (N/m^2)
T	torque (N m), tension (N)
T_a	alternating torque (N m)
T_m	mid-range torque (N m)
U	strain energy (J)
W_i	mass (kg) or weight of node i (N)
x	coordinate (m)
y	static deflection (m)
y_i	static deflection of W_i (m)
σ_a	amplitude of alternating stress (N/m^2)
σ_e	endurance limit of the item (N/m^2)
σ_e'	endurance limit of test specimen (N/m^2)
σ_m	mid-range steady component of stress (N/m^2)
σ_{max}	maximum stress (N/m^2)
σ_{min}	minimum stress (N/m^2)
σ_{uts}	ultimate tensile strength (N/m^2)
σ_y	yield strength (N/m^2)
σ_a'	alternating von Mises stress (N/m^2)
σ_m'	mean von Mises stress (N/m^2)
σ_{max}'	max von Mises stress (N/m^2)
τ_a	alternating stress (N/m^2)
τ_m	mid-range stress (N/m^2)
ω	frequency
ω_c	critical speed (rad/s)

7.1 Introduction to shaft design

The objective of this chapter is to introduce the concepts and principles of shaft design. Specific attention is given to the arrangement of machine elements and features on a shaft, the connection of shafts, determining the deflection of shafts and

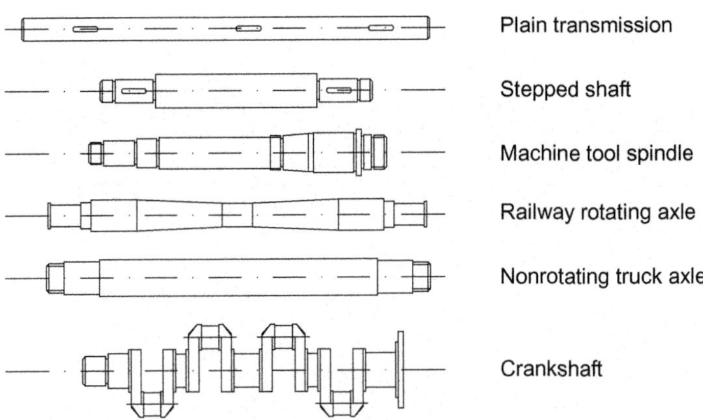

Fig. 7.1 Typical shaft arrangements.
Based on Reshetov, D.N., 1978. Machine Design. Mir Publishers, 1978.

Fig. 7.2 An example of a machine shaft.

critical speeds as well as specifying shaft dimensions for strength and fluctuating load integrity. An overall shaft design procedure is presented including consideration of bearing and component mounting and shaft dynamics for transmission shafting.

The term shaft usually refers to a component of circular cross section that rotates and transmits power from a driving device, such as a motor or engine, through a machine. Shafts can carry gears, pulleys and sprockets to transmit rotary motion and power via mating gears, belts and chains. Alternatively, a shaft may simply connect to another via a mechanical or magnetic flux coupling. A shaft can be stationary and support a rotating member such as the short shafts that support the nondriven wheels of automobiles often referred to as spindles. Some common shaft arrangements are illustrated in Fig. 7.1 and some examples are shown in Figs 7.2–7.4.

Fig. 7.3 Example of gear shafts.
Image courtesy of Daimler AG.

Fig. 7.4 An example of an automotive crank shaft.
Image courtesy of Daimler AG.

Shaft design considerations include the following:

(1) size and spacing of components (as on a general assembly drawing), tolerances,
(2) material selection, material treatments,
(3) deflection and rigidity,
    bending deflection,
    torsional deflection,
    slope at bearings,
    shear deflection,

(4) stress and strength,
static strength,
fatigue,
reliability,
(5) frequency response,
(6) manufacturing constraints.

Shafts typically consist of a series of stepped diameters accommodating bearing mounts and providing shoulders for locating devices such as gears, sprockets and pulleys to butt up against and keys are often used to prevent rotation, relative to the shaft, of these 'added' components. A typical arrangement illustrating the use of constant diameter sections and shoulders is illustrated in Fig. 7.5 for a transmission shaft supporting a gear and pulley wheel.

Shafts must be designed so that deflections are within acceptable levels. Too much deflection can, for example, degrade gear performance, and cause noise and vibration. The maximum allowable deflection of a shaft is usually determined by limitations set on the critical speed, minimum deflections required for gear operation and bearing requirements. In general, deflections should not cause mating gear teeth to separate more than about 0.13 mm and the slope of the gear axes should not exceed about 0.03 degrees. The deflection of the journal section of a shaft across a plain bearing should be small in comparison with the oil film thickness. The critical speed (see Section 7.4) is the rotational speed at which dynamic forces acting on the system cause it to vibrate at its natural frequency. Operation of a shaft at this speed can cause resonance and significant vibrations that can severely damage a machine and is therefore a key consideration in the design of a rotating machine. Torsional and lateral deflection both contribute to lower critical speed. In addition, shaft angular deflection at

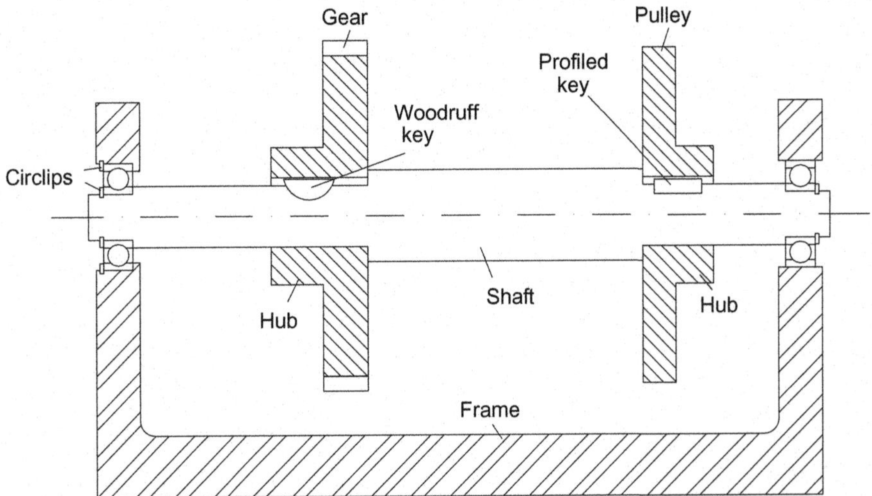

Fig. 7.5 Typical shaft arrangement incorporating constant diameter sections and shoulders for locating added components.

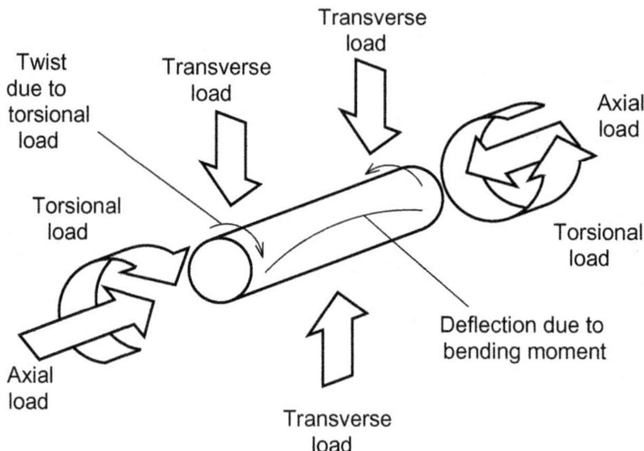

Fig. 7.6 Typical machine shaft loading and deflection
Based on Beswarick, J., 1994a. Shaft for strength and rigidity. In: Hurst, K. (Ed.), Rotary Power Transmission Design. McGraw Hill, 135–141.

rolling element bearings should not exceed 0.04 degree, with the exception being self-aligning rolling element bearings.

Shafts can be subjected to a variety of combinations of axial, bending and torsional loads (see Fig. 7.6) which may fluctuate or vary with time. Typically, a rotating shaft transmitting power is subjected to a constant torque together with a completely reversed bending load, producing a mean torsional stress and an alternating bending stress respectively.

As indicated previously, shafts should be designed to avoid operation at, or near, critical speeds. This is usually achieved by the provision of sufficient lateral rigidity so that the lowest critical speed is significantly above the range of operation. If torsional fluctuations are present (e.g. engine crankshafts, camshafts, compressors), the torsional natural frequencies of the shaft must be significantly different to the torsional input frequency. This can be achieved by providing sufficient torsional stiffness so that the shaft's lowest natural frequency is much higher than the highest torsional input frequency.

Rotating shafts must generally be supported by bearings. For simplicity of manufacture, it is desirable to use just two sets of bearings. If more bearings are required, precise alignment of the bearings is necessary. Provision for thrust load capability and axial location of the shaft is normally supplied by just one thrust bearing taking thrust in each direction. It is important that the structural members supporting the shaft bearings are sufficiently strong and rigid.

The following list outlines a shaft design procedure for a shaft experiencing constant loading. The flow charts given in Figs 7.7 and 7.8 can be used to guide and facilitate design for shaft strength and rigidity and fluctuating load capability. Pyrhonen

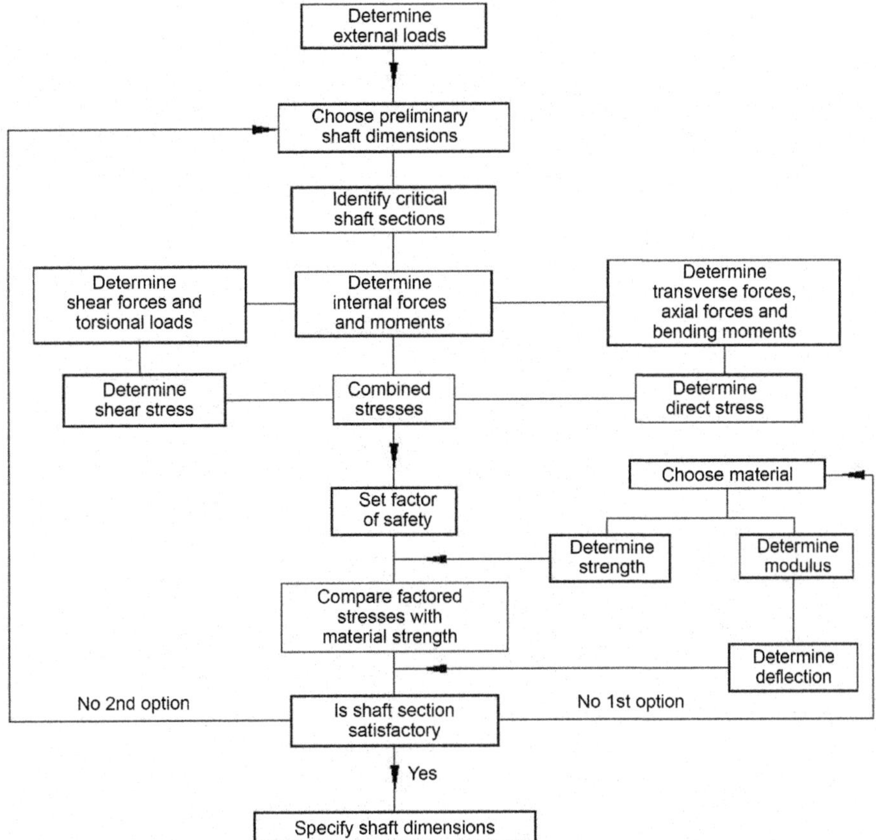

Fig. 7.7 Design procedure flow chart for shaft strength and rigidity.
Based on Beswarick, J., 1994a. Shaft for strength and rigidity. In: Hurst, K. (Ed.), Rotary Power Transmission Design. McGraw Hill, 135–141.

et al. (2008) provides an overview of shaft design with specific reference to electrical machine design.

(1) Determine the shaft rotational speed.
(2) Determine the power or torque to be transmitted by the shaft.
(3) Determine the dimensions of the power-transmitting devices and other components to be mounted on the shaft and
(4) Specify an axial location along the shaft for each device and component. As a general principle keep shafts as short as possible. If possible locate stress raisers away from highly stressed regions of the shaft.
(5) Specify the locations of the bearings to support the shaft.

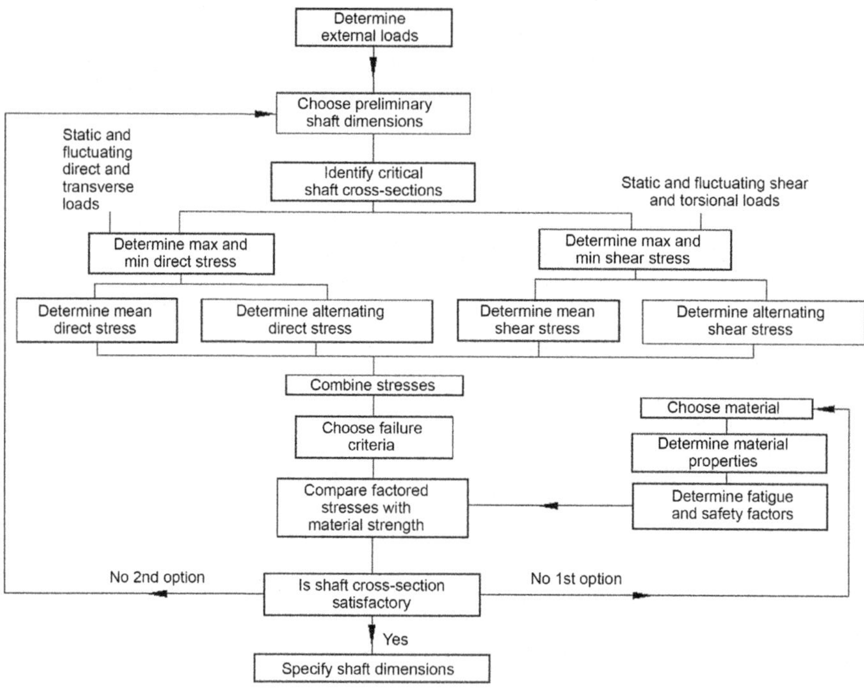

Fig. 7.8 Design procedure flow chart for a shaft with fluctuating loading.
Based on Beswarick, J., 1994b. Shaft with fluctuating load. In: Hurst, K. (Ed.), Rotary Power Transmission Design. McGraw Hill, 142–148.

(6) Propose a general layout for the shaft geometry considering how each component will be located axially and how power transmission will take place. As a principle, use generous fillet radii.
(7) Determine the magnitude of the torques throughout the shaft.
(8) Determine the forces exerted on the shaft.
(9) Produce shearing force and bending moment diagrams so that the distribution of bending moments in the shaft can be determined.
(10) Select a material for the shaft and specify any heat treatments, etc.
(11) Determine an appropriate design stress taking into account the type of loading (whether smooth, shock, repeated and reversed).
(12) Analyse all the critical points on the shaft and determine the minimum acceptable diameter at each point to ensure safe design.
(13) Determine the deflections of the shaft at critical locations and estimate the critical frequencies.
(14) Specify the final dimensions of the shaft. This is best achieved using a detailed manufacturing drawing to a recognised standard (see, for example, the Manual of British Standards in Engineering and Drawing Design) and the drawing should include all the information required to ensure the desired quality. Typically, this will include material specifications, dimensions and tolerances (bilateral, runout, data, etc., see Chapter 19), surface finishes, material treatments and inspection procedures.

The following general principles should be observed in shaft design.

- Keep shafts as short as possible with the bearings close to applied loads. This will reduce shaft deflection and bending moments and increase critical speeds.
- If possible, locate stress raisers away from highly stressed regions of the shaft. Use generous fillet radii and smooth surface finishes and consider using local surface strengthening processes such as shot-peening and cold-rolling.
- If weight is critical use hollow shafts.

An overview of shaft-hub connection methods is given in Section 7.2, shaft-to-shaft connection methods in Section 7.3 and the determination of critical speeds in Section 7.4. In Section 7.5, analytical methods for determining the diameter in the design of transmission shafts are introduced.

7.2 Shaft-hub connection

Power-transmitting components such as gears, pulleys and sprockets need to be mounted on shafts securely and located axially with respect to mating components. In addition, a method of transmitting torque between the shaft and the component must be supplied. The portion of the component in contact with the shaft is called the hub and can be attached to, or driven by, the shaft by keys, pins, setscrews, press and shrink fits, splines and taper bushes. Table 7.1 identifies the merits of various connection

Table 7.1 Merits of various shaft-hub connections

	Pin	**Grub screw**	**Clamp**	**Press fit**	**Shrink fit**	**Spline**	**Key**	**Taper bush**
High-torque capacity	x	x	✓	x	✓	✓	✓	✓
High axial loads	✓	x	✓	x	✓	x	x	✓
Axially compact	x	x	x	✓	✓	✓	✓	✓
Axial location provision	✓	✓	✓	✓	✓	x	x	✓
Easy hub replacement	x	✓	✓	x	x	✓	✓	✓
Fatigue	x	x	✓	✓	✓	x	x	✓
Accurate angular positioning	✓	x	x	x	x	✓	✓	(✓)
Easy position adjustment	x	✓	✓	x	x	x	x	✓

After Hurst, K., 1994. Shaft/hub connections. In Hurst, K. (Ed.) Rotary Power Transmission Design. McGraw Hill, 55-65.

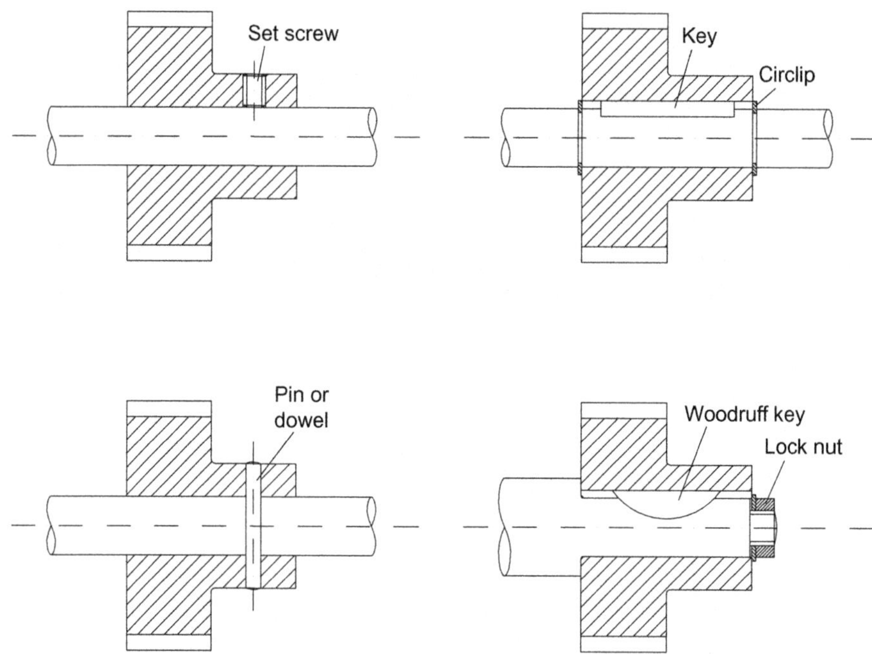

Fig. 7.9 Alternative methods of shaft-hub connection.

methods. Alternatively, the component can be formed as an integral part of a shaft as, for example, the cam on an automotive camshaft.

Fig. 7.9 illustrates the practical implementation of several shaft-hub connection methods. Gears, for example, can be gripped axially between a shoulder on a shaft and a spacer with the torque transmitted through a key. Various configurations of keys exist including square, flat and round keys as shown in Figs 7.10 and 7.11. The grooves in the shaft and hub into which the key fits are called keyways or key seats. A simpler and less expensive method for transmitting light loads is to use pins and various pin types are illustrated in Fig. 7.12. An inexpensive method of providing axial location of hubs and bearings on shafts is to use circlips as shown in Figs 7.5 and 7.13. One of the simplest hub-shaft attachments is to use an interference fit, where the hub bore is slightly smaller than the shaft diameter. Assembly is achieved by press fitting, or thermal expansion of the outer ring by heating and thermal contraction of the inner by use of liquid nitrogen. The design of interference fits is covered in greater detail in Sec. 19.2.2. Mating splines, as shown in Figs 7.14 and 7.15, comprise teeth cut into both the shaft and the hub and provide one of the strongest methods of transmitting torque. Both splines and keys can be designed to allow axial sliding along the shaft.

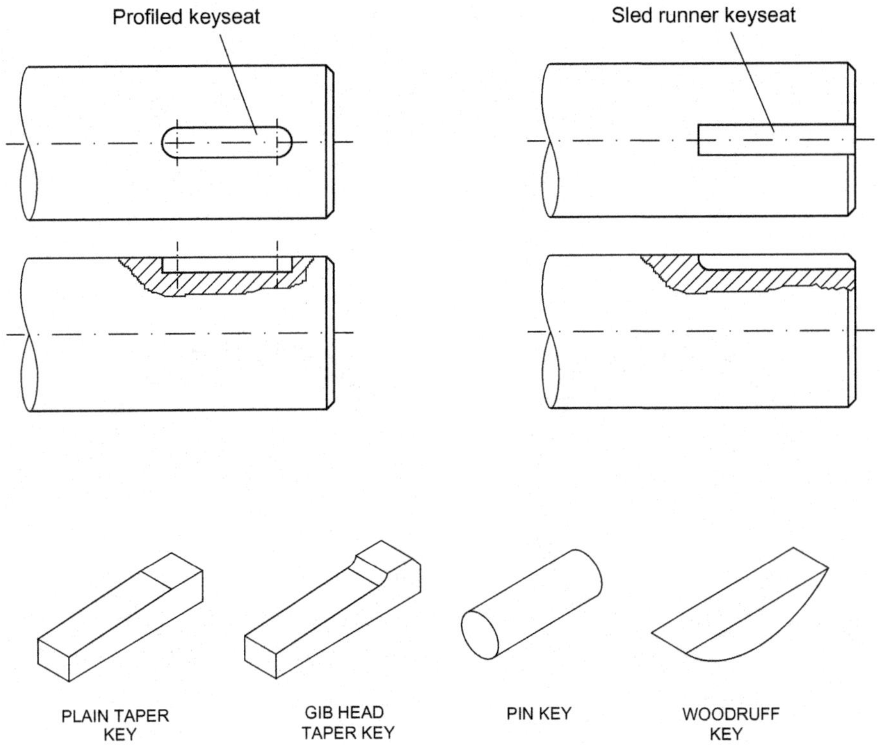

Profiled keyseat

Sled runner keyseat

PLAIN TAPER KEY

GIB HEAD TAPER KEY

PIN KEY

WOODRUFF KEY

Fig. 7.10 Keyways and keys for torque transmission and component location.

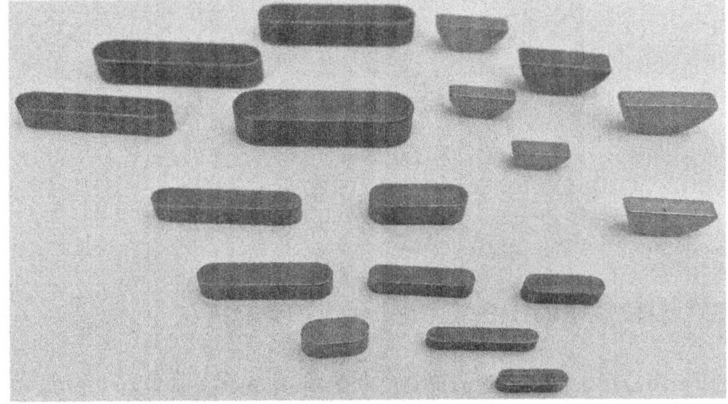

Fig. 7.11 Various keys for torque transmission and component location.

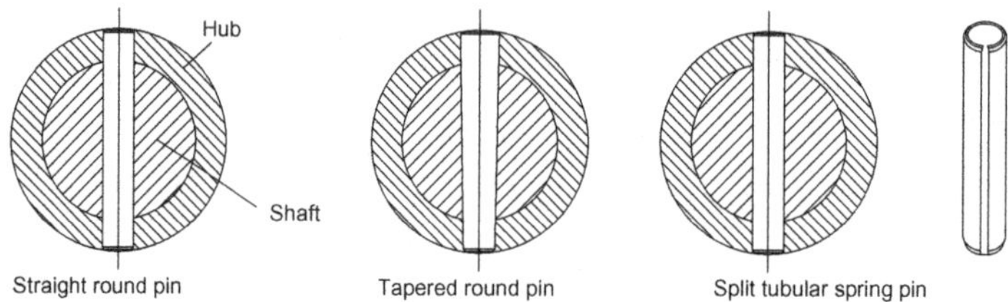

Fig. 7.12 Pins for torque transmission and component location.

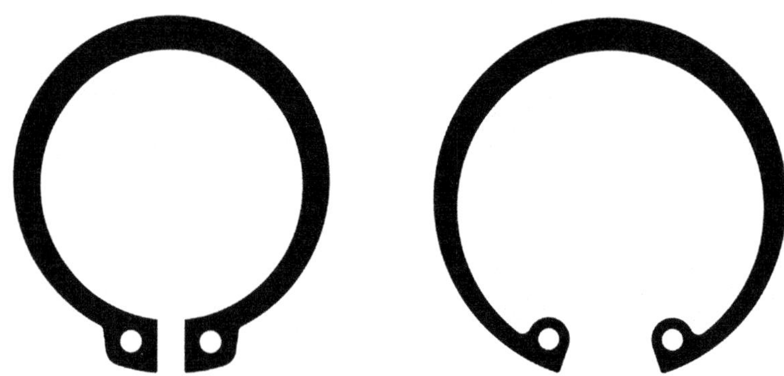

Fig. 7.13 Snap rings or circlips.

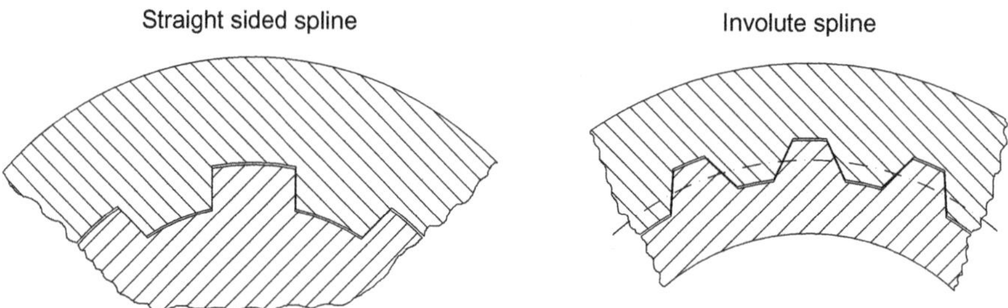

Fig. 7.14 Splines.

7.3 Shaft-shaft connection—couplings

In order to transmit power from one shaft to another, a coupling or clutch can be used (for clutches see Chapter 13). There are two general types of coupling, rigid and flexible. Rigid couplings are designed to connect two shafts together so that no relative motion occurs between them (see Fig. 7.16). Rigid couplings are suitable when precise alignment of two shafts is required. If significant radial or axial misalignment occurs high stresses may result which can lead to early failure. Flexible couplings (see Fig. 7.17) are designed to transmit torque whilst permitting some axial, radial

Fig. 7.15 Shaft splines.

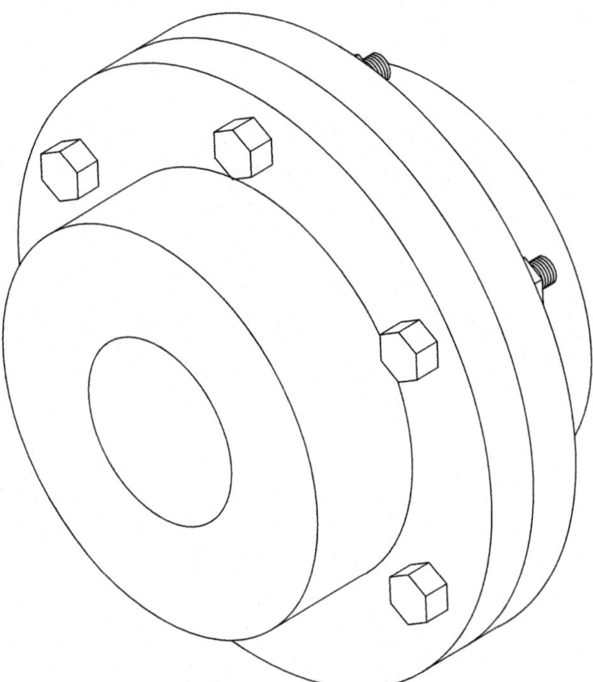

Fig. 7.16 Rigid coupling.

Fig. 7.17 Flexible couplings.
Photograph courtesy of Cross and Morse.

Fig. 7.18 Universal joints.

and angular misalignment. Many forms of flexible coupling are available (e.g. see manufacturers' catalogues such as Turboflex and Fenner). Each coupling is designed to transmit a given limiting torque. Generally, flexible couplings are able to tolerate up to ± 3 degree of angular misalignment and up to 0.75 mm parallel misalignment depending on their design. If more misalignment is required a universal joint can be used (sees Figs 7.18 and 7.19). Couplings are considered in detail by Neale et al. (1998), Piotrowski (2006), and Schlater (2011).

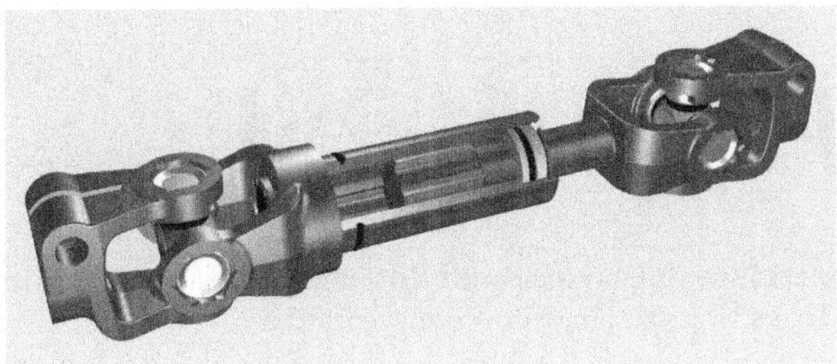

Fig. 7.19 Universal joints.
Image courtesy of Pailton Engineering Limited.

7.4 Critical speeds and shaft deflection

The centre of mass of a rotating system (for example, a mid-mounted disc on a shaft supported by bearings at each end) will never coincide with the centre of rotation due to manufacturing and operational constraints. As the shaft rotational speed is increased, the centrifugal force acting at the centre of mass tends to bow the shaft. The more the shaft bows the greater the eccentricity and the greater the centrifugal force. Below the lowest critical speed of rotation, the centrifugal and elastic forces balance at a finite value of shaft deflection. At the critical speed equilibrium theoretically requires infinite deflection of the centre of mass although realistically bearing damping, internal hysteresis and windage causes equilibrium to occur at a finite displacement. This displacement can be large enough to break a shaft, damage bearings and cause destructive machine vibration and therefore deflections need to be determined along the shaft and the consequences evaluated.

The critical speed of rotation is the same as the lateral frequency of vibration, which is induced when rotation is stopped and the centre displaced laterally and suddenly released (i.e. the same as the frequency you would obtain if you 'wacked' the stationary shaft with a hammer and monitored the frequency it vibrated at). For all shafts, except for the single concentrated mass shaft, critical speeds also occur at higher frequencies.

At the first critical speed, the shaft will bend into the simplest possible shape, and at the second critical speed it will bend into the next simplest shape. For example, the shapes the shaft will bend into at the first two critical speeds (or modes) for an end-supported shaft with two masses are illustrated in Fig. 7.20.

In certain circumstances, the fundamental frequency of a shaft system cannot be made higher than the shaft design speed. If the shaft can be accelerated rapidly through and beyond the first resonant critical frequency, before the vibrations have a chance to build up in amplitude, then the system can be run at speeds higher than the natural frequency. This is the case with steam and gas turbines where the size of the turbomachinery and generators give low natural frequency but must be run at high speed due to efficiency considerations. As a general design principle maintaining the

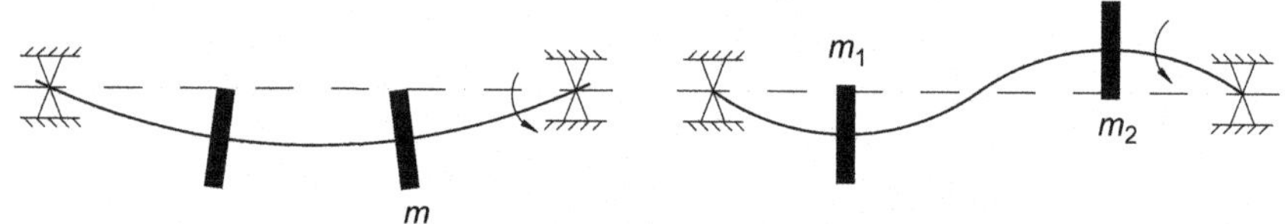

Fig. 7.20 Shaft shapes for a simply supported shaft with two masses at the first and second critical speeds. Masses large in comparison with shaft mass.

operating speed of a shaft below half the shaft whirl critical frequency is normally acceptable.

A complete analysis of the natural frequencies of a shaft can be performed using a finite element analysis package, such as ANSYS, called a 'nodal analysis'. This can give a large number of natural frequencies in three dimensions from the fundamental upwards. This is the sensible and easiest approach for complex systems but a quick estimate for a simplified system can be undertaken for design purposes as outlined in this section.

The critical speed of a shaft with a single mass attached can be approximated by

$$\omega_c = \sqrt{g/y} \tag{7.1}$$

where

ω_c = critical angular velocity (rad/s),
y = static deflection at the location of the mass (m),
g = acceleration due to gravity (m/s^2).

The first critical speed of a shaft carrying several concentrated masses is approximated by the Rayleigh-Ritz equation (see Eq. 7.2). The dynamic deflections of a shaft are generally unknown. Rayleigh showed that an estimate of the deflection curve is suitable provided it represents the maximum deflection and the boundary conditions. The static deflection curve due to the shaft's own weight and the weight of any attached components gives a suitable estimate. Note that external loads are not considered in this analysis only those due to gravitation. The resulting calculation gives a value higher than the actual natural frequency by a few percent.

$$\omega_c = \sqrt{\frac{g\sum_{i=1}^{n} W_i y_i}{\sum_{i=1}^{n} W_i y_i^2}} \tag{7.2}$$

where

ω_c = critical angular velocity (rad/s),
W_i = mass or weight of node i (kg or N),
y_i = static deflection of W_i (m).

Alternatively, the Dunkerley equation can be used to estimate the critical speed:

$$\frac{1}{\omega_c^2} = \frac{1}{\omega_1^2} + \frac{1}{\omega_2^2} + \frac{1}{\omega_3^2} + \ldots \tag{7.3}$$

where ω_1 is the critical speed if only mass number 1 is present, etc.

Both the Rayleigh-Ritz and the Dunkerley equations are approximations to the first natural frequency of vibration, which is assumed nearly equal to the critical speed. The Dunkerley equation tends to underestimate and the Rayleigh-Ritz equation to overestimate the critical frequency.

Often we need to know the second critical speed. For a two mass system, as illustrated in Fig. 7.20, approximate values for the first two critical speeds can be found by solving the frequency equation given in Eq. (7.4).

$$\frac{1}{\omega^4} - (a_{11}m_1 + a_{22}m_2)\frac{1}{\omega^2} + (a_{11}a_{22} - a_{12}a_{21})m_1m_2 = 0 \tag{7.4}$$

where a_{11}, a_{12}, etc., are the influence coefficients. These correspond to the deflection of the shaft at the locations of the loads as a result of 1-N loads. The first subscript refers to location of the deflection, the second to the location of the 1-N force. For example, the influence coefficients for a simply supported shaft with two loads, as illustrated in Fig. 7.21, are listed as following:

a_{11} is the deflection at the location of mass 1, that would be caused by a 1-N weight at the location of mass 1.

a_{21} is the deflection at the location of mass 2, that would be caused by a 1-N weight at the location of mass 1.

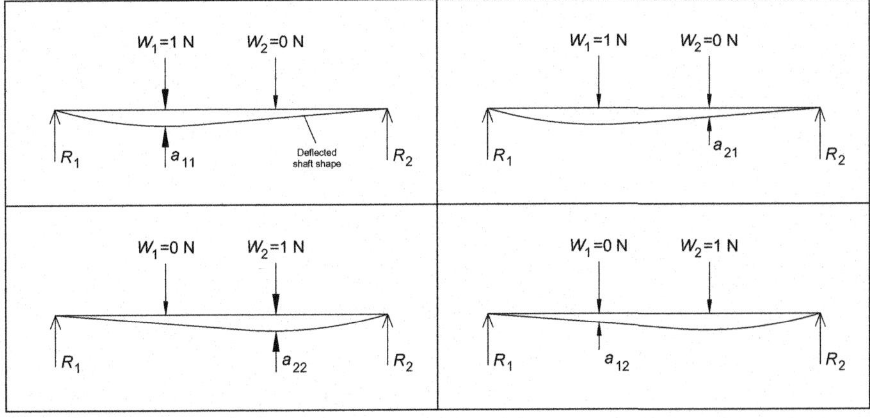

Fig. 7.21 Example of influence coefficient definition for a simply supported shaft with two concentrated loads.

a_{22} is the deflection at the location of mass 2, that would be caused by a 1-N weight at the location of mass 2.
a_{12} is the deflection at the location of mass 1, that would be caused by a 1-N weight at the location of mass 2.

Note that values for a_{np} and a_{pn} are equal by the principle of reciprocity; for example $a_{12}=a_{21}$.

For a multimass system, the frequency equation can be obtained by equating the following determinate to zero:

$$\begin{vmatrix} \left(a_{11}m_1 - \frac{1}{\omega^2}\right) & a_{12}m_2 & a_{13}m_3 & \cdots \\ a_{21}m_1 & \left(a_{22}m_2 - \frac{1}{\omega^2}\right) & a_{23}m_3 & \cdots \\ a_{31}m_1 & a_{32}m_2 & \left(a_{33}m_3 - \frac{1}{\omega^2}\right) & \cdots \\ \cdots & \cdots & \cdots & \cdots \end{vmatrix} = 0 \tag{7.5}$$

It should be noted that lateral vibration requires an external source of energy. For example, vibrations can be transferred from another part of a machine and the shaft will vibrate in one or more lateral planes regardless of whether the shaft is rotating. Shaft whirl is a self-excited vibration caused by the shaft's rotation acting on an eccentric mass.

These analysis techniques for calculating the critical frequency require the determination of the shaft deflection. Section 7.4.1 introduces the Macaulay method which is suitable for calculating the deflection of a constant diameter shaft and Section 7.4.2 introduces the strain energy method which is suitable for more complex shafts with stepped diameters.

7.4.1 Macaulay's Method for Calculating the Deflection of Beams

Macaulay's method can be used to determine the deflection of a constant cross-section shaft. The general rules for this method are given in the following. Having calculated the deflections of a shaft, this information can then be used to determine critical frequencies.

- **(i)** Take an origin at the left-hand side of the beam.
- **(ii)** Express the bending moment at a suitable section XX in the beam to include the effect of all the loads.
- **(iii)** Uniformly distributed loads must be made to extend to the right-hand end of the beam. Use negative loads to compensate.
- **(iv)** Put in square brackets all functions of length other than those involving single powers of x.
- **(v)** Integrate as a whole any term in square brackets.
- **(vi)** When evaluating the moment, slope or deflection, neglect the square brackets terms when they become negative.

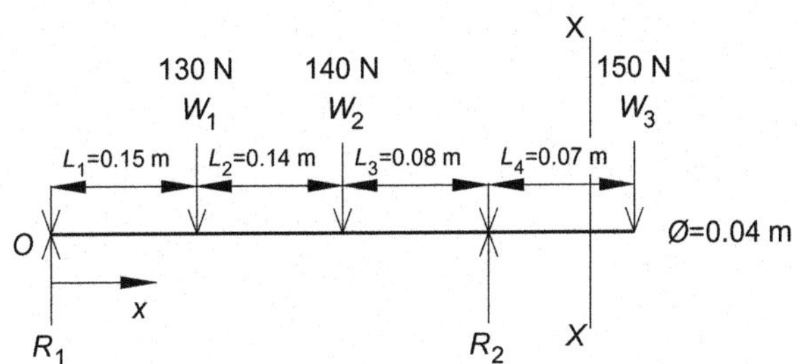

Fig. 7.22 Machine shaft example.

(vii) In the moment equation, express concentrated moments in the form $M_1[x-a]^0$ where M_1 is the concentrated moment and $x-a$ is its point of application relative to the section XX.

Macaulay's method for calculating the deflection of beams and shafts is useful, in that, it is relatively simple to use and easily programmed.

Example 7.1. As part of the preliminary design of a machine shaft a check is to be undertaken to determine that the critical speed is significantly higher than the design speed of 7000 rpm. The components can be represented by three point masses as shown in Fig. 7.22. Assume the bearings are stiff and act as simple supports. The shaft diameter is 40 mm and the material is steel with a Young's modulus of $200 \times 10^9 \text{ N/m}^2$.

Solution

Macaulay's method is used to determine the shaft deflections:
Resolving vertically, $R_1+R_2=W_1+W_2+W_3$.
Clockwise moments about O:

$$W_1L_1+W_2(L_1+L_2)-R_2(L_1+L_2+L_3)+W_3(L_1+L_2+L_3+L_4)=0.$$

Hence, $R_2=\frac{W_1L_1+W_2(L_1+L_2)+W_3(L_1+L_2+L_3+L_4)}{L_1+L_2+L_3}$

Calculating the moment at XX,

$$M_{XX}=-R_1x+W_1[x-L_1]+W_2[x-(L_1+L_2)]-R_2[x-(L_1+L_2+L_3)] \tag{7.6}$$

The relationship for the deflection y of a beam subjected to a bending moment M is given by Eq. (7.7).

$$EI\frac{d^2y}{dx^2}=M \tag{7.7}$$

where

E is the modulus of elasticity or Young's modulus (N/m^2),
I is the second moment of area (m^4),
Y = deflection (m),

x is the distance from the end of the beam to the location at which the deflection is to be determined (m) and.
M = moment (N m).

Eq. (7.7) can be integrated once to find the slope dy/dx and twice to find the deflection y.

Integrating Eq. (7.6) to find the slope:

$$EI\frac{dy}{dx} = -R_1\frac{x^2}{2} + \frac{W_1}{2}[x - L_1]^2 + \frac{W_2}{2}[x - (L_1 + L_2)]^2 - \frac{R_2}{2}[x - (L_1 + L_2 + L_3)]^2 + A.$$

Integrating again to find the deflection equation gives the following:

$$EIy = -R_1\frac{x^3}{6} + \frac{W_1}{6}[x - L_1]^3 + \frac{W_2}{6}[x - (L_1 + L_2)]^3 \\ - \frac{R_2}{6}[x - (L_1 + L_2 + L_3)]^3 + Ax + B.$$

Note that Macaulay's method requires that terms within square brackets be ignored when the sign of the bracket goes negative.

It is now necessary to substitute boundary conditions to find the constants of integration. In the case of a shaft, the deflection at the bearings may be known in which case these can be used as boundary conditions.

Assuming that the deflection at the bearing is zero then substituting $y=0$ at $x=0$ into the equation for deflection above gives $B=0$.

At $x = L_1 + L_2 + L_3$, $y=0$.

$$0 = -\frac{R_1}{6}(L_1 + L_2 + L_3)^3 + \frac{W_1}{6}(L_2 + L_3)^3 + \frac{W_2}{6}(L_3)^3 + A(L_1 + L_2 + L_3).$$

Hence, $A = \dfrac{\frac{R_1}{6}(L_1 + L_2 + L_3)^3 - \frac{W_1}{6}(L_2 + L_3)^3 - \frac{W_2}{6}(L_3)^3}{L_1 + L_2 + L_3}$.

The second moment of area for a solid round shaft is given by

$$I = \frac{\pi d^4}{64}$$

so $I = \frac{\pi 0.04^4}{64} = 1.2566 \times 10^{-7}\,\text{m}^4$.

With $W_1 = 130\,\text{N}$, $W_2 = 140\,\text{N}$, $W_3 = 150\,\text{N}$, $\phi = 0.04\,\text{m}$, and $E = 200 \times 10^9\,\text{N/m}^2$, substitution of these values into the previous equations gives the following:

$R_1 = 79.19\,\text{N}$,
$R_2 = 340.8\,\text{N}$,
$A = 1.151\,\text{N m}^2$.

At $x = 0.15\,\text{m}$, $y = 5.097 \times 10^{-6}\,\text{m}$.
At $x = 0.29\,\text{m}$, $y = 2.839 \times 10^{-6}\,\text{m}$.
At $x = 0.44\,\text{m}$, $y = -1.199 \times 10^{-6}\,\text{m}$.

Use *absolute* values for the displacement in the Rayleigh-Ritz equation:

$$\omega_c =$$

$$\sqrt{\frac{9.81\left(\left(130\times 5.097\times 10^{-6}\right)+\left(140\times 2.839\times 10^{-6}\right)+\left(150\times \left|-1.199\times 10^{-6}\right|\right)\right)}{\left(130\times \left(5.097\times 10^{-6}\right)^2\right)+\left(140\times \left(2.839\times 10^{-6}\right)^2\right)+\left(150\times \left(\left|-1.199\times 10^{-6}\right|\right)^2\right)}}$$

$$= \sqrt{\frac{9.81\times 1.240\times 10^{-3}}{4.722\times 10^{-9}}} = 1605\,\text{rad/s}.$$

So the critical speed is $1605 \times 2\pi/60 = 15{,}530\,\text{rpm}$.

Example 7.2. An air compressor consists of four compressor discs mounted on a steel shaft 50 mm apart as shown in Fig. 7.23. The shaft is simply supported at either end. Each compressor wheel mass is 18 kg. Calculate the deflection at each wheel using Macaulay's method and calculate, using the Rayleigh-Ritz equation, the first critical frequency of the shaft. The shaft outer diameter is 0.05 m, the inner diameter is 0.03 m.

Solution

$$-R_1x + W_1[x - L_1] + W_2[x-(L_1+L_2)] + W_3[x-(L_1+L_2+L_3)] + W_4[x-(L_1+L_2+L_3+L_4)] = EI\frac{d^2y}{dx^2}.$$

$$-\frac{R_1x^3}{6} + \frac{W_1}{6}[x-L_1]^3 + \frac{W_2}{6}[x-(L_1+L_2)]^3 + \frac{W_3}{6}[x-(L_1+L_2+L_3)]^3 + \frac{W_4}{6}[x-(L_1+L_2+L_3+L_4)]^3 + Ax + B = EIy.$$

Boundary conditions:

$y = 0$ at $x = 0$,
$y = 0$ at $x = L_1 + L_2 + L_3 + L_4 + L_5 = L$.

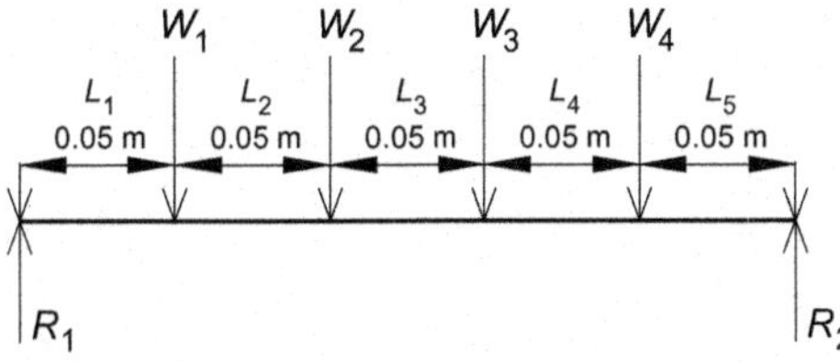

Fig. 7.23 Air compressor shaft.

Hence, $B = 0$.

$$A = \frac{1}{L}\left(\frac{R_1 L^3}{6} - \frac{W_1}{6}(L - L_1)^3 - \frac{W_2}{6}(L_3 + L_4 + L_5)^3 - \frac{W_3}{6}(L_4 + L_5)^3 - \frac{W_4}{6}L_5^3\right).$$

$$R_1 + R_2 = W_1 + W_2 + W_3 + W_4.$$

Moments about R_1:

$$W_1 L_1 + W_2(L_1 + L_2) + W_3(L_1 + L_2 + L_3) + W_4(L_1 + L_2 + L_3 + L_4) - R_2 L = 0.$$

Hence,

$$A = 2.20725,$$

$$R_2 = 353.16\text{N},$$

$$R_1 = 353.16\text{N},$$

The second moment of area, $I = 2.67 \times 10^{-7}\,\text{m}^4$, and Young's modulus, $E = 200 \times 10^9\,\text{N/m}^2$

$$y_{w1} = 1.9286726 \times 10^{-6}\,\text{m},$$

$$y_{w2} = 3.0996525 \times 10^{-6}\,\text{m},$$

$$y_{w3} = 3.0996525 \times 10^{-6}\,\text{m},$$

$$y_{w4} = 1.9286726 \times 10^{-6}\,\text{m},$$

$$\omega = 1923.8\,\text{rad/s} = 18{,}371\,\text{rpm}.$$

Example 7.3. This example explores the influence on the critical frequency of adding an overhang load to a beam.

(i) Determine the critical frequency for a 0.4 m long steel shaft running on bearings with a single point load as illustrated in Fig. 7.24. Assume that the bearings are rigid and act as simple supports. Ignore the self-weight of the shaft. Take Young's modulus of elasticity as $200 \times 10^9\,\text{N/m}^2$.

(ii) Calculate the critical frequency if an overhang load of 150 N is added at a distance of 0.15 m from the right-hand bearing as shown in Fig. 7.25.

Solution

(i) Resolving vertically: $R_1 + R_2 = \text{W}_1$.

Clockwise moments about O: $375 \times 0.25 - R_2 0.4 = 0$.

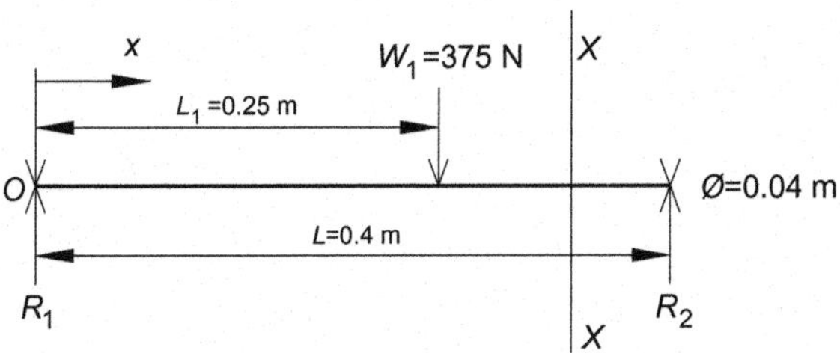

Fig. 7.24 Simple shaft with single concentrated mass.

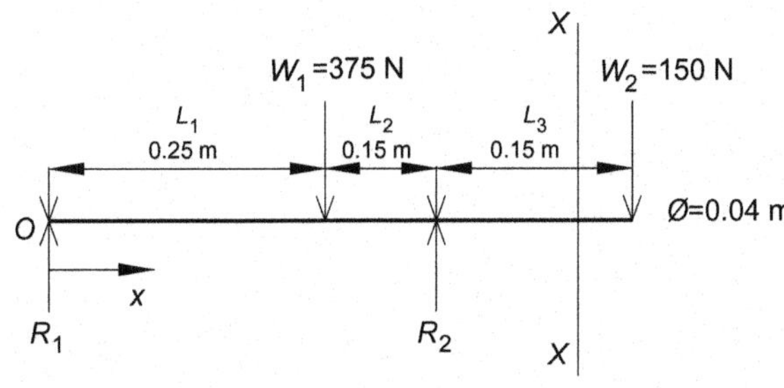

Fig. 7.25 Simple shaft with a single concentrated mass between bearing and overhang mass.

Hence, $R_2=234.375\,\text{N}$, $R_1=140.625\,\text{N}$.

$$EI\frac{d^2y}{dx^2}=-R_1x+W_1[x-L_1].$$

$$EI\frac{dy}{dx}=-R_1\frac{x^2}{2}+\frac{W_1}{2}[x-L_1]^2+A.$$

$$EIy=-R_1\frac{x^3}{6}+\frac{W_1}{6}[x-L_1]^3+Ax+B.$$

Boundary conditions: assuming that the deflection of the shaft is zero at the bearings, then substituting $y=0$ and $x=0$ into the previous equation gives $B=0$.

At $x=L$, $y=0$. Hence, $A=\frac{1}{L}\left(R_1\frac{L^2}{6}-\frac{W_1}{6}(L-L_1)^3\right)=\frac{1}{0.4}(1.5-0.2109)=3.22266$.

For a solid circular shaft, the second moment of area, $I=\frac{\pi d^4}{64}=\frac{\pi 0.04^4}{64}=1.2566\times10^{-7}\,\text{m}^4$.

$E=200\times10^9\,\text{N/m}^2$.

$$EIy=-140.625\times\frac{0.25^3}{6}+(3.22266\times0.25).$$

$$y=1.7485\times10^{-5}\,\text{m}.$$

Application of Macaulay's method to the geometry given in Fig. 7.24 yields $y_1=1.74853\times10^{-5}\,\text{m}$. So from Eq. (7.1),

$$\omega_c = \sqrt{\frac{g}{y_1}} = 749.03\,\text{rad/s} = 7152\,\text{rpm}.$$

(ii) $EI\frac{d^2y}{dx^2} = -R_1x + W_1[x - L_1] - R_2[x - (L_1 + L_2)]$

$$EI\frac{dy}{dx} = -R_1\frac{x^2}{2} + \frac{W_1}{2}[x - L_1]^2 - \frac{R_2}{2}[x - (L_1 + L_2)]^2 + A$$

$$EIy = -R_1\frac{x^3}{6} + \frac{W_1}{6}[x - L_1]^3 - \frac{R_2}{6}[x - (L_1 + L_2)]^3 + Ax + B$$

Boundary conditions: assuming negligible deflection at the bearings, at $x=0$, $y=0$. Hence, $B=0$.

At $x=L_1+L_2$, $y=0$. Hence, $A = \frac{1}{L_1+L_2}\left(\frac{R_1}{6}(L_1+L_2)^3 - \frac{W_1}{6}L_2^3\right)$.

Resolving vertically, $R_1+R_2=W_1+W_2$.
Moments about O: $W_1L_1+W_2(L_1+L_2+L_3)-R_2(L_1+L_2)=0$.

Hence, $R_2 = \frac{1}{L^1+L_2}(W_1L_1 + W_2(L_1+L_2+L_3))$, $R_1=W_1+W_2-R_2$, $R_1=84.375\,\text{N}$, $R_2=440.625\,\text{N}$.
Substitution for A gives $A=1.722656\,\text{N}\,\text{m}^2$.
Evaluating y at $x=L_1$ gives $y=8.3929381 \times 10^{-6}\,\text{m}$.
Evaluating y at $x=L_1+L_2+L_3$ gives $y=1.8884145 \times 10^{-6}\,\text{m}$.
Hence,

$$\omega_c = \sqrt{\frac{9.81\left(3.147\times10^{-3} + 2.8326\times10^{-4}\right)}{2.64155\times10^{-8} + 5.34061\times10^{-10}}} =$$

$$\sqrt{\frac{9.81\times3.4306\times10^{-3}}{2.695\times10^{-8}}} = 1117.5\,\text{rad/s}.$$

$$\omega_c = 10671\,\text{rpm}.$$

Comparison of the results from (i) and (ii) shows that the addition of an overhang weight can have a significant effect in raising the critical frequency from 7150 rpm to 10,670 rpm in this case. This technique of adding overhang masses to increase the critical speed is sometimes used in plant machinery.

Example 7.4.

(i) Calculate the first two critical speeds for the loadings on a steel shaft indicated in Fig. 7.26. Assume the mass of the shaft can be ignored for the purpose of this calculation and that the bearings can be considered to be stiff simple supports. The internal and external diameters of the shaft are 0.03 m and 0.05 m, respectively. Take Young's modulus as 200 GN/m^2.

(ii) The design speed for the shaft is nominally 5000 rpm; what would be the implication of the bearing data given in Table 7.2 on the shaft operation assuming the bearings allow flexibility only in directions perpendicular to the shaft axis?

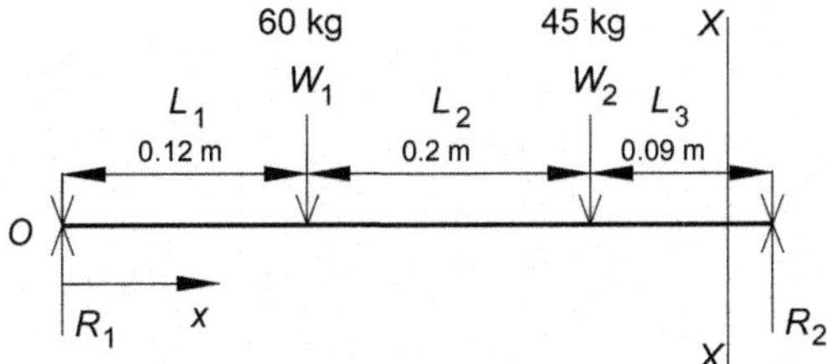

Fig. 7.26 Simple shaft.

Table 7.2 Example bearing data for deflection versus load

Load on bearing (N)	Deflection at bearing (mm)
50	0.003
100	0.009
200	0.010
300	0.0115
400	0.0126
500	0.0136
1000	0.0179
2000	0.0245

Solution

(i) Applying Macaulay's method to find the influence coefficients: Firstly calculating the deflections at L_1 and L_2 with $W_1 = 1$ N and $W_2 = 0$ N to give a_{11} and a_{21}. Then calculating the deflections at L_1 and L_2 with $W_1 = 0$ N and $W_2 = 1$ N to give a_{12} and a_{22}.

$$EIy = -R_1\frac{x^3}{6} + \frac{W_1}{6}[x - L_1]^3 + \frac{W_2}{6}[x - (L_1 + L_2)]^3 + Ax + B.$$

With $W_1 = 1$N, $W_2 = 0$N, $R_1 = 0.707317$N, $R_2 = 0.292683$N, $A = 9.90244 \times 10^{-3}$ Nm^2 ($B = 0$).

Hence, $a_{11} = 1.84355 \times 10^{-8}$ m, $a_{21} = 1.19688 \times 10^{-8}$m.

With $W_1 = 0$N, $W_2 = 1$N, $R_1 = 0.2195122$N, $R_2 = 0.7804878$N, $A = 5.853659 \times 10^{-3}$Nm^2.

Hence, $a_{22} = 1.26264 \times 10^{-8}$m, $a_{12} = 1.19688 \times 10^{-8}$m.

For this example, $a_{11} = 1.84355 \times 10^{-8}$m, $a_{21} = 1.19688 \times 10^{-8}$m, $a_{12} = 1.19688 \times 10^{-8}$m, $a_{22} = 1.26264 \times 10^{-8}$m.

Solving $\frac{1}{\omega^4} - (a_{11}m_1 + a_{22}m_2)\frac{1}{\omega^2} + (a_{11}a_{22} - a_{12}a_{21})m_1m_2 = 0.$ by multiplying through by ω^4 and solving as a quadratic:

$$\omega^2 = \frac{-b \pm \sqrt{b^2 - 4ac}}{2a},$$

with

$$a = (a_{11}a_{22} - a_{12}a_{21})m_1m_2,$$
$$b = -(a_{11}m_1 + a_{22}m_2),$$
$$c = 1,$$

gives $\omega_{c1} = 812.5$ rad/s (7759 rpm) and $\omega_{c2} = 2503.4$ rad/s (23,905 rpm).

(ii) The deflection at the bearings should also be taken into account in determining the critical frequency of the shaft. The reaction at each of the bearings is approximately 500 N ($R_1 = 513.2$ N and $R_2 = 516.8$ N). The deflection at each bearing can be assumed roughly equal (so from Table 7.2 $y_{\text{bearing}} = 0.0136$ mm). This deflection can be simply added to the static deflection of the shaft due to bending by superposition. If the deflections were not equal, similar triangles could be used to calculate the deflection at the mass locations.

The influence coefficients can also be used to determine the deflection under the actual loads applied. $W_1 = 60 \times 9.81 = 588.6$ N, $W_2 = 45 \times 9.81 = 441.45$ N:

$$y_{1\,\text{Load}} = W_1 a_{11} + W_2 a_{12} = 1.613 \times 10^{-5}\,\text{m},$$

$$y_{2\,\text{Load}} = W_1 a_{21} + W_2 a_{22} = 1.26 \times 10^{-5}\,\text{m},$$

$$y_{1\,\text{Total}} = 1.61 \times 10^{-5} + 1.36 \times 10^{-5} = 2.97 \times 10^{-5}\,\text{m},$$

$$y_{2\,\text{Total}} = 1.26 \times 10^{-5} + 1.36 \times 10^{-5} = 2.62 \times 10^{-5}\,\text{m}.$$

These deflections give (using the Rayleigh-Ritz equation) a first critical frequency of

$$\omega_c = \sqrt{\frac{9.81 \times 2.96 \times 10^{-3}}{8.38 \times 10^{-8}}} = 588.6\,\text{rad/s} = 5620\,\text{rpm}.$$

This is very close to the design speed. Different bearings or a stiffer shaft would be advisable.

Example 7.5. A 22.63 kg compressor impeller wheel is driven by a 13.56 kg turbine mounted on a common shaft (see Fig. 7.27) manufactured from steel with Young's modulus $E = 207\,\text{GN/m}^2$. The design speed is 10,000 rpm. Determine the nominal shaft diameter so that the first critical speed is 12,000 rpm giving a safety margin of 2000 rpm. Use Rayleigh's equation and assume rigid bearings and a massless shaft.

Solution

$W_1 = 9.81 \times 22.63 = 222\,\text{N}, W_2 = 9.81 \times 13.56 = 133\,\text{N}.$

Moments about the left-hand bearing (Fig. 7.28):

$$222 \times 0.254 - R_R \times 0.508 + 133 \times 0.762 = 0$$

$$R_R = 310.5\,\text{N}.$$

Resolving vertically, $R_L + R_R = 222 + 133$.

$$R_L = 44.5\,\text{N}.$$

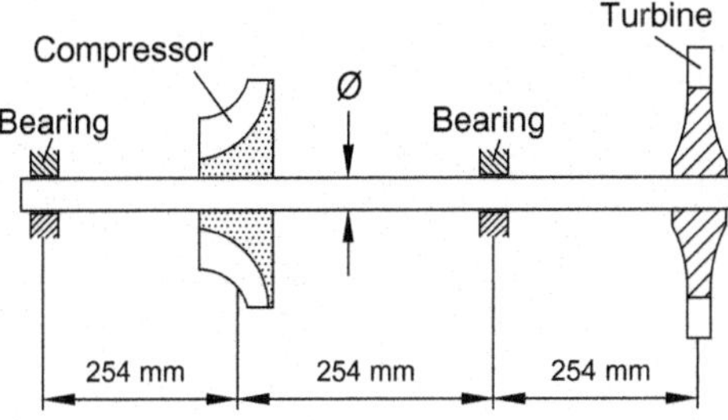

Fig. 7.27 Turbine shaft example.

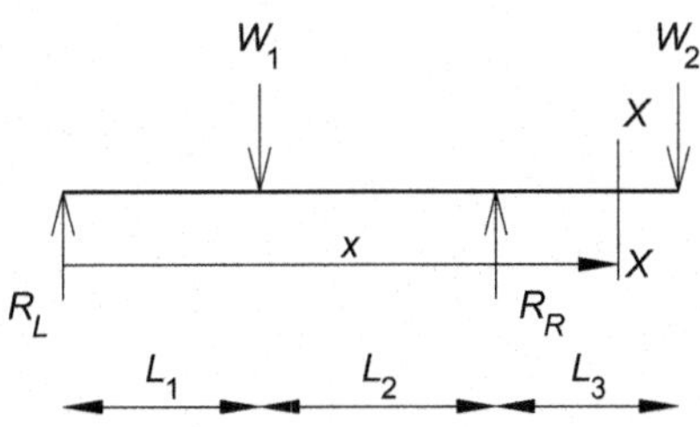

Fig. 7.28 Principal Loads.

Using Macaulay's method to determine the slope and deflection,

$$M_{xx} = -R_L x + W_1[x - L_1] - R_R[x - (L_1 + L_2)]$$

$$EI\frac{dy}{dx} = -R_L\frac{x^2}{2} + \frac{W_1}{2}[x - L_1]^2 - \frac{R_R}{2}[x - (L_1 + L_2)]^2 + A$$

$$EIy = -R_L\frac{x^3}{6} + \frac{W_1}{6}[x - L_1]^3 - \frac{R_R}{6}[x - (L_1 + L_2)]^3 + Ax + B$$

Boundary conditions:

At $x=0$, $y=0$ (left-hand bearing), so from the previous equation, $B=0$.

At $x=L_1+L_2$, $y=0$ (right-hand bearing).

$$0 = -R_L\frac{(L_1 + L_2)^3}{6} + \frac{W_1}{6}L_2^3 + A(L_1 + L_2)$$

$$A = \frac{R_L\frac{(L_1 + L_2)^3}{6} - \frac{W_1}{6}L_2^3}{(L_1 + L_2)}$$

$$A = \frac{44.5(0.508)^3/6 - 222(0.254)^3/6}{0.508} = 0.7204$$

y_1 = deflection at location of load W_1.
y_2 = deflection at location of load W_2.

$$y_1 = \frac{1}{EI}\left(-\frac{44.5 \times 0.254^3}{6} + 0.7204 \times 0.254\right) = \frac{0.06144}{EI}$$

$$y_2 = \frac{1}{EI}\left(-\frac{44.5 \times 0.762^3}{6} + \frac{222}{6}(0.762 - 0.254)^3 - \frac{310.5}{6}(0.254)^3 + 0.7204 \times 0.762\right) = \frac{1.270}{EI}$$

Let the first critical speed = 12,000 rpm = 12,000 × 2π/60 rad/s = 1257 rad/s,

$$1257 = \sqrt{\frac{9.81\left(222 \times \frac{0.01644}{EI} + 133 \times \frac{1.270}{EI}\right)}{222\left(\frac{0.01644}{EI}\right)^2 + 133\left(\frac{1.270}{EI}\right)^2}} = \sqrt{EI \times 9.81 \times \frac{172.5}{214.6}} = \sqrt{EI \times 7.89}$$

$$1257^2 = EI \times 7.89 = \frac{\pi d^4}{64} \times 207 \times 10^9 \times 7.89$$

$$d = \left(1.971 \times 10^{-5}\right)^{0.25} = 0.06663 \text{ m}$$

The nominal diameter is therefore $d \approx 67$ mm.

7.4.2 Castigliano's Theorem for Calculating Shaft Deflections

The strain energy, U, in a straight beam subjected to a bending moment M is

$$U = \int \frac{M^2 dx}{2EI} \tag{7.8}$$

Castigliano's strain energy equation can be applied to problems involving shafts of nonconstant section to calculate deflections, and hence critical speeds.

Example 7.6. Determine the deflection of the stepped shaft illustrated in Fig. 7.29 under a single concentrated load.

Solution

The total strain energy due to bending is $U = \int \frac{M^2 dx}{2EI} = U_1 + U_2$
where,

U_1 = the energy from $x = 0$ to L_1,
U_2 = the energy from $x = L_1$ to L.

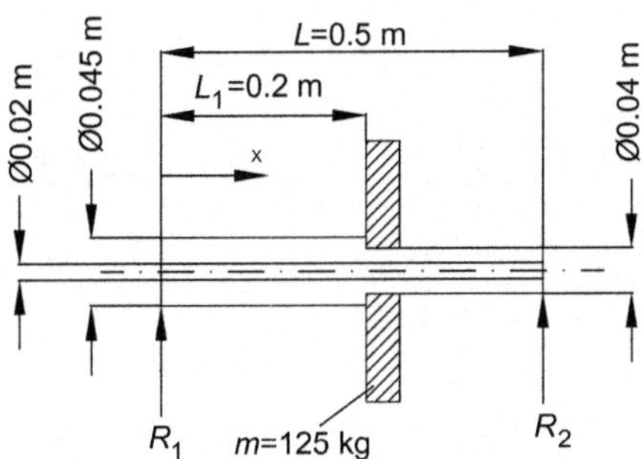

Fig. 7.29 Stepped shaft with single load.

(In general, consider the strain energy for each section of a beam, i.e. between loads and between any change of shaft section.)

Resolving vertical loads, $R_1 + R_2 = W$.

Taking moments about the left-hand bearing,

$$R_2 = \frac{WL_1}{L}.$$

Hence, $R_1 = W\left(1 - \frac{L_1}{L}\right)$.

Splitting the beam between $x=0$ and L_1, the moment can be expressed as R_1x.

$$U_1 = \int_0^{L_1} \frac{(R_1x)^2}{2EI_1}dx = \left[\frac{(W(1-(L_1/L)))^2x^3}{6EI_1}\right]_0^{L_1} = \frac{(W(1-(L_1/L)))^2L_1^3}{6EI_1}.$$

Splitting the beam between $x=L_1$ and L, the moment can be expressed as $R_2(L-x)$.

$$U_2 = \int_{L_1}^{L} \frac{(R_2(L-x))^2}{2EI_2}dx = \left[\frac{(WL_1/L)^2(L^2x - Lx^2 + x^3/3)}{2EI_2}\right]_{L_1}^{L} =$$

$$\frac{W^2L_1^2}{2EI_2L^2}\left(L^3 - L^3 + \frac{L^3}{3} - L^2L_1 + LL_1^2 - \frac{L_1^3}{3}\right).$$

$$U = U_1 + U_2$$
$$= \frac{(W(1-(L_1/L)))^2L_1^3}{6EI_1} + \frac{W^2L_1^2}{2EI_2L^2}\left(L^3 - L^3 + \frac{L^3}{3} - L^2L_1 + LL_1^2 - \frac{L_1^3}{3}\right).$$

Differentiating the previous expression with respect to W gives the deflection at W:

$$\frac{\partial U}{\partial W} = \frac{L_1^3}{6EI_1}\left(2W - \frac{4WL_1}{L} + \frac{2WL_1^2}{L^2}\right) + \frac{2WL_1^2}{2EI_2L^2}\left(\frac{L^3}{3} - L^2L_1 + LL_1^2 - \frac{L_1^3}{3}\right). \quad (7.9)$$

The second moment of areas for the two sections of the shaft are given by

$$I_1 = \frac{\pi\left(d_o^4 - d_i^4\right)}{64} = \frac{\pi\left(0.045^4 - 0.02^4\right)}{64} = 1.9343 \times 10^{-7}\,\text{m}^4.$$

$$I_2 = \frac{\pi\left(0.04^4 - 0.02^4\right)}{64} = 1.1781 \times 10^{-7}\,\text{m}^4.$$

Substitution of $W = 125 \times 9.81\,\text{N}, L_1 = 0.2\,\text{m}, L = 0.5\,\text{m}, E = 200 \times 10^9\,\text{N/m}^2, I_1$ and I_2 into Eq. (7.9) gives a deflection $\partial U/\partial W = 1.05372 \times 10^{-4}\,\text{m}$ and hence a first critical frequency using the approximation of Eq. (7.1) of

$$\sqrt{\frac{9.81}{1.05372 \times 10^{-4}}} = 305.121\,\text{rad/s} = 2913\,\text{rpm}.$$

Example 7.7. Using Castigliano's theorem for calculating deflections, determine the critical frequency of the steel shaft shown in Fig. 7.30. Neglect the shaft mass.

Solution

- Looking at the shaft from the left,
  section (I) $M = R_L\,x$,
  section (II) $M = R_L\,x$,
  section (III) $M = R_R\,(L - x)$.

$$R_L + R_R = W.$$

$$WL_2 - R_R L = 0.$$

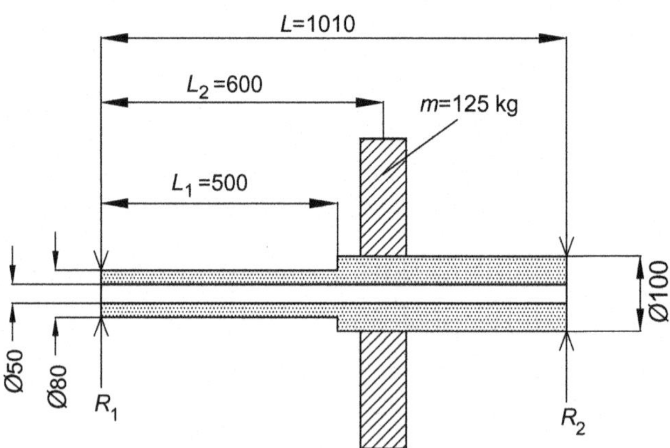

Fig. 7.30 Simple stepped shaft and load (dimensions in mm).

$$R_R = WL_2/L.$$

$$R_L = W\left(1-\frac{L_2}{L}\right).$$

By definition,

$$U = \int \frac{M^2 dx}{2EI}$$

$$U_1 = \int_0^{L_1} \frac{R_L^2 x^2}{2EI_1} dx = \left[\frac{W^2\left(1-\frac{L_2}{L}\right)^2 \frac{x^3}{3}}{2EI_1}\right]_0^{L_1} = \frac{W^2\left(1-\frac{L_2}{L}\right)^2 \frac{L_1^3}{3}}{2EI_1}.$$

$$\frac{\partial U_1}{\partial W} = \frac{2W\left(1-\frac{L_2}{L}\right)^2 \frac{L_1^3}{3}}{2EI_1}.$$

$$U_2 = \int_{L_1}^{L_2} \frac{R_L^2 x^2}{2EI_2} dx = \left[\frac{W^2}{2EI_2}\left(1-\frac{L_2}{L}\right)^2 \frac{x^3}{3}\right]_{L_1}^{L_2} = \frac{W^2}{2EI_2}\left(1-\frac{L_2}{L}\right)^2\left(\frac{L_2^3}{3}-\frac{L_1^3}{3}\right)$$

$$\frac{\partial U_2}{\partial W} = \frac{2W}{2EI_2}\left(1-\frac{L_2}{L}\right)^2\left(\frac{L_2^3}{3}-\frac{L_1^3}{3}\right).$$

$I_3 = I_2$ (same section)

$$U_3 = \int \frac{W^2 (L_2/L)^2}{2EI_3} dx$$

$$= \left[\frac{W^2 (L_2/L)^2 \left(L^2 x - x^2 L + \frac{x^3}{3}\right)}{2EI_3}\right]_{L_2}^{L} = \frac{W^2}{2EI_3}\left(\frac{L_2}{L}\right)^2\left(L^3 - L^3 + \frac{L^3}{3} - L^2 L_2 + L_2^2 L - \frac{L_2^3}{3}\right)$$

$$\frac{\partial U_3}{\partial W} = \frac{W}{EI_3}\left(\frac{L_2}{L}\right)^2\left(\frac{L^3}{3} - L^2 L_2 + L_2^2 L - \frac{L_2^3}{3}\right).$$

Substituting values for the loads, lengths, Young's modulus and second moments of area,

$$\frac{\partial U_1}{\partial W} = 2.4708039 \times 10^{-5}$$

$$\frac{\partial U_2}{\partial W} = 6.6596749 \times 10^{-6}$$

$$\frac{\partial U_3}{\partial W} = 1.0801836 \times 10^{-5}$$

$$\frac{\partial U}{\partial W} = \frac{\partial U_1}{\partial W} + \frac{\partial U_2}{\partial W} + \frac{\partial U_3}{\partial W}$$

Hence, $\frac{\partial U}{\partial W} = 4.2169551 \times 10^{-5}$

The first critical frequency using the approximation of Eq. (7.1) is.

$$\omega_c = 482.3 \text{rad/s},$$

$$\omega_c = 4605.8 \text{rpm}.$$

7.5 Analysis of transmission shafting

According to Fuchs and Stephens (1980), between 50% and 90% of all mechanical failures are fatigue failures. This challenges the engineer to consider the possibility of fatigue failure at the design stage. Fig. 7.31 shows the characteristic variation of fatigue strength for steel with the number of stress cycles. For low-strength steels, a 'levelling off' occurs in the graph between 10^6 and 10^7 cycles under noncorrosive conditions and regardless of the number of stress cycles; beyond this, the component will not fail due to fatigue. The value of stress corresponding to this levelling off is called the endurance stress or the fatigue limit.

A rotating shaft will experience fluctuating loads, with any given location on a circumference going in and out of compression with each rotation. A key requirement in shaft design is therefore the selection of diameters in order to avoid failure due to fatigue as well to avoid yielding. For many shafts, the loads are a combination of

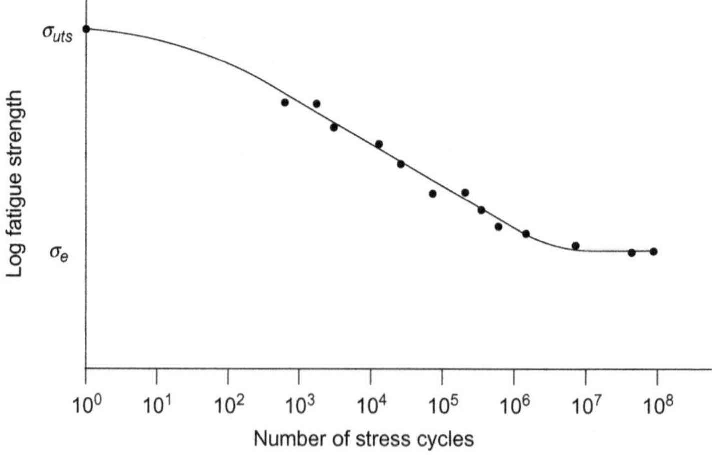

Fig. 7.31 A typical strength cycle diagram for various steels.

bending and torsion. In order to design solid or hollow rotating shafts under combined cyclic bending and steady torsional loading for limited life, a variety of approaches are possible. This section outlines the modelling of fluctuating forces in a shaft and criteria that have been developed in order to provide increased confidence in a design in order to avoid failure. A series of equations that allow for either the calculation of a factor of safety for a given design, or the diameter at a location of interest, have been developed. One or more of these can be used to assess or develop a design.

Fluctuating forces in machinery, e.g. Fig. 7.32, can be characterised by a waveform with a mid-range steady component and an amplitude as follows:

$$F_m = \frac{F_{max} + F_{min}}{2} \tag{7.10}$$

$$F_a = \left| \frac{F_{max} - F_{min}}{2} \right| \tag{7.11}$$

where

F_m = mid-range steady component of force (N),
F_{max} = maximum force (N),
F_{min} = minimum force (N) and.
F_a = amplitude of alternating force (N).

For a fluctuating stress waveform, mid-range and amplitude stress components for a location of interest are given by

$$\sigma_m = \frac{\sigma_{max} + \sigma_{min}}{2} \tag{7.12}$$

$$\sigma_a = \left| \frac{\sigma_{max} - \sigma_{min}}{2} \right| \tag{7.13}$$

where

σ_m = mid-range steady component of stress (N/m^2),
σ_{max} = maximum stress (N/m^2),
σ_{min} = minimum stress (N/m^2) and.
σ_a = amplitude of alternating stress (N/m^2).

A stress ratio, R, and an amplitude ratio, A, can be defined by Eqs (7.14) and (7.15), respectively:

$$R = \frac{\sigma_{\min}}{\sigma_{\max}} \tag{7.14}$$

$$A = \frac{\sigma_a}{\sigma_m} \tag{7.15}$$

A fatigue graph of stress amplitude, σ_a, versus mean stress, σ_m, is given in Fig. 7.33 showing various failure criteria. The lines represent zones of safe and unsafe

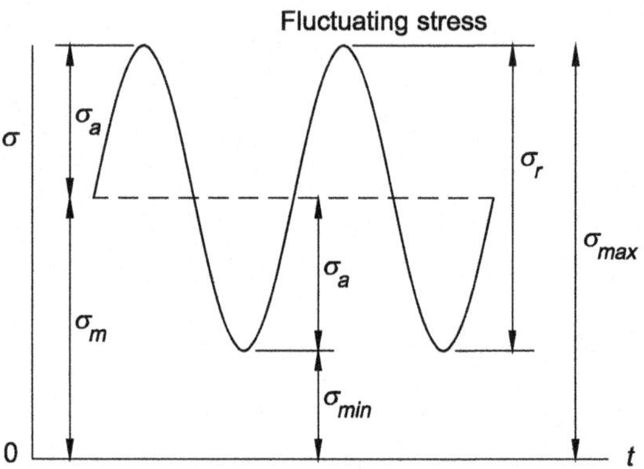

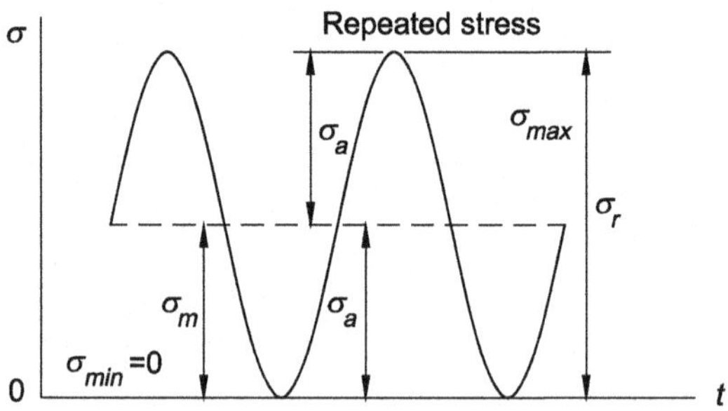

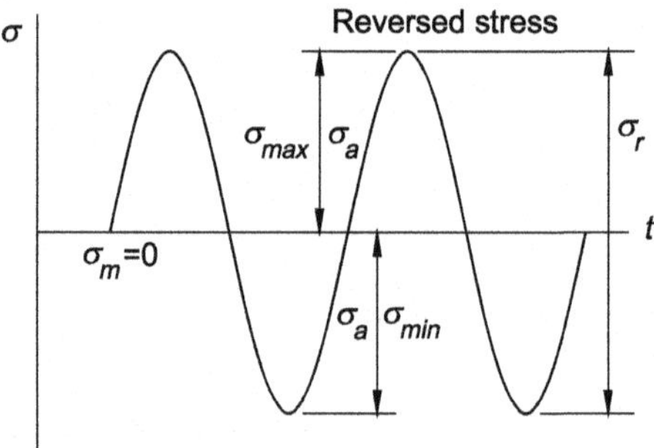

Fig. 7.32 Fluctuating load.

operation. Operation to the right-hand side of any of the criterion lines identified provides an indication of likelihood of failure.

The relationship between S_a, S_e, S_m and S_y for the Soderberg line is

$$\frac{S_a}{S_e} + \frac{S_m}{S_y} = 1 \tag{7.16}$$

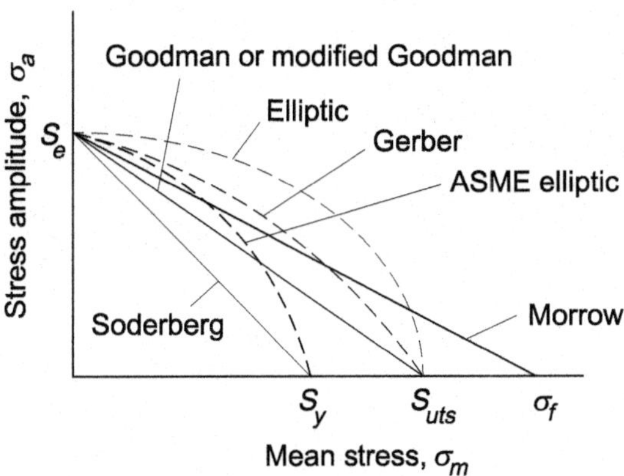

Fig. 7.33 Fatigue graph of σ_a versus σ_m showing various failure criteria.

where

S_a = alternating strength (N/m^2),
S_e = endurance limit (N/m^2),
S_m = mid-range strength (N/m^2) and.
S_y = yield stress (N/m^2).

The relationship between S_a, S_e, S_m and S_{ut} for the Goodman line is

$$\frac{S_a}{S_e} + \frac{S_m}{S_{ut}} = 1 \tag{7.17}$$

where, S_{ut} = ultimate strength (N/m^2).

For the Gerber line, the relationship between S_a, S_e, S_m and S_{ut} is

$$\frac{S_a}{S_e} + \left(\frac{S_m}{S_{ut}}\right)^2 = 1 \tag{7.18}$$

For the ASME elliptic, the relationship between S_a, S_e, S_m and S_{ut} is

$$\left(\frac{S_a}{S_e}\right)^2 + \left(\frac{S_m}{S_{ut}}\right)^2 = 1 \tag{7.19}$$

The yield line

$$S_a + S_m = S_y \tag{7.20}$$

By substituting $n\sigma_a = S_a$ and $n\sigma_m = S_m$, where n is a factor of safety, in Eq. (7.16), gives

$$\frac{\sigma_a}{S_e} + \frac{\sigma_m}{S_y} = \frac{1}{n} \tag{7.21}$$

Similarly, the modified Goodman line is given by

$$\frac{\sigma_a}{S_e} + \frac{\sigma_m}{S_{ut}} = \frac{1}{n} \tag{7.22}$$

And for the Gerber line,

$$\frac{n\sigma_a}{S_e} + \left(\frac{n\sigma_m}{S_{ut}}\right)^2 = 1 \tag{7.23}$$

The ASME elliptic is given by

$$\left(\frac{n\sigma_a}{S_e}\right)^2 + \left(\frac{n\sigma_m}{S_{ut}}\right)^2 = 1 \tag{7.24}$$

The yield line is given by

$$\sigma_a + \sigma_m = \frac{S_y}{n} \tag{7.25}$$

If we ignore axial loads, which tend to be small at critical locations where bending and torsion are significant, then the fluctuating stresses due to bending and torsion can be represented by

$$\sigma_a = K_f \frac{M_a c}{I} \tag{7.26}$$

$$\sigma_m = K_f \frac{M_m c}{I} \tag{7.27}$$

$$\tau_a = K_{fs} \frac{T_a c}{J} \tag{7.28}$$

$$\tau_m = K_{fs} \frac{T_m c}{J} \tag{7.29}$$

where

K_f = fatigue stress concentration factor for bending,
M_a = alternating bending moment (N m),
K_{fs} = fatigue stress concentration factor for torsion,
M_m = mid-range bending moment (N m),
c = distance from the neutral axis (m),
I = second moment of area (m^4),
τ_a = alternating stress (N/m^2),
T_a = alternating torque (N m),
J = polar second moment of area (m^4),

τ_m = mid-range stress (N/m^2) and.
T_m = mid-range torque (N m).

K_f and K_{fs} are fatigue stress concentration factors that account for the lessened sensitivity of some materials to the presence of notches.

For a rotating solid shaft, rotating about its central axis, $I=\pi d^4/64$, $J=\pi d^4/32$, $c=d/2$ giving

$$\sigma_a = K_f \frac{32M_a}{\pi d^3} \tag{7.30}$$

$$\sigma_m = K_f \frac{32M_m}{\pi d^3} \tag{7.31}$$

$$\tau_a = K_{fs} \frac{16T_a}{\pi d^3} \tag{7.32}$$

$$\tau_m = K_{fs} \frac{16T_m}{\pi d^3} \tag{7.33}$$

From consideration of distortion energy (DE), which predicts that a structure is safe provided the maximum value of the DE per unit volume is less than the DE per unit volume that causes yielding in a tensile test, von Mises stresses, σ', for a shaft with bending and torsional shear stresses and axial stress, take the form

$$\sigma' = \left(\sigma_x^2 + 3\tau_{xy}^2\right)^{0.5} \tag{7.34}$$

Neglecting axial loads,

$$\sigma_a' = \left(\sigma_a^2 + 3\tau_a^2\right)^{0.5} = \left[\left(\frac{32K_f M_a}{\pi d^3}\right)^2 + 3\left(\frac{16K_{fs}T_a}{\pi d^3}\right)^2\right]^{0.5} \tag{7.35}$$

$$\sigma_m' = \left(\sigma_m^2 + 3\tau_m^2\right)^{0.5} = \left[\left(\frac{32K_f M_m}{\pi d^3}\right)^2 + 3\left(\frac{16K_{fs}T_m}{\pi d^3}\right)^2\right]^{0.5} \tag{7.36}$$

where

σ_a' = alternating von Mises stress (N/m^2),
σ_m' = mean von Mises stress (N/m^2).

These stresses can be determined using one of the fatigue failure criteria on the modified Goodman diagram.

For example, for the modified Goodman line,

$$\frac{1}{n} = \frac{\sigma_a'}{S_e} + \frac{\sigma_m'}{S_{ut}} \tag{7.37}$$

Substitutions for σ_a' and σ_m' gives

$$\frac{1}{n}=\frac{16}{\pi d^3}\left\{\frac{1}{S_e}\left[4\left(K_f M_a\right)^2+3\left(K_{fs}T_a\right)^2\right]^{0.5}+\frac{1}{S_{ut}}\left[4\left(K_f M_m\right)^2+3\left(K_{fs}T_m\right)^2\right]^{0.5}\right\} \tag{7.38}$$

Rearranging to give the equation for a diameter,

$$d=\left(\frac{16n}{\pi}\left\{\frac{1}{S_e}\left[4\left(K_f M_a\right)^2+3\left(K_{fs}T_a\right)^2\right]^{0.5}+\frac{1}{S_{ut}}\left[4\left(K_f M_m\right)^2+3\left(K_{fs}T_m\right)^2\right]^{0.5}\right\}\right)^{1/3} \tag{7.39}$$

Similar equations for the factor of safety and diameter can be developed for each of the common failure criteria as follows:

For the modified DE Goodman criteria,

$$\frac{1}{n}=\frac{16}{\pi d^3}\left\{\frac{1}{S_e}\left[4\left(K_f M_a\right)^2+3\left(K_{fs}T_a\right)^2\right]^{0.5}+\frac{1}{S_{ut}}\left[4\left(K_f M_m\right)^2+3\left(K_{fs}T_m\right)^2\right]^{0.5}\right\} \tag{7.40}$$

$$d=\left(\frac{16n}{\pi}\left\{\frac{1}{S_e}\left[4\left(K_f M_a\right)^2+3\left(K_{fs}T_a\right)^2\right]^{0.5}+\frac{1}{S_{ut}}\left[4\left(K_f M_m\right)^2+3\left(K_{fs}T_m\right)^2\right]^{0.5}\right\}\right)^{1/3} \tag{7.41}$$

For DE Gerber,

$$\frac{1}{n}=\frac{8\sqrt{4\left(K_f M_a\right)^2+3\left(K_{fs}T_a\right)^2}}{\pi d^3 S_e}\left\{1+\left[1+\left(\frac{2S_e\sqrt{4\left(K_f M_m\right)^2+3\left(K_{fs}T_m\right)^2}}{S_{ut}\sqrt{4\left(K_f M_a\right)^2+3\left(K_{fs}T_a\right)^2}}\right)^2\right]^{0.5}\right\} \tag{7.42}$$

$$d=\left(\frac{8n\sqrt{4\left(K_f M_a\right)^2+3\left(K_{fs}T_a\right)^2}}{\pi S_e}\left\{1+\left[1+\left(\frac{2S_e\sqrt{4\left(K_f M_m\right)^2+3\left(K_{fs}T_m\right)^2}}{S_{ut}\sqrt{4\left(K_f M_a\right)^2+3\left(K_{fs}T_a\right)^2}}\right)^2\right]^{0.5}\right\}\right)^{1/3} \tag{7.43}$$

For the DE ASME Elliptic,

$$\frac{1}{n}=\frac{16}{\pi d^3}\left[4\left(\frac{K_f M_a}{S_e}\right)^2+3\left(\frac{K_{fs}T_a}{S_e}\right)^2+4\left(\frac{K_f M_m}{S_y}\right)^2+3\left(\frac{K_{fs}T_m}{S_y}\right)^2\right]^{0.5} \tag{7.44}$$

$$d = \left\{ \frac{16n}{\pi} \left[4\left(\frac{K_f M_a}{S_e}\right)^2 + 3\left(\frac{K_{fs} T_a}{S_e}\right)^2 + 4\left(\frac{K_f M_m}{S_y}\right)^2 + 3\left(\frac{K_{fs} T_m}{S_y}\right)^2 \right]^{0.5} \right\}^{1/3} \tag{7.45}$$

For the DE Soderberg,

$$\frac{1}{n} = \frac{16}{\pi d^3} \left\{ \frac{1}{S_e} \left[4(K_f M_a)^2 + 3(K_{fs} T_a)^2 \right]^{0.5} + \frac{1}{S_{yt}} \left[4(K_f M_m)^2 + 3(K_{fs} T_m)^2 \right]^{0.5} \right\} \tag{7.46}$$

$$d = \left(\frac{16n}{\pi} \left\{ \frac{1}{S_e} \left[4(K_f M_a)^2 + 3(K_{fs} T_a)^2 \right]^{0.5} + \frac{1}{S_{yt}} \left[4(K_f M_m)^2 + 3(K_{fs} T_m)^2 \right]^{0.5} \right\} \right)^{1/3} \tag{7.47}$$

For many ductile materials, the yield strength for tension and compression can be taken as equal, $S_{yt} = S_{yc} = S_y$.

For a rotating shaft with constant bending and torsion, $M_m = 0$, $T_a = 0$, allowing simplification of Eqs (7.40)–(7.47).

For the DE Goodman criteria, with constant bending and torsion:

$$\frac{1}{n} = \frac{16}{\pi d^3} \left\{ \frac{1}{S_e} \left[4(K_f M_a)^2 \right]^{0.5} + \frac{1}{S_{ut}} \left[3(K_{fs} T_m)^2 \right]^{0.5} \right\} \tag{7.48}$$

$$d = \left(\frac{16n}{\pi} \left\{ \frac{1}{S_e} \left[4(K_f M_a)^2 \right]^{0.5} + \frac{1}{S_{ut}} \left[3(K_{fs} T_m)^2 \right]^{0.5} \right\} \right)^{1/3} \tag{7.49}$$

For DE Gerber, with constant bending and torsion:

$$\frac{1}{n} = \frac{8\sqrt{(K_f M_a)^2}}{\pi d^3 S_e} \left\{ 1 + \left[1 + \left(\frac{2S_e \sqrt{3(K_{fs} T_m)^2}}{S_{ut} \sqrt{(K_f M_a)^2}} \right)^2 \right]^{0.5} \right\} \tag{7.50}$$

$$d = \left(\frac{8n\sqrt{4(K_f M_a)^2}}{\pi S_e} \left\{ 1 + \left[1 + \left(\frac{2S_e \sqrt{3(K_{fs} T_m)^2}}{S_{ut} \sqrt{4(K_f M_a)^2}} \right)^2 \right]^{0.5} \right\} \right)^{1/3} \tag{7.51}$$

For the DE ASME Elliptic, with constant bending and torsion:

$$\frac{1}{n} = \frac{16}{\pi d^3} \left[4\left(\frac{K_f M_a}{S_e}\right)^2 + 3\left(\frac{K_{fs} T_m}{S_y}\right)^2 \right]^{0.5} \tag{7.52}$$

$$d=\left\{\frac{16n}{\pi}\left[4\left(\frac{K_fM_a}{S_e}\right)^2+3\left(\frac{K_{fs}T_m}{S_y}\right)^2\right]^{0.5}\right\}^{1/3} \tag{7.53}$$

For the DE Soderberg, with constant bending and torsion:

$$\frac{1}{n}=\frac{16}{\pi d^3}\left\{\frac{1}{S_e}\left[4\left(K_fM_a\right)^2\right]^{0.5}+\frac{1}{S_{yt}}\left[3\left(K_{fs}T_m\right)^2\right]^{0.5}\right\} \tag{7.54}$$

$$d=\left(\frac{16n}{\pi}\left\{\frac{1}{S_e}\left[4\left(K_fM_a\right)^2\right]^{0.5}+\frac{1}{S_{yt}}\left[3\left(K_{fs}T_m\right)^2\right]^{0.5}\right\}\right)^{1/3} \tag{7.55}$$

The modified Goodman and Gerber criteria do not address failure due to yield, and a von Mises stress can be used to check for this:

$$\begin{aligned}\sigma'_{\max}&=\sqrt{(\sigma_m+\sigma_a)^2+3(\tau_m+\tau_a)^2}\\&=\left[\left(\frac{32K_f(M_m+M_a)}{\pi d^3}\right)^2+3\left(\frac{16K_{fs}(T_m+T_a)}{\pi d^3}\right)^2\right]^{0.5}\end{aligned} \tag{7.56}$$

$$n_y=\frac{S_y}{\sigma'_{\max}} \tag{7.57}$$

If the diameters and layout for a shaft along with the loading are known, then a value for the endurance limit can be determined along with the various modifying and stress concentration factors (S_e, S_{ut}, K_f, K_{fs}, k_a, k_b, k_c, k_d, k_e, k_g). The factor of safety estimated by any, or all, of the DE Goodman, DE Gerber, DE ASME Elliptic, DE Soderberg criteria and yielding factor of safety n_y can then be determined. Similarly, if the nominal diameter is known for a shaft, the equations for diameter can be used to provide an estimate for the small diameter on shoulders. Alternatively, if a factor of safety is known, or can be proposed, then a diameter for a shaft or critical location along a shaft can be determined.

Example 7.8. A shaft of diameter 40 mm experiences an alternating moment of 165 N m and a mean torque of 132 N m. The endurance strength is 85 MPa, the ultimate tensile strength is 560 MPa and the yield strength is 440 MPa. The fatigue stress concentration factors for bending and torsion are 1.65 and 1.52, respectively. Determine the factor of safety according to the DE modified Goodman, DE Gerber, DE ASME Elliptic and DE Soderberg criteria, respectively.

Solution

For the modified DE Goodman criteria,

$$\frac{1}{n}=\frac{16}{\pi d^3}\left\{\frac{1}{S_e}\left[4\left(K_fM_a\right)^2+3\left(K_{fs}T_a\right)^2\right]^{0.5}+\frac{1}{S_{ut}}\left[4\left(K_fM_m\right)^2+3\left(K_{fs}T_m\right)^2\right]^{0.5}\right\}$$

$$\left[4\left(K_f M_a\right)^2 + 3(K_{fs} T_a)^2\right]^{0.5} = 544.5$$

$$\left[4\left(K_f M_m\right)^2 + 3(K_{fs} T_m)^2\right]^{0.5} = 347.5$$

$$\frac{1}{n} = \frac{16}{\pi\left(0.04^3\right)}\left\{\frac{1}{85 \times 10^6} \times 544.5 + \frac{1}{560 \times 10^6} \times 347.5\right\} = 0.5591$$

$$n = 1.788.$$

For DE Gerber,

$$\frac{1}{n} = \frac{8\sqrt{4\left(K_f M_a\right)^2 + 3(K_{fs} T_a)^2}}{\pi d^3 S_e}\left\{1 + \left[1 + \left(\frac{2S_e\sqrt{4\left(K_f M_m\right)^2 + 3(K_{fs} T_m)^2}}{S_{ut}\sqrt{4\left(K_f M_a\right)^2 + 3(K_{fs} T_a)^2}}\right)^2\right]^{0.5}\right\}$$

$$\left[4\left(K_f M_a\right)^2 + 3(K_{fs} T_a)^2\right]^{0.5} = 544.5$$

$$\left[4\left(K_f M_m\right)^2 + 3(K_{fs} T_m)^2\right]^{0.5} = 347.5$$

$$\frac{1}{n} = \frac{8 \times 544.5}{\pi\left(0.04^3\right) \times 85 \times 10^6}\left\{1 + \left[1 + \left(\frac{2 \times 85 \times 10^6 \times 347.5}{560 \times 10^6 \times 544.5}\right)^2\right]^{0.5}\right\} = 0.5145$$

$$n = 1.944.$$

For the DE ASME Elliptic,

$$\frac{1}{n} = \frac{16}{\pi d^3}\left[4\left(\frac{K_f M_a}{S_e}\right)^2 + 3\left(\frac{K_{fs} T_a}{S_e}\right)^2 + 4\left(\frac{K_f M_m}{S_y}\right)^2 + 3\left(\frac{K_{fs} T_m}{S_y}\right)^2\right]^{0.5}$$

$$4\left(\frac{K_f M_a}{S_e}\right)^2 = 4\left(\frac{1.65 \times 165}{85 \times 10^6}\right)^2 = 4.104 \times 10^{-11}$$

$$3\left(\frac{K_{fs} T_m}{S_y}\right)^2 = 3\left(\frac{1.52 \times 132}{440 \times 10^6}\right)^2 = 6.238 \times 10^{-13}$$

$$\frac{1}{n} = \frac{16}{\pi\left(0.04^3\right)}\left[4.104 \times 10^{-11} + 6.238 \times 10^{-13}\right]^{0.5} = 0.5136$$

$$n = 1.947.$$

For the DE Soderberg,

$$\frac{1}{n}=\frac{16}{\pi d^3}\left\{\frac{1}{S_e}\left[4\left(K_f M_a\right)^2+3\left(K_{fs} T_a\right)^2\right]^{0.5}+\frac{1}{S_{yt}}\left[4\left(K_f M_m\right)^2+3\left(K_{fs} T_m\right)^2\right]^{0.5}\right\}$$

$$\left[4\left(K_f M_a\right)^2+3\left(K_{fs} T_a\right)^2\right]^{0.5}=544.5$$

$$\left[4\left(K_f M_m\right)^2+3\left(K_{fs} T_m\right)^2\right]^{0.5}=347.5$$

$$\frac{1}{n}=\frac{16}{\pi\left(0.04^3\right)}\left\{\frac{544.5}{85\times 10^6}+\frac{347.5}{440\times 10^6}\right\}=0.5726$$

$$n=1.746.$$

Example 7.9. A factor of safety of 2 is required for a shaft experiencing an alternating moment of 205 N m and a mean torque of 145 N m. The endurance strength is 192 MPa, the ultimate tensile strength is 620 MPa and the yield strength is 480 MPa. The fatigue stress concentration factors for bending and torsion are 1.55 and 1.38, respectively. Determine the nominal shaft diameter according to the DE modified Goodman, DE Gerber, DE ASME Elliptic and DE Soderberg criteria, respectively.

Solution

DE Modified Goodman,

$$d=\left(\frac{16n}{\pi}\left\{\frac{1}{S_e}\left[4\left(K_f M_a\right)^2+3\left(K_{fs} T_a\right)^2\right]^{0.5}+\frac{1}{S_{ut}}\left[4\left(K_f M_m\right)^2+3\left(K_{fs} T_m\right)^2\right]^{0.5}\right\}\right)^{1/3}$$

$$\left[4\left(K_f M_a\right)^2+3\left(K_{fs} T_a\right)^2\right]^{0.5}=635.5$$

$$\left[4\left(K_f M_m\right)^2+3\left(K_{fs} T_m\right)^2\right]^{0.5}=346.6$$

$$d=\left(\frac{16\times 2}{\pi}\left\{\frac{1}{192\times 10^6}[635.5]^{0.5}+\frac{1}{620\times 10^6}[346.6]^{0.5}\right\}\right)^{1/3}=0.03403\,\text{m}$$

DE Gerber:

$$d=\left(\frac{8n\sqrt{4\left(K_f M_a\right)^2+3\left(K_{fs} T_a\right)^2}}{\pi S_e}\left\{1+\left[1+\left(\frac{2S_e\sqrt{4\left(K_f M_m\right)^2+3\left(K_{fs} T_m\right)^2}}{S_{ut}\sqrt{4\left(K_f M_a\right)^2+3\left(K_{fs} T_a\right)^2}}\right)^2\right]^{0.5}\right\}\right)^{1/3}$$

$$\left[4\left(K_f M_a\right)^2 + 3(K_{fs} T_a)^2\right]^{0.5} = 635.5$$

$$\left[4\left(K_f M_m\right)^2 + 3(K_{fs} T_m)^2\right]^{0.5} = 346.6$$

$$d = \left(\frac{8 \times 2 \times 635.5}{\pi \times 192 \times 10^6}\left\{1 + \left[1 + \left(\frac{2 \times 192 \times 10^6 \times 346.6}{620 \times 10^6 \times 635.5}\right)^2\right]^{0.5}\right\}\right)^{1/3} = 0.03260\,\text{m}$$

DE ASME Elliptic:

$$d = \left\{\frac{16n}{\pi}\left[4\left(\frac{K_f M_a}{S_e}\right)^2 + 3\left(\frac{K_{fs} T_a}{S_e}\right)^2 + 4\left(\frac{K_f M_m}{S_y}\right)^2 + 3\left(\frac{K_{fs} T_m}{S_y}\right)^2\right]^{0.5}\right\}^{1/3}$$

$$4\left(\frac{K_f M_a}{S_e}\right)^2 = 4\left(\frac{1.55 \times 205}{192 \times 10^6}\right)^2 = 1.096 \times 10^{-11}$$

$$3\left(\frac{K_{fs} T_m}{S_y}\right)^2 = 3\left(\frac{1.38 \times 145}{480 \times 10^6}\right)^2 = 5.214 \times 10^{-13}$$

$$d = \left\{\frac{16 \times 2}{\pi}\left[1.096 \times 10^{-11} + 5.214 \times 10^{-13}\right]^{0.5}\right\}^{1/3} = 0.03256\,\text{m}$$

DE Soderberg:

$$d = \left(\frac{16n}{\pi}\left\{\frac{1}{S_e}\left[4(K_f M_a)^2 + 3(K_{fs} T_a)^2\right]^{0.5} + \frac{1}{S_{yt}}\left[4(K_f M_m)^2 + 3(K_{fs} T_m)^2\right]^{0.5}\right\}\right)^{1/3}$$

$$\left[4(K_f M_a)^2 + 3(K_{fs} T_a)^2\right]^{0.5} = 635.5$$

$$\left[4(K_f M_m)^2 + 3(K_{fs} T_m)^2\right]^{0.5} = 346.6$$

$$d = \left(\frac{16 \times 2}{\pi}\left\{\frac{1}{192 \times 10^6} \times 635.5 + \frac{1}{480 \times 10^6} \times 346.6\right\}\right)^{1/3} = 0.03450\,\text{m}$$

The ASME design code for the design of transmission shafting provided an approach but was suspended in 1995 by the ASME as improved understanding and more sophisticated methodologies have become available allowing more precise modelling. The ASME procedure aimed to ensure that the shaft is properly sized to provide adequate service life but the designer must still check that the shaft is stiff enough to limit deflections of power transfer elements such as gears and to minimise misalignment through seals and bearings. In addition, the shaft's stiffness must be such that it avoids unwanted vibrations through the running range. The ASME equation for determining the diameter for a solid shaft is given by

$$d = \left[\frac{32n_s}{\pi}\sqrt{\left(\frac{M}{\sigma_e}\right)^2 + \frac{3}{4}\left(\frac{T}{\sigma_y}\right)^2}\right]^{1/3} \tag{7.58}$$

where

d = diameter (m),
n_s = factor of safety,
M = bending moment (N m),
σ_e = endurance limit of the item (N/m^2),
T = torque (N m),
σ_y = yield strength (N/m^2).

As noted, the ASME design code has now been suspended, but this equation is included here due to its relative simplicity, and usefulness on occasion to provide a starting estimate for a shaft diameter. It should be noted that Eq. (7.58) tends to underestimate the diameter required.

There is usually some uncertainty regarding what level a component will actually be loaded to, how strong a material is and how accurate the modelling methods are. Factors of safety are frequently used to account for these uncertainties. The value of a factor of safety is usually based on experience of what has given acceptable performance in the past. The level is also a function of the consequences of component failure and the cost of providing an increased safety factor. As a guide typical values for the factor of safety based on strength recommended by Vidosek (1957) are as follow:

- 1.25–1.5 for reliable materials under controlled conditions subjected to loads and stresses known with certainty;
- 1.5–2 for well-known materials under reasonably constant environmental conditions subjected to known loads and stresses;
- 2–2.5 for average materials subjected to known loads and stresses;
- 2.5–3 for less well-known materials under average conditions of load, stress and environment;
- 3–4 for untried materials under average conditions of load, stress and environment; and
- 3 to 4 for well-known materials under uncertain conditions of load, stress and environment.

The endurance limit, σ_e, for a mechanical element can be estimated by Eq. (7.59) (after Marin, 1962). Here, a series of modifying factors are applied to the endurance limit of a test specimen for various effects such as size, load and temperature.

$$\sigma_e = k_a k_b k_c k_d k_e k_g \sigma'_e \tag{7.59}$$

where

k_a = surface factor,
k_b = size factor,
k_c = reliability factor,
k_d = temperature factor,
k_e = duty cycle factor,
k_g = miscellaneous effects factor,
σ_e = endurance limit of test specimen (N/m^2).

If the stress at the location under consideration is greater than σ_e, then the component will fail eventually due to fatigue (i.e. the component has a limited life).

Mischke (1987) has determined the following approximate relationships between the endurance limit of test specimens and the ultimate tensile strength of the material (for steels only):

$$\sigma_e' = 0.504\sigma_{uts} \text{ for } \sigma_{uts} \leq 1400\,\text{MPa} \tag{7.60}$$

$$\sigma_e' = 700\,\text{MPa for } \sigma_{uts} \geq 1400\,\text{MPa} \tag{7.61}$$

The surface finish factor is given by Eq. (7.62).

$$k_a = a\sigma_{uts}^b \tag{7.62}$$

Values for a and b can be found in Table 7.3.

The size factor k_b can be calculated from Eq. (7.63) or (7.64) (Kuguel (1969)).

$$\text{For } d < 50\,\text{mm}, k_b = \left(\frac{d}{7.62}\right)^{-0.1133} \tag{7.63}$$

$$\text{For } d > 50\,\text{mm}, k_b = 1.85\text{d}^{-0.19} \tag{7.64}$$

The reliability factor k_c is given in Table 7.4 as a function of the nominal reliability desired.

For temperatures between −57°C and 204°C, the temperature factor, k_d, can be taken as 1. The ASME standard documents values to use that are outside of this range.

Table 7.3 Surface finish factors Noll and Lipson (1946)

Surface finish	***a* (MPa)**	***b***
Ground	1.58	−0.085
M/c or cold-drawn	4.51	−0.265
Hot-rolled	57.7	−0.718
Forged	272.0	−0.995

Table 7.4 Reliability factors for use in the ASME transmission shaft equation

Shaft nominal reliability	k_c
0.5	1.0
0.9	0.897
0.95	0.868
0.99	0.814
0.999	0.753

ANSI/ASME B106.1M-1985.

The duty cycle factor k_e is used to account for cycle loading experienced by the shaft such as stops and starts, transient overloads, shock loading etc. and requires prototype fatigue testing for its quantification. k_e is taken as 1 in the examples presented here.

K_f or K_{fs} are the component fatigue stress concentration factors for bending and torsion, respectively, used to account for stress concentrations regions such as notches, holes, keyways and shoulders, which are given by

$$K_f = 1 + q(K_t - 1) \tag{7.65}$$

or

$$K_{fs} = 1 + q_{shear}(K_{ts} - 1) \tag{7.66}$$

where

q = notch sensitivity,
q_{shear} = notch sensitivity for reversed torsion,
K_t = geometric stress concentration factor for bending,
K_{ts} = geometric stress concentration factor for shear.

Some values for the notch sensitivity and typical geometric stress concentration factors are given in Figs 7.34–7.39 and Table 7.5; see also Zappalorto et al. (2011), Savruk and Kazberuk (2006), Noda et al. (1995), Peterson (1959, 1974). For the case of a keyway, setting K_t and K_{ts} equal to 2.14 and 3, respectively, provides a reasonable starting point (Budynas and Nisbett, 2011), prior to more detailed analysis or testing data. If there is uncertainty on the value to use for notch sensitivity, a conservative assumption is to set $K_f = K_t$, and $K_{fs} = K_{ts}$.

The miscellaneous factor k_g is used to account for residual stresses, heat treatment, corrosion, surface coatings, vibration, environment and unusual loadings. k_g is taken as 1 here.

In general, the following principles should be used when designing to avoid fatigue failures:

- Calculations should allow an appropriate safety factor, particularly where stress concentrations occur. (e.g. keyways, notches and change of section).
- Provide generous radii at changes of section and introduce stress relief grooves, etc.

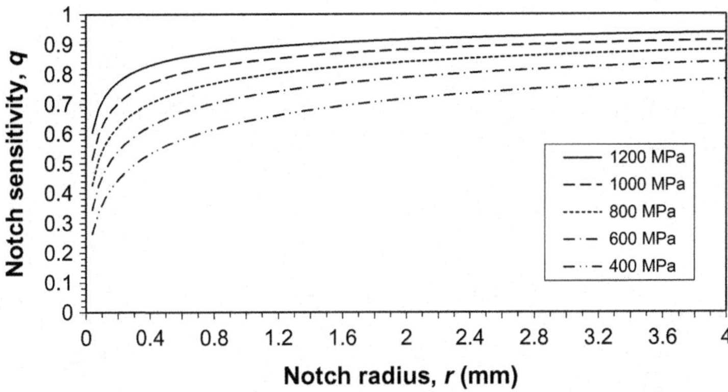

Fig. 7.34 Notch sensitivity index versus notch radius for a range of steels subjected to reversed bending or reversed axial loads.
Data from Kuhn, P., Hardrath, H.F., 1952. An engineering method for estimating notch size effect on fatigue tests of steel. National Advisory Committee for Aeronautics, NACA TN2805.

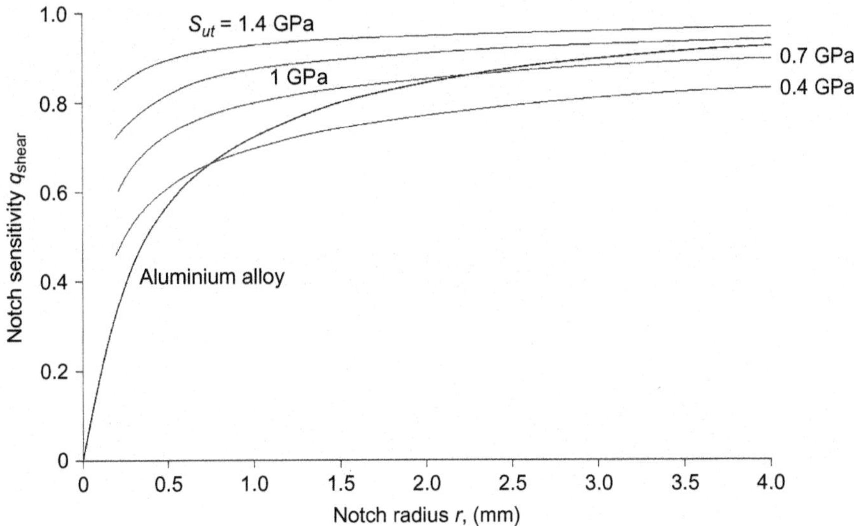

Fig. 7.35 Notch sensitivity index versus notch radius for a range of steels subjected to reversed shear.
Data from Budynas, R.G., Nisbett, J.K., 2011. Shigley's Mechanical Engineering Design, 9th ed. McGraw Hill, New York.

- Choose materials if possible that have limiting fatigue stresses, e.g. most steels.
- Provide for suitable forms of surface treatment, e.g. shot-peening, work-hardening and nitriding. Avoid treatments which introduce residual tensile stresses such as electro-plating.
- Specify fine surface finishes.
- Avoid corrosive conditions.
- Stress relieving should be used where possible, particularly for welded structures.

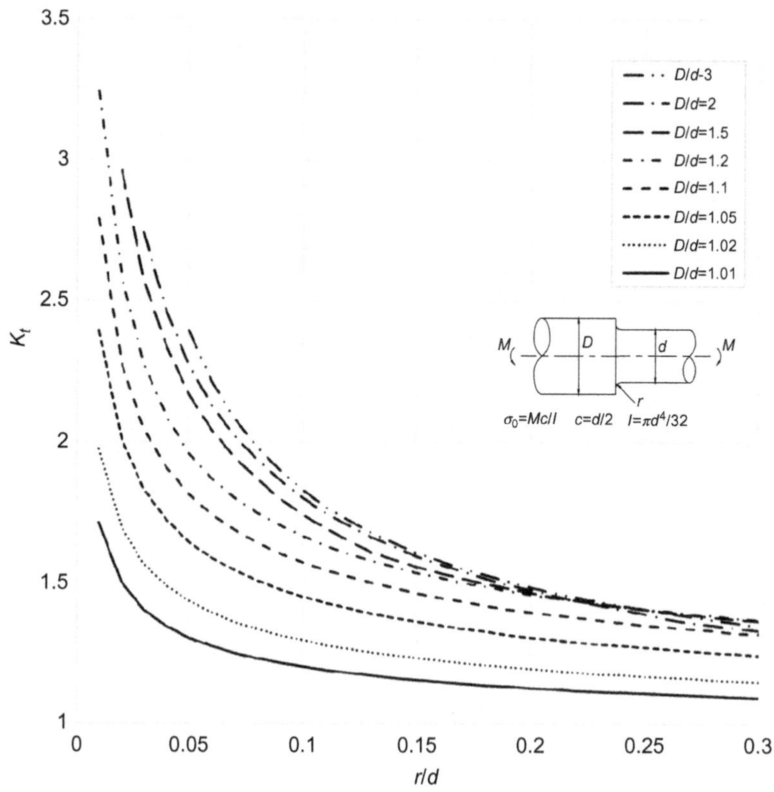

Fig. 7.36 Stress concentration factors for a shaft with a fillet subjected to bending. Data from Peterson, R.E., 1974. Stress Concentration Factors. Wiley.

Any appropriate material can be selected for a shaft. The requirements to resist high loads and high-cycle fatigue often dictate consideration of steel. The coding system used for steels in the UK is given in British Standard 970. The standard uses a six digit designation; for example, 070M20.

The first three digits of the designation denote the family of steels to which the alloy belongs.

000–199 Carbon and carbon-manganese steels. The first three figures indicate one hundred times the mean manganese content.
200–240 Free-cutting steels where the second and third digits represent one hundred times the minimum or mean sulphur content.
250 Silicon-manganese spring steels.
300–399 Stainless, heat resisting and valve steels.
500–999 Alloy steels in groups of ten or multiples of groups of ten according to alloy type.

The fourth digit of the designation is a letter: either A, H, M or S.

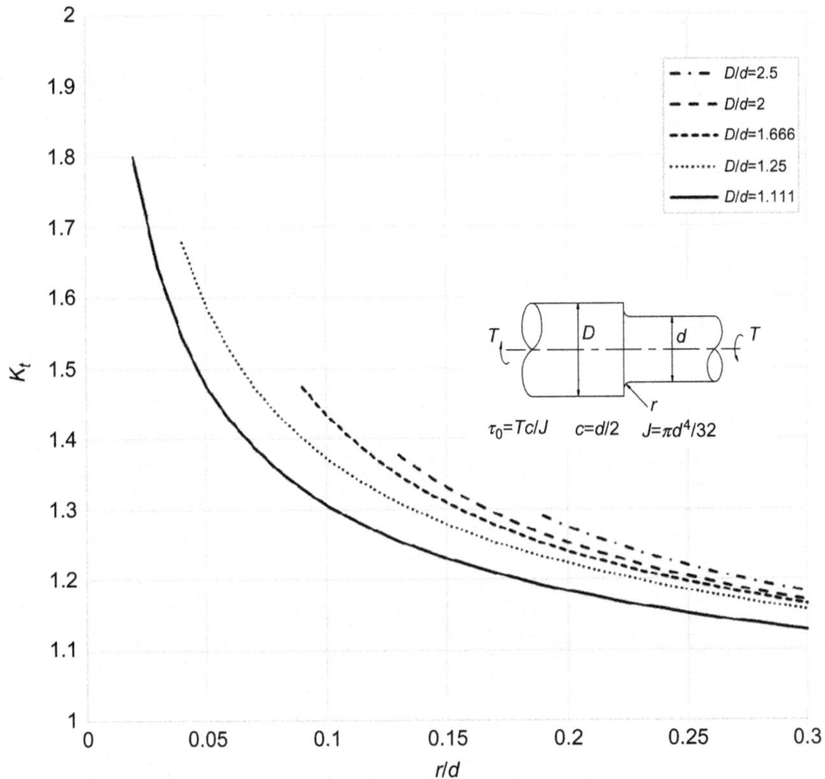

Fig. 7.37 Stress concentration factors for a shaft with a fillet subjected to torsion. Data from Peterson, R.E., 1974. Stress Concentration Factors. Wiley.

'A' denotes that the steel will be supplied to close limits of chemical composition,
'H' denotes a hardenability requirement for the material,
'M' denotes specified mechanical properties,
'S' denotes a stainless steel.

The fifth and sixth digits of the designation represent one hundred times the mean carbon content of the steel. For example, 080A40 denotes a carbon-manganese steel, supplied to close limits of chemical composition, containing 0.7%–0.9% manganese and 0.4% carbon.

Tables 7.6 and 7.7 list some steel types and typical material properties values.

The choice of steel for a particular application can sometimes be a bewildering experience. Within the current British Standard alone for steels (BS970), there are several hundred steel specifications. In practice, a relatively few steels are used for the majority of applications and some of the popular specifications are listed in the following.

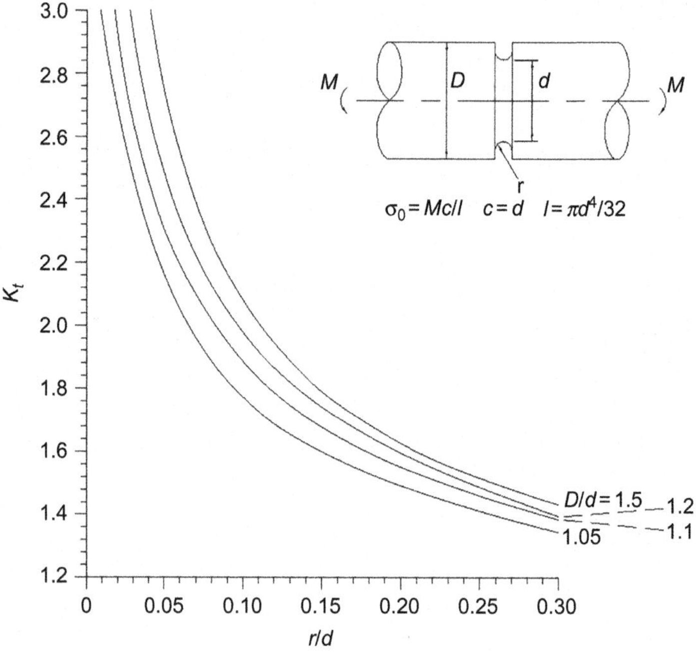

Fig. 7.38 Stress concentration factors for a shaft with a groove subjected to bending. Data from Peterson, R.E., 1974. Stress Concentration Factors. Wiley.

Fig. 7.39 Stress concentration factors for a shaft with a transverse hole subjected to bending. Data from Peterson, R.E., 1974. Stress Concentration Factors. Wiley.

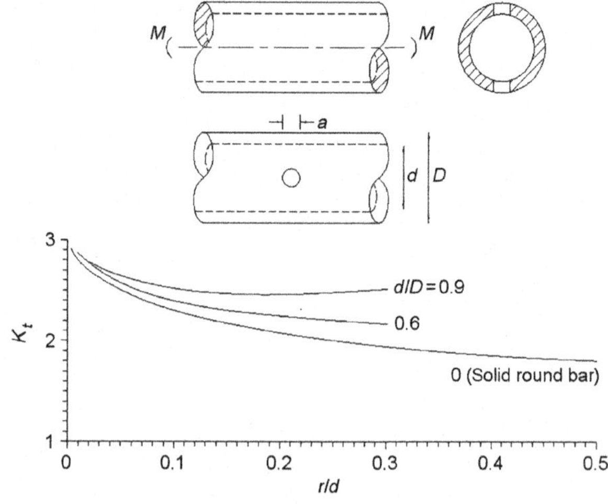

Table 7.5 Fatigue stress concentration factors, k_f for keyways in steel shafts. Juvinall (1967), Juvinall and Marshek (2012)

Steel	k_f			
	Bending	Torsion	Bending	Torsion
Annealed <200 BHN	1.3	1.3	1.6	1.3
Quenched and drawn >200 BHN	1.6	1.6	2.0	1.6

Steels can be divided into seven principal groupings:

(1) Low-carbon free-cutting steels. These are the most popular type of steel for the production of turned components where machinability and surface finish are important. Applications include automotive and general engineering. The principal specification is 230M07.

(2) Low-carbon steels or mild steels. These are used for lightly stressed components, welding, bending, forming and general engineering applications. Some of the popular specifications are 040A10, 045M10, 070M20, 080A15 and 080M15.

(3) Carbon and carbon manganese case hardening steels. These steels are suitable for components that require a wear resisting surface and tough core. Specifications include 045A10, 045M10, 080M15, 210M15 and 214M15.

(4) Medium carbon and carbon-manganese steels. These offer greater strength than mild steels and respond to heat treatment. Tensile strengths in the range of 700–1000 MPa can be attained. Applications include gears, racks, pinions, shafts, rollers, bolts and nuts. Specifications include 080M30, 080M40, 080A42, 080M50, 070M55 and 150M36.

(5) Alloy case-hardening steels. These are used when a hard wear-resisting surface is required but because of the alloying elements superior mechanical properties can be attained in comparison with carbon and carbon manganese case hardening steels. Typical applications include gears, cams, rolled and transmission components. Types include 635M15, 655M13, 665M17, 805M20 and 832M13.

(6) Alloy direct hardening steels. These steels include alloying elements such as Ni, Cr, Mo and V and are used for applications where high strength and shock resistance are important. Types include 605M36, 708M40, 709M40, 817M40 and 826M40.

(7) Stainless steels. There are three types of stainless steels: martensitic, ferritic and austenitic.

Table 7.6 Guide to the ultimate tensile and yield strengths for selected steels

BS970 designation	σ_{uts} (MN m^{-2})	σ_y (MN m^{-2})	Old EN code
230M07	475		1A
080M15	450[a]	330[a]	32C
070M20	560[a]	440[a]	3B
210M15	490		32M
214M15	720		202
080M30	620[a]	480[a]	6
080M40	660[a]	530[a]	8
635M15	770 min		351
655M13	1000 min		36
665M17	770 min		34
605M36	700–850[b]	540[b]	16
709M40	700–1225[b]	540–955[b]	19
817M40	850–1000[b]	700[b]	24
410S21	700–850[c]	525[c]	56A
431S29	850–1000	680	57
403S17	420 min	280	
430S17	430 min	280	
304S15	480		58E
316S11	490		
303S31	510		58M
303S42	510		58AM

[a]Hot-rolled and cold-drawn.
[b]Hardened and tempered and cold-drawn.
[c]Tempered.

Table 7.7 General physical properties

Steel	ρ (kgm^{-3})	E (GPa)	G (GPa)	ν	α (K^{-1})
080A15	7859	211.665	82.046	0.291	11.7×10^{-6}
080M40	7844	211.583	82.625	0.287	11.2×10^{-6}
708M40	7830	210.810	83.398	0.279	12.3×10^{-6}
303S—	7930	193.050	–	–	17.2×10^{-6}

Martensitic stainless steels can be hardened and tempered to give tensile strengths in the range 550–1000 MNm^{-2}. Applications include fasteners, valves, shafts, spindles, cutlery and surgical instruments. Specifications include 410S21, 420S29, 420S45, 431S29, 416S21, 416S41, 416S37 and 441S49.

Ferritic stainless steels are common in strip and sheet form and applications include domestic and automotive trim, catering equipment and exhaust systems. They have good ductility and are easily formed. Specifications include 403S17 and 430S17.

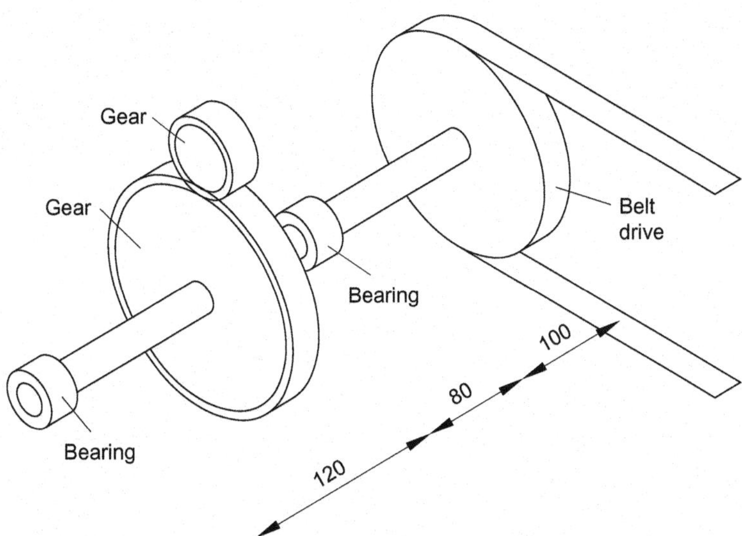

Fig. 7.40 Transmission drive shaft.

Austenitic stainless steels offer the highest resistance to corrosion and applications include the food, chemical, gas and oil industries as well as medical equipment and domestic appliances. Specifications include 302S31, 304S15, 316S11, 316S31, 320S31, 321S31, 303S31, 325S31, 303S42 and 326S36.

Example 7.10. Determine a sensible minimum nominal diameter for the drive shaft illustrated in Fig. 7.40 consisting of a mid-mounted spur gear and overhung pulley wheel. The shaft is to be manufactured using 817M40 hot-rolled alloy steel with $\sigma_{uts}=1000\,\text{MPa}$, $\sigma_y=770\,\text{MPa}$ and Brinell Hardness approximately 220 BHN. The radius of the fillets at the gear and pulley shoulders is 3 mm. The power to be transmitted is 8 kW at 900 rpm. The pitch circle diameter of the 20 degree pressure angle spur gear is 192 mm and the pulley diameter is 250 mm. The masses of the gear and pulley are 8 kg and 10 kg, respectively. The ratio of belt tensions should be taken as 2.5. Profiled keys are used to transmit torque through the gear and pulley. A shaft nominal reliability of 90% is desired. A nominal diameter of 32 mm can be assumed for the shaft.

Solution

In order to estimate the diameter for a transmission shaft, using the failure criteria listed in Eqs (7.48)–(7.55), the maximum combination of torque and bending moment need to be determined. A sensible approach is to determine the overall bending moment diagram for the shaft, as this information may also be of use in other design calculations such as calculating the shaft deflection.

The loading on the shaft can be resolved into both horizontal and vertical planes and must be considered in determining the resulting bending moments on the shaft. Fig. 7.41 shows the combined loadings on the shaft from the gear forces, the gear's

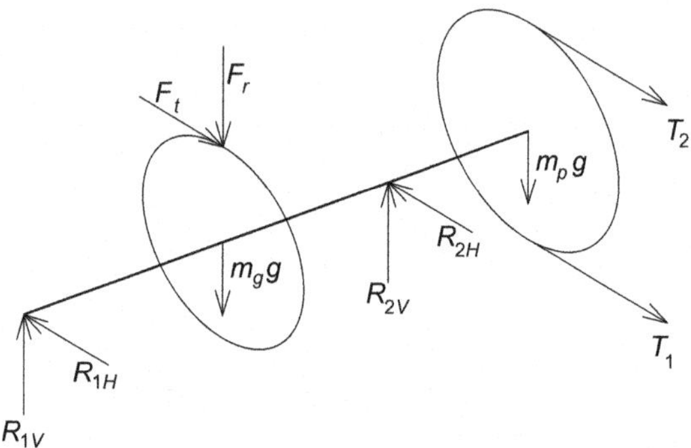

Fig. 7.41 Shaft loading diagram.

mass, the belt tensions and the pulley's mass. The mass of the shaft itself has been ignored. The tension on a pulley belt is tighter on the 'pulling' side than on the 'slack' side and the relationship of these tensions is normally given as a ratio, which in this case is 2.5:1.

The power transmitted through the shaft is 8000 W.

The torque is given by Torque $= \frac{\text{Power}}{\omega} = \frac{8000}{(2\pi/60) \times 900} = 84.9\,\text{N m}$.

Belts are introduced in more detail in Chapter 12. Here, only the tensions will be considered as necessary for the development of the solution.

The ratio of belt tensions is $T_1/T_2 = 2.5$, so $T_1 = 2.5\,T_2$.

The torque on the pulley in terms of the belt tensions is given by.

$$\text{Torque} = T_1 r_p - T_2 r_p = 2.5\,T_2 r_p - T_2 r_p = 1.5\,T_2 r_p.$$

The belt tensions are $T_2 = 84.9/(1.5 \times 0.125) = 452.8\,\text{N}$,

$$T_1 = 2.5\,T_2 = 1131.8\text{N}.$$

Forces on spur gears are considered in detail in Chapter 8. Here, the relevant relationships will be stated as necessary and used in the development of the solution. A spur gear experiences both radial and tangential loading. The tangential load is given by F_t = torque/pitch circle radius.

$$F_t = \frac{\text{Torque}}{r_g} = \frac{84.9}{0.192/2} = 884.4\,\text{N}.$$

The radial load is given by $F_r = F_t \tan\phi$, where ϕ is the pressure angle.

$$F_r = F_t \tan\phi = 884.4 \times \tan 20 = 321.9\text{N}.$$

Note that when determining the horizontal and vertical bending moment diagrams both transmitted and gravitational loads should be included.

The vertical loads on the shaft are illustrated in Fig. 7.42.

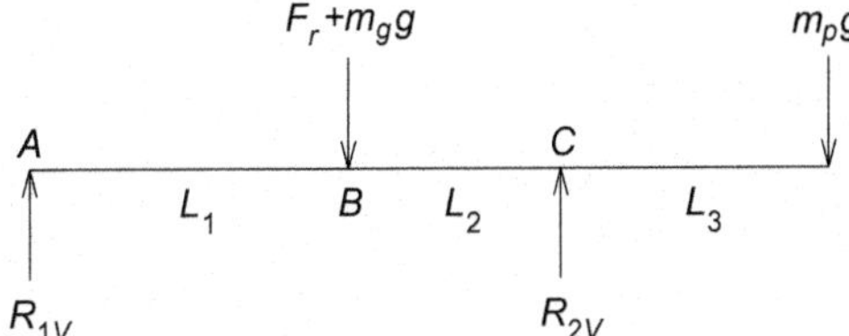

Fig. 7.42 Vertical loading diagram.

Moments about A:

$$\left(F_r + m_g g\right)L_1 - R_{2V}(L_1 + L_2) + m_p g(L_1 + L_2 + L_3) = 0.$$

$$R_{2V} = \frac{\left(F_r + m_g g\right)L_1 + m_p g(L_1 + L_2 + L_3)}{L_1 + L_2}$$
$$= \frac{(321.9 + (8 \times 9.81))0.12 + (10 \times 9.81)0.3}{0.12 + 0.08} = 387.4\,\text{N}.$$

Resolving vertical forces, $R_{1V}+R_{2V}=F_r+m_g g+m_p g$.

Hence, $R_{1V}=321.9+78.5+98.1-387.4=111.1\,\text{N}$.

The horizontal loads on the shaft are illustrated in Fig. 7.43.

$$T = \text{total tension} = T_1 + T_2 = 1131.8 + 452.8 = 1585\,\text{N}.$$

Moments about A: $F_t L_1 - R_{2H}(L_1+L_2)+T(L_1+L_2+L_3)=0$.

Hence,

$$R_{2H} = \frac{F_t L_1 + T(L_1 + L_2 + L_3)}{L_1 + L_2} = \frac{(884.4 \times 0.12) + (1585 \times 0.3)}{0.2} = 2908\,\text{N}.$$

Resolving horizontal forces: $R_{1H}+R_{2H}=F_t+T$.

Hence, $R_{1H}=884.4+1585-2908=-438.5\,\text{N}$.

The bending moment diagrams can now be determined, Figs 7.44 and 7.45. You may wish to recall that the bending moment is the algebraic sum of the moments of the external forces to one side of the section about an axis through the section.

The vertical bending moments at B and C are given by.

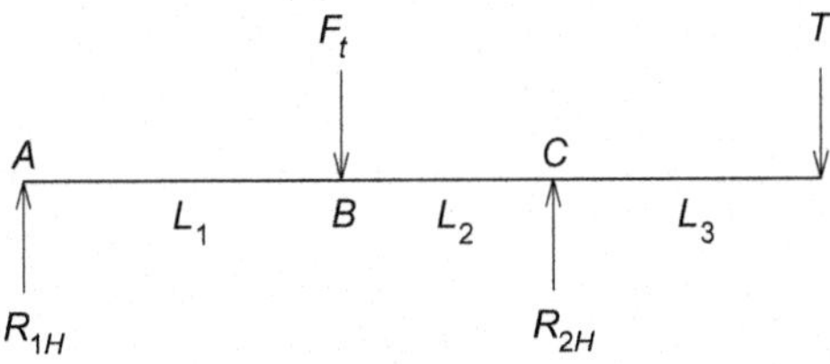

Fig. 7.43 Horizontal loading diagram.

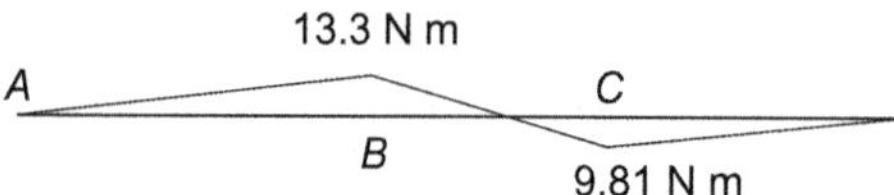

Fig. 7.44 Vertical bending moment diagram.

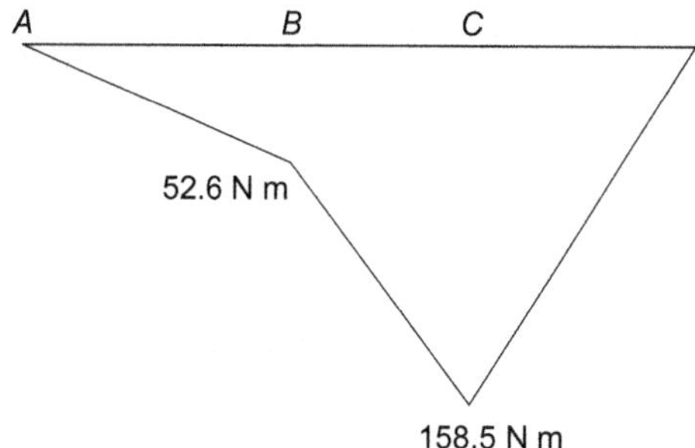

Fig. 7.45 Horizontal bending moment diagram.

$$M_{BV} = R_{1V}L_1 = 111.1 \times 0.12 = 13.3\text{Nm},$$

$$\begin{aligned} M_{CV} &= R_{1V}(L_1 + L_2) - \left(\left(F_r + m_g g\right)L_2\right) \\ &= 111.1(0.12 + 0.08) - (321.9 + 8 \times 9.81)0.08 = -9.810\text{Nm}. \end{aligned}$$

The horizontal bending moments at B and C are given by.

$$M_{BH} = R_{1H}L_1 = -438.5 \times 0.12 = -52.62\text{Nm},$$

$$M_{CH} = R_{1H}(L_1 + L_2) - (F_t L_2) = -438.5 \times 0.2 - 884.4 \times 0.08 = -158.5\text{Nm}.$$

The resultant bending moment diagram, Fig. 7.46, can be determined by calculating the resultant bending moments at each point.

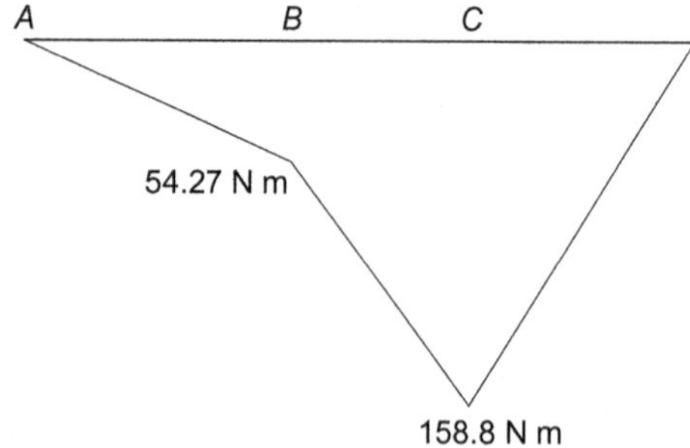

Fig. 7.46 Resultant bending moment diagram.

$$|M_B| = \sqrt{(13.3)^2 + (-52.62)^2} = 54.27\,\text{N m.},$$

$$|M_C| = \sqrt{(-9.81)^2 + (-158.5)^2} = 158.8\,\text{N m.}$$

From Fig. 7.46, it can be seen that the maximum bending moment occurs at the right-hand bearing. The value of the resultant bending moment here should be used in the estimation of a suitable diameter for the design.

The next task is to determine the endurance limit of the shaft and the modifying factors. The shaft material is hot-rolled steel and the endurance limit of the test specimen if unknown can be estimated using $\sigma_e' = 0.504\sigma_{uts}$. The ultimate tensile strength for 817M40 is 1000MPa and σ_y is 770MPa. So $\sigma_e' = 0.504 \times 1000 = 504$ MPa.

The material is hot-rolled so from Eq. (7.14) and Table 7.3, $k_a = a\sigma_{uts}^b = 57.7\,(1000)^{-0.718} = 0.4047$.

Assuming that the diameter will be about 38mm, the size factor can be estimated using the relevant equation for k_b, in this case for a diameter under 50mm, $k_b = (d/7.62)^{-0.1133} = (32/7.62)^{-0.1133} = 0.8499$. If the shaft is significantly different in size to 32mm, then the calculation should be repeated until convergence between the assumed and the final calculated value for the shaft diameter is achieved.

The desired nominal reliability is 90%, so $k_c = 0.897$.

The operating temperature is not stated so a value of temperature factor of $k_d = 1$ is assumed. The duty cycle factor is assumed to be $k_e = 1$.

The fillet radius at the shoulders is 3mm. The ratio of diameters $D/d = (3+3+32)/32 = 38/32 = 1.188$. $r/d = 3/32 = 0.09375$. So from Fig. 7.36, the geometric stress concentration factor $k_t = 1.7$ and from Fig. 7.37, $k_{ts} = 1.35$.

From Fig. 7.34, the notch sensitivity index for a 1000MPa strength material with notch radius of 3mm is approximately $q = 0.9$.

$$K_f = 1 + q(k_t - 1) = 1 + 0.9(1.7 - 1) = 1.63.$$

From Fig. 7.35, the notch sensitivity index for materials in reversed torsion, for a 1000MPa strength material with notch radius of 3mm is $q_{\text{shear}} = 0.92$.

$$K_{fs} = 1 + q_{\text{shear}}(k_{ts} - 1) = 1 + 0.92(1.35 - 1) = 1.322.$$

The miscellaneous factor is taken as $k_g = 1$.

The endurance limit can now be calculated from $\sigma_e = k_a k_b k_c k_d k_e k_g \sigma_e'$.

$$\sigma_e = 0.4047 \times 0.8499 \times 0.897 \times 1 \times 1 \times 504 = 155.5\,\text{MPa}.$$

As a well-known material has been selected subject to known loads, the factor of safety can be taken as $n_s = 2$. The diameter can now be calculated using the DE Modified Goodman, DE Gerber, DE ASME Elliptic and Soderberg criteria.

DE Modified Goodman:

$$d=\left(\frac{16n}{\pi}\left\{\frac{1}{S_e}\left[4\left(K_f M_a\right)^2+3(K_{fs}T_a)^2\right]^{0.5}+\frac{1}{S_{ut}}\left[4\left(K_f M_m\right)^2+3(K_{fs}T_m)^2\right]^{0.5}\right\}\right)^{1/3}$$

$$\left[4\left(K_f M_a\right)^2+3(K_{fs}T_a)^2\right]^{0.5}=517.7$$

$$\left[4\left(K_f M_m\right)^2+3(K_{fs}T_m)^2\right]^{0.5}=194.4$$

$$d=\left(\frac{16\times 2}{\pi}\left\{\frac{1}{155.5\times 10^6}[517.7]^{0.5}+\frac{1}{1000\times 10^6}[194.4]^{0.5}\right\}\right)^{1/3}=0.03298\text{ m}$$

DE Gerber:

$$d=\left(\frac{8n\sqrt{4\left(K_f M_a\right)^2+3(K_{fs}T_a)^2}}{\pi S_e}\left\{1+\left[1+\left(\frac{2S_e\sqrt{4\left(K_f M_m\right)^2+3(K_{fs}T_m)^2}}{S_{ut}\sqrt{4\left(K_f M_a\right)^2+3(K_{fs}T_a)^2}}\right)^2\right]^{0.5}\right\}\right)^{1/3}$$

$$\left[4\left(K_f M_a\right)^2+3(K_{fs}T_a)^2\right]^{0.5}=517.7$$

$$\left[4\left(K_f M_m\right)^2+3(K_{fs}T_m)^2\right]^{0.5}=194.4$$

$$d=\left(\frac{8\times 2\times 517.7}{\pi\times 155.5\times 10^6}\left\{1+\left[1+\left(\frac{2\times 155.5\times 10^6\times 194.4}{1000\times 10^6\times 517.7}\right)^2\right]^{0.5}\right\}\right)^{1/3}=0.03240\,m$$

DE ASME Elliptic:

$$d=\left\{\frac{16n}{\pi}\left[4\left(\frac{K_f M_a}{S_e}\right)^2+3\left(\frac{K_{fs}T_a}{S_e}\right)^2+4\left(\frac{K_f M_m}{S_y}\right)^2+3\left(\frac{K_{fs}T_m}{S_y}\right)^2\right]^{0.5}\right\}^{1/3}$$

$$4\left(\frac{K_f M_a}{S_e}\right)^2=4\left(\frac{1.63\times 158.8}{155.5\times 10^6}\right)^2=1.108\times 10^{-11}$$

$$3\left(\frac{K_{fs}T_m}{S_y}\right)^2=3\left(\frac{1.322\times 84.9}{770\times 10^6}\right)^2=6.374\times 10^{-14}$$

$$d=\left\{\frac{16\times 2}{\pi}\left[1.108\times 10^{-11}+6.374\times 10^{-14}\right]^{0.5}\right\}^{1/3}=0.03240\,\text{m}$$

DE Soderberg:

$$d=\left(\frac{16n}{\pi}\left\{\frac{1}{S_e}\left[4\left(K_fM_a\right)^2+3\left(K_{fs}T_a\right)^2\right]^{0.5}+\frac{1}{S_{yt}}\left[4\left(K_fM_m\right)^2+3\left(K_{fs}T_m\right)^2\right]^{0.5}\right\}\right)^{1/3}$$

$$\left[4\left(K_fM_a\right)^2+3\left(K_{fs}T_a\right)^2\right]^{0.5}=517.7$$

$$\left[4\left(K_fM_m\right)^2+3\left(K_{fs}T_m\right)^2\right]^{0.5}=194.4$$

$$d=\left(\frac{16\times 2}{\pi}\left\{\frac{1}{155.5\times 10^6}\times 517.7+\frac{1}{770\times 10^6}\times 194.4\right\}\right)^{1/3}=0.03316\,\text{m}$$

These value are close to the assumed value used to evaluate the size and geometric stress factors so further iteration is not necessary. Comparison of the values obtained for the different criteria indicates similar values around 32–33 mm. For manufacturing convenience, it may be necessary to modify this diameter to the nearest standard size as used within the company or taking into account materials readily available from suppliers; in this case, a diameter of 33, 34 or 35 mm may be a pragmatic choice. Standard sizes can be found in texts such as the Machinery's Handbook. An advantage of using standard sizes is that standard stock bearings can be selected to fit.

Example 7.11. Determine the diameter of the drive shaft for a chain conveyor, which has the loading parameters illustrated in Fig. 7.47. A roller chain sprocket of 500 mm pitch diameter, weighing 90 kg will be mid-mounted between two bearings. A 400 mm, 125 kg roller chain sprocket will be mounted overhung. The drive shaft

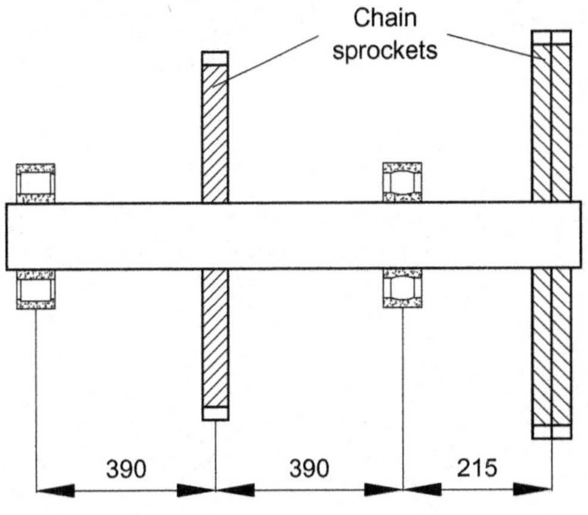

Fig. 7.47 Chain conveyor drive shaft (dimensions in mm).

is to be manufactured using cold-drawn 070 M20 steel. Operating temperatures are not expected to exceed 65°C and the operating environment is noncorrosive. The shaft is to be designed for nonlimited life of $>10^8$ cycles with a 90% survival rate. The shaft will carry a steady driving torque of 1600 N m and rotate at 36 rpm. A sled runner keyway will be used for the overhung pulley and a profile keyway for the mid-mounted pulley. A nominal diameter of 85 mm can be assumed for the shaft. (Example adapted from ASME B106.1M—1985).

Solution

Figs 7.48–7.51 illustrate the principal loads and bending moment diagrams.

$$070\text{M}20\ \sigma_{uts} = 560\,\text{MPa}, \sigma_y = 440\,\text{MPa}.$$

$$\text{Torque } 1600\,\text{N m} = P \times d,$$

$$P_1 = 1600/0.25 = 6400\,\text{N},$$

$$P_2 = 1600/0.2 = 8000\,\text{N},$$

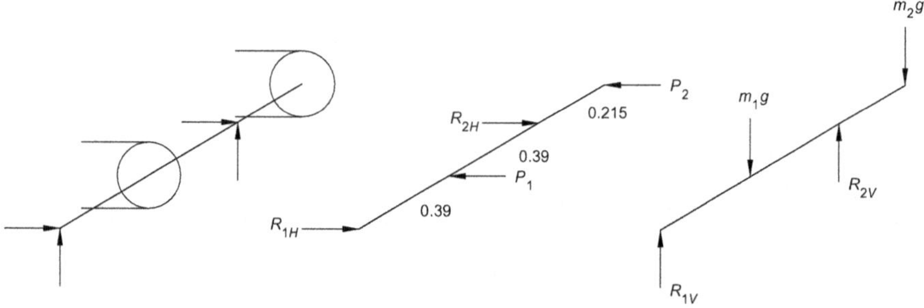

Fig. 7.48 Principal loads.

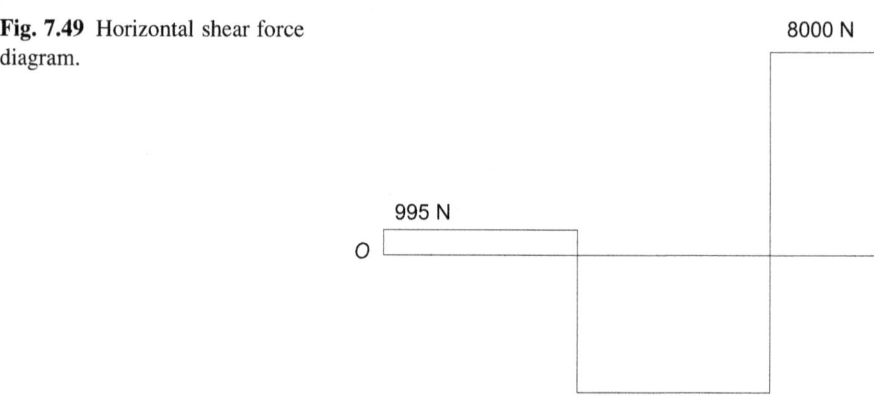

Fig. 7.49 Horizontal shear force diagram.

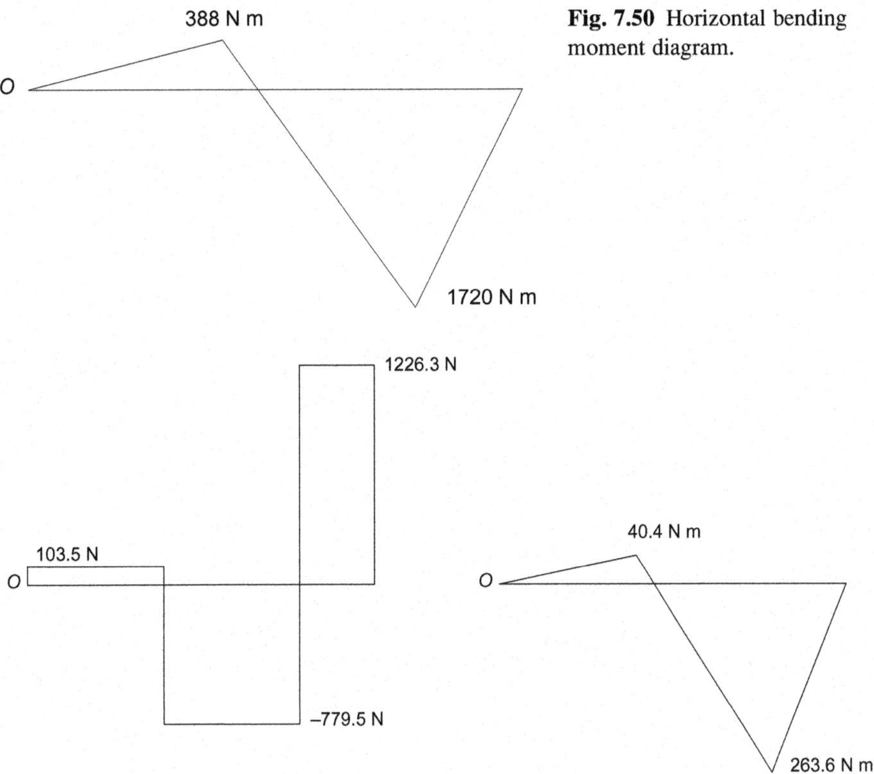

Fig. 7.50 Horizontal bending moment diagram.

Fig. 7.51 Vertical shear force diagram and bending moment diagram.

Moments about R_1,

$$6400 \times 0.39 - R_{2H} \times 0.78 + 8000 \times 0.995 = 0.$$

$$R_{2H} = 13,405.13\text{N}.$$

Resolving horizontal loads:

$$R_{1H} + R_{2H} = P_1 + P_2,$$

$$R_{1H} = 6400 + 8000 - 13,405.13 = 994.87\text{N}.$$

Horizontal Bending Moment Diagram:

At Pulley 1, $995 \times 0.39 = 388\,\text{N m}$,
At Pulley 2, $8000 \times 0.215 = 1720\,\text{N m}$.

Vertical Loads and bending moments:

$$R_{1V} + R_{2V} = 882.9 + 1226.3.$$

Moments about R_1:

$$882.9 \times 0.39 - R_{2V} \times 0.78 + 1226.3 \times 0.995 = 0.$$

$$R_{2V} = 2005.7\text{N},$$

$$R_{1V} = 103.45\text{N}.$$

Vertical SFD Vertical BMD:

Vertical moment at pulley 1: $103.45 \times 0.39 = 40.4\,\text{N m}$.
Vertical moment at pulley 2: $1226.3 \times 0.215 = 263.6\,\text{N m}$.

Resultant maximum bending moment:

$$M_{\max} = \left(263.6^2 + 1720^2\right)^{0.5} = 1740\text{N m}.$$

Factors for the Marin equation:

$$k_a = 4.51{\sigma_{uts}}^{-0.265} = 0.8432.$$

$$k_b = 1.85d^{-0.19}.\ \text{Assume}\ d = 85\text{mm, so}\ k_b = 0.7954.$$

$$k_c = 0.897.$$

$$k_d = 1.$$

$$k_e = 1.$$

Taking $K_t = 2.14$ and $K_{ts} = 3$ for the key seat (the bending moment at the profiled keyway region is far greater. Also in the absence of information about the hardness assume worse case value). Hence $K_f = 2.14$ and $K_{fs} = 3$

$$k_g = 1,$$

$${\sigma_e}' = 0.504\sigma_{uts} = 0.504 \times 560 = 282.2\text{MPa},$$

$$\sigma_e = 0.8432 \times 0.7954 \times 0.897 \times 1 \times 1 \times 282.1 = 169.8\text{MPa}.$$

As a well-known material has been selected subject to known loads the factor of safety can be taken as $n_s = 2$. The diameter can now be calculated using the DE Modified Goodman, DE Gerber, DE ASME Elliptic and Soderberg criteria.

DE Modified Goodman:

$$d = \left(\frac{16n}{\pi}\left\{\frac{1}{S_e}\left[4(K_f M_a)^2 + 3(K_{fs}T_a)^2\right]^{0.5} + \frac{1}{S_{ut}}\left[4(K_f M_m)^2 + 3(K_{fs}T_m)^2\right]^{0.5}\right\}\right)^{1/3}$$

$$\left[4(K_fM_a)^2+3(K_{fs}T_a)^2\right]^{0.5}=7447$$

$$\left[4(K_fM_m)^2+3(K_{fs}T_m)^2\right]^{0.5}=8314$$

$$d=\left(\frac{16\times 2}{\pi}\left\{\frac{1}{169.8\times 10^6}[7447]^{0.5}+\frac{1}{560\times 10^6}[8314]^{0.5}\right\}\right)^{1/3}=0.08425\text{ m}$$

DE Gerber:

$$d=\left(\frac{8n\sqrt{4(K_fM_a)^2+3(K_{fs}T_a)^2}}{\pi S_e}\left\{1+\left[1+\left(\frac{2S_e\sqrt{4(K_fM_m)^2+3(K_{fs}T_m)^2}}{S_{ut}\sqrt{4(K_fM_a)^2+3(K_{fs}T_a)^2}}\right)^2\right]^{0.5}\right\}\right)^{1/3}$$

$$\left[4(K_fM_a)^2+3(K_{fs}T_a)^2\right]^{0.5}=7447$$

$$\left[4(K_fM_m)^2+3(K_{fs}T_m)^2\right]^{0.5}=8314$$

$$d=\left(\frac{8\times 2\times 7447}{\pi\times 169.8\times 10^6}\left\{1+\left[1+\left(\frac{2\times 169.8\times 10^6\times 8314}{560\times 10^6\times 7447}\right)^2\right]^{0.5}\right\}\right)^{1/3}$$

$$=0.07901\text{ m}$$

DE ASME Elliptic:

$$d=\left\{\frac{16n}{\pi}\left[4\left(\frac{K_fM_a}{S_e}\right)^2+3\left(\frac{K_{fs}T_a}{S_e}\right)^2+4\left(\frac{K_fM_m}{S_y}\right)^2+3\left(\frac{K_{fs}T_m}{S_y}\right)^2\right]^{0.5}\right\}^{1/3}$$

$$4\left(\frac{K_fM_a}{S_e}\right)^2=4\left(\frac{2\times 1740}{169.8\times 10^6}\right)^2=1.924\times 10^{-9}$$

$$3\left(\frac{K_{fs}T_m}{S_y}\right)^2=3\left(\frac{3\times 1600}{560\times 10^6}\right)^2=3.570\times 10^{-10}$$

$$d=\left\{\frac{16\times 2}{\pi}\left[1.924\times 10^{-9}+3.570\times 10^{-10}\right]^{0.5}\right\}^{1/3}=0.07865\text{ m}$$

DE Soderberg:

$$d=\left(\frac{16n}{\pi}\left\{\frac{1}{S_e}\left[4(K_fM_a)^2+3(K_{fs}T_a)^2\right]^{0.5}+\frac{1}{S_{yt}}\left[4(K_fM_m)^2+3(K_{fs}T_m)^2\right]^{0.5}\right\}\right)^{1/3}$$

$$\left[4(K_fM_a)^2+3(K_{fs}T_a)^2\right]^{0.5}=7447$$

$$\left[4(K_fM_m)^2+3(K_{fs}T_m)^2\right]^{0.5}=8314$$

$$d=\left(\frac{16\times 2}{\pi}\left\{\frac{1}{155.5\times 10^6}\times 7447+\frac{1}{560\times 10^6}\times 8314\right\}\right)^{1/3}=0.08614\ \text{m}$$

(Note if d assumed for k_b is significantly different from the result, repeat calculation until reasonable convergence).

In this case, each of the criteria is indicating a shaft of between 78 and 86 mm. A conservative option could be to select a shaft of 86 mm.

Example 7.12. A shaft is required for the chain drive illustrated in Fig. 7.52. The power transmitted is 12 kW at 100 rpm. The masses of the sprockets are 5 and 16.3 kg with respective diameters of 254.6 and 460.8 mm. The material selected for

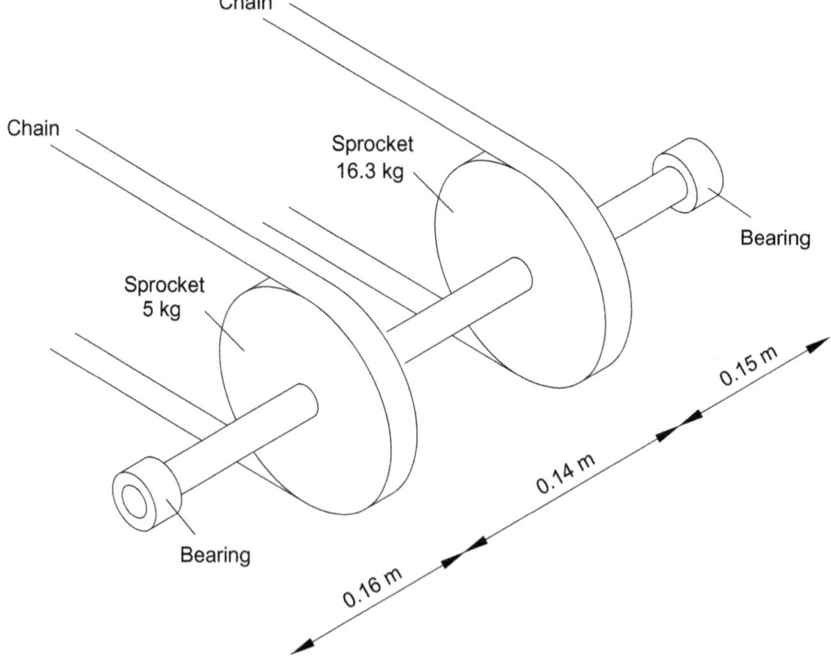

Fig. 7.52 Chain drive.

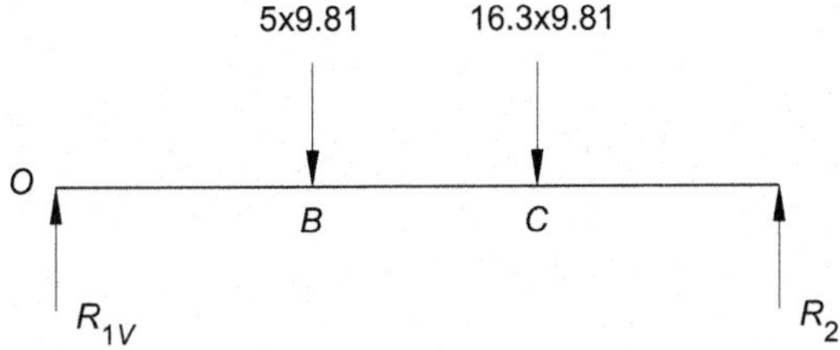

Fig. 7.53 Vertical loading.

the shaft is quenched and drawn 817M40 with $\sigma_{uts} = 850\,\text{MPa}$ and $\sigma_y = 700\,\text{MPa}$. The torque is transmitted between the shaft and sprockets by means of keys in profiled keyways. Determine the minimum diameter for the shaft for the design of transmission shafting if the desired reliability is 99%. Use an initial estimate for the shaft diameter of 68 mm.

Solution

Vertical loading, Fig. 7.53

Moments about O:

$$5 \times 9.81 \times 0.16 + 16.3 \times 9.81 \times 0.3 - R_{2V} \times 0.56 = 0,$$

$$R_{2V} = 124.0\text{N}.$$

Resolving vertically,

$$R_{1V} + R_{2V} = W_1 + W_{2,}$$

$$R_{1V} = 49.05 + 159.9 - 124 = 84.91\,\text{N},$$

$$T = \frac{12000}{100 \times (2\pi/60)} = 1146\,\text{N m}$$

$$P_1 = T/r_{1,}$$

$$P_2 = T/r_{2,}$$

$$P_1 = \frac{1146}{0.2546/2} = 9002\,\text{N}$$

$$P_2 = \frac{1146}{0.4608/2} = 4974\,\text{N}$$

Moments about O, Fig. 7.54, gives

$$-9002 \times 0.16 - 4974 \times 0.3 + 0.45 R_{2H} = 0,$$

$$R_{2H} = 6517\text{N},$$

Fig. 7.54 Horizontal loading.

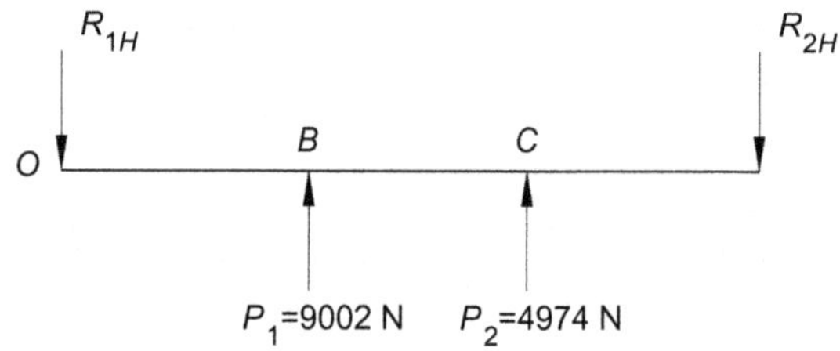

$$R_{1H} + R_{2H} = 9002 + 4974,$$

$$R_{1H} = 7459\text{N}.$$

Vertical bending moments:

$$M_{BV} = L_1 \times R_{1V} = 84.91 \times 0.16 = 13.59\text{Nm},$$

$$M_{CV} = R_{1V} \times 0.3 - 5 \times 9.81 \times 0.14 = 18.61\text{Nm}.$$

Horizontal bending moments

$$M_{BH} = R_{1H} \times L_1 = 7459 \times 0.16 = 1193\text{Nm},$$

$$M_{CH} = R_{1H} \times (L_1 + L_2) - 9002 \times 0.14 = 7459 \times 0.3 - 9002 \times 0.14 = 977.4\text{Nm}$$

$$|M_{\max}| = \sqrt{13.59^2 + 1193^2} = 1193 \text{ N m}$$

The vertical contribution was insignificant and could have been ignored.

$$T_{\max} = 1146\text{Nm}.$$

$$\sigma_e = k_a k_b k_c k_d k_e k_g \sigma'_e$$

$$\sigma'_e = 0.504 \times 850 = 428.4 \text{ MPa}$$

$$k_a = 4.51 \sigma_{uts}^{-0.265} = 0.7549$$

$$k_b = 1.85\text{d}^{-0.19}. \text{ Assume } d = 68\text{mm, so } k_b = 0.8145,$$

$$k_c = 0.814 \text{ for } 99\% \text{reliability},$$

$$k_d = 1 \text{ (no special factors assumed)},$$

$$k_e = 1 \text{ (no special factors assumed)}.$$

Assuming K_t=2.14 and K_{ts}=3 to account for the keyway. Hence, K_f=2.14 and K_{sf}=3.

$k_g = 1$ (no special factors assumed),

$\sigma_e = 0.7549 \times 0.8298 \times 0.814 \times 1 \times 1 \times 428.4 = 218.4\text{MPa}.$

As a well-known material has been selected subject to known loads the factor of safety can be taken as $n_s = 2$. The diameter can now be calculated using the DE Modified Goodman, DE Gerber, DE ASME Elliptic and Soderberg criteria.

DE Modified Goodman:

$$d = \left(\frac{16n}{\pi}\left\{\frac{1}{S_e}\left[4(K_f M_a)^2 + 3(K_{fs}T_a)^2\right]^{0.5} + \frac{1}{S_{ut}}\left[4(K_f M_m)^2 + 3(K_{fs}T_m)^2\right]^{0.5}\right\}\right)^{1/3}$$

$$\left[4(K_f M_a)^2 + 3(K_{fs}T_a)^2\right]^{0.5} = 5106$$

$$\left[4(K_f M_m)^2 + 3(K_{fs}T_m)^2\right]^{0.5} = 5955$$

$$d = \left(\frac{16 \times 2}{\pi}\left\{\frac{1}{218.4 \times 10^6}[5106]^{0.5} + \frac{1}{850 \times 10^6}[5955]^{0.5}\right\}\right)^{1/3} = 0.06764\text{ m}$$

DE Gerber:

$$d = \left(\frac{8n\sqrt{4(K_f M_a)^2 + 3(K_{fs}T_a)^2}}{\pi S_e}\left\{1 + \left[1 + \left(\frac{2S_e\sqrt{4(K_f M_m)^2 + 3(K_{fs}T_m)^2}}{S_{ut}\sqrt{4(K_f M_a)^2 + 3(K_{fs}T_a)^2}}\right)^2\right]^{0.5}\right\}\right)^{1/3}$$

$$\left[4(K_f M_a)^2 + 3(K_{fs}T_a)^2\right]^{0.5} = 5106$$

$$\left[4(K_f M_m)^2 + 3(K_{fs}T_m)^2\right]^{0.5} = 5955$$

$$d = \left(\frac{8 \times 2 \times 5106}{\pi \times 218.4 \times 10^6}\left\{1 + \left[1 + \left(\frac{2 \times 218.4 \times 10^6 \times 5955}{850 \times 10^6 \times 5106}\right)^2\right]^{0.5}\right\}\right)^{1/3} = 0.06365\text{ m}$$

DE ASME Elliptic:

$$d = \left\{\frac{16n}{\pi}\left[4\left(\frac{K_f M_a}{S_e}\right)^2 + 3\left(\frac{K_{fs}T_a}{S_e}\right)^2 + 4\left(\frac{K_f M_m}{S_y}\right)^2 + 3\left(\frac{K_{fs}T_m}{S_y}\right)^2\right]^{0.5}\right\}^{1/3}$$

$$4\left(\frac{K_f M_a}{S_e}\right)^2 = 4\left(\frac{2.14 \times 1193}{218.4 \times 10^6}\right)^2 = 5.463 \times 10^{-10}$$

$$3\left(\frac{K_{fs}T_m}{S_y}\right)^2 = 3\left(\frac{3\times 1146}{700\times 10^6}\right)^2 = 7.237\times 10^{-11}$$

$$d = \left\{\frac{16\times 2}{\pi}\left[5.463\times 10^{-10} + 7.237\times 10^{-11}\right]^{0.5}\right\}^{1/3} = 0.06328\,\text{m}$$

DE Soderberg:

$$d = \left(\frac{16n}{\pi}\left\{\frac{1}{S_e}\left[4(K_fM_a)^2 + 3(K_{fs}T_a)^2\right]^{0.5} + \frac{1}{S_{yt}}\left[4(K_fM_m)^2 + 3(K_{fs}T_m)^2\right]^{0.5}\right\}\right)^{1/3}$$

$$\left[4(K_fM_a)^2 + 3(K_{fs}T_a)^2\right]^{0.5} = 5106$$

$$\left[4(K_fM_m)^2 + 3(K_{fs}T_m)^2\right]^{0.5} = 5955$$

$$d = \left(\frac{16\times 2}{\pi}\left\{\frac{1}{218.4\times 10^6}\times 5106 + \frac{1}{700\times 10^6}\times 5955\right\}\right)^{1/3} = 0.06873\,\text{m}$$

(Note if d assumed for k_b is significantly different from the result, repeat calculation until reasonable convergence). In this case, the diameters are between 63 and 69 mm indicating a diameter of 69 or 70 mm could be suitable.

7.6 Detailed design case study

A small and medium enterprise (SME) company has investigated the potential market for a small electric motor-driven air compressor. The market analysis has confirmed the requirement for an 80 m^3/h, 7 bar air supply driven by a 5 kW electric motor. Preliminary design has already resulted in the selection of an electric motor, with a synchronous speed of 2850 rpm, and the piston and cylinder and associated components for a reciprocating compressor. The compressor design speed is 1500 rpm. High reliability is required. At this stage of the design task, some uncertainty exists on loads and material quality. A schematic for the compressor concept is given in Fig. 7.55. The task here is to explore the detailed design of the shafts for this compressor if a geared transmission has been selected. Various alternatives might be possible for the transmission, based on the distance between the motor and the final drive and some of these are illustrated in Fig. 7.56, including a chain, belt or gear drive as well as direct drive. Here it is assumed that the gear drive provides the most suitable arrangement and the detailed design for this option is explored.

For a reduction ratio of 1.9, a single pair of gears is possible. Spur, helical, bevel and worm gears are all possible. Helical gears are generally quieter and more compact

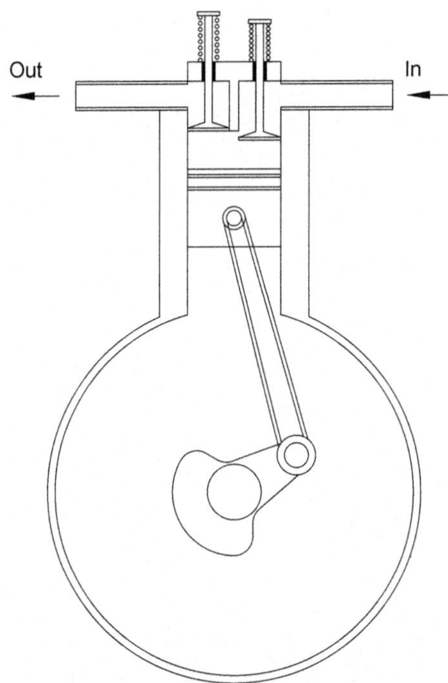

Fig. 7.55 Schematic for a 7 bar, 80 m^3/h reciprocating compressor.

than spur gears but the thrust load they generate must be accommodated, bevel gears provide drive through right angles, which is not a requirement here and worm gears are too inefficient. Spur gears appear to be the most appropriate selection for the design.

Given that this design must be capable of operating for 2 h a day, 200 days a year with a warrantee life of 18 months, detailed calculations based on the AGMA (American Gear Manufacturers Association) standards can be made for the gears. The size of the gears is unknown at this stage. The results of calculations for a range of modules and face widths are given in Table 7.8 (see Childs, 2004).

Examination of Table 7.8 indicates that all the proposals have high factors of safety for bending. The highest threat to failure for all the proposals is due to wear on the pinion. The proposal with the highest factor of safety for contact stress on the pinion and acceptable values for the other parameters is $m=3$. This design proposal has therefore been selected for the transmission.

The next step in the design is to size the shaft. This design experiences bending moments and torque so it is important to model these in the analysis. Fig. 7.57 gives an approximate arrangement for the shaft illustrating proposed dimensions, shoulders, steps and key ways. In order to determine the shaft diameter using Eqs. 7.48–7.55, it is necessary to know the torque and bending moment and endurance limit for the material.

The transmitted load from the gears is 558 N.

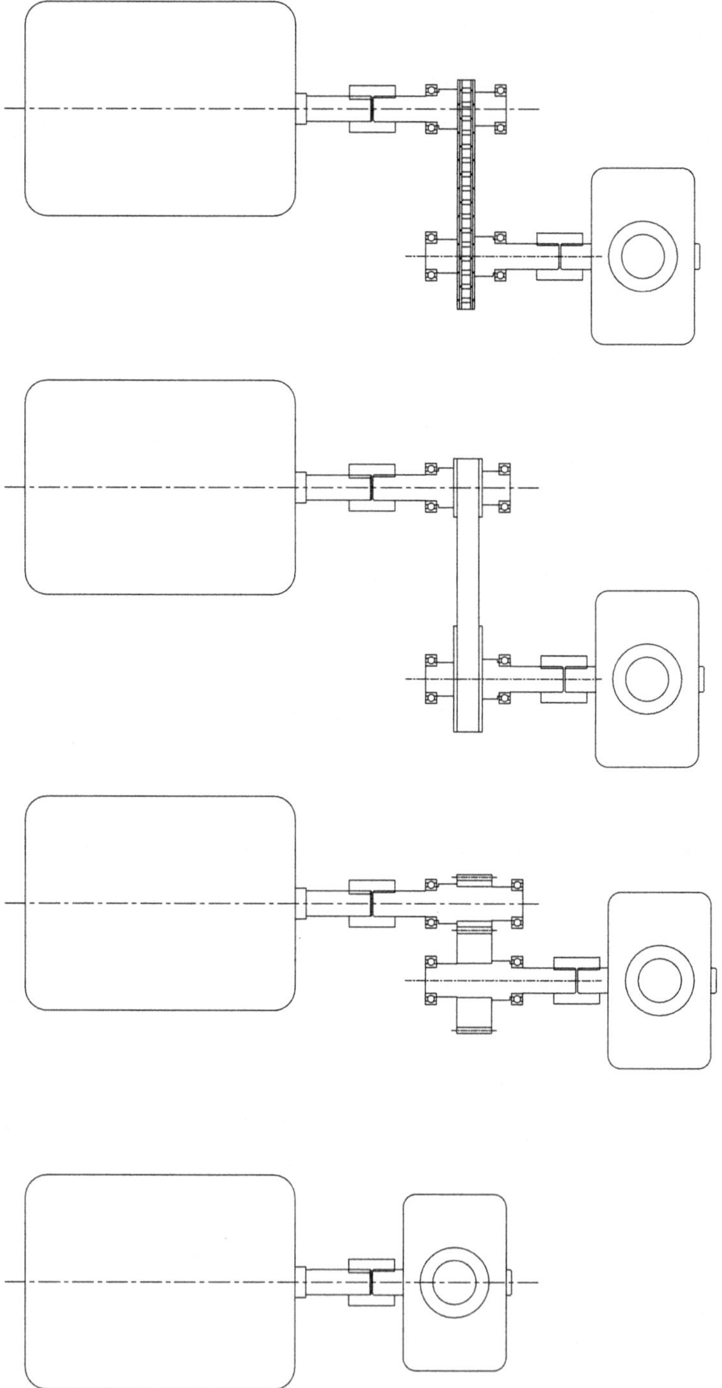

Fig. 7.56 Conceptual arrangements for the transmission.

Table 7.8 AGMA bending and contact safety factors for a variety of modules and face widths

Module	$m=2$	$m=2$	$m=2.5$	$m=3$
N_P	20	20	20	20
N_G	38	38	38	38
d_P (mm)	40	40	50	60
d_G (mm)	76	76	95	114
V (m/s)	5.97	5.97	7.46	8.95
F (mm)	20	50	30	40
W_t (N)	838	838	670	558
H_{BP}	240	240	240	240
H_{BG}	200	200	200	200
S_{FP}	1.6	3.7	3.6	6.6
S_{FG}	1.7	3.9	3.8	7.0
S_{HP}	0.76	1.1	1.1	1.5
S_{HG}	0.96	1.4	1.4	1.9

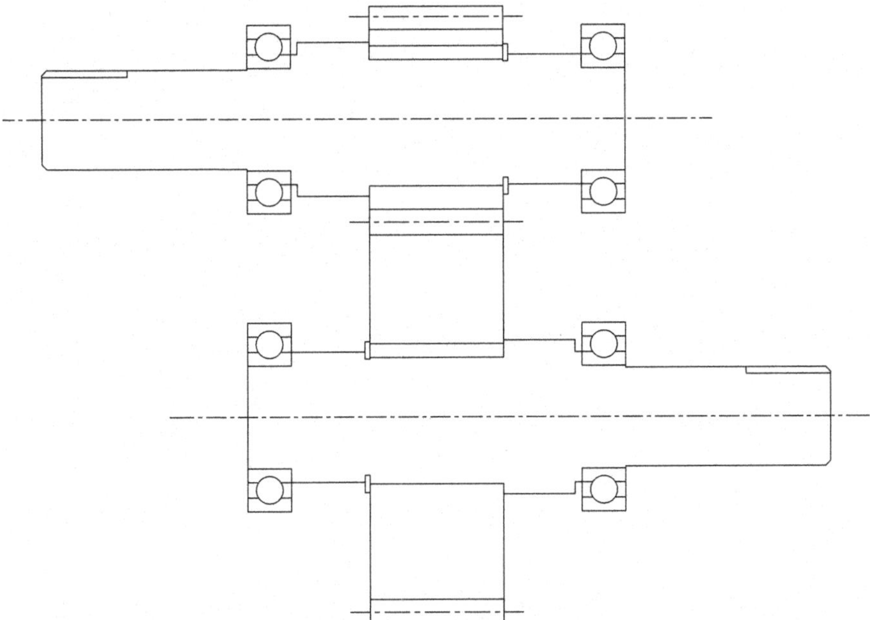

Fig. 7.57 General arrangement proposal for the pinion and gear shafts.

For a spur gear, the radial load is given by

$$W_r = W_t \tan\phi = 558 \tan 20 = 203\ \text{N}$$

The mass of the pinion is approximately 0.46 kg. The mass of the shaft is approximately 0.57 kg. Combining these values gives an approximate vertical load due to gravity of 10.1 N.

The mass of the gear is approximately 1.66 kg. The mass of the shaft is approximately 0.57 kg. Combining these values gives an approximate vertical load due to gravity of 21.9 N.

These various loads and the geometry proposed in Fig. 7.57 are used to develop the shearing force and bending moment diagrams given in Fig. 7.58. The maximum resultant bending moments for the pinion and gear shafts are 15.1 Nm and 15.4 Nm, respectively.

The torque transmitted by the pinion shaft is given by

$$T = \frac{\text{power}}{\omega} = \frac{5000}{2850 \times \frac{2\pi}{60}} = 16.8\ \text{Nm}$$

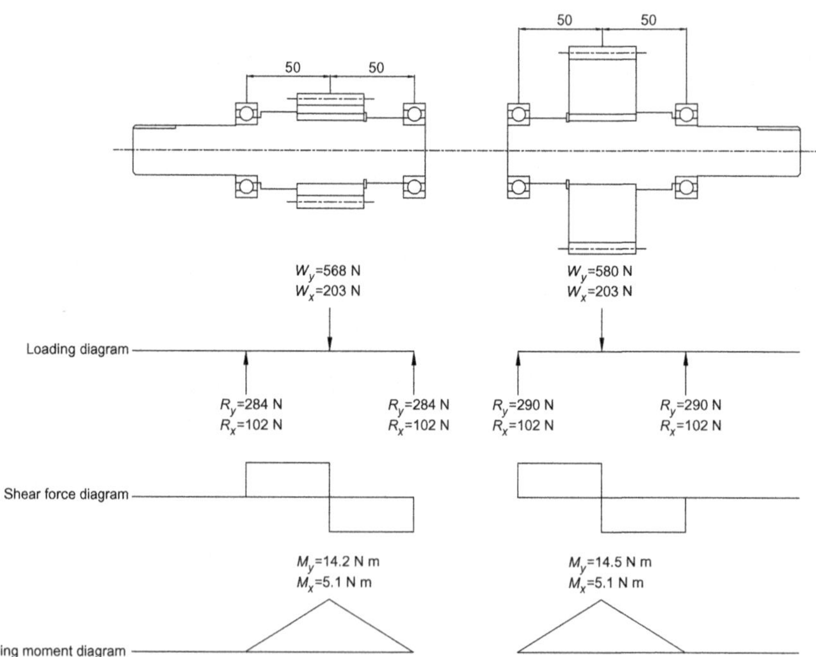

Fig. 7.58 Shear and bending moment diagrams for the proposed shafts.

The torque transmitted by the gear shaft is given by

$$T = \frac{\text{power}}{\omega} = \frac{5000}{1500 \times \frac{2\pi}{60}} = 31.8\,\text{N m}$$

An estimate of the shaft diameter needs to be made in order to evaluate the endurance limit. Here, $d = 30$ mm will be assumed. An 070M20 steel has been proposed for the shaft. The ultimate tensile and yield strengths for this material are 560 MPa and 440 MPa respectively. Taking $k_a = 0.8432$, $k_b = 0.8562$, $k_c = 0.753$ (99.9%), $k_d = 1$, $k_e = 1$, $k_g = 1$, and $\sigma_e' = 282.2$ MPa gives $\sigma_e = 153.4$ MPa. Also taking $K_t = 3.2$, $K_{ts} = 5.1$, $\underline{n} = 4$.

The DE Goodman, DE Gerber, DE ASME Elliptic, DE Soderberg criteria give shaft diameters of 26.32 mm, 24.55 mm, 24.42 mm and 27.01 mm, respectively, for the pinion shaft and 28.56 mm, 26.53 mm, 26.42 mm and 29.66 mm, respectively, for the gear shaft. It should be noted that the diameter determined is particularly sensitive to the surface finish and the fatigue stress concentration factors. Given the uncertainty associated with the design, a diameter of 30 mm is proposed for both the pinion and the gear shafts.

It is necessary to check the critical frequency, deflection and slope characteristics of the shaft. The deflection of the shaft can be estimated by considering the shaft to be a simply supported beam. The deflection at mid-span of a simply supported beam with a mid-mounted load is given by

$$y = -\frac{WL^3}{48EI}$$

Taking the total mass of the gear and the shaft located at mid-span the deflection is

$$y = -\frac{29.1 \times 0.1^3}{48 \times 200 \times 10^9 \pi 0.03^4/64} = 7.624 \times 10^{-8}\ \text{m}$$

This is a very small deflection indicating that slope of the shaft, at the bearings, and deflection, are unlikely to be a problem in this design.

Substitution in the Rayleigh-Ritz equation gives a first critical frequency of

$$\omega_c = \sqrt{g/y} = \sqrt{9.81/7.624 \times 10^{-8}} = 11340\,\text{rad/s}$$

This represents a rotating speed of 108,000 rpm, which is an order of magnitude higher than the operating speed. Even accounting for deflection in the bearings the critical speed will still be significantly higher than the design speed of the gearbox. This is not surprising given the stocky nature of the shaft with it small span.

Deep groove ball bearings are proposed for the transmission as these bearings provide support for the radial load and axial alignment. The design life for the bearings is

1.02×10^8 revolutions. The maximum load on the bearings is approximately 310 N. The basic dynamic load rating required is given by

$$C = 310\left(\frac{1.02 \times 10^8}{1 \times 10^6}\right)^{1/3} = 1448\,\text{N}$$

A 30 mm bore bearing with a basic dynamic load rating greater than this value and a speed capability with grease filled bearings of >2850 rpm would be suitable. The bearings from a number of specialist bearing manufacturers should be considered. From Table 6.6, ISO designation 6006 would be suitable with a considerable margin of safety. It is worth considering worse case usage. If the machine was operated continuously for the warrantee period and other components did not fail, the design life would be

$$2850 \times 60 \times 24 \times 365 \times 1.5 = 2.25 \times 10^9 \text{ revolutions}$$

Using this figure, the required basic dynamic load rating is 3798 N.

Selecting a low-carbon, cold-drawn steel for the key with a yield strength of 350 MPa, assuming a square key of 4 × 4 mm and setting $N = 4$ for this application, the necessary key length is given by

$$L = \frac{4TN}{DW\sigma_y} = \frac{4 \times 31.8 \times 4}{0.03 \times 0.004 \times 350 \times 10^6} = 0.0121\,\text{m}$$

This length can readily be accommodated in the shaft and mating components. Because it is comparatively short, it would be possible to use the same type of key for the gears and couplings. This would have the advantage of a reduced parts inventory, simplifying tooling and economy of scale.

An almost finished engineering drawing for the pinion shaft is given in Fig. 7.59. The drawing includes the results of a considerable number of decisions. The calculations outlined so far in this case study have resulted in the specification of some of the major dimensions. However, geometric tolerances need to be specified on certain features in order to control features such as location, form, profile, orientation and runout, and ensure adequate assembly and operation of the machine. In addition, particular surface finishes must be specified on critical regions such as mating surfaces and areas experiencing reversed bending. A typical surface finish for a ground surface is 0.4 μm and that for turning or milling is 1.6 μm. Fig. 7.59 can be examined by the reader and each dimension and feature scrutinised.

The similarity in diameter required for the pinion shaft and gear shaft provides an opportunity to use a common design for both shafts. This has the advantage of reduced inventory, setting up costs and economies of scale. This approach has been adopted for this design.

The options for the transmission support and enclosure realistically include fabricating, moulding or casting using metal or plastic. The temperature of operation indicates that a plastic material is a possibility provided rigidity, strength and cost

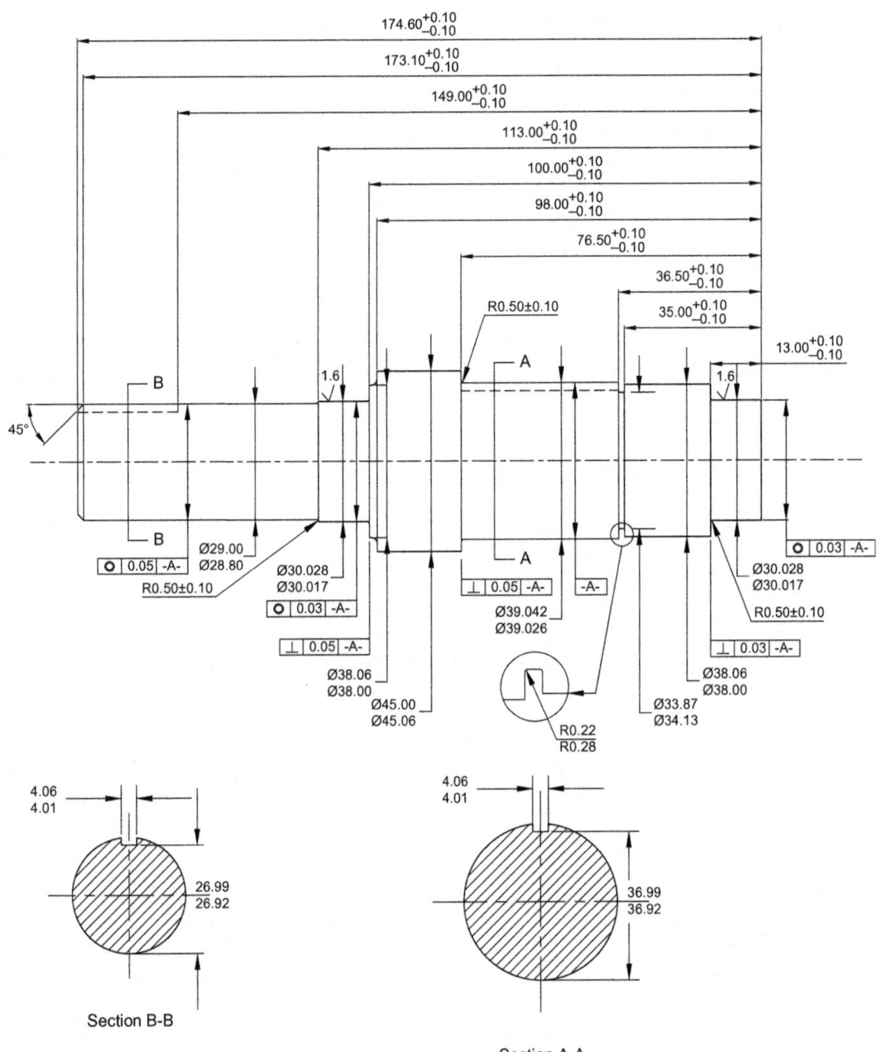

Fig. 7.59 Pinion shaft.

requirements are met. Metals are however an obvious choice. In order to provide rigid support, use of a two-part casting assembly is explored in Fig. 7.60. Turned inserts are proposed for the stationary bearing mountings. Although this solution appears adequate, fabrication and use of a plastic moulding should also be explored to identify whether they can meet the specification and provide a cost advantage.

Engineering drawings, similar to that given in Fig. 7.59, should also be produced for all the components in the transmission. Here, however, just a general arrangement for the overall transmission is given in Fig. 7.61.

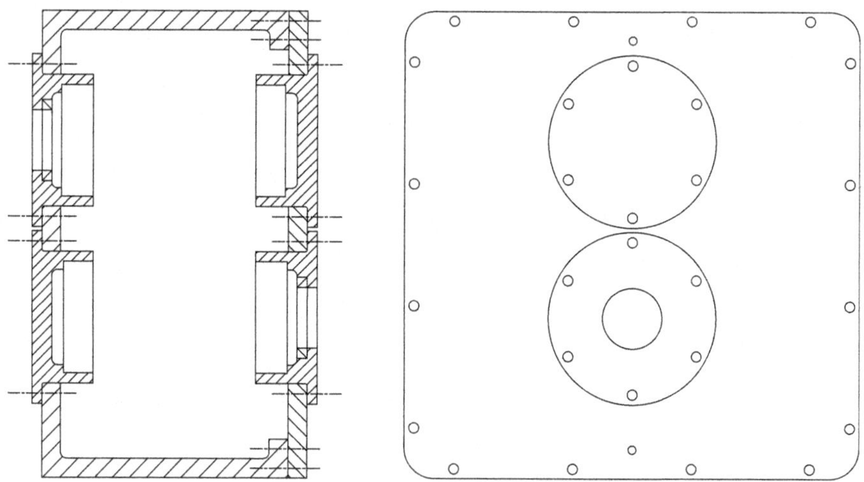

Fig. 7.60 Casing.

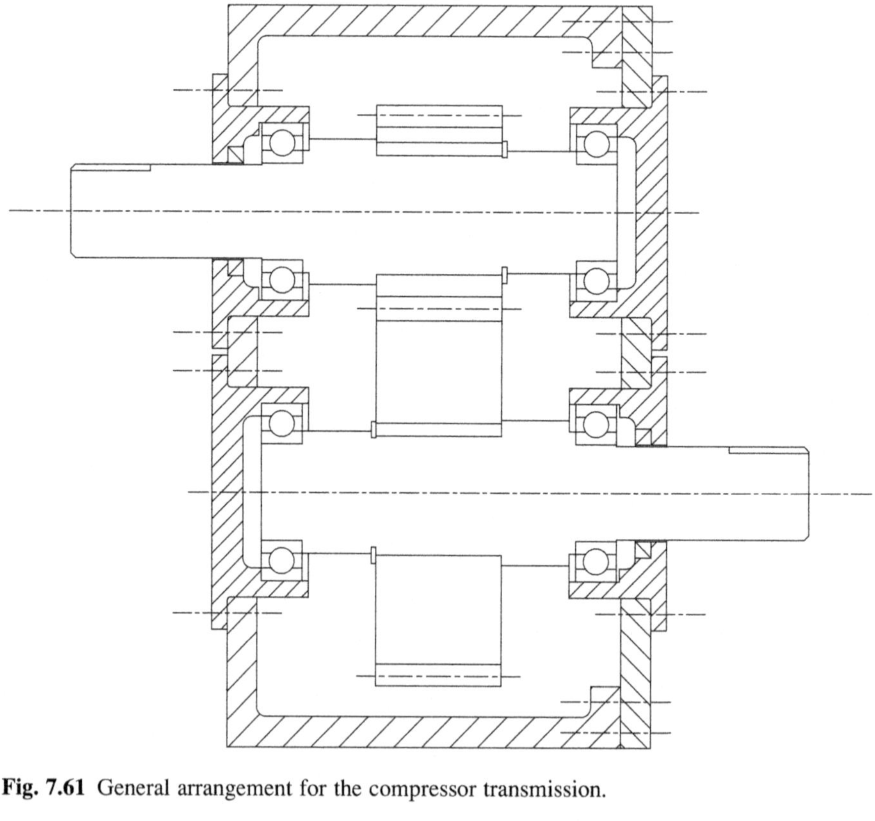

Fig. 7.61 General arrangement for the compressor transmission.

Although the design developed is probably far from optimal, an estimate of the works cost price should be developed in order to identify whether the proposal is likely to meet the specification in terms of price and which aspects are responsible for determining the majority of the costs. These can then be assessed and the use of expensive features justified or, if possible, removed from the design altogether. Estimating the cost of a product at the design stage is notoriously difficult. In the design proposal presented here estimates for the costs of stock items such as the gears, bearings, keys and fasteners can be obtained by examination of price lists if published or by contacting the Original Equipment Manufacturer (OEM) or stock supplier concerned and obtaining a quotation. Ensure that the quotation provides the appropriate discount for the quantities concerned. For a production run of 1000 units per year, it is likely that the transmissions will be manufactured in batches. In this case, a quotation for the necessary batch size would seem appropriate so that variations in demand will not tie up company capital unnecessarily. Estimates for the cost of the castings will need to be obtained by means of quotations from casting companies. In this design, machining of the shafts and finish machining of the castings need to be performed either in-house or bought in. Obvious quotations from machine shops can be obtained to evaluate the possibility of buying the shafts directly. For in-house machining, reliance needs to be placed on experience or use of standard estimates for machining times. Standard estimates for a wide range of machining processes are outlined in Swift and Booker (2013).

From Table 7.9, taking the materials overhead as 130% and the labour overhead as 225%, the works cost price is given by

Table 7.9 Estimates for component costs based on a batch run of 100 gearboxes

Item	Number	Time	Unit cost (Euro)	Total cost (Euro)	Source of estimate
Pinion	1 off		12	12	Stock gear manufacturer
Gear	1 off		18	18	Stock gear manufacturer
Keys	6 off		0.5	3	Stock supplier
Shaft material	2 off		10	20	Metal supplier
Bearings	4 off		10	40	Stock supplier
Split rings	2 off		2	4	Stock supplier
Main casting	1 off		60	60	Quotation from casting specialist
Casting cover	1 off		20	20	Quotation from casting specialist

Continued

Table 7.9 Continued

Item	Number	Time	Unit cost (Euro)	Total cost (Euro)	Source of estimate
Casting seal	1 off		2	2	O ring cording supplier
Bearing mountings material	4 off		5	20	Stock supplier
Shaft seals	2 off		8	16	Seal manufacturer
Couplings	2 off		20	40	Coupling manufacturer
Fasteners	20 off		0.25	5	Fastener stock supplier
Gear Hardening	Batch of 200		100	1 per transmission	Specialist service
Materials subtotal				261	
Key way machining	6 off	2 min each	60/h	12	Machine shop
Turning	2 off	15 min/ shaft	60/h	30	In-house estimate
Grinding	2 off	10 min/ shaft	60/h	20	In-house estimate
Casting machining	1 off	30 min	60/h	30	In-house estimate
Bearing mountings machining	4 off	10 min/ mounting	60/h	40	In-house estimate
Assembly	1 off	30 min	40/h	20	In-house estimate
Packaging	1off	4 min	4	4	In-house estimate
Labour subtotal				156	

$$\text{works cost price} = (261 \times 1.3) + (156 \times 2.25) = 690.3\ \text{Euros}$$

The works cost price for the transmission system is close to the target price of 600 Euros. With careful control of procurement costs and in-house manufacturing costs, it should be possible to get this total cost well below the 600 Euro target. As such the design is likely to meet the specification.

A proposal for the compressor transmission has been developed. The design appears adequate in terms of function and compliance with the specification, although the works cost price has exceeded the target. It should however be possible to refine the design of the shaft and casing and bearing mounts to reduce costs and cheaper quotations for the stock items may be available. It is also possible that one of the alternative concepts, such as the belt drive, could be considerably cheaper to produce.

7.7 Conclusions

Shaft design involves consideration of the layout of features and components to be mounted on the shaft, specific dimensions and allowable tolerances, materials, deflection, frequency response, life and manufacturing constraints. This chapter has introduced the use of steps and shoulders and miscellaneous devices to locate components, methods to calculate the deflection of a shaft and its critical speeds, and methods to determine the minimum safe diameter for shaft experiencing torque and bending for a given life.

References

Budynas, R.G., Nisbett, J.K., 2011. Shigley's Mechanical Engineering Design, 9th ed. McGraw Hill, New York.

Childs, P.R.N., 2004. Mechanical Design, 2nd ed. Butterworth Heinemann.

Fuchs, H.O., Stephens, R.I., 1980. Metal Fatigue in Engineering. Wiley, New York.

Juvinall, R.C., 1967. Engineering Considerations of Stress, Strain and Strength. McGraw Hill, New York.

Juvinall, R.C., Marshek, K.M., 2012. Fundamentals of Machine Component Design, fifth ed. Wiley, Chichester.

Kuguel, R., 1969. A relation between theoretical stress concentration factor and fatigue notch factor deduced from the concept of highly stressed volume. Proc. ASTM 61, 732–748.

Marin, J., 1962. Mechanical Behaviour of Engineering Materials. Prentice Hall, Englewood Cliffs, NJ.

Mischke, C.R., 1987. Prediction of stochastic endurance strength. Trans. ASME. J. Vib. Acoust. Stress Reliab. Des. 113–122.

Neale, M.J., Needham, P., Horrell, R., 1998. Couplings and Shaft Alignment. Professional Engineering Publishing, Bury St Edmunds.

Noda, N.A., Sera, M., Takase, Y., 1995. Stress concentration factors for round and flat test specimens with notches. Int. J. Fatigue 17, 163–178.

Noll, C.G., Lipson, C., 1946. Allowable working stresses. Soc. Exp. Stress Anal.. 3.

Peterson, R.E., 1959. Notch sensitivity. Metal Fatigue 293–306.

Peterson, R.E., 1974. Stress Concentration Factors. Wiley, New York.

Piotrowski, J., 2006. Shaft Alignment Handbook, third ed. CRC Press, Boca Raton.

Pyrhonen, J., Jokinen, T., Hrabovcova, V., 2008. Design of Rotating Electrical Machines. Wiley-Blackwell, Chichester.

Savruk, M.P., Kazberuk, A., 2006. Relationship between the stress intensity and stress concentration factors for sharp and rounded notches. Mater. Sci. 42, 725–738.

Sclater, N., 2011. Mechanisms and Mechanical Devices Sourcebook, fifth ed. McGraw-Hill, New York.

Swift, K.G., Booker, J.D., 2013. Manufacturing Process Selection Handbook. Butterworth Heinemann.

Vidosek, J.P., 1957. Machine Design Projects. Ronald Press, New York.

Zappalorto, M., Berto, F., Lazzarin, P., 2011. Practical expressions for the notch stress concentration factors of round bars under torsion. Int. Fatigue J. 33, 382–395.

Standards

ANSI/ASME B106.1M-1985. 'Design of Transmission Shafting'.

BSI, 1984. 'Manual of British Standards in engineering drawing and design'. BSI Hutchinson.

British Standards Institution. 'BS 4235: Part 2: 1977. Specification for metric keys and keyways: Woodruff keys and keyways'.

British Standards Institution. 'BS 4235: Part 1: 1992. Specification for metric keys and keyways. Parallel and taper keys'.

British Standards Institution. 'BS 3550: 1963. Specification for involute splines'.

British Standards Institute. BS 970: Part 1:1996. Specification for wrought steels for mechanical and allied engineering purposes. General inspection and testing procedures and specific requirements for carbon, carbon manganese, alloy and stainless steels.

Websites

At the time of going to press the world-wide-web contained useful information relating to this chapter at the following sites:

www.allpowertrans.com.
www.ameridrives.com.
www.couplingcorp.com.
www.heli-cal.com.
www.magnaloy.com.
www.mayrcorp.com.
www.peterstubs.com.
www.rw-america.com.
www.servometer.com.
www.zero-max.com.

Further reading

Arola, D., Williams, C.L., 2002. Estimating the fatigue stress concentration factor of machined surfaces. Int. J. Fatigue 24, 923–930.

Beswarick, J., 1994a. Shaft for strength and rigidity. In: Hurst, K. (Ed.), Rotary Power Transmission Design. McGraw Hill, London, pp. 135–141.

Beswarick, J., 1994b. Shaft with fluctuating load. In: Hurst, K. (Ed.), Rotary Power Transmission Design. McGraw Hill, London, pp. 142–148.

Hurst, K., 1994. Shaft/hub connections. In: Hurst, K. (Ed.), Rotary Power Transmission Design. McGraw Hill, London, pp. 55–65.

Kuhn, P., Hardrath, H.F., 1952. An engineering method for estimating notch size effect on fatigue tests of steel. National Advisory Committee for Aeronautics, NACA TN2805.

Macreadys, 1995. Standard Stock Range of Quality Steels and Specifications. Glynwed Steels Ltd.
Mischke, C.R., 1996. Shafts. In: Shigley, J.E., Mischke, C.R. (Eds.), Standard Handbook of Machine Design. McGraw Hill, New York.
Mott, R.L., 1992. Machine Elements in Mechanical Design. Merrill, New York.
Oberg, E., 2012. Machinery's Handbook 29. Industrial Press Inc, New York.
Reshetov, D.N., 1978. Machine Design. Mir Publishers, Moscow.
Sines, G., Waisman, J.L., 1959. Metal Fatigue. McGraw Hill, New York.

Gears

8

Chapter Outline

Nomenclature

Generally, preferred SI units have been stated.

a addendum (mm)
b dedendum (mm)
c clearance (mm)
d pitch diameter (mm)
d_G pitch diameter of the gear (mm)
d_P pitch diameter of the pinion (mm)
d_{ring} pitch diameter of the ring gear (mm)
d_{sun} pitch diameter of the sun gear (mm)
e train value
F face width (mm)
F_r radial force component (N)
F_t tangential force component (N)
h_k working depth (mm)
h_t whole depth (mm)
H power (kW)
K_v velocity factor
m module (mm)
n rotational speed (rpm)

Mechanical Design Engineering Handbook. https://doi.org/10.1016/B978-0-08-102367-9.00008-1

n_P rotational speeds of the pinion (rpm)
n_G rotational speeds of the gear (rpm)
N number of teeth
N_G number of teeth in the gear
N_P number of teeth in the pinion or planet gear
N_R number of teeth in annulus gear
N_s number of teeth in sun gear
O/D outside diameter (mm)
p circular pitch (mm)
PCD pitch circle diameter (mm)
P_d diametral pitch (teeth per inch)
r pitch radius (mm)
r_f fillet radius of basic rack (mm)
t tooth thickness (mm)
t_o width of top land (mm)
V velocity (m/s)
W_t transmitted load (N)
Y Lewis form factor
ϕ pressure angle (°)
σ bending stress (N/m^2)
σ_P permissible stress (MPa)
σ_{uts} ultimate tensile strength (MPa)
ω angular velocity (rad/s)
ω_G angular velocities of the gear (rad/s)
ω_P angular velocities of the pinion (rad/s)

8.1 Introduction

Gears are toothed cylindrical wheels used for transmitting mechanical power from one rotating shaft to another. Several types of gears are in common use and are available as stock items from original equipment suppliers worldwide. This chapter introduces various types of gears and gear transmission and details the design, specification and selection of spur gears in particular, based consideration of failure due to bending using the Lewis equation. Detailed procedures to calculate the stress in spur and helical, bevel, worm and wheel gears are given in Chapters 9, 10 and 11.

The principal functions of a transmission include

- To transmit power from one location to another—a familiar example is found in a bicycle transmitting power from the crank to the wheel.
- To match the torque and speed of a driving and a driven machine component—an example is a car transmission where the driving torque of an internal combustion engine (ICE) is ill-matched to the demand and gears can be used to alter the torque-speed characteristic of the engine. This is discussed further later in this section and also in Section 8.3.1.
- To change the direction of a rotating component. An example is the transmission axle in a rear-wheel-drive vehicle where the axle transmits power from a longitudinal prop-shaft to transverse drive shafts
- To synchronise the motion of one rotating component with another.

These general functions are illustrated for some typical applications in Fig. 8.1.

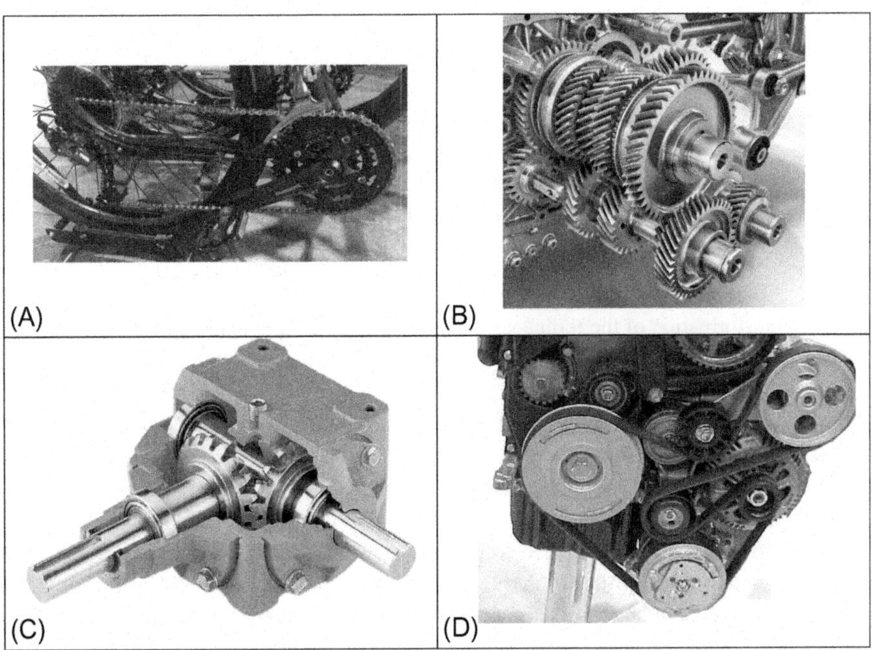

Fig. 8.1 Examples of principal functions of a transmission (A) transmission of power from one location to another, (B) matching the torque and speed of a driving and a driven machine component (Image courtesy of Daimler AG), (C) to change direction (Image courtesy Boston Gear - Altra Industrial Motion) and (D) to synchronise motion.

Many power-producing machines, or prime movers, such as ICEs, industrial gas turbine engines (GTEs) and electric motors produce power in the form of rotary motion. The operating characteristics of prime movers vary according to their type and size. Typical examples are illustrated in Fig. 8.2 for an ICE; an alternating current electric motor and a series wound direct current electric motor. These characteristics may not match those of a given application. For example, the ICE gives little torque at the low speeds associated with starting. If the load and driving characteristics are not well matched, for example, if the starting torque is low, then a prime mover may not

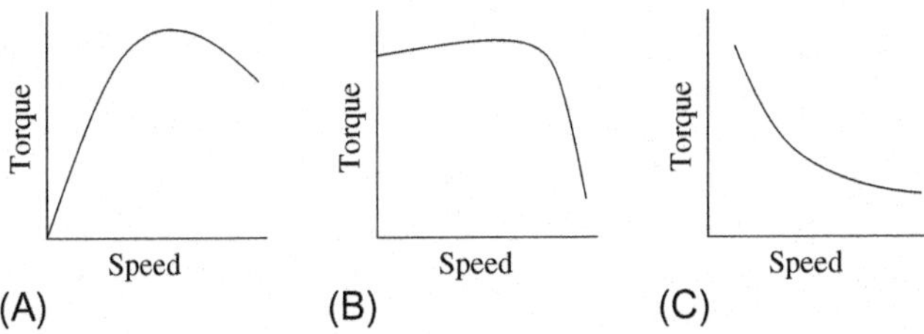

Fig. 8.2 Typical torque-speed characteristics: (A) ICE, (B) three-phase induction alternating current motor and (C) series wound direct current motor.

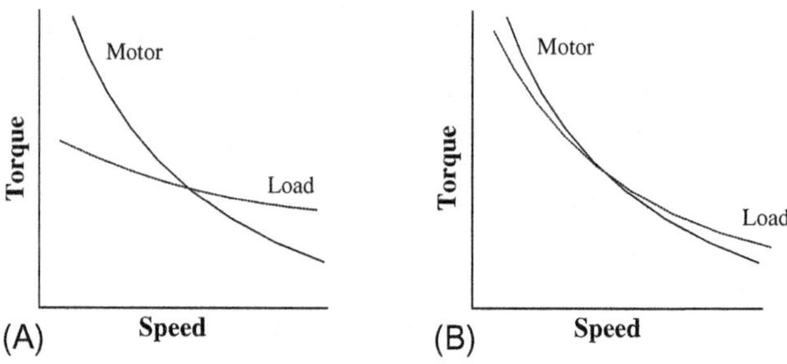

Fig. 8.3 Transformation of the torque-speed characteristic using a reduction gear ratio: (A) mismatched system, (B) transformed system.

enough power to get the load to its desired running speed or it may even stall or not start. It is for this reason that a car engine requires a clutch to decouple the engine from the load during starting and a gearbox to transform the characteristic of the prime mover to match that of the load.

A change of speed can be used to transform the torque-speed characteristic of a prime mover to a useful output characteristic. An example is the use of a reduction gearbox to reduce the speed and increase the torque seen by a load as illustrated in Fig. 8.3. Recalling the relationship between torque, angular velocity and power,

$$\text{Torque} = \frac{\text{Power}}{\text{Angular velocity}}. \tag{8.1}$$

Examination of Eq. (8.1) indicates that as the speed reduces the torque rises for a given power. So if increased torque is required, then a reduction gearbox, which decreases the speed of the output shaft relative to the input shaft, can be used to increase the torque delivered by the output shaft of the gearbox.

When transmitting power from a source to the required point of application, a series of devices are available including belts, pulleys, chains, hydraulic and electrical systems, and gears. Generally, if the distances of power transmission are large, gears are not suitable and chains or belts can be considered (see Chapter 12). However, when a compact, efficient or high-speed drive is required gear trains offer a competitive and suitable solution.

Gears, belts and chains are all available from specialist suppliers. The manufacture of gears requires specialist machinery and the stresses in the region of contact between gears in mesh can be exceptionally high requiring the use of specialist materials (Davis, 2005). Similarly, the design of roller chain and associated sprockets involves high contact stresses, and in addition the design of an articulated joint and associated bearings. For these reasons, combined with financial expediency, leaving the cost of specialist manufacture infrastructure to other companies, gear, chain and belt manufacture is usually best left to the specialist and the task of the machine designer becomes that of selecting and specifying appropriate components from the specialist stock suppliers and original equipment manufacturers. A general introduction to gears

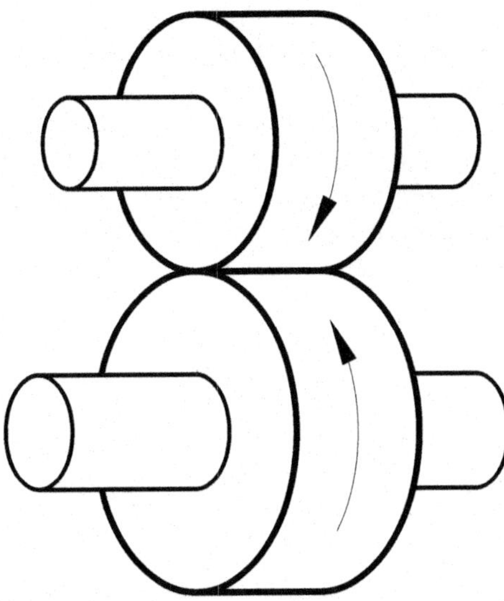

Fig. 8.4 Disc or roller drives.

is considered in this chapter with Chapters 9, 10 and 11 providing further detailed consideration of calculating the stress in spur and helical gears, bevel gears and worm and wheels, respectively. The design and selection of belts and chains are considered in Chapter 12.

Simplistically, a speed change can be achieved by running discs of different diameter together, Fig. 8.4, or alternatively cones for turning corners as well, Fig. 8.5. However, the torque capacity of disc or cone drives is limited by the frictional properties of

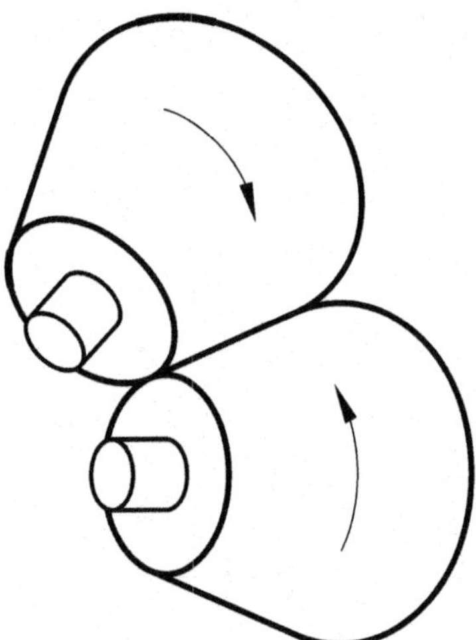

Fig. 8.5 Cone drives.

the surfaces. The addition of teeth to the surfaces of the discs or cones makes the drive positive, ensuring synchronisation, and substantially increases the torque capacity.

Early historical examples of gear trains consisted of teeth or pegs located on discs as illustrated in Figs. 8.6 and 8.7. The disadvantage of these simple teeth is that the velocity ratio is not constant and changes as the teeth go through the meshing cycle causing noise and vibration problems at elevated speeds. The solution to this problem can be achieved by using a profile on the gear teeth, which gives a constant velocity ratio throughout the meshing cycle. Several different geometrical forms can be used but the full-depth, involute form is primarily used in current professional engineering practice.

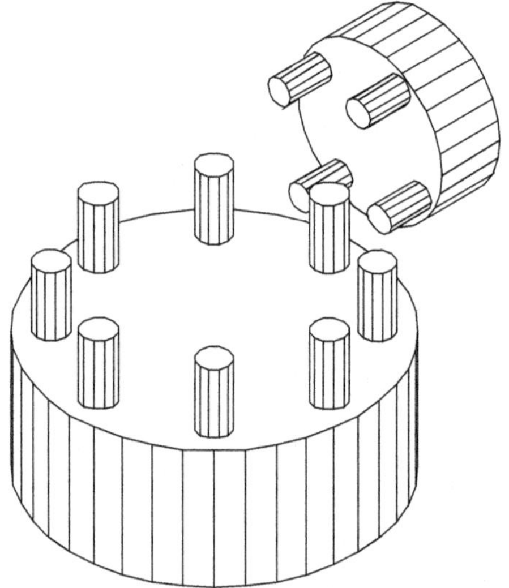

Fig. 8.6 Primitive teeth gears.

Fig. 8.7 Primitive gears.

Gears can be divided into several broad classifications as listed and illustrated in figures

(1) Parallel axis gears:

(a) Spur gears (see Fig. 8.8),

(b) Helical gears (see Figs. 8.9 and 8.10) and

(c) Internal gears.

Fig. 8.8 Spur gears.
Photograph courtesy of Hinchliffe Precision Components Ltd.

Fig. 8.9 Helical gears.
Photograph courtesy of Hinchliffe Precision Components Ltd.

Fig. 8.10 Double helical gears.

(2) Nonparallel, coplanar gears (intersecting axes):
  (a) Bevel gears (see Fig. 8.11),
  (b) Face gears and
  (c) Conical involute gearing.
(3) Nonparallel, noncoplanar gears (nonintersecting axes):
  (a) Crossed axis helicals (see Fig. 8.12),
  (b) Cylindrical worm gearing (see Fig. 8.13),
  (c) Single enveloping worm gearing,
  (d) Double enveloping worm gearing,
  (e) Hypoid gears,
  (f) Spiroid and helicon gearing and
  (g) Face gears (off centre).
(4) Special gear types:
  (a) Square and rectangular gears,
  (b) Elliptical gears (see Litvin, 2014),
  (c) Scroll gears and
  (d) Multiple sector gears (Litvin, 2014).

Spur gears (Fig. 8.8) are the cheapest of all types for parallel shaft applications. Their straight teeth allow running engagement or disengagement using sliding shaft and clutch mechanisms. Typical applications of spur gears include manual and automatic motor vehicle gearboxes, machine tool drives, conveyor systems, electric motor gearboxes, timing mechanisms and power tool drives. The majority of power gears are manufactured from hardened and case hardened steel. Other materials used include iron, brass and bronze, and polymers such as polyamide (e.g., nylon), polyacetal (e.g., Delrin) and Synthetic Resin Bonded Fabric (SRBF, e.g., Tufnol). Typical

Fig. 8.11 Bevel gears.

Fig. 8.12 Crossed axis helical gears.
Photograph courtesy of Hinchliffe Precision Components Ltd.

Fig. 8.13 Worm gears.
Photograph courtesy of Hinchliffe Precision Components Ltd.

material matches are listed in Table 8.1. A harder material may sometimes be used for a pinion, than that for the mating gear wheel, as the pinion teeth will experience more use than those on the gear wheel. Steel gears are often supplied in unhardened form to allow for grub screws, keyways or splines to be added before hardening and finishing.

A helical gear is a cylindrical gear whose tooth traces are helixes, as shown in Fig. 8.9. Common helix angles are 15–30 degrees. Helical gears are typically used for heavy-duty high-speed (>3500 rpm) power transmission, turbine drives, locomotive gearboxes and machine tool drives. Helical gears are generally more expensive than spur gears. Noise levels are lower than for spur gears because teeth in mesh make point contact rather than line contact. The forces arising from meshing of helical gears can be resolved into three components loads, radial, tangential and axial (axial loads are often called thrust loads). Bearings used to support the gear on a shaft must be able to withstand the axial or thrust force component. A solution to this problem is the use of double helical gears, as illustrated in Fig. 8.10. These eliminate the need for thrust bearings because the axial force components cancel each other. Whilst spur gears are generally cheaper, a comparable helical gear will be smaller.

Table 8.1 Typical material matches for gears and pinions

Gear material	Pinion material
Cast iron	Cast iron
Cast iron	Carbon steel
Carbon steel	Alloy steel
Alloy steel	Alloy steel
Alloy steel	Case hardened steel

Bevel gears have teeth cut on conical blanks (see Fig. 8.11) and a gear pair can connect nonparallel intersecting shafts. Bevel gears are used for motor transmission differential drives, valve control and mechanical instruments.

A worm gear is a cylindrical helical gear with one or more threads (see Fig. 8.13). A worm wheel is a cylindrical gear with flanks cut in such a way as to ensure contact with the flanks of the worm gear. Worm gears are used for steering in vehicles, winch blocks, low-speed gearboxes, rotary tables and remote valve control. Worm gear sets are capable of high-speed reduction and high load applications where nonparallel, noninteracting shafts are used. The 90 degree configuration is most common. Frictional heat generation is high in worm gears, so continuous lubrication is required and provision for heat dissipation must be made. The high level of friction associated with worm and wheel sets can be used to produce a self-braking system so that when the input torque is removed the gear set does not rotate. This principle is used in some winches.

Some gear axes can be allowed to rotate about others. In such cases, the gear trains are called planetary or epicyclic (see Fig. 8.14). Planetary trains always include a sun gear, a planet carrier or arm, and one or more planet gears.

All gear mechanisms and gear trains demand continuous lubrication, which must be pressure-fed for high-speed gears in order to counteract centrifugal effects on the oil. Plastic gears, made from for example nylon, can be used in certain applications and have the advantage that there is no need for lubrication but are only suitable for low-speed applications. Plastic gears can reduce noise levels significantly.

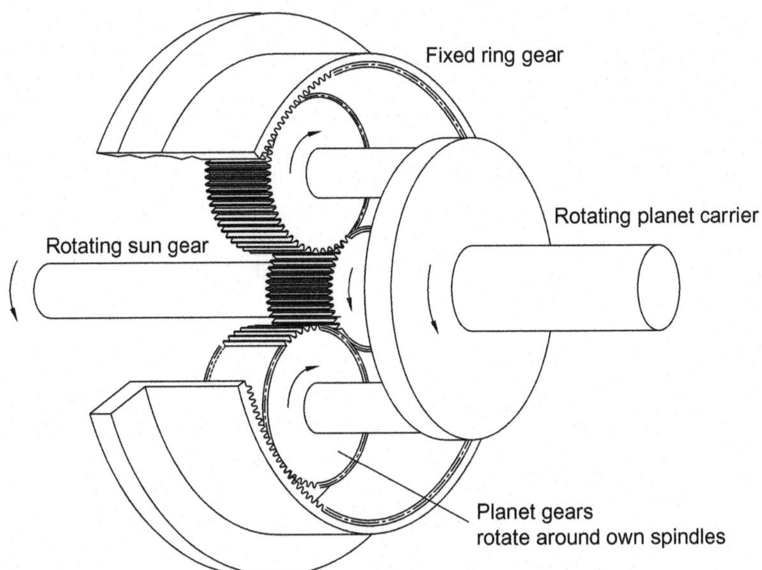

Fig. 8.14 Schematic of an epicyclic gearbox.

Table 8.2 Useful range of gear ratios

Gear	Ratio range	Pitch line velocity (m/s)	Efficiency (%)
Spur	1:1–6:1	25	98–99
Helical	1:1–10:1	50	98–99
Double helical	1:1–15:1	150	98–99
Straight bevel	3:2–5:1	20	93–97
Spiral bevel	3:2–4:1		95–99
Worm	5:1–75:1	30	20–98
Hypoid	10:1–200:1		80–95
Crossed helical	1:1–6:1	30	70–98

Generally, the pinion of a pair of gears should have the largest number of teeth consistent with adequate resistance to failure by bending stress in the teeth. For a given diameter, the larger the number of teeth the finer the pitch and consequently the weaker they are and the greater the liability to fracture. Table 8.2 lists the range of gear ratios and performance characteristics typically achievable.

When the number of teeth selected for a gear wheel is equal to the product of an integer and the number of teeth in the mating pinion, then the same tooth on the gear wheel will touch the same tooth on the pinion in each revolution of the gear wheel. If there is no common factor between the numbers of teeth on the gear wheel and pinion, then each tooth on the gear wheel will touch each tooth on the pinion in regular succession with the frequency of contact between a particular pair of teeth being the speed of the pinion divided by the number of teeth on the gear wheel. The avoidance of an integer ratio between the number of teeth on the gear wheel and the pinion can be achieved by the addition of an extra tooth to the gear wheel, provided that there is no operational need for an exact velocity ratio. This extra tooth is called a hunting tooth and can have advantages for wear equalisation (see Tuplin (1962) for a fuller discussion).

Various definitions used for describing gear geometry are illustrated in Fig. 8.15 and listed in the following. For a pair of meshing gears, the smaller gear is called the 'pinion', the larger is called the 'gear wheel' or simply the 'gear'.

Pitch circle: This is a theoretical circle on which calculations are based. Its diameter is called the pitch diameter.

$$d = mN \tag{8.2}$$

where d is the pitch diameter (mm), m is the module (mm) and N is the number of teeth. Care must be taken to distinguish the module from the unit symbol for a metre.

Circular pitch: This is the distance from a point on one tooth to the corresponding point on the adjacent tooth measured along the pitch circle.

$$p = \pi m = \frac{\pi d}{N} \tag{8.3}$$

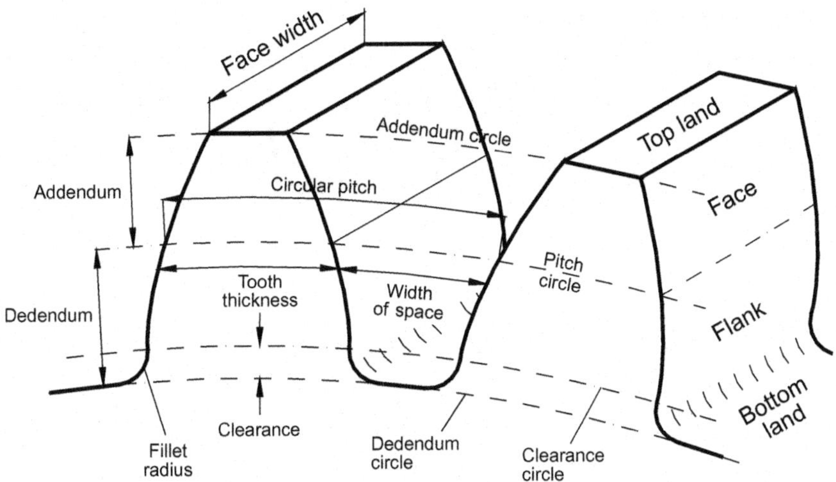

Fig. 8.15 Spur gear schematic showing principal terminology.

where p is the circular pitch (mm), m the module, d the pitch diameter (mm) and N the number if teeth.

Module: This is the ratio of the pitch diameter to the number of teeth. The unit of the module should be millimetres (mm). The module is defined by the ratio of pitch diameter and number of teeth. Typically, the height of a tooth is about 2.25 times the module. Various modules are illustrated in Fig. 8.16.

$$m = \frac{d}{N}. \tag{8.4}$$

Addendum, a: This is the radial distance from the pitch circle to the outside of the tooth.

Dedendum, b: This is the radial distance from the pitch circle to the bottom land.

Backlash: The amount by which the width of a tooth space exceeds the thickness of the engaging tooth measured on the pitch circle.

Prior to use of the metric module, the term diametral pitch was commonly used. The diametral pitch is the ratio of the number of teeth in the gear to the pitch diameter, $P_d = N/d$ (usually in US/English units only, i.e., teeth per inch (tpi)). To convert from diametral pitch P_d (tpi) to the module m (in mm), $m = 25.4/P_d$ can be used.

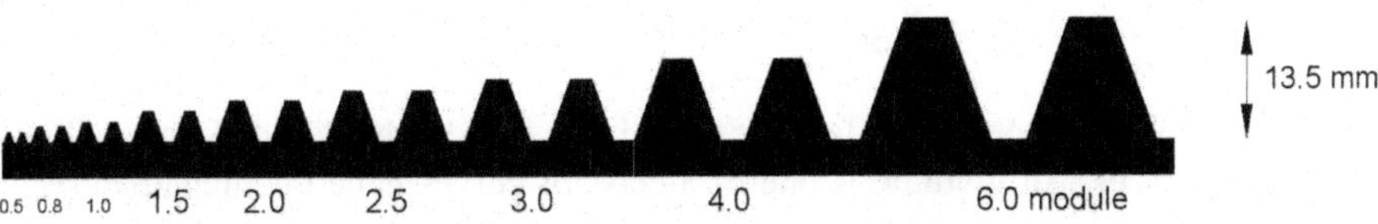

Fig. 8.16 Gear tooth size as a function of the module. If the height of the module 6 tooth from the bottom land to the top land is 13.5 mm, then this diagram has been printed to a 1:1 scale.

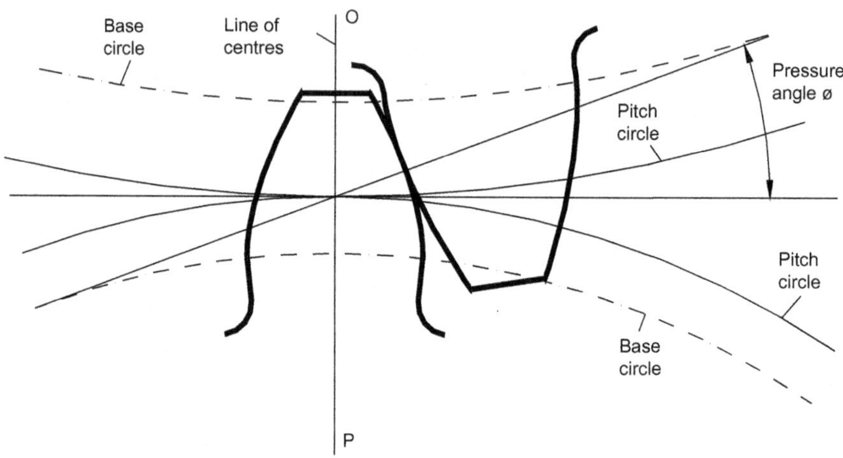

Fig. 8.17 Schematic showing the pressure line and pressure angle.

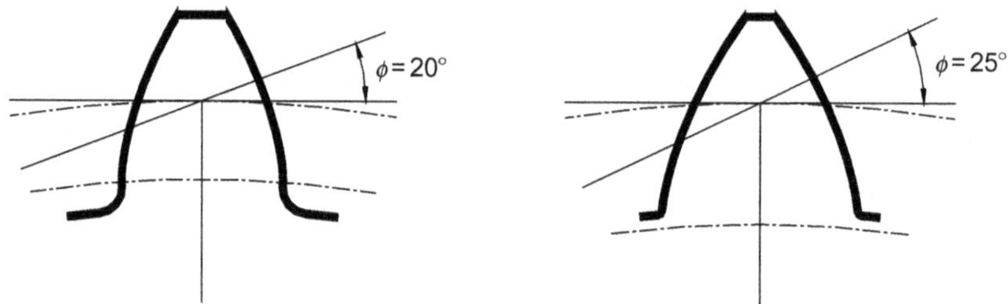

Fig. 8.18 Full depth involute form for varying pressure angles.

Fig. 8.17 shows the line of centres OP connecting the rotation axes of a pair of meshing gears. The angle ϕ is called the pressure angle. The pressure line (also called the generating line or line of action) is defined by the pressure angle. The resultant force vector between a pair of operating meshing gears acts along this line. The actual shape or form of the gear teeth depends on the pressure angle chosen. Fig. 8.18 shows two gear teeth of differing pressure angle for a gear with identical number of teeth, pitch and pitch diameter. Current standard pressure angles are 20 and 25 degree with the 20 degree form most widely available.

8.2 Construction of gear tooth profiles

The most widely used tooth form for spur gears is the full-depth involute form, as illustrated in Fig. 8.19. An involute is one of a class of curves called conjugate curves. When two involute form gear teeth are in mesh, there is a constant velocity ratio between them. From the moment of initial contact to the moment of disengagement,

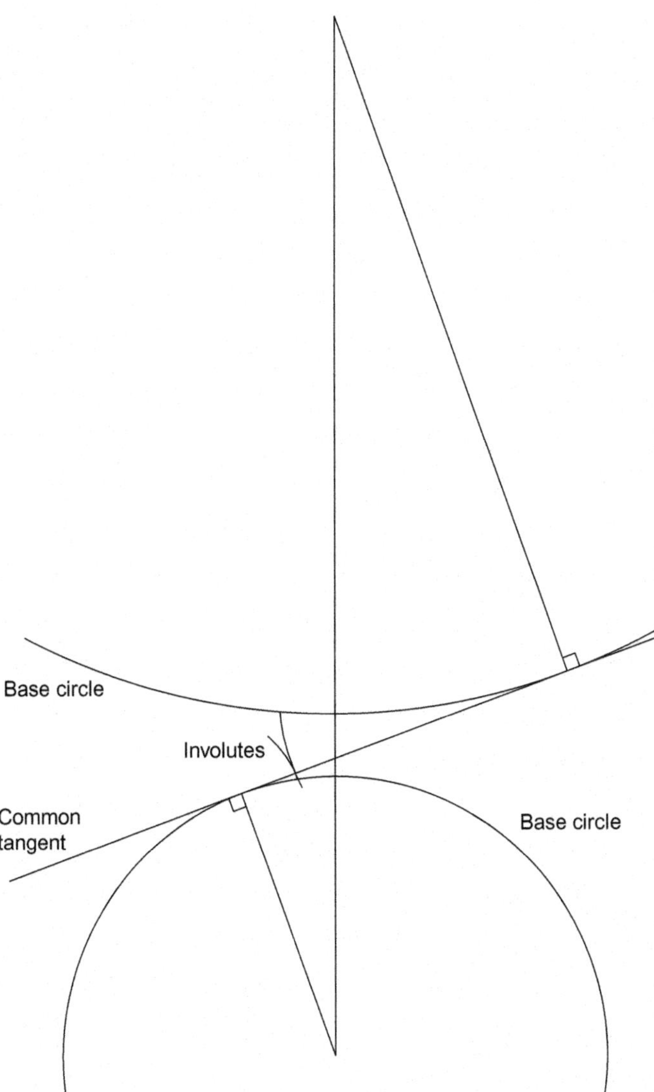

Fig. 8.19 Schematic of the involute form.

the speed of the gear is in constant proportion to the speed of the pinion. The resulting action of the gears is very smooth. If the velocity ratio were not constant, there would be accelerations and decelerations during the engagement and disengagement, causing vibration noise and potentially damaging torsional oscillations.

The layout and geometry for a pair of meshing spur gears can be determined by the procedure set out in the following. This procedure assumes access to a computer-aided design (CAD) drawing package. It should be noted that gears are commonly available as standard items from specialist manufacturers and suppliers and rarely need not to be designed from scratch.

(1) Calculate the pitch diameter and draw pitch circles tangential to each other (see Fig. 8.20).

$$d = mN,$$

$$d_p = mN_p \text{ and}$$

$$d_g = mN_g.$$

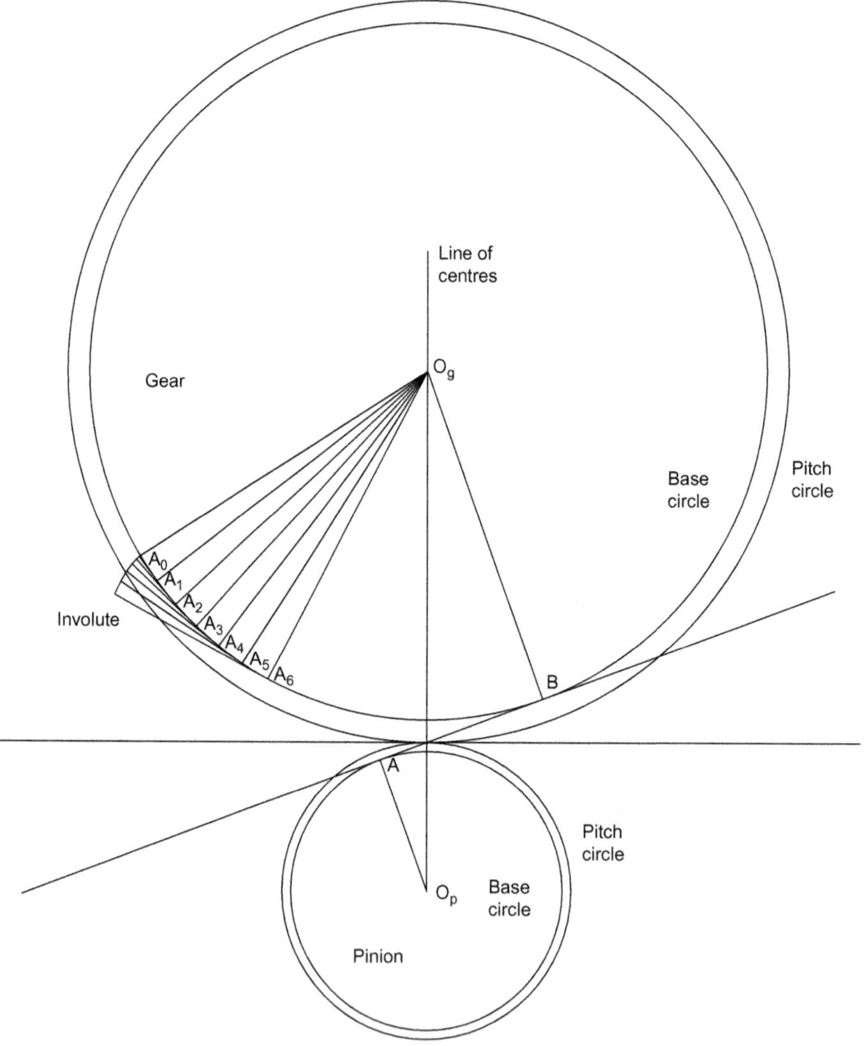

Fig. 8.20 Construction of gear geometry.

In the example shown, the module has been selected as $m=2.5$, the number of teeth in the pinion 20, and in the gear 50. So $d_p=2.5\times 20=50\,\text{mm}$ and $d_g=2.5\times\ 50\ =125\,\text{mm}$.

(2) Draw a line perpendicular to the line of centres through the pitch point (this is the point of tangency of the pitch circles). Draw the pressure line at an angle equal to the pressure angle from the perpendicular. It is called the pressure line because the resultant tooth force is along this line during meshing. Here the pressure angle is 20 degree.

(3) Construct perpendiculars O_pA and O_gB to the pressure line through the centres of each gear. The radial distance of each of these lines are the radii of the base circles of the pinion and gear, respectively. Draw the base circles.

(4) Draw an involute curve on each base circle. This is illustrated on the gear. First divide the base circle in equal parts $A_0, A_1, A_2, A_3, A_4, A_5, \ldots, A_n$. Construct radial lines O_gA_0, O_gA_1, $O_gA_2, O_gA_3, \ldots, O_gA_n$. Construct perpendiculars to these radial lines. The involute begins at A_0. The second point is obtained by measuring off the distance A_0A_1 on the perpendicular through A_1. The next point is found by measuring off twice the distance A_0A_1 on the perpendicular through A_2 and so on. The curve constructed through these points is the involute for the gear. The involute for the pinion is constructed in the same way on the base circle of the pinion.

(5) Calculate the circular pitch, $p=\pi m$.

The width of the teeth and the width of the spaces are equal to half the circular pitch. Mark these distances off on the pitch circles. (Here, $p=\pi\times 2.5$).

(6) Draw the addendum and dedendum circles for the pinion and gear (see Fig. 8.21). Here, a tooth system (see Section 8.4 and Table 8.3) has been selected with

$$a=m,$$

$$b=1.25m.$$

(7) Mirror the involute profile about a line constructed using a distance half the tooth width along the pitch circle and the gear centre. Using a polar array, generate all of the teeth for the gear. Construct the root fillets as appropriate. Construct the tooth top and bottom lands.

Gears can be purchased as standard items from specialist manufacturers and the exact geometry for the blade teeth is not necessary for engineering drawings and general design purposes (unless you are designer gear cutters/forming tools). However, the tooth geometry is necessary for detailed stress/reliability and dynamic analysis.

8.3 Gear trains

A gear train is one or more pairs of gears operating together to transmit power. Fig. 8.22 shows examples of a simple gear train, a compound gear train where more than one gear is located on a shaft to provide a compact design, and a reverted gear train where the input and output shafts are coaxial.

When two gears are in mesh, their pitch circles roll on each other without slippage. If r_1=pitch radius of gear 1, r_2=pitch radius of gear 2, ω_1=angular velocity of gear 1, ω_2=angular velocity of gear 2 then the pitch line velocity is given by

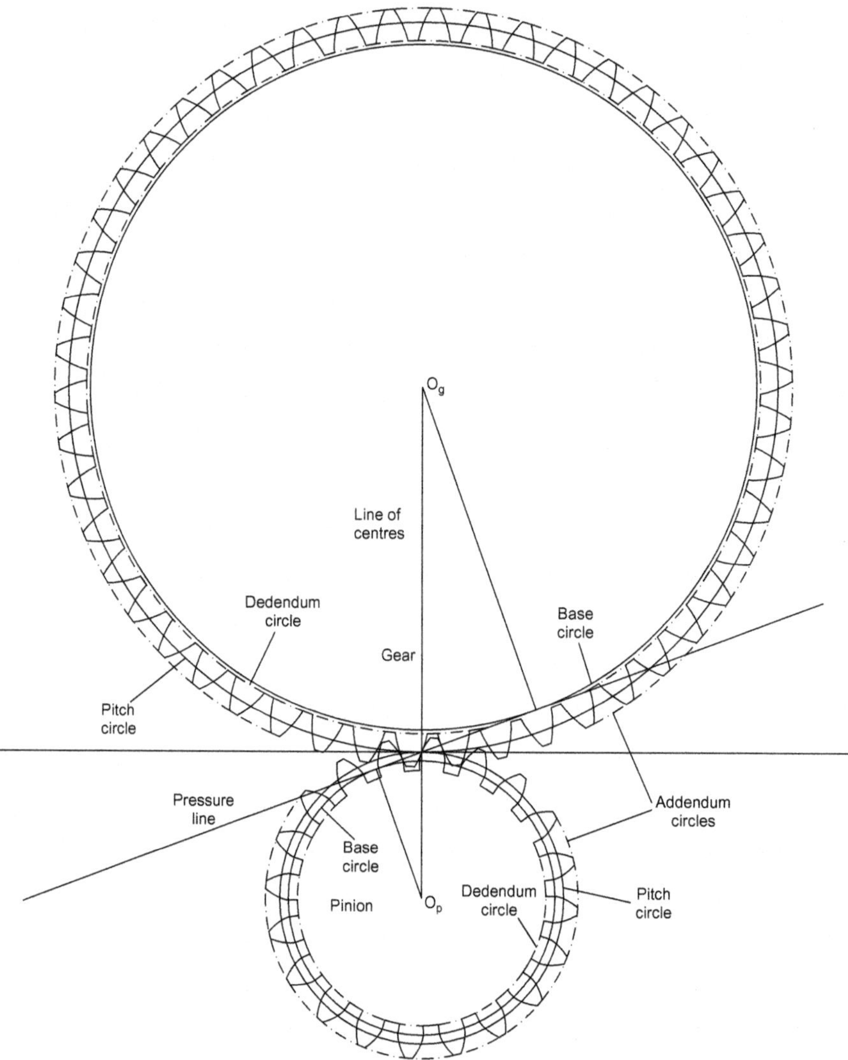

Fig. 8.21 Layout of a pair of meshing spur gears.

$$V = |r_1\omega_1| = |r_2\omega_2|. \tag{8.5}$$

The modulus mathematical symbol is applied here to remove information concerning the direction of rotation.

The velocity ratio is

$$\left|\frac{\omega_1}{\omega_2}\right| = \frac{r_2}{r_1}, \tag{8.6}$$

Table 8.3 Tooth dimension formulas for $\phi=20$ and 25 degrees

Addendum	$a=m$
Dedendum	$b=1.25m$
Working depth	$h_k=2m$
Whole depth (min)	$h_t=2.25m$
Tooth thickness	$t=\pi m/2$
Fillet radius of basic rack	$r_f=0.3m$
Clearance (min)	$c=0.25m$
Clearance for shaved or ground teeth	$c=0.35m$
Minimum number of pinion teeth $\phi=20$ degree	$N_P=18$
Minimum number of pinion teeth $\phi=25$ degree	$N_P=12$
Minimum number of teeth per pair $\phi=20$ degree	$N_P+N_G=36$
Minimum number of teeth per pair $\phi=25$ degree	$N_P+N_G=24$
Width of top land (min)	$t_o=0.25m$

Note: In this table, 'm' stands for module.

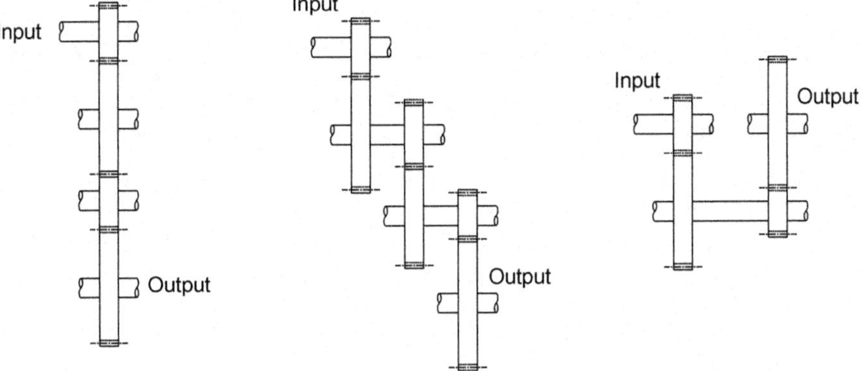

Fig. 8.22 Some example gear trains: (A) Simple gear train, (B) compound gear train, (C) reverted compound gear train.

and can be defined in any of the following ways:

$$\frac{\omega_P}{\omega_G}=\frac{n_P}{n_G}=-\frac{N_G}{N_P}=-\frac{d_G}{d_P}, \tag{8.7}$$

where

ω_P and ω_G are the angular velocities of the pinion and gear, respectively (rad/s);
n_P and n_G are the rotational speeds of the pinion and gear, respectively (rpm);
N_P and N_G are the number of teeth in the pinion and gear, respectively; and.
d_P and d_G are the pitch diameter of the pinion and gear, respectively (mm).

Consider a pinion 1, driving a gear 2. The speed of the driven gear is as follows:

$$n_2 = \left|\frac{N_1}{N_2}n_1\right| = \left|\frac{d_1}{d_2}n_1\right|, \tag{8.8}$$

where n = revolutions or rpm, N = number of teeth, d = pitch diameter.

Eq. (8.8) applies to any gear set (spur, helical, bevel or worm). For spur and parallel helical gears, the convention for direction is positive for anticlockwise rotation, i.e.,

$$n_2 = -\frac{N_1}{N_2}n_1. \tag{8.9}$$

Example 8.1. For the gear pair illustrated in Fig. 8.23, determine the output speed if the pinion has 22 teeth and is being driven at 2650 rpm clockwise. The gear wheel has 72 teeth.

Solution

$$n_2 = -\frac{N_1}{N_2}n_1,$$

$$n_2 = -\frac{22}{72} \times -2650 = 809.7 \text{ rpm}.$$

For this configuration and input conditions, the output shaft will rotate at 809.7 rpm anticlockwise.

Example 8.2. Consider the gear train shown in Fig. 8.24. Calculate the speed of gear five.

Solution

$$n_2 = -\frac{N_1}{N_2}n_1,$$

$$n_3 = -\frac{N_2}{N_3}n_2,$$

$n_4 = n_3$ (on the same shaft)

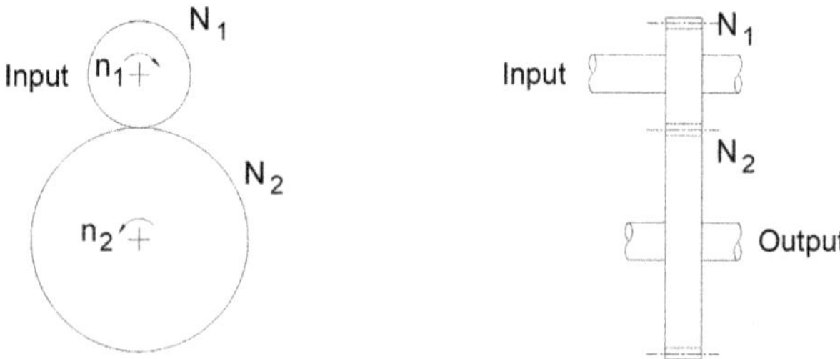

Fig. 8.23 Pinion and gear.

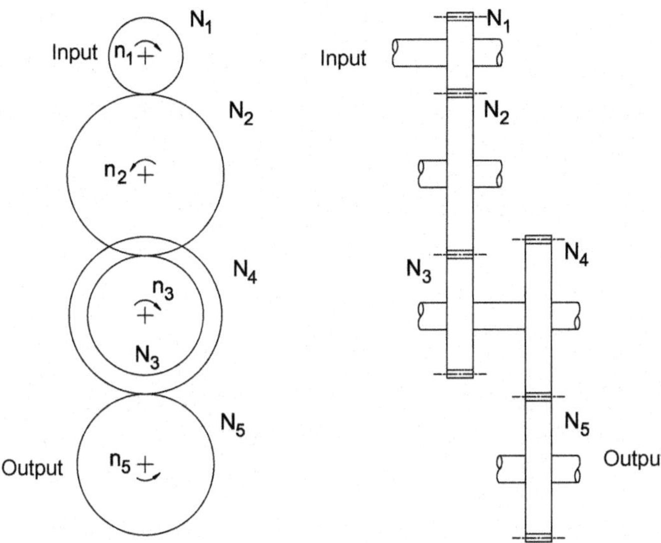

Fig. 8.24 Example gear train.

$$n_5 = -\frac{N_4}{N_5} n_4,$$

$$n_5 = -\frac{N_4 N_2 N_1}{N_5 N_3 N_2} n_1.$$

Example 8.3. For the double reduction, gear train shown in Fig. 8.25, if the input speed is 1750 rpm in a clockwise direction what is the output speed?

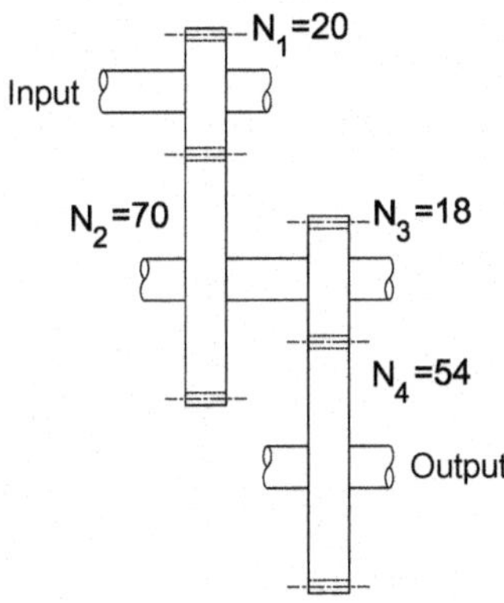

Fig. 8.25 Double reduction gear train.

Solution

$$n_2 = -\frac{N_1}{N_2} n_1,$$

$$n_3 = n_2 \text{ (on the same shaft)}$$

$$n_4 = -\frac{N_3}{N_4} n_3,$$

$$n_4 = \frac{N_3 N_1}{N_4 N_2} n_1 = \left(\frac{18}{54}\right)\left(\frac{20}{70}\right)(-1750) = -166.7\,\text{rpm}.$$

Example 8.4. For the double reduction gear train with an idler shown in Fig. 8.26, if the input speed is 1750 rpm in a clockwise direction what is the output speed?

Solution

$$n_5 = -\frac{N_4}{N_5} n_4,$$

$$n_4 = -\frac{N_3}{N_4} n_3,$$

$$n_3 = n_2,$$

$$n_2 = -\frac{N_1}{N_2} n_1,$$

$$n_5 = -\frac{N_4 N_3 N_1}{N_5 N_4 N_2} n_1 = -\left(\frac{22}{54}\right)\left(\frac{18}{22}\right)\left(\frac{20}{70}\right)(-1750) = 166.7\,\text{rpm}.$$

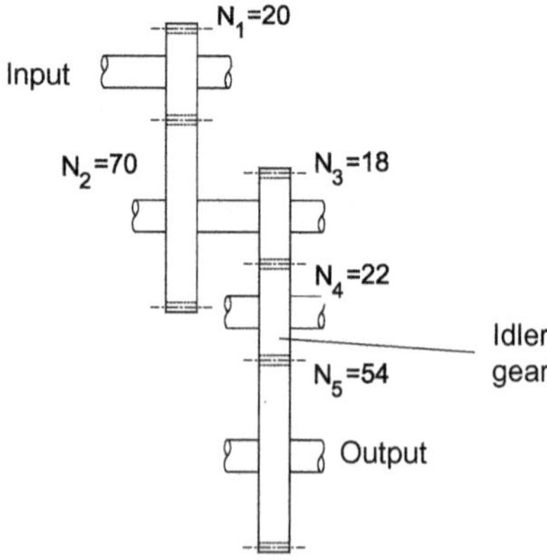

Fig. 8.26 Double reduction gear with idler.

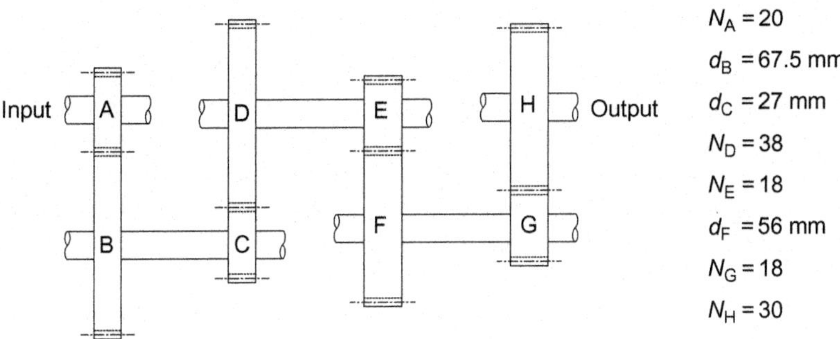

Fig. 8.27 Gear train.

Notice that the presence of the idler gear has caused the gear train output to reverse direction but has not altered the gear ratio in comparison to the previous example.

Example 8.5. For the gear train illustrated in Fig. 8.27, determine the output speed and direction of rotation, if the input shaft rotates at 1490 rpm clockwise. Gears A–D have a module of 1.5 and gears E–H a module of 2.

Solution

$d = mN$, $N = d/m$, $N_B = 67.5/1.5 = 45$, $N_C = 27/1.5 = 18$, $N_F = 56/2 = 28$.

$$n_H = \frac{N_G}{N_H} n_G,$$

$$n_G = n_F,$$

$$n_F = \frac{N_E}{N_F} n_E,$$

$$n_E = n_D,$$

$$n_D = \frac{N_C}{N_D} n_C.$$

$$n_C = n_B,$$

$$n_B = \frac{N_A}{N_B} n_A,$$

$$n_H = \frac{N_G}{N_H} \times \frac{N_E}{N_F} \times \frac{N_C}{N_D} \times \frac{N_A}{N_B} n_A = \frac{18}{30} \times \frac{18}{28} \times \frac{18}{38} \times \frac{20}{45} \times -1490 = -121 \text{ rpm}$$

Gear	A	B	C	D	E	F	G	H
Direction	CW	ACW	ACW	CW	CW	ACW	ACW	CW

CW = clockwise, ACW = anticlockwise.

8.3.1 Manually shifted automotive transmissions

The torque-speed characteristic of a petrol ICE delivers low torque at starting and at low speeds. This does not match the load, which, to enable a car to pull away from stationary, demands a high torque. The torque from an ICE can be increased by means of gearing. Similarly, at other load conditions such as high acceleration, hills and cruising the torque-speed characteristics of the engine may not be compatible with the load. A multiratio gearbox is the traditional solution to this problem (see, for example, Heisler (1999), Stokes (1992) and Naunheimer et al. (2010)).

A typical manually shifted gearbox for a passenger car is illustrated in Fig. 8.28. This has five forward speeds and reverse. The basic elements of a manually shifted transmission are a single or multiplate clutch for engaging and disengaging the power from the load, a variable ration transmission unit with permanent mesh gears and a gear shift mechanism and lever. Clutches are considered more fully in Chapter 13 and their function here is to interrupt the power flow. The demands of the automobile product mean that harshness in gear changes is unacceptable to the majority of customers. A method of overcoming shocks from gears crashing into each other as they are engaged is to hold the gears in permanent mesh but not necessarily transmitting any power. This is illustrated for gears 2, 3 and 4 which although in mesh in Fig. 8.28A do not transmit torque because one of each of the pairs of gears concerned is free to rotate relative to its shaft. In order to transmit torque, a spline is moved on the shaft causing the gear to rotate at the same speed as the shaft and therefore it is able to transmit power. This is the principle of the synchromesh illustrated in Fig. 8.29. Depending on the gear train desired, the gear lever shift mechanism moves the synchromesh splines to engage the appropriate gears. (See Heisler (1999) and SAE (2012) for automotive transmissions.)

8.3.2 Epicyclic gear trains

Fig. 8.30 illustrates an example of an epicyclic gear train. Epicyclic gear trains include a sun gear, a planet carrier and one or more planetary gears. The planetary gears may also mesh with an internal ring gear. The use of a planet gear between the sun and ring gear increases the number of teeth and hence the torque and a significant gear ratio can be obtained within a compact unit, in comparison to standard parallel axis gear trains. It should be noted that the number of planets used does not alter the gear ratio but does serve to share the load if required in order to deliver acceptable stress levels. Epicyclic gear trains can be assembled with spur, helical or bevel gearing.

Epicyclic gearboxes are available as complete units from specialist manufactures or integrated with a gear motor. Depending on the application, it may be necessary to design an epicyclic gearbox from scratch, sourcing the individual components from stock suppliers. For simple epicyclic gear trains, gear ratios of between 2:1 and 12:1 are readily achievable. Two or more gear trains can be arranged in series in order to provide much higher overall gear ratios, typically ranging from 5:1 to 120:1, with efficiencies varying widely from 25% to 97% (e.g., see Pennestri and Freudenstein (1993)).

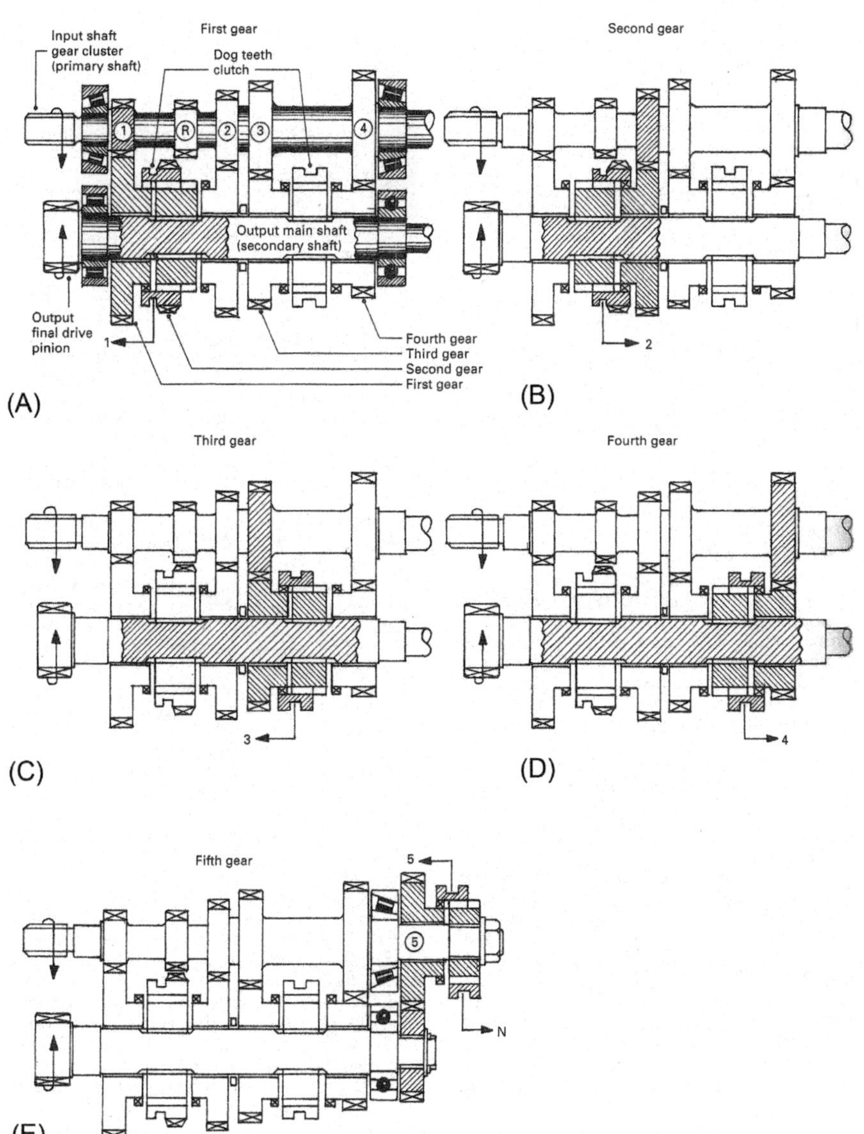

Fig. 8.28 A five speed and reverse single-stage constant mesh transmission.
Figure reproduced from Heisler, H., 1999. Vehicle and Engine Technology, second ed. Butterworth Heinemann.

The range of possible configurations is extensive and basic possible configurations as considered by Levai (1969) are illustrated in Fig. 8.31, although it should be noted that this represents an innovation space and new configurations are emerging periodically. The basic configurations can be connected to produce a larger train having more degrees of freedom.

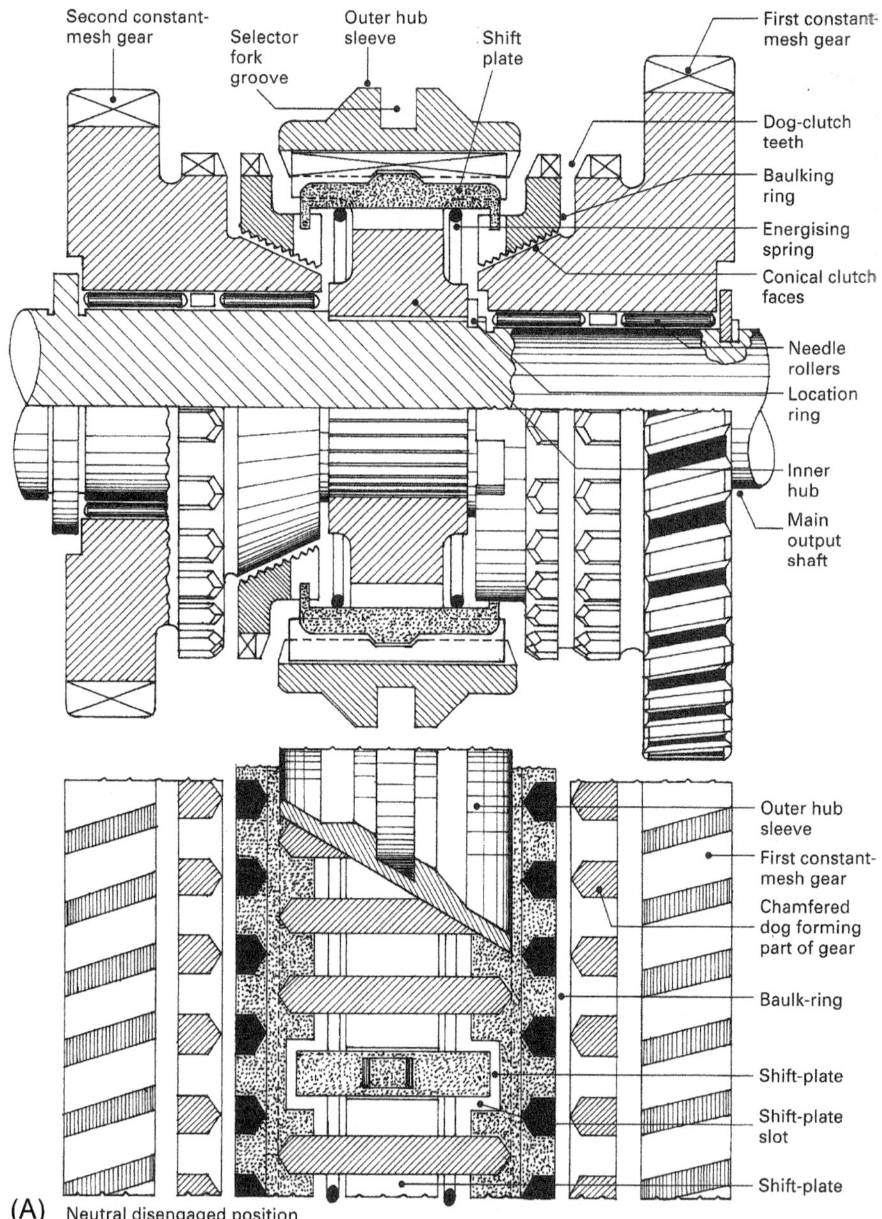

Fig. 8.29 Positive baulk-ring synchromesh.
Figure reproduced from Heisler, H., 1999. Vehicle and Engine Technology, second ed. Butterworth Heinemann.

(Continued)

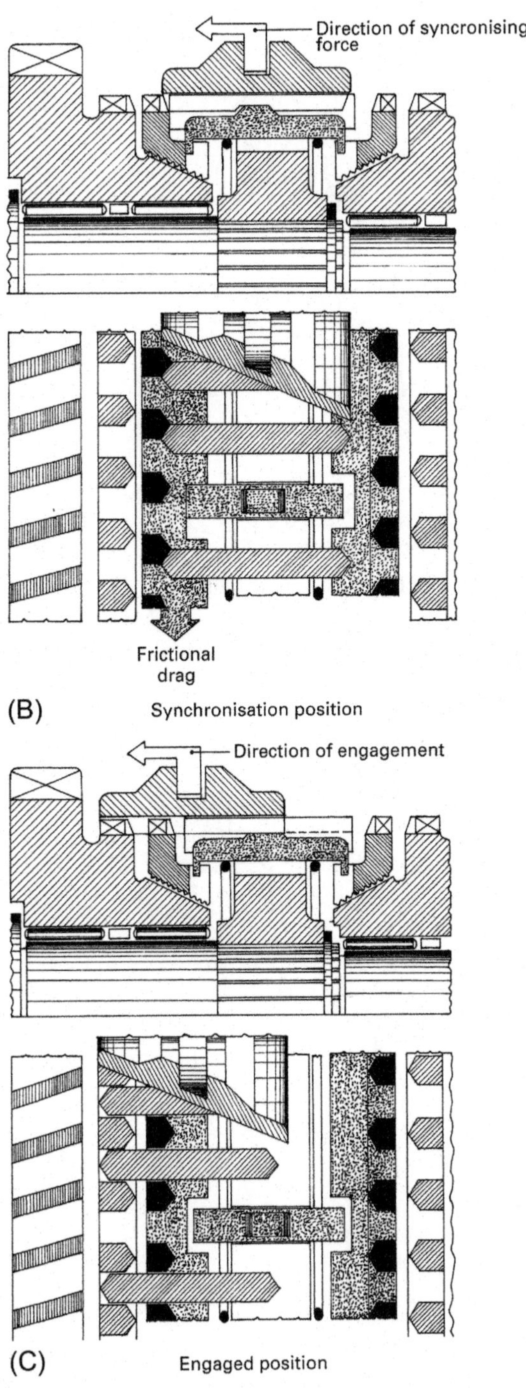

Fig. 8.29, Cont'd

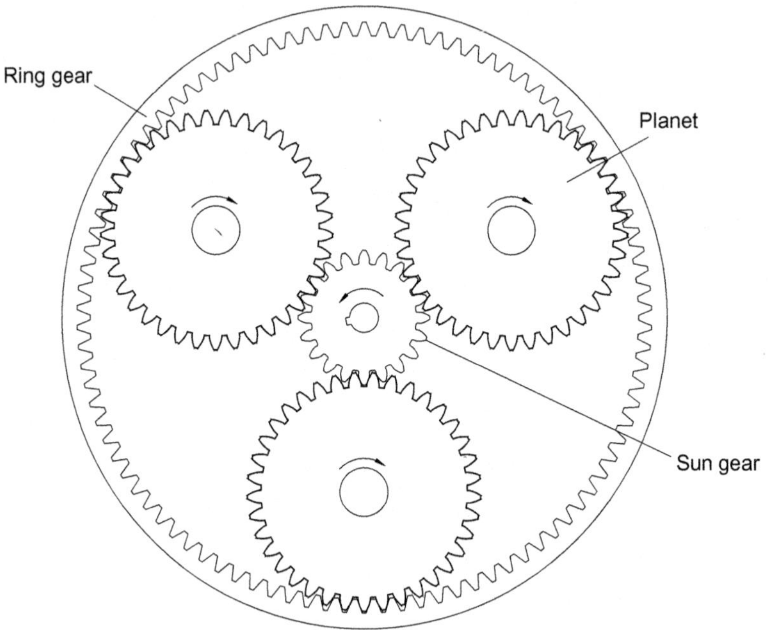

Fig. 8.30 An example of a simple epicyclic gear train.

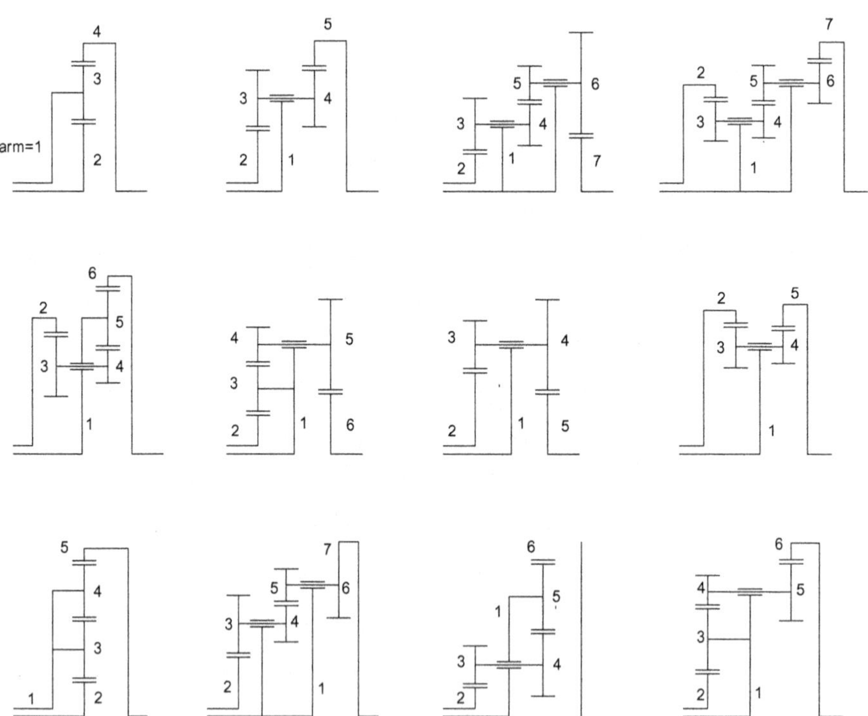

Fig. 8.31 Types of epicyclic gear train as categorised by Levai (1969).

Epicyclic gear trains have two degrees of freedom and in order to determine their motion two inputs must be defined such as the speed of both the sun and the ring gear (see for example, Jelaska (2012) and Radzevich (2016)). For the example shown in Fig. 8.13, the rotational speed of the sun gear is given by

$$n_{\text{sun}} = n_{\text{arm}} + n_{\text{sun/arm}}. \tag{8.10}$$

If n_F represents the speed of the first gear in the train and n_L the speed of the last gear in the train,

$$n_{F/\text{arm}} = n_F - n_{\text{arm}}, \tag{8.11}$$

$$n_{L/\text{arm}} = n_L - n_{\text{arm}}. \tag{8.12}$$

Dividing Eq. (8.12) by (8.11) gives

$$\frac{n_{L/\text{arm}}}{n_{F/\text{arm}}} = \frac{n_L - n_{\text{arm}}}{n_F - n_{\text{arm}}} = e = \pm \frac{\text{Product of driving teeth numbers}}{\text{Product of driven teeth numbers}} = \frac{N_S N_P}{N_P N_R}, \tag{8.13}$$

where e is the train value. For spur and helical gears, the train value is positive if the last gear rotates in the same direction as the first.

Eq. (8.13) can be rearranged to give the reduction ratio defined as $n_{\text{sun}}/n_{\text{arm}}$ which for an epicyclic gearbox, with the ring fixed stationary, $n_L = 0$, $n_F = n_{\text{sun}}$, is given by

$$\frac{n_{\text{sun}}}{n_{\text{arm}}} = \frac{N_R}{N_S} + 1. \tag{8.14}$$

For calculations of stress, it may also be necessary to determine the speed of the planetary gears. Velocity difference equations can be used for this. For the gear train illustrated in Fig. 8.13,

$$n_{\text{planet/arm}} = n_{\text{planet}} - n_{\text{arm}}, \tag{8.15}$$

$$n_{\text{sun/arm}} = n_{\text{sun}} - n_{\text{arm}}. \tag{8.16}$$

From knowledge of n_{arm}, N_P and N_S, these equations can be solved to determine n_P.

Note that if equi-spacing of the planetary gears is desired, the ratio given in Eq. (8.17) must be an integer. Equi-spacing of planets is desirable to produce balanced forces on the pinion and ring gear.

$$\frac{N_S + N_R}{\text{Number of planets}} = \text{Integer}. \tag{8.17}$$

For a star arrangement, Fig. 8.32, where the sun is the driver, the ring is the driven element and the planetary carrier is fixed, then the train value is given by

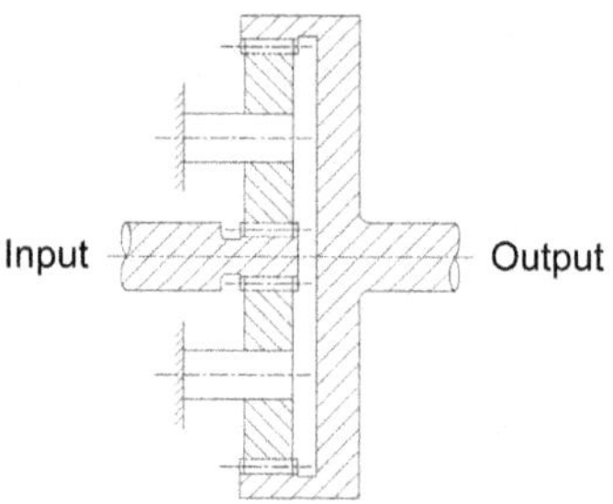

Fig. 8.32 Star arrangement.

$$e = \left(\frac{N_S}{N_P}\right)\left(-\frac{N_P}{N_R}\right) = -\frac{N_S}{N_R}, \tag{8.18}$$

$n_{arm} = 0,$

$$e = \frac{n_A - n_{arm}}{n_S - n_{arm}} = \frac{n_A}{n_S}. \tag{8.19}$$

Hence,

$$\frac{n_A}{n_S} = -\frac{N_S}{N_R}. \tag{8.20}$$

In the solar arrangement, Fig. 8.33, the sun gear is fixed and the annulus gear, and planet carrier rotate.

$$e = \left(-\frac{N_S}{N_P}\right)\left(\frac{N_P}{N_R}\right) = -\frac{N_S}{N_R}, \tag{8.21}$$

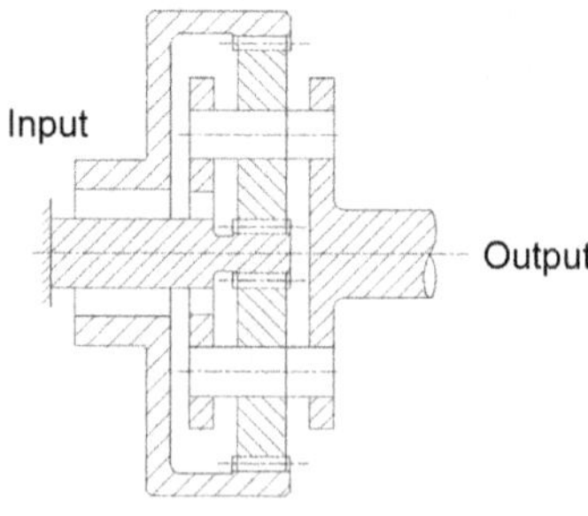

Fig. 8.33 Solar arrangement.

$n_S = 0,$

$$e = \frac{n_R - n_{arm}}{n_S - n_{arm}} = \frac{n_R - n_{arm}}{-n_{arm}}, \tag{8.22}$$

$$\frac{n_R}{n_{arm}} = 1 + \frac{N_S}{N_R}. \tag{8.23}$$

Example 8.6. For the epicyclic gearbox illustrated in Fig. 8.34, determine the speed and direction of the final drive and also the speed and direction of the planetary gears. The teeth numbers of the sun, planets and ring gear are 20, 30 and 80, respectively. The speed and direction of the sun gear is 1000 rpm clockwise and the ring gear is held stationary.

Solution

From Eq. (8.14),

$$n_{arm} = \frac{n_{sun}}{(80/20) + 1} = \frac{-1000}{5} = -200 \text{ rpm}.$$

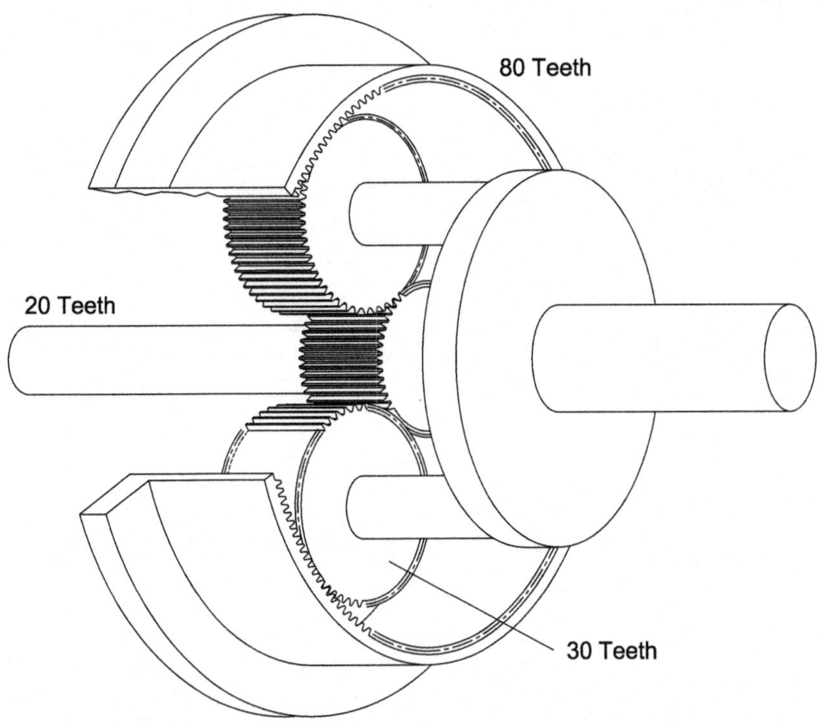

Fig. 8.34 Epicyclic gear train.

The speed of the final drive is 200 rpm clockwise.

The reduction ratio for the gearbox is given by $n_{sun}/n_{arm} = 1000/200 = 5$.

To determine the speed of the planets, use Eqs. (8.15) and (8.16).

The planets and sun are in mesh, so

$$\frac{n_{planet}/n_{arm}}{n_{sun}/n_{arm}} = -\frac{N_S}{N_P},$$

$$\frac{n_{planet} - n_{arm}}{n_{sun} - n_{arm}} = -\frac{N_S}{N_P},$$

$$\frac{n_{planet} - (-200)}{-1000 - (-200)} = -\frac{20}{30},$$

$$n_{planet} = -\frac{20}{30} \times (-800) - 200 = 333 \text{ rpm}.$$

The speed of rotation of the planetary gears is 333 rpm counterclockwise.

Example 8.7. An epicyclic gearbox is required to transmit 12 kW from a shaft rotating at 3000 rpm. The desired output speed is approximately 500 rpm. Using a module 3 hardened steel gear, determine the number of teeth in each gear for a single-stage reduction gearbox.

Solution

Ratio sun/arm = 3000/500 = 6

$$\frac{n_{sun}}{n_{arm}} = \frac{3000}{500} = 6,$$

$$\frac{n_L - n_{arm}}{n_F - n_{arm}} = -\frac{N_S N_P}{N_P N_R}.$$

Set $N_s = 18$ (minimum normally recommended).

$$\frac{0 - 500}{3000 - 500} = -\frac{18}{N_R},$$

$N_R = 90$.

With $m = 3$ does this fit in?

$d_R = mN = 3 \times 90 = 270$ mm,

$d_S = mN = 3 \times 18 = 54$ mm,

This leaves 270/2–54/2 = 108 mm for the planetary gears' diameter.

With $m = 3$, $N_P = 36$, i.e., $d = mN = 3 \times 36 = 108$ mm. This fits. Design OK so far.

Does this combination give even spacing using say three planets?

$$\frac{N_S + N_R}{\text{Number of planets}} = \frac{18 + 90}{3} = \text{Integer} = 3.$$

Integer result: even spacing. Design OK.

A check on stresses although not asked for here would confirm the validity of the design using a module of 3.

8.4 Tooth systems

Tooth systems are standards that define the geometric proportions of gear teeth. Table 8.3 lists the basic tooth dimensions for full-depth teeth with pressure angles of 20 and 25 degrees.

Table 8.4 lists preferred values for the module, m, which are used to minimise gear cutting tool requirements, and Table 8.5 lists the preferred standard gear teeth numbers.

The failure of gears can principally be attributed to tooth breakage, surface failure and scuffing and scoring of the teeth surfaces, and failure of lubrication (Ku, 1976; Höhn and Michaelis, 2004). Section 8.5 outlines the forces involved on spur gear teeth and Section 8.6 defines methods for estimating gear stresses and the preliminary selection of standard stock gears.

8.5 Force analysis

Fig. 8.35 shows the forces involved for two spur gears in mesh. The force acting at the pressure angle ϕ can be subdivided into two components: The tangential component F_t and the radial component F_r. The radial component serves no useful purpose. The tangential component F_t transmits the load from one gear to the other. If W_t is defined as the transmitted load, $W_t = F_t$. The transmitted load is related to the power transmitted through the gears by the equation:

$$W_t = \frac{P}{V}, \tag{8.24}$$

where

W_t = transmitted load (N),
P = power (W), and
V = pitch line velocity (m/s).

Alternatively, the pitch line velocity can be defined by

$$W_t = \frac{60 \times 10^3 H}{\pi d n}, \tag{8.25}$$

where

Table 8.4 Preferred values for the module, m

m	0.5	0.8	1	1.25	1.5	2	2.5	3	4	5	6	8	10	12	16	20	25	32	40	50

Table 8.5 Preferred standard gear teeth numbers

12	**13**	**14**	**15**	**16**	**18**	**20**	**22**	**24**	**25**	**28**	**30**
32	34	38	40	45	50	54	60	64	70	72	75
80	84	90	96	100	120	140	150	180	200	220	250

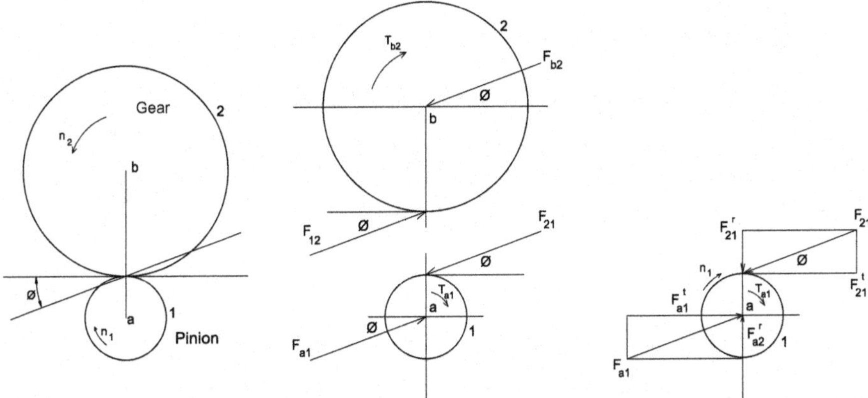

Fig. 8.35 Forces for two spur gears in mesh.

W_t = transmitted load (kN),
H = power (kW),
d = pitch diameter (mm), and
n = speed (rpm).

8.5.1 Introduction to gear stresses

Gears experience two principal types of stress, bending stress at the root of the teeth due to the transmitted load and contact stresses on the flank of the teeth due to repeated impact, or sustained contact, of one tooth surface against another. A simple method of calculating bending stresses is presented in Section 8.5.2 and this is utilised within a gear selection procedure given in Section 8.6. However, this methodology, whilst suitable for determining the geometry for a gear train at the conceptual phase is inadequate for designing a gear train for highly stressed applications or where life of the components is important. For the determination of life, critical parts national and international standards for gears are available such as the AGMA (American Gear Manufacturers Association), BS (British Standards) and ISO (International Organization for Standardization) standards for gears. These are considered in Chapter 9, for spur and helical gears, and the AGMA equations recommended for bending and contact stress for spur gears are introduced.

8.5.2 Bending stresses

The calculation of bending stress in gear teeth can be based on the Lewis formula

$$\sigma = \frac{W_t}{FmY}, \tag{8.26}$$

where

W_t = transmitted load (N),
F = face width (m or mm),
m = module (m or mm) and
Y = the Lewis form factor and can be found from Table 8.6.

When teeth mesh, the load is delivered to the teeth with some degree of impact. The velocity factor is used to account for this and is given, in the case of cut or milled profile gears, by the Barth equation (8.27)

$$K_v = \frac{6.1}{6.1 + V}, \tag{8.27}$$

Table 8.6 Values for the Lewis form factor *Y* defined for two different tooth standards (Mitchener and Mabie, 1982)

N Number of teeth	*Y* $\phi = 20$ degree $a = 0.8m$ $b = m$	*Y* $\phi = 20$ degree $a = m$ $b = 1.25m$
12	0.33512	0.22960
13	0.34827	0.24317
14	0.35985	0.25530
15	0.37013	0.26622
16	0.37931	0.27610
17	0.38757	0.28508
18	0.39502	0.29327
19	0.40179	0.30078
20	0.40797	0.30769
21	0.41363	0.31406
22	0.41883	0.31997
24	0.42806	0.33056
26	0.43601	0.33979
28	0.44294	0.34790
30	0.44902	0.35510
34	0.45920	0.36731
38	0.46740	0.37727
45	0.47846	0.39093
50	0.48458	0.39860
60	0.49391	0.41047
75	0.50345	0.42283
100	0.51321	0.43574
150	0.52321	0.44930
300	0.53348	0.46364
Rack	0.54406	0.47897

Note: a = addendum, b = dedendum, ϕ = pressure angle, m = module.

where V is the pitch line velocity which is given by

$$V = \frac{d}{2} \times 10^{-3} n \frac{2\pi}{60}, \tag{8.28}$$

where d is in mm and n in rpm.

Introducing the velocity factor into the Lewis equation gives the following:

$$\sigma = \frac{W_t}{K_v F m Y}. \tag{8.29}$$

Eq. (8.29) forms the basis of a simple approach to the calculation of bending stresses in gears.

Example 8.8. A 20 degree full-depth spur pinion is to transmit 1.25 kW at 850 rpm. The pinion has 18 teeth. Determine the Lewis bending stress if the module is 2 and the face width is 25 mm.

Solution

Calculating the pinion pitch diameter: $d_P = mN_P = 2 \times 18 = 36\,\text{mm}$.

Calculating the pitch line velocity:

$$V = \frac{d_P}{2} \times 10^{-3} \times n \frac{2\pi}{60} = \frac{0.036}{2} \times 850 \times 0.1047 = 1.602\,\text{m/s}.$$

Calculating the velocity factor,

$$K_v = \frac{6.1}{6.1 + V} = \frac{6.1}{6.1 + 1.602} = 0.7920.$$

Calculating the transmitted load,

$$W_t = \frac{\text{Power}}{V} = \frac{1250}{1.602} = 780.2\,\text{N}.$$

From Table 8.6 for $N_P = 18$, the Lewis form factor $Y = 0.29327$.

The Lewis equation for bending stress gives the following:

$$\sigma = \frac{W_t}{K_v F m Y} = \frac{780.2}{0.792 \times 0.025 \times 0.002 \times 0.29327} =$$

$$67.18 \times 10^6\ \text{N/m}^2 = 67.18\ \text{MPa}.$$

8.6 Simple gear selection procedure

The Lewis formula in the form given by Eq. (8.29) ($\sigma = W_t/(K_v F m Y)$) can be used in a provisional spur gear selection procedure for a given transmission power, input and output speeds. The procedure is outlined in the following list and illustrated in Table 8.7.

(1) Select the number of teeth for the pinion and the gear to give the required gear ratio (observe the guidelines presented in Table 8.2 for maximum gear ratios). Note that the minimum number of teeth permissible when using a pressure angle of 20 degree is 18 (Table 8.3). Use either the standard teeth numbers as listed in Table 8.5 or as listed in a stock gear catalogue.

(2) Select a material. This will be limited to those listed in the stock gear catalogues.

(3) Select a module, m from Table 8.4 or as listed in a stock gear catalogue (see Tables 8.8–8.11 which give examples of a selection of stock gears available.

(4) Calculate the pitch diameter, $d = mN$.

(5) Calculate the pitch line velocity, $V = (d/2) \times n \times (2\pi/60)$. Ensure this does not exceed the guidelines given in Table 8.2.

(6) Calculate the dynamic factor, $K_v = 6.1/(6.1 + V)$.

(7) Calculate the transmitted load, $W_t = \text{Power}/V$.

(8) Calculate an acceptable face width using the Lewis formula in the form,

$$F = \frac{W_t}{K_v m Y \sigma_p}. \tag{8.30}$$

The Lewis form factor, Y, can be obtained from Table 8.6.

The permissible bending stress, σ_p, can be taken as σ_{uts}/factor of safety, where the factor of safety is set by experience but may range from 2 to 5. Alternatively use values of σ_p as listed for the appropriate material in a stock gear catalogue. Certain plastics are suitable for use as gear materials in application where low-weight, low-friction, high corrosion resistance, low wear and quiet operation are beneficial. The strength of plastic is usually significantly lower than that of metals. Plastics are often formed using a filler to improve strength, wear, impact resistance, temperature performance as well as other properties and it is therefore difficult to regulate or standardise properties for plastics and these need to be obtained instead from the manufacturer or a traceable testing laboratory. Plastics used for gears include acrylonitrile-butadiene-styrene (ABS), acetal, nylon, polycarbonate, polyester, polyurethane and styrene-acrylonitrile (SAN). Values of permissible bending stress for a few gear materials are listed in Table 8.12.

The design procedure consists of proposing teeth numbers for the gear and pinion, selecting a suitable material, selecting a module, calculating the various parameters as listed resulting in a value for the face width. If the face width is greater than those available in the stock gear catalogue or if the pitch line velocity is too high, repeat the process for a different module. If this does not provide a sensible solution try a different material, etc. The process can be optimised taking cost and other

Table 8.7 Spur gear outline selection procedure based on bending

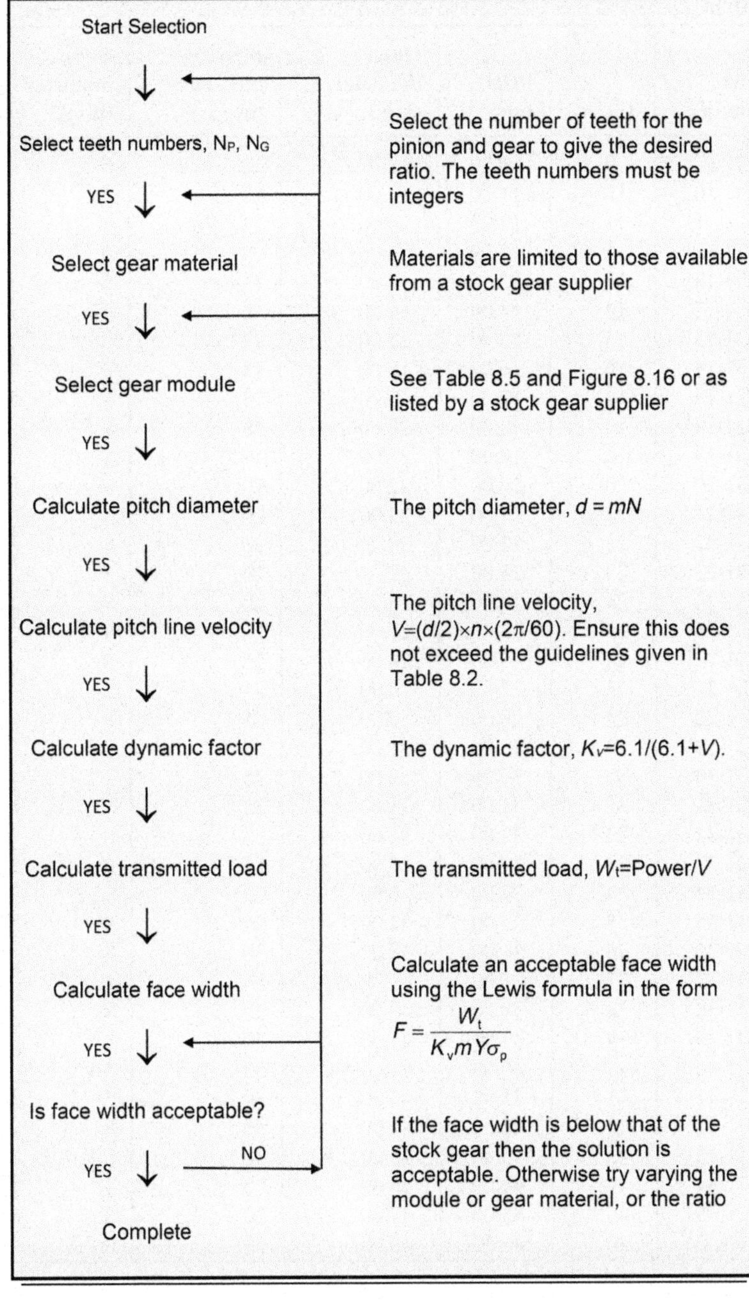
Start Selection
Select teeth numbers, N_P, N_G
Select the number of teeth for the pinion and gear to give the desired ratio. The teeth numbers must be integers
YES
Select gear material
Materials are limited to those available from a stock gear supplier
YES
Select gear module
See Table 8.5 and Figure 8.16 or as listed by a stock gear supplier
YES
Calculate pitch diameter
The pitch diameter, $d = mN$
YES
Calculate pitch line velocity
The pitch line velocity, $V=(d/2)\times n\times(2\pi/60)$. Ensure this does not exceed the guidelines given in Table 8.2.
YES
Calculate dynamic factor
The dynamic factor, $K_v=6.1/(6.1+V)$.
YES
Calculate transmitted load
The transmitted load, W_t=Power/V
YES
Calculate face width
Calculate an acceptable face width using the Lewis formula in the form
$F = \frac{W_t}{K_v m Y \sigma_p}$
YES
Is face width acceptable?
If the face width is below that of the stock gear then the solution is acceptable. Otherwise try varying the module or gear material, or the ratio
YES
NO
Complete

Table 8.8 Spur gears; 1.0 module; heavy duty steel 817M40, 655M13; face width 15 mm

Part number	Teeth	PCD (mm)	Outer diameter (mm)	Boss diameter (mm)	Bore diameter (mm)
SG1-9	9	10.00	12.00	12	6
SG1-10	10	11.00	13.00	13	6
SG1-11	11	12.00	14.00	14	6
SG1-12	12	12.00	14.00	14	6
SG1-13	13	13.00	15.00	15	6
SG1-14	14	14.00	16.00	16	6
SG1-15	15	15.00	17.00	17	6
SG1-16	16	16.00	18.00	18	8
SG1-17	17	17.00	19.00	18	8
SG1-18	18	18.00	20.00	18	8
SG1-19	19	19.00	21.00	18	8
SG1-20	20	20.00	22.00	20	8
SG1-21	21	21.00	23.00	20	8
SG1-22	22	22.00	24.00	20	8
SG1-23	23	23.00	25.00	20	8
SG1-24	24	24.00	26.00	20	8
SG1-25	25	25.00	27.00	20	8
SG1-26	26	26.00	28.00	25	8
SG1-27	27	27.00	29.00	25	8
SG1-28	28	28.00	30.00	25	8
SG1-29	29	29.00	31.00	25	8
SG1-30	30	30.00	32.00	25	8
SG1-31	31	31.00	33.00	25	8
SG1-32	32	32.00	34.00	25	8
SG1-33	33	33.00	35.00	25	8
SG1-34	34	34.00	36.00	25	8
SG1-35	35	35.00	37.00	25	8
SG1-36	36	36.00	38.00	30	8
SG1-37	37	37.00	39.00	30	8
SG1-38	38	38.00	40.00	30	8
SG1-39	39	39.00	41.00	30	8
SG1-40	40	40.00	42.00	30	8
SG1-41	41	41.00	43.00	30	8
SG1-42	42	42.00	44.00	30	8
SG1-43	43	43.00	45.00	30	8
SG1-44	44	44.00	46.00	30	8
SG1-45	45	45.00	47.00	30	8
SG1-46	46	46.00	48.00	30	8

Table 8.8 Continued

Part number	Teeth	PCD (mm)	Outer diameter (mm)	Boss diameter (mm)	Bore diameter (mm)
SG1-47	47	47.00	49.00	30	8
SG1-48	48	48.00	50.00	30	8
SG1-49	49	49.00	51.00	30	8
SG1-50	50	50.00	52.00	35	10
SG1-51	51	51.00	53.00	35	10
SG1-52	52	52.00	54.00	35	10
SG1-53	53	53.00	55.00	35	10
SG1-54	54	54.00	56.00	35	10
SG1-55	55	55.00	57.00	35	10
SG1-56	56	56.00	58.00	35	10
SG1-57	57	57.00	59.00	35	10
SG1-58	58	58.00	60.00	35	10
SG1-59	59	59.00	61.00	35	10
SG1-60	60	60.00	62.00	35	10
SG1-61	61	61.00	63.00	35	10
SG1-62	62	62.00	64.00	35	10
SG1-63	63	63.00	65.00	35	10
SG1-64	64	64.00	66.00	35	10
SG1-65	65	65.00	67.00	35	10
SG1-66	66	66.00	68.00	35	10
SG1-68	68	68.00	70.00	35	10
SG1-70	70	70.00	72.00	35	10
SG1-72	72	72.00	74.00	35	10
SG1-74	74	74.00	76.00	45	10
SG1-76	76	76.00	78.00	45	10
SG1-78	78	78.00	80.00	45	10
SG1-80	80	80.00	82.00	45	10
SG1-84	84	84.00	86.00	45	10
SG1-88	88	88.00	90.00	45	10
SG1-90	90	90.00	92.00	45	10
SG1-96	96	96.00	98.00	45	10
SG1-100	100	100.00	102.00	45	10
SG1-112	112	112.00	114.00	45	10
SG1-120	120	120.00	122.00	45	10
SG1-130	130	130.00	132.00	45	10
SG1-150	150	150.00	152.00	45	10

Table 8.9 Spur gears; 1.5 module; heavy duty steel 817M40, 655M13; face width 20mm

Part number	Teeth	PCD (mm)	Outer diameter (mm)	Boss diameter (mm)	Bore diameter (mm)
SG1.5-9	9	15.00	18.00	18	8
SG1.5-10	10	16.50	19.50	19.5	8
SG1.5-11	11	18.00	21.00	21	8
SG1.5-12	12	18.00	21.00	21	8
SG1.5-13	13	19.50	22.50	20	8
SG1.5-14	14	21.00	24.00	20	8
SG1.5-15	15	22.50	25.50	20	8
SG1.5-16	16	24.00	27.00	25	10
SG1.5-17	17	25.50	28.50	25	10
SG1.5-18	18	27.00	30.00	25	10
SG1.5-19	19	28.50	31.50	25	10
SG1.5-20	20	30.00	33.00	25	10
SG1.5-21	21	31.50	34.50	25	10
SG1.5-22	22	33.00	36.00	30	10
SG1.5-23	23	34.50	37.50	30	10
SG1.5-24	24	36.00	39.00	30	10
SG1.5-25	25	37.50	40.50	30	10
SG1.5-26	26	39.00	42.00	30	10
SG1.5-27	27	40.50	43.50	30	10
SG1.5-28	28	42.00	45.00	30	10
SG1.5-29	29	43.50	46.50	30	10
SG1.5-30	30	45.00	48.00	30	10
SG1.5-31	31	46.50	49.50	30	10
SG1.5-32	32	48.00	51.00	30	10
SG1.5-33	33	49.50	52.50	30	10
SG1.5-34	34	51.00	54.00	30	10
SG1.5-35	35	52.50	55.50	50	15
SG1.5-36	36	54.00	57.00	50	15
SG1.5-37	37	55.50	58.50	50	15
SG1.5-38	38	57.00	60.00	50	15
SG1.5-39	39	58.50	61.50	50	15
SG1.5-40	40	60.00	63.00	50	15
SG1.5-41	41	61.50	64.50	50	15
SG1.5-42	42	63.00	66.00	50	15
SG1.5-43	43	64.50	67.50	50	15
SG1.5-44	44	66.00	69.00	50	15
SG1.5-45	45	67.50	70.50	50	15
SG1.5-46	46	69.00	72.00	50	15

Table 8.9 Continued

Part number	Teeth	PCD (mm)	Outer diameter (mm)	Boss diameter (mm)	Bore diameter (mm)
SG1.5-47	47	70.50	73.50	50	15
SG1.5-48	48	72.00	75.00	50	15
SG1.5-49	49	73.50	76.50	50	15
SG1.5-50	50	75.00	78.00	50	15
SG1.5-51	51	76.50	79.50	50	15
SG1.5-52	52	78.00	81.00	50	15
SG1.5-53	53	79.50	82.50	60	15
SG1.5-54	54	81.00	84.00	60	15
SG1.5-55	55	82.50	85.50	60	15
SG1.5-56	56	84.00	87.00	60	15
SG1.5-57	57	85.50	88.50	60	15
SG1.5-58	58	87.00	90.00	60	15
SG1.5-59	59	88.50	91.50	60	15
SG1.5-60	60	90.00	93.00	60	15
SG1.5-62	62	93.00	96.00	60	15
SG1.5-64	64	96.00	99.00	60	15
SG1.5-65	65	97.50	100.50	60	15
SG1.5-66	66	99.00	102.00	60	15
SG1.5-68	68	102.00	105.00	60	15
SG1.5-70	70	105.00	108.00	60	15
SG1.5-71	71	106.50	109.50	60	15
SG1.5-72	72	108.00	111.00	60	15
SG1.5-73	73	109.50	112.50	60	15
SG1.5-74	74	111.00	114.00	60	15
SG1.5-75	75	112.50	115.50	60	15
SG1.5-76	76	114.00	117.00	60	15
SG1.5-78	78	117.00	120.00	75	15
SG1.5-80	80	120.00	123.00	75	15
SG1.5-86	86	129.00	132.00	75	15
SG1.5-90	90	135.00	138.00	75	15
SG1.5-96	96	144.00	147.00	75	15
SG1.5-98	98	147.00	150.00	75	15
SG1.5-100	100	150.00	153.00	75	15
SG1.5-105	105	157.50	160.50	75	15
SG1.5-110	110	165.00	168.00	75	15
SG1.5-115	115	175.50	178.50	75	15
SG1.5-120	120	180.00	183.00	75	15

Table 8.10 Spur gears; 2.0 module; heavy duty steel 817M40, 655M13; Face width 25 mm

Part number	Teeth	PCD (mm)	Outer diameter (mm)	Boss diameter (mm)	Bore diameter (mm)
SG2-9	9	20.00	24.00	24	12
SG2-10	10	22.00	26.00	26	12
SG2-11	11	24.00	28.00	28	12
SG2-12	12	24.00	28.00	28	12
SG2-13	13	26.00	30.00	30	12
SG2-14	14	28.00	32.00	30	12
SG2-15	15	30.00	34.00	30	12
SG2-16	16	32.00	36.00	30	12
SG2-17	17	34.00	38.00	35	15
SG2-18	18	36.00	40.00	35	15
SG2-19	19	38.00	42.00	35	15
SG2-20	20	40.00	44.00	35	15
SG2-21	21	42.00	46.00	35	15
SG2-22	22	44.00	48.00	35	15
SG2-23	23	46.00	50.00	35	15
SG2-24	24	48.00	52.00	35	15
SG2-25	25	50.00	54.00	35	15
SG2-26	26	52.00	56.00	35	15
SG2-27	27	54.00	58.00	50	20
SG2-28	28	56.00	60.00	50	20
SG2-29	29	58.00	62.00	50	20
SG2-30	30	60.00	64.00	50	20
SG2-31	31	62.00	66.00	50	20
SG2-32	32	64.00	68.00	50	20
SG2-33	33	66.00	70.00	50	20
SG2-34	34	68.00	72.00	50	20
SG2-35	35	70.00	74.00	50	20
SG2-36	36	72.00	76.00	50	20
SG2-37	37	74.00	78.00	50	20
SG2-38	38	76.00	80.00	50	20
SG2-39	39	78.00	82.00	50	20
SG2-40	40	80.00	84.00	50	20
SG2-41	41	82.00	86.00	50	20
SG2-42	42	84.00	88.00	50	20
SG2-43	43	86.00	90.00	60	20
SG2-44	44	88.00	92.00	60	20
SG2-45	45	90.00	94.00	60	20
SG2-46	46	92.00	96.00	60	20
SG2-47	47	94.00	98.00	60	20
SG2-48	48	96.00	100.00	60	20
SG2-49	49	98.00	102.00	60	20
SG2-50	50	100.00	104.00	60	20
SG2-51	51	102.00	106.00	60	20

Table 8.10 Continued

Part number	Teeth	PCD (mm)	Outer diameter (mm)	Boss diameter (mm)	Bore diameter (mm)
SG2-52	52	104.00	108.00	60	20
SG2-53	53	106.00	110.00	60	20
SG2-54	54	108.00	112.00	60	20
SG2-55	55	110.00	114.00	60	20
SG2-56	56	112.00	116.00	60	20
SG2-57	57	114.00	118.00	60	20
SG2-58	58	116.00	120.00	60	20
SG2-59	59	118.00	122.00	60	20
SG2-60	60	120.00	124.00	60	20
SG2-62	62	124.00	128.00	75	20
SG2-64	64	128.00	132.00	75	20
SG2-65	65	130.00	134.00	75	20
SG2-66	66	132.00	136.00	75	20
SG2-68	68	136.00	140.00	75	20
SG2-70	70	140.00	144.00	75	20
SG2-71	71	142.00	146.00	75	20
SG2-72	72	144.00	148.00	75	20
SG2-73	73	146.00	150.00	75	20
SG2-74	74	148.00	152.00	75	20
SG2-75	75	150.00	154.00	75	20
SG2-76	76	152.00	156.00	75	20
SG2-78	78	156.00	160.00	100	20
SG2-80	80	160.00	164.00	100	20
SG2-86	86	172.00	176.00	100	20
SG2-90	90	180.00	184.00	100	20
SG2-96	96	192.00	196.00	100	20
SG2-98	98	196.00	200.00	100	20
SG2-100	100	200.00	204.00	100	20
SG2-105	105	210.00	214.00	100	20
SG2-110	110	220.00	224.00	100	20
SG2-115	115	230.00	234.00	100	20
SG2-120	120	240.00	244.00	100	20

performance criteria into account if necessary and can be programmed. As a rule of thumb, the face width for spur and helical gears is generally in the range $9m \leq F \leq 14m$, where m is the module, or $F \leq 0.6d_P$ for straight spur gears.

Example 8.9. A 20 degree full-depth spur pinion is required to transmit 1.8 kW at a speed of 1100 rpm. If the pinion has 18 teeth and is manufactured from heavy duty 817M40 steel, select a suitable gear from the limited choice given in Tables 8.8–8.11, specifying the module and face width based on the Lewis formula.

Table 8.11 Spur gears; 3.0 module; heavy duty steel 817M40, 655M13; Face width 35 mm

Part number	Teeth	PCD (mm)	Outer diameter (mm)	Boss diameter (mm)	Bore diameter (mm)
SG3-9	9	30.00	36.00	36	15
SG3-10	10	33.00	39.00	39	15
SG3-11	11	36.00	42.00	42	15
SG3-12	12	36.00	42.00	42	20
SG3-13	13	39.00	45.00	45	20
SG3-14	14	42.00	48.00	45	20
SG3-15	15	45.00	51.00	45	20
SG3-16	16	48.00	54.00	45	20
SG3-17	17	51.00	57.00	45	20
SG3-18	18	54.00	60.00	45	20
SG3-19	19	57.00	63.00	45	20
SG3-20	20	60.00	66.00	60	25
SG3-21	21	63.00	69.00	60	25
SG3-22	22	66.00	72.00	60	25
SG3-23	23	69.00	75.00	60	25
SG3-24	24	72.00	78.00	60	25
SG3-25	25	75.00	81.00	60	25
SG3-26	26	78.00	84.00	60	25
SG3-27	27	81.00	87.00	60	25
SG3-28	28	84.00	90.00	60	25
SG3-29	29	87.00	93.00	60	25
SG3-30	30	90.00	96.00	60	25
SG3-31	31	93.00	99.00	60	25
SG3-32	32	96.00	102.00	60	25
SG3-33	33	99.00	105.00	60	25
SG3-34	34	102.00	108.00	75	25
SG3-35	35	105.00	111.00	75	25
SG3-36	36	108.00	114.00	75	25
SG3-37	37	111.00	117.00	75	25
SG3-38	38	114.00	120.00	75	25
SG3-39	39	117.00	123.00	75	25
SG3-40	40	120.00	126.00	75	25
SG3-41	41	123.00	129.00	75	25
SG3-42	42	126.00	132.00	75	25
SG3-43	43	129.00	135.00	75	25
SG3-44	44	132.00	138.00	100	25
SG3-45	45	135.00	141.00	100	25
SG3-46	46	138.00	144.00	100	25
SG3-47	47	141.00	147.00	100	25
SG3-48	48	144.00	150.00	100	25
SG3-49	49	147.00	153.00	100	25
SG3-50	50	150.00	156.00	100	25

Table 8.11 Continued

Part number	Teeth	PCD (mm)	Outer diameter (mm)	Boss diameter (mm)	Bore diameter (mm)
SG3-51	51	153.00	159.00	100	25
SG3-52	52	156.00	162.00	100	25
SG3-53	53	159.00	165.00	100	25
SG3-54	54	162.00	168.00	100	25
SG3-55	55	165.00	171.00	100	25
SG3-56	56	168.00	174.00	100	25
SG3-57	57	171.00	177.00	127	25
SG3-58	58	174.00	180.00	127	25
SG3-59	59	177.00	183.00	127	25
SG3-60	60	180.00	186.00	127	25
SG3-62	62	186.00	192.00	127	25
SG3-63	63	189.00	195.00	127	25
SG3-64	64	192.00	198.00	127	25
SG3-65	65	195.00	201.00	127	25
SG3-68	68	204.00	210.00	127	25
SG3-70	70	210.00	216.00	150	30
SG3-72	72	216.00	222.00	150	30
SG3-75	75	225.00	231.00	150	30
SG3-76	76	228.00	234.00	150	30
SG3-78	78	234.00	240.00	150	30
SG3-80	80	240.00	246.00	150	30
SG3-82	82	246.00	252.00	150	30
SG3-84	84	252.00	258.00	150	30
SG3-86	86	258.00	264.00	150	30
SG3-90	90	270.00	276.00	150	30
SG3-92	92	276.00	282.00	150	30
SG3-94	94	282.00	288.00	150	30
SG3-95	95	285.00	291.00	150	30
SG3-96	96	288.00	294.00	150	30

Note: PCD = pitch circle diameter, O/D = outer diameter.
Data adapted from HPC Gears Ltd.

Solution

Power = 1.8 kW

$N_P = 18$

Try $m = 1$

Assuming a hardened material will be used to improve wear resistance, $\sigma_p = 183$ MPa.

$Y_{18} = 0.29327$,

$n_P = 1100$ rpm,

Table 8.12 Permissible bending stresses for various commonly used gear materials

Material	Treatment	σ_{UTS} (MPa)	Permissible bending stress σ_P (MPa)
Nylon		65 (20°C)	27
Tufnol		110	31
080M40		540	131
080M40	Induction hardened	540	117
817M40		772	221
817M40	Induction hardened	772	183
045M10		494	117
045M10	Case hardened	494	276
655M13	Case hardened	849	345

$d_P = mN = 0.001 \times 18 = 0.018\,\text{m},$

$$V = \frac{d_P}{2} \times 10^{-3} \times n\frac{2\pi}{60} = \frac{0.018}{2} \times 1100 \times 0.1047 = 1.037\ \text{m/s},$$

$$K_v = \frac{6.1}{6.1+V} = \frac{6.1}{6.1+1.037} = 0.8548,$$

$$W_t = \frac{\text{Power}}{V} = \frac{1800}{1.037} = 1737\ \text{N},$$

$$F = \frac{W_t}{K_v m Y \sigma} = \frac{1737}{0.8548 \times 0.001 \times 0.29327 \times 183 \times 10^6} = 0.0379\ \text{m}.$$

This value of face width is greater than the stock value available as listed in Table 8.8 for a module 1 gear. Therefore, an alternative design needs to be considered.

Trying $m = 1.5\,\text{mm}$,
$d_P = mN = 0.0015 \times 18 = 0.027\,\text{m}$,
$V = 1.555\,\text{m/s}$,
$K_v = 0.7969$,
$W_t = 1158\,\text{N}$,

$$F = \frac{W_t}{K_v m Y \sigma} = \frac{1158}{0.7969 \times 0.0015 \times 0.29327 \times 183 \times 10^6} = 0.01805\ \text{m}.$$

This is less than the $F=20$ mm available for the 1.5 module stock gears listed in Table 8.9. This gear is therefore likely to be acceptable in terms of bending stress capability.

The gear specification is therefore: $N_P=18$, $m=1.5$ mm, $F=20$ mm, 817M40 induction hardened.

Example 8.10. A gearbox is required to transmit 18 kW from a shaft rotating at 2650 rpm. The desired output speed is approximately 12,000 rpm. For space limitation and standardisation reasons a double step-up gearbox is requested with equal ratios. Using the limited selection of gears presented in Tables 8.8–8.11 select suitable gears for the gear wheels and pinions.

Solution

The overall ratio = 12,000/2650 = 4.528.

First-stage ratio $=\sqrt{4.528}=2.128$.

This could be achieved using a gear with 38 teeth and pinion with 18 teeth (ratio = 38/18 = 2.11).

The gear materials listed in Tables 8.8–8.11 are 817M40 and 655M13 steels.

From Table 8.12, the 655M13 is the stronger steel and this is selected for this example prior to a more detailed consideration.

For 655M13 case hardened steel gears, the permissible stress $\sigma_p=345$ MPa.

Calculations for gear 1: $Y_{38}=0.37727$, $n=2650$ rpm.

m	1.5	2.0
$d=mN$ (mm)	57	76
$V=\frac{d}{2}\times 10^{-3}\times n\frac{2\pi}{60}$ (m/s)	7.9	10.5
$W_t=\frac{\text{Power}}{V}$ (N)	2276	1707
$K_v=\frac{6.1}{6.1+V}$	0.4357	0.3675
$F=\frac{W_t}{K_v m Y \sigma_p}$ (m)	0.027	0.018

$m=1.5$ gives a face width greater than the catalogue value of 20 mm, so try $m=2$.

$m=2$ gives a face width less than the catalogue value of 25 mm, so OK.

Calculations for pinion 1: $Y_{18}=0.29327$, $n=5594$ rpm.

(No need to do calculations for $m=1.5$ because it has now been rejected.)

m	2.0
d (mm)	36
V (m/s)	10.5
W_t (N)	1707
K_v	0.3676
F (m)	0.023

$m=1.5$ gives a face width greater than the catalogue value of 20 mm, so try $m=2$.

$m=2$ gives a face width less than the catalogue value of 25 mm, so OK.

Calculations for gear 2: $Y_{38}=0.37727$, $n=5594$ rpm.

m	2.0
d (mm)	76
V (m/s)	22.26
W_t (N)	808.6
K_v	0.215
F (m)	0.0144

$m=2$ gives value for face width lower than catalogue specification, so the design is OK.

m	2.0
d (mm)	36
V (m/s)	22.26
W_t (N)	808.6
K_v	0.215
F (m)	0.0186

Calculations for pinion 2: $Y_{18}=0.29327$, $n=11{,}810$ rpm.
$m=2$ gives value for face width lower than catalogue specification, so the design is OK.

A schematic for the overall design is given in Fig. 8.36.

Example 8.11. A gearbox for a cordless hand tool, Fig. 8.37, is required to transmit 5 W from an electric motor running at 3000 rpm to an output drive running at approximately 75 rpm. The specification for the gearbox is that it should be reliable, efficient, maintenance free for the life of the device, light and fit within an enclosure of 50 mm diameter. The length is not critical. The tolerance on the output speed acceptable is ±10%. Specify an appropriate gear train including values for the module, teeth numbers and gear materials.

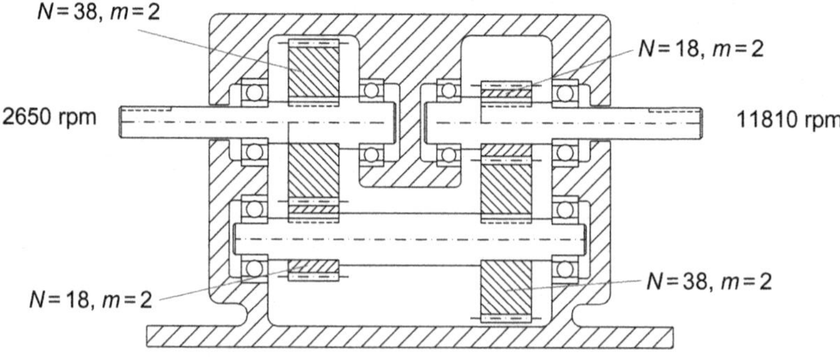

Fig. 8.36 Step up gearbox solution schematic.

Fig. 8.37 Cordless hand-tool moulding, electric motor and gearbox.

Solution

The reduction ratio 3000/75 is quite large and could be achieved by a compound gear train, a worm and wheel or a two-stage reduction epicyclic. The restriction on the maximum diameter precludes the use of compound gear trains, especially if the drive input is axial as would be expected from an electric motor. The high efficiency demand rules out the use of a worm and wheel.

We want an overall ratio of 40 (=3000/75). This could be achieved in a two-stage reduction epicyclic gearbox as illustrated in Fig. 8.38. For identical stages, the gear ratio of each stage would be $\sqrt{40}=6.3$ ($6.3\times 6.3\approx 39.7$).

Try $N_s=18$, $N_R=72$. This gives a reduction ratio per stage of $N_R/N_s+1=72/18+1=5$. Overall reduction ratio of the two-stage gear train $=5\times 5=25$ No good.

Try $N_s=12$, $N_R=72$. This gives a reduction ratio of 7. Overall reduction ratio $7\times 7=49$ No good.

Try $N_s=12$, $N_R=63$. This gives a reduction ratio of 6.25. Overall reduction ratio $6.25\times 6.25=39.1$ OK.

Can this fit in the space?

Propose a module. Say $m=0.5$.

With $N_S=12$, $d_{\text{sun}}=mN=0.5\times 12=6\,\text{mm}$.

With $N_R=63$, $d_{\text{ring}}=mN=0.5\times 63=31.5\,\text{mm}$.

Room available for planet gear is $31.5/2–6/2=12.75\,\text{mm}$. With a module of 0.5, there are no gears with this diameter. Impossible solution.

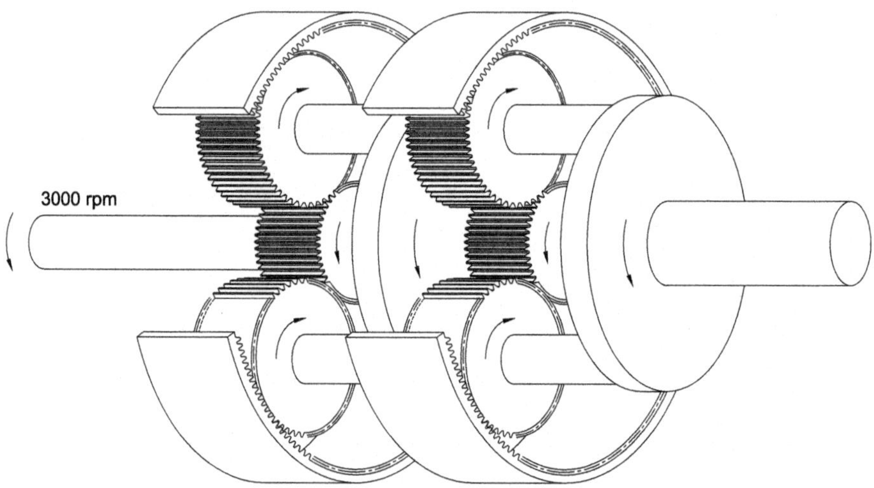

Fig. 8.38 Two stage reduction epicyclic gear train.

Try $N_s = 12$, $N_R = 66$. This gives a reduction ratio of 6.5. Reduction ratio $6.5 \times 6.5 = 42.3$.

This would give an output speed of 71 rpm.

This is within the tolerance allowable on the output speed. OK.

Can this fit in the space?

Propose a module. Say m = 0.5.

With $N_S = 12$, $d_{\text{sun}} = mN = 0.5 \times 12 = 6\,\text{mm}$.

With $N_R = 66$, $d_{\text{ring}} = mN = 0.5 \times 66 = 33\,\text{mm}$.

Room available for planet gear is 33/2–6/2 = 13.5 mm.

With a module of 0.5, the planetary gears would have 13.5/0.5 = 27 teeth. $N_p = 27$.

Proposed design: $N_s = 12$, $N_p = 27$, $N_R = 66$, module 0.5.

What face width should the gears have? For a module of 0.5 HPC Gears Ltd. offers standard 0.5 module gears with a face width of 5 mm available in Tufnol, delrin, brass and 045M10 steel.

The Lewis formula can be used as an approximate check to determine whether the gears with a face width of 5 mm manufactured from a particular material are strong enough.

Analysing the input sun gear

$$V_s = \frac{6 \times 10^{-3}}{2} \times 3000 \times \frac{2\pi}{60} = 0.942\,\text{m/s},$$

$$K_v = \frac{6.1}{6.1 + V} = \frac{6.1}{6.1 + 0.942} = 0.866,$$

$$W_t = \frac{\text{Power}}{V} = \frac{5}{0.942} = 5.3\,\text{N}.$$

From Table 8.6, $Y_{12} = 0.2296$.

Try Tufnol, $\sigma_p = 31\,\text{MPa}$.

$$F = \frac{W_t}{K_v m Y \sigma_p} = \frac{5.3}{0.866 \times 0.0005 \times 0.2296 \times 31 \times 10^6} = 1.72 \times 10^{-3}\,\text{m}$$

This face width of 1.7 mm is less than the 5 mm available, indicating that the gear will be more than strong enough. If this figure had been >5 mm, then another material such as steel or brass could be considered.

Analysing the planets:

$$n_{\text{arm}} = 3000/6.5 = 461.5\,\text{rpm}.$$

To determine the speed of the planets, use Eqs. (8.14) and (8.15)

$$\frac{n_P - n_{\text{arm}}}{n_S - n_{\text{arm}}} = -\frac{N_S}{N_P},$$

$$\frac{n_P - (-461.5)}{-3000 - (-461.5)} = -\frac{12}{27},$$

$$n_P = -\frac{12}{27} \times (-2538) - 461.5 = 666.7\,\text{rpm},$$

$$V_p = \frac{13.5 \times 10^{-3}}{2} \times 666.7 \times \frac{2\pi}{60} = 0.47\ \text{m/s},$$

$$K_v = \frac{6.1}{6.1 + V} = \frac{6.1}{6.1 + 0.47} = 0.928.$$

The power is divided between three planets, so

$$W_t = \frac{\text{Power}}{V} = \frac{5/3}{0.47} = 3.55\ \text{N}.$$

From Table 8.6, $Y_{27} = 0.34385$.

Try Tufnol, $\sigma_p = 31\,\text{MPa}$.

$$F = \frac{W_t}{K_v m Y \sigma_p} = \frac{3.55}{0.928 \times 0.0005 \times 0.34385 \times 31 \times 10^6} = 0.72 \times 10^{-3}\,\text{m}$$

This face width of 0.72 mm is less than the 5 mm available, indicating that the gear will be more than strong enough.

Analysing the ring gear, the stress on the ring gear will be less than on the planets so it does not necessarily need to be analysed at this stage.

Analysing the stage 2 sun gear ($n_{sun} = n_{planet\ carrier}$).

$$V_s = \frac{6 \times 10^{-3}}{2} \times 461.5 \times \frac{2\pi}{60} = 0.145\ \text{m/s}$$

$$K_v = \frac{6.1}{6.1 + V} = \frac{6.1}{6.1 + 0.145} = 0.98,$$

$$W_t = \frac{\text{Power}}{V} = \frac{5}{0.145} = 34.5\ \text{N}.$$

From Table 8.6, $Y_{12} = 0.2296$.

Try Tufnol, $\sigma_p = 31\,\text{MPa}$.

$$F = \frac{W_t}{K_v m Y \sigma_p} = \frac{34.5}{0.98 \times 0.0005 \times 0.2296 \times 31 \times 10^6} = 0.0099\ \text{m}.$$

This is larger than the available standard size (5 mm). So try a stronger material.

Using 045M10 steel, $\sigma_p = 117\,\text{MPa}$.

$$F = \frac{W_t}{K_v m Y \sigma_p} = \frac{34.5}{0.98 \times 0.0005 \times 0.2296 \times 117 \times 10^6} = 2.62 \times 10^{-3}\,\text{m}.$$

This face width of 2.62 mm is less than the 5 mm available, indicating that the gear will be more than strong enough. So the sun gear must be manufactured from 045M10 steel. Manufacturing expediency might dictate that the planet carrier should also be made from the same steel.

Analysing the planets,

$$\frac{n_P - (-71)}{-461.5 - (-71)} = -\frac{12}{27},$$

$$n_P = -\frac{12}{27} \times (-390.5) - 71 = 102.6\ \text{rpm},$$

$$V_p = \frac{13.5 \times 10^{-3}}{2} \times 102.6 \times \frac{2\pi}{60} = 0.0725\ \text{m/s},$$

and

$$K_v = \frac{6.1}{6.1 + V} = \frac{6.1}{6.1 + 0.0725} = 0.99.$$

The power is divided between three planets, so,

$$W_t = \frac{\text{Power}}{V} = \frac{5/3}{0.0725} = 23.0\ \text{N},$$

$Y_{27} = 0.34385.$

Try Tufnol, $\sigma_p = 31\,\text{MPa}$.

$$F = \frac{W_t}{K_v m Y \sigma_p} = \frac{23.0}{0.99 \times 0.0005 \times 0.34285 \times 31 \times 10^6} = 4.4 \times 10^{-3}\ \text{m}$$

This face width of 4.4 mm is less than the 5 mm available, indicating that the gear will be more than strong enough.

Design summary:

Stage 1: module 0.5, $F = 5\,\text{mm}$, $N_s = 12$ (Tufnol), $N_p = 27$ (three planets, Tufnol), $N_R = 66$ (Tufnol).

Stage 2: module 0.5, $F = 5\,\text{mm}$, $N_s = 12$ (steel), $N_p = 27$ (three planets, Tufnol), $N_R = 66$ (Tufnol).

8.7 Condition monitoring

A key consideration in running a gear set is ensuring its on-going effective operation. Many approaches have been developed managing gear 'health' ranging from periodic or real-time inspection, monitoring temperatures and vibration to an in-depth analysis of data feeds from diverse measurements (see Zakrajsek et al., 1993; Jardine et al., 2006; Randall, 2004; Bartelmus, 2001; Feng et al., 2012; Feng and Zuo, 2012; Lei et al., 2014). The vibration of a feature on gearbox can be measured readily and its spectrum analysed. Significant experience from operation of multiple gearsets has resulted in the development of a knowledge base for a wide range of indications or vibration signatures of gear faults and onset of failure (see, for example, Wang and McFadden, 1993a,b; Lebold et al., 2000; Dalpiaz et al., 2000; Jena et al., 2013, 2014; Yan et al., 2014).

8.8 Conclusions

For transmissions where compact size, high efficiency or high speed are required gears offer a competitive solution in comparison to other types of drive such as belts and chains (see Chapter 12). This chapter has introduced a wide range of types of gears and calculations enabling simple gear trains to be analysed. A preliminary selection procedure for spur gears based on the Lewis formula for bending stresses has been presented. This serves as a starting point in proposing the details for a gear train but more sophisticated analysis is necessary if a robust gearbox is required. Analysis for determining bending and contact stresses and associated safety factors based on national standards for spur and helical gears is outlined in Chapter 9, bevel gears in Chapter 10 and worm and wheels in Chapter 11.

References

Bartelmus, W., 2001. Mathematical modelling and computer simulations as an aid to gearbox diagnostics. Mech. Syst. Signal Process. 15, 855–871.

Dalpiaz, G., Rivola, A., Rubini, R., 2000. Effectiveness and sensitivity of vibration processing techniques for local fault detection in gears. Mech. Syst. Signal Process. 14, 387–412.

Davis, J.R., 2005. Gear Materials, Properties, and Manufacture. SAE.

Feng, Z.P., Zuo, M.J., 2012. Vibration signal models for fault diagnosis of planetary gearboxes. J. Sound Vib. 331, 4919–4939.

Feng, Z.P., Liang, M., Zhang, Y., Hou, S.M., 2012. Fault diagnosis for wind turbine planetary gearboxes via demodulation analysis based on ensemble empirical mode decomposition and energy separation. Renew. Energy 47, 112–126.

Heisler, H., 1999. Vehicle and Engine Technology, second ed. Butterworth Heinemann.

Höhn, B.R., Michaelis, K., 2004. Influence of oil temperature on gear failures. Tribol. Int. 37, 103–109.

Jardine, A.K., Lin, D., Banjevic, D., 2006. A review on machinery diagnostics and prognostics implementing condition-based maintenance. Mech. Syst. Signal Process. 20, 1483–1510.

Jelaska, D.T., 2012. Gears and Gear Drives. Wiley.

Jena, D.P., Panigrahi, S.N., Kumar, R., 2013. Gear fault identification and localization using analytic wavelet transform of vibration signal. Measurement 46, 1115–1124.

Jena, D.P., Sahoo, S., Panigrahi, S.N., 2014. Gear fault diagnosis using active noise cancellation and adaptive wavelet transform. Measurement 47, 356–372.

Ku, P.M., 1976. Gear failure modes—importance of lubrication and mechanics. ASLe Transactions 19, 239–249.

Lebold, M., McClintic, K., Campbell, R., Byington, C., Maynard, K., 2000. Review of vibration analysis methods for gearbox diagnostics and prognostics. In: Proceedings of the 54th meeting of the society for machinery failure prevention technology.634, p. 16.

Lei, Y.G., Lin, J., Zuo, M.J., He, Z.J., 2014. Condition monitoring and fault diagnosis of planetary gearboxes: a review. Measurement 48, 292–305.

Levai, Z., 1969. Structure and analysis of planetary gear trains. J. Mech. 3, 131–148.

Litvin, F.L., 2014. Noncircular Gears: Design and Generation. Cambridge University Press.

Mitchener, R.G., Mabie, H.H., 1982. The determination of the Lewis form factor and the AGMA geometry factor J for external spur gear teeth. ASME J. Mech. Design 104, 148–158.

Naunheimer, H., Bertsche, B., Ryborz, J., Novak, W., 2010. Automotive Transmissions: Fundamentals, Selection, Design and Application, second ed. Springer.

Pennestri, E., Freudenstein, F., 1993. The mechanical efficiency of epicyclic gear trains. J. Mech. Des. 115, 645–651.

Radzevich, S.P., 2016. Dudley's Handbook of Practical Gear Design and Manufacture, third ed. CRC Press.

Randall, R.B., 2004. Detection and diagnosis of incipient bearing failure in helicopter gearboxes. Eng. Fail. Anal. 11, 177–190.

SAE, 2012. Design Practices: Passenger Car Automatic Transmissions. SAE.

Stokes, A., 1992. Manual Gearbox Design. Elsevier.

Tuplin, W.A., 1962. Gear Design. The Machinery Publishing Co.

Wang, W.J., McFadden, P.D., 1993a. Early detection of gear failure by vibration analysis. I. Calculation of the time-frequency distribution. Mech. Syst. Signal Process. 7, 193–203.

Wang, W.J., McFadden, P.D., 1993b. Early detection of gear failure by vibration analysis. II. Interpretation of the time-frequency distribution using image processing techniques. Mech. Syst. Signal Process. 7, 205–215.

Yan, R., Gao, R.X., Chen, X., 2014. Wavelets for fault diagnosis of rotary machines: a review with applications. Signal Process. 96, 1–15.

Zakrajsek, J.J., Townsend, D.P. and Decker, H.J. An analysis of gear fault detection methods as applied to pitting fatigue failure data, NASA, NASA-E-7470, 1993.

Standards

British Standards Institution. BS436: Part 1: 1967 Specification for spur and helical gears. Part 1 Basic rack form, pitches and accuracy (diametral pitch series).

British Standards Institution. BS436: Part 2: 1970 Specification for spur and helical gears. Part 2 Basic rack form, modules and accuracy (1 to 50 metric module).

British Standards Institution. BS2519: Part 1, 1976. Glossary for gears. Geometrical definitions.

British Standards Institution. BS2519: Part 2, 1976. Glossary for gears. Notation.

British Standards Institution. BS 6168: 1987. Specification for non-metallic spur gears.

British Standards Institution. BS ISO 9083:2001 Calculation of load capacity of spur and helical gears. Application to marine gears.

Websites

At the time of going to press the world-wide-web contained useful information relating to this chapter at the following sites.

moorecustomgear.com.
www.bga.org.uk.
www.cross-morse.co.uk.
www.davall.co.uk.
www.drgears.com.
www.gearing-hk.com.
www.geartechnology.com.
www.hpcgears.com.
www.omnigearandmachine.com.
www.penfold.co.nz.
www.powertransmission.com.
www.precipart.com.
www.qtcgears.com.
www.rapidgear.com.
www.reliance.co.uk.
www.shanthigears.com.
www.textronpt.com.
www.textronpt.com.
www.ugaco.com.
www.wgear.com.

Further reading

David Brown Special Products Ltd. David Brown basic gear book. 1995.
Hofmann, D.A., Kohler, H.K., Munro, R.G., 1991. Gear Technology Teaching Pack. British Gear Association.
HPC Gears Ltd. Gears catalogue. http://www.hpcgears.com/ (Accessed 27 August 2018).
Townsend, D.P., 1992. Dudley's Gear Handbook. McGraw Hill.

Spur and helical gear stressing

9

Chapter Outline

Nomenclature

Generally, preferred SI units have been stated.

A	used in evaluating the dynamic factor
A'	used in evaluating the hardness ratio factor
B	used in evaluating the dynamic factor
C_e	used in evaluating the load distribution factor K_H
C_f	surface condition factor for pitting resistance
C_H	hardness ratio factor
C_{ma}	used in evaluating the load distribution factor K_H
C_{mc}	used in evaluating the load distribution factor K_H
C_P	elastic coefficient ((MPa)$^{0.5}$)
C_{pf}	used in evaluating the load distribution factor K_H
C_{pm}	used in evaluating the load distribution factor K_H
d	pitch diameter (m)
d_G	pitch diameter of the gear (m)
d_P	pitch diameter of the pinion (m)
E	Young's modulus (N/m^2)
E_G	Young's modulus for the gear (N/m^2)
E_P	Young's modulus for the pinion (N/m^2)
F	face width (m)
h_t	gear tooth whole depth (m)
H	power
H_B	Brinell hardness
H_{BG}	Brinell hardness of the gear
H_{BP}	Brinell hardness of the pinion
I	geometry factor for pitting resistance

Mechanical Design Engineering Handbook. https://doi.org/10.1016/B978-0-08-102367-9.00009-3

J	geometry factor for bending strength
K_B	rim thickness factor
K_H	load distribution factor
K_m	load distribution factor
K_O	overload factor
K_R	reliability factor
K_s	size factor
K_T	temperature factor
K_v'	velocity factor (Eq. 9.4)
K_V'	dynamic factor (Eq. 9.22)
m	module (mm or m)
m_n	normal module (mm or m)
m_N	load sharing ratio
m_G	speed ratio (mm or m)
n	speed (rpm)
N	number of teeth
N_G	number of teeth in the gear
N_P	number of teeth in the pinion
P_d	transverse diametral pitch (in.)
p_n	normal circular pitch (m)
p_N	base pitch (m)
p_t	transverse circular pitch (m)
Q_v	quality control numbers, transmission accuracy level number
r_1	radius of curvature (m)
r_2	radius of curvature (m)
S	distance (m)
S_c	allowable contact stress (lb/in.^{2})
S_F	AGMA factor of safety
S_H	AGMA factor of safety
S_{HG}	AGMA factor of safety for the gear
S_{HP}	AGMA factor of safety for the pinion
S_t	allowable bending stress (lb/in.^{2})
S_1	distance (m)
t_R	rim thickness below the tooth (m)
V	velocity (m/s)
W	resultant load (N)
W_a	axial load (N)
W_r	radial load (N)
W_t	transmitted load (N)
Y	Lewis form factor
Y_J	geometry factor for bending strength
Y_N	stress cycle factor for bending stress
Y_Z	reliability factor
Y_θ	temperature factor
Z	length of the line of action in the transverse plane (m)
Z_e	elastic coefficient (MPa)$^{0.5}$
Z_I	geometry factor for pitting resistance
Z_N	stress cycle life factor
Z_R	surface condition factor for pitting resistance

Z_W hardness ratio factor
ϕ pressure angle (degree)
ϕ_t transverse pressure angle (degree)
ν_G Poisson's ratio for the gear
ν_P Poisson's ratio for the pinion
σ bending stress (N/m^2)
σ_{all} allowable bending stress (N/m^2)
$\sigma_{c,all}$ allowable contact stress (N/m^2)
σ_G bending stress for the gear (N/m^2)
σ_P bending stress for the pinion (N/m^2)
σ_c contact stress (N/m^2)
σ_{FP} allowable bending stress (MPa)
σ_{HP} allowable contact stress (MPa)
σ_p permissible stress (N/m^2)
ψ helix angle at the standard pitch diameter (degree)

9.1 Introduction

Although suitable for initial selection of a gear, the Lewis equation introduced in Chapter 8 is inadequate for detailed design calculations where life of the machine is critical or where the gears are highly loaded. National and international standards are available detailing the calculation of bending and contact stresses for gears. This chapter outlines the American Gear Manufacturers Association (AGMA) standards for the calculation of bending and contact stresses for spur and helical gears.

The Lewis equation provides a useful starting point allowing the configuration of a gear set that is to be developed. The Lewis equation takes account of the bending stress on a gear. Gears however can fail to excessive bending stress or wear. Wear occurs as the teeth move in and out of contact with each other with accompanying local deformation of the gear teeth surfaces in the region of contact. The stresses resulting from the surface deformation are known as contact stresses (Fernandes and McDuling, 1997; Hassan, 2009; Li and Kahraman, 2014; Patil et al., 2014). If the stresses are too high then material failure can take the form of loss of material from the surfaces, known as pitting. Pitting is a surface fatigue failure due to too many repetitions of high contact stresses (Fig. 9.1). Wear can also occur due to scuffing (Fig. 9.2), scoring

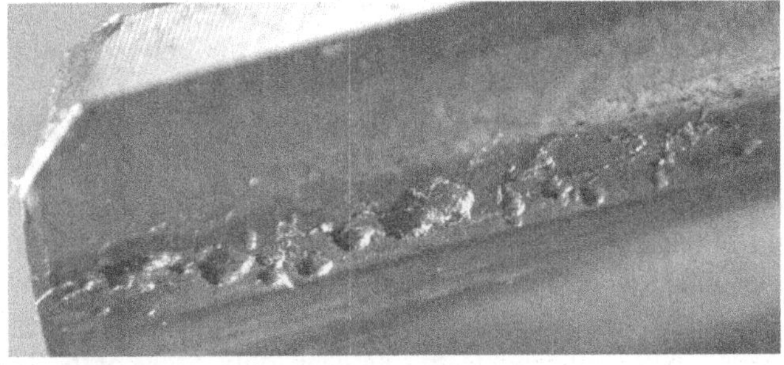

Fig. 9.1 Pitting damage to a helical gear tooth (Batsch, 2016).

Fig. 9.2 Scuffing gear failure on the flank of a test gear due to poor extreme-pressure (EP) gear oil properties (Michalczewski et al., 2013).

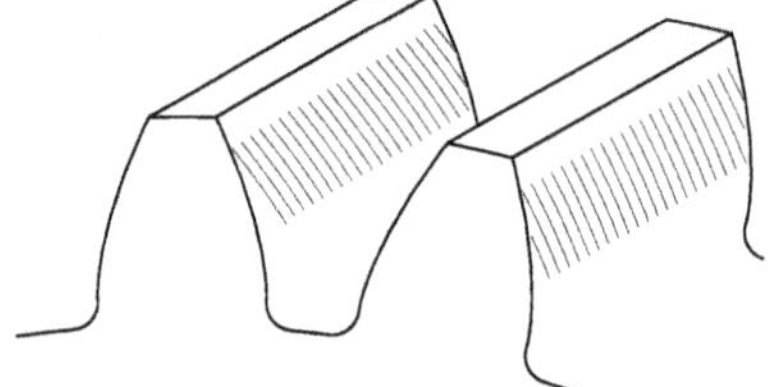

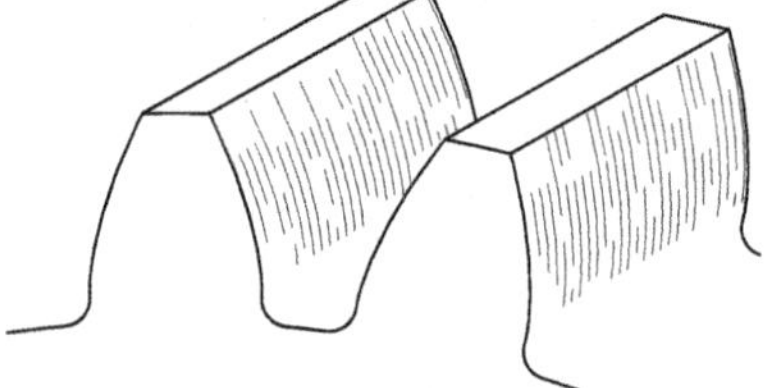

Fig. 9.3 Gear failure due to scoring.

(Fig. 9.3), failure of lubrication, abrasion, or the presence of a foreign material such as grit or powder.

The modelling of gear stresses has been the subject of extensive research. The analysis is complex because of the dynamic nature of the loads and the very high stresses involved for some gear designs. Whilst techniques such as finite element analysis (Barbieri et al., 2014; Chen and Tsay, 2002; Jyothirmai et al., 2014; Rao and Muthuveerappan, 1993) allow detailed modelling of a specific gear geometry, the activity can be labour intensive in terms of setting up the model and uncertainty often remains concerning the validity of boundary conditions and material properties. An alternative to finite element analysis is the use of procedural approaches such as those outlined in standards developed by the American Gear Manufacturers Association (AGMA), the British Standards Institution (BSI) and the International Organisation for Standardisation (ISO).

This chapter concentrates on analysing whether failure due to bending and contact stresses in a gear set is likely. In Section 9.2, a relatively simple relationship for contact stresses is outlined prior to the introduction of AGMA equations for bending and contact stresses in Section 9.3. The AGMA equations rely on the evaluation of a significant number of factors and geometrical parameters, which depend on the specific geometry and materials concerned. In this chapter the basic approach is outlined for 20 degree pressure angle, full depth teeth spur gears only. Some of the detailed charts and equations given in the source standards have been omitted and the reader is encouraged to view the most recent standards as necessary. A general strategy for gear design is outlined in Section 9.4.

9.2 Wear failure

As well as failure due to bending stresses in gears (Fernandes, 1996), the failure due to wear on the surface of gear teeth should be considered. Possible surface failures are pitting, which is a surface fatigue failure due to many repetitions of high contact stresses, scoring due to failure of lubrication and abrasion due to the presence of foreign particles.

The surface compressive, Hertzian or contact stress for a gear can be modelled by:

$$\sigma_c = -C_p \left[\frac{K_v' W_t}{F \cos\phi} \left(\frac{1}{r_1} + \frac{1}{r_2} \right) \right]^{0.5} \tag{9.1}$$

where

C_p is an elastic coefficient,
r_1 and r_2 are radii of curvature,
$K_v{}'$ is the velocity factor,
W_t is the transmitted load (N),
F is the face width (m),
ϕ is the pressure angle (degree)

If the units of C_P are in $\sqrt{\text{MPa}}$, W_t in N, r_1, r_2 in meters, then Eq. (9.1) gives the contact stress σ_c in kPa.

The radii of curvature are given by

$$r_1 = \frac{d_P \sin\phi}{2} \tag{9.2}$$

$$r_2 = \frac{d_G \sin\phi}{2} \tag{9.3}$$

where d_P and d_G are the pitch diameters of the pinion and gear respectively.

The velocity factor $K_v{}'$ for cut or milled profile gears is given by

$$K_v' = \frac{6.1 + V}{6.1} \tag{9.4}$$

where V is the pitch line velocity (m/s).

Note that this definition for the velocity factor, now used in the latest AGMA standards is the reciprocal of that previously defined. For this reason, following the notation suggested by Mischke in Shigley and Mischke (2001), a superscipt $'$ has been added to allow it to be distinguished from previous definitions.

Table 9.1 Values of the elastic coefficient C_p ($\sqrt{\text{MPa}}$) (AGMA 218.01)

Pinion material	E_{pinion} (GPa)	Gear material					
		Steel	Malleable iron	Nodular iron	Cast iron	Aluminium bronze	Tin bronze
Steel	200	191	181	179	174	162	158
Malleable iron	170	181	174	172	168	158	154
Nodular iron	170	179	172	170	166	156	152
Cast iron	150	174	168	166	163	154	149
Aluminium bronze	120	162	158	156	154	145	141
Tin bronze	110	158	154	152	149	141	137

The elastic coefficient C_P ($\sqrt{\text{MPa}}$)) can be calculated from Eq. (9.5) or obtained from Table 9.1.

$$C_p = \left[\frac{1}{\pi \left(\frac{1-\nu_P^2}{E_P} + \frac{1-\nu_G^2}{E_G} \right)} \right]^{0.5} \tag{9.5}$$

where

ν_P is Poisson's ratio for the pinion,
ν_G is Poisson's ratio for the gear,
E_P is Young's modulus for the pinion (MPa),
E_G is Young's modulus for the gear (MPa).

Example 9.1. A speed reducer has a 22 tooth spur pinion made of steel, driving a 60 tooth gear made of cast iron. The transmitted power is 10 kW. The pinion speed is 1200 rpm, module 4 and face width 50 mm. Determine the contact stress.

Solution

$N_P = 22$; steel pinion, $N_G = 60$; cast iron gear. $n = 1200$ rpm, $m = 4$, $F = 50$ mm, $H = 10$ kW.

$d_P = mN_P = 4 \times 22 = 88$ mm, $d_G = 4 \times 60 = 240$ mm.

$$V = \frac{0.088}{2} \times 1200 \times \frac{2\pi}{60} = 5.529 \text{ m/s}.$$

$$K_v' = \frac{6.1 + 5.529}{6.1} = 1.906.$$

$$W_t = \frac{10000}{5.529} = 1809 \text{ N}.$$

From Table 9.1 for a steel pinion and cast iron gear, $C_P = 174\ \sqrt{\text{MPa}}$.

From Eqs (9.2) and (9.3),

$$r_1 = \frac{88 \times 10^{-3} \sin 20}{2} = 0.01505 \text{ m},$$

$$r_2 = \frac{240 \times 10^{-3} \sin 20}{2} = 0.04104 \text{ m}.$$

From Eq. (9.1),

$$\sigma_c = -174 \left[\frac{1.906 \times 1809}{0.05 \cos 20} (90.81) \right]^{0.5} = -174 \times 2581 \text{ kPa} = -449{,}200 \text{ kPa}$$

$$\cong -449 \text{ MPa}.$$

9.3 AGMA equations for bending and contact stress

The calculation of bending and contact stresses in spur and helical gears can be determined using standardised methods presented by the British Standards Institution, the International Organisation for Standardisation (ISO), the Deutsches Institut für Normung (DIN), and the American Gear Manufacturers Association (AGMA). The AGMA standards have recently been reaffirmed (ANSI/AGMA 2101-D04) and are widely used and have therefore been selected for presentation here. The procedures make extensive use of a series of geometry and design factors, which can be determined from design charts and tables.

The AGMA formula for bending stress for spur gears is given in SI units by:

$$\sigma = W_t K_O K_V' K_s \frac{1}{Fm} \frac{K_H K_B}{Y_J} \tag{9.6}$$

or US customary units by:

$$\sigma = W_t K_O K_V' K_s \frac{P_d}{F} \frac{K_m K_B}{J} \tag{9.7}$$

where

σ is the bending stress (N/m^2),
W_t is the transmitted tangential load (N),
K_O is the overload factor,
K_v' is the dynamic factor,

K_s is the size factor,
P_d is the transverse diametral pitch (in.),
F is the face width (m or in.),
m is the normal metric module for spur gears and the transverse metric module, m_t, for helical gears (with the unit of meters),
m_t is the transverse metric module (with the unit of meters),
K_H or K_m is the load distribution factor,
K_B is the rim thickness factor,
Y_J or J is the geometry factor for bending strength.

The transverse metric module for a helical gear is given by

$$m_t = \frac{m_n}{\cos\psi} \tag{9.8}$$

where

ψ is the helix angle at the standard pitch diameter (degree),
m_n is the normal module (m or mm).

The circular pitch for helical gears is measured in two ways. The transverse circular pitch, p_t, is measured along the pitch circle, and the normal circular pitch, p_n, is measured normal to the helix of the gear. For a helical gear the normal and transverse circular pitch are given by Eqs (9.9) and (9.10), respectively

$$p_n = \pi m_n \tag{9.9}$$

$$p_t = \pi m_t \tag{9.10}$$

The pitch circle diameter for a helical gear is given by

$$d = m_t N = \frac{m_n}{\cos\psi} N \tag{9.11}$$

The transverse pressure angle, ϕ_t, is given by

$$\phi_t = \tan^{-1}\left(\frac{\tan\phi_n}{\cos\psi}\right) \tag{9.12}$$

The radial, axial and resultant loads, W_r, W_a, W, for a helical gear are given respectively by

$$W_r = W_t \tan\phi_t \tag{9.13}$$

$$W_a = W_t \tan\psi \tag{9.14}$$

$$W = \frac{W_t}{\cos\phi_n \cos\psi} \tag{9.15}$$

The AGMA equation for pitting resistance is given in SI units by:

$$\sigma_c = Z_e \left(W_t K_O K_V' K_s \frac{K_H}{Fd} \frac{Z_R}{Z_I} \right)^{0.5} \tag{9.16}$$

or in US customary units by:

$$\sigma_c = C_p \left(W_t K_O K_V' K_s \frac{K_m}{Fd} \frac{C_f}{I} \right)^{0.5} \tag{9.17}$$

where

σ_c is the absolute value of contact stress (kPa),
Z_e is an elastic coefficient $(\mathrm{MPa})^{0.5}$,
W_t is the transmitted tangential load (N),
K_O is the overload factor,
K_v' is the dynamic factor,
K_s is the size factor,
K_H or K_m is the load distribution factor,
F is the face width (m),
d is the operating pitch diameter of the pinion (m),
Z_R or C_f is the surface condition factor for pitting resistance,
Z_I or I is the geometry factor for pitting resistance.

For Eq. (9.16), if W_t is in N, F in m, d in m, C_P in $\sqrt{\mathrm{MPa}}$ then the units of the contact stress σ_c are in kPa.

Based on safe working practices, the AGMA has defined allowable stress equations for gears for the allowable bending stress and for the allowable contact stress. The AGMA equation for determining a safe value for the allowable bending stress in SI units is given by:

$$\sigma_{all} = \frac{\sigma_{FP} Y_N}{S_F Y_\theta Y_Z} \tag{9.18}$$

or US customary units by:

$$\sigma_{all} = \frac{S_t Y_N}{S_F K_T K_R} \tag{9.19}$$

where

S_t or σ_{FP} is the allowable bending stress (lb/in.^{2} or MPa),
Y_N is the stress cycle factor for bending stress,
S_F is the AGMA factor of safety,
Y_θ or K_T is the temperature factor,
Y_Z or K_R is the reliability factor.

The AGMA equation for determining a safe value for the allowable contact stress in SI units is given by:

$$\sigma_{c,all} = \frac{\sigma_{HP}}{S_H}\frac{Z_N Z_W}{Y_\theta Y_Z} \tag{9.20}$$

or in US customary units by:

$$\sigma_{c,all} = \frac{S_c}{S_H}\frac{Z_N C_H}{K_T K_R} \tag{9.21}$$

where

S_c or σ_{HP} is the allowable contact stress (lb/in.^{2} or MPa),
Z_N is the stress cycle life factor,
Z_W or C_H is a hardness ratio factor,
Y_θ or K_T is the temperature factor,
Y_Z or K_R is the reliability factor,
S_H is the AGMA factor of safety.

The purpose of the overload factor, K_O, is to account for externally applied loads in excess of the nominal tangential load W_t. A typical example is the variation in torque from an internal combustion engine. Values for K_O tend to be derived from extensive experience of a particular application. Values for the overload factor can be estimated from Table 9.2. Within this text the value for the overload factor will generally be assumed to be unity.

The dynamic factor, K_v', is used to account for the effect of tooth spacing and profile errors, the magnitude of the pitch line velocity, the inertia and stiffness of the rotating components and the transmitted load per unit face width. The AGMA has defined a set of quality control numbers which can be taken as equal to the transmission accuracy level number, Q_V, to quantify these parameters (AGMA 390.01). Classes $3 \le Q_V \le 7$ include most commercial quality gears and are generally suitable for applications such as agricultural and plant machinery, presses and conveyors, classes of $8 \le Q_V \le 12$ are of precision quality and are generally suitable for power tools, washing machines and automotive transmissions. Generally the higher the peripheral speed and the smaller the specific tooth load, the more accurate the gear manufacture must be and a gear with a higher quality control number should be used. The dynamic factor K_v' can be determined from

$$K_v' = \left(\frac{A + \sqrt{200V}}{A}\right)^B \tag{9.22}$$

where

$$B = \frac{(12 - Q_V)^{2/3}}{4} \tag{9.23}$$

and

$$A = 50 + 56(1 - B) \tag{9.24}$$

Table 9.2 Estimate for overload factors

	Driven machine			
Driving machine	**Uniform. For example continuous duty generator sets.**	**Light Shock. For example fans, low speed centrifugal pumps, variable duty generators, uniformly loaded conveyors, positive displacement pumps**	**Moderate Shock. For example high speed centrifugal pumps, reciprocating pumps and compressors, heavy duty conveyors, machine tools drives and saws.**	**Heavy Shock. For example punch presses and crushers.**
Uniform. For example electric motors or constant speed gas turbines.	1	1.25	1.5	1.75
Light shock. For example, variable speed drives.	1.2	1.4	1.75	2.25
Moderate shock. For example multicylinder engines.	1.3	1.7	2	2.75

Data from Mott, R.L., 1998. Machine Elements in Mechanical Design. third ed. Prentice Hall.

The size factor K_s is usually taken as unity. Higher values can be used if there is any nonuniformity in the material properties due to size.

The load distribution factor K_H is used to account for nonuniform distribution of load across the line of contact. Ideally gears should be sited at a location where there is zero slope on the shaft under the load. This is normally at mid-span.

For $F/d \leq 2$, gears mounted between bearings, $F \leq 1.016\,\text{m}$, and contact across the full width of the narrowest member,

$$K_H = K_m = 1 + C_{mc}\left(C_{pf}C_{pm} + C_{ma}C_e\right) \tag{9.25}$$

For uncrowned teeth $C_{mc} = 1$ and for crowned teeth $C_{mc} = 0.8$.

If $F \leq 0.0254\,\text{m}$ then,

$$C_{pf} = \frac{F}{10d} - 0.025 \tag{9.26}$$

If $0.0254\,\text{m} < F \leq 0.4318\,\text{m}$ then,

$$C_{pf} = \frac{F}{10d} - 0.0375 + 0.4921F \tag{9.27}$$

If $0.4318\,\text{m} < F \leq 1.016\,\text{m}$ then,

$$C_{pf} = \frac{F}{10d} + 0.815F - 0.3534F^2 \tag{9.28}$$

For values of $F/10d < 0.05$ then set $F/10d = 0.05$ in Eqs (9.26)–(9.28).

For a straddle mounted pinion with $S_1/S < 0.175$ use $C_{pm} = 1$.

For a straddle mounted pinion with $S_1/S \geq 0.175$ use $C_{pm} = 1.1$.

The definition for S_1 and S is given in Fig. 9.4.

$$C_{ma} = A + BF + CF^2 \tag{9.29}$$

where the coefficients A, B and C can be found from Table 9.3.

For gearing adjusted at assembly or if compatibility is improved by lapping or both, $C_e = 0.8$. For all other conditions take $C_e = 1$.

If the rim thickness is insufficient to provide full support for the bearing then this can be accounted for by the rim thickness factor K_B. This is necessary when the ratio $t_R/h_t < 1.2$. In the examples given here it will be assumed that $t_R/h_t \geq 1.2$ in which case $K_B = 1$.

The bending strength geometry factor, Y_J, for 20 degree pressure angle full depth teeth spur gears can be obtained from Fig. 9.5. The bending strength geometry factor,

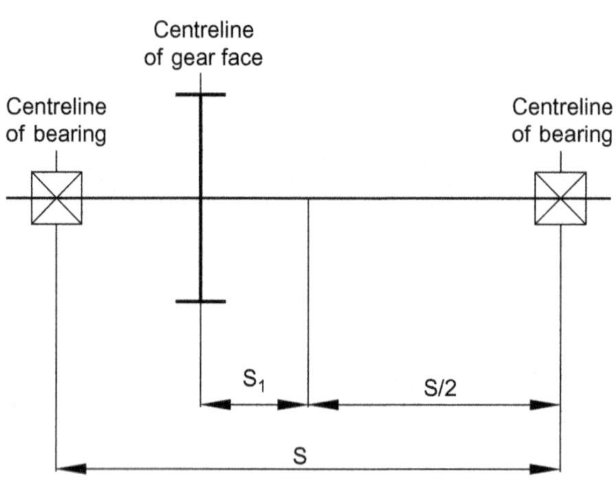

Fig. 9.4 Definition of distances S_1 and S used for evaluating C_{pm}. ANSI/AGMA 2001/C95.

Table 9.3 Coefficients used in determining C_{ma} (Eq. 9.29)

Condition	A	B	C
Open gearing	0.247	0.657	−0.119
Commercial enclosed gear units	0.127	0.622	−0.0144
Precision enclosed gear units	0.0675	0.504	−0.144
Extra-precision enclosed gear units	0.00360	0.402	−0.127

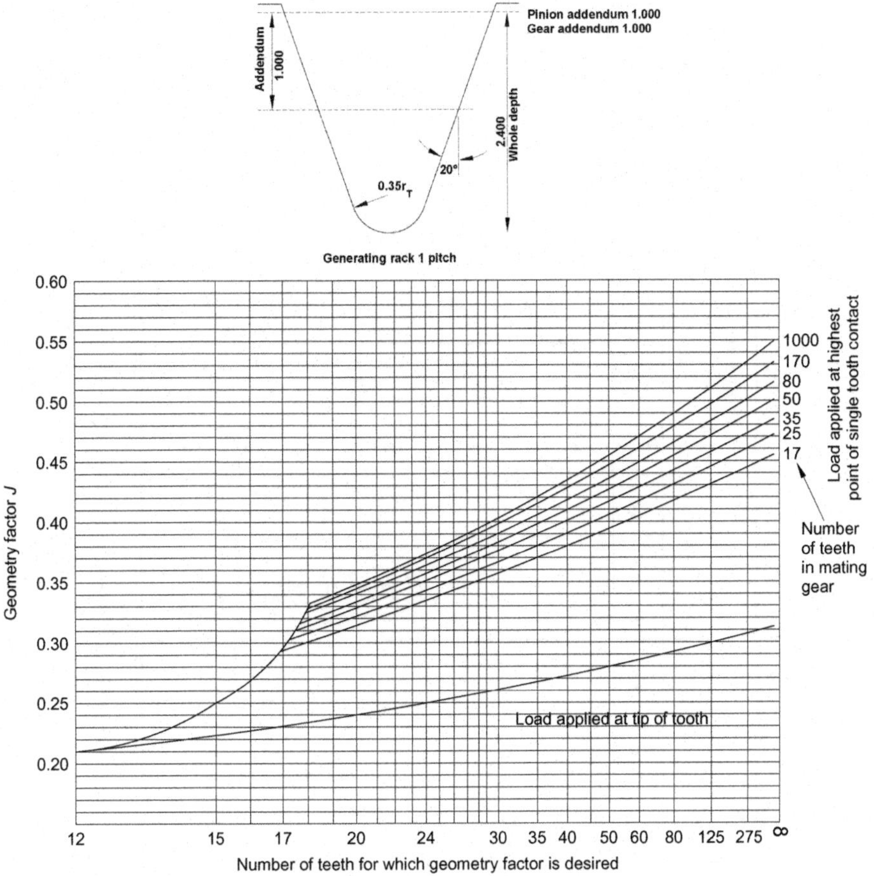

Fig. 9.5 Bending strength geometry factor Y_J or J for 20 degree pressure angle full depth teeth spur gears.

Y_J, for helical gears are given in AGMA Information Sheet 908-B89. A selection of these are given in Tables 9.4–9.8.

The surface condition factor Z_R is used to account for surface finish, residual stress and work hardening. A value for Z_R greater than unity is suggested when a detrimental surface finish is known to exist.

Table 9.4 AGMA bending geometry factor, J, and pitting resistance geometry factor I

Gear teeth		Pinion teeth															
		12		14		17		21		26		35		55		135	
		P	G	P	G	P	G	P	G	P	G	P	G	P	G	P	G
12	I																
	J	U	U														
14	I																
	J	U	U	U	U												
17	I																
	J	U	U	U	U	U	U										
21	I							0.078									
	J	U	U	U	U	U	U	0.24	0.24								
26	I							0.084		0.079							
	J	U	U	U	U	U	U	0.24	0.25	0.25	0.25						
35	I							0.091		0.088		0.080					
	J	U	U	U	U	U	U	0.24	0.26	0.25	0.26	0.26	0.26				
55	I							0.102		0.101		0.095		0.080			
	J	U	U	U	U	U	U	0.24	0.28	0.25	0.28	0.26	0.28	0.28	0.28		
135	I							0.118		0.121		0.120		0.112		0.080	
	J	U	U	U	U	U	U	0.24	0.29	0.25	0.29	0.26	0.29	0.28	0.29	0.29	0.29

Pressure angle = 20 degree, helix angle = 0 degree, full depth teeth with tip loading. AGMA 908-B89.

Table 9.5 AGMA bending geometry factor, *J*, and pitting resistance geometry factor *I*

Gear teeth		Pinion teeth 12		14		17		21		26		35		55		135	
		P	**G**	**P**	**G**	**P**	**G**	**P**	**G**	**P**	**G**	**P**	**G**	**P**	**G**	**P**	**G**
12	*I*																
	J	U	U														
14	*I*																
	J	U	U	U	U												
17	*I*					0.0124											
	J	U	U	U	U	0.43	0.43										
21	*I*					0.139											
	J	U	U	U	U	0.44	0.46	0.47	0.47								
26	*I*					0.154		0.143		0.132							
	J	U	U	U	U	0.45	0.49	0.48	0.50	0.50	0.50						
35	*I*					0.175											
	J	U	U	U	U	0.46	0.52	0.49	0.53	0.51	0.53	0.54	0.54				
55	*I*					0.204											
	J	U	U	U	U	0.47	0.55	0.50	0.56	0.53	0.57	0.56	0.58	0.59	0.59		
135	*I*					0.244		0.241		0.237		0.229		0.209		0.151	
	J	U	U	U	U	0.48	0.59	0.51	0.60	0.54	0.61	0.57	0.62	0.61	0.64	0.65	0.65

Pressure angle = 20 degree, helix angle = 15 degree, full depth teeth with tip loading. AGMA 908-B89.

Table 9.6 AGMA bending geometry factor, J, and pitting resistance geometry factor I

Gear teeth		Pinion teeth															
		12		14		17		21		26		35		55		135	
		P	G	P	G	P	G	P	G	P	G	P	G	P	G	P	G
12	I																
	J	U	U														
14	I																
	J	U	U	U	U												
17	I					0.125											
	J	U	U	U	U	0.44	0.44										
21	I					0.140		0.129									
	J	U	U	U	U	0.45	0.46	0.47	0.47								
26	I					0.156		0.145		0.133							
	J	U	U	U	U	0.45	0.49	0.48	0.49	0.50	0.50						
35	I					0.177		0.167		0.155		0.138					
	J	U	U	U	U	0.46	0.51	0.49	0.52	0.51	0.53	0.54	0.54				
55	I					0.205		0.197		0.188		0.172		0.144			
	J	U	U	U	U	0.47	0.54	0.50	0.55	0.52	0.56	0.55	0.57	0.58	0.58		
135	I					0.245		0.242		0.238		0.229		0.209		0.151	
	J	U	U	U	U	0.48	0.58	0.51	0.59	0.54	0.60	0.57	0.61	0.60	0.62	0.64	0.64

Pressure angle = 20 degree, helix angle = 20 degree, full depth teeth with tip loading. AGMA 908-B89.

Table 9.7 AGMA bending geometry factor, J, and pitting resistance geometry factor I

Gear teeth		Pinion teeth															
		12		14		17		21		26		35		55		135	
		P	G	P	G	P	G	P	G	P	G	P	G	P	G	P	G
12	I																
	J	U	U														
14	I			0.123													
	J	U	U	0.40	0.40												
17	I			0.137		0.126											
	J	U	U	0.41	0.43	0.43	0.43										
21	I			0.152		0.142		0.130									
	J	U	U	0.41	0.45	0.44	0.45	0.46	0.46								
26	I			0.167		0.157		0.146		0.134							
	J	U	U	0.42	0.47	0.44	0.47	0.47	0.48	0.49	0.49						
35	I			0.187		0.178		0.168		0.156		0.138					
	J	U	U	0.43	0.49	0.45	0.50	0.48	0.50	0.50	0.51	0.52	0.52				
55	I			0.213		0.207		0.199		0.189		0.173		0.144			
	J	U	U	0.44	0.52	0.46	0.52	0.49	0.53	0.51	0.54	0.53	0.55	0.56	0.56		
135	I			0.248		0.247		0.244		0.239		0.230		0.210		0.151	
	J	U	U	0.45	0.55	0.47	0.56	0.50	0.56	0.52	0.57	0.54	0.58	0.57	0.59	0.61	0.61

Pressure angle = 20 degree, helix angle = 25 degree, full depth teeth with tip loading. AGMA 908-B89.

Table 9.8 AGMA bending geometry factor, J, and pitting resistance geometry factor I

Gear teeth		Pinion teeth															
		12		14		17		21		26		35		55		135	
		P	**G**	**P**	**G**	**P**	**G**	**P**	**G**	**P**	**G**	**P**	**G**	**P**	**G**	**P**	**G**
12	I																
	J	U	U														
14	I			0.125													
	J	U	U	0.39	0.39												
17	I			0.139		0.128											
	J	U	U	0.39	0.41	0.41	0.41										
21	I			0.154		0.144		0.132									
	J	U	U	0.40	0.43	0.42	0.43	0.44	0.44								
26	I			0.169		0.159		0.148		0.135							
	J	U	U	0.41	0.44	0.43	0.45	0.45	0.46	0.46	0.46						
35	I			0.189		0.180		0.170		0.158		0.139					
	J	U	U	0.41	0.46	0.43	0.47	0.45	0.48	0.47	0.48	0.49	0.49				
55	I			0.215		0.208		0.200		0.190		0.174		0.145			
	J	U	U	0.42	0.49	0.44	0.49	0.0.46	0.50	0.48	0.50	0.50	0.51	0.52	0.52		
135	I			0.250		0.248		0.245		0.240		0.231		0.210		0.151	
	J	U	U	0.43	0.51	0.45	0.52	0.47	0.53	0.49	0.53	0.51	0.54	0.53	0.55	0.56	0.56

Pressure angle = 20 degree, helix angle = 30 degree, full depth teeth with tip loading. AGMA 908-B89.

For spur and helical gears the surface strength geometry factor, Z_I, I, can be determined from Eq. (9.22) or (9.23).

$$\text{For external gears}: Z_I = I = \frac{\cos\phi_t \sin\phi_t}{2m_N}\frac{m_G}{m_G+1} \tag{9.30}$$

$$\text{For internal gears}: Z_I = I = \frac{\cos\phi_t \sin\phi_t}{2m_N}\frac{m_G}{m_G-1} \tag{9.31}$$

where $m_N = 1$ for spur gears. (ϕ_t is the transverse pressure angle used for helical gears. Replace ϕ_t by ϕ for spur gears.) For helical gears the load sharing ratio can be taken as

$$m_N = p_N/0.95Z \tag{9.32}$$

where p_N is the base pitch and Z is the length of the line of action in the transverse plane. The line of action in the transverse plane can be obtained from Eq. (9.33) (Shigley and Uicker Jr, 1980).

$$Z = \left[(r_P + a)^2 - r_{bP}^2\right]^{0.5} + \left[(r_G + a)^2 - r_{bG}^2\right]^{0.5} - (r_P + r_G)\sin\phi_t \tag{9.33}$$

where, depending on the application, $a = m_n$, $r_{bP} = r_p\cos\phi_t$, $r_{bG} = r_G\cos\phi_t$.

The speed ratio m_G is

$$m_G = \frac{N_G}{N_P} \tag{9.34}$$

The elastic coefficient Z_e or C_p can be found from Table 9.1 or Eq. (9.5).

The life factors Y_N and Z_N are used to modify the AGMA strengths for lives other than 10^7 cycles. The life required for a design is a critical decision. Consideration should be given to the maximum number of cycles that a machine in service will see. Obviously different customers may use machines in different ways. A common approach is to consider 95% of the machines in use. This allows for exclusion of extreme cases. Take for instance a washing machine. Market analysis would be required to identify what the maximum use of the machine would be. The 95 percentile excludes the tail of extreme users who might only rarely use their machines or use them 24h a day. Such high use may well be excluded from warrantee claims in any case. An estimate of use can also be made. It is conceivable that a washing machine in a family household might be used twice a day, every day of the year. Assuming that the washing process involves operation at 30rpm for 1h and 800rpm for 15min, then the total number of cycles in a year is given by $365 \times 2 \times ((30 \times 60) + (800 \times 15)) = 10.074 \times 10^6$ or about 10^7 cycles. If a one-year warranty is to be considered then the number of cycles to be used in calculations should be 10^7. The life factor Y_N can be determined from Fig. 9.6 and Z_N from Fig. 9.7. Note that if the pinion and gear have different teeth numbers then they will experience different numbers of cycles.

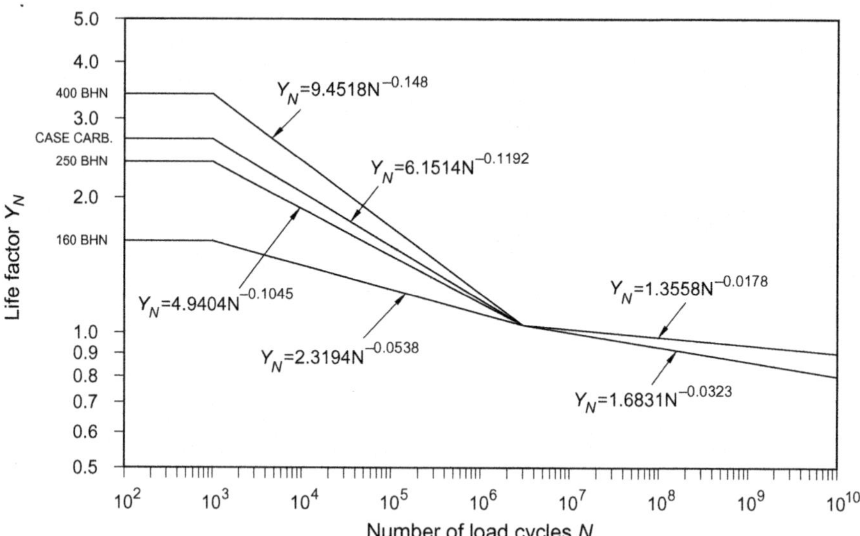

Fig. 9.6 Bending strength life factor Y_N.
Data from ANSI/AGMA 2001-C95.

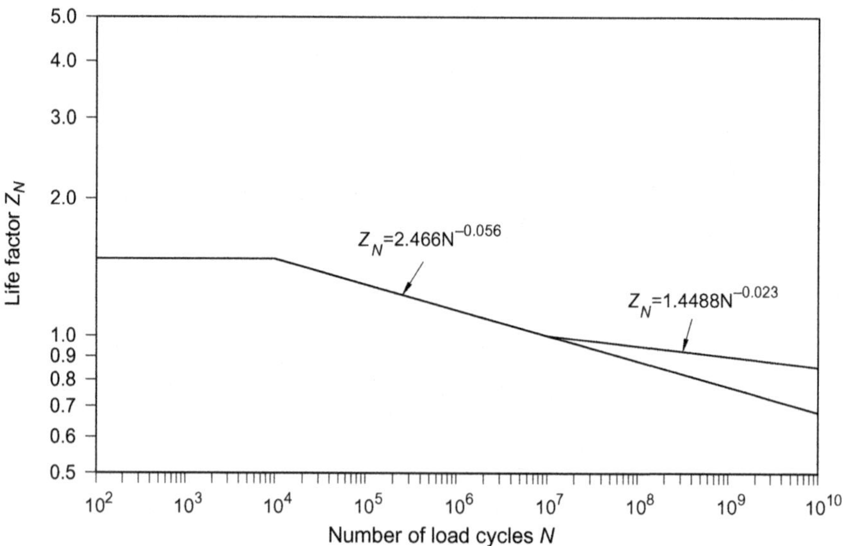

Fig. 9.7 Pitting resistance life factor Z_N.
Data from ANSI/AGMA 2001-C95.

Table 9.9 AGMA reliability factor Y_Z or K_R

Reliability	Y_Z, K_R
0.5	0.70
0.90	0.85
0.99	1.00
0.999	1.25
0.9999	1.50

After ANSI/AGMA 2001-C95

For oil or gear temperatures up to 120°C the temperature factor Y_θ can be taken as unity. Operation at temperatures above this value are not recommended due to degradation of the lubricant and use of a heat exchanger to reduce the lubricant temperature is advised in such circumstances. If the temperature exceeds 120°C the temperature factor Y_θ needs to be increased above unity.

The reliability factor, K_R or Y_Z, is used to account for the statistical distributions of failure of material by fatigue. Depending on the level of reliability desired the factor Y_Z should be adjusted accordingly as identified in Table 9.4. For values other than those listed in Table 9.9 use logarithmic interpolation.

The allowable bending stress for through hardened steels for grade 1 components is given by

$$\sigma_{FP} = 0.533H_B + 88.3 \text{ MPa} \tag{9.35}$$

and for grade 2:

$$\sigma_{FP} = 0.703H_B + 113 \text{ MPa} \tag{9.36}$$

The allowable bending stress for nitrided through hardened steels for grade 1 components is given by

$$\sigma_{FP} = 0.568H_B + 83.8 \text{ MPa} \tag{9.37}$$

and for grade 2:

$$\sigma_{FP} = 0.749H_B + 110 \text{ MPa} \tag{9.38}$$

Values for the allowable bending stress can be found from Eq. (9.35)–(9.38) for through hardened and nitrided through hardened steels and from Table 9.10 for iron and bronze gears. Alternatively an estimate for σ_{FP} can be obtained from Eq. (9.39).

$$\sigma_{FP} \approx 20.55 + 1.176H_B - 9.584 \times 10^{-4}H_B^2 \tag{9.39}$$

where σ_{FP} is the allowable bending stress in MPa.

Table 9.10 Repeatedly applied allowable bending stress and allowable contact stress for a selection of iron and bronze gear materials at 10^7 cycles and 99% reliability. ANSI/AGMA 2001/C95

Material	Designation	Heat treatment	Typical minimum surface hardness	Allowable bending stress (MPa)	Allowable contact stress (MPa)
ASTM A48 grey cast iron	Class 20	As cast	–	34	345 to 414
	Class 30	As cast	174 H_B	59	448 to 517
	Class 40	As cast	201 H_B	90	517 to 586
ASTM A536 ductile (nodular) iron	Grade 60-40-18	Annealed	140 H_B	152 to 228	530 to 634
	Grade 80-55-06	Quenched and tempered	179 H_B	152 to 228	530 to 634
	Grade 100-70-03	Quenched and tempered	229 H_B	186 to 276	634 to 772
	Grade 120-90-02	Quenched and tempered	269 H_B	214 to 303	710 to 869
Bronze		Sand cast	$\sigma_{uts\ min} = 275$ MPa	39	207
	ASTM B-148 alloy 954	Heat treated	$\sigma_{uts\ min} = 620$ MPa	163	448

The allowable contact stress, σ_{HP}, assuming unidirectional loading, 10^7 cycles and 99% reliability for grade 1 through hardened steel gears is given by

$$\sigma_{HP} = 2.22H_B + 200 \quad (9.40)$$

and for grade 2 through hardened steel gears by

$$\sigma_{HP} = 2.41H_B + 237 \quad (9.41)$$

Values for the allowable contact stress can be found from Eqs (9.40) and (9.41) for grade 1 and 2 through hardened steels and from Table 9.10 for iron and bronze gears. Alternatively an estimate for σ_{HP} can be obtained from Eq. (9.42).

$$\sigma_{HP} \approx 182.7 + 2.382H_B \quad (9.42)$$

where σ_{HP} is the allowable contact stress in MPa.

The hardness ratio factor is used only for the gear. Its purpose is to adjust the surface strengths to account for different hardness of the gear and pinion. The hardness ratio factor for a gear is given by:

$$Z_W = C_H = 1 + A'(m_G - 1) \tag{9.43}$$

where

$$A' = 8.98 \times 10^{-3}\left(\frac{H_{BP}}{H_{BG}}\right) - 8.29 \times 10^{-3} \tag{9.44}$$

for

$$1.2 \leq \frac{H_{BP}}{H_{BG}} \leq 1.7 \tag{9.45}$$

where H_{BP} and H_{BG} are the Brinell hardness of the pinion and gear respectively. $m_G = N_G/N_P$.

Example 9.2. A gear drive proposal consists of a 20 degree pressure angle spur pinion with 18 teeth driving a 50 tooth gear. The module of the gears is 2.5 and the face width is 30 mm. The pinion runs at 1425 rpm. The gear set is to transmit 3 kW and the load and drive can be considered smooth. The gear material proposed is grade 1 steel and the Brinell hardness of the pinion and gear is 240 and 200 respectively. The gears are to be manufactured to number 6 AGMA quality standard and the teeth are uncrowned. The gears are straddle mounted within an enclosed unit. Assume a pinion life of 10^8 cycles and a reliability of 90%. Calculate the AGMA bending and contact stresses and the corresponding factors of safety for both the pinion and the gear.

Solution

The pinion diameter, $d_P = 18 \times 2.5 = 45$ mm.

The gear diameter, $d_G = 50 \times 2.5 = 125$ mm.

The speed ratio (Eq. 9.34) is given by:

$$m_G = \frac{N_G}{N_P} = \frac{50}{18} = 2.778$$

Assume the overload factor, $K_O = 1$.

$$V = \omega r = 1425 \times \frac{2\pi}{60} \times \frac{0.045}{2} = 3.358 \text{m/s}$$

$$W_t = \frac{\text{Power}}{V} = \frac{3000}{3.358} = 893.5 \text{ N}$$

From Eqs (9.23) and (9.24),

$$B = \frac{(12 - Q_v)^{2/3}}{4} = \frac{(12 - 6)^{2/3}}{4} = 0.8255$$

$$A = 50 + 56(1 - B) = 50 + 56(1 - 0.8255) = 59.77$$

The dynamic factor (Eq. 9.22) is given by:

$$K_v' = \left(\frac{A + \sqrt{200V}}{A}\right)^B = \left(\frac{59.77 + \sqrt{200 \times 3.358}}{59.77}\right)^{0.8255} = 1.346$$

Assume the size factor $K_s = 1$.

The teeth are uncrowned, so $C_{mc} = 1$.

$F = 30\,\text{mm}$, so from Eq. (9.27),

$$C_{pf} = \frac{F}{10d} - 0.0375 + 0.4921F = \frac{0.03}{10 \times 0.045} - 0.0375 + 0.4921 \times 0.03$$
$$= 0.04392$$

$C_{pm} = 1$.

$C_e = 1$.

For commercially enclosed units, from Eq. (9.29),

$$C_{ma} = 0.127 + 0.622F - 0.01442F^2 = 0.0127 + 0.622 \times 0.03 - 0.01442 \times 0.03^2$$
$$= 0.1456$$

The load distribution factor K_H (Eq. 9.25), is given by

$$K_H = 1 + C_{mc}\left(C_{pf}C_{pm} + C_{ma}C_e\right) = 1 + 1(0.04392 \times 1 + 0.1456 \times 1) = 1.190$$

The repeatedly applied bending strength stress cycle factor for the pinion, Y_{NP}, from Fig. 9.6 taking the worst case line for 10^8 cycles is given by

$$Y_{NP} = 1.6831\left(1 \times 10^8\right)^{-0.0323} = 0.9284$$

The gear does comparatively less cycles and the repeatedly applied bending strength stress cycle factor for the gear, Y_{NG}, is given by

$$Y_{NG} = 1.6831\left(\frac{1 \times 10^8}{m_G}\right)^{-0.0323} = 1.6831\left(\frac{1 \times 10^8}{2.778}\right)^{-0.0323} = 0.9284$$

The pitting resistance stress cycle factor for the pinion, Z_{NP}, from Fig. 9.7 taking the worst case line for 10^8 cycles is given by

$$Z_{NP} = 2.466\,N^{-0.056} = 2.466\left(1 \times 10^8\right)^{-0.056} = 0.879$$

The gear does comparatively less cycles and the pitting resistance stress cycle factor for the gear, Z_{NG}, is given by

$$Z_{NG} = 2.466\left(\frac{1 \times 10^8}{m_G}\right)^{-0.056} = 2.466\left(\frac{1 \times 10^8}{2.778}\right)^{-0.056} = 0.9308$$

From Table 9.9, the reliability factor $Y_Z = 0.85$ for a reliability of 90%.

Assume the temperature factor $Y_\theta = 1$.

Assume the surface condition factor $Z_R = 1$.

For a steel pinion in mesh with a steel gear the elastic coefficient, $Z_e = 191\ (\text{MPa})^{0.5}$ from Table 9.1.

The allowable bending stress for through hardened grade 1 steel, in the case of the pinion from Eq. (9.35) is given by:

$$\sigma_{FP} = 0.533 H_B + 88.3 = 0.533 \times 240 + 88.3 = 216\,\text{MPa}$$

For the gear,

$$\sigma_{FG} = 0.533 \times 200 + 88.3 = 195\ \text{MPa}$$

The allowable contact stress for through hardened steel gears, in the case of the pinion (Eq. 9.40) is given by

$$\sigma_{HP} = 2.22 H_B + 200 = 2.22 \times 240 + 200 = 733\ \text{MPa}$$

For the gear,

$$\sigma_{HG} = 2.22 \times 200 + 200 = 644\ \text{MPa}$$

From Eq. (9.44),

$$A' = 8.98 \times 10^{-3}\left(\frac{H_{BP}}{H_{BG}}\right) - 8.29 \times 10^{-3} = 8.98 \times 10^{-3}\left(\frac{240}{200}\right) - 8.29 \times 10^{-3}$$
$$= 2.486 \times 10^{-3}$$

The hardness ratio factor, Z_W (Eq. 9.43) is given by

$$Z_W = 1 + A'(m_G - 1) = 1 + 2.486 \times 10^{-3}(2.778 - 1) = 1.004$$

From Fig. 9.5, the bending strength geometry factors $Y_{JP} = 0.316$ and $Y_{JG} = 0.396$.
The geometry factor Z_I (Eq. 9.30) is given by

$$Z_I = \frac{\cos\phi_t \sin\phi_t}{2m_N} \frac{m_G}{m_G + 1} = \frac{\cos 20 \ \sin 20}{2} \frac{2.778}{2.778 + 1} = 0.1182$$

The AGMA bending stress (Eq. 9.6) is given by

$$\sigma = W_t K_O K'_V K_s \frac{1}{Fm} \frac{K_H K_B}{Y_J}$$

Substitution for the factors and geometrical variables gives, for the pinion,

$$\sigma_P = 893.5 \times 1 \times 1.346 \times 1 \times \frac{1}{0.03 \times 0.0025} \times \frac{1.190 \times 1}{0.316} = 60.39 \times 10^6 \text{ Pa}$$

For the gear,

$$\sigma_G = 893.5 \times 1 \times 1.346 \times 1 \times \frac{1}{0.03 \times 0.0025} \times \frac{1.190 \times 1}{0.396} = 48.18 \times 10^6 \text{ Pa}$$

The allowable bending stress (Eq. 9.18) is given by

$$\sigma_{all} = \frac{\sigma_{FP} Y_N}{S_F Y_\theta Y_Z}$$

Setting $S_F = 1$, and determining the allowable bending stress for the pinion gives,

$$\sigma_{all} = \frac{216 \times 0.9284}{1 \times 1 \times 0.85} = 236 \text{ MPa}$$

The factor of safety for the pinion, S_{FP}, is given by

$$S_{FP} = \frac{236}{60.4} = 3.91$$

Setting $S_F = 1$, and determining the allowable bending stress for the gear gives,

$$\sigma_{all} = \frac{195 \times 0.9583}{1 \times 1 \times 0.85} = 220 \text{ MPa}$$

The factor of safety for the gear, S_{FG}, is given by

$$S_{FG} = \frac{220}{48.2} = 4.56$$

The values for S_{FP} and S_{FG} are >1 indicating that the design seems acceptable at this stage.

The contact stress (Eq. 9.16) is given by

$$\sigma_c = Z_e\left(W_t K_O K_V' K_s \frac{K_H}{Fd}\frac{Z_R}{Z_I}\right)^{0.5}$$

For the pinion,

$$\sigma_c = 191\sqrt{893.5\times 1\times 1.346\times 1\times \frac{1.190}{0.045\times 0.03}\times \frac{1}{0.1182}}\ \text{kPa}$$
$$= 191\times 2994\ \text{kPa} = 572\ \text{MPa}$$

For the gear,

$$\sigma_c = 191\sqrt{893.5\times 1\times 1.346\times 1\times \frac{1.190}{0.125\times 0.03}\times \frac{1}{0.1182}}\ \text{kPa}$$
$$= 191\times 1796\ \text{kPa} = 342\ \text{MPa}$$

The allowable contact stress (Eq. 9.20) is given by,

$$\sigma_{c,all} = \frac{\sigma_{HP}}{S_H}\frac{Z_N Z_W}{Y_\theta Y_Z}$$

For the pinion, omitting Z_W and setting $S_H = 1$ gives,

$$\sigma_{c,all} = \frac{\sigma_{HP}}{S_H}\frac{Z_N}{Y_\theta Y_Z} = \frac{732.8}{1}\times\frac{0.879}{1\times 0.85} = 758\ \text{MPa}$$

$$S_{HP} = \frac{758}{572} = 1.33$$

For the gear,

$$\sigma_{c,all} = \frac{644}{1}\times\frac{0.9308\times 1.004}{1\times 0.85} = 708\ \text{MPa}$$

$$S_{HG} = \frac{708}{342} = 2.06$$

The values for S_{HP} and S_{HG} are >1 indicating that the design seems acceptable at this stage.

Care needs to be taken when comparing values for the factors of safety for bending and contact stress. S_F should be compared with S_H^2 if the teeth are uncrowned or S_F

with S_H^3 if the teeth are crowned. The exponent linearises the safety factor for contact stress with respect to the transmitted load.

In Example 9.2, $S_{FP}=3.91$ and $S_{HP}=1.33$. Comparing $S_{FP}=3.91$ with $S_{HP}^2=1.77$ shows that the risk of failure for the pinion is more likely due to wear. Similarly for the gear, comparing $S_{FP}=4.56$ with $S_{HG}^2=4.24$ shows that the risk of failure is evenly balanced between bending and wear. In conclusion failure due to wear is most likely to cause retirement of the gear set.

Example 9.3. A gear drive proposal consists of a 20 degree pressure angle helical pinion with 17 teeth and a right hand helix angle of 30 degree, driving a 55 tooth helical gear. The normal module of the gears is 2.5 and the face width is 30 mm. The pinion runs at 1740 rpm. The gear set is to transmit 2.2 kW and the load and drive can be considered smooth. The gear material proposed is grade 1 steel and the Brinell hardness of the pinion and gear is 240 and 200 respectively. The gears are to be manufactured to number 6 AGMA quality standard and the teeth are uncrowned. The gears are straddle mounted within an enclosed unit. Assume a pinion life of 10^8 cycles and a reliability of 90%. Calculate the AGMA bending and contact stresses and the corresponding factors of safety for both the pinion and the gear.

Solution

The speed ratio (Eq. 9.34) is given by:

$$m_G=\frac{N_G}{N_P}=\frac{55}{17}=3.235$$

The transverse module is given by

$$m_t=\frac{m_n}{\cos\psi}=\frac{2.5}{\cos 30}=2.887$$

Assume the overload factor, $K_O=1$.

Assume the size factor, $K_s=1$.

The pitch diameter for the helical pinion is given by $d=m_tN$

$$d_P=m_tN_P=2.887\times 17=49.07\text{ mm}$$

And for the gear

$$d_G=m_tN_G=2.887\times 55=158.8\text{ mm}$$

$$V=\omega r=1740\times\frac{2\pi}{60}\times\frac{0.04907}{2}=4.471\text{ m/s}$$

$$W_t=\frac{\text{Power}}{V}=\frac{2200}{4.471}=492.1\text{ N}$$

The transverse pressure angle is given by:

$$\phi_t = \tan^{-1}\left(\frac{\tan\phi_n}{\cos\psi}\right) = \tan^{-1}\left(\frac{\tan 20}{\cos 30}\right) = 22.80^\circ$$

The radial, axial and resultant loads are given respectively by

$$W_r = W_t \tan\phi_t = 492.1 \tan 20 = 206.8 \text{ N}$$

$$W_a = W_t \tan\psi = 492.1 \tan 30 = 284.1 \text{ N}$$

$$W = \frac{W_t}{\cos\phi_n \cos\psi} = \frac{492.1}{\cos 20 \cos 30} = 604.6 \text{ N}$$

From Eqs (9.23) and (9.24)

$$B = \frac{(12 - Q_v)^{2/3}}{4} = \frac{(12-6)^{2/3}}{4} = 0.8255$$

$$A = 50 + 56(1 - B) = 50 + 56(1 - 0.8255) = 59.77$$

The dynamic factor (Eq. 9.22) is given by:

$$K'_V = \left(\frac{A + \sqrt{200\,V}}{A}\right)^B = \left(\frac{59.77 + \sqrt{200 \times 4.471}}{59.77}\right)^{0.8255} = 1.398$$

The teeth are uncrowned, so $C_{mc} = 1$.

$F = 30\,\text{mm}$, so from Eq. (9.27),

$$C_{pf} = \frac{F}{10d} - 0.0375 + 0.4921F = \frac{0.03}{10 \times 0.04908} - 0.0375 + 0.4921 \times 0.03$$
$$= 0.03839$$

$C_{pm} = 1$.

$C_e = 1$.

For commercially enclosed units, from Eq. (9.29),

$$C_{ma} = 0.127 + 0.622F - 0.01442F^2 = 0.0127 + 0.622 \times 0.03 - 0.01442 \times 0.03^2$$
$$= 0.1456$$

The load distribution factor K_H (Eq. 9.25) is given by

$$K_H = 1 + C_{mc}\left(C_{pf}C_{pm} + C_{ma}C_e\right) = 1 + 1(0.03839 \times 1 + 0.1456 \times 1) = 1.184$$

The repeatedly applied bending strength stress cycle factor for the pinion, Y_{NP}, from Fig. 9.6 taking the worst case line for 10^8 cycles is given by

$$Y_{NP} = 1.6831\left(1 \times 10^8\right)^{-0.0323} = 0.9283$$

The gear does comparatively less cycles and the repeatedly applied bending strength stress cycle factor for the gear, Y_{NG}, is given by

$$Y_{NG} = 1.6831\left(\frac{1 \times 10^8}{m_G}\right)^{-0.0323} = 1.6831\left(\frac{1 \times 10^8}{3.325}\right)^{-0.0323} = 0.9642$$

The pitting resistance stress cycle factor for the pinion, Z_{NP}, from Fig. 9.7 taking the worst case line for 10^8 cycles is given by

$$Z_{NP} = 2.466\ N^{-0.056} = 2.466\left(1 \times 10^8\right)^{-0.056} = 0.879$$

The gear does comparatively less cycles and the pitting resistance stress cycle factor for the gear, Z_{NG}, is given by

$$Z_{NG} = 2.466\left(\frac{1 \times 10^8}{m_G}\right)^{-0.056} = 2.466\left(\frac{1 \times 10^8}{3.325}\right)^{-0.056} = 0.9642$$

From Table 9.9, the reliability factor $Y_Z = 0.85$ for a reliability of 90%.

Assume the temperature factor $Y_\theta = 1$.

Assume the surface condition factor $Z_R = 1$.

For a steel pinion in mesh with a steel gear the elastic coefficient, $Z_e = 191\ (\text{MPa})^{0.5}$ from Table 9.1.

The allowable bending stress for through hardened grade 1 steel, in the case of the pinion from Eq. (9.35) is given by:

$$\sigma_{FP} = 0.533H_B + 88.3 = 0.533 \times 240 + 88.3 = 216\ \text{MPa}$$

For the gear,

$$\sigma_{FG} = 0.533 \times 200 + 88.3 = 195\ \text{MPa}$$

The allowable contact stress for through hardened steel gears, in the case of the pinion (Eq. 9.40) is given by

$$\sigma_{HP} = 2.22H_B + 200 = 2.22 \times 240 + 200 = 733\ \text{MPa}$$

For the gear,

$$\sigma_{HG} = 2.22 \times 200 + 200 = 644\ \text{MPa}$$

From Eq. (9.44),

$$A' = 8.98 \times 10^{-3}\left(\frac{H_{BP}}{H_{BG}}\right) - 8.29 \times 10^{-3} = 8.98 \times 10^{-3}\left(\frac{240}{200}\right) - 8.29 \times 10^{-3}$$
$$= 2.486 \times 10^{-3}$$

The hardness ratio factor, Z_W (Eq. 9.43) is given by

$$Z_W = 1 + A'(m_G - 1) = 1 + 2.486 \times 10^{-3}(3.235 - 1) = 1.006$$

From Table 9.8, the bending strength geometry factors $Y_{JP} = 0.44$ and $Y_{JG} = 0.49$.
The load sharing ratio, m_N (Eq. 9.32) is given by

$$m_N = \frac{p_N}{0.95Z}$$

The length of the line of action in the transverse plane (Eq. 9.33) is given by

$$Z = \left[(r_P + a)^2 - r_{bP}^2\right]^{0.5} + \left[(r_G + a)^2 - r_{bG}^2\right]^{0.5} - (r_P + r_G)\sin\varphi_t$$

$Z = 11.28\,\text{mm}$.
Hence $m_N = 0.6889$.
The geometry factor Z_I (Eq. 9.30) is given by

$$Z_I = \frac{\cos\phi_t \sin\phi_t}{2m_N}\frac{m_G}{m_G + 1} = \frac{\cos 22.80\ \sin 22.80}{2 \times 0.6889}\frac{3.235}{3.235 + 1} = 0.1980$$

The AGMA bending stress (Eq. 9.6) is given by

$$\sigma = W_t K_O K_V' K_s \frac{1}{Fm}\frac{K_H K_B}{Y_J}$$

Substitution for the factors and geometrical variables gives, for the pinion,

$$\sigma_P = 492.1 \times 1 \times 1.398 \times 1 \times \frac{1}{0.03 \times 0.002887} \times \frac{1.184 \times 1}{0.44} = 21.05 \times 10^6\ \text{Pa}$$

For the gear,

$$\sigma_G = 492.1 \times 1 \times 1.398 \times 1 \times \frac{1}{0.03 \times 0.002887} \times \frac{1.184 \times 1}{0.49} = 19.05 \times 10^6\,\text{Pa}$$

The allowable bending stress (Eq. 9.18) is given by

$$\sigma_{all} = \frac{\sigma_{FP} Y_N}{S_F Y_\theta Y_Z}$$

Setting $S_F = 1$, and determining the allowable bending stress for the pinion gives,

$$\sigma_{all} = \frac{216 \times 0.9283}{1 \times 1 \times 0.85} = 236\,\text{MPa}$$

The factor of safety for the pinion, S_{FP}, is given by

$$S_{FP} = \frac{236}{21.05} = 11.2$$

Setting $S_F = 1$, and determining the allowable bending stress for the gear gives,

$$\sigma_{all} = \frac{195 \times 0.9642}{1 \times 1 \times 0.85} = 221\,\text{MPa}$$

The factor of safety for the gear, S_{FG}, is given by

$$S_{FG} = \frac{221}{19.06} = 11.6$$

The values for S_{FP} and S_{FG} are >1 indicating that the design seems acceptable at this stage.

The contact stress (Eq. 9.16) is given by

$$\sigma_c = Z_e \left(W_t K_O K'_V K_s \frac{K_H}{F d} \frac{Z_R}{Z_I} \right)^{0.5}$$

For the pinion,

$$\sigma_c = 191 \sqrt{492.1 \times 1 \times 1.398 \times 1 \times \frac{1.184}{0.04907 \times 0.03} \times \frac{1}{0.1980}}\,\text{kPa} = 316.8\ \text{MPa}$$

For the gear,

$$\sigma_c = 191 \sqrt{492.1 \times 1 \times 1.398 \times 1 \times \frac{1.184}{0.1588 \times 0.03} \times \frac{1}{0.1980}}\,\text{kPa} = 318.1\ \text{MPa}$$

The allowable contact stress (Eq. 9.20) is given by,

$$\sigma_{c,all} = \frac{\sigma_{HP}}{S_H} \frac{Z_N Z_W}{Y_\theta Y_Z}$$

For the pinion, omitting Z_W and setting $S_H = 1$ gives,

$$\sigma_{c,all} = \frac{\sigma_{HP}}{S_H}\frac{Z_N Z_W}{Y_\theta Y_Z} = \frac{732}{1} \times \frac{0.8790 \times 1.006}{1 \times 0.85} = 762 \text{ MPa}$$

$$S_{HP} = \frac{762}{318.1} = 2.39$$

For the gear,

$$\sigma_{c,all} = \frac{644}{1} \times \frac{0.9387 \times 1.006}{1 \times 0.85} = 715 \text{ MPa}$$

$$S_{HG} = \frac{715}{318.1} = 2.24$$

The values for S_{HP} and S_{HG} are >1 indicating that the design seems acceptable at this stage.

A general approach for the calculation of AGMA bending and contact stresses for spur and helical gears is given in the flow charts in Figs. 9.8 and 9.9.

9.4 Gear selection procedure

Fig. 9.10 illustrates a general selection procedure for gears. Normally the required rotational speeds for the gear and pinion are known from the design requirements, as is the power the drive must transmit. A designer has to decide on the type of gears to be used, their arrangements on the shafts, materials, heat treatments, gear geometry: number of teeth, module, tooth form, face width, quality. There is no one solution to a gear design requirement. Several 'good' or optimum designs are possible. The designer's judgement and creativity are important aspects. Whilst Fig. 9.10 shows the general selection procedure, the steps given below outline a general calculation procedure for a pinion and gear.

Select the number of teeth for the pinion and the gear to give the required gear ratio.

Propose a material for the pinion and gear.

Select a module, m. Fig. 8.16 can be used as a visual guide to the size of gear teeth for a given module.

Calculate the pitch diameters.

$$d_P = mN_P, \quad d_G = mN_G$$

Calculate the pitch line velocity, $V = \omega r$.

Calculate the transmitted load, $W_t = \dfrac{\text{Power}}{V}$.

Calculate the pinion diameter, $d_P=mN$

↓

Calculate the gear diameter, $d_G=mN$

↓

Determine the speed ration m_G

↓

The speed ratio, Eq. (9.34), is given by:

↓

Assume the overload factor, $K_O=1$.

↓

Calculate $V = \omega r$

↓

Calculate $W_t = \frac{Power}{V}$

↓

Determine the factors A and B from Eq. (9.15) and (9.16),

↓

Calculate the dynamic factor, K_V', Eq. (9.22)

↓

Assume the size factor $K_s=1$.

↓

Determine C_{mc}, C_{pf}, C_e, C_{ma}

↓

Calculate the load distribution factor K_H, Eq. (9.25)

↓

Determine the repeatedly applied bending strength stress cycle factor for the pinion, Y_{NP}, from Fig. 9.6

↓

Determine Y_{NG}

↓

Determine the pitting resistance stress cycle factor for the pinion, Z_{NP}, from Fig. 9.7

↓

Determine Z_{NG}

↓

Determine the reliability factor Y_Z

↓

Assume the temperature factor $Y_\theta=1$.

↓

Assume the surface condition factor $Z_R=1$.

↓

Determine the elastic coefficient, Z_e

↓

Determine the allowable bending stresses, σ_{FP}, σ_{FG}

↓

Determine the allowable contact stresses, σ_{HP}, σ_{HG}

↓

Calculate A' from Eq. (9.44),

↓

Calculate the hardness ratio factor, Z_W, Eq. (9.43)

↓

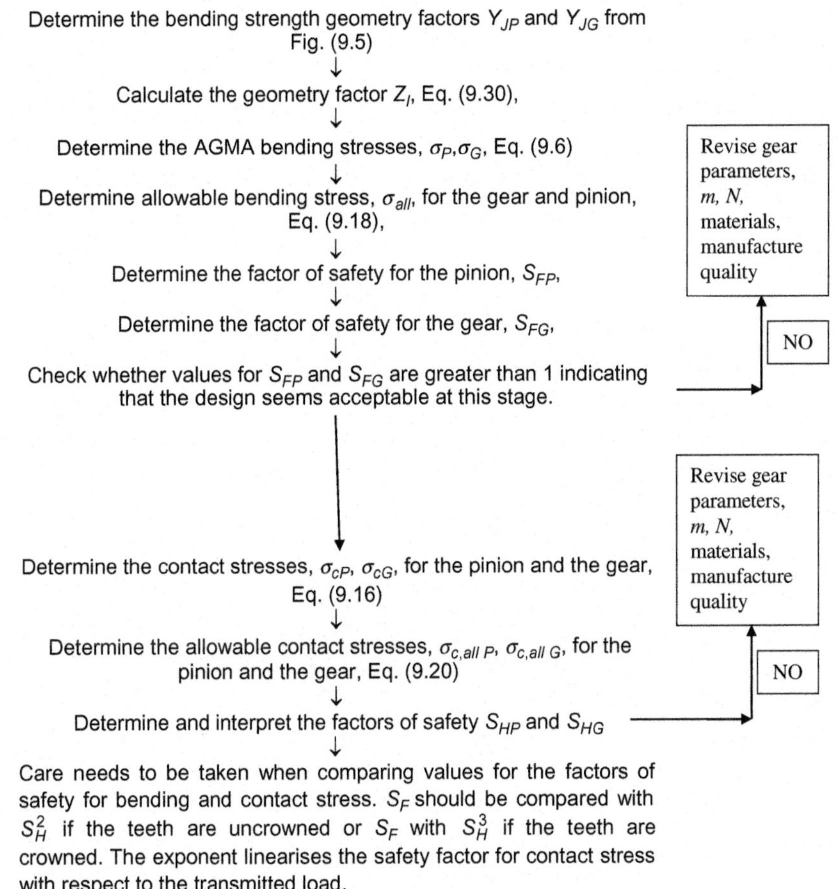

Determine the bending strength geometry factors Y_{JP} and Y_{JG} from Fig. (9.5)
Calculate the geometry factor Z_I, Eq. (9.30),
Determine the AGMA bending stresses, σ_P,σ_G, Eq. (9.6)
Determine allowable bending stress, σ_{all}, for the gear and pinion, Eq. (9.18),
Determine the factor of safety for the pinion, S_{FP},
Determine the factor of safety for the gear, S_{FG},
Check whether values for S_{FP} and S_{FG} are greater than 1 indicating that the design seems acceptable at this stage.
Revise gear parameters, m, N, materials, manufacture quality
NO
Determine the contact stresses, σ_{cP}, σ_{cG}, for the pinion and the gear, Eq. (9.16)
Determine the allowable contact stresses, $\sigma_{c,all\,P}$, $\sigma_{c,all\,G}$, for the pinion and the gear, Eq. (9.20)
Determine and interpret the factors of safety S_{HP} and S_{HG}
Revise gear parameters, m, N, materials, manufacture quality
NO
Care needs to be taken when comparing values for the factors of safety for bending and contact stress. S_F should be compared with S_H^2 if the teeth are uncrowned or S_F with S_H^3 if the teeth are crowned. The exponent linearises the safety factor for contact stress with respect to the transmitted load.

Fig. 9.8 General procedure for the calculation of AGMA bending and contact stresses for a spur gear.

Specify the face width or calculate using the procedure outlined in Section 8.6 from:

$$F = \frac{W_t}{K_v m Y \sigma_p}$$

As a general rule spur gears should be designed with a face width between three and five times the circular pitch.

Check the proposed gears for bending and contact stress. Determine the AGMA factors K_O, K_v', K_H, Y_N, Z_N, Z_W, Z_I, Y_J.

Calculate the speed ratio, m_G, Eq. (9.34)

↓

Calculate the transverse module

↓

Assume the overload factor, $K_O=1$.

↓

Assume the size factor, $K_s=1$.

↓

Calculate the pitch diameters for the helical pinion and gear, $d=m_tN$

↓

Calculate V

↓

Determine W_t

↓

Calculate the transverse pressure angle

↓

Calculate the radial, axial and resultant loads, W_r, W_a, W

↓

Determine the factors B and A from Eq. (9.23) and (9.24),

↓

Calculate the dynamic factor, K_V' Eq. (9.22)

↓

Determine C_{mc}, C_{pf}, C_e, C_{ma}

↓

Calculate the load distribution factor K_H

↓

Determine the repeatedly applied bending strength stress cycle factor for the pinion, Y_{NP}, from Fig. (9.6)

↓

Determine Y_{NG}

↓

Determine the pitting resistance stress cycle factor for the pinion, Z_{NP}, from Fig. (9.7)

↓

Determine Z_{NG}

↓

Determine the reliability factor Y_Z

↓

Assume the temperature factor $Y_\theta=1$.

↓

Assume the surface condition factor $Z_R=1$.

↓

Determine the elastic coefficient, Z_e

↓

Determine the allowable bending stresses, σ_{FP}, σ_{FG}, Eq. (9.35)

↓

Determine the allowable contact stresses, σ_{HP}, σ_{HG}, Eq. (9.40)

↓

Determine A' from Eq. (9.44),

↓

Calculate the hardness ratio factor, Z_W, Eq. (9.43)

↓

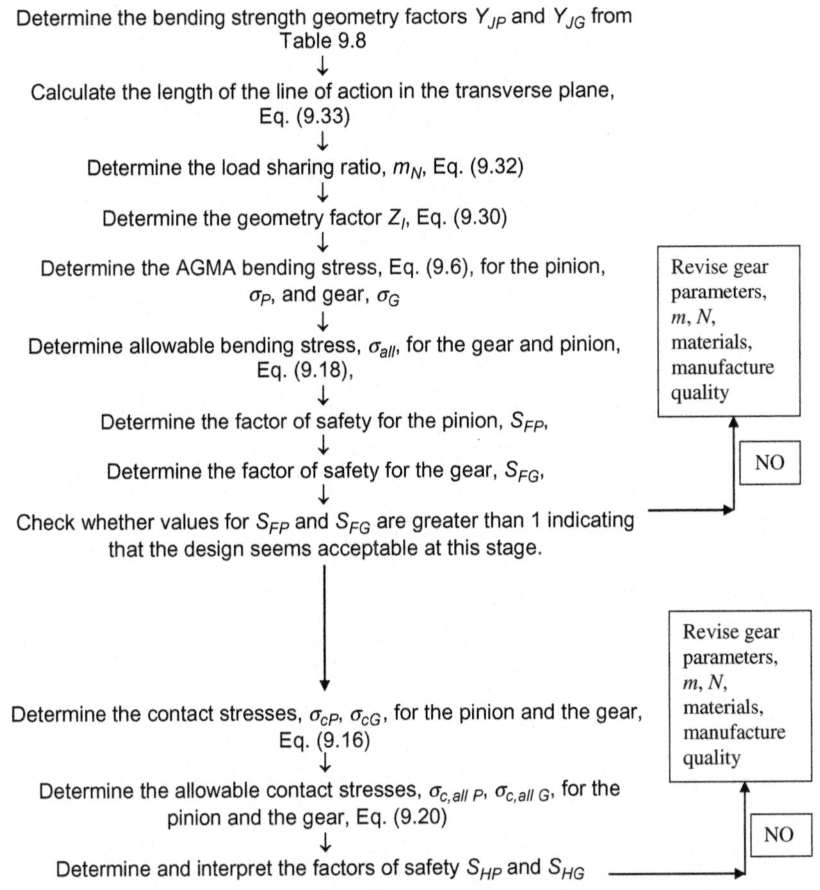

Fig. 9.9 General procedure for the calculation of AGMA bending and contact stresses for a helical gear.

Calculate the AGMA bending stress (Eq. 9.6), $\sigma = W_t K_O K'_V K_s \frac{1}{Fm}\frac{K_H K_B}{Y_J}$.

Calculate the AGMA contact stress (Eq. 9.16), $\sigma_c = Z_e\left(W_t K_O K'_V K_s \frac{K_H}{Fd}\frac{Z_R}{Z_I}\right)^{0.5}$.

Calculate the allowable bending stress (Eq. 9.18), $\sigma_{all} = \frac{\sigma_{FP} Y_N}{S_F Y_\theta Y_Z}$.

Calculate the allowable contact stress (Eq. 9.20), $\sigma_{c,all} = \frac{\sigma_{HP}}{S_H}\frac{Z_N Z_W}{Y_\theta Y_Z}$.

If values for the factors of safety S_F and S_H are unsatisfactory, either change initial geometry or change materials.

This procedure can be reordered say if a particular gear material is to be used and stress is to be limited to a certain value, or if a particular factor of safety is required and so on.

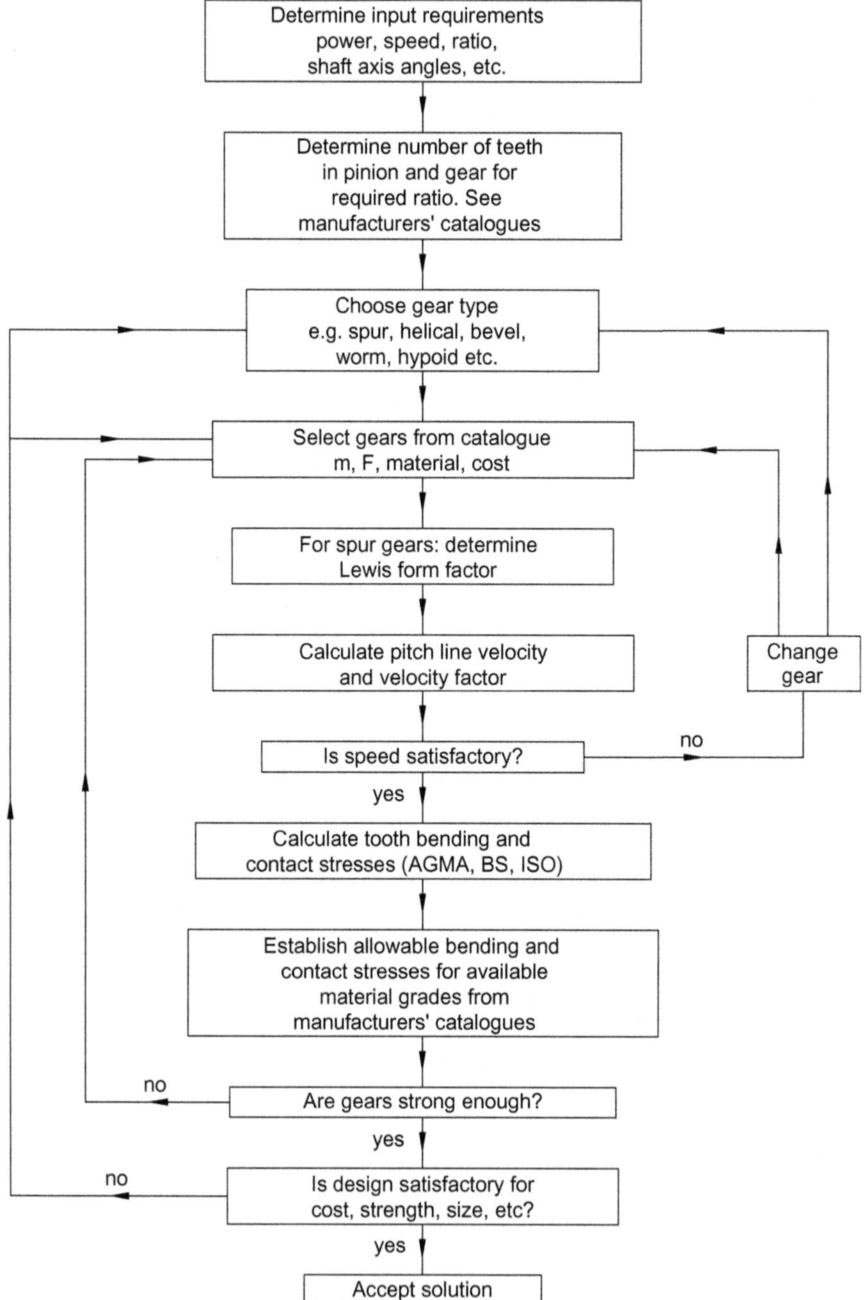

Fig. 9.10 General selection procedure for gears.

9.5 Conclusions

Gears can fail due to excessive bending stresses or by wear. Wear can take the form of pitting of the surface due to excessive contact stresses, scoring due to failure of lubrication or abrasion due to the presence of foreign particulates. This chapter has introduced the AGMA equations for modelling bending and contact stresses in spur gears. The calculations involved in evaluating the AGMA bending and contact stresses are lengthy and rely on a large number of factors and geometrical parameters. These can be readily embedded in a programme or spreadsheet calculation procedure if required.

References

Barbieri, M., Zippo, A., Pellicano, F., 2014. Adaptive grid-size finite element modeling of helical gear pairs. Mech. Mach. Theory 82, 17–32.

Batsch, M., 2016. Comparative fatigue testing of gears with involute and convexo-concave teeth profiles. Adv. Manuf. Sci. Technol. 40, 5–25.

Chen, Y.C., Tsay, C.B., 2002. Stress analysis of a helical gear set with localized bearing contact. Finite Elem. Anal. Des. 38, 707–723.

Fernandes, P.J.L., 1996. Tooth bending fatigue failures in gears. Eng. Fail. Anal. 3, 219–225.

Fernandes, P.J.L., McDuling, C., 1997. Surface contact fatigue failures in gears. Eng. Fail. Anal. 4, 99–107.

Hassan, A.R., 2009. Contact stress analysis of spur gear teeth pair. World Academy of Science. Eng. Technol. 58, 597–602.

Jyothirmai, S., Ramesh, R., Swarnalatha, T., Renuka, D., 2014. A finite element approach to bending, contact and fatigue stress distribution in helical gear systems. Proc. Mater. Sci. 6, 907–918.

Li, S., Kahraman, A., 2014. A micro-pitting model for spur gear contacts. Int. J. Fatigue 59, 224–233.

Michalczewski, R., Kalbarczyk, M., Michalak, M., Piekoszewski, W., Szczerek, M., Tuszynski, W., Wulczynski, J., 2013. New scuffing test methods for the determination of the scuffing resistance of coated gears. Chapter 6, In: Gegner, J. (Ed.), Tribology—Fundamentals and Advancements.

Patil, S.S., Karuppanan, S., Atanasovska, I., Wahab, A.A., 2014. Contact stress analysis of helical gear pairs, including frictional coefficients. Int. J. Mech. Sci. 85, 205–211.

Rao, C.R.M., Muthuveerappan, G., 1993. Finite element modelling and stress analysis of helical gear teeth. Comput. Struct. 49, 1095–1106.

Shigley, J.E., Mischke, C.R., 2001. Mechanical Engineering Design, sixth ed. McGraw Hill.

Shigley, J.E., Uicker Jr., J.J., 1980. Theory of Machines and Mechanisms. McGraw Hill.

Standards

AGMA 908-B89. Geometry factors for determining the pitting resistance and bending strength of spur, helical and herringbone gear teeth. AGMA Information Sheet.

ANSI/AGMA 2001-C95. Fundamental rating factors and calculation methods for involute spur and helical gear teeth. American Gear Manufacturers Association.

ANSI/AGMA 2001-D04. Fundamental rating factors and calculation methods for involute spur and helical gear teeth.

ANSI/AGMA 2101-D04. Fundamental rating factors and calculation methods for involute spur and helical gear teeth.

BS 2519-2:1976, ISO 701:1976.Glossary for gears. Notation.

BS 436-5:1997, ISO 1328-2:1997. Spur and helical gears. Definitions and allowable values of deviations relevant to radial composite deviations and runout information.

BS 6168:1987. Specification for non-metallic spur gears.

BS ISO 6336-1:2006. Calculation of load capacity of spur and helical gears. Basic principles, introduction and general influence factors.

BS ISO 6336-2:2006. Calculation of load capacity of spur and helical gears. Calculation of surface durability (pitting).

BS ISO 6336-3:2006. Calculation of load capacity of spur and helical gears. Calculation of tooth bending strength. Corrigendum 1.

BS ISO 6336-5:2003. Calculation of load capacity of spur and helical gears. Strength and quality of materials.

BS ISO 6336-6:2006. Calculation of load capacity of spur and helical gears. Calculation of service life under variable load.

BS ISO 9084:2000. Calculation of load capacity of spur and helical gears. Application to high speed gears and gears of similar requirements.

PD 6457:1970. Guide to the application of addendum modification to involute spur and helical gears.

Websites

At the time of going to press the world wide web contained useful information relating to this chapter at the following sites

www.agma.org.

bsol.bsigroup.com.

www.esdu.com.

www.iso.org.

Further reading

Mott, R.L., 1998. Machine Elements in Mechanical Design, third ed. Prentice Hall.

Bevel gears

10

Chapter Outline

Nomenclature

Generally, preferred SI units have been stated.

d	pitch diameter (mm)
d_G	pitch diameter for the pinion (mm)
d_P	pitch diameter for the gear (mm)
d_{e1}	pinion outer pitch diameter (mm)
E_1	Young's modulus for the pinion (MPa)
E_2	Young's modulus for the gear (MPa)
F	net face width (mm)
F_1'	effective face width of the pinion (mm)
F_2'	effective face width of the gear (mm)
g_c	length of line contact (mm)
H_{B1}	minimum Brinell hardness for the pinion (HB, MPa)
H_{B2}	minimum Brinell hardness for the gear (HB, MPa)
K_θ	temperature factor
K_A	overload factor
$K_{H\beta}$	load distribution factor
K_{mb}	load distribution modifier
K_v	dynamic factor
L	orthogonal length from the gear back cone to the axis (mm)
m	module
m_{et}	metric outer traverse module (mm)
m_{et}	outer traverse module (mm)
m_G	gear ratio
m_{mt}	mean traverse module (mm)
n_1	pinion speed (rpm)
n_P	pinion speed (rpm)

Mechanical Design Engineering Handbook. https://doi.org/10.1016/B978-0-08-102367-9.00010-X

n_G	gear speed (rpm)
N	number of teeth
N_G	number of teeth in the gear
N_P	number of teeth in the pinion
P	power (kW)
P_{ay}	allowable transmitted power (kW)
P_{az}	allowable transmitted power (kW)
Q_v	transmission accuracy number
R_{a1}	pinion surface roughness (μm)
r_{av}	pitch radius at the midpoint of the tooth for the gear under consideration (m)
r_{c0}	cutter radius (mm)
R_m	mean cone distance (mm)
r_{mpt1}	mean transverse pitch radius for the pinion (mm)
r_{mpt2}	mean transverse pitch radius for the gear (mm)
r_{myo1}	mean transverse radius to the point of load application for the pinion (mm)
r_{myo2}	mean transverse radius to the point of load application for the gear (mm)
S_F	bending safety factor
S_H	contact safety factor
T	peak operating gear temperature (K)
T_q	torque (N m)
T_{q1}	operating pinion torque (N m)
v_{et}	pitch line velocity at the outside pitch diameter (m/s)
W_a	axial force (N)
W_r	radial force (N)
W_t	transmitted load (N)
Y_β	tooth lengthwise curvature factor
Y_i	inertia ratio for gears with low contact ratio
Y_J	bending strength geometry factor
Y_{K1}	tooth form factor for the pinion
Y_{K2}	tooth form factor for the gear
Y_{NT}	stress cycle factor
Y_x	size factor
Y_Z	reliability factor
Z_E	elastic coefficient ($\sqrt{\text{MPa}}$)
Z_i	inertia factor for gears with a low contact ratio
Z_I	pitting resistance geometry factor
Z_{NT}	stress cycle factor
Z_W	hardness ratio factor
Z_x	size factor
Z_{xc}	crowning factor
Z_Z	reliability factor
α	pitch angle (degree)
α_G	pitch angle for the gear (degree)
α_n	normal pressure angle (degree)
α_P	pitch angle for the pinion (degree)
β_m	mean spiral angle (degree)

ε_{NI}	load sharing ratio
ε_{NJ}	load sharing ratio
ϕ	pressure angle (degree)
ρ_{y0}	relative radius of profile curvature (mm)
ν_1	Poisson's ratio for the pinion
ν_2	Poisson's ratio for the gear
$\sigma_{F\,\lim}$	allowable bending stress (N/mm^2, MPa)
σ_F	bending stress at the root of the tooth (N/mm^2, MPa)
σ_{FP}	permissible bending stress (N/mm^2, MPa)
$\sigma_{H\,\lim}$	allowable contact stress (N/mm^2, MPa)
σ_H	contact stress (N/mm^2, MPa)
σ_{HP}	permissible contact stress (N/mm^2, MPa)
ω	angular velocity (rad/s)
ψ	spiral angle (degree)

10.1 Introduction

Bevel gears have a conical form and can be used to transmit rotational power through shafts that are typically at an angle of 90° to each other. This functionality is useful in a wide range of applications from cordless hand-tools to automotive transmissions and outboard motors where the prime-mover location is not coaxial with the driven shaft. This chapter introduces bevel gears and the ANSI/AGMA 2003-B97 Standard that provides a conservative means for estimating the bending and contact stress in straight, zerol and spiral bevel gears and comparing the merits of different design proposals.

Bevel gears have teeth cut on conical blanks (Figs. 10.1 and 10.2) and a gear pair can connect nonparallel intersecting shafts. Bevel gears are used for motor transmission differential drives, valve control and mechanical instruments. A variety of tooth forms are possible including:

- straight bevel gears;
- spiral bevel gears;
- zerol bevel gears.

Straight bevel gears (Fig. 10.1) have, as implied in their name, a straight tooth form cut parallel to the cone axis, which if extended would pass through a point of intersection on the shaft axis. Bevel gears are normally mounted on shafts that are at 90 degrees to each other, but bevel gears can be designed for other angles. Straight bevel gears can be noisy, due to the abrupt line or flank contact between teeth as they mesh. This abrupt meshing means that there can be a high impact stress if the teeth are used under heavy loads or at high speeds which can lead to the possibility of tooth breakages and gear failure (Park, 2003; Siddiqui et al., 2013; Zakrajsek et al., 1994). Straight bevel gears are usually only suitable for speeds up to 5 m/s.

Fig. 10.1 Straight cut bevel gears.

Spiral bevel gears (Fig. 10.2) have curved teeth that are formed along a spiral angle to the cone axis. The most common spiral angle is 35 degree with a pressure angle of 20 degree. The advantage of spiral bevel gears over straight teeth is that the gears engage more gradually with contact commencing at one end of the tooth which increases until there is contact across the whole length of the tooth. This enables a smoother transmission of power and reduces the risk of tooth breakage. As a result of the smoother operation, spiral bevel gears tend to be quieter and can have a smaller diameter for the same load capacity than straight bevel gears. Spiral bevel gears are recommended for pitch line speeds in the range from 5 m/s to 40 m/s. Higher speeds, in the region of 50 m/s, are possible with high precision finished gears.

Zerol bevel gears have a tooth form that is curved but with a zero spiral angle (Fig. 10.3). They represent an intermediate category between straight and spiral bevel gears. Zerol gears are generally smoother and quieter than straight bevel gears but they do produce side loads.

Fig. 10.2 Spiral bevel gear.
Image courtesy Boston Gear—Altra Industrial Motion.

Fig. 10.3 Zero bevel gear for electrical power tool, Crown gear 37 T × Pinion gear 6 T.
(Figure courtesy Bevel Gear Co Ltd).

In general a maximum reduction ratio of 10:1 is recommended for a bevel gear set. A maximum step ratio of 5:1 is recommended when using a bevel gear set for increasing the speed of a transmission. Bevel gears tend to be made as a paired set and are in general not interchangeable.

The mountings for bevel gears need to be rigid to maintain the displacement of the gears within specified limits. To align bevel gears, attention should be given to the accuracy of the mountings and careful design and fitting of keys and keyways.

The design procedure for a bevel gear set can be approached in a number of ways, ranging from consideration of the bending and contact stresses using a methodology based on a standard (e.g. ISO or AGMA), to optimisation and finite element analysis (Lin and Fong, 2015; Padmanabhan et al., 2011).

If the power and speed are known then the procedure can be approached by proposing values for the tooth system, module, teeth numbers, face width, gear material and case hardness, quality number, mounting arrangement and associated precision. These factors need to be considered in accordance with the required or desired reliability, design life, space available and cost. The process involves determining the AGMA contact and bending stresses, comparing these to the respective permitted stresses for the proposed gears and assessing the suitability of the safety factors. The process is typically iterative involving an initial proposal and then reconsideration of a series of factors. Alternatively the power rating, based on the AGMA equations for contact and bending stress, can be determined for a given proposal for the input speed, tooth system, module, teeth numbers, face width, gear material and case hardness, quality number, mounting arrangement, reliability, design life.

10.2 Force analysis

Usual practice for bevel gears is to approximate the tangential or transmitted load by the load that would occur if all the forces were concentrated at the mid-point of the tooth.

$$W_t = \frac{T_q}{r_{av}} \tag{10.1}$$

where

W_t is the transmitted load (N);
r_{av} is the pitch radius at the mid-point of the tooth for the gear or pinion under consideration (m);
T_q is the torque (N m).

The resultant force has three components, W_a, W_t and W_r

$$W_r = W_t \tan\phi \cos\alpha \tag{10.2}$$

$$W_a = W_t \tan\phi \sin\alpha \tag{10.3}$$

where

W_a is the axial force (N);
W_t is the tangential force or transmitted load (N);
W_r is the radial force (N);
ϕ is the pressure angle (degree);
α is the pitch cone angle (degree).

For the pitch cone angle (Fig. 10.4) the relevant angle for the pinion, α_P, or the gear, α_G, should be substituted, depending on which element is under analysis.

$$\alpha_G = \tan^{-1}\left(\frac{N_G}{N_P}\right) \tag{10.4}$$

$$\alpha_P = 90 - \alpha_G \tag{10.5}$$

where

N_G is the number of teeth in the gear;
N_P is the number of teeth in the pinion.

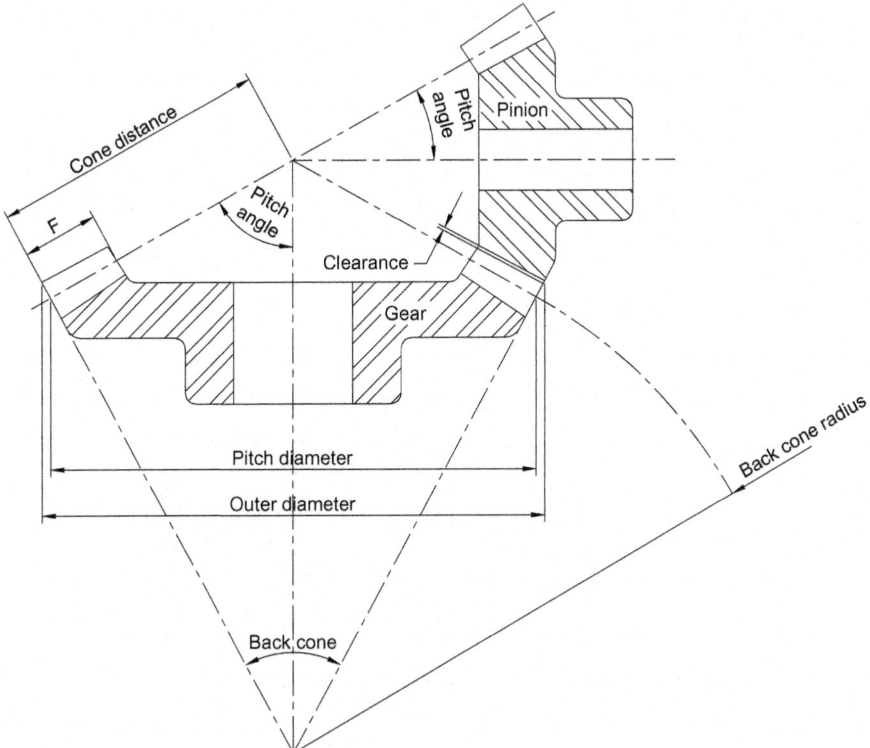

Fig. 10.4 Bevel gear nomenclature.

For a spiral bevel gear,

$$W_r = \frac{W_t}{\cos\psi}(\tan\phi_n \cos\alpha \pm \sin\psi \sin\alpha) \tag{10.6}$$

$$W_a = \frac{W_t}{\cos\psi}(\tan\phi_n \sin\alpha \mp \sin\psi \cos\alpha) \tag{10.7}$$

The hand of a spiral bevel gear is defined so that right hand teeth incline away from the axis in a clockwise direction looking at the small end. As with helical gears, the hand of a gear is opposite to that of its mate with the hand of the gear pair defined as that of the pinion. The upper signs in Eqs (10.6) and (10.7) are used for a driving pinion with a right hand spiral rotating clockwise viewed from its large end, or for a driving pinion with a left hand spiral rotating counter-clockwise viewed from its large end. The lower signs are used for the corresponding opposite directions.

10.3 Stress analysis

Bevel gear sets are often mounted with one of the gears outboard of the bearings (Fig. 10.5). As a result shaft deflections can be more significant than for a mid-mounted gear. To compound the challenge in modelling gear stresses, the teeth are tapered. To achieve perfect line contact through the cone centre, the teeth would need to bend more at the larger end of the cone than at the smaller end. Such conditions dictate proportionally greater loads at the large end. In general, as a result of the varying load, it is desirable to have a fairly short face width for bevel gears. In general the face width, F, is limited to a maximum of $L/3$ where

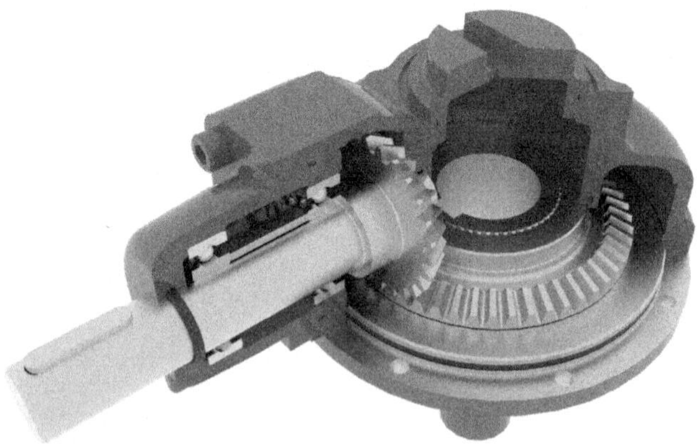

Fig. 10.5 Outboard bevel gear arrangement (Image courtesy of Rotork).

$$L = \frac{d_P}{2\sin\alpha_P} = \frac{d_G}{2\sin\alpha_G} \quad (10.8)$$

where L is the orthogonal length from the gear back cone to the axis (mm) and is also known as the back cone distance.

The ANSI/AGMA 2003-B97 (2003) Standard provides a conservative means for estimating the bending and contact stress in straight, zerol and spiral bevel gears and comparing the merits of different design proposals.

Use of the standard requires determination of a significant number of factors. The information provided in this section provides a summary of these.

Bevel gear teeth can be formed by cutting, hard-finish cutting or grinding (Deng et al., 2011; Jelaska, 2012; Litvin et al., 2006). There are several methods for cutting the teeth, dependent upon the volume of parts to be produced, the desired quality and the availability of equipment. Hard-finish cutting and grinding of the bevel gear teeth after heat treatment can produce machine elements of superior quality with good profile and root fillet finishes, tooth position and concentricity. Precision finish, as used in defining the American Gear Manufacturers Association (AGMA) standard, refers to a secondary machine finish operation which includes grinding, skiving or hard-cut finishing.

The teeth of most bevel gears are crowned in profile and lengthwise directions during the manufacturing to accommodate deflection and tooth variations. Crowning involves modifying the surface profile so that the teeth have convex surfaces along the face width. Under design load, the contact should spread over the tooth flank without any concentration of the pattern at the edges.

As most bevel gears are made from carburised case-hardened steel, the factors incorporated in the AGMA standard are principally based on tests performed on gears made using this heat treating process.

The dynamic response of the system results in additional gear tooth loads due to the characteristic relative motions of the driver and the driven equipment. The overload factor, K_A, is used to account for the operating characteristics of the driving and driven equipment.

Failure of gear teeth by breakage can be caused by severe instantaneous overloads, excessive pitting (Sekercioglu and Kovan, 2007; Siddiqui et al., 2013), case crushing or bending fatigue. Assuming that the gears have been designed to avoid pitting and case crushing, the formulas for bending strength rating can be used. The strength ratings determined by the AGMA standard are based on cantilever beam theory modified to consider:

- the compressive stress at the tooth roots caused by the radial component of the tooth load;
- nonuniform moment distribution of the load resulting from the inclined contact lines on the teeth of spiral bevel gears;
- stress concentration at the tooth root fillet;
- load sharing between adjacent contacting teeth;
- the lack of smoothness due to low contact ratio.

The capacity of a gear tooth is determined by considering the following values.

(a) The calculated stress number based on: the geometry of the gear tooth; manufacturing tolerances; the rigidity of the gear blank, bearings, housing and foundation; the operating torque.
(b) The permissible stress number based on the allowable stress and the effect of the working conditions under which the gears operate.
(c) The maximum power that the gear is capable of transmitting at the operating speeds and loads and under the working conditions. The maximum power is based on the allowable stress.

The fundamental ANSI/AGMA 2003-B97 contact stress for pitting resistance of bevel gear teeth can be calculated by:

$$\sigma_H = Z_E\sqrt{\frac{2000T_{q1}}{Fd_{e1}^2 Z_I}K_A K_v K_{H\beta} Z_x Z_{xc}} \tag{10.9}$$

where

σ_H is the contact stress (N/mm^2, MPa);
Z_E is the elastic coefficient ($\sqrt{\text{MPa}}$);
Z_x is the size factor;
T_{q1} is the operating pinion torque (N m);
K_A is the overload factor;
K_v is the dynamic factor;
F is the net face width (mm);
d_{e1} is the pinion outer pitch diameter (mm);
$K_{H\beta}$ is the load distribution factor;
Z_{xc} is the crowning factor;
Z_I is the pitting resistance geometry factor.

It should be noted that Eq. (10.9) has lengths expressed in terms of mm and produces a result in MPa. The various factors required for Eq. (10.9) are considered in this section and their use illustrated in two examples.

The permissible ANSI/AGMA 2003-B97 contact stress is given by

$$\sigma_{HP} = \frac{\sigma_{H\lim} Z_{NT} Z_W}{S_H K_\theta Z_Z} \tag{10.10}$$

where

σ_{HP} is the permissible contact stress (N/mm^2, MPa);
$\sigma_{H\ \lim}$ is the allowable contact stress (N/mm^2, MPa);
Z_{NT} is the stress cycle factor;
S_H is the contact safety factor;
Z_W is hardness ratio factor;
K_θ is the temperature factor;
Z_Z is the reliability factor.

The various factors required for Eq. (10.10) are considered in this section and their use illustrated in the two examples.

The calculated contact stress must be less than or equal to the permissible contact stress, that is $\sigma_H \leq \sigma_{HP}$.

The rated power at a specific speed to resist pitting of the gear teeth, according to ANSI/AGMA 2003-B97, is given by

$$P_{az} = \frac{n_1 F}{1.91 \times 10^7} \frac{Z_I}{K_v K_{H\beta} K_A Z_x Z_{xc}} \left(\frac{\sigma_{H\lim} d_{e1} Z_{NI} Z_W}{S_H Z_E K_\theta Z_Z} \right)^2 \tag{10.11}$$

where

P_{az} is the allowable transmitted power (kW);
n_1 is the pinion speed (rpm).

The values of $\sigma_{H\,\lim}$ and Z_{NT} may be different for the pinion and the gear. The rating should be based on the machine element that has the lowest value of $\sigma_{H\,\lim}$.

The bending stress of the gear teeth according to ANSI/AGMA 2003-B97 is given by

$$\sigma_F = \frac{2000 T_{q1}}{F d_{e1}} \frac{K_A K_v}{m_{et}} \frac{Y_x K_{H\beta}}{Y_\beta Y_J} \tag{10.12}$$

where

σ_F is the bending stress at the root of the tooth (N/mm^2, MPa);
m_{et} is the metric outer transverse module (mm);
Y_x is the size factor;
Y_β is the tooth lengthwise curvature factor;
Y_J is the bending strength geometry factor.

It should be noted that Eq. (10.12) has lengths expressed in terms of mm and produces a result in MPa.

The permissible bending stress according to ANSI/AGMA 2003-B97 is given by

$$\sigma_{FP} = \frac{\sigma_{F\lim} Y_{NT}}{S_F K_\theta Y_z} \tag{10.13}$$

where

σ_{FP} is the bending stress (N/mm^2, MPa);
$\sigma_{F\,\lim}$ is the allowable bending stress (N/mm^2, MPa);
S_F is the bending safety factor;
Y_{NT} is the stress cycle factor;
Y_Z is the reliability factor.

The tensile bending stress must be less than or equal to the permissible bending stress, that is $\sigma_F \leq \sigma_{FP}$.

The rated power to resist teeth breakage, according to ANSI/AGMA 2003-B97, is given by

$$P_{ay} = \frac{n_1 F}{1.91 \times 10^7} \frac{Y_J Y_\beta}{Y_x K_{H\beta} K_A K_v} \frac{\sigma_{F\lim} d_{e1} m_{et}}{1} \frac{Y_{NT}}{K_\theta Y_z S_F} \quad (10.14)$$

where P_{ay} is the allowable transmitted power (kW).

The values for the stress cycle factor, K_L, may be different for the pinion and the gear and the rating should therefore be based on the lower value of the product $\sigma_{F\ \lim} Y_{NT}$.

In most gear applications, load is not constant. To obtain a nominal value for the operating torque the following equation can be used based on the value of the power and speed of operation that will be in effect over significant periods of time and under the most severe operating conditions.

$$T_{q1} = \frac{9550P}{n_1} \quad (10.15)$$

where

P is the power (kW).
n_1 is the pinion speed (rpm).

The overload factor, K_A, makes allowance for any externally applied loads in excess of the nominal tangential load. The overload factor is typically between 1 and 2.25 and depends on the characteristic motion of the driven and driving machinery. Factors can only be established with confidence after considerable field experience is gained for a particular application. Table 10.1, however, can be used to provide an estimate for the overload factor, based on a reasonable assessment of the prime mover and driven machine characteristics.

A safety factor can be used to account for uncertainties in the design analysis, material properties and manufacturing quality as well as the consequences of failure.

Table 10.1 Estimated overload factors, K_A, based on the driving and driven machine characteristics

	Driven machine characteristics			
Prime mover characteristics	**Uniform**	**Light shocks**	**Medium shocks**	**Heavy shocks**
Uniform	1.00	1.25	1.50	1.75 or higher
Light shocks	1.10	1.35	1.60	1.85 or higher
Medium shocks	1.25	1.50	1.75	2.00 or higher
Heavy shocks	1.50	1.75	2.00	2.25 or higher

Data sourced from ANSI/AGMA 2003-B97. American National Standard. Rating the pitting resistance and bending strength of generated straight bevel, zerol bevel and spiral bevel gear teeth. AGMA, 2003.

Care needs to be taken when comparing values for the factors of safety for bending and contact stress. As indicated previously in Chapter 9 for spur and helical gears, S_F should be compared with S_H^2 if the teeth are uncrowned or S_F with S_H^3 if the teeth are crowned. The exponent linearises the safety factor for contact stress with respect to the transmitted load.

The dynamic factor accounts for the effects of gear tooth quality and the associated variation of stress with speed and load. The AGMA standards use a quality number rating factor, Q_V, that varies according to the expected variation of component pitch and profile variations in a production run. High-accuracy gearing requires less derating than low-accuracy gearing. Heavily loaded gearing requires less derating than lightly loaded gearing. The dynamic factor, K_V, thus decreases with improving gear quality. In the absence of specific knowledge of the dynamic loads, then estimates for the approximate dynamic factor, K_v, can be obtained using Fig. 10.6, or by means of the curve fit given in Eq. (10.16).

$$K_v = \left(\frac{A}{A + \sqrt{200 v_{et}}}\right)^{-B} \tag{10.16}$$

The constants A and B and the pitch line velocity, v_{et}, can be determined by Eqs (10.17)–(10.19).

$$A = 50 + 56(1 - B) \tag{10.17}$$

$$B = 0.25(12 - Q_v)^{0.667} \tag{10.18}$$

$$v_{et} = 5.236 \times 10^{-5} d_1 n_1 \tag{10.19}$$

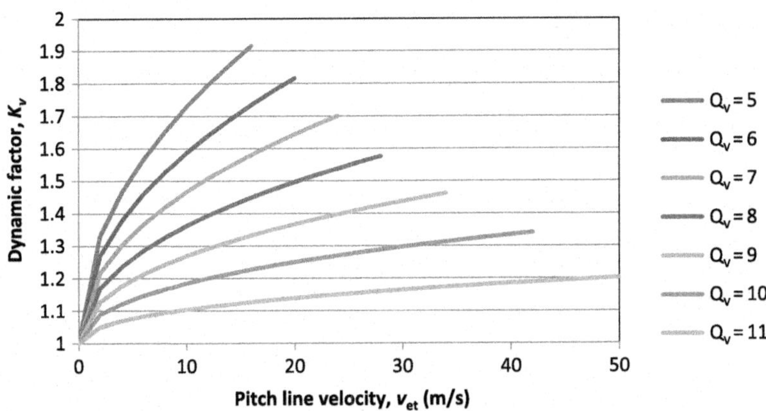

Fig. 10.6 Dynamic factor, K_v, as a function of pitch line velocity and quality number.

where

Q_v is the transmission accuracy number;
v_{et} is the pitch line velocity at the outside pitch diameter (m/s).

The maximum recommended pitch line velocity for a given transmission accuracy number can be determined from

$$v_{et\max} = \frac{[A + (Q_v - 3)]^2}{200} \tag{10.20}$$

The size factor accounts for nonuniformity of material properties and depends on tooth size, diameter, ratio of tooth size to diameter, face width, area of the stress pattern and material characteristics.

The size factor for pitting resistance can be determined from Fig. 10.7.

For $F < 12.7\,\text{mm}$, $Z_x = 0.5$

For $12.7\,\text{mm} \leq F \leq 114.3\,\text{mm}$,

$$Z_x = 0.00492F + 0.4375 \tag{10.21}$$

where Z_x is size factor for pitting resistance.

For $F \leq 12.7$, $Z_x = 0.5$. For $F \geq 114.5\,\text{mm}$, $Z_x = 1$.

The size factor for the bending strength of bevel gears can be determined using Fig. 10.8.

For $m_{et} < 1.6$, $Y_x = 0.5$.

For $1.6 \leq m_{et} \leq 50$ the size factor for the bending strength, Y_x, is given by

$$Y_x = 0.4867 + 0.008399m_{et} \tag{10.22}$$

The load distribution factor, $K_{H\beta}$, is used to modify the rating formulas to account for the nonuniform load along the tooth length. Fig. 10.9 can be used to estimate the load distribution factor for gears with properly crowned teeth.

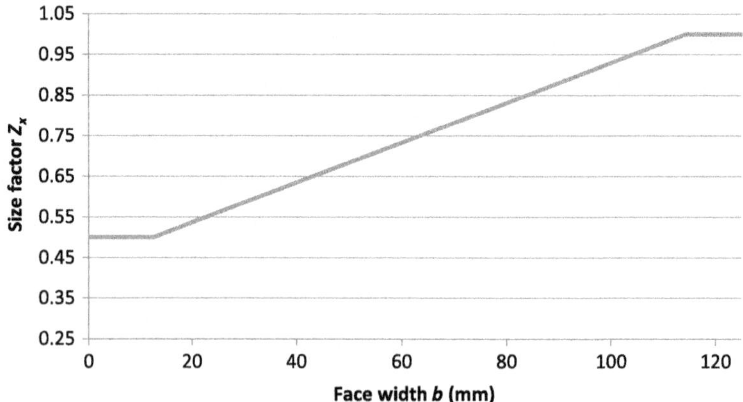

Fig. 10.7 Size factor for pitting resistance, Z_x.

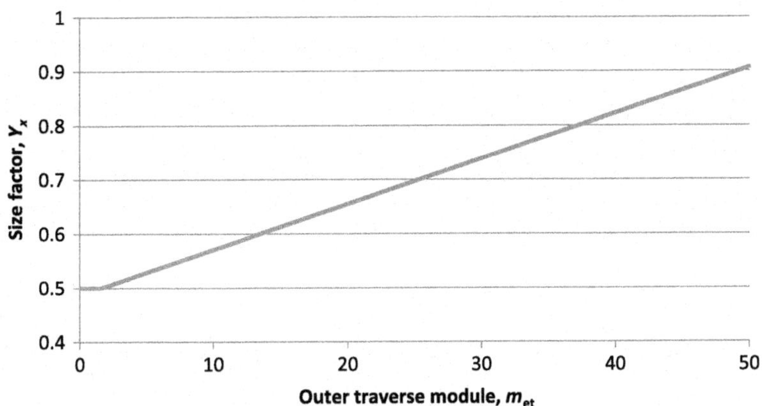

Fig. 10.8 Size factor for bending strength, Y_x.

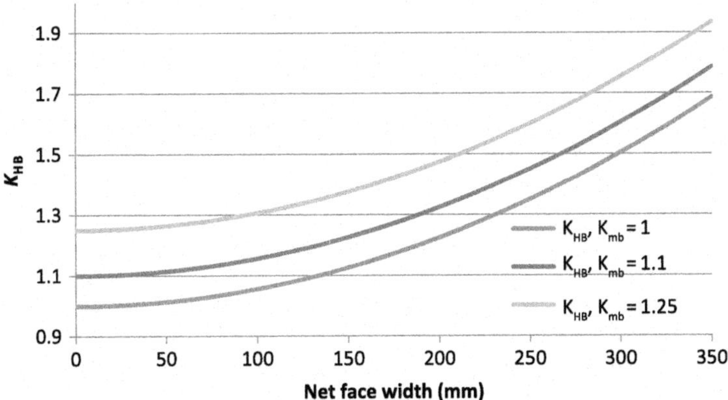

Fig. 10.9 Load distribution factor, $K_{H\beta}$, for $F \leq 356\,\text{mm}$.

The load distribution factor, $K_{H\beta}$, for $F \leq 356\,\text{mm}$, can be calculated from

$$K_{H\beta} = K_{mb} + 5.6 \times 10^{-6} F^2 \tag{10.23}$$

where K_{mb} is the load distribution modifier. In general straddle mounting with bearings on both sides of the gear is preferable for providing the best support for the gear. This arrangement is however often impossible to arrange in practice for intersecting shafts. Typically, in most applications, the gear is straddle mounted and the pinion is cantilevered.

$K_{mb} = 1$ for both gear and pinion straddle mounted.
$K_{mb} = 1.1$ if one member is straddle mounted.
$K_{mb} = 1.2$ if neither the gear nor the pinion is straddle mounted.

To accommodate deflection of the mountings, the bevel teeth are crowned in the lengthwise direction during the manufacture process. To account for the variation in contact pattern the crowning factor, Z_{xc}, can be used.

For properly crowned teeth, $Z_{xc}=1$.
For large or noncrowned teeth, $Z_{xc}=2$ or higher.

The lengthwise curvature factor for bending strength, Y_β, is given, for spiral bevel gears, by

$$Y_\beta = 0.211\left(\frac{r_{c0}}{R_m}\right)^q + 0.789 \tag{10.24}$$

where

r_{c0} is the cutter radius (mm);
R_m is the mean cone distance (mm).

$$q = \frac{0.279}{\log_{10}(\sin\beta_m)} \tag{10.25}$$

β_m is the mean spiral angle (degree).

For straight bevel, zerol bevel and skew bevel gears, $Y_\beta=1$.

The geometry factor, Z_I, is used to evaluate the effects of the gear tooth form on the gear tooth stresses. It accounts for the relative radius of curvature of the mating tooth surfaces, and the load sharing between adjacent pairs of teeth at the point on the tooth surfaces where the calculated contact pressure will reach its maximum value.

$$Z_I = \frac{g_c \rho_{yo} \cos\beta_m \cos\alpha_n}{b d_{e1} Z_i \varepsilon_{NI}} \frac{m_{mt}}{m_{et}} \tag{10.26}$$

where

g_c is the length of line contact (mm):
ρ_{y0} is the relative radius of profile curvature (mm);
α_n is the normal pressure angle (degree);
Z_i is an inertia factor for gears with a low contact ratio;
ε_{NI} is the load sharing ratio;
m_{et} is the outer transverse module (mm);
m_{mt} is the mean transverse module (mm).

The geometry factor, Y_J, is used to evaluate the shape of the tooth, the position at which the most damaging load is applied, the stress concentration due to the geometric shape of the root fillet, the sharing of load between adjacent pairs of teeth, the tooth thickness balance between the gear and mating pinion, the effective face width due to lengthwise crowning of the teeth, and the buttressing effect of an extended face width on one member of the pair. Both tangential (bending) and radial (compressive) components of the tooth load are included.

As spiral bevel gear teeth are not symmetrical in the lengthwise direction, the stresses will be different on the concave and convex sides of the teeth. Normal practice, for machines driven in one direction, is to calculate the stresses on the concave side of the pinion and the convex side of the gear tooth as these are the usual driving surfaces. For bidirectional operation it would be necessary to calculate the stresses for both sides for both the gear and the pinion.

The bending strength geometry factors, Y_{J1} and Y_{J2}, for the pinion and gear respectively, can be calculated from

$$Y_{J1} = \frac{Y_{K1}}{\varepsilon_{NJ} Y_i} \frac{r_{myo1}}{r_{mpt1}} \frac{F_1'}{F} \frac{m_{mt}}{m_{et}} \tag{10.27}$$

$$Y_{J2} = \frac{Y_{K2}}{\varepsilon_{NJ} Y_i} \frac{r_{myo2}}{r_{mpt2}} \frac{F_2'}{F} \frac{m_{mt}}{m_{et}} \tag{10.28}$$

where

Y_{K1} is a tooth form factor for the pinion;
Y_{K2} is a tooth form factor for the gear;
ε_{NJ} is the load sharing ratio;
Y_i is the inertia ratio for gears with low contact ratio;
r_{myo1} is the mean transverse radius to the point of load application for the pinion (mm);
r_{myo2} is the mean transverse radius to the point of load application for the gear (mm);
r_{mpt1} is the mean transverse pitch radius for the pinion (mm);
r_{mpt2} is the mean transverse pitch radius for the gear (mm);
F'_1 is the effective face width of the pinion (mm);
F'_2 is the effective face width of the gear (mm);
F is the face width (mm).

Alternatively the geometry factors Z_I and Y_J can be determined from graphs presented in the ANSI/AGMA 2003-B97 standard. A selection of these are reproduced in Figs. 10.10–10.15. It should be noted that the geometry factor varies as a function of the spiral angle and shaft angle. Figs. 10.10 and 10.11 are for a straight bevel gear with 20 degree pressure angle and 90 degree shaft angle. Figs. 10.12 and 10.13 are for a 20 degree pressure angle, 35 degree spiral angle and 90 degree shaft angle. Figs. 10.14 and 10.15 are for a 20 degree pressure angle, 25 degree spiral angle and 90 degree shaft angle.

The stress cycle factors Z_{NT} and Y_{NT} account for the number of cycles of the gears in operation during the design life. A stress cycle factor is used in the equations for the permissible contact and bending stress. The stress cycle factors for pitting resistance and bending strength can be determined using Figs. 10.16 and 10.17, respectively.

The stress cycle factor for pitting resistance,
$Z_{NT} = 2$ for $10^3 \le n_L \le 10^4$
$Z_{NT} = 3.4822 n_L^{-0.0602}$ for $10^4 \le n_L \le 10^{10}$
The stress cycle factor for bending strength,
$Y_{NT} = 2.7$ for $10^2 \le n_L \le 10^3$

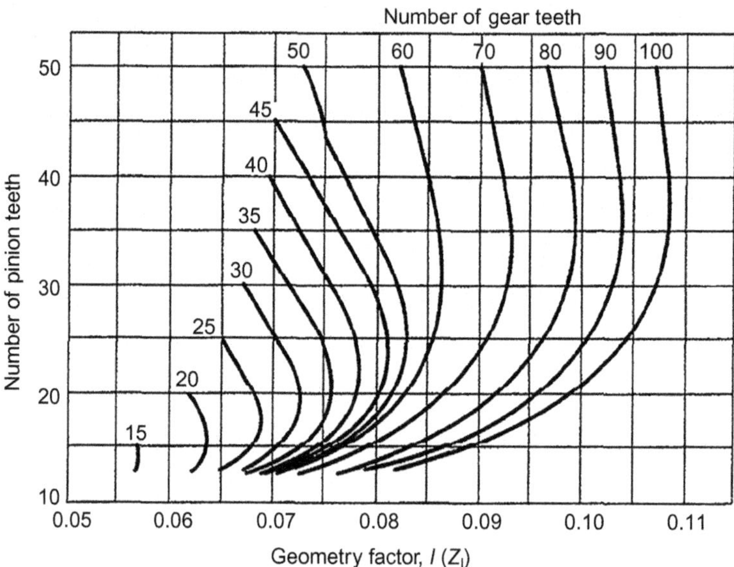

Fig. 10.10 Geometry factor Z_I for coniflex straight bevel gears with 20 degree pressure angle and 90 degree shaft angle.
Image courtesy: ANSI/AGMA 2003-B97. American National Standard. Rating the pitting resistance and bending strength of generated straight bevel, zerol bevel and spiral bevel gear teeth. AGMA, 2003.

$Y_{NT} = 6.1514 n_L^{-0.1182}$ for $10^3 \leq n_L \leq 3 \times 10^6$
$Y_{NT} = 1.6831 n_L^{-0.0323}$ for $3 \times 10^6 \leq n_L \leq 10^{10}$ for general conditions
$Y_{NT} = 1.3558 n_L^{-0.0323}$ for $3 \times 10^6 \leq n_L \leq 10^{10}$ for critical conditions

The hardness ratio factor for pitting resistance is a function of the gear ratio and the relative hardness of the materials used for the gear and the pinion. The hardness ratio factor for pitting resistance can be calculated using Eq. (10.29) and is plotted in Fig. 10.18.

$$Z_W = 1 + B_1 \left(\frac{N_G}{N_P} - 1 \right) \tag{10.29}$$

where

$$B_1 = 0.00898 \left(\frac{H_{B1}}{H_{B2}} \right) - 0.00829 \tag{10.30}$$

Eqs (10.29) and (10.30) are valid for the ratio of hardnesses for the pinion and gear in the range $1.2 \leq H_{B1}/H_{B2} \leq 1.7$, where

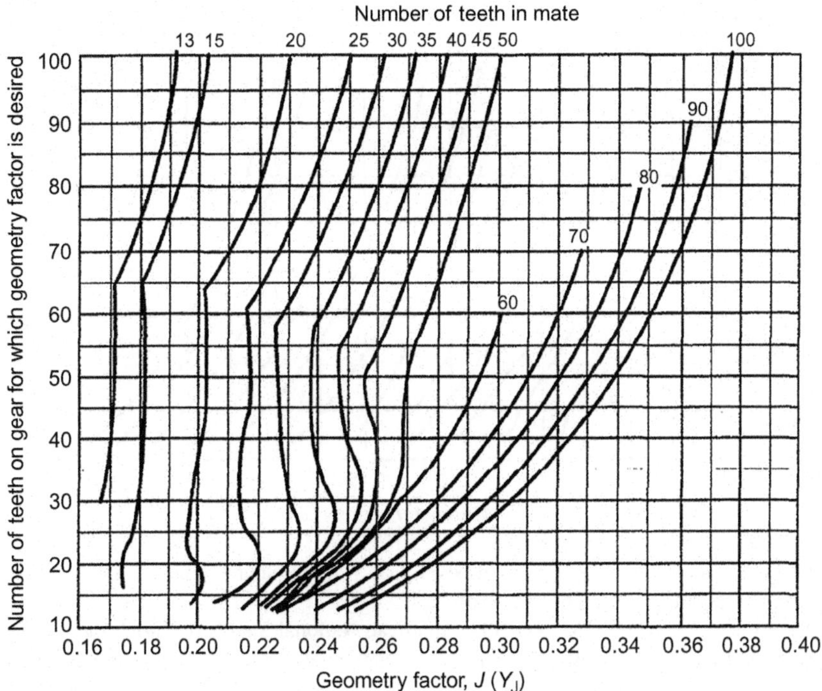

Fig. 10.11 Geometry factor Y_J for coniflex straight bevel gears with 20 degree pressure angle and 90 degree shaft angle.
Image courtesy: ANSI/AGMA 2003-B97. American National Standard. Rating the pitting resistance and bending strength of generated straight bevel, zerol bevel and spiral bevel gear teeth. AGMA, 2003.

H_{B1} is the minimum Brinell hardness for the pinion (HB, MPa);
H_{B2} is the minimum Brinell hardness for the gear (HB, MPa).

It should be noted that the units of hardness are the same as for stress but are seldom used.

For a surface hardened pinion (48 HRC or harder) run with a through hardened gear (180 to 400 HB), a work hardening effect takes place and the hardness ratio factor for pitting resistance can be calculated using Eq. (10.31).

$$Z_W = 1 + B_2(450 - H_{B2}) \tag{10.31}$$

where

$$B_2 = 0.00075e^{-0.52R_{a1}} \tag{10.32}$$

where R_{a1} is the pinion surface roughness (μm).

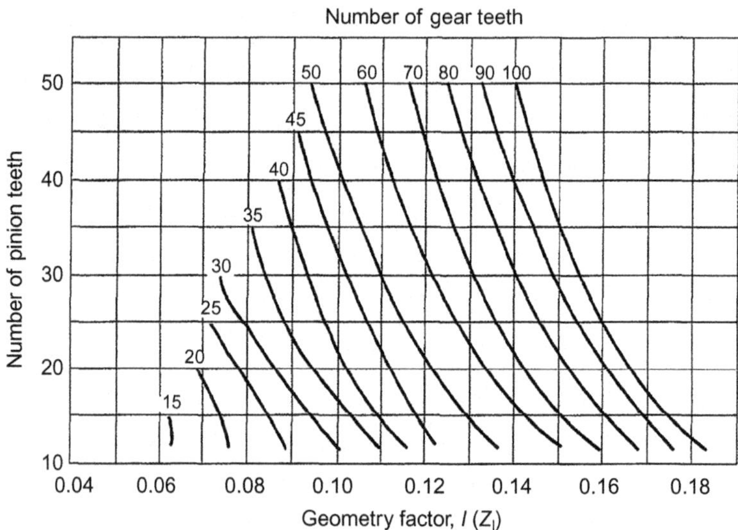

Fig. 10.12 Geometry factor Z_I for spiral bevel gears with 20 degree pressure angle and 35 degree spiral angle and 90 degree shaft angle.
Image courtesy: ANSI/AGMA 2003-B97. American National Standard. Rating the pitting resistance and bending strength of generated straight bevel, zerol bevel and spiral bevel gear teeth. AGMA, 2003.

Under normal conditions the temperature factor, K_θ, can be taken as $K_\theta = 1$. For high temperature operation the temperature factor can be determined using Eq. (10.33).

$$K_\theta = \frac{T}{393} \tag{10.33}$$

where T is the peak operating gear temperature (K).

The reliability factors Z_Z and Y_Z account for the effects of the statistical failure in material testing that is assumed to follow a normal distribution. If less than one failure in 100 is acceptable (99% reliability) then $Z_Z = Y_Z = 1$. For other requirements the values summarised in Table 10.2 can be used.

The elastic coefficient for pitting resistance, Z_E, is a function of Poisson's ratio and the Young's modulus and can be determined using Eq. (10.34) or from Table 10.3.

$$Z_E = \sqrt{\frac{1}{\pi\left(\frac{1-\nu_1^2}{E_1} + \frac{1-\nu_2^2}{E_2}\right)}} \tag{10.34}$$

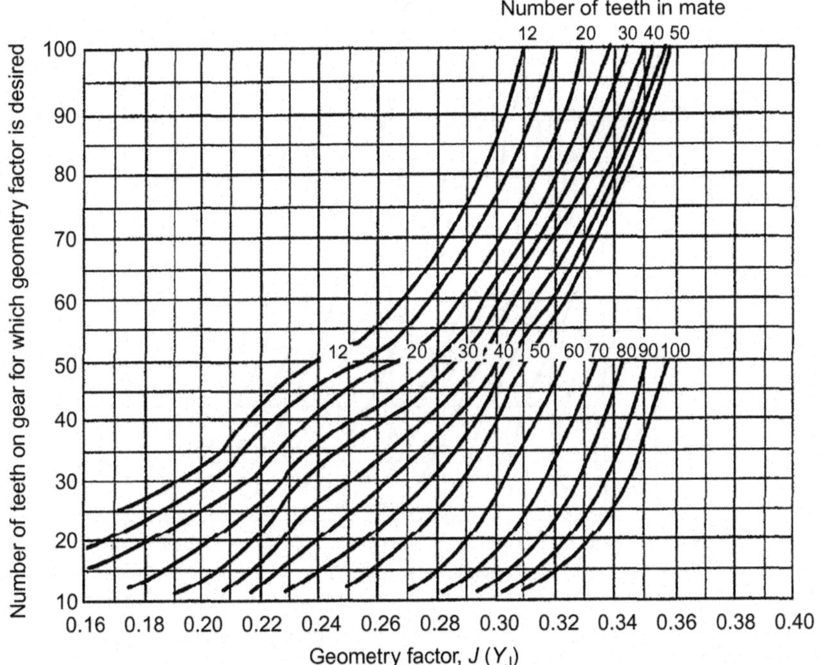

Fig. 10.13 Geometry factor Y_J for spiral bevel gears with 20 degree pressure angle and 35 degree spiral angle and 90 degree shaft angle.
Image courtesy: ANSI/AGMA 2003-B97. American National Standard. Rating the pitting resistance and bending strength of generated straight bevel, zerol bevel and spiral bevel gear teeth. AGMA, 2003.

where

Z_E is the elastic coefficient for pitting resistance (0);
ν_1 is the Poisson's ratio for the pinion;
ν_2 is the Poisson's ratio for the gear;
E_1 is the Young's modulus for the pinion (MPa);
E_2 is the Young's modulus for the gear (MPa).

The allowable stresses for pitting resistance and bending strength, $\sigma_{H\ \mathrm{lim}}$ and $\sigma_{F\ \mathrm{lim}}$, respectively, for unity values of the overload and life factors, need to be determined by field experience and laboratory trials. Testing of bevel gears has been principally on carburised case hardened steels and the available data for the allowable contact stress for other materials is limited or based on estimates from data for other types of gear.

The AGMA standards have established allowable stress numbers for a range of material types and grade, defined by specific quality control requirements (see ANSI/AGMA 2001-D04, 2001a). The allowable bending stress for through hardened steels for grade 1 quality components can be approximated by

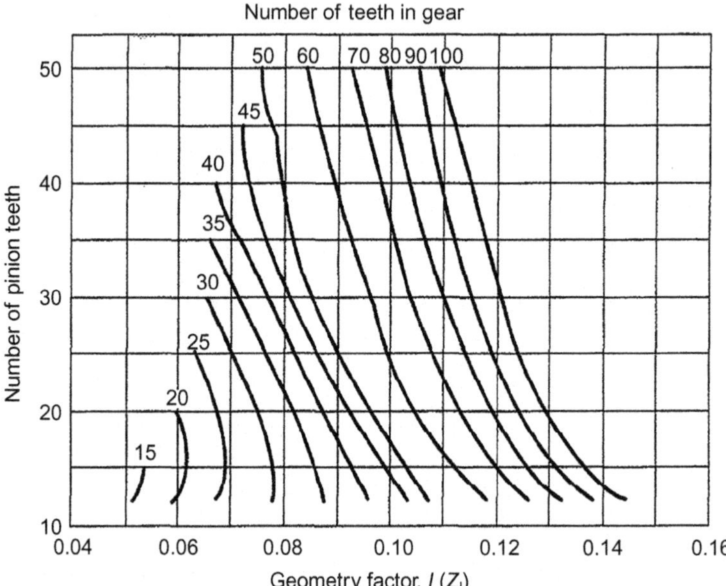

Fig. 10.14 Geometry factor Z_I for spiral bevel gears with 20 degree pressure angle and 25 degree spiral angle and 90 degree shaft angle.
Image courtesy: ANSI/AGMA 2003-B97. American National Standard. Rating the pitting resistance and bending strength of generated straight bevel, zerol bevel and spiral bevel gear teeth. AGMA, 2003.

$$\sigma_{F\lim} = 0.533 H_B + 88.3 \text{ MPa} \tag{10.35}$$

and for grade 2:

$$\sigma_{F\lim} = 0.703 H_B + 113 \text{ MPa} \tag{10.36}$$

The allowable bending stress for nitrided through hardened steels for grade 1 quality components can be approximated by

$$\sigma_{F\lim} = 0.568 H_B + 83.8 \text{ MPa} \tag{10.37}$$

and for grade 2:

$$\sigma_{F\lim} = 0.749 H_B + 110 \text{ MPa} \tag{10.38}$$

Values for the allowable bending stress can be found from Eqs (10.35)–(10.38) for through hardened and nitrided through hardened steels or using the values given in Table 10.4. Alternatively an estimate for $\sigma_{F\ \lim}$ can be obtained from Eq. (10.39).

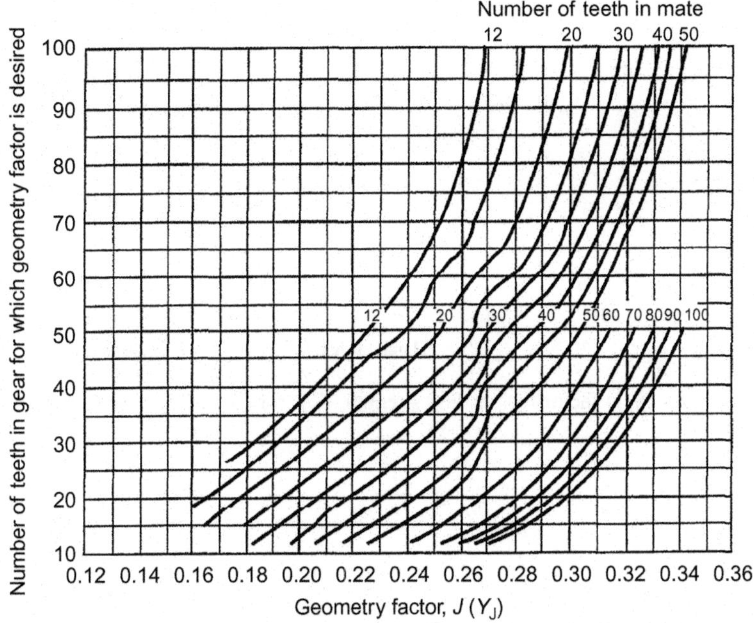

Fig. 10.15 Geometry factor Y_J for spiral bevel gears with 20 degree pressure angle and 25 degree spiral angle and 90 degree shaft angle.
Image courtesy: ANSI/AGMA 2003-B97. American National Standard. Rating the pitting resistance and bending strength of generated straight bevel, zerol bevel and spiral bevel gear teeth. AGMA, 2003.

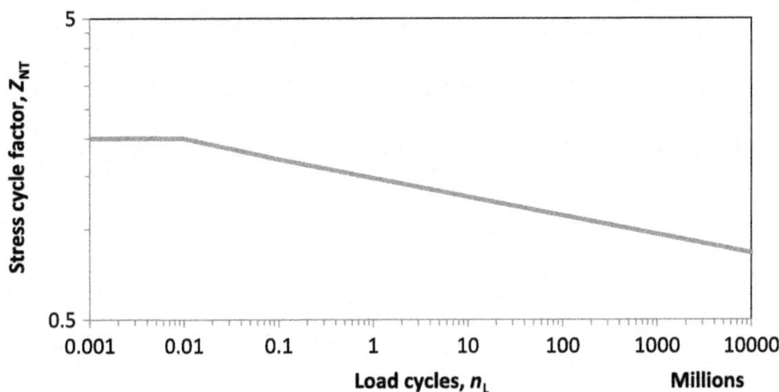

Fig. 10.16 Stress cycle factor for pitting resistance, Z_{NT}, for carburised case hardened steel bevel gears.

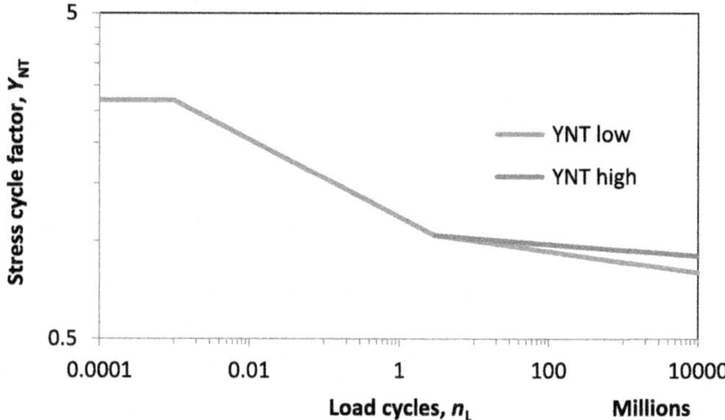

Fig. 10.17 Stress cycle factor for bending strength, Y_{NT}, for carburised case hardened steel bevel gears.

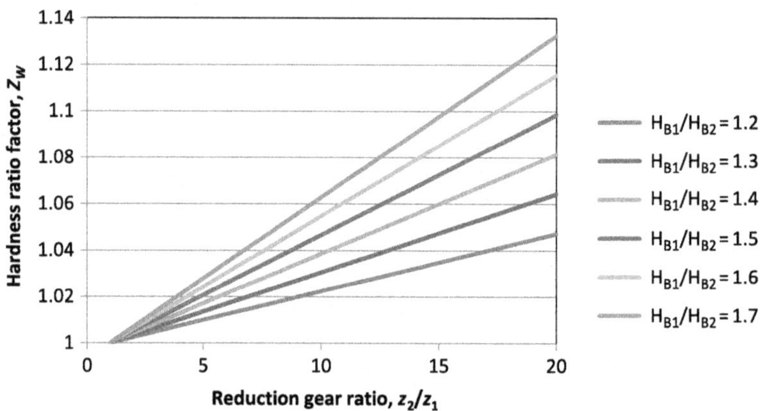

Fig. 10.18 Hardness ratio factor, Z_W, for a through hardened pinion and gear

$$\sigma_{F\lim} \approx 20.55 + 1.176H_B - 9.584 \times 10^{-4}H_B^2 \tag{10.39}$$

where $\sigma_{F\ \lim}$ is the allowable bending stress in MPa.

The allowable contact stress, $\sigma_{H\ \lim}$, assuming unidirectional loading, 10^7 cycles and 99% reliability for grade 1 through hardened steel gears can be approximated by

$$\sigma_{H\lim} = 2.22H_B + 200 \tag{10.40}$$

and for grade 2 through hardened steel gears by

$$\sigma_{H\lim} = 2.41H_B + 237 \tag{10.41}$$

Table 10.2 Reliability factors

Failure requirement	Z_Z	Y_Z
Fewer than one failure in 10,000	1.22	1.50
Fewer than one failure in 1000	1.12	1.25
Fewer than one failure in 100	1.00	1.00
Fewer than one failure in 10	0.92	0.85
Fewer than one failure in 2	0.84	0.70

After ANSI/AGMA 2003-B97. American National Standard. Rating the pitting resistance and bending strength of generated straight bevel, zerol bevel and spiral bevel gear teeth. AGMA, 2003.

Table 10.3 Approximate values for the elastic coefficient Z_E ($\sqrt{\text{MPa}}$) (AGMA 218.01, 2001b)

		Gear material					
Pinion material	**E_{pinion} (GPa)**	**Steel**	**Malleable iron**	**Nodular iron**	**Cast iron**	**Aluminium bronze**	**Tin bronze**
Steel	200	191	181	179	174	162	158
Malleable iron	170	181	174	172	168	158	154
Nodular iron	170	179	172	170	166	156	152
Cast iron	150	174	168	166	163	154	149
Aluminium bronze	120	162	158	156	154	145	141
Tin bronze	110	158	154	152	149	141	137

Table 10.4 Allowable bending stress number for steel gears

Material	Heat treatment	Minimum surface hardness	$\sigma_{F\ lim}$ (MPa)	$\sigma_{F\ lim}$ (MPa)	$\sigma_{F\ lim}$ (MPa)
			Grade 1	Grade 2	Grade 3
Steel	Through hardened		See Eq. (10.35)	See Eq. (10.36)	
	Flame or induction hardened				
	Unhardened roots	50 HRC	85	95	
	Hardened roots		154		
	Carburised and case hardened		205	240	275

Continued

Table 10.4 Continued

Material	Heat treatment	Minimum surface hardness	$\sigma_{F\ lim}$ (MPa)	$\sigma_{F\ lim}$ (MPa)	$\sigma_{F\ lim}$ (MPa)
AISI 4140	Nitrided	84.5 HR15N		150	
Nitralloy	Nitrided	90.0 HR15N		165	

Table 10.5 Allowable contact stress number for steel gears

Material	Heat treatment	Minimum surface hardness	$\sigma_{H\ lim}$ (MPa)	$\sigma_{H\ lim}$ (MPa)	$\sigma_{H\ lim}$ (MPa)
			Grade 1	Grade 2	Grade 3
Steel	Through hardened		See Eq. (10.40)	See Eq. (10.41)	
	Flame or induction hardened	50 HRC	1210	1310	
	Carburised and case hardened		1380	1550	1720
AISI 4140	Nitrided	84.5 HR15N		1000	
Nitralloy 135M	Nitrided	90.9 HR15N		1100	

Data sourced from ANSI/AGMA 2003-B97. American National Standard. Rating the pitting resistance and bending strength of generated straight bevel, zerol bevel and spiral bevel gear teeth. AGMA, 2003.

Values for the allowable contact stress can be found from Eqs (10.40) and (10.41) for grade 1 and 2 through hardened steels or using the values given in Table 10.5. Alternatively an estimate for $\sigma_{H\ \text{lim}}$ can be obtained from Eq. (10.42).

$$\sigma_{H\text{lim}} \approx 182.7 + 2.382 H_B \tag{10.42}$$

where $\sigma_{H\ \text{lim}}$ is the allowable contact stress in MPa.

10.4 Calculation procedure summary

The flow chart illustrated in Fig. 10.19 provides an indication of the procedure for determining the contact and bending stresses for a bevel gear set, based on the AGMA standard.

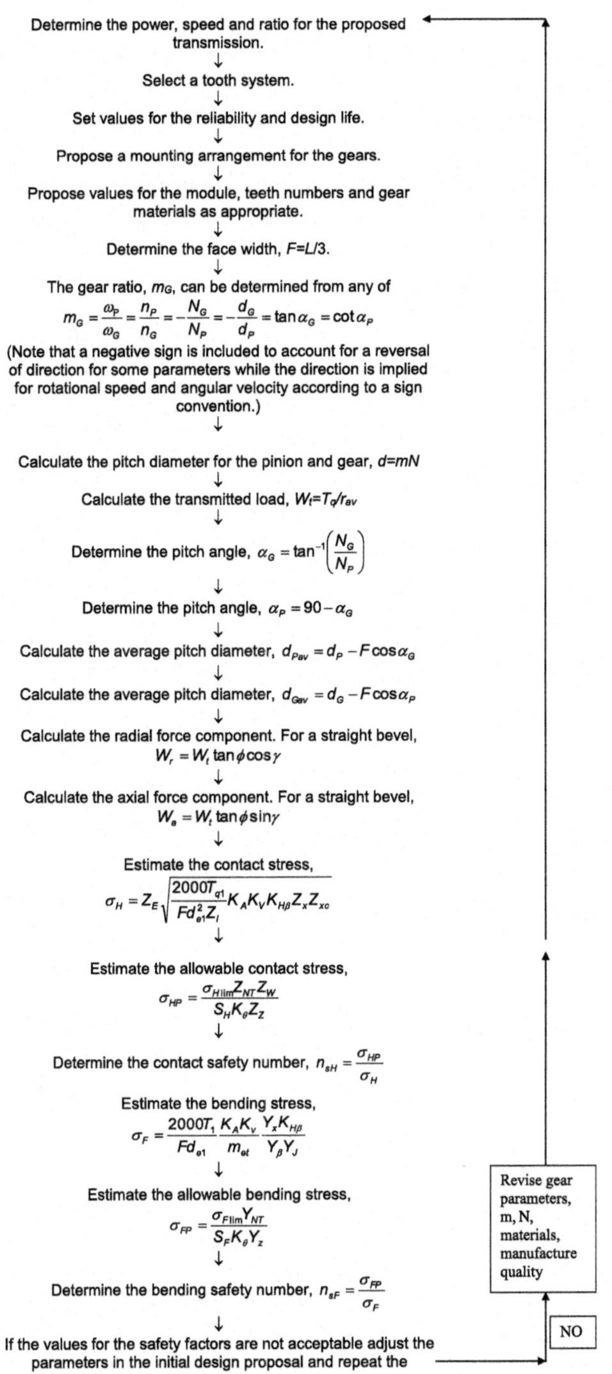

Fig. 10.19 Procedure for determining the contact and bending stresses for a bevel gear set, based on the AGMA standard.

Example 10.1. Determine the bending and contact stresses and associated safety factors for a straight bevel gear set comprising a 25 tooth bevel and a 35 tooth gear with a module of 3 mm. The pinion speed is 3000 rpm and the transmission power is 8.5 kW. The pinion and gear are manufactured from Grade 1 steel through hardened to 280 HB and 240 HB respectively with a quality number of 8 and the teeth are crowned. The gear set is driven by an electric motor and the driven machinery has a uniform driving characteristic. The gear is straddle mounted and the pinion is cantilevered.

Solution

The initial values for the design are:

$N_P = 25$; $N_G = 25$; $m = 3$ mm; $n_P = 3000$ rpm, Power$=3$ kW, pinion material Grade 1 through-hardened steel, $\text{HB}_1 = 280$ HB, $\text{HB}_2 = 240$ HB.

The torque, T_q, is given by

$$T_q = \frac{\text{Power}}{\omega} = \frac{8500}{(2\pi/60)3000} = 27.06 \text{ N m}$$

The pinion and gear pitch diameters are

$$d_P = mN_P = 3 \times 25 = 75 \text{ mm}$$

$$d_G = mN_G = 3 \times 35 = 105 \text{ mm}$$

The gear and pinion pitch angles are

$$\alpha_G = \tan^{-1}\left(\frac{N_G}{N_P}\right) = \tan^{-1}\left(\frac{35}{25}\right) = 54.46^o$$

$$\alpha_P = 90 - \alpha_G = 35.54^o$$

The back cone distance, L, is given by

$$L = \frac{d_P}{2\sin\alpha_P} = \frac{75}{2\sin 35.54} = 64.52 \text{ mm}$$

The face width, F, is normally L/3, so

$$F = \frac{64.52}{3} = 21.51 \text{ mm}$$

From Fig. 10.10, the pitting resistance geometry factor for a 20 degree pressure angle straight bevel gears, for a pinion with 25 teeth and a gear with 35 teeth is $Z_I = 0.075$.

From Table 10.1, for smooth or uniform driver and driven characteristics, the overload factor, K_A, is $K_A = 1$.

The pitch line velocity is

$$V_{et} = \omega\frac{d}{2} = \left(\frac{2\pi}{60} \times 3000\right)\frac{75 \times 10^{-3}}{2} = 11.78 \text{ m/s}$$

$$B = 0.25(12 - Q_v)^{0.667} = 0.25(12 - 8)^{0.667} = 0.6303$$

$$A = 50 + 56(1 - B) = 50 + 56(1 - 0.6303) = 70.71$$

The dynamic factor is given by

$$K_v = \left(\frac{A}{A + \sqrt{200 v_{et}}}\right)^{-B} = \left(\frac{70.71}{70.71 + \sqrt{200 \times 11.78}}\right)^{-0.6303} = 1.39$$

The maximum recommended pitch line velocity is

$$v_{et\max} = \frac{[A + (Q_v - 3)]^2}{200} = \frac{[70.71 + (8 - 3)]^2}{200} = 28.66 \text{ m/s}$$

As $v_{et} < v_{et\ max}$, the pitch line velocity is acceptable.

As one of the gears is straddle mounted, the load distribution modifier, $K_{mb} = 1.1$.

The load distribution factor, $K_{H\beta}$, is given by

$$K_{H\beta} = K_{mb} + 5.6 \times 10^{-6} F^2 = 1.1 + 5.6 \times 10^{-6}(21.51)^2 = 1.103$$

The size factor for pitting resistance, Z_x, is given by

$$Z_x = 0.00492F + 0.4375 = (0.00492 \times 21.51) + 0.4375 = 0.5433$$

For properly crowned teeth, $Z_{xc} = 1$.

For a steel pinion and a steel gear (Table 10.3) $Z_E = 191\sqrt{MPa}$

From Eq. (10.9), the AGMA contact stress for pitting resistance, for the pinion, is given by,

$$\sigma_H = Z_E \sqrt{\frac{2000 T_{q1}}{F d_{e1}^2 Z_I} K_A K_V K_{H\beta} Z_x Z_{xc}}$$

$$= 191 \sqrt{\frac{2000 \times 27.06}{21.51 \times 75^2 \times 0.076} \times 1 \times 1.39 \times 1.103 \times 0.5433 \times 1}$$

$$\sigma_H = 422.9 \text{ MPa}$$

From Eq. (10.9), the AGMA contact stress for pitting resistance, for the gear, is given by,

$$\sigma_H = Z_E \sqrt{\frac{2000 T_{q1}}{F d_{e1}^2 Z_I} K_A K_V K_{H\beta} Z_x Z_{xc}}$$

$$= 191 \sqrt{\frac{2000 \times 27.06}{21.51 \times 105^2 \times 0.076} \times 1 \times 1.39 \times 1.103 \times 0.5433 \times 1}$$

$$\sigma_H = 302.0 \text{ MPa}$$

If the number of cycles for the gear set is taken as 1×10^9, the stress cycle factor, Z_{NT}, is given by

$$Z_{NT} = 3.4822 n_L^{-0.0602} = 3.4822\left(1 \times 10^9\right)^{-0.0602} = 1.000$$

$$B_1 = 0.00898\left(\frac{H_{B1}}{H_{B2}}\right) - 0.00829 = 0.00898\left(\frac{280}{240}\right) - 0.00829 = 0.002187$$

The hardness ratio factor, Z_W, is given by

$$Z_W = 1 + B_1\left(\frac{N_G}{N_P} - 1\right) = 1 + 0.002187\left(\frac{35}{25} - 1\right) = 1.001$$

The temperature factor can be taken as $K_\theta = 1$.

The reliability factor for 99% reliability can be taken as $Z_Z = 1$.

The allowable contact stress for grade 1 through hardened steel can be estimated from Eq. (10.40).

For the pinion, $\sigma_{H\lim} = 2.22H_B + 200 = (2.22 \times 280) + 200 = 821.6$ MPa

For the gear, $\sigma_{H\lim} = 2.22H_B + 200 = (2.22 \times 240) + 200 = 732.8$ MPa

Setting the contact safety factor, $S_H = 1$, to enable an estimate for the permissible AGMA contact stress to be calculated, then for the pinion,

$$\sigma_{HP} = \frac{\sigma_{H\lim} Z_{NT} Z_W}{S_H K_\theta Z_Z} = \frac{821.6 \times 1 \times 1.001}{1 \times 1 \times 1} = 822.4 \text{ MPa}$$

For the gear,

$$\sigma_{HP} = \frac{\sigma_{H\lim} Z_{NT} Z_W}{S_H K_\theta Z_Z} = \frac{732.8 \times 1 \times 1.001}{1 \times 1 \times 1} = 733.5 \text{ MPa}$$

The ratio of the contact stress to the permissible contact stress for the pinion is,

$$N_{sH} = \frac{\sigma_{HP}}{\sigma_H} = \frac{822.4}{422.9} = 1.9$$

The ratio of the contact stress to the permissible contact stress for the gear is,

$$N_{sH} = \frac{\sigma_{HP}}{\sigma_H} = \frac{733.5}{302.0} = 2.4$$

The relevant factors to enable the AGMA bending stress to be evaluated can now be established.

From Fig. 10.11, the bending strength geometry factor for the pinion, is $Y_J = 0.245$.

From Fig. 10.11, the bending strength geometry factor for the gear, is $Y_J = 0.213$. The size factor Y_x, is given by

$$Y_x = 0.4867 + 0.008399 m_{et} = 0.4867 + (0.008399 \times 3) = 0.5119$$

The tooth lengthwise curvature factor, Y_β, can be taken as $Y_\beta = 1$.

The AGMA bending stress for the pinion is given by

$$\sigma_F = \frac{2000 T_{q1}}{F d_{e1}} \frac{K_A K_v}{m_{et}} \frac{Y_x K_{H\beta}}{Y_\beta Y_J} = \frac{2000 \times 27.06}{21.51 \times 75} \times \frac{1 \times 1.39}{3} \times \frac{0.5119 \times 1.103}{1 \times 0.245}$$
$$= 35.81 \text{ MPa}$$

The AGMA bending stress for the gear is given by

$$\sigma_F = \frac{2000 T_{q1}}{F d_{e1}} \frac{K_A K_v}{m_{et}} \frac{Y_x K_{H\beta}}{Y_\beta Y_J} = \frac{2000 \times 27.06}{21.51 \times 105} \times \frac{1 \times 1.39}{3} \times \frac{0.5119 \times 1.103}{1 \times 0.213}$$
$$= 29.42 \text{ MPa}$$

The allowable bending stress for grade 1 through hardened steel can be estimated from Eq. 10.35.

For the pinion,

$$\sigma_{F\lim} = 0.533 H_B + 88.3 = (0.533 \times 280) + 88.3 = 237.5 \text{ MPa}$$

For the gear,

$$\sigma_{F\lim} = 0.533 H_B + 88.3 = (0.533 \times 240) + 88.3 = 216.2 \text{ MPa}$$

The stress cycle factor for bending strength, Y_{NT}, can be calculated from

$$Y_{NT} = 1.6831 n_L^{-0.0323} = 1.6831 \left(1 \times 10^9\right)^{-0.0323} = 0.8618$$

The reliability factor for 99% reliability can be taken as $Y_Z = 1$.

Setting the contact safety factor, $S_F = 1$, to enable an estimate for the permissible AGMA bending stress to be calculated, then for the pinion,

$$\sigma_{FP} = \frac{\sigma_{F\lim} Y_{NT}}{S_F K_\theta Y_z} = \frac{237.5 \times 0.8618}{1 \times 1 \times 1} = 204.7 \text{ MPa}$$

For the gear

$$\sigma_{FP} = \frac{\sigma_{F\lim} Y_{NT}}{S_F K_\theta Y_z} = \frac{216.2 \times 0.8618}{1 \times 1 \times 1} = 186.3 \text{ MPa}$$

The ratio of the bending stress to the permissible bending stress for the pinion is,

$$N_{sF} = \frac{\sigma_{FP}}{\sigma_F} = \frac{204.7}{35.81} = 5.7$$

The ratio of the bending stress to the permissible bending stress for the gear is,

$$N_{sF} = \frac{\sigma_{FP}}{\sigma_F} = \frac{186.3}{29.42} = 6.3$$

Comment: The safety factors for bending are significantly higher than those for contact resistance, let alone comparing N_{sF} to N_{sH}^2 or N_{sH}^3. This suggests the design proposal is overly conservative in bending and that alternative initial values could result in a cheaper design.

Example 10.2. A bevel gear set has been proposed for a right angle drive from an electrical motor running at 14,000 rpm. The power being transmitted is 900 W from the electric motor and the driven machine is expected to experience medium shocks. The design proposed has a spiral bevel pinion with 15 teeth and the gear has 32 teeth with a 20 degree pressure angle and a 35 degree spiral angle. The module selected is $m = 1.5$ mm. A Grade 1 steel through hardened with a hardness of 280 HB for the pinion and 180 HB for the gear is proposed. The quality number for manufacture is 8 and the teeth are crowned. Determine the bending and contact stresses and associated safety factors and comment on the suitability of the proposed gear-set.

Solution

The initial values for the design are:

$N_P = 15$; $N_G = 25$; $m = 1.5$ mm; $n_P = 14{,}000$ rpm, Power $= 0.9$ kW, pinion material Grade 1 through-hardened steel, $HB_1 = 280$ HB, $HB_2 = 180$ HB.

The torque, T_q, is given by

$$T_q = \frac{\text{Power}}{\omega} = \frac{900}{(2\pi/60)14000} = 0.6139 \text{ N m}$$

The pinion and gear pitch diameters are

$$d_P = mN_P = 1.5 \times 15 = 22.5 \text{ mm}$$

$$d_G = mN_G = 1.5 \times 32 = 48 \text{ mm}$$

The gear and pinion pitch angles are

$$\alpha_G = \tan^{-1}\left(\frac{N_G}{N_P}\right) = \tan^{-1}\left(\frac{32}{15}\right) = 64.89^o$$

$$\alpha_P = 90 - \alpha_G = 25.11^o$$

The back cone distance, L, is given by

$$L = \frac{d_P}{2\sin\alpha_P} = \frac{22.5}{2\sin 25.11} = 26.51 \text{ mm}$$

The face width, F, is normally L/3, so

$$F = \frac{26.51}{3} = 8.835\,\text{mm}$$

From Fig. 10.12, the pitting resistance geometry factor for a 20 degree pressure angle and 35 degree spiral angle and 90 degree shaft angle spiral bevel gears, for a pinion with 15 teeth and a gear with 32 teeth is $Z_I = 0.097$.

From Table 10.1, for smooth or uniform driver and medium shock driven characteristics, the overload factor, K_A, is $K_A = 1.5$.

The pitch line velocity is

$$V_{et} = \omega\frac{d}{2} = \left(\frac{2\pi}{60} \times 14000\right)\frac{22.5 \times 10^{-3}}{2} = 16.49\ \text{m/s}$$

$$B = 0.25(12 - Q_v)^{0.667} = 0.25(12 - 8)^{0.667} = 0.6303$$

$$A = 50 + 56(1 - B) = 50 + 56(1 - 0.6303) = 70.71$$

The dynamic factor is given by

$$K_v = \left(\frac{A}{A + \sqrt{200 v_{et}}}\right)^{-B} = \left(\frac{70.71}{70.71 + \sqrt{200 \times 16.49}}\right)^{-0.6303} = 1.455$$

The maximum recommended pitch line velocity is

$$v_{etmax} = \frac{[A + (Q_v - 3)]^2}{200} = \frac{[70.71 + (8 - 3)]^2}{200} = 28.66\ \text{m/s}$$

As $v_{et} < v_{et\ max}$, the pitch line velocity is acceptable.

As one of the gears is straddle mounted, the load distribution modifier, $K_{mb} = 1.1$.

The load distribution factor, $K_{H\beta}$, is given by

$$K_{H\beta} = K_{mb} + 5.6 \times 10^{-6}F^2 = 1.1 + 5.6 \times 10^{-6}(8.835)^2 = 1.100$$

The size factor for pitting resistance, Z_x, is given by

$$Z_x = 0.00492F + 0.4375 = (0.00492 \times 8.835) + 0.4375 = 0.4810$$

For properly crowned teeth, $Z_{xc} = 1$.

For a steel pinion and a steel gear (Table 10.3) $Z_E = 191\sqrt{MPa}$

From Eq. (10.9), the AGMA contact stress for pitting resistance, for the pinion, is given by,

$$\sigma_H = Z_E\sqrt{\frac{2000T_{q1}}{Fd_{e1}^2 Z_I}K_A K_V K_{H\beta} Z_x Z_{xc}}$$

$$= 191\sqrt{\frac{2000 \times 0.6139}{8.835 \times 22.5^2 \times 0.097} \times 1.5 \times 1.455 \times 1.1 \times 0.481 \times 1}$$

$\sigma_H = 345.3$ MPa

From Eq. (10.9), the AGMA contact stress for pitting resistance, for the gear, is given by,

$$\sigma_H = Z_E\sqrt{\frac{2000T_{q1}}{Fd_{e1}^2 Z_I}K_A K_V K_{H\beta} Z_x Z_{xc}}$$

$$= 191\sqrt{\frac{2000 \times 0.6139}{8.835 \times 48^2 \times 0.097} \times 1.5 \times 1.455 \times 1.1 \times 0.481 \times 1}$$

$$\sigma_H = 161.8 \text{ MPa}$$

If the number of cycles for the gear set is taken as 1×10^9, the stress cycle factor, Z_{NT}, is given by

$$Z_{NT} = 3.4822 n_L^{-0.0602} = 3.4822\left(1 \times 10^9\right)^{-0.0602} = 1.000$$

$$B_1 = 0.00898\left(\frac{H_{B1}}{H_{B2}}\right) - 0.00829 = 0.00898\left(\frac{280}{180}\right) - 0.00829 = 0.005679$$

The hardness ratio factor, Z_W, is given by

$$Z_W = 1 + B_1\left(\frac{N_G}{N_P} - 1\right) = 1 + 0.005679\left(\frac{32}{15} - 1\right) = 1.006$$

The temperature factor can be taken as $K_\theta = 1$.

The reliability factor for 99% reliability can be taken as $Z_Z = 1$.

The allowable contact stress for grade 1 through hardened steel can be estimated from Eq. (10.40).

For the pinion, $\sigma_{H\lim} = 2.22H_B + 200 = (2.22 \times 280) + 200 = 821.6$ MPa

For the gear, $\sigma_{H\lim} = 2.22H_B + 200 = (2.22 \times 180) + 200 = 599.6$ MPa

Setting the contact safety factor, $S_H = 1$, to enable an estimate for the permissible AGMA contact stress to be calculated, then for the pinion,

$$\sigma_{HP} = \frac{\sigma_{H\lim} Z_{NT} Z_W}{S_H K_\theta Z_Z} = \frac{821.6 \times 1 \times 1.006}{1 \times 1 \times 1} = 827.0 \text{ MPa}$$

For the gear,

$$\sigma_{HP} = \frac{\sigma_{H\lim} Z_{NT} Z_W}{S_H K_\theta Z_Z} = \frac{599.6 \times 1 \times 1.006}{1 \times 1 \times 1} = 603.5 \text{ MPa}$$

The ratio of the contact stress to the permissible contact stress for the pinion is,

$$N_{sH} = \frac{\sigma_{HP}}{\sigma_H} = \frac{827.0}{345.3} = 2.4$$

The ratio of the contact stress to the permissible contact stress for the gear is,

$$N_{sH} = \frac{\sigma_{HP}}{\sigma_H} = \frac{603.5}{161.8} = 3.7$$

The relevant factors to enable the AGMA bending stress to be evaluated can now be established.

From Fig. 10.13, the bending strength geometry factor for the pinion, is $Y_J = 0.21$.

From Fig. 10.13, the bending strength geometry factor for the gear, is $Y_J = 0.205$.

The size factor Y_x, is given by

$$Y_x = 0.4867 + 0.008399 m_{et} = 0.4867 + (0.008399 \times 1.5) = 0.4993$$

The tooth lengthwise curvature factor, Y_β, can be taken as $Y_\beta = 1$.

The AGMA bending stress for the pinion is given by

$$\sigma_F = \frac{2000 T_{q1}}{F d_{e1}} \frac{K_A K_v}{m_{et}} \frac{Y_x K_{H\beta}}{Y_\beta Y_J} = \frac{2000 \times 0.6139}{8.835 \times 22.5} \times \frac{1.5 \times 1.455}{1.5} \times \frac{0.4993 \times 1.1}{1 \times 0.21}$$
$$= 23.50 \text{ MPa}$$

The AGMA bending stress for the gear is given by

$$\sigma_F = \frac{2000 T_{q1}}{F d_{e1}} \frac{K_A K_v}{m_{et}} \frac{Y_x K_{H\beta}}{Y_\beta Y_J} = \frac{2000 \times 0.6139}{8.835 \times 48} \times \frac{1.5 \times 1.455}{1.5} \times \frac{0.4993 \times 1.1}{1 \times 0.205}$$
$$= 11.29 \text{ MPa}$$

The allowable bending stress for grade 1 through hardened steel can be estimated from Eq. (10.35).

For the pinion,

$$\sigma_{F\lim} = 0.533 H_B + 88.3 = (0.533 \times 280) + 88.3 = 237.5 \text{ MPa}$$

For the gear,

$$\sigma_{F\lim} = 0.533 H_B + 88.3 = (0.533 \times 180) + 88.3 = 184.2 \text{ MPa}$$

The stress cycle factor for bending strength, Y_{NT}, can be calculated from

$$Y_{NT} = 1.6831 n_L^{-0.0323} = 1.6831\left(1 \times 10^9\right)^{-0.0323} = 0.8618$$

The reliability factor for 99% reliability can be taken as $Y_Z = 1$.

Setting the contact safety factor, $S_F = 1$, to enable an estimate for the permissible AGMA bending stress to be calculated, then for the pinion,

$$\sigma_{FP} = \frac{\sigma_{F\lim} Y_{NT}}{S_F K_\theta Y_z} = \frac{237.5 \times 0.8618}{1 \times 1 \times 1} = 204.7 \text{ MPa}$$

For the gear

$$\sigma_{FP} = \frac{\sigma_{F\lim} Y_{NT}}{S_F K_\theta Y_z} = \frac{184.2 \times 0.8618}{1 \times 1 \times 1} = 158.8 \text{ MPa}$$

The ratio of the bending stress to the permissible bending stress for the pinion is,

$$N_{sF} = \frac{\sigma_{FP}}{\sigma_F} = \frac{204.7}{23.5} = 8.7$$

The ratio of the bending stress to the permissible bending stress for the gear is,

$$N_{sF} = \frac{\sigma_{FP}}{\sigma_F} = \frac{158.8}{11.29} = 14$$

Comment: The safety factors for bending are significantly higher than those for contact resistance, let alone comparing N_{sF} to N_{sH}^2 or N_{sH}^3. This suggests the design proposal is overly conservative in bending and that alternative initial values could result in a cheaper design.

10.5 Conclusions

Bevel gears can be used to transmit power through a 90 degree angle. The efficiency of transmission can be high, up to 99%, and this form of gearing can be found in a wide range of applications from machine tools to differential drives. The AGMA equations presented provide a conservative means for assessing the bending and contact stresses.

References

American Gear Manufacturers Association. ANSI/AGMA 2001-D04. Fundamental rating factors and calculation methods for involute spur and helical gear teeth. 2001a.

American Gear Manufacturers Association. ANSI/AGMA 218.01. 2001b.

ANSI/AGMA 2003-B97. American National Standard. Rating the pitting resistance and bending strength of generated straight bevel, zerol bevel and spiral bevel gear teeth. AGMA, 2003.

Deng, X., Hua, L., Han, X., Song, Y., 2011. Numerical and experimental investigation of cold rotary forging of a 20CrMnTi alloy spur bevel gear. Mater. Des. 32, 1376–1389.

Jelaska, D.T., 2012. Gears and Gear Drives. Wiley.

Lin, C.H., Fong, Z.H., 2015. Numerical tooth contact analysis of a bevel gear set by using measured tooth geometry data. Mech. Mach. Theory 84, 1–24.

Litvin, F.L., Fuentes, A., Hayasaka, K., 2006. Design manufacture, stress analysis, and experimental tests of low-noise high endurance spiral bevel gears. Mech. Mach. Theory 41, 83–118.

Padmanabhan, S., Srinivasa Raman, V., Asokan, P., Arunachalam, S., Page, T., 2011. Design optimisation of bevel gear pair. Int. J. Des. Eng 4, 364–393.

Park, M., 2003. Failure analysis of an accessory bevel gear installed on a J69 turbojet engine. Eng. Fail. Anal. 10, 371–382.

Sekercioglu, T., Kovan, V., 2007. Pitting failure of truck spiral bevel gear. Eng. Fail. Anal. 14 (4), 614–619.

Siddiqui, N.A., Deen, K.M., Khan, M.Z., Ahmad, R., 2013. Investigating the failure of bevel gears in an aircraft engine. Case Stud. Eng. Failure Anal. 1, 24–31.

Zakrajsek, J.J., Handschuh, R.F. and Decker, H.J. Application of fault detection techniques to spiral bevel gear fatigue data, 1994.

Websites

www.agma.org/.
www.bga.org.uk/.
www.motionco.co.uk/?gclid=CMDP3p_bzaYCFUhO4QodpCr7iw.
www.arrowgear.com/products/spiral_bevel_gears.htm.
www.qtcgears.com/rfq/bevelgears.htm.
www.spiralbevel.com/.
www.gears-manufacturers.com/spiral-bevel-gears.html.
www.amarillogear.com/spiralbevel.html.
www.bostongear.com/products/open/miter.html.
www.gearshub.com/bevel-gears.html.
www.roymech.co.uk/Useful_Tables/Drive/Bevel_Gears.html.
www.derangear.com/spiral_gear.htm.
www.geartechnology.com/issues/.
www.bevelgeartw.com.

Further reading

Shigley, J.E., Mischke, C.R., 1986. Gearing. McGraw Hill.

Worm gears

11

Chapter Outline

Nomenclature

Generally, preferred SI units have been stated.

a	addendum (mm)
b	dedendum (mm)
c	clearance (mm)
C	centre distance (mm)
C_m	ratio correction factor
C_s	materials factor
C_v	velocity factor
d	worm pitch diameter (mm)
d_g	mean diameter of the gear (mm)
d_G	pitch diameter of the worm gear (mm)
d_m	mean diameter of the worm gear (mm)
d_{max}	maximum pitch diameter of the worm (mm)
d_{min}	minimum pitch diameter of the worm (mm)
d_o	worm outer diameter (mm)
d_r	worm root diameter (mm)
d_W	pitch diameter of the worm (mm)
D_m	mean gear diameter (mm)
D_o	worm gear outer diameter (mm)
D_r	worm gear root diameter (mm)
D_t	worm gear throat diameter (mm)
f	coefficient of friction
F	effective face width (mm)
F_G	worm gear face width (mm)
F_W	worm face width (mm)

Mechanical Design Engineering Handbook. https://doi.org/10.1016/B978-0-08-102367-9.00011-1

h_t full depth of worm thread (mm)
l_{pitch} worm axial pitch (mm)
L lead (mm)
m_G gear ratio
n rotational speed of the worm (rpm)
N_G number of teeth on the worm gear
N_{min} minimum number of worm gear teeth
N_W number of threads on the worm
P_{input} rated input power (kW)
P_{loss} lost power (kW)
P_{output} rated output power (kW)
R_a roughness (μm)
T_q torque (N m)
V_G pitch line velocity of the gear (m/s)
V_s sliding velocity (m/s)
V_t sliding velocity at the mean worm diameter (m/s)
V_W pitch line velocity of the worm (m/s)
W resultant force (N)
W_f friction force (N).
W_{Gr} radial force component acting against the gear (N)
W_{Gt} tangential force component acting against the gear (N)
W_t tangential load on the worm gear (N)
W_{tg} worm gear tangential force (N)
W_{Wa} axial force component acting against the gear (N)
W_{Wa} axial force component acting against the worm (N)
W_{Wr} radial force component acting against the worm (N)
W_{Wt} tangential force component acting against the worm (N)
W_x force component in the tangential direction on the worm (N)
W_y force component in the radial direction on the worm (N)
W_z force component in the axial direction on the worm (N)
ϕ_n normal pressure angle of the worm thread at the mean diameter (degree)
η efficiency
λ lead angle (degree)
ψ helix angle (degree)
ψ_G helix angle (degree)
ψ_W worm helix angle (degree)

11.1 Introduction

Worm and wheel gears are widely used for non-parallel, non-intersecting, right angle gear drive system applications where a high transmission gearing ratio is required. In comparison to other gear, belt and chain transmission elements, worm and wheel gear sets tend to offer a more compact solution. In certain configurations a worm and wheel gear set can provide sufficiently high friction to be self-locking which can be a desirable feature if a defined position is required for a gear train if it is not braked or unpowered. This chapter provides an overview of worm and wheels and outlines a selection procedure.

A worm gear is a cylindrical helical gear with one or more threads and resembles a screw thread. A worm wheel or worm gear is a cylindrical gear with flanks cut in such a way as to ensure contact with the flanks of the worm gear. The worm wheel is analogous to a nut that fits on the screw thread of the worm. If the worm is restrained axially within its housing, then if the worm is rotated, the worm gear will also rotate. Typical forms for worms and worm gears are shown in Fig. 11.1.

In a worm and wheel gear-set rotary power can transmitted between nonparallel and nonintersecting shafts. A worm and wheel gear-set is typically used when the speed ratio of the two shafts is high, say three or more.

Worm and wheel gear-sets are used for steering gear, winch blocks (Fig. 11.2) low speed gearboxes, rotary tables and remote valve control. Worm and wheel gear-sets are capable of high-speed reduction and high load applications where nonparallel, noninteracting shafts are used (Merritt, 1935; Radzevich, 2016). The 90 degree configuration is most common, although other angles are possible. Frictional heat generation is high in worm gears because of the high sliding velocities, so continuous lubrication is required and provision for heat dissipation must be made.

The direction of rotation of the worm wheel depends on the direction of rotation of the worm and on whether the worm teeth have a right or a left hand thread. The direction of rotation for a worm and wheel gear-sets is illustrated is Fig. 11.3.

Worms usually have just one tooth and can therefore produce gearing ratios as high as the number of teeth on the gear wheel. Herein lies the principal merit of worm and wheel gear-sets. In comparison to other gear sets which are typically limited to a gear ratio of up to 10:1, worm and wheel gear-sets can achieve gear ratios of up to 360:1, although most manufacturers quote ranges between 3:1 and 100:1. Ratios above 30:1 generally have one thread on the worm, while ratios below 30:1 tend to have a worm with multiple threads (sometimes referred to as starts).

Fig. 11.1 Worm and wheel gear sets.

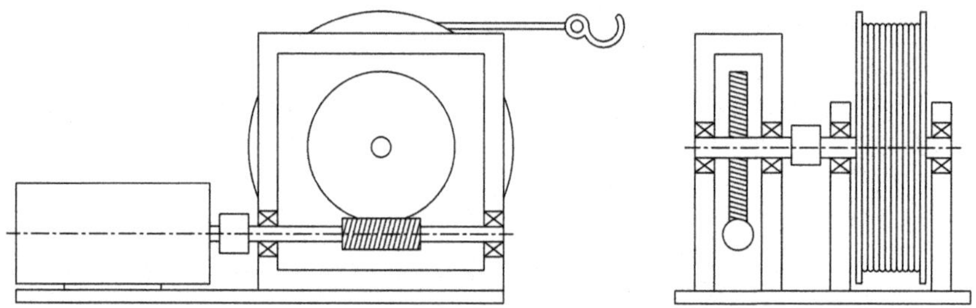

Fig. 11.2 Possible outline winch configuration incorporating a worm and wheel gear-set.

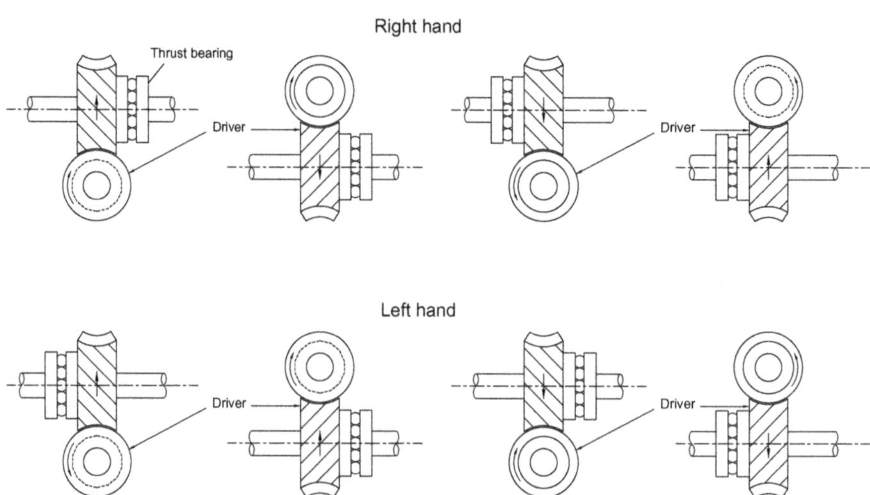

Fig. 11.3 Rotation and hand relations for worm and wheel gear-sets.
After *Boston Gear Division*.

The gear ratio for a worm and wheel gear-set is given by

$$m_G = \frac{N_G}{N_W} \tag{11.1}$$

where

m_G = gear ratio;
N_G = number of teeth in the worm gear;
N_W = number of threads in the worm.

A particular merit of worm and wheel gear-sets is their ability to self-lock. If a worm set is self-locking it will not back drive and any torque applied to the worm gear will not rotate the worm. A self-locking worm and wheel gear-set can only be driven

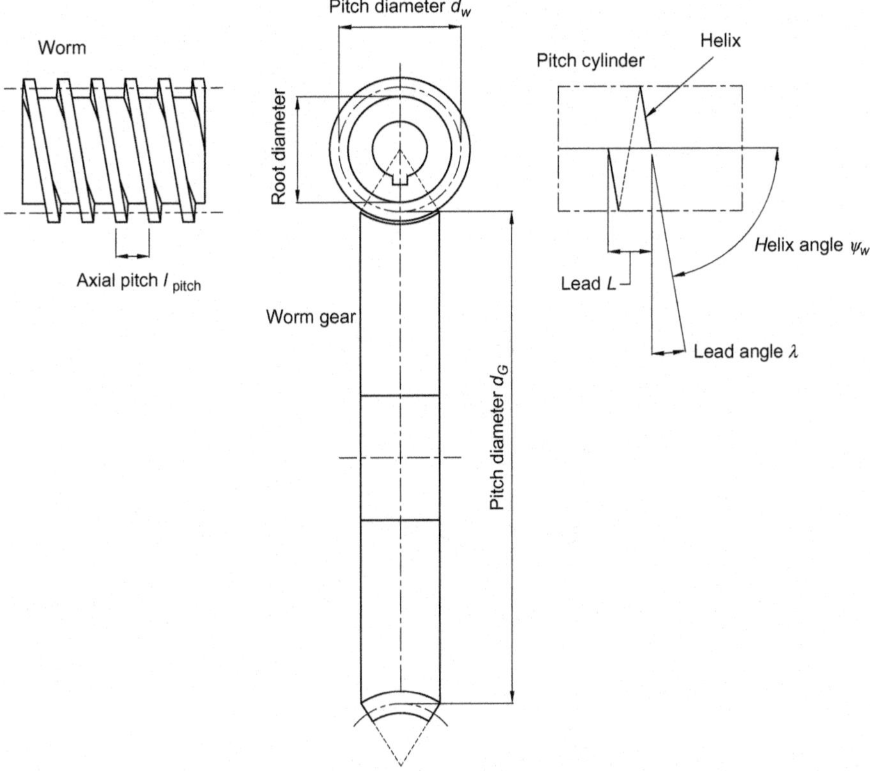

Fig. 11.4 Nomenclature for a single enveloping worm and wheel gear-set.

forward by rotation of the worm. This principle can be exploited in lifting equipment to hold a load or in applications where rapid braking is required. Whether a worm and wheel gear set will be self-locking depends on frictional contact between the worm and the worm wheel flanks.

There are two types of worm and wheel gear-sets, depending on whether the teeth of one or both wrap around each other.

- Single enveloping worm and wheel gear-sets, see Fig. 11.4.
- Double enveloping worm and wheel gear-sets, see Fig. 11.5.

As the worm rotates through the worm gear, lines of contact roll or progress from the tip to the root of the worm gear teeth. At any instant in time there may be two or three teeth in contact and transmitting power as illustrated in Fig. 11.6.

Some of the key geometric features and dimensions for a worm gear are illustrated in Fig. 11.7.

The helix angle on a worm is usually high and that on the worm wheel low. Normal convention is to define a lead angle, λ, on the worm and a helix angle, ψ_G, on the

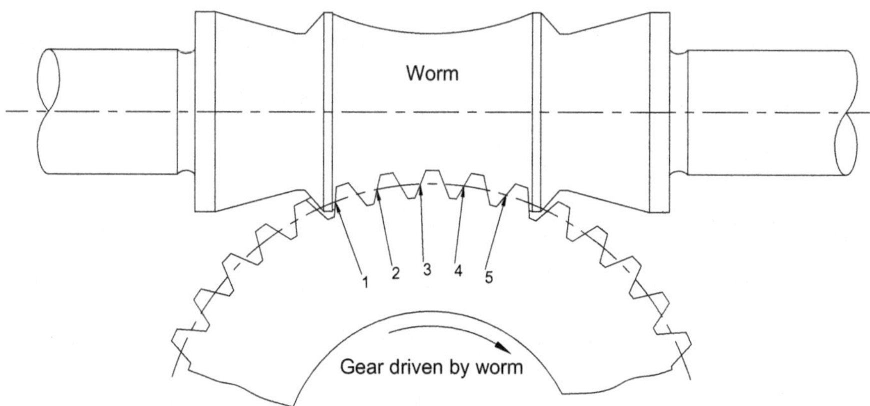

Fig. 11.5 Double enveloping worm and wheel gear-set.

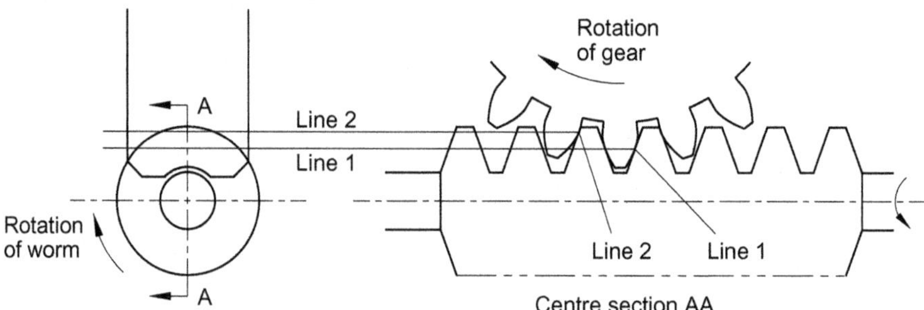

Fig. 11.6 Lines of contact for a worm and wheel gear-set.

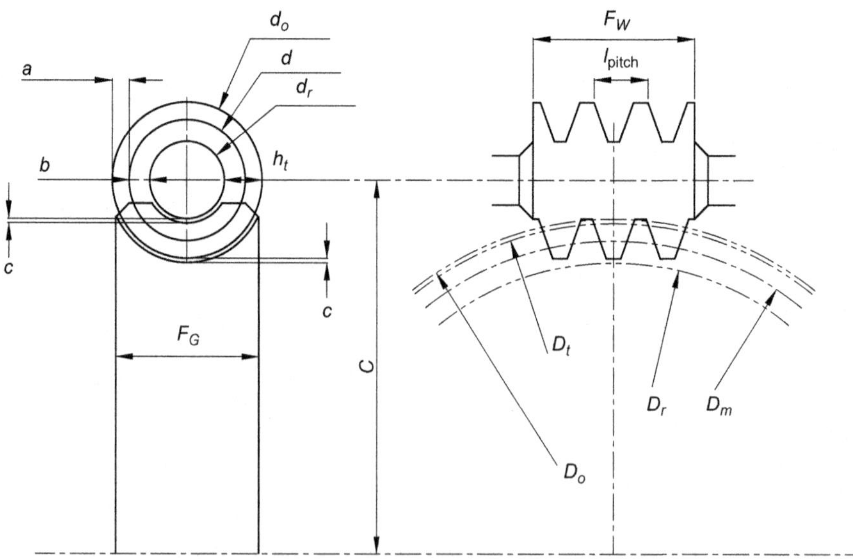

Fig. 11.7 Worm gear dimension.

worm gear. For a 90 degree configuration, $\lambda = \psi_G$. The distance that a point on the mating worm gear moves axially in one revolution of the worm is called the lead, L.

The following relationships apply to the lead, L and lead angle λ

$$L = l_{\text{pitch}} N_W = \frac{\pi d_G N_W}{N_G} \tag{11.2}$$

$$\tan\lambda = \frac{L}{\pi d_W} \tag{11.3}$$

where

L = lead (mm);
l_{pitch} = worm axial pitch (mm);
N_W = number of teeth on the worm;
d_G = pitch diameter of the worm gear (mm);
N_G = number of teeth on the worm gear;
λ = lead angle (degree);
d_W = pitch diameter of the worm (mm).

The worm lead angle and the worm helix angle, ψ_W, are related by $\lambda = 90$ degree $- \psi_W$.

The lead angle will vary from the root to the outside diameter of the worm as indicated in Fig. 11.8. Generally self-locking occurs for lead angles below 6 degree. However care is necessary in relying solely on self-locking to brake or sustain a load as vibration has been known to result in a reduction of the frictional contact between

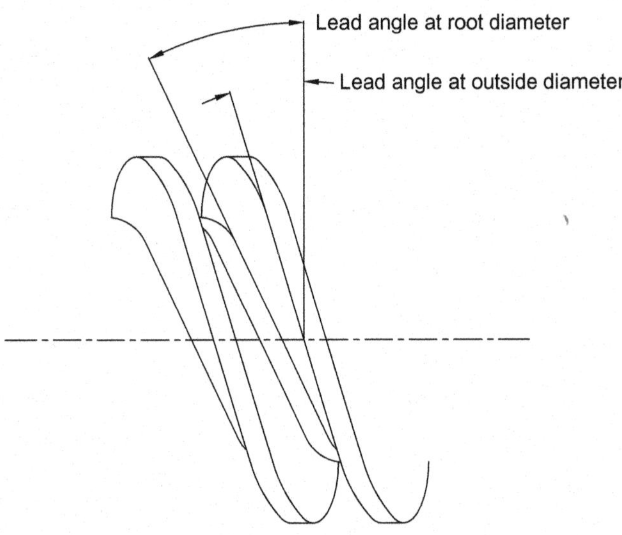

Fig. 11.8 Variation of the lead angle on a worm gear.
Based on AGMA.

the worm and wheel for lead angles below 6 degree and resulting movement or failure of the device.

The axial pitch of the worm and the transverse circular pitch of the wheel will be equal for a 90 degree set configuration.

$$d = mN \tag{11.4}$$

The worm can have any pitch diameter, as this is not related to the number of teeth. General guidance for optimum power capacity indicates that the pitch diameter, d, of the worm should fall in the following ranges (AGMA 6022-C93):

$$\frac{C^{0.875}}{1.6} \leq d_{max} \leq \frac{C^{0.875}}{1.07} \tag{11.5}$$

$$\frac{C^{0.875}}{3} \leq d_{min} \leq \frac{C^{0.875}}{2} \tag{11.6}$$

where

C = centre distance (mm);
d = worm pitch diameter (mm).
d_{max} = maximum worm pitch diameter (mm).
d_{min} = minimum worm pitch diameter (mm).

Dudley (1984) recommends

$$d \approx \frac{C^{0.875}}{2.2} \tag{11.7}$$

The pitch diameter of the worm gear, d_G, is related to the centre distance C and the pitch diameter of the worm, by:

$$d_G = 2C - d \tag{11.8}$$

The addendum, a, and dedendum, b, are given by:

$$a = 0.3183 l_{pitch} \tag{11.9}$$

$$b = 0.3683 l_{pitch} \tag{11.10}$$

The face width of a worm gear (Fig. 11.7) is limited by the worm diameter. The ANSI/AGMA 6034-B92 recommendation for the minimum face width, for a pitch exceeding 4.06 mm, is given by

$$F_G = 0.67d \tag{11.11}$$

The tooth forms for worm and wheel gear sets are not involutes. They are manufactured as matched sets. The worm is subject to high stresses and is normally made using a hardened steel such as AISI 1020, 1117, 8620, 4320 hardened to HRC 58-62 or a medium carbon steel such as AISI 4140 or 4150 induction or flame hardened to a case of HRC 58-62 (Norton, 2006). They are typically ground or polished to a roughness of $R_a = 0.4$ μm. The worm gear needs to be of softer material that is compliant enough to run-in and conform to the worm under the high sliding running conditions. Sand-cast or forged bronze is commonly used. Cast iron and polymers are sometimes used for lightly loaded, low speed applications.

An analysis of the forces associated a worm and wheel gear-set can be undertaken readily (e.g. see Dudás, 2005; Litvin and Kin, 1992) and this is outlined in Section 11.2. Such information is critical to enable suitable bearings to be selected for both shafts. Worm and wheel gear-sets tend to fail due to pitting and wear (see Dudley, 1984; Maitra, 1994; Radzevich, 2016). The AGMA power ratings based on wear and pitting resistance are presented in Section 11.3 and an associated design procedure in Section 11.4.

11.2 Force analysis

The force exerted on a worm by a gear is illustrated in Fig. 11.9 where, for the time being, friction has been neglected. The resultant force W will have three components:

$$W_x = W \cos\phi_n \sin\lambda \tag{11.12}$$

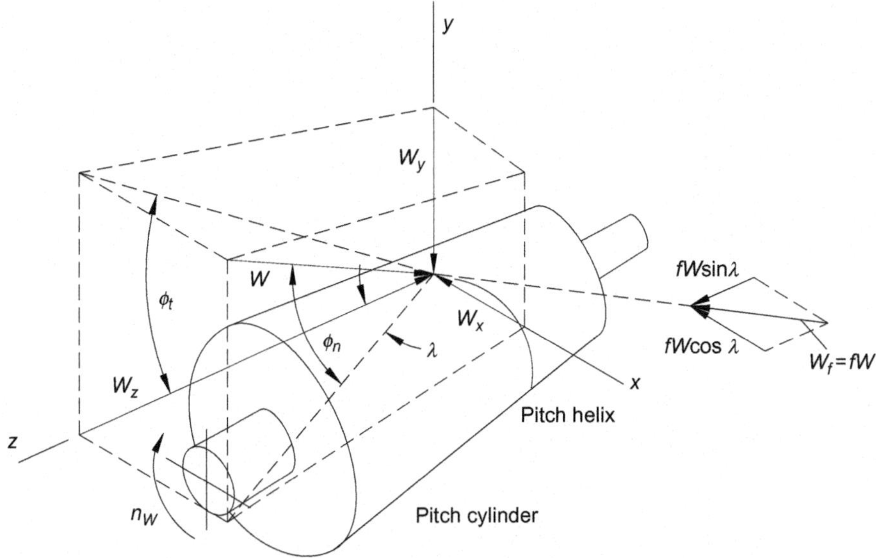

Fig. 11.9 Pitch cylinder of the worm, showing the forces exerted on the worm by the worm gear.
Based on Shigley, J.E., 1986. Mechanical Engineering Design, first metric ed. McGraw Hill.

$$W_y = W \sin \phi_n \tag{11.13}$$

$$W_z = W \cos \phi_n \cos \lambda \tag{11.14}$$

where

W_x = force component in the tangential direction on the worm (N);
W_y = force component in the radial direction on the worm (N);
W_z = force component in the axial direction on the worm (N);
λ = lead angle (°);
ϕ_n = normal pressure angle of the worm thread at the mean diameter (°).

Standard pressure angles for worm and wheel gear-sets are 14.5, 17.5, 20, 22.5, 25, 27.5 and 30 degree. The higher the pressure, the higher the tooth strength, albeit at the expense of higher friction, bearing loads and bending stresses in the worm.

The minimum number of worm gear teeth, N_{min}, as a function of the pressure angle is listed in Table 11.1.

As the forces on the worm and worm gear are equal and opposite, the tangential, radial and axial forces are given by

$$W_{Wt} = -W_{Ga} = W_x \tag{11.15}$$

$$W_{Wr} = -W_{Gr} = W_y \tag{11.16}$$

$$W_{Wa} = -W_{Gt} = W_z \tag{11.17}$$

where

W_{Wt} = tangential force component acting against the worm (N);
W_{Wr} = radial force component acting against the worm (N);
W_{Wa} = axial force component acting against the worm (N);
W_{Gt} = tangential force component acting against the gear (N);
W_{Gr} = radial force component acting against the gear (N);
W_{Wa} = axial force component acting against the gear (N).

Table 11.1 Suggested minimum number of teeth for the worm (AGMA)

Pressure angle (degree)	N_{min}
14.5	40
17.5	27
20	21
22.5	17
25	14
27.5	12
30	10

Introducing a coefficient of friction, f, to account for the sliding motion experienced in the motion between a worm thread and the wheel teeth surfaces,

$$W_x = W(\cos\phi_n \sin\lambda + f\cos\lambda) \tag{11.18}$$

$$W_y = W\sin\phi_n \tag{11.19}$$

$$W_z = W(\cos\phi_n \cos\lambda - f\sin\lambda) \tag{11.20}$$

From Eqs (11.15)–(11.18),

$$W_f = fW = \frac{fW_{Gt}}{f\sin\lambda - \cos\phi_n \cos\lambda} \tag{11.21}$$

$$W_{Wt} = W_{Gt}\frac{\cos\phi_n \sin\lambda + f\cos\lambda}{f\sin\lambda - \cos\phi_n \cos\lambda} \tag{11.22}$$

The efficiency of a worm and wheel gear-set can be defined by

$$\eta = \frac{W_{Wt,\text{ without friction}}}{W_{Wt,\text{ with friction}}} \tag{11.23}$$

or

$$\eta = \frac{\cos\phi_n - f\tan\lambda}{\cos\phi_n + f\cot\lambda} \tag{11.24}$$

A typical value for the coefficient of friction for worm gears is $f \approx 0.05$. The variation of efficiency with helix angle is given in Table 11.2.

Experiments have shown that efficiency for a worm and wheel gear-set is a function of the sliding velocity. Taking V_G as the pitch line velocity of the gear, V_W as the pitch line velocity of the worm, the sliding velocity, V_s, is given by vector addition:

Table 11.2 Worm and wheel efficiency (taking $f = 0.05$)

ψ	λ	F	ϕ_n	η
1	1	0.05	20	24.7
2.5	2.5	0.05	20	45.0
5	5	0.05	20	61.9
7.5	7.5	0.05	20	70.7
10	10	0.05	20	76.1
15	15	0.05	20	82.2
20	20	0.05	20	85.6
25	25	0.05	20	87.5
30	30	0.05	20	88.7

$$V_W = V_G + V_s \tag{11.25}$$

or

$$V_s = \frac{V_W}{\cos\lambda} \tag{11.26}$$

where

V_G = pitch line velocity of the gear (m/s);
V_W = pitch line velocity of the worm (m/s);
V_S = sliding velocity (m/s).

11.3 AGMA equations

Worm sets are generally rated by their capacity to handle a particular level of input power, output power or allowable torque at a particular speed for the input or output shaft. The AGMA power rating is based on pitting and wear resistance as this is the usual failure mode for worm sets. The AGMA rating (ANSI/AGMA 6034-B92) is based on 10h of continuous operation under a uniform load.

The input-power rating, P_{input}, is given by

$$P_{\text{input}} = P_{\text{output}} + P_{\text{loss}} \tag{11.27}$$

where P_{loss} is the power lost due to friction in the mesh (kW).

The output power is given by

$$P_{\text{output}} = \frac{nW_{tg}d_g}{1.91 \times 10^7 m_G} \tag{11.28}$$

where

n = rotational speed of the worm (rpm);
W_{tg} = worm gear tangential force (N);
P_{output} = output power (kW);
m_G = gear ratio;
d_g = mean gear diameter (mm).

The power lost is given by

$$P_{\text{loss}} = \frac{V_t W_f}{1000} \tag{11.29}$$

where

P_{loss} = lost power (kW);
V_t = sliding velocity at the mean worm diameter (m/s);
W_f = friction force (N).

The AGMA tangential load on a worm gear is given by

$$W_t = \frac{C_s C_m C_v d_g^{0.8} F}{75.948} \tag{11.30}$$

where

C_s = materials factor;
d_g = mean diameter of the gear (mm);
F = effective face width (mm);
C_m = ratio correction factor;
C_v = velocity factor.

The friction force can be determined by

$$W_f = \frac{fW_t}{\cos\lambda \cos\phi_n} \tag{11.31}$$

where

f = coefficient of friction;
W_t = tangential load on the worm gear tooth (N);
λ = lead angle (degree);
ϕ_n = normal pressure angle of the worm thread at the mean diameter (degree).

The sliding velocity at the mean worm diameter can be determined by

$$V_t = \frac{nd_m}{19098\cos\lambda} \tag{11.32}$$

where

n = rotational speed of the worm (rpm);
d_m = mean worm diameter (mm).

Values for the ratio correction factor, the velocity factor and materials factors can be found from tables provided in the ANSI/AGMA 6034-B92 standard. The following equations for the ratio correction factor, the velocity factor and materials factors provide approximations to the values given in the tables.

The ratio correction factor C_m is a function of the gear ratio, m_G.

For $3 \le m_G \le 20$

$$C_m = 0.02\left(-m_G^2 + 40m_G - 76\right)^{0.5} + 0.46 \tag{11.33}$$

For $20 \le m_G \le 76$

$$C_m = 0.0107\left(-m_G^2 + 56m_G + 5145\right)^{0.5} \tag{11.34}$$

For $m_G > 76$

$$C_m = 1.1483 - 0.00658 m_G \tag{11.35}$$

For $0 \le V_t \le 3.556\,\text{m/s}$

$$C_v = 0.659 e^{-0.2165 V_t} \tag{11.36}$$

For $3.556 \le V_t \le 15.24\,\text{m/s}$

$$C_v = 0.652 V_t^{-0.571} \tag{11.37}$$

For $V_t > 15.24\,\text{m/s}$

$$C_v = 1.098 V_t^{-0.774} \tag{11.38}$$

For $V_t = 0$, take $f = 0.15$.

For $0 \le V_t \le 0.0508\,\text{m/s}$

$$f = 0.124 e^{\left(-2.233 V_t^{0.645}\right)} \tag{11.39}$$

For $V_t > 0.0508\,\text{m/s}$

$$f = 0.103 e^{\left(-1.185 V_t^{0.45}\right)} \tag{11.40}$$

The materials factor C_s depends on the method of casting.

For $C \le 76.2\,\text{mm}$, an initial estimate for the materials factor can be obtained from

$$C_s = 720 + 0.000633 C^3 \tag{11.41}$$

This value can be compared to the values obtained for the relevant means of casting as indicated below and the smaller value used.

For sand cast gears then:
for $d_m < 63.5\,\text{mm}$, $C_s = 1000$:
for $d_m > 63.5\,\text{mm}$,

$$C_s = 1859.104 - 476.5454 \log_{10} d_m \tag{11.42}$$

For chill cast bronze gears then:
for $d_m < 203.2\,\text{mm}$,
$C_s = 1000$.
for $d_m > 203.2\,\text{mm}$

$$C_s = 2052.011 - 455.8259 \log_{10} d_m \tag{11.43}$$

For centrifugally cast gears then:
for $d_m < 635\,\text{mm}$,
$C_s = 1000$.
for $d_m > 635\,\text{mm}$,

$$C_s = 1503.811 - 179.7503 \log_{10} d_m \tag{11.44}$$

The efficiency, in percent, for worm gearing is given by

$$\eta = \frac{P_{\text{output}}}{P_{\text{input}}} \times 100 \tag{11.45}$$

Substituting for the output power,

$$\eta = \frac{n W_t d_m}{1.91 \times 10^7 m_G P_{\text{input}}} \times 100 \tag{11.46}$$

where

P_{output} = rated output power (kW);
P_{input} = rated input power (kW);
n = rotational speed of the worm (rpm);
W_t = tangential load on the wormgear (N);
d_m = mean diameter of the gear (mm);
m_G = gear ratio.

11.4 Design procedure

An outline design procedure for a worm and wheel gear-set using the AGMA equations is given in Fig. 11.10.

Example 11.1. Develop a design for a worm and wheel gear-set. The prime mover is an electric motor running at 1470 rpm. A reduction ratio of 50:1 is required. The peak torque required is 850 N m. Assume sand cast gears.

Solution

Input speed = 1470 rpm. Ratio 50:1.

Output = 29.4 rpm.

Sand cast gears.

If a single start worm is specified then a 50 tooth worm gear will be needed to give a ratio of 50:1.

Fifty teeth is above the minimum recommended (Table 11.1).

An estimate for the centre distance needs to be made. If the centre distance between the worm and wheel is taken as 140 mm, then from Eqs (11.5) and (11.6), the pitch diameter range is found to be between 20.72 mm and 70.55 mm. A mid-value of 50 mm is selected here.

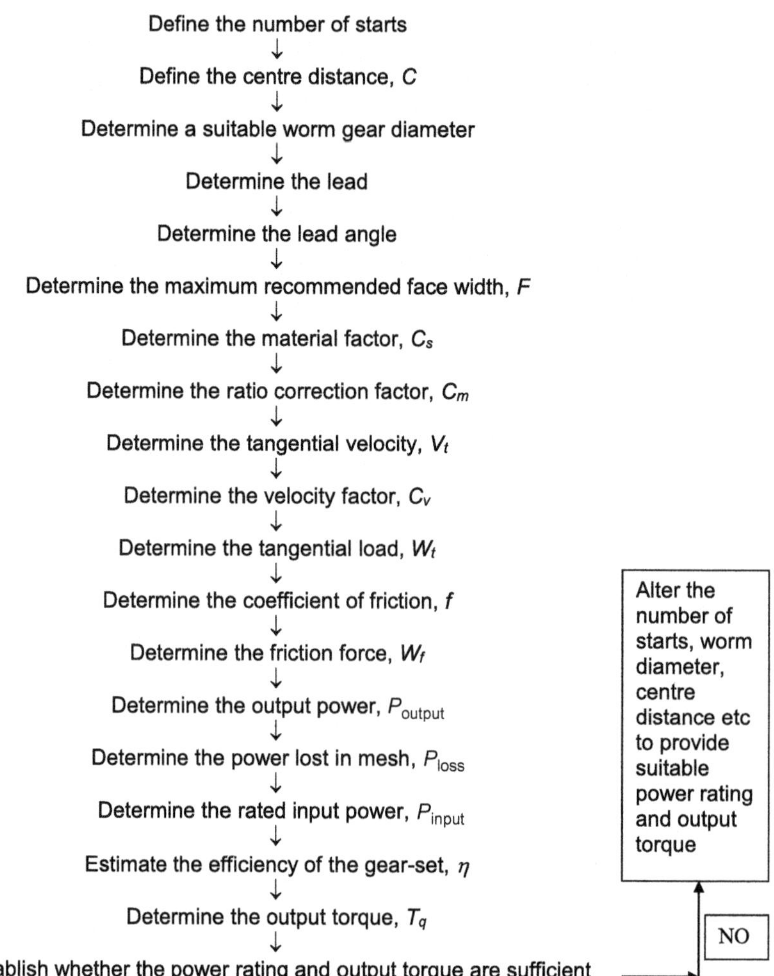
Define the number of starts
Define the centre distance, C
Determine a suitable worm gear diameter
Determine the lead
Determine the lead angle
Determine the maximum recommended face width, F
Determine the material factor, Cs
Determine the ratio correction factor, Cm
Determine the tangential velocity, Vt
Determine the velocity factor, Cv
Determine the tangential load, Wt
Determine the coefficient of friction, f
Determine the friction force, Wf
Determine the output power, Poutput
Determine the power lost in mesh, Ploss
Determine the rated input power, Pinput
Estimate the efficiency of the gear-set, η
Determine the output torque, Tq
Establish whether the power rating and output torque are sufficient for the application.
NO
Alter the number of starts, worm diameter, centre distance etc to provide suitable power rating and output torque

Fig. 11.10 Outline design procedure for a worm and wheel gear set using the AGMA equations.

From Eq. (11.8),
$d_G = 2C - d = (2 \times 140) - 50 = 230.0\,\text{mm}$.
The lead angle, from Eq. (11.2), is

$$L = \frac{\pi d_G N_W}{N_G} = \pi \times 230.0 \times \frac{1}{50} = 14.45\,\text{mm}$$

The lead angle, from Eq. (11.3), is

$$\lambda = \tan^{-1}\left(\frac{L}{\pi d_w}\right) = \tan^{-1}\left(\frac{14.45}{\pi \times 50}\right) = 5.256^{\circ}$$

Generally self-locking occurs for lead angles below 6 degrees. This is less than 6 degrees so the worm set will be self-locking.

The face width can be determined from Eq. (11.11).

$F_G = 0.67d = 0.67 \times 50 = 33.5\,\text{mm}$.

The materials factor for sand cast gears is given by Eq. (11.42), if the mean worm diameter is >63.5 mm. Here the mean worm diameter is 50 mm so $C_s = 1000$.

$m_G = 50$.

From Eq. (11.34),

$C_m = 0.7896$.

The tangential velocity at the mean worm diameter can be determined from Eq. (11.32),

$$V_t = \frac{nd}{19098 \cos\lambda} = \frac{1470 \times 50.0}{19098 \cos 5.256} = 3.865 \text{ m/s}$$

The velocity factor, from Eq. (11.37), is

$C_v = 0.3013$.

From Eq. (11.30), the tangential load is given by

$$W_t = \frac{1000 \times 0.7896 \times 0.3013 \times 230^{0.8} \times 33.50}{75.948} = 8134 \text{ N}$$

The coefficient of friction is given by Eq. (11.40)

$$\mu = 0.103 e^{\left(-1.185 V_t^{0.45}\right)} = 0.103 e^{\left(-1.185 \times 3.865^{0.45}\right)} = 0.01167$$

The friction force is given by

$$W_f = \frac{f W_t}{\cos\lambda \cos\phi} = \frac{0.01167 \times 8134}{\cos 5.256 \cos 20} = 101.5 \text{ N}$$

The rated output power (Eq. 11.28) is

$$P_{\text{output}} = \frac{n W_{tg} d_g}{1.91 \times 10^7 m_G} = \frac{1470 \times 8134 \times 230}{1.91 \times 10^7 \times 50} = 2.880 \text{ kW}$$

The lost power (Eq. 11.29) is

$$P_{\text{loss}} = \frac{V_t W_f}{1000} = \frac{3.865 \times 101.5}{1000} = 0.392 \text{ kW}$$

The input power rating is $2.880 + 0.392 = 3.272\,\text{kW}$.

The efficiency of the gear set is $2.880/3.272 = 0.881 = 88.1\%$.

The output torque is given by

$$T_q = W_t \frac{d}{2} = 8134 \times \frac{0.230}{2} = 935.4 \text{ N m}$$

This torque exceeds the requirement, suggesting that the design is suitable. If the value was below that required, then an alternative centre distance could be explored and, if necessary, the number of starts could be increased.

11.5 Conclusions

Worm and wheel gears are usually used for nonparallel, nonintersecting, right angle gear drive system applications, where a high gear ratio is required. They can also be used for medium speed reductions. The worm is generally the driving member. The high transmission ratio leads to a compact solution for many applications in comparison to other types of gearing. For certain arrangements, self-locking is possible and this can provide an attribute for some applications where a set position is desirable if the drive train is not braked or powered.

References

Books and papers

Dudás, I., 2005. The Theory and Practice of Worm Gear Drives. Butterworth-Heinemann.

Dudley, D.W., 1984. Handbook of Practical Gear Design. McGraw Hill.

Litvin, F.L., Kin, V., 1992. Computerized simulation of meshing and bearing contact for single-enveloping worm-gear drives. J. Mech. Des. 114, 313–316.

Maitra, G.M., 1994. Handbook of Gear Design, second ed. Tata McGraw Hill.

Merritt, H.E., 1935. Worm Gear Performance. Proc. Inst. Mech. Eng. 129, 127–194.

Norton, R.L., 2006. Machine Design, third ed. Pearson.

Radzevich, S.P., 2016. Dudley's Handbook of Practical Gear Design and Manufacture, third ed. CRC Press.

Standards

AGMA Design manual for cylindrical wormgearing. ANSI/AGMA Standard 6022-C93. Reaffirmed 2008.

AGMA Practice for enclosed cylindrical wormgear speed reducers and gearmotors. ANSI/AGMA Standard 6034-B92. Reaffirmed 2005.

BS 721-1:1963. Specification for worm gearing. Imperial units.

BS 721-2:1983. Specification for worm gearing. Metric units.

BS ISO TR 10828:1997. Worm gears. Geometry of worm profiles.

PD ISO/TR 14521:2010. Gears. Calculation of load capacity of wormgears.

Websites

At the time of going to press the world-wide-web contained useful information relating to this chapter at the following sites.

www.ashokaengineering.com/.
www.bandhgears.co.uk.
www.bellgears.co.uk.
www.bostongear.com/products/open/worms.html.
www.brentwingearcompany.co.uk.
www.davall.co.uk.
www.delroyd.com.
www.gearcutting.com.
www.gearmanufacturer.net/.
www.girard-transmissions.com.
www.hewitt-topham.co.uk/.
www.hopwoodgear.com.
www.hpcgears.com.
www.huco.com.
www.mmestrygears.com.
www.muffettgears.co.uk/.
www.qtcgears.com/.
www.rarodriguez.co.uk.
www.traceygear.com.
www.wmberg.com.

Further reading

Shigley, J.E., 1986. Mechanical Engineering Design, first metric ed. McGraw Hill.

Shigley, J.E., Mischke, C.R., Budynas, R.G., 2004. Mechanical Engineering Design, seventh ed. McGraw Hill.

Townsend, D.P., 1992. Dudley's Gear Handbook, second ed. McGraw Hill.

Belt and chain drives

12

Chapter Outline

Nomenclature

Generally, preferred SI units have been stated.

C	centre distance (m)
D	pitch diameter (m)
f_1	application factor
f_2	tooth factor
L	number of pitches
N_1	number of teeth in the driving sprocket
N_2	number of teeth in the driven sprocket
p	chain pitch (m)
R_1	radius of driving pulley or sprocket (m)
R_2	radius of driven pulley or sprocket (m)
$V_{pitchline}$	pitchline velocity (m/s)
θ	angle of contact (rad)
ω_1	angular velocity of driving pulley or sprocket (rad/s)
ω_2	angular velocity of driven pulley or sprocket (rad/s)

12.1 Introduction

Belt and chain drives are used to transmit power from one rotational drive to another. A belt is a flexible power transmission element that runs tightly on a set of pulleys. A chain drive consists of a series of pin-connected links that run on a set of sprockets. This chapter introduces various types of belt and chain drives and presents selection procedures for wedge, synchronous and flat belts and also for roller chains.

Mechanical Design Engineering Handbook. https://doi.org/10.1016/B978-0-08-102367-9.00012-3

Belt and chain drives consist of flexible elements running on either pulleys or sprockets as illustrated in Fig. 12.1 with practical examples shown in Figs 12.2 and 12.3. The purpose of a belt or chain drive is to transmit power from one rotating shaft to another.

The speed ratio between the driving and driven shaft is dependent on the ratio of the pulley or sprocket diameters as is given by

$$V_{\text{pitchline}} = \omega_1 R_1 = \omega_2 R_2 \tag{12.1}$$

$$\text{Angular velocity ratio} = \frac{\omega_1}{\omega_2} = \frac{R_2}{R_1} \tag{12.2}$$

where $V_{\text{pitchline}}$ = pitchline velocity (m/s), ω_1 = angular velocity of driving pulley or sprocket (rad/s), ω_2 = angular velocity of driven pulley or sprocket (rad/s), R_1 = radius of driving pulley or sprocket (m) and R_2 = radius of driven pulley or sprocket (m).

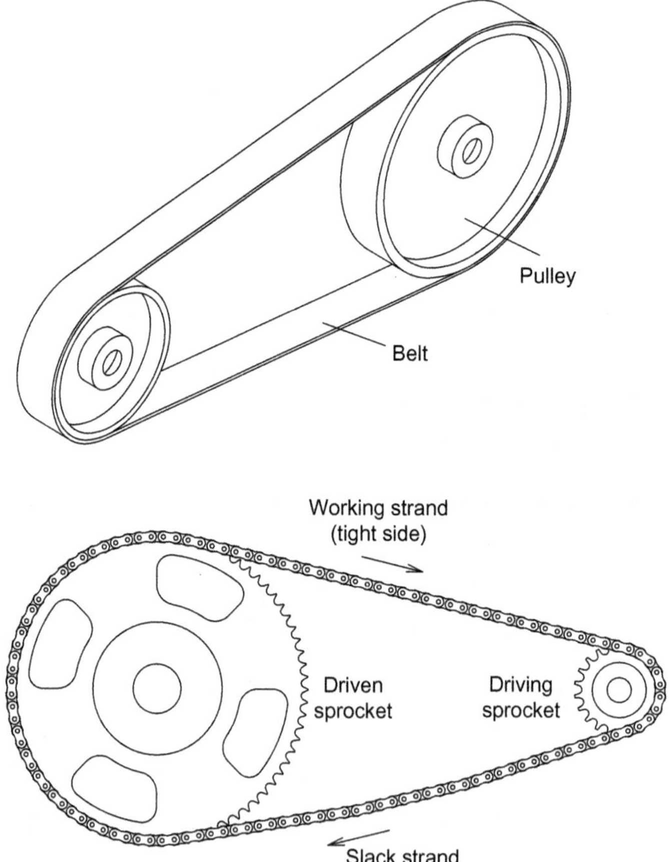

Fig. 12.1 Belt drive and chain drive.

Fig. 12.2 The Strida bicycle featuring a belt drive.
Image courtesy of Mark Sanders, MAS Design Ltd.

Fig. 12.3 A motorbike chain drive.

A belt drive transmits power between shafts by using a belt to connect pulleys on the shafts by means of frictional contact or mechanical interference. A chain consists of a series of links connected by pins. The chain is designed to mesh with corresponding teeth on sprockets located on both the driving and the driven shafts.

Power transmission between shafts can be achieved by a variety of means including belt, chain and gear drives and their use should be compared for suitability and

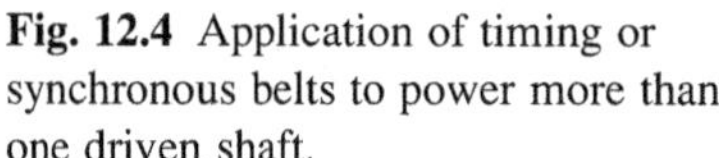

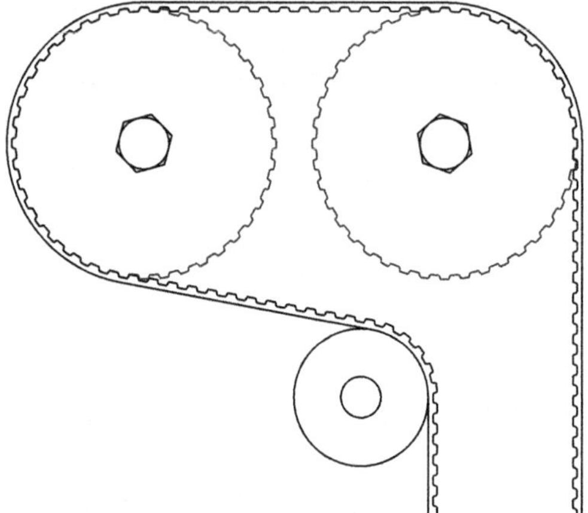

Fig. 12.4 Application of timing or synchronous belts to power more than one driven shaft.

optimisation for any given application. In addition to power transmission, constant speed ratio or synchronisation of the angular position of the driving and driven shaft may be critical to operation. This can be achieved by means of gears, chains or special toothed belts called synchronous or timing belts, see Fig. 12.4.

Both belt and chain drives can transmit power between shafts that are widely separated giving designers and engineers greater scope for control over machine layout. In comparison to gears, they do not require as much precision in the location of centre distances. Both belt and chain drives are often cheaper than the equivalent gear drive. Belts and chains are generally complimentary covering a range of operational requirements.

In general, belt drives are used when the rotational tangential velocity component of the pulleys are of the order of 10–60 m/s. At lower speeds, the tension in the belt becomes too high for typical belt sections. At higher speeds, centrifugal forces throw the belts off the pulleys reducing the torque capacity and dynamic phenomena reduce the effectiveness and life of the belt drive. Chain drives are typically used at lower speeds and consequently higher torques than belts. Recall that for a rotating machine torque is proportional to power/ω. As the angular velocity reduces, for a given power, the torque increases.

Belt drives can have numerous advantages over gear and chain drives including easy installation, low maintenance, high reliability, adaptability to nonparallel drive and high-transmission speeds. The principal disadvantages of belt drives are their limited power transmission capacity and limited speed ratio capability. Belt drives are less compact than either gear or chain drives and are susceptible to changes in environmental conditions such as contamination with lubricants. In addition, vibration and shock loading can damage belts.

Chains are usually more compact than belt drives for a given speed ratio and power capacity. Chain drives are generally more economical than the equivalent gear drive and are usually competitive with belt drives. Chains are inherently stronger than belt drives due to the use of steels in their manufacture and can therefore support higher tension and transmit greater power. The disadvantages of chain drives are limited speed ratios and power transmission capability and also safety issues. Chain drives can break and get thrown off the sprockets with large forces and high speeds. Guards should be provided for chain drives and some belt drives to prevent damage caused by a broken chain or belt and also to prevent careless access to the chain or belt drive.

12.2 Belt drives

As mentioned previously, a belt drive consists of a flexible element that runs on a set of pulleys mounted on the shafts. Belt drives can be used to simply transmit power between one shaft and another with the speed of the driving and driven shaft equal. In this case, the pulley diameters would be equal. Alternatively, the driven shaft velocity can be decreased by using a bigger diameter pulley on the driven shaft than on the driving shaft. It is also possible to use belt drives to step up or increase the speed of a driven shaft but this is a less common application.

There are various types of belt drive configurations including flat, round, V, wedge and synchronous belt drives, each with their individual merits. The cross sections of various belts are illustrated in Fig. 12.5. Most belts are manufactured from rubber or polymer-based materials.

Power is transmitted by means of friction in the case of flat, round, wedge and V belts and by a combination of friction and positive mechanical interference in the case of synchronous belt drives. When the driven shaft rotates, friction between the pulley and the belt causes the belt to grip the pulley increasing the tension in the side near to the point of first rotational contact. The tensile force exerts a tangential force, and associated torque, on the driven pulley. The opposite side of the belt is also under tension but to a lesser extent and is referred to as the slack side.

A frequent application of belt drives is to reduce the speed output from electric motors which typically run at specific synchronous speeds which are high in comparison to the desired application drive speed. Because of their good 'twistability', belt drives are well suited to applications where the rotating shafts are in different planes. Some of the standard layouts are shown in Fig. 12.6. Belts are installed by moving the shafts closer together, slipping the belt over the pulleys and then moving the shafts back into their operating locations.

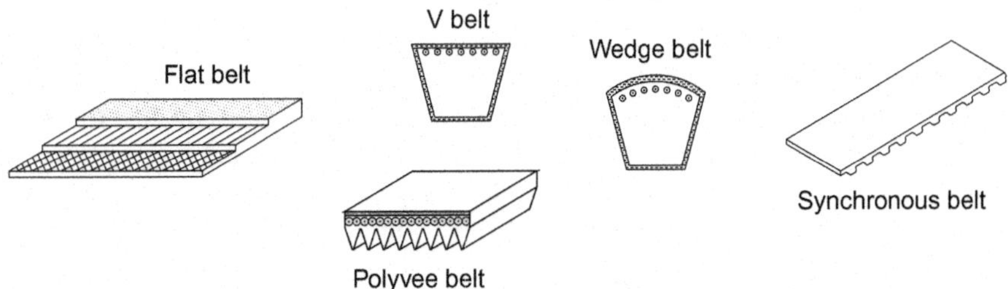

Fig. 12.5 Various belt cross sections.

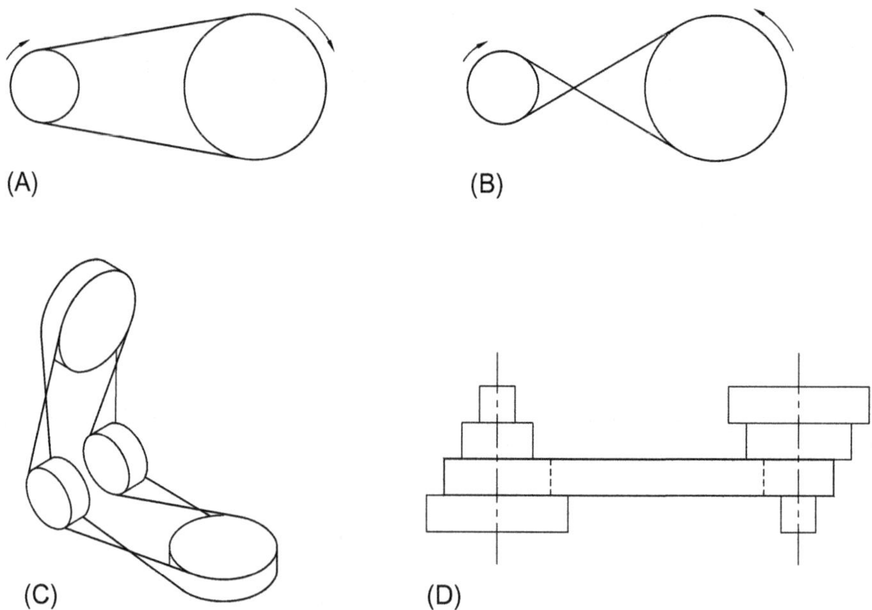

Fig. 12.6 Example belt drive configurations (A) nonreversing open belt; (B) reversing crossed belt; (C) twisted belt drive and (D) variable speed drive.

Flat belts that have high strength, can be used for large speed ratios (>8:1), have a low pulley cost, give low noise levels and are good at absorbing torsional vibration. The belts are typically made from multiple plies with each layer serving a special purpose. A typical three ply belt consists of a friction ply made from synthetic rubber, polyurethane or chrome leather, a tension ply made from polyamide strips or polyester cord and an outer skin made from polyamide fabric, chrome leather or an elastomer. The corresponding pulleys are made from cast iron or polymer materials and are relatively smooth to limit wear. The driving force is limited by the friction between the belt and the pulley. Applications include manufacturing tools, saw mills, textile machinery, food-processing machines, multiple spindle drives, pumps and compressors.

The most widely used type of belt in industrial and automotive applications is the V belt or wedge belt, Fig. 12.7. It is familiar from its automotive application where it is used to connect the crankshaft to accessory drives such as the alternator, water pump and cooling fan. It is also used for general engineering purposes from domestic appliances to heavy duty rolling machines. The V or wedge shape causes the belt to wedge into the corresponding groove in the pulley increasing friction and torque capacity. Multiple belts are commonly used so that a cheaper small cross-sectional area belt can be used to transmit more power. Note that long centre distances are not recommended for V or wedge belts.

Synchronous belts, also called timing belts, have teeth which mesh with corresponding teeth on the pulleys. This mechanical interference or positive contact

Fig. 12.7 V belt drive.

between the pulleys and the belt provides angular synchronisation between the driving and the driven shafts and ensures a constant speed ratio. Synchronous belts combine the advantages of normal friction belt drives with the capability of synchronous drive (see, Perneder and Osbourne, 2012). The meshing of the belt and pulleys is critical to their effective operation. The teeth of the advancing belt must mesh correctly with the corresponding grooves on the pulley wheels and remain in mesh throughout the arc of contact. To achieve this, the pitch of the belts and the pulleys must correspond exactly. A disadvantage of synchronous belts can be the noise generated by compression of air between the teeth especially at high speeds.

Table 12.1 lists the comparative merits of various belt drives. Fig. 12.8 provides a guide for the selection of belt type.

It is necessary to manage the tension in a belt drive. This can be achieved by means of setting the distance between pulley wheels to a specified distance as defined by the manufacturer or through analysis. It may however not be practical or convenient to assemble all the components under tension or running conditions may mandate control

Table 12.1 Comparison of belt performance

Parameter	Flat belt drive	V belt drive	Wedge belt drive	Synchronous belt drive
Optimum efficiency (%)	98	80	86	98
Maximum speed (m/s)	70	30	40	50
Minimum pulley diameter (mm)	40	67	60	16
Maximum speed ratio	20	7	8	9
Optimum tension ratio	2.5	5	5	–

Data from Hamilton, P., 1994. Belt drives. In: Hurst, K. (Ed.), Rotary Power Transmission Design. McGraw Hill.

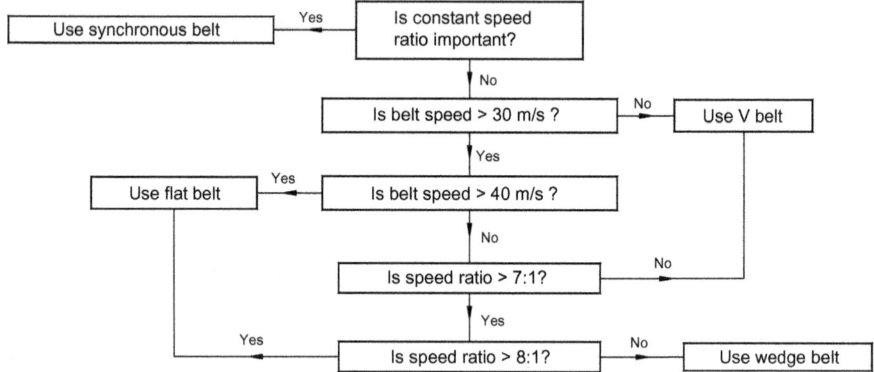

Fig. 12.8 Procedure for the selection of belt type.
After Hamilton, P., 1994. Belt drives. In: Hurst, K. (Ed.), Rotary Power Transmission Design. McGraw Hill.

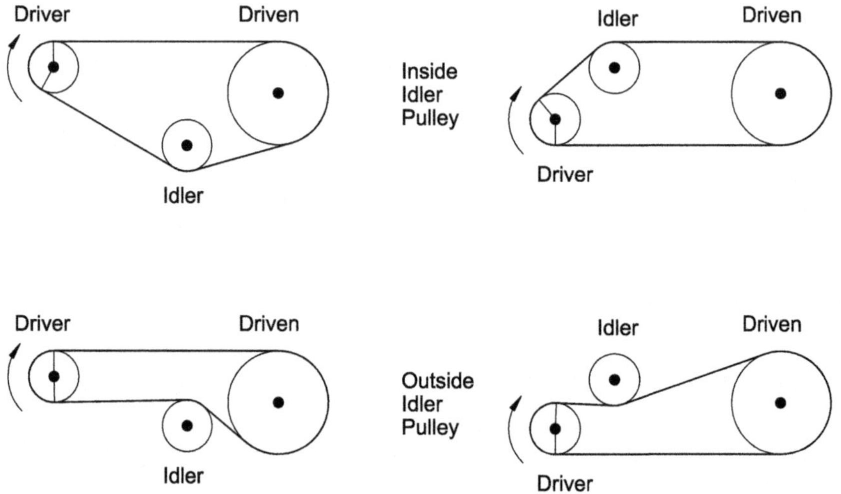

Fig. 12.9 Belt tensioning examples.

of the tension. In these circumstances, a belt tensioner can be used. A belt tensioner can take up slack in a belt so that the belt does not slip on the pulley wheels; improve drive performance (see, Ulsoy et al., 1985); enable use of a longer belt; change the direction of the belt; enable easier installation and servicing. Some examples of the use of belt tensioning are illustrated in Fig. 12.9.

12.2.1 Belt selection

Two approaches are presented here for the selection of a belt drive. Use can be made of the design procedures and accompanying charts provided by most belt drive manufacturers. The use of charts for belt selection is illustrated by considering a wedge belt drive in Section 12.2.2 and for synchronous belts in Section 12.2.3. Alternatively, use can be made of fundamental relationships for the belt tensions and torque transmission and a belt selected based on its maximum permissible tensile stress. These equations are illustrated by an example using a flat belt drive in Section 12.2.4.

12.2.2 Wedge belt selection

The typical design and selection procedure for use in conjunction with power-speed rating charts supplied by commercial companies is outlined for a wedge belt, as follows and illustrated in Fig. 12.10.)

(1) Define the operating conditions.	These include the nominal power to be transmitted, the rotational speeds of the shafts, and any space, layout or other constraints such as environmental conditions.
(2) Determine the service factor.	Service factors are used to downrate the power transmission capability listed by belt suppliers to account for the differences between practical applications and test conditions. Table 12.2 lists typical values for service factors.
(3) Calculate the design power.	This is the product of the nominal power and the service factor.
(4) Select the belt type.	To assist in the selection of which type of belt drive to use the procedure given in Fig. 12.8 can be used as a guideline.
(5) Select a belt.	Using a manufacturer's rating chart, select a specific belt for the design power and speed.
(6) Select the pulley diameters.	Pulleys are normally available in standard sizes. Choose the smallest sizes available such that the speed ratio is acceptable.
(7) Set the centre distance.	This is dependent on the application. As a general guideline, the centre distance should be greater than the diameter of the larger pulley.
(8) Determine the belt length.	Note that belts are usually manufactured in standard lengths. So some iteration around the design parameters may be necessary to arrive at a satisfactory compromise.
(9) Apply power correction factors.	These are used to compensate for speed ratio and the belt geometry and are provided within belt manufacturer's design guides.
(10) Determine the allowable power per belt (or per belt width for flat belts).	This is a function of the belt dimensions and is available from manufacturer's design guides.
(11) Determine the number of belts.	The number of belts is given by dividing the design power by the allowable power per belt and rounding up to the nearest integer.

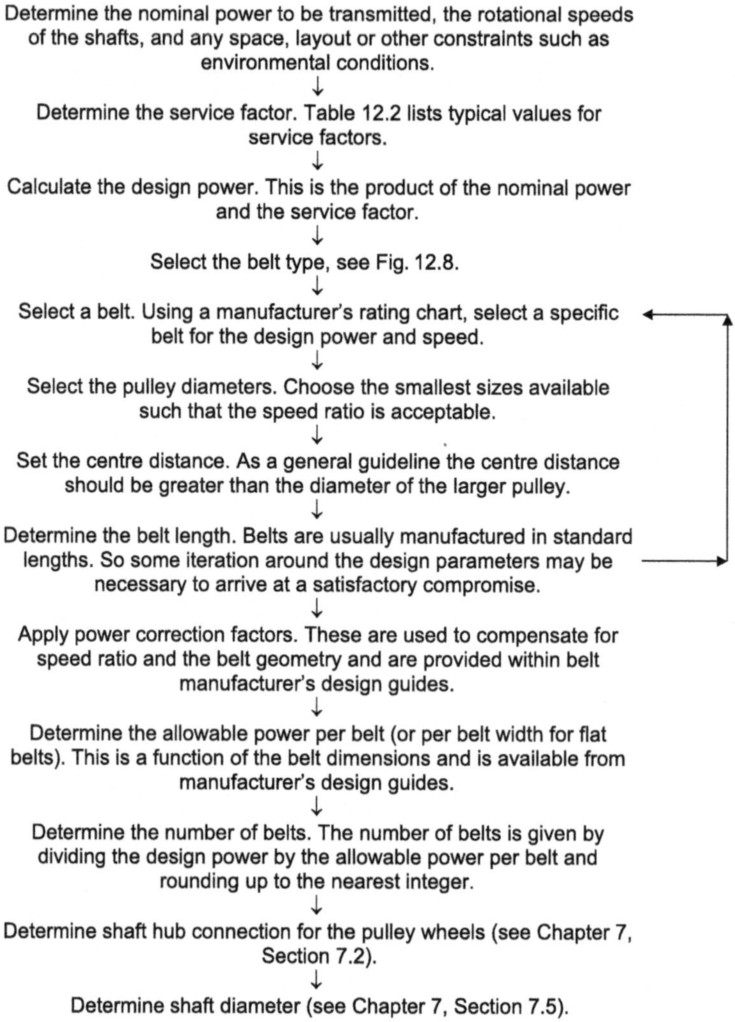

Determine the nominal power to be transmitted, the rotational speeds of the shafts, and any space, layout or other constraints such as environmental conditions.

↓

Determine the service factor. Table 12.2 lists typical values for service factors.

↓

Calculate the design power. This is the product of the nominal power and the service factor.

↓

Select the belt type, see Fig. 12.8.

↓

Select a belt. Using a manufacturer's rating chart, select a specific belt for the design power and speed.

↓

Select the pulley diameters. Choose the smallest sizes available such that the speed ratio is acceptable.

↓

Set the centre distance. As a general guideline the centre distance should be greater than the diameter of the larger pulley.

↓

Determine the belt length. Belts are usually manufactured in standard lengths. So some iteration around the design parameters may be necessary to arrive at a satisfactory compromise.

↓

Apply power correction factors. These are used to compensate for speed ratio and the belt geometry and are provided within belt manufacturer's design guides.

↓

Determine the allowable power per belt (or per belt width for flat belts). This is a function of the belt dimensions and is available from manufacturer's design guides.

↓

Determine the number of belts. The number of belts is given by dividing the design power by the allowable power per belt and rounding up to the nearest integer.

↓

Determine shaft hub connection for the pulley wheels (see Chapter 7, Section 7.2).

↓

Determine shaft diameter (see Chapter 7, Section 7.5).

Fig. 12.10 Typical wedge belt design procedure.

Fig. 12.11 shows a design chart for power rating versus speed and Tables 12.3–12.6 list values for the minimum pulley diameter, centre distance and power ratings for wedge belts. The use of these in conjunction with the procedure given is illustrated by the following examples (Examples 12.1 and 12.2).

Example 12.1. Select a wedge belt and determine the pulley diameters for a reciprocating compressor driven by a 28-kW two-cylinder diesel engine. The engine speed is 1500 rpm and the compressor speed is 950 rpm. The proposed distance between the engine and compressor shaft centres is approximately 1.5 m. The system is expected to be used for <10 h per day.

Solution

The speed ratio is $\frac{1500}{950} = 1.58$.

From Table 12.2 for 'heavy starts', 'heavy duty' and <10 h of operation per day, the service factor is 1.4.

The design power $= 28 \times 1.4 = 39.2$ kW.

From Fig. 12.11, the combination of power equal to 39.2 kW and the speed equal to 1500 rpm is found to be within the envelope curve and range defined as suitable for SPB belt drives. The combination is also with the envelope range suitable for SPA belts. As the power is near the upper range of the SPA curve, an SPB belt is selected here, although the choice is marginal.

From Table 12.3 for a design power of 39.2 kW and driving shaft speed of 1500 rpm, an approximate minimum pulley diameter of 160 mm is suitable.

From Table 12.4, the actual pulley diameters can be selected by tracing down the left hand side to the speed ratio and reading across for the pitch diameters of the driving and driven pulleys. The minimum diameter from Table 12.3 can be used as a guideline when there is a choice.

From Table 12.4, for a speed ratio of 1.58, suitable pitch diameters are $D_1 = 224$ mm and $D_2 = 355$ mm. The nearest centre distance listed to the 1.5 m desired is 1.544 m and tracing the column upwards defines the belt length as 4000 mm and by tracing the shading upwards the arc-length correction factor is 1.05.

From Table 12.5, the rated power per belt for $n_1 = 1500$ rpm and $D_1 = 224$ mm is 15.97 kW/belt.

From Table 12.6, the additional power per belt accounting for the speed ratio is 1.11 kW.

The corrected value for the power per belt is $(15.97 + 1.11) \times 1.05 = 17.93$ kW/belt.

The total number of belts is therefore $\frac{39.2}{17.93} = 2.19$.

As a partial belt cannot be used round this number up to the nearest integer, thus, three SPB4000 belts should be used running on sprockets of 224 mm and 315 mm pitch diameter with a centre distance of 1.544 m.

Example 12.2. A wedge belt drive is required to transmit 18.5 kW from an electric motor running at 1455 rpm to a uniformly loaded conveyor running at 400 rpm. The desired centre distance is 1.4 m and expected use is 15 h a day. Select a suitable belt, or belts, and determine the pulley diameters.

Solution

18.5 kW, $n_1 = 1455$ rpm, $n_2 = 400$ rpm. $C \approx 1.4$ m.

The speed ratio is 1455/400 = 3.6375.

From Table 12.2, the service factor is 1.1.

Design power $= 18.5 \times 1.1 = 20.35$ kW.

From Fig. 12.11, an SPB belt is suitable.

From Table 12.3, the minimum pulley diameter is approximately 112 mm.

From Table 12.4, selecting the closest ratio as 3.57, $D_1 = 224$ mm, $D_2 = 800$ mm, $C = 1.416$ m (SPB4500). The combined arc and belt length correction factor is 1.0.

From Table 12.5, the rated power per belt is approximately 15.47 kW. From Table 12.6, the additional power per belt is 1.21 kW.

Table 12.2 Service Factors

	Type of prime mover					
	Soft starts. Electric motors: AC—star delta start, DC—shunt wound. Internal combustion (IC) engines with four or more cylinders. Prime movers fitted with centrifugal clutches, dry or fluid couplings or electronic soft start drives.			**Heavy starts. Electric motors: AC—star delta start, DC—shunt wound. IC engines with less than four cylinders. Prime movers not fitted with soft start drives.**		
	Duty (hours per day)					
Types of driven machine	**<10**	**10–16**	**>16**	**<10**	**10–16**	**>16**
Light duty, e.g. agitators (uniform density), blowers, exhausters and fans (up to 7.5 kW), centrifugal compressors, rotodynamic pumps, uniformly loaded belt conveyors.	1.0	1.1	1.2	1.1	1.2	1.3
Medium duty, e.g. agitators (variable density) blowers exhausters and fans (over 7.5 kW), rotary compressors and pumps (other than centrifugal), nonuniformly loaded conveyors, generators, machine tools, printing machinery, sawmill machinery.	1.1	1.2	1.3	1.2	1.3	1.4
Heavy duty, e.g. brick machinery, bucket elevators, reciprocating compressors and pumps, heavy duty conveyors, hoists, pulverisers, punches, presses, quarry plant, textile machinery.	1.2	1.3	1.4	1.4	1.5	1.6
Extra heavy duty, e.g. crushers.	1.3	1.4	1.5	1.5	1.6	1.8

After Fenner Power Transmission UK. *Belt Drives Design Manual.*

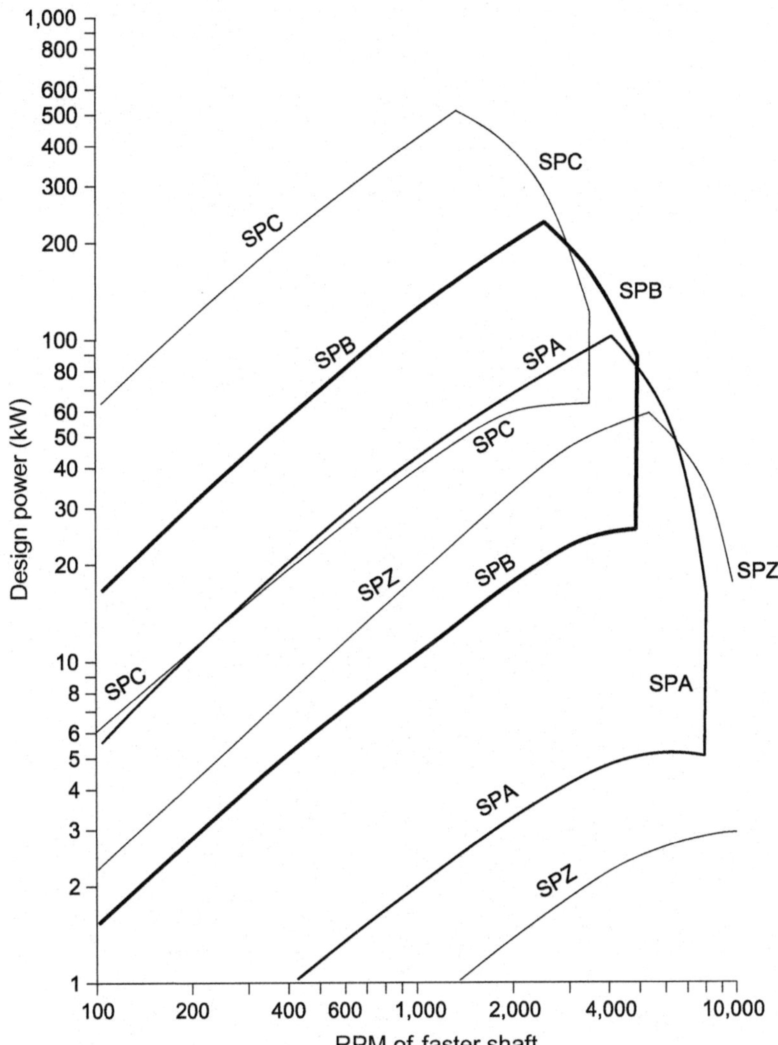

Fig. 12.11 Selection chart for wedge belts.
Courtesy of Fenner Drives, UK.

The corrected power per belt is $(15.47 + 1.21) \times 1.0 = 16.68\,\text{kW}$.

$$\text{Number of belts required} = \frac{20.35}{16.68} = 1.22$$

Rounding up, use two SPB4500 belts with pulley diameters of 224 and 800 mm and centre distance 1.416 m.

Table 12.3 Minimum recommended pulley diameters for wedge belt drives

Speed of faster shaft (rpm)	Minimum pulley diameters (mm) Design power (kW)																			
	<1	**3.0**	**4.0**	**5.0**	**7.5**	**10.0**	**15.0**	**20.0**	**25**	**30**	**40**	**50**	**60**	**75**	**90**	**110**	**130**	**150**	**200**	**250**
500	56	90	100	112	125	140	180	200	212	236	250	280	280	315	375	400	450	475	500	560
600	56	85	90	100	112	125	140	180	200	212	224	250	265	280	300	335	375	400	475	500
720	56	80	85	90	100	106	132	150	160	170	200	236	250	265	280	300	335	375	450	500
960	56	75	80	85	95	100	112	132	150	180	180	200	224	250	280	280	300	335	400	450
1200	56	71	80	80	95	95	106	118	132	150	160	180	200	236	236	250	265	300	335	355
1440	56	63	75	80	85	85	100	112	125	140	160	170	190	212	236	236	250	280	315	335
1800	56	63	71	75	80	85	95	106	112	125	150	160	170	190	212	224	236	265	300	335
2880	56	60	67	67	80	80	85	90	100	112	125	140	160	170	180	212	224	236	–	–

Courtesy of Fenner Power Transmission UK.

Table 12.4 Centre distances for selected SPB wedge belts

Combined arc and belt length correction factor					0.85		0.90		0.95		1.00			1.05			1.10		1.15	
	Pitch diameter of pulleys		Power per belt (kW)		Belt Length for SPB belt type															
Speed ratio	Driver	Driven	1440 rpm	960 rpm	1250	1400	1800	2000	2240	2500	2800	3150	3550	4000	4500	5000	5600	6300	7100	8000
1.27	315	400	24.56	17.91	-	-	-	436	557	687	837	1013	1213	1438	1688	1938	2238	2588	2988	3438
1.27	118	150	6.37	4.58	414	489	689	789	-	-	-	-	-	-	-	-	-	-	-	-
1.28	125	160	7.24	5.18	401	476	676	776	-	-	-	-	-	-	-	-	-	-	-	-
1.29	140	180	7.69	5.54	373	448	648	748	869	998	1148	1342	1524	1749	1999	2249	2549	2899	3299	3749
1.29	132	170	8.10	5.78	387	462	663	763	-	-	-	-	-	-	-	-	-	-	-	-
1.53	118	180	6.51	4.67	390	465	665	765	-	-	-	-	-	-	-	-	-	-	-	-
1.56	180	280	12.03	8.61	259	335	536	637	757	887	1038	1213	1413	1638	1888	2138	2438	2788	3188	3638
1.56	160	250	9.95	7.13	300	375	576	676	797	927	1077	1252	1452	1677	1927	2178	2478	2828	3228	3678
1.57	150	236	8.89	6.37	319	394	595	696	816	946	1096	1271	1471	1696	1946	2196	2496	2847	3247	3697
1.57	200	315	14.05	10.05	-	290	492	593	713	844	994	1169	1369	1594	1845	2095	2395	2745	3145	3595
1.58	224	355	16.54	11.86	-	-	440	541	662	793	943	1118	1319	1544	1794	2044	2344	2694	3095	3545
1.59	315	500	24.81	18.09	-	-	-	-	471	603	754	930	1131	1357	1607	1858	2158	2508	2908	3359
1.60	125	200	7.49	5.35	368	443	644	744	-	-	-	-	-	-	-	-	-	-	-	-
1.60	140	224	7.95	5.71	336	412	613	713	833	963	1113	1288	1489	1714	1964	2214	2514	2864	3264	3714
1.60	250	400	19.02	13.68	-	-	382	484	605	736	886	1062	1262	1488	1738	1988	2288	2368	3039	3489
1.87	190	355	13.17	9.42	-	-	465	566	687	818	968	1144	1344	1570	1820	2070	2371	2721	3121	3571
1.89	212	400	15.37	11.00	-	-	409	511	633	764	915	1090	1291	1516	1767	2017	2317	2668	3068	3518
1.89	125	236	7.49	5.35	337	413	614	714	-	-	-	-	-	-	-	-	-	-	-	-
1.89	112	212	5.88	4.23	367	443	644	744	-	-	-	-	-	-	-	-	-	-	-	-
1.89	132	250	8.36	5.95	320	396	597	697	-	-	-	-	-	-	-	-	-	-	-	-
2.09	170	355	11.21	8.02	-	-	479	580	702	833	983	1159	1360	1585	1835	2086	2386	2736	3136	3586
2.10	150	315	9.11	6.53	-	324	528	629	750	881	1031	1207	1407	1633	1883	2133	2433	2784	3184	3634
2.11	190	400	13.26	9.48	-	-	424	526	648	780	931	1107	1307	1533	1784	2034	2334	2685	3085	3535
2.11	112	236	5.97	4.30	346	422	624	724	-	-	-	-	-	-	-	-	-	-	-	-
2.12	118	250	6.72	4.81	429	406	607	708	-	-	-	-	-	-	-	-	-	-	-	-

Continued

Table 12.4 Continued

2.23	224	500	16.64	11.92	-	-	-	408	534	667	820	997	1198	1425	1676	1926	2227	2578	2978	3429
2.24	125	280	7.59	5.42	297	374	577	677	-	-	-	-	-	-	-	-	-	-	-	-
2.25	140	315	8.04	5.78	252	331	535	637	758	888	1039	1214	1415	1640	1891	2141	2441	2791	3191	3642
2.25	280	630	21.87	15.81	-	-	-	-	364	505	662	842	1046	1273	1525	1777	2078	2429	2830	3281
2.35	170	400	11.21	8.02	-	-	437	540	663	794	945	1121	1322	1548	1799	2049	2350	2700	3100	3550
2.54	315	800	24.81	18.15	-	-	-	-	-	-	-	654	865	1097	1353	1606	1909	2261	2663	3115
2.63	190	500	13.26	9.48	-	-	-	430	557	691	844	1021	1223	1450	1701	1952	2253	2603	3004	3455
2.67	150	400	9.11	6.53	-	-	451	554	677	808	960	1136	1337	1563	1814	2064	2365	2715	3116	3566
2.67	118	315	6.72	4.81	267	346	551	652	-	-	-	-	-	-	-	-	-	-	-	-
2.67	236	630	17.79	12.77	-	-	-	-	390	533	692	873	1077	1305	1557	1809	2111	2462	2863	3314
3.39	236	800	17.86	12.82	-	-	-	-	-	-	-	705	918	1152	1408	1662	1966	2319	2722	3174
3.50	180	630	12.31	8.80	-	-	-	-	425	569	729	911	1116	1345	1598	1850	2152	2504	2905	3356
3.57	112	400	6.04	4.34	-	-	476	580	-	-	-	-	-	-	-	-	-	-	-	-
3.57	140	500	8.11	5.82	-	-	351	462	590	725	879	1057	1259	1486	1738	1989	2290	2641	3042	3493
3.57	224	800	16.71	11.97	-	-	-	-	-	-	-	713	926	1160	1416	1671	1975	2328	2731	3183
3.57	280	1000	21.94	15.85	-	-	-	-	-	-	-	-	673	925	1190	1450	1758	2114	2519	2973
3.71	170	630	11.28	8.06	-	-	-	-	431	576	736	918	1123	1352	1605	1857	2159	2511	2913	3364
3.77	212	800	15.53	11.11	-	-	-	-	-	-	-	720	934	1168	1425	1679	1983	2337	2739	3192
3.79	132	500	8.52	6.06	-	-	356	467	-	-	-	-	-	-	-	-	-	-	-	-
3.94	160	630	10.23	7.32	-	-	-	-	437	582	742	925	1130	1359	1612	1865	2167	2519	2920	3371
4.72	212	1000	15.53	11.11	-	-	-	-	-	-	-	-	714	968	1235	1496	1805	2162	2568	3022
5.00	160	800	10.23	7.32	-	-	-	-	-	-	554	753	968	1203	1461	1716	2021	2374	2778	3230
5.00	200	1000	14.34	10.25	-	-	-	-	-	-	-	-	722	976	1243	1504	1813	2171	2576	3031
5.26	190	1000	13.33	9.53	-	-	-	-	-	-	-	-	728	982	1250	1511	1820	2178	2584	3038
5.33	150	800	9.18	6.57	-	-	-	-	-	-	-	759	975	1210	1468	1723	2028	2382	2785	3238

Note: This is only a partial selection from a typical catalogue relevant to the worked examples.
Courtesy of Fenner Power Transmission UK.

Table 12.5 Power ratings for SPB wedge belts

rpm of faster shaft	Rated power (kW) per belt for small pulley pitch diameter (mm)												
	140	**150**	**160**	**170**	**180**	**190**	**200**	**212**	**224**	**236**	**250**	**280**	**315**
100	0.73	0.82	0.92	1.01	1.10	1.20	1.29	1.40	1.51	1.62	1.74	2.01	2.33
200	1.33	1.51	1.69	1.87	2.05	2.22	2.40	2.61	2.82	3.02	3.26	3.78	4.37
300	1.89	2.15	2.41	2.67	2.93	3.18	3.44	3.74	4.04	4.35	4.70	5.44	6.30
400	2.42	2.76	3.09	3.43	3.77	4.10	4.43	4.83	5.22	5.61	6.07	7.04	8.15
500	2.92	3.33	3.75	4.16	4.57	4.98	5.39	5.87	6.36	6.84	7.39	8.58	9.94
600	3.40	3.89	4.38	4.87	5.35	5.83	6.31	6.89	7.45	8.02	8.67	10.06	11.66
700	3.86	4.43	4.99	5.55	6.11	6.66	7.21	7.87	8.52	9.17	9.92	11.50	13.32
720	3.95	4.53	5.11	5.69	6.26	6.82	7.39	8.06	8.73	9.39	10.16	11.79	13.65
800	4.31	4.95	5.59	6.22	6.84	7.47	8.08	8.82	9.55	10.28	11.12	12.90	14.93
900	4.75	5.46	6.16	6.86	7.56	8.25	8.93	9.75	10.56	11.36	12.29	14.25	16.47
960	5.00	5.75	6.50	7.24	7.98	8.71	9.43	10.29	11.15	11.99	12.97	15.03	17.37
1000	5.17	5.95	6.72	7.49	8.25	9.01	9.76	10.65	11.53	12.41	13.42	15.55	17.96
1100	5.58	6.42	7.27	8.10	8.93	9.75	10.56	11.52	12.48	13.43	14.52	16.80	19.39
1200	5.97	6.89	7.79	8.69	9.58	10.46	11.34	12.37	13.40	14.41	15.57	18.01	20.75
1300	6.36	7.34	8.31	9.27	10.22	11.16	12.09	13.19	14.28	15.36	16.59	19.17	22.05
1400	6.73	7.77	8.81	9.83	10.84	11.84	12.82	13.99	15.14	16.27	17.57	20.28	23.28
1440	6.88	7.95	9.00	10.05	11.08	12.10	13.11	14.30	15.47	16.63	17.96	20.70	23.75
1500	7.09	8.20	9.29	10.37	11.44	12.49	13.53	14.76	15.97	17.15	18.51	21.33	24.43
1600	7.44	8.61	9.76	10.90	12.02	13.12	14.21	15.50	16.76	18.00	19.41	22.33	25.51
1700	7.78	9.01	10.21	11.40	12.58	13.73	14.87	16.21	17.52	18.81	20.27	23.27	26.51
1800	8.11	9.39	10.65	11.90	13.12	14.32	15.50	16.89	18.25	19.58	21.08	24.15	27.43
1900	8.43	9.76	11.08	12.37	13.64	14.88	16.11	17.54	18.94	20.31	21.85	24.97	28.27
2000	8.73	10.12	11.48	12.82	14.14	15.43	16.69	18.16	19.60	20.99	22.57	25.72	29.01
2100	9.02	10.46	11.88	13.26	14.62	15.94	17.24	18.75	20.22	21.64	23.23	26.41	29.67
2200	9.31	10.79	12.25	13.68	15.07	16.44	17.76	19.31	20.80	22.24	23.85	27.03	30.22

Continued

Table 12.5 Continued

rpm of faster shaft	Rated power (kW) per belt for small pulley pitch diameter (mm)												
	140	**150**	**160**	**170**	**180**	**190**	**200**	**212**	**224**	**236**	**250**	**280**	**315**
2300	9.57	11.11	12.61	14.08	15.51	16.90	18.26	19.83	21.35	22.80	24.42	27.57	30.68
2400	9.83	11.41	12.95	14.46	15.92	17.34	18.72	20.32	21.85	23.31	24.93	28.05	31.04
2500	10.08	11.70	13.28	14.82	16.31	17.76	19.16	20.77	22.31	23.78	25.38	28.44	–
2600	10.31	11.97	13.59	15.16	16.68	18.14	19.56	21.19	22.73	24.19	25.78	28.76	–
2700	10.53	12.23	13.88	15.47	17.02	18.50	19.93	21.56	23.11	24.56	26.12	28.99	–
2800	10.73	12.47	14.15	15.77	17.33	18.83	20.27	21.90	23.44	24.87	26.40	–	–
2880	10.89	12.65	14.35	15.99	17.57	19.07	20.51	22.14	23.67	25.08	26.57	–	–
2900	10.93	12.69	14.40	16.04	17.62	19.13	20.57	22.20	23.72	25.12	26.61	–	–
3000	11.10	12.90	14.63	16.30	17.89	19.40	20.84	22.46	23.96	25.33	26.76	–	–

Courtesy of Fenner Power Transmission UK.

Table 12.6 Additional power increment per SPB belt

Speed of faster shaft (rpm)	Additional power (kW) per belt for speed ratio									
	1.00–1.01	**1.02–1.05**	**1.06–1.11**	**1.12–1.18**	**1.19–1.26**	**1.27–1.38**	**1.39–1.57**	**1.58–1.94**	**1.95–3.38**	**3.39 and over**
100	0.00	0.01	0.02	0.04	0.04	0.06	0.07	0.07	0.08	0.08
200	0.00	0.01	0.04	0.07	0.09	0.11	0.13	0.15	0.16	0.17
300	0.00	0.02	0.06	0.10	0.14	0.17	0.20	0.22	0.24	0.25
400	0.00	0.03	0.07	0.13	0.19	0.22	0.26	0.29	0.32	0.34
500	0.00	0.04	0.09	0.17	0.23	0.28	0.33	0.37	0.40	0.43
600	0.00	0.04	0.12	0.20	0.28	0.34	0.40	0.45	0.48	0.51
700	0.00	0.05	0.13	0.24	0.33	0.39	0.46	0.52	0.57	0.59
720	0.00	0.05	0.14	0.25	0.33	0.41	0.48	0.54	0.59	0.62
800	0.00	0.06	0.16	0.28	0.37	0.45	0.53	0.60	0.65	0.69
900	0.00	0.07	0.18	0.31	0.42	0.51	0.60	0.66	0.72	0.77
960	0.00	0.07	0.19	0.32	0.44	0.54	0.62	0.70	0.77	0.81
1000	0.00	0.07	0.19	0.34	0.46	0.56	0.66	0.74	0.81	0.86
1100	0.00	0.08	0.22	0.37	0.51	0.62	0.72	0.81	0.89	094
1200	0.00	0.09	0.23	0.41	0.56	0.68	0.79	0.89	0.97	1.03
1300	0.00	0.09	0.25	0.44	0.60	0.73	0.86	0.96	1.05	1.11
1400	0.00	0.10	0.28	0.48	0.65	0.79	0.93	1.04	1.13	1.20
1440	0.00	0.10	0.28	0.48	0.66	0.79	0.94	1.06	1.15	1.21
1500	0.00	0.10	0.29	0.51	0.69	0.84	0.99	1.11	1.21	1.28
1600	0.00	0.11	0.31	0.54	0.75	0.90	1.05	1.19	1.29	1.37
1700	0.00	0.12	0.34	0.58	0.79	0.95	1.12	1.26	1.37	1.45
1800	0.00	0.13	0.35	0.61	0.84	1.01	1.19	1.34	1.45	1.54
1900	0.00	0.13	0.37	0.65	0.88	1.07	1.25	1.41	1.54	1.63
2000	0.00	0.14	0.39	0.68	0.93	1.13	1.32	1.48	1.62	1.71
2100	0.00	0.15	0.41	0.72	0.98	1.18	1.39	1.56	1.69	1.79

Continued

Table 12.6 Continued

Speed of faster shaft (rpm)	Additional power (kW) per belt for speed ratio									
	1.00–1.01	**1.02–1.05**	**1.06–1.11**	**1.12–1.18**	**1.19–1.26**	**1.27–1.38**	**1.39–1.57**	**1.58–1.94**	**1.95–3.38**	**3.39 and over**
2200	0.00	0.16	0.43	0.75	1.02	1.24	1.45	1.63	1.78	1.88
2300	0.00	0.16	0.45	0.78	1.07	1.29	1.51	1.71	1.86	1.97
2400	0.00	0.17	0.47	0.82	1.11	1.35	1.58	1.78	1.94	2.05
2500	0.00	0.18	0.49	0.85	1.16	1.41	1.65	1.86	2.02	2.14
2600	0.00	0.19	0.51	0.89	1.21	1.46	1.72	1.92	2.10	2.22
2700	0.00	0.19	0.53	0.92	1.25	1.52	1.78	1.99	2.18	2.31
2800	0.00	0.20	0.54	0.95	1.29	1.57	1.84	2.07	2.26	2.39
2880	0.00	0.20	0.56	0.97	1.32	1.60	1.88	2.11	2.31	2.44
2900	0.00	0.21	0.57	0.99	1.34	1.63	1.91	2.15	2.34	2.48
3000	0.00	0.22	0.59	1.02	1.39	1.69	1.98	2.23	2.42	2.57

Courtesy of Fenner Power Transmission UK.

12.2.3 Synchronous belts

Synchronous belt drives have teeth on the belt and corresponding grooves on the pulley wheels. The meshing teeth provide positive angular location and hence there is normally no relative motion or slipping between the two elements in mesh giving a constant speed ratio between the driving and driven shafts. Synchronous belts can therefore be used for applications such as automatic machinery where a definite motion sequence or indexing is necessary. A key consideration in belt selection and design is the material concerned and this topic remains an ongoing area of development (see, e.g. Iizuka et al., 1994; Dalgarno et al., 1994; as well as the various belt manufacturers).

The principal dimensions of a synchronous belt drive are the number of grooves, the pitch and the width. The belt pitch is the distance between two adjacent tooth centres measured on the pitch line of the belt. The belt pitch length is the total length of the belt measured along the pitch line.

The selection procedure for synchronous belts is specific to the belt manufacturer concerned, although there are similarities. Here, a procedure based on TDP3 belt drives provided by Fenner is presented in the following list, and Fig. 12.12, and the associated figures and tables, see Fig. 12.13 and Tables 12.7–12.19.

(1) Define the principal requirements.	These include the nominal power to be transmitted; the rotational speeds of the shafts; the power capability and starting characteristics of the prime mover and power absorbed by the driven machine; any space, layout or other constraints such as the centre distance and shaft diameters; environmental conditions.
(2) Determine the service factor.	Service factors are used to downrate the power transmission capability listed by belt suppliers to account for the differences between practical applications and test conditions. Table 12.13 lists typical values for service factors.
(3) Calculate the design power.	This is the product of the nominal power and the service factor.
(4) Belt pitch	Use the belt pitch selection chart, Fig. 12.13, to identify a suitable pitch for the drive. The chart defines envelopes of suitability based on the intersection between the smaller pulley speed and design power to identify the belt pitch.
(5) Speed ratio	This is the ratio of rotational speed of the faster shaft over the rotational speed of the slower shaft.
(6) Pulley selection	The charts given in Tables 12.14–12.19 can be used for the particular pitch concerned to identify the number of grooves for the driving and driven pulleys.
(7) Belt length and centre distance	Using the charts given in Tables 12.14–12.19, read across the row for the selected number of grooves for the driving and driven pulleys and select the centre distance that is closest to that required.

(8) Power rating and belt width	For the chosen belt pitch, select the appropriate power rating table, Table 12.7 or 12.10. Locate the smaller pulley groove number speed combination and note the power rating. Multiply this rating by the belt length factor, Table 12.8 or 12.11. Divide the design power by the length corrected power rating to give the require belt width factor. Refer to Tables 12.9 and 12.12 and select the belt width with a factor equal to or greater than that required.

Determine the nominal power to be transmitted; the rotational speeds of the shafts; the power capability and starting characteristics of the prime mover and power absorbed by the driven machine; any space, layout or other constraints such as the centre distance and shaft diameters; environmental conditions.

↓

Determine the service factor. Table 12.13 lists typical values for service factors.

↓

Calculate the design power. This is the product of the nominal power and the service factor.

↓

Use the belt pitch selection chart, Fig. 12.13, to identify a suitable pitch for the drive. The chart defines envelopes of suitability based on the intersection between the smaller pulley speed and design power to identify the belt pitch.

↓

Determine the speed ratio. This is the ratio of rotational speed of the faster shaft over the rotational speed of the slower shaft.

↓

Pulley selection. The charts given in Tables 12.14–12.19 can be used for the particular pitch concerned to identify the number of grooves for the driving and driven pulleys.

↓

Belt length and centre distance. Using the charts given in Tables 12.14–12.19, read across the row for the selected number of grooves for the driving and driven pulleys and select the centre distance that is closest to that required.

↓

Power rating and belt width. For the chosen belt pitch select the appropriate power rating table, Table 12.7 or 12.10. Locate the smaller pulley groove number speed combination and note the power rating. Multiply this rating by the belt length factor, Table 12.8 or 12.11. Divide the design power by the length corrected power rating to give the require belt width factor. Refer to Tables 12.9 and 12.12 and select the belt width with a factor equal to or greater than that required.

↓

Determine shaft hub connection for the pulley wheels (see Chapter 7, Section 7.2)

↓

Determine shaft diameter (see Chapter 7, Section 7.5)

Fig. 12.12 Example selection procedure for synchronous belts. *Note*: Refer to a specific manufacturer's guidance for a specific brand of synchronous belt.

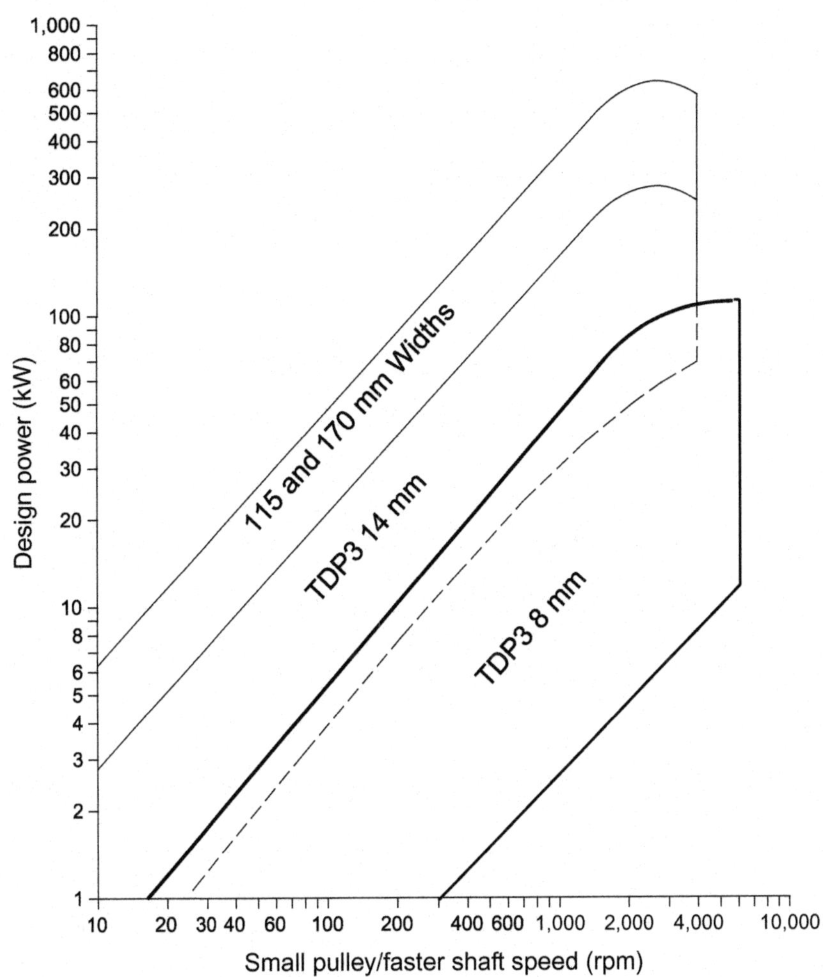

Fig. 12.13 Belt pitch selection guide.
Chart courtesy of Fenner.

Example 12.3. A synchronous belt is required for a gear pump that is driven by a 1470 rpm AC electric motor. The pump rotational speed required is 720 rpm within ±15 rpm. The motor has been matched to the pump power requirements and has a rating of 50 kW. The centre distance required between the motor axis and the gear pump drive axis is 700–750 mm. The pump shaft diameter is 70 mm. The pump operation required is continuous.

Solution

Assuming continuous operation, and medium duty characteristics, from Table 12.13, the service factor is 1.7.

Table 12.7 Power ratings (kW) for 20 mm wide 8MXP belts

Rev./min of small pulley	Number of grooves														
	22	**24**	**26**	**28**	**30**	**32**	**34**	**36**	**38**	**40**	**44**	**48**	**56**	**64**	**72**
100	0.43	0.50	0.57	0.65	0.73	0.81	0.89	0.97	1.06	1.14	1.30	1.45	1.71	1.88	1.94
200	0.83	0.97	1.11	1.26	1.41	1.57	1.72	1.88	2.04	2.19	2.50	2.79	3.27	3.59	3.71
300	1.23	1.43	1.64	1.85	2.07	2.30	2.53	2.76	2.99	3.21	3.66	4.07	4.76	5.22	5.37
400	1.61	1.88	2.15	2.43	2.72	3.01	3.31	3.61	3.91	4.20	4.78	5.31	6.21	6.78	6.97
500	1.99	2.32	2.65	3.00	3.35	3.71	4.08	4.44	4.81	5.17	5.87	6.52	7.60	8.29	8.51
600	2.37	2.75	3.14	3.55	3.97	4.40	4.83	5.26	5.69	6.11	6.93	7.69	8.96	9.76	9.99
720	2.82	3.27	3.73	4.22	4.71	5.21	5.73	6.23	6.74	7.24	8.20	9.09	10.57	11.49	11.74
800	3.10	3.60	4.11	4.64	5.19	5.74	6.29	6.85	7.40	7.95	8.99	9.96	11.56	12.54	12.79
960	3.65	4.23	4.83	5.45	6.08	6.72	7.37	8.01	8.65	9.28	10.50	11.62	13.44	14.54	14.80
1000	3.82	4.42	5.05	5.70	6.36	7.03	7.71	8.38	9.05	9.71	10.97	12.13	14.02	15.16	15.41
1200	4.52	5.23	5.97	6.73	7.50	8.29	9.08	9.86	10.64	11.40	12.86	14.20	16.36	17.63	17.85
1450	5.38	6.21	7.08	7.98	8.89	9.81	10.73	11.64	12.55	13.43	15.12	16.66	19.11	20.51	20.67
1600	5.57	6.43	7.33	8.25	9.18	10.12	11.07	12.00	12.93	13.83	15.55	17.12	19.58	20.95	21.07
1800	6.54	7.55	8.59	9.66	10.75	11.84	12.94	14.02	15.09	16.13	18.10	19.89	22.67	24.17	24.21
2000	7.18	8.29	9.43	10.59	11.77	12.96	14.15	15.32	16.47	17.59	19.71	21.62	24.56	26.08	26.02
2500	8.74	10.07	11.43	12.82	14.22	15.62	17.01	18.39	19.73	21.02	23.46	25.60	28.82	30.30	29.91
2850	9.79	11.26	12.76	14.29	15.83	17.37	18.89	20.38	21.83	23.24	25.84	28.11	31.42	32.79	32.10
3000	10.23	11.76	13.32	14.91	16.50	18.09	19.66	21.20	22.70	24.14	26.80	29.12	32.44	33.74	32.91
3500	11.65	13.36	15.10	16.87	18.63	20.38	22.10	23.78	25.40	26.94	29.77	32.18	35.46	36.43	35.06
4000	13.00	14.88	16.78	18.70	20.61	22.49	24.34	26.12	27.83	29.45	32.37	34.79	37.88	38.39	36.37
4500	14.28	16.31	18.36	20.41	22.44	24.44	26.37	28.23	30.00	31.66	34.61	36.98	39.71	39.63	36.86
5000	15.50	17.66	19.83	22.00	24.13	26.21	28.21	30.12	31.92	33.59	36.49	38.73	40.98	40.16	36.55
5500	16.65	18.93	21.21	23.47	25.68	27.81	29.86	31.79	33.58	35.23	38.02	40.06	41.68	40.00	35.44
6000	17.74	20.12	22.49	24.82	27.08	29.26	31.31	33.24	35.00	36.60	39.21	40.98	41.82	39.16	33.56

Courtesy of Fenner.

Table 12.8 Belt length correction factors for 20mm wide 8MXP belts

Belt length (mm)	384–600	640–880	960–1200	1280–1760	1800–4400
Length factor	0.8	0.9	1.0	1.1	1.2

Courtesy of Fenner.

Table 12.9 Belt width factors for 20mm wide 8MXP belts

Belt width (mm)	20	30	50	85
Width factor	1.00	1.58	2.73	4.76

Courtesy of Fenner.

The design power is given by $1.7 \times 50 = 85$ kW.

The belt pitch selection chart, Fig. 12.13, for a power of 85kW and a small pulley speed of 1470rpm, indicates that a TDP3 14-mm pitch belt drive would be suitable.

The speed ratio is 1470/720 = 2.042.

From Table 12.18, the corresponding chart for a 14mm pitch belt with a speed ratio of 2.05, the nearest listed to the desired speed ratio of 2.05, the corresponding number of grooves on the driving and driven pulleys are 44 and 90, respectively.

A speed ratio of 2.05 gives an output speed of 717rpm. This is within the ±15 rpm stated requirement.

From Table 12.18, for the 2.05 ratio, 44 and 90 groove pulley combination, a belt length of 2450mm gives a centre distance of 749mm. This is within the centre distance range requirement.

Table 12.10 gives a power rating of 69.77 for a pulley with 44 grooves running at 1450rpm, for a 40mm belt.

From Table 12.11, the belt length factor is 1.00.

The ratio of the design power to the power rating is 85/69.77 = 1.22. From Table 12.12, the next larger standard width factor is 1.44, with a corresponding belt width of 55mm. A belt width of 55mm should therefore be used.

The pulley specification is therefore as follows.

Motor pulley: 44-14M-55 HTD pulley, corresponding to a 44 groove pulley with a 55-mm belt width.

Pump pulley: 90-14M-55 HTD pulley, corresponding to a 90 groove pulley with a 55-mm belt width.

Belt: 14MXP-2450-55 Torque Drive Plus 3 belt, corresponding to a 2450-mm long, 55-mm wide belt.

A check also needs to be made that the bores for the pulleys are compatible with the motor and pump shaft diameters.

Table 12.10 Power ratings (kW) for 40 mm wide 14MXP belts

Rev/min	Number of grooves										
	28	**29**	**30**	**32**	**34**	**36**	**38**	**40**	**44**	**48**	**56**	**64**
10	0.44	0.47	0.50	0.55	0.60	0.65	0.69	0.74	0.88	0.92	1.10	1.29
20	0.85	0.90	0.96	1.06	1.15	1.24	1.33	1.42	1.69	1.78	2.13	2.48
50	1.99	2.12	2.24	2.48	2.71	2.92	3.14	3.35	3.98	4.20	5.03	5.87
100	3.77	4.03	4.27	4.73	5.16	5.58	5.99	6.40	7.60	8.00	9.59	11.18
200	7.09	7.58	8.04	8.91	9.72	10.52	11.29	12.06	14.30	15.06	18.02	20.95
300	10.19	10.91	11.57	12.82	14.00	15.13	16.25	17.34	20.55	21.61	25.78	29.88
400	13.13	14.06	14.92	16.52	18.04	19.50	20.92	22.31	26.40	27.73	32.98	38.10
500	15.93	17.06	18.11	20.05	21.88	23.64	25.34	27.02	31.89	33.48	39.68	45.67
600	18.61	19.93	21.16	23.42	25.54	27.58	29.55	31.48	37.06	38.87	45.91	52.63
720	21.78	23.34	24.77	27.40	29.87	32.22	34.50	36.72	43.13	45.20	53.17	60.71
800	23.65	25.33	26.88	29.73	32.38	34.90	37.34	39.71	46.51	48.69	57.04	64.84
960	27.20	29.20	30.90	34.20	37.20	40.00	42.70	45.33	52.90	55.30	64.30	72.60
1000	28.30	30.30	32.14	35.50	38.60	41.55	44.38	47.11	54.64	57.29	66.51	74.85
1200	32.59	34.88	36.97	40.78	44.28	47.57	50.70	53.71	62.11	64.73	74.38	82.77
1450	37.49	40.10	42.46	46.74	50.62	54.24	57.65	60.90	69.77	72.47	82.07	89.83
1600	40.20	42.97	45.47	49.98	54.04	57.81	61.34	64.66	73.62	76.30	85.57	92.58
1800	43.54	46.50	49.72	53.93	58.18	62.08	65.69	69.07	77.83	80.50	88.95	94.56
2000	46.59	49.71	52.51	57.46	61.84	65.80	69.43	72.78	81.29	83.66	90.90	94.63
2200	49.35	52.61	55.51	60.59	65.02	68.99	72.57	75.81	83.74	85.81	91.44	92.82
2500	52.95	56.35	59.33	64.49	68.89	72.72	76.09	79.04	85.59	87.06	89.47	86.40
2850	56.46	59.94	62.96	68.05	72.25	75.77	78.71	81.12	85.46	85.99	83.65	
3000	57.70	61.20	64.20	69.21	73.26	76.58	79.26	81.37	84.54	84.58	79.82	
3500	60.83	64.23	67.06	71.55	74.87	77.27	78.85	79.68	77.95	76.02		
4000	62.41	65.54	68.03	71.64	73.86	74.94	75.01	74.12	66.07			

Courtesy of Fenner.

Table 12.11 Belt length correction factors for 40 mm wide 14MXP belts

Belt length (mm)	966–1190	1400–1610	1778–1890	2100–2450	2590–3360	3500–6860
Length factor	0.80	0.90	0.95	1.00	1.05	1.10

Courtesy of Fenner.

Table 12.12 Belt width factors for 40 mm wide 14MXP belts

Belt width (mm)	40	55	85	115	170
Width factor	1.00	1.44	2.31	3.18	4.78

Courtesy of Fenner.

12.2.4 *Flat belt drives*

Flat belts, Fig. 12.14, can be used for high-speed low-power applications, transmission of power across significant centre distances as in some agricultural machinery and in conveyer systems. Flat belts can be made from a wide range of materials with urethane, neoprene, chlorosulfonated polyethylene (CSPE), silicone and Ethyene Propylene Diene Monomer (EPDM) common. Flat belts are often reinforced with textiles, fibres and metal wires with a range of surface coatings applied to improve performance. In general, flat belt drive performance is enhanced by belts with high flexibility, low mass and surfaces with high coefficient of friction. In order to prevent a flay belt from running off its pulleys, the pulleys can be crowned or guides can be used as shown in Fig. 12.15.

For the simple belt drive configuration shown in Fig. 12.16, the angles of contact between the belt and the pulleys are given by

$$\theta_d = \pi - 2\sin^{-1}\frac{D-d}{2C} \tag{12.3}$$

$$\theta_D = \pi + 2\sin^{-1}\frac{D-d}{2C} \tag{12.4}$$

where d = the diameter of the small pulley (m), D = the diameter of the large pulley (m), C = the distance between the pulley centres (m), θ_d = the angle of contact between the belt and the small pulley (rad) and θ_D = the angle of contact between the belt and the large pulley (rad).

The length of the belt can be obtained by summing the arc lengths of contact and the spanned distances and is given by

$$L = \sqrt{4C^2 - (D-d)^2} + \frac{1}{2}(D\theta_D + d\theta_d) \tag{12.5}$$

Table 12.13 Synchronous belt drive service factors

Types of driven machine	Types of prime mover					
Special cases For speed increasing drives of: 1.00–1.24—no additional factor 1.25–1.74–add 0.1 1.75–2.49–add 0.2 2.50–3.49–add 0.3 3.50 and greater—add 0.4 Seasonal/intermittent use—subtract 0.2 Idler pulley used on drive—add 0.2	'Soft' starts			'Heavy' starts		
	AC electric motors – star/delta start – synchronous – split wound – inverter control	**DC electric motors** – shunt wound – stepper motors		**AC electric motors** – DOL start – single phase – slip ring	**DC electric motors** – series wound – compound – servo motors	
	IC engines with 4 or more cylinders. Prime movers with centrifugal clutches or fluid couplings.			**IC engines with <4 cylinders**		
	Hours per day duty			**Hours per day duty**		
	<10	**16 ≥ hours > 10**	**>16**	**<10**	**16 ≥ Hours > 10**	**>16**
Light duty Agitators (uniform density); bakery machinery: dough mixers, blowers except positive displacement; centrifugal pumps and compressors; belt conveyors (uniformly loaded); exhausters; fans up to 7.5 kW; paper machinery: agitators; calendars; dryers; printing machinery: linotype machines, cutters, folders; screens: drum; conical; woodworking machinery: lathes, band saws	1.2	1.4	1.6	1.6	1.8	2.0

Medium duty Agitators and mixers (variable density); belt conveyors (not uniformly loaded); brick and clay machinery, augers, mixers, granulators; fans over 7.5 kW; generators; line shafts; laundry machinery; punches, presses, shears; printing machinery: presses, newspaper, rotary embossing, flat bed magazine; pumps: positive displacement, rotary; screens, vibrating; machine tools	1.3	1.5	1.7	1.7	1.9	2.1
Heavy duty Blowers, positive displacement; bucket elevators; centrifuges; conveyors: drag, pan, screw; paper machinery: beaters, Jordans, mash pumps, pulpers; pumps, piston; pulverisers; woodworking machinery; textile machinery; exiters	1.5	1.7	1.9	1.9	2.1	2.3
Extra heavy duty Brick machinery; pug mills; compressors, piston; crushers: gyratory, jaw roll; hoists; mills: ball, rod, tube, rubber; rubber machinery: calendars, extruders, mills	1.7	1.9	2.1	2.1	2.3	2.5

Courtesy of Fenner.

Table 12.14 Fenner Torque Drive Plus 8MXP and 8M drives (centre distance, mm)

			Belt pitch length (mm)																			
	Number of grooves		480	560	600	640	720	800	880	960	1040	1120	1200	1280	1440	1600	1760	1800	2000	2400	2600	2800	
Speed Ratio	Driving Pulley	Driven Pulley	60 teeth	70 teeth	75 teeth	80 teeth	90 teeth	100 teeth	110 teeth	120 teeth	130 teeth	140 teeth	150 teeth	160 teeth	180 teeth	200 teeth	220 teeth	225 teeth	250 teeth	300 teeth	325 teeth	350 teeth	Speed Ratio
1.00	24	24	144	184	204	224	264	304	344	384	424	464	504	544	624	704	784	804	904	1104	1204	1304	1.00
1.00	26	26	136	176	196	216	256	296	336	376	416	456	496	536	616	696	776	796	896	1096	1196	1296	1.00
1.00	28	28	128	168	188	208	248	288	328	368	408	448	488	528	608	688	768	788	888	1088	1188	1288	1.00
1.00	30	30	120	160	180	200	240	280	320	360	400	440	480	520	600	680	760	780	880	1080	1180	1280	1.00
1.00	32	32	112	152	172	192	232	272	312	352	392	432	472	512	592	672	752	772	872	1072	1172	1272	1.00
1.00	34	34	104	144	164	184	224	264	304	344	384	424	464	504	584	664	744	764	864	1064	1164	1264	1.00
1.00	36	36	–	136	156	176	216	256	296	336	376	416	456	496	576	656	736	756	856	1056	1156	1256	1.00
1.00	38	38	–	128	148	168	208	248	288	328	368	408	448	488	568	648	728	748	848	1048	1148	1248	1.00
1.00	40	40	–	120	140	160	200	240	280	320	360	400	440	480	560	640	720	740	840	1040	1140	1240	1.00
1.00	44	44	–	–	–	144	184	224	264	304	344	384	424	464	544	624	704	724	824	1024	1124	1224	1.00
1.00	48	48	–	–	–	–	168	208	248	288	328	368	408	448	528	608	688	708	808	1008	1108	1208	1.00
1.00	56	56	–	–	–	–	–	176	216	256	296	336	376	416	496	576	656	676	776	976	1076	1176	1.00
1.00	64	64	–	–	–	–	–	–	184	224	264	304	344	384	464	544	624	644	744	944	1044	1144	1.00
1.00	72	72	–	–	–	–	–	–	–	–	232	272	312	352	432	512	592	612	712	912	1012	1112	1.00
1.00	80	80	–	–	–	–	–	–	–	–	–	240	280	320	400	480	560	580	680	880	980	1080	1.00
1.05	38	40	–	124	144	164	204	244	284	324	364	404	444	484	564	644	724	744	844	1044	1144	1244	1.05
1.06	36	38	–	132	152	172	212	252	292	332	372	412	452	492	572	652	732	752	852	1052	1152	1252	1.06
1.06	34	36	–	140	160	180	220	260	300	340	380	420	460	500	580	660	740	760	860	1060	1160	1260	1.06
1.06	32	34	108	148	168	188	228	268	308	348	388	428	468	508	588	668	748	768	868	1068	1168	1268	1.06
1.07	30	32	116	156	176	196	236	276	316	356	396	436	476	516	596	676	756	776	876	1076	1176	1276	1.07
1.07	28	30	124	164	184	204	244	284	324	364	404	444	484	524	604	684	764	784	884	1084	1184	1284	1.07
1.08	26	28	132	172	192	212	252	292	332	372	412	452	492	532	612	692	772	792	892	1092	1192	1292	1.08
1.08	24	26	140	180	200	220	260	300	340	380	420	460	500	540	620	700	780	800	900	1100	1200	1300	1.08
1.09	44	48	–	–	–	136	176	216	256	296	336	376	416	456	536	616	696	716	816	1016	1116	1216	1.09
1.10	40	44	–	–	132	152	192	232	272	312	352	392	432	472	552	632	712	732	832	1032	1132	1232	1.10
1.11	36	40	–	128	148	168	208	248	288	328	368	408	448	488	568	648	728	748	848	1048	1148	1248	1.11
1.11	72	80	–	–	–	–	–	–	–	–	216	256	296	336	416	496	576	596	696	896	996	1096	1.11
1.12	34	38	–	136	156	176	216	256	296	336	376	416	456	496	576	656	736	756	856	1056	1156	1256	1.12
1.13	32	36	104	144	164	184	224	264	304	344	384	424	464	504	584	664	744	764	864	1064	1164	1264	1.13
1.13	64	72	–	–	–	–	–	–	–	208	248	288	328	368	448	528	608	628	728	928	1028	1128	1.13
1.13	80	90	–	–	–	–	–	–	–	–	–	–	260	300	380	460	540	560	660	860	960	1060	1.13

1.13	30	34	112	152	172	192	232	272	312	352	392	432	472	512	592	672	752	772	872	1072	1172	1272	1.13
1.14	28	32	120	160	180	200	240	280	320	360	400	440	480	520	600	680	760	780	880	1080	1180	1280	1.14
1.14	56	64	–	–	–	–	–	–	200	240	280	320	360	400	480	560	640	660	760	960	1060	1160	1.14
1.15	26	30	128	168	188	208	248	288	328	368	408	448	488	528	608	688	768	788	888	1088	1188	1288	1.15
1.16	38	44	–	–	136	156	196	236	276	316	356	396	436	476	556	636	716	736	836	1036	1136	1236	1.16
1.17	24	28	136	176	196	216	256	296	336	376	416	456	496	536	616	696	776	796	896	1096	1196	1296	1.17
1.17	48	56	–	–	–	–	152	192	232	272	312	352	392	432	512	592	672	692	792	992	1092	1192	1.17
1.18	34	40	–	132	152	172	212	252	292	332	372	412	452	492	572	652	732	752	852	1052	1152	1252	1.18
1.19	32	38	–	140	160	180	220	260	300	340	380	420	460	500	580	660	740	760	860	1060	1160	1260	1.19
1.20	30	36	108	148	168	188	228	268	308	348	388	428	468	508	588	668	748	768	868	1068	1168	1268	1.20
1.20	40	48	–	–	–	144	184	224	264	304	344	384	424	464	544	624	704	724	824	1024	1124	1224	1.20
1.21	28	34	116	156	176	196	236	276	316	356	396	436	476	516	596	676	756	776	876	1076	1176	1276	1.21
1.22	36	44	–	120	140	160	200	240	280	320	360	400	440	480	560	640	720	740	840	1040	1140	1240	1.22
1.23	26	32	124	164	184	204	244	284	324	364	404	444	484	524	604	684	764	784	884	1084	1184	1284	1.23
1.25	24	30	132	172	192	212	252	292	332	372	412	452	492	532	612	692	772	792	892	1092	1192	1292	1.25
1.25	32	40	–	136	156	176	216	256	296	336	376	416	456	496	576	656	736	756	856	1056	1156	1256	1.25
1.25	64	80	–	–	–	–	–	–	–	–	231	271	311	351	432	512	592	612	712	912	1012	1112	1.25
1.25	72	90	–	–	–	–	–	–	–	–	–	235	275	315	395	475	556	576	676	876	976	1076	1.25
1.26	38	48	–	–	127	147	188	228	268	308	348	388	428	468	548	628	708	728	828	1028	1128	1228	1.26
1.27	30	38	103	144	164	184	224	264	304	344	384	424	464	504	584	664	744	764	864	1064	1164	1264	1.27
1.27	44	56	–	–	–	–	159	199	240	280	320	360	400	440	520	600	680	700	800	1000	1100	1200	1.27
1.29	28	36	112	152	172	192	232	272	312	352	392	432	472	512	592	672	752	772	872	1072	1172	1272	1.29
1.29	56	72	–	–	–	–	–	–	183	223	263	303	343	383	464	544	624	644	744	944	1044	1144	1.29
1.29	34	44	–	123	143	164	204	244	284	324	364	404	444	484	564	644	724	744	844	1044	1144	1244	1.29
1.31	26	34	120	160	180	200	240	280	320	360	400	440	480	520	600	680	760	780	880	1080	1180	1280	1.31
1.33	24	32	128	168	188	208	248	288	328	368	408	448	488	528	608	688	768	788	888	1088	1188	1288	1.33
1.33	30	40	–	139	159	180	220	260	300	340	380	420	460	500	580	660	740	760	860	1060	1160	1260	1.33
1.33	36	48	–	–	131	151	191	231	272	312	352	392	432	472	552	632	712	732	832	1032	1132	1232	1.33
1.33	48	64	–	–	–	–	–	175	215	255	295	335	375	416	496	576	656	676	776	976	1076	1176	1.33
1.36	28	38	107	147	168	188	228	268	308	348	388	428	468	508	588	668	748	768	868	1068	1168	1268	1.36
1.38	32	44	–	127	147	167	207	248	288	328	368	408	448	488	568	648	728	748	848	1048	1148	1248	1.38
1.38	26	36	115	155	176	196	236	276	316	356	396	436	476	516	596	676	756	776	876	1076	1176	1276	1.38
1.40	40	56	–	–	–	–	167	207	247	287	327	367	407	448	528	608	688	708	808	1008	1108	1208	1.40
1.40	80	112	–	–	–	–	–	–	–	–	–	–	–	–	334	414	494	514	615	815	915	1015	1.40

Centre distance (mm).
Courtesy of Fenner.

Table 12.15 Fenner Torque Drive Plus 8MXP and 8M drives (centre distance, mm)

	Number of grooves		Belt pitch length (mm)																				
			480	560	600	640	720	800	880	960	1040	1120	1200	1280	1440	1600	1760	1800	2000	2400	2600	2800	
Speed ratio	Driving pulley	Driven pulley	60 teeth	70 teeth	75 teeth	80 teeth	90 teeth	100 teeth	110 teeth	120 teeth	130 teeth	140 teeth	150 teeth	160 teeth	180 teeth	200 teeth	220 teeth	225 teeth	250 teeth	300 teeth	325 teeth	350 teeth	Speed ratio
1.41	64	90	–	–	–	–	–	–	–	–	–	250	290	330	411	491	571	591	691	891	991	1091	1.41
1.41	34	48	–	–	135	155	195	235	275	315	356	396	436	476	556	636	716	736	836	1036	1136	1236	1.41
1.42	24	34	123	164	184	204	244	284	324	364	404	444	484	524	604	684	764	784	884	1084	1184	1284	1.42
1.43	28	40	103	143	163	183	223	264	304	344	384	424	464	504	584	664	744	764	864	1064	1164	1264	1.43
1.43	56	80	–	–	–	–	–	–	–	206	246	286	327	367	447	527	607	627	727	927	1028	1128	1.43
1.45	44	64	–	–	–	–	–	182	223	263	303	343	383	423	503	583	664	684	784	984	1084	1184	1.45
1.46	26	38	111	151	171	191	231	272	312	352	392	432	472	512	592	672	752	772	872	1072	1172	1272	1.46
1.47	30	44	–	131	151	171	211	251	291	332	372	412	452	492	572	652	732	752	852	1052	1152	1252	1.47
1.47	38	56	–	–	–	–	170	211	251	291	331	371	411	451	532	612	692	712	812	1012	1112	1212	1.47
1.50	24	36	119	159	179	199	240	280	320	360	400	440	480	520	600	680	760	780	880	1080	1180	1280	1.50
1.50	32	48	–	118	139	159	199	239	279	319	359	399	440	480	560	640	720	740	840	1040	1140	1240	1.50
1.50	48	72	–	–	–	–	–	–	198	238	278	319	359	399	479	559	639	659	759	960	1060	1160	1.50
1.54	26	40	107	147	167	187	227	267	307	348	388	428	468	508	588	668	748	768	868	1068	1168	1268	1.54
1.56	36	56	–	–	–	134	174	214	255	295	335	375	415	455	535	615	696	716	816	1016	1116	1216	1.56
1.56	72	112	–	–	–	–	–	–	–	–	–	–	–	267	348	429	509	530	630	830	931	1031	1.56
1.57	28	44	–	134	155	175	215	255	295	335	375	416	456	496	576	656	736	756	856	1956	1156	1256	1.57
1.58	24	38	115	155	175	195	235	275	315	356	396	436	476	516	596	676	756	776	876	1076	1176	1276	1.58
1.60	30	48	–	122	142	162	203	243	283	323	363	403	443	483	564	644	724	744	844	1044	1144	1244	1.60
1.60	40	64	–	–	–	–	149	190	230	270	310	351	391	431	511	591	671	691	791	992	1092	1192	1.60
1.61	56	90	–	–	–	–	–	–	–	–	224	264	305	345	426	506	586	606	707	907	1007	1107	1.61
1.64	44	72	–	–	–	–	–	164	205	245	286	326	366	406	487	567	647	667	767	967	1067	1167	1.64
1.65	34	56	–	–	–	137	178	218	258	299	339	379	419	459	539	619	699	719	820	1020	1120	1220	1.65
1.67	24	40	110	151	171	191	231	271	311	351	391	432	472	512	592	672	752	772	872	1072	1172	1272	1.67
1.67	48	80	–	–	–	–	–	–	179	220	261	301	342	382	462	542	623	643	743	943	1043	1143	1.67
1.68	38	64	–	–	–	–	152	193	234	274	314	354	395	435	515	595	675	695	795	995	1095	1196	1.68
1.69	26	44	–	138	158	179	219	259	299	339	379	419	459	499	580	660	740	760	860	1060	1160	1260	1.69
1.71	28	48	–	125	146	166	206	247	287	327	367	407	447	487	567	647	728	748	848	1048	1148	1248	1.71
1.75	32	56	–	–	–	141	181	222	262	302	343	383	423	463	543	623	703	723	823	1024	1124	1224	1.75
1.75	64	112	–	–	–	–	–	–	–	–	–	–	240	281	363	444	524	545	645	846	946	1046	1.75
1.78	36	64	–	–	–	–	156	197	237	278	318	358	398	439	519	599	679	699	799	999	1099	1199	1.78
1.80	40	72	–	–	–	–	–	171	212	253	293	334	374	414	494	575	655	675	775	975	1075	1175	1.80
1.80	80	144	–	–	–	–	–	–	–	–	–	–	–	–	–	342	424	445	546	748	848	948	1.80
1.82	44	80	–	–	–	–	–	–	186	227	268	309	349	389	470	550	630	650	751	951	1051	1151	1.82

1.83	24	44	–	142	162	182	223	263	303	343	383	423	463	503	583	664	744	764	864	1064	1164	1264	1.83
1.85	26	48	–	129	149	170	210	250	291	331	371	411	451	491	571	651	731	751	852	1052	1152	1252	1.85
1.87	30	56	–	–	–	144	185	226	266	306	346	387	427	467	547	627	707	727	827	1027	1128	1228	1.87
1.88	48	90	–	–	–	–	–	–	–	197	238	279	320	360	441	521	602	622	722	922	1023	1123	1.88
1.88	34	64	–	–	–	–	159	200	241	281	322	362	402	442	523	603	683	703	803	1003	1103	1203	1.88
1.89	38	72	–	–	–	–	–	175	216	256	297	337	378	418	498	578	659	679	779	979	1079	1179	1.89
2.00	24	48	–	132	153	173	214	254	294	335	375	415	455	495	575	655	735	755	855	1056	1156	1256	2.00
2.00	28	56	–	–	127	148	189	229	270	310	350	390	431	471	551	631	711	731	831	1031	1131	1231	2.00
2.00	32	64	–	–	–	–	163	204	245	285	325	366	406	446	526	607	687	707	807	1007	1107	1207	2.00
2.00	36	72	–	–	–	–	–	178	219	260	301	341	381	422	502	582	662	682	783	983	1083	1183	2.00
2.00	40	80	–	–	–	–	–	–	193	234	275	316	356	397	477	558	638	658	758	959	1059	1159	2.00
2.00	56	112	–	–	–	–	–	–	–	–	–	–	254	295	377	458	539	559	660	861	961	1062	2.00
2.00	72	144	–	–	–	–	–	–	–	–	–	–	–	–	–	356	438	459	561	762	863	964	2.00
2.05	44	90	–	–	–	–	–	–	–	204	245	286	327	367	448	529	609	629	730	930	1030	1130	2.05
2.10	80	168	–	–	–	–	–	–	–	–	–	–	–	–	–	–	367	388	491	695	796	897	2.10
2.11	38	80	–	–	–	–	–	–	197	238	279	320	360	400	481	561	642	662	762	963	1063	1163	2.11
2.12	34	72	–	–	–	–	–	182	223	264	304	345	385	425	506	586	666	686	787	987	1087	1187	2.12
2.13	30	64	–	–	–	–	166	207	248	289	329	369	410	450	530	610	691	711	811	1011	1111	1211	2.13
2.15	26	56	–	–	130	151	192	233	273	314	354	394	434	474	555	635	715	735	835	1035	1135	1235	2.15
2.22	36	80	–	–	–	–	–	–	200	242	282	323	364	404	485	565	646	666	766	966	1067	1167	2.22
2.25	32	72	–	–	–	–	–	185	226	267	308	348	389	429	509	590	670	690	790	991	1091	1191	2.25
2.25	40	90	–	–	–	–	–	–	–	210	252	293	334	375	456	536	617	637	737	938	1038	1138	2.25
2.25	64	144	–	–	–	–	–	–	–	–	–	–	–	–	286	370	453	473	575	777	878	979	2.25
2.29	28	64	–	–	–	–	170	211	252	292	333	373	413	454	534	614	694	715	815	1015	1115	1215	2.29
2.33	24	56	–	–	134	155	196	236	277	317	358	398	438	478	559	639	719	739	839	1039	1139	1239	2.33
2.33	48	112	–	–	–	–	–	–	–	–	–	225	268	309	392	473	554	574	675	876	977	1077	2.33
2.33	72	168	–	–	–	–	–	–	–	–	–	–	–	–	–	–	380	401	505	709	811	912	2.33
2.35	34	80	–	–	–	–	–	161	204	245	286	327	367	408	488	569	649	669	770	970	1070	1171	2.35
2.37	38	90	–	–	–	–	–	–	–	214	255	297	338	378	459	540	620	641	741	942	1042	1142	2.37
2.40	30	72	–	–	–	–	146	188	230	271	311	352	392	433	513	594	674	694	794	995	1095	1195	2.40
2.40	80	192	–	–	–	–	–	–	–	–	–	–	–	–	–	–	–	–	432	640	742	844	2.40
2.46	26	64	–	–	–	131	173	215	255	296	337	377	417	457	538	618	698	718	819	1019	1119	1219	2.46

Centre distance (mm).
Courtesy of Fenner.

Table 12.16 Fenner Torque Drive Plus 8MXP and 8M drives (centre distance, mm)

			Belt pitch length (mm)																				
	Number of grooves		480	560	600	640	720	800	880	960	1040	1120	1200	1280	1440	1600	1760	1800	2000	2400	2600	2800	
Speed ratio	Driving pulley	Driven pulley	60 teeth	70 teeth	75 teeth	80 teeth	90 teeth	100 teeth	110 teeth	120 teeth	130 teeth	140 teeth	150 teeth	160 teeth	180 teeth	200 teeth	220 teeth	225 teeth	250 teeth	300 teeth	325 teeth	350 teeth	Speed ratio
2.50	32	80	–	–	–	–	–	165	207	248	290	330	371	411	492	573	653	673	774	974	1074	1174	2.50
2.50	36	90	–	–	–	–	–	–	–	217	259	300	341	382	463	544	624	644	745	946	1046	1146	2.50
2.55	44	112	–	–	–	–	–	–	–	–	–	232	274	316	399	480	561	582	683	884	984	1085	2.55
2.57	28	72	–	–	–	–	150	192	233	274	315	356	396	436	517	597	678	698	798	998	1099	1199	2.57
2.57	56	144	–	–	–	–	–	–	–	–	–	–	–	–	299	384	467	487	589	792	893	994	2.57
2.63	64	168	–	–	–	–	–	–	–	–	–	–	–	–	–	–	394	415	519	724	825	927	2.63
2.65	34	90	–	–	–	–	–	–	178	220	262	304	345	385	467	547	628	648	749	949	1050	1150	2.65
2.67	24	64	–	–	–	134	177	218	259	300	340	381	421	461	542	622	702	722	822	1023	1123	1223	2.65
2.67	30	80	–	–	–	–	–	168	210	252	293	334	375	415	496	576	657	677	777	978	1078	1178	2.67
2.67	72	192	–	–	–	–	–	–	–	–	–	–	–	–	–	–	–	–	446	654	757	858	2.67
2.77	26	72	–	–	–	–	153	195	237	278	319	359	400	440	521	601	681	702	802	1002	1102	1203	2.77
2.80	40	112	–	–	–	–	–	–	–	–	–	238	281	323	406	487	569	589	690	891	992	1092	2.80
2.81	32	90	–	–	–	–	–	–	181	224	266	307	348	389	470	551	632	652	752	953	1053	1154	2.81
2.86	28	80	–	–	–	–	–	171	214	255	297	338	378	419	500	580	661	681	781	982	1082	1182	2.86
2.95	38	112	–	–	–	–	–	–	–	–	–	242	284	326	409	491	572	593	694	895	996	1096	2.95
3.00	24	72	–	–	–	–	156	199	240	281	322	363	403	444	524	605	685	705	806	1006	1106	1206	3.00
3.00	30	90	–	–	–	–	–	–	184	227	269	311	352	393	474	555	635	656	756	957	1057	1157	3.00
3.00	48	144	–	–	–	–	–	–	–	–	–	–	–	–	312	397	480	501	604	807	908	1009	3.00
3.00	56	168	–	–	–	–	–	–	–	–	–	–	–	–	–	320	407	428	533	738	840	941	3.00
3.00	64	192	–	–	–	–	–	–	–	–	–	–	–	–	–	–	–	350	459	668	771	873	3.00
3.08	26	80	–	–	–	–	–	174	217	259	300	341	382	422	503	584	664	685	785	986	1086	1186	3.08
3.11	36	112	–	–	–	–	–	–	–	–	–	245	288	330	413	495	576	596	697	899	999	1100	3.11
3.21	28	90	–	–	–	–	–	–	187	230	273	314	355	396	477	558	639	659	760	961	1061	1161	3.21
3.27	44	144	–	–	–	–	–	–	–	–	–	–	–	–	319	404	487	508	611	814	915	1016	3.27
3.29	34	112	–	–	–	–	–	–	–	–	204	248	291	333	416	498	579	600	701	903	1003	1104	3.29
3.33	24	80	–	–	–	–	–	178	220	262	304	345	385	426	507	588	668	688	789	989	1090	1190	3.33
3.43	56	192	–	–	–	–	–	–	–	–	–	–	–	–	–	–	340	363	472	682	785	887	3.43
3.46	26	90	–	–	–	–	–	–	191	234	276	318	359	400	481	562	643	663	764	985	1065	1165	3.46
3.50	32	112	–	–	–	–	–	–	–	–	207	251	294	337	420	502	583	603	705	906	1007	1107	3.50

3.50	48	168	–	–	–	–	–	–	–	–	–	–	–	–	–	333	420	442	547	752	854	956	3.50
3.60	40	144	–	–	–	–	–	–	–	–	–	–	–	–	325	411	494	515	618	821	922	1023	3.60
3.73	30	112	–	–	–	–	–	–	–	–	210	255	298	340	423	505	587	607	708	910	1011	1111	3.73
3.75	24	90	–	–	–	–	–	–	194	237	279	321	362	403	485	566	647	667	767	968	1069	1169	3.75
3.79	38	144	–	–	–	–	–	–	–	–	–	–	–	–	328	414	498	518	621	825	926	1027	3.79
3.82	44	168	–	–	–	–	–	–	–	–	–	–	–	–	–	339	427	448	553	760	862	963	3.82
4.00	28	112	–	–	–	–	–	–	–	–	213	258	301	343	427	509	590	611	712	914	1014	1115	4.00
4.00	36	144	–	–	–	–	–	–	–	–	–	–	–	–	331	417	501	522	625	829	930	1031	4.00
4.00	48	192	–	–	–	–	–	–	–	–	–	–	–	–	–	–	352	375	485	696	799	901	4.00
4.20	40	168	–	–	–	–	–	–	–	–	–	–	–	–	–	346	433	455	560	767	869	970	4.20
4.24	34	144	–	–	–	–	–	–	–	–	–	–	–	244	335	421	505	525	628	832	933	1035	4.24
4.31	26	112	–	–	–	–	–	–	–	–	216	261	304	347	430	512	594	614	716	917	1018	1119	4.31
4.36	44	192	–	–	–	–	–	–	–	–	–	–	–	–	–	–	358	381	492	703	806	908	4.36
4.42	38	168	–	–	–	–	–	–	–	–	–	–	–	–	–	349	437	459	564	770	872	974	4.42
4.50	32	144	–	–	–	–	–	–	–	–	–	–	–	247	338	424	508	529	632	836	937	1038	4.50
4.67	24	112	–	–	–	–	–	–	–	–	219	264	308	350	434	516	597	618	719	921	1022	1122	4.67
4.67	36	168	–	–	–	–	–	–	–	–	–	–	–	–	–	352	440	461	567	774	876	978	4.67
4.80	30	144	–	–	–	–	–	–	–	–	–	–	–	250	341	427	511	532	635	839	941	1042	4.80
4.80	40	192	–	–	–	–	–	–	–	–	–	–	–	–	–	–	365	388	498	710	813	916	4.90
4.94	34	168	–	–	–	–	–	–	–	–	–	–	–	–	–	355	443	465	570	777	879	981	4.94
5.05	38	192	–	–	–	–	–	–	–	–	–	–	–	–	–	–	368	391	502	713	816	919	5.05
5.14	28	144	–	–	–	–	–	–	–	–	–	–	–	253	344	431	515	536	639	843	944	1046	5.14
5.25	32	168	–	–	–	–	–	–	–	–	–	–	–	–	–	358	446	468	574	781	883	995	5.25
5.33	36	192	–	–	–	–	–	–	–	–	–	–	–	–	–	–	371	394	505	716	820	923	5.33
5.54	26	144	–	–	–	–	–	–	–	–	–	–	–	256	348	434	518	539	642	847	948	1049	5.54
5.60	30	168	–	–	–	–	–	–	–	–	–	–	–	–	–	361	450	471	577	784	887	988	5.60
5.65	34	192	–	–	–	=	–	–	–	–	–	–	–	–	–	–	374	397	508	720	823	926	5.65
6.00	24	144	–	–	–	–	–	–	–	–	–	–	–	259	351	437	522	542	646	850	952	1053	6.00
6.00	28	168	–	–	–	–	–	–	–	–	–	–	–	–	269	364	453	475	581	785	890	992	6.00
6.00	32	192	–	–	–	–	–	–	–	–	–	–	–	–	–	–	377	400	511	723	827	930	6.00
6.40	30	192	–	–	–	–	–	–	–	–	–	–	–	–	–	–	380	403	515	727	830	933	6.40
6.46	26	168	–	–	–	–	–	–	–	–	–	–	–	–	272	368	478	456	584	791	894	996	6.46
6.86	28	192	–	–	–	–	–	–	–	–	–	–	–	–	–	–	383	406	519	730	834	937	6.86

Courtesy of Fenner.

Table 12.17 Fenner Torque Drive Plus 14MXP and 14M drives (centre distance, mm)

	Number of grooves		Belt pitch length (mm)																
			966	1190	1400	1610	1778	1890	2100	2310	2450	2590	2800	3150	3500	3850	4326	4578	
Speed ratio	Driving pulley	Driven pulley	69 teeth	85 teeth	100 teeth	115 teeth	127 teeth	135 teeth	150 teeth	165 teeth	175 teeth	185 teeth	200 teeth	225 teeth	250 teeth	275 teeth	309 teeth	327 teeth	Speed ratio
1.00	28	28	287	399	504	609	693	749	854	959	1029	1099	1204	1379	1554	1729	1967	2093	1.00
1.00	29	29	280	392	497	602	686	742	847	952	1022	1092	1197	1372	1547	1722	1960	2086	1.00
1.00	30	30	273	385	490	595	679	735	840	945	1015	1085	1190	1365	1540	1715	1953	2079	1.00
1.00	32	32	259	371	476	581	665	721	826	931	1001	1071	1176	1351	1526	1701	1939	2065	1.00
1.00	34	34	245	357	462	567	651	707	812	917	987	1057	1162	1337	1512	1687	1925	2051	1.00
1.00	36	36	231	343	448	553	637	693	798	903	973	1043	1148	1323	1498	1673	1911	2037	1.00
1.00	38	38	217	329	434	539	623	679	784	889	959	1029	1134	1309	1484	1659	1897	2023	1.00
1.00	40	40	203	315	420	525	609	665	770	875	945	1015	1120	1295	1470	1645	1883	2009	1.00
1.00	44	44	–	287	392	497	581	637	742	847	917	987	1092	1267	1442	1617	1855	1981	1.00
1.00	48	48	–	259	364	469	553	609	714	819	889	959	1064	1239	1414	1589	1827	1953	1.00
1.00	56	56	–	–	308	413	497	553	658	763	833	903	1008	1183	1358	1533	1771	1897	1.00
1.03	29	30	276	388	493	598	682	738	843	948	1018	1088	1193	1368	1543	1718	1956	2082	1.00
1.04	28	29	283	395	500	605	689	745	850	955	1025	1095	1200	1375	1550	1725	1963	2089	1.04
1.05	38	40	210	322	427	532	616	672	777	882	952	1022	1127	1302	1477	1652	1890	2016	1.05
1.06	36	38	224	336	441	546	630	686	791	896	966	1036	1141	1316	1491	1666	1904	2030	1.06
1.06	34	36	238	350	455	560	644	700	805	910	980	1050	1155	1330	1505	1680	1918	2044	1.06
1.06	32	34	252	364	469	574	658	714	819	924	994	1064	1169	1344	1519	1694	1932	2058	1.06
1.07	30	32	266	378	483	588	672	728	833	938	1008	1078	1183	1358	1533	1708	1946	2072	1.07
1.07	28	30	280	392	497	602	686	742	847	952	1022	1092	1197	1372	1547	1722	1960	2086	1.07
1.09	44	48	–	273	378	483	567	623	728	833	903	973	1078	1253	1428	1603	1841	1967	1.09
1.10	40	44	–	301	406	511	595	651	756	861	931	1001	1106	1281	1456	1631	1869	1995	1.10
1.10	29	32	269	381	486	591	675	731	836	941	1011	1081	1186	1361	1536	1711	1949	2075	1.10
1.11	36	40	217	329	434	539	623	679	784	889	959	1029	1134	1309	1484	1659	1897	2023	1.11
1.12	34	38	231	343	448	553	637	693	798	903	973	1043	1148	1323	1498	1673	1911	2037	1.12
1.13	32	36	245	357	462	567	651	707	812	917	987	1057	1162	1337	1512	1687	1925	2051	1.13

1.13	30	34	259	371	476	581	665	721	826	931	1001	1071	1176	1351	1526	1701	1939	2065	1.13
1.14	28	32	273	385	490	595	679	735	840	945	1015	1085	1190	1365	1540	1715	1953	2079	1.14
1.14	56	64	–	–	–	385	469	525	630	735	805	875	980	1155	1330	1505	1743	1869	1.14
1.16	38	44	–	308	413	518	602	658	763	868	938	1008	1113	1288	1463	1638	1876	2002	1.16
1.17	48	56	–	–	336	441	525	581	686	791	861	931	1036	1211	1386	1561	1799	1925	1.17
1.17	29	34	262	374	479	584	668	724	829	934	1004	1074	1179	1354	1529	1704	1942	2068	1.17
1.18	34	40	224	336	441	546	630	686	791	896	966	1036	1141	1316	1491	1666	1904	2030	1.18
1.19	32	38	238	350	455	560	644	700	805	910	980	1050	1155	1330	1505	1680	1918	2044	1.19
1.20	30	36	252	364	469	574	658	714	819	924	994	1064	1169	1344	1519	1694	1932	2058	1.20
1.20	40	48	–	286	392	497	581	637	742	847	917	987	1092	1267	1442	1617	1855	1981	1.20
1.21	28	34	266	378	483	588	672	728	833	938	1008	1078	1183	1358	1533	1708	1946	2072	1.21
1.22	36	44	202	314	420	525	609	665	770	875	945	1015	1120	1295	1470	1645	1883	2009	1.22
1.24	29	36	255	367	472	577	661	717	822	927	997	1067	1172	1347	1522	1697	1935	2061	1.24
1.25	32	40	230	343	448	553	637	693	798	903	973	1043	1148	1323	1498	1673	1911	2037	1.25
1.26	38	48	–	293	398	504	588	644	749	854	924	994	1099	1274	1449	1624	1862	1988	1.26
1.27	30	38	244	357	462	567	651	707	812	917	987	1057	1162	1337	1512	1687	1925	2051	1.27
1.27	44	56	–	244	349	454	538	594	699	805	875	945	1050	1225	1400	1575	1813	1939	1.27
1.29	28	36	258	371	476	581	665	721	826	931	1001	1071	1176	1351	1526	1701	1939	2065	1.29
1.29	56	72	–	–	–	355	440	496	601	706	776	846	951	1126	1302	1477	1715	1841	1.29
1.29	34	44	209	321	426	532	616	672	777	882	952	1022	1127	1302	1477	1652	1890	2016	1.29
1.31	29	38	248	360	465	570	654	710	815	920	990	1060	1165	1340	1515	1690	1928	2054	1.31
1.33	30	40	237	349	454	560	644	700	805	910	980	1050	1155	1330	1505	1680	1918	2044	1.33
1.33	36	48	–	300	405	510	594	650	756	861	931	1001	1106	1281	1456	1631	1869	1995	1.33
1.33	48	64	–	–	306	411	496	552	657	762	832	902	1007	1182	1358	1533	1771	1897	1.33
1.36	28	38	251	363	468	574	658	714	819	924	994	1064	1169	1344	1519	1694	1932	2058	1.36
1.38	32	44	215	328	433	538	622	678	784	889	959	1029	1134	1309	1484	1659	1897	2023	1.38
1.38	29	40	240	353	458	563	647	703	808	913	983	1053	1158	1333	1508	1683	1921	2047	1.38
1.40	40	56	–	257	362	468	552	608	713	818	888	958	1063	1238	1414	1589	1827	1953	1.40
1.41	34	48	–	306	412	517	601	657	762	867	937	1008	1113	1288	1463	1638	1876	2002	1.41
1.43	28	40	244	356	461	566	650	706	812	917	987	1057	1162	1337	1512	1687	1925	2051	1.43

Centre distance (mm).
Courtesy of Fenner.

Table 12.18 Fenner Torque Drive Plus 14MXP and 14M drives (centre distance, mm)

	Number of grooves		Belt pitch length (mm)																
			966	1190	1400	1610	1778	1890	2100	2310	2450	2590	2800	3150	3500	3850	4326	4578	
Speed Ratio	Driving Pulley	Driven Pulley	69 teeth	85 teeth	100 teeth	115 teeth	127 teeth	135 teeth	150 teeth	165 teeth	175 teeth	185 teeth	200 teeth	225 teeth	250 teeth	275 teeth	309 teeth	327 teeth	Speed Ratio
1.43	56	80	–	–	–	325	410	466	571	677	747	817	922	1098	1273	1448	1686	1812	1.43
1.45	44	64	–	–	319	425	509	565	671	776	846	916	1021	1196	1371	1546	1784	1910	1.45
1.47	30	44	222	335	440	545	629	685	790	895	965	1036	1141	1316	1491	1666	1904	2030	1.47
1.47	38	56	–	263	369	474	559	615	720	825	895	965	1070	1245	1420	1595	1834	1960	1.47
1.50	32	48	200	313	418	524	608	664	769	874	944	1014	1119	1295	1470	1645	1883	2009	1.50
1.50	48	72	–	–	–	381	466	522	628	733	803	873	979	1154	1329	1504	1742	1868	1.50
1.52	29	44	225	338	443	548	633	689	794	899	969	1039	1144	1319	1494	1669	1907	2033	1.52
1.56	36	56	–	269	375	481	565	621	727	832	902	972	1077	1252	1427	1602	1840	1966	1.56
1.57	28	44	228	341	447	552	636	692	797	902	972	1042	1147	1323	1498	1673	1911	2037	1.57
1.60	30	48	206	319	425	530	615	671	776	881	951	1021	1126	1301	1476	1652	1890	2016	1.60
1.60	40	64	–	–	332	438	522	579	684	789	859	929	1035	1210	1385	1560	1798	1924	1.60
1.61	56	90	–	–	–	–	370	427	534	640	710	780	886	1061	1237	1412	1650	1776	1.61
1.64	44	72	–	–	287	394	479	535	641	746	817	887	992	1167	1343	1518	1756	1882	1.64
1.65	34	56	–	276	382	488	572	628	733	839	909	979	1084	1259	1434	1609	1847	1973	1.65
1.66	29	48	209	323	428	534	618	674	779	884	955	1025	1130	1305	1480	1655	1893	2019	1.66
1.67	48	80	–	–	–	350	435	492	598	703	774	844	949	1125	1300	1475	1714	1840	1.67
1.68	38	64	–	–	338	444	529	585	691	796	866	936	1041	1217	1392	1567	1805	1931	1.68
1.71	28	48	212	326	432	537	621	678	783	888	958	1028	1133	1308	1483	1658	1896	2023	1.71
1.75	32	56	–	282	388	494	579	635	740	845	915	986	1091	1266	1441	1616	1854	1980	1.75
1.78	36	64	–	–	344	451	535	592	697	803	873	943	1048	1223	1399	1574	1812	1938	1.78
1.80	40	72	–	–	300	407	492	548	654	760	830	900	1005	1181	1356	1531	1770	1896	1.80
1.82	44	80	–	–	–	362	448	505	611	717	787	857	963	1138	1314	1489	1727	1853	1.82
1.87	30	56	–	288	395	501	585	641	747	852	922	992	1097	1273	1448	1623	1861	1987	1.87
1.88	48	90	–	–	–	–	395	452	559	665	736	807	912	1088	1264	1439	1677	1804	1.88
1.88	34	64	–	243	351	457	542	598	704	809	879	950	1055	1230	1405	1581	1819	1945	1.88

1.89	38	72	–	–	306	413	498	555	661	766	837	907	1012	1188	1356	1538	1776	1902	1.89
1.93	29	56	–	291	389	504	588	645	750	855	926	996	1101	1276	1451	1626	1865	1991	1.93
2.00	28	56	–	294	401	507	592	648	753	859	929	999	1104	1279	1455	1630	1868	1994	2.00
2.00	32	64	–	249	357	464	548	605	710	816	886	956	1062	1237	1412	1587	1826	1952	2.00
2.00	36	72	–	–	312	419	505	561	667	773	843	913	1019	1194	1370	1545	1783	1909	2.00
2.00	40	80	–	–	–	374	460	517	624	730	800	870	976	1152	1327	1502	1741	1867	2.00
2.00	56	112	–	–	–	–	–	–	444	553	625	696	802	979	1155	1331	1570	1696	2.00
2.05	44	90	–	–	–	320	407	465	572	678	749	820	925	1101	1277	1452	1691	1817	2.05
2.11	38	80	–	–	–	380	467	524	630	736	807	877	983	1158	1334	1509	1747	1874	2.11
2.12	34	72	–	–	318	426	511	568	674	779	850	920	1026	1201	1376	1552	1790	1916	2.12
2.13	30	64	–	255	363	470	555	611	717	823	893	963	1068	1244	1419	1594	1832	1959	2.13
2.21	29	64	–	258	366	473	558	615	720	826	896	966	1072	1247	1422	1598	1836	1962	2.21
2.22	36	80	–	–	–	387	473	530	636	743	813	884	989	1165	1340	1516	1754	1880	2.22
2.25	32	72	–	–	324	432	517	574	680	786	856	927	1032	1208	1383	1558	1797	1923	2.25
2.25	40	90	–	–	–	331	419	477	584	691	762	833	938	1114	1290	1466	1704	1831	2.25
2.29	28	64	–	261	369	476	561	618	724	829	899	970	1075	1250	1426	1601	1839	1965	2.29
2.33	48	112	–	–	–	–	–	–	468	577	649	721	828	1005	1181	1358	1597	1723	2.33
2.35	34	80	–	–	282	393	479	536	643	749	820	890	996	1172	1347	1523	1761	1887	2.35
2.37	38	90	–	–	–	337	425	483	591	697	768	839	945	1121	1297	1472	1711	1837	2.37
2.40	30	72	–	–	330	438	524	580	687	792	863	933	1039	1214	1390	1565	1804	1930	2.40
2.48	29	72	–	–	333	441	527	584	690	796	866	937	1042	1218	1393	1569	1807	1933	2.48
2.50	32	80	–	–	288	399	485	542	649	755	826	897	1002	1178	1354	1529	1758	1894	2.50
2.50	36	90	–	–	–	343	431	489	597	704	775	845	951	1128	1303	1479	1718	1844	2.50
2.55	44	112	–	–	–	–	–	368	480	590	662	733	840	1018	1194	1371	1610	1736	2.55
2.57	28	72	–	–	336	444	530	587	693	799	869	940	1045	1221	1397	1572	1810	1937	2.57
2.57	56	144	–	–	–	–	–	–	–	–	485	561	671	852	1031	1209	1450	1577	2.57
2.65	34	90	–	–	–	349	437	495	603	710	781	852	958	1134	1310	1486	1724	1851	2.65
2.67	30	80	–	–	294	405	491	549	656	762	833	903	1009	1185	1360	1536	1775	1901	2.67
2.76	29	80	–	–	297	408	494	552	659	765	836	906	1012	1188	1364	1539	1778	1904	2.76
2.80	40	112	–	–	–	–	–	379	492	602	674	746	853	1031	1207	1384	1623	1750	2.80

Centre distance (mm)
Courtesy of Fenner.

Table 12.19 Fenner Torque Drive Plus 14MXP and 14M drives (centre distance, mm)

	Number of grooves		Belt pitch length (mm)																
			966	1190	1400	1610	1778	1890	2100	2310	2450	2590	2800	3150	3500	3850	4326	4578	
Speed ratio	Driving pulley	Driven pulley	69 teeth	85 teeth	100 teeth	115 teeth	127 teeth	135 teeth	150 teeth	165 teeth	175 teeth	185 teeth	200 teeth	225 teeth	250 teeth	275 teeth	309 teeth	327 teeth	Speed Ratio
2.81	32	90	–	–	–	354	443	501	609	716	787	858	964	1141	1317	1492	1731	1858	2.81
2.86	28	80	–	–	300	411	498	555	662	768	839	910	1015	1191	1367	1543	1781	1907	2.86
2.95	38	112	–	–	–	–	–	385	498	608	680	752	859	1037	1214	1390	1630	1756	2.95
3.00	30	90	–	–	–	360	449	507	615	723	794	865	971	1147	1323	1499	1738	1864	3.00
3.00	48	144	–	–	–	–	–	–	–	–	508	584	695	877	1056	1234	1475	1603	3.00
3.00	56	168	–	–	–	–	–	–	–	–	–	–	560	749	933	1113	1356	1484	3.00
3.10	29	90	–	–	–	363	452	510	619	726	797	868	974	1150	1327	1502	1741	1868	3.10
3.11	36	112	–	–	–	–	–	390	504	614	686	758	865	1043	1220	1397	1636	1763	3.11
3.21	28	90	–	–	–	366	455	513	622	729	800	871	977	1154	1330	1506	1745	1871	3.21
3.27	44	144	–	–	–	–	–	–	–	441	519	595	707	889	1069	1247	1488	1616	3.27
3.29	34	112	–	–	–	–	–	396	509	620	692	764	872	1050	1227	1403	1643	1769	3.29
3.43	56	192	–	–	–	–	–	–	–	–	–	–	—	635	826	1012	1259	1388	3.43
3.50	32	112	–	–	–	–	–	401	515	626	698	770	878	1056	1233	1410	1649	1776	3.50
3.50	48	168	–	–	–	–	–	–	–	–	–	–	583	773	957	1138	1381	1509	3.50
3.60	40	144	–	–	–	–	–	–	–	452	530	607	719	901	1081	1260	1501	1629	3.60
3.73	30	112	–	–	–	–	343	407	521	632	704	777	884	1062	1240	1416	1656	1783	3.73
3.79	38	144	–	–	–	–	–	–	–	457	536	612	725	907	1087	1266	1507	1635	3.79
3.82	44	168	–	–	–	–	–	–	–	–	–	–	594	784	969	1150	1394	1522	3.82
3.86	29	112	–	–	–	–	346	410	524	635	707	780	887	1065	1243	1419	1659	1786	3.86
4.00	28	112	–	–	–	–	349	413	527	638	710	783	890	1069	1246	1423	1662	1789	4.00
4.00	36	144	–	–	–	–	–	–	–	462	542	618	730	913	1094	1272	1514	1641	4.00
4.00	48	192	–	–	–	–	–	–	–	–	–	–	—	654	849	1035	1283	1413	4.00
4.20	40	168	–	–	–	–	–	–	–	–	–	–	605	796	981	1162	1406	1534	4.20

4.24	34	144	–	–	–	–	–	–	–	488	547	624	736	919	1100	1279	1520	1648	4.24
4.36	44	192	–	–	–	–	–	–	–	–	–	–	—	668	861	1047	1295	1425	4.36
4.42	38	168	–	–	–	–	–	–	–	–	–	488	610	802	986	1168	1412	1541	4.42
4.50	32	144	–	–	–	–	–	–	–	473	553	630	742	925	1106	1285	1527	1654	4.50
4.67	36	168	–	–	–	–	–	–	–	–	–	493	616	807	992	1174	1419	1547	4.67
4.80	30	144	–	–	–	–	–	–	–	479	558	635	748	931	1112	1291	1533	1661	4.80
4.80	40	192	–	–	–	–	–	–	–	–	–	–	—	678	872	1059	1307	1437	4.80
4.94	34	168	–	–	–	–	–	–	–	–	–	499	621	813	998	1180	1425	1553	4.94
4.97	29	144	–	–	–	–	–	–	–	481	561	638	751	934	1115	1294	1536	1664	4.97
5.05	38	192	–	–	–	–	–	–	–	–	–	–	—	684	878	1065	1313	1443	5.05
5.14	28	144	–	–	–	–	–	–	–	484	564	641	754	937	1118	1297	1539	1667	5.14
5.25	32	168	–	–	–	–	–	–	–	–	–	504	627	819	1004	1186	1431	1560	5.25
5.33	36	192	–	–	–	–	–	–	–	–	–	–	—	689	884	1071	1319	1449	5.33
5.60	30	168	–	–	–	–	–	–	–	–	–	509	632	825	1010	1192	1437	1566	5.60
5.65	34	192	–	–	–	–	–	–	–	–	–	–	—	695	889	1076	1325	1455	5.65
5.79	29	168	–	–	–	–	–	–	–	–	–	512	635	828	1013	1195	1440	1569	5.79
6.00	28	168	–	–	–	–	–	–	–	–	–	514	638	830	1016	1198	1443	1572	6.00
6.00	32	192	–	–	–	–	–	–	–	–	–	–	—	700	895	1082	1331	1462	6.00
6.40	30	192	–	–	–	–	–	–	–	–	–	–	—	706	901	1088	1337	1468	6.40
6.62	29	192	–	–	–	–	–	–	–	–	–	–	—	708	904	1091	1340	1471	6.62
6.86	28	192	–	–	–	–	–	–	–	–	–	–	—	711	906	1094	1343	1474	6.86

Centre distance (mm).
Courtesy of Fenner.

Fig. 12.14 Flat belt drive.

Fig. 12.15 Crowned and guided pulley wheels for flat belt drives.

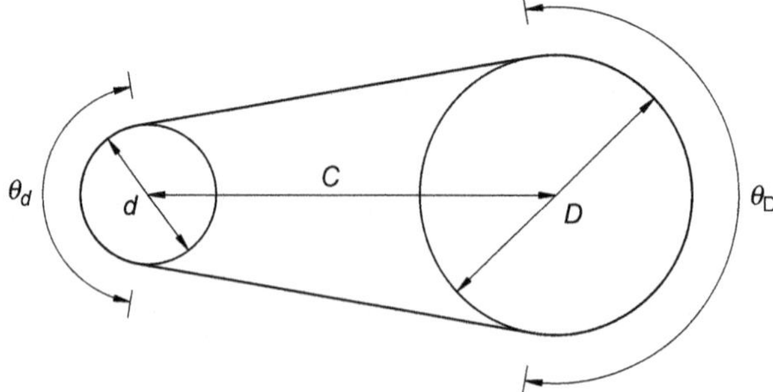

Fig. 12.16 Belt drive geometry definition.

The power transmitted by a belt drive is given by

$$\text{Power} = (F_1 - F_2)V \tag{12.6}$$

where F_1 = belt tension in the tight side (N), F_2 = belt tension in the slack side (N), V = belt speed (ms^{-1}).

The torque is given by

$$\text{Torque} = (F_1 - F_2)r \tag{12.7}$$

Assuming that the friction is uniform throughout the arc of contact and ignoring centrifugal effects, the ratio of the tensions in the belts can be modelled by Eytlewein's formula:

$$\frac{F_1}{F_2} = e^{\mu\theta} \tag{12.8}$$

where μ = the coefficient of friction, θ = the angle of contact (rad), usually taken as the angle for the smaller pulley.

Centrifugal forces acting on the belt along the arcs of contact reduce the surface pressure (Firbank, 1970). The centrifugal force is given by

$$F_c = \rho V^2 A = mV^2 \tag{12.9}$$

where ρ = the density of the belt material (kg/m^3), A = cross-sectional area of the belt (m^2), m = the mass per unit length of the belt (kg/m).

The centrifugal force acts on both the tight and the slack sides of the belt and Eytlewein's formula can be modelled to model the effect:

$$\frac{F_1 - F_c}{F_2 - F_c} = e^{\mu\theta} \tag{12.10}$$

The maximum allowable tension, $F_{1,\max}$, in the tight side of a belt depends on the allowable stress of the belt material, $\sigma_{\max}$, see Table 12.20.

$$F_{1,\max} = \sigma \max A \tag{12.11}$$

The required cross-sectional area for a belt drive can be found from

$$A = \frac{F_1 - F_2}{\sigma_1 - \sigma_2} \tag{12.12}$$

Example 12.4. A fan is belt driven by an electric motor running at 1500 rpm. The pulley diameters for the fan and motor are 500 mm and 355 mm, respectively. A flat belt has been selected with a width of 100 mm, thickness 3.5 mm, coefficient

Table 12.20 Guideline values for the maximum permissible stress for high-performance flat belts

Multiply structure			
Friction surface coating	**Core**	**Top surface**	**Maximum permissible stress, σ_{max} (MN/m^2)**
Elastomer	Polyamide sheet	Polyamide fabric	8.3–19.3
Elastomer	Polyamide sheet	Elastomer	6.6–13.7
Chrome leather	Polyamide sheet	None	6.3–11.4
Chrome leather	Polyamide sheet	Polyamide fabric	5.7–14.7
Chrome leather	Polyamide sheet	Chrome leather	4–8
Elastomer	Polyester cord	Elastomer	Up to 21.8
Chrome leather	Polyester cord	Polyamide fabric	5.2–12
Chrome leather	Polyester cord	Chrome leather	3.1–8

of friction 0.8, density 1100 kg/m^3 and permissible stress 11 MN/m^2. The centre distance is 1500 mm. Determine the power capacity of the belt.

Solution

The arcs of contact for the driving and driven pulleys are.

$$\theta_d = \pi - 2\sin^{-1}\frac{D-d}{2C} = \pi - 2\sin^{-1}\frac{500-355}{2\times 1500} = 3.045 \text{ rad}$$

$$\theta_D = \pi + 2\sin^{-1}\frac{D-d}{2C} = \pi + 2\sin^{-1}\frac{500-355}{2\times 1500} = 3.238 \text{ rad}$$

The maximum tension in the tight side is given as a function of the maximum permissible stress in the belt by

$$F_1 = \sigma_{max}A = 11\times 10^6\left(3.5\times 10^{-3}\times 100\times 10^{-3}\right) = 3850 \text{ N}$$

The belt velocity

$$V = 1500\times\frac{2\pi}{60}\times\frac{0.355}{2} = 27.88 \text{ m/s}$$

The mass per unit length

$$m = \rho A = 1100\times 0.1\times 0.0035 = 0.385\, kg/m$$

The centrifugal loads given by

$$F_c = 0.385(27.88)^2 = 299.3 \text{ N}$$

Using Eq. (12.10),

$$\frac{F_1 - F_c}{F_2 - F_c} = e^{\mu\theta} = \frac{3850 - 299.3}{F_2 - 299.3} = e^{0.8\times 3.045}$$

$$\Rightarrow F_2 = 610.0 \text{ N}$$

The power capacity is given by $(F_1 - F_2)V = (3850\text{–}610)27.88 = 90.3\,\text{kW}$.

Example 12.5. A flat belt is required to transmit 22 kW from a 250 mm diameter pulley running at 1450 rpm to a 355 mm diameter pulley. The coefficient of friction can be taken as 0.7, the density of the belt is 1100 kg/m^3 and the maximum permissible stress is 7 MPa. The distance between the shaft centres is 1.8 m. The proposed belt is 3.5 mm thick. Calculate the width required.

Solution

22 kW, $D_1 = 250$ mm, $D_2 = 355$ mm, $n_1 = 1450$ rpm. $\mu = 0.7$, $\rho = 1100\,\text{kg/m}^3$, $\sigma_{max} = 7$ MPa, $C = 1.8$ m, $t = 3.5$ mm.

$$\theta_d = \pi - 2\sin^{-1}\frac{355 - 250}{2\times 1800} = 3.083$$

$$F_1 = \sigma_{max}\,A = \sigma_{max}\,tw = 7\times 10^6 \times 3.5\times 10^{-3} w = 24,\,500w$$

$$F_c = \rho A V^2 = \rho t w \times \left(1450 \times \frac{2\pi}{60} \times \frac{0.25}{2}\right)^2$$

$$= w \times 1100 \times 3.5 \times 10^{-3} \times \left(1450 \times \frac{2\pi}{60} \times \frac{0.25}{2}\right)^2 = 1387w$$

Using Eq. (12.10),

$$\frac{F_1 - 1387w}{F_2 - 1387w} = e^{2.1581} = 8.655$$

$$F_1 - 1387w = 8.655(F_2 - 1387w)$$

$$F_1 - F_2 = 22{,}000/18.98 = 1159$$

$$F_2 = F_1 - 1159 = 24500w - 1159$$

$$\frac{24500w - 1387w}{8.655} = 24500w - 1159 - 1387w$$

$$w = 0.0567 \text{ m.}\quad w = 56.7 \text{ mm.}\quad \text{Use } w = 60 \text{ mm}$$

12.3 Chain drives

A chain is a power transmission device consisting of a series of pin-connected links as illustrated in Fig. 12.17. The chain transmits power between two rotating shafts by meshing with toothed sprockets as shown in Fig. 12.18. A sprocket is shown for simple chain in Fig. 12.19, the cross section of a sprocket for the case of triplex chain is illustrated in Fig. 12.20 and an application of triplex chain in Fig. 12.21.

Chain drives are usually manufactured using high-strength steel and for this reason are capable of transmitting high torque. Chain drives are complimentary and competitive with belt drives serving the function of transmitting a wide range of powers for shaft speeds up to about 6000 rpm. At higher speeds, the cyclic impact between the chain links and the sprocket teeth, high noise and difficulties in providing lubrication, limit the application of chain drives. Table 12.21 shows a comparison of chain, belt and gear attributes. Chain drives are principally used for power transmission, conveyors and for supporting or lifting loads (see, American Chain Association, 2005). Applications range from motorcycle and bicycle transmissions, automotive

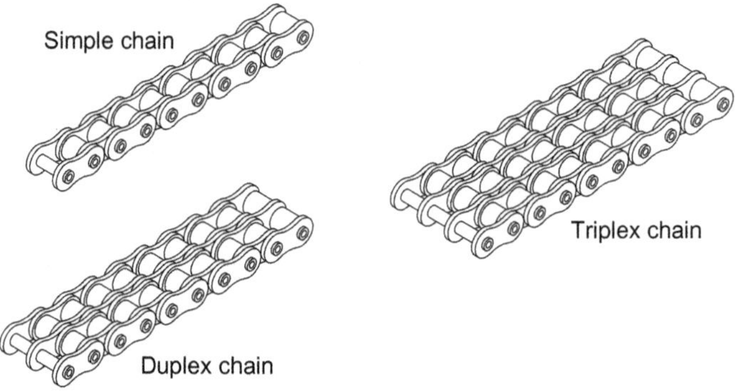

Fig. 12.17 Roller chain.

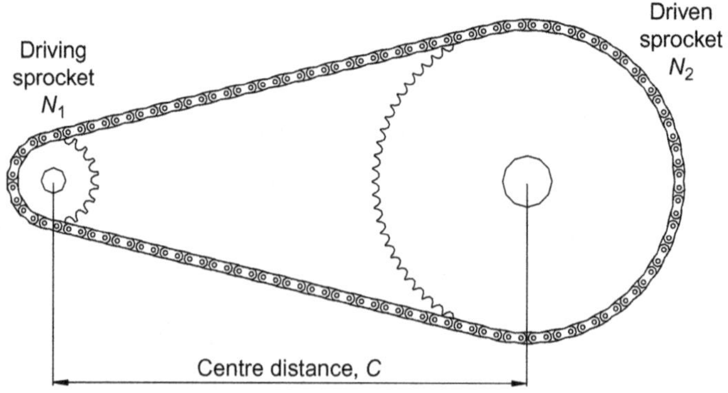

Fig. 12.18 Simple chain drive.

Fig. 12.19 Chain drive sprocket.

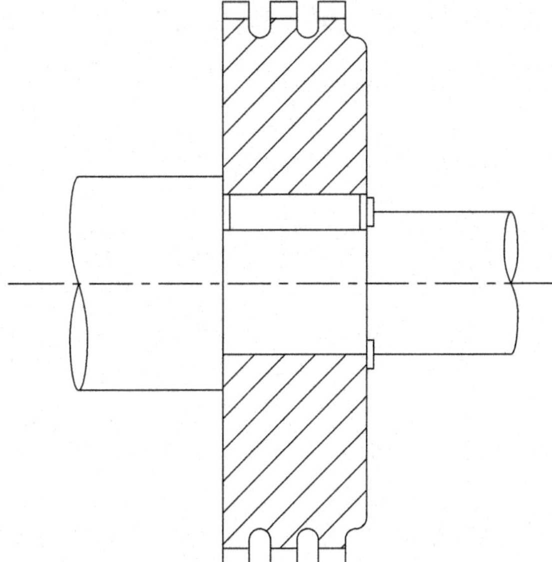

Fig. 12.20 Triplex sprocket.

camshafts drives (although synchronous belts have replaced chains in some automotive applications), machine tools and aerospace drives such as the thruster nozzles for the classic Harrier aircraft, to conveyors and packaging machinery. Efficiencies of up to 98.9%, where necessary ratios of up to 9:1 and power transmission of several

Fig. 12.21 Use of triplex chain for a dynamometer.

Table 12.21 Comparison of chain, belt and gear performance

Feature	Chain drive	Belt drive	Gear drive
Efficiency	A	A	A
Positive drive	A	A[a]	A
Large centre distance	A	A	C
Wear resistance	A	B	A
Multiple drives	A	A	C
Heat resistance	A	C	B
Chemical resistance	A	C	B
Oil resistance	A	C	A
Power range	A	B	A
Speed range	C	A	A
Ease of maintenance	A	B	C
Environment	A	C	A

A = excellent, B = good, C = poor.
[a]Using a synchronous belt drive.

hundred kilowatts can be achieved. Chain drives are typically used with reduction ratios of up to three to one giving high efficiency and ratios of up to five giving reasonable efficiency.

The range of chain drives is extensive as illustrated in Fig. 12.22. The most common type is roller chain, which is used for high-power transmission and conveyor applications. The selection and specification of this type of chain is introduced in Section 12.3.1. Roller chain is made up of a series of links. Each chain consists of side plates, pins, bushes and rollers as shown in Fig. 12.23. The chain runs on toothed sprockets and as the teeth of the sprockets engage with the rollers, rolling motion occurs and the chain articulates onto the sprocket. Roller chain is classified by its pitch. This is the distance between the corresponding points of adjacent links. Roller chain is available in stainless steel and nylon (for hygienic applications) as well as a series of other materials for specialist applications.

Conveyor chain is specially designed for use in materials handling and conveyor equipment and is characterised by its long pitch, large roller diameters and high-tensile strength (Jagtap et al., 2014). The applications of conveyor chain demand that

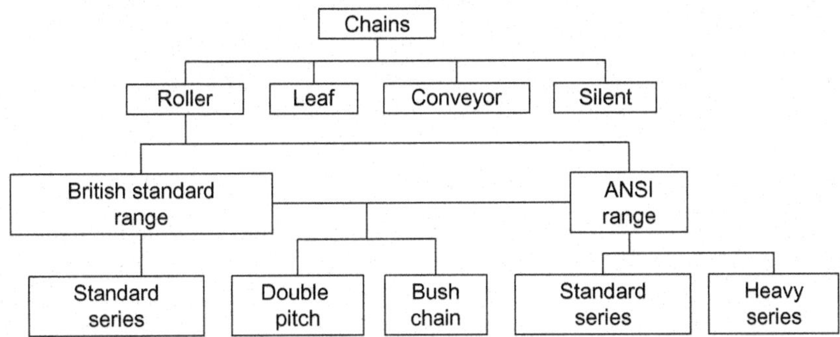

Fig. 12.22 Chain types.

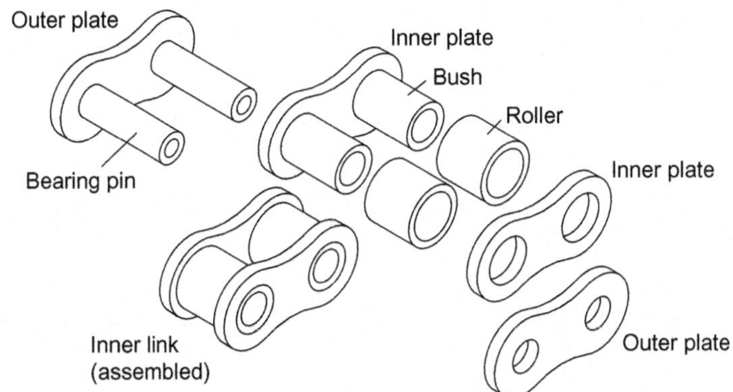

Fig. 12.23 Roller chain components.

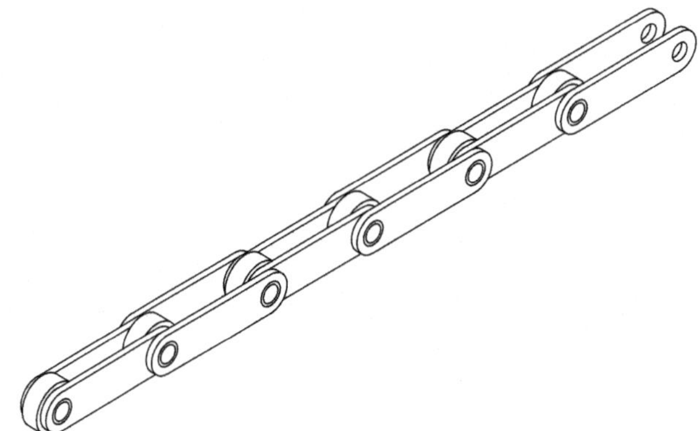

Fig. 12.24 Conveyor chain.

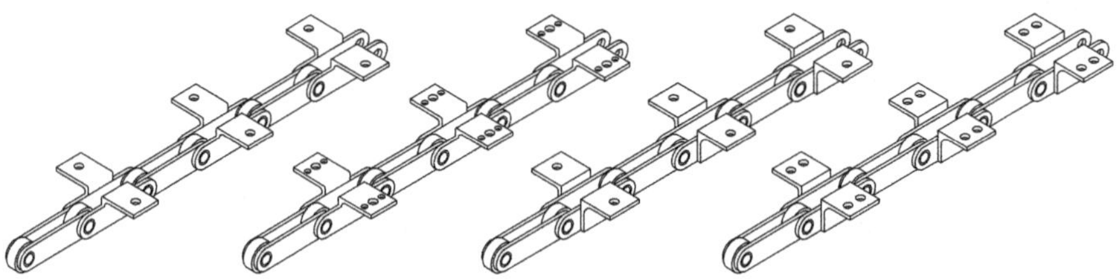

Fig. 12.25 Conveyer chain showing a selection of side plate features.

it is capable of pulling significant loads usually along a straight line at relatively low speeds. The sides of conveyor chains often incorporate special features to aid connection to conveyor components as illustrated in Figs. 12.24 and 12.25.

Leaf chain consists of a series of pin-connected side plates as illustrated in Fig. 12.26. It is generally used for load balancing applications, for example Fig. 12.27, and is essentially a special form of transmission chain.

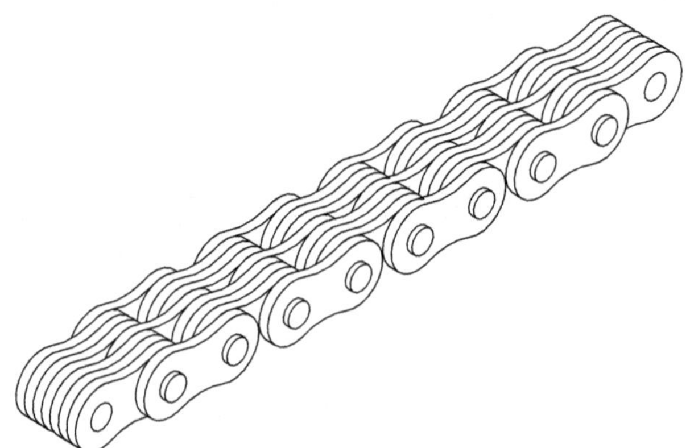

Fig. 12.26 Leaf chain.

Fig. 12.27 Leaf chain used for load balancing in a forklift truck.

Silent chain (also called inverted tooth chain) has the teeth formed in the link plates as shown in Figs. 12.28 and 12.29. The chain consists of alternately mounted links so that the chain can articulate onto the mating sprocket teeth. Silent chains can operate at higher speeds than comparatively sized roller chain and, as the name suggests, more quietly. Silent chain is used in a wide variety of applications from power transmission in conveyors to driving internal combustions (IC) engine camshafts. The dynamic behaviour of silent chain can be improved by fine-tuning the meshing angles (see, Meng et al., 2007; Cali et al., 2016) and silent chain represents an area of ongoing technical development and opportunity for improvement.

The direction of rotation relative to the chain layout is important to the effective operation of a chain drive. Figure 12.30 shows some examples of recommended and acceptable practice for chain drive layouts. The idler sprockets shown are used to engage the chain in the slack region for chain drives with large centre distances or when the drive is vertical. Chain tensioners can be used to ensure the correct chain tension, see Figs. 12.31 and 12.32, for example.

Chain design is based on ensuring that the power transmission capacity is within limits for three modes of failure: fatigue, impact and galling. Chains are designed so that the maximum tensile stress is below the fatigue endurance limit for finite life of the material. Failure would nevertheless eventually occur but it would be due to wear, not fatigue. In service, failures due to wear can be eliminated by inspection and replacement intervals.

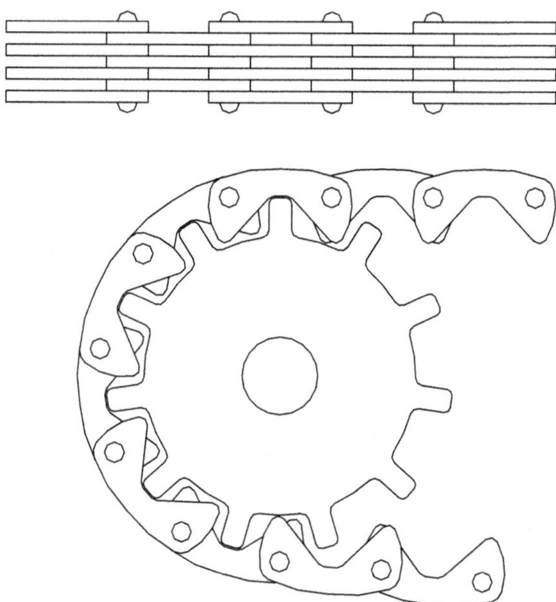

Fig. 12.28 Silent chain.

Fig. 12.29 Silent chain example.

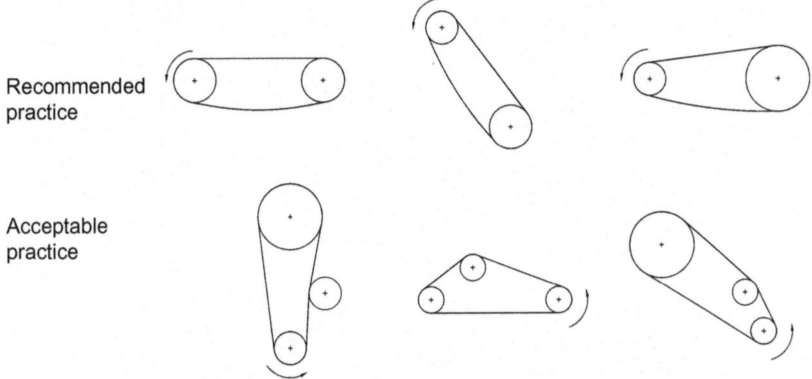

Fig. 12.30 Examples of typical chain drive layouts.

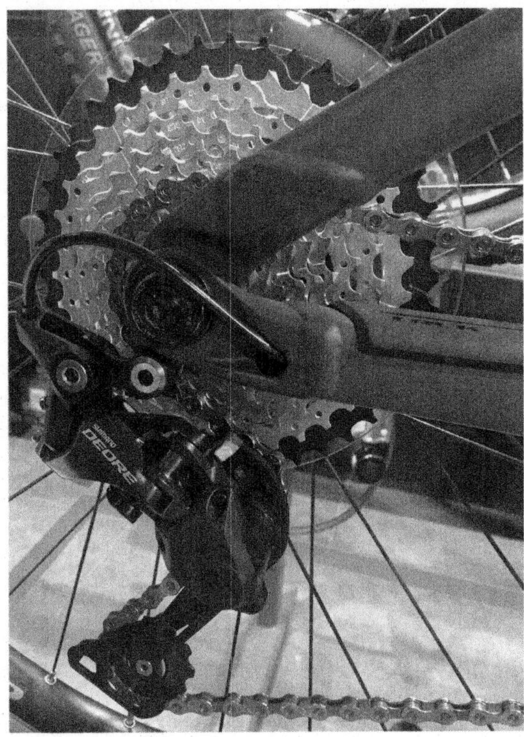

Fig. 12.31 Classic example of a chain tensioner.

Fig. 12.32 Chain tensioner for a cam shaft.
Image courtesy of CC-BY-SA-3.0.

When the chain rollers mesh with the sprocket teeth, an impact occurs and a Hertz contact stress occurs, similar to that found for gear meshing. The power rating charts for chain drives limit the selection of the drive so that these modes of failure should not occur assuming proper installation, operation and lubrication.

Details of the selection procedure for standard roller chains are outlined in Section 12.3.1. For details of chain geometry, the reader is referred to the standards listed in the references and for details of selection and design procedures for other chain types to manufacturers' catalogues.

12.3.1 Roller chain selection

Once the use of a chain drive has been shown to be preferable to other forms of drive, the type of chain to be used can be selected from the range available as illustrated in Fig. 12.22. The next step is the design of the chain drive layout and selection of the standard components available from chain manufacturers. The method outlined here is for roller chains. Procedures for the selection of other chain types can be found in manufacturers' catalogues.

The method is based upon the use of power rating charts for the chain drive, which ensure 15,000 h operation assuming proper installation, operation and lubrication. The steps for the method are itemised as follow and illustrated in Fig. 12.33.

1. Determine the power to be transmitted.
2. Determine the speeds of the driving and driven shafts.
3. Determine the characteristics of the driving and driven shaft, for example type of running, whether smooth or shock loadings, etc.
4. Set the approximate centre distance. This should normally be in the range of 30–50 times the chain pitch.

Start selection

Determine the power	Determine the power to be transmitted.
OK	
Determine the shafts speeds	Determine the speeds of the driving and driven shafts.
OK	
Determine the driving and driven shaft characteristics	Determine the characteristics of the driving and driven shaft, for example, whether smooth or shock loadings.
OK	
Set the approximate centre distance	This should normally be in the range of 30–50 times the chain pitch.
OK	
Select the speed ratio	This is dependent on the standard sizes available as listed in Table 12.22. Ideally sprockets should have a minimum of 19 teeth. For high speed drives subjected to transient loads the minimum number of teeth is 25. Note that the maximum number of teeth should not exceed 114.
OK	
Establish the application and tooth factors	Establish the application and tooth factors: Table 12.23 gives values for the application factor f_1. The tooth factor is given by $f_2=19/N_1$ assuming that the selection rating charts are based on a 19 tooth sprocket.
OK	
Calculate the selection power	Calculate the selection power. Selection power = Power $\times f_1 \times f_2$.
OK	

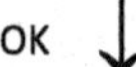

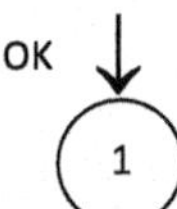

Fig. 12.33 See the legend on next page.

(Continued)

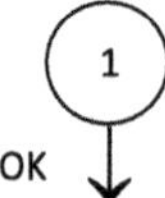

Select the chain drive pitch	Use power speed rating charts as supplied by chain manufacturers (e.g. Fig. 12.34). The smallest pitch of a simple chain should be used as this normally gives the most economical drive. If the power requirement at a given speed is beyond the capacity of a single strand of chain then the use of a multistrand chain, such as duplex (two strands), triplex (three strands) and up to decuplex (ten strands) for the ANSI range, permits higher power to be transmitted and can be considered.

Calculate the chain length	The chain length as a function of the number of pitches is given in Equation 12.13. Note that the value for the length should be rounded up to the nearest even integer. $L=\frac{N_1+N_2}{2}+\frac{2C}{p}+\left(\frac{N_2-N_1}{2\pi}\right)^2\frac{p}{C}$

Calculate the exact centre distance	The exact centre distance can be calculated using Eq. (12.14). $C=\frac{p}{8}\left[2L-N_2-N_1+\sqrt{(2L-N_2-N_1)^2-\frac{\pi}{3.88}(N_2-N_1)^2}\right]$

Specify the lubrication method	See Fig. 12.35 for some options for chain lubrication.

Complete	

Fig. 12.33, Cont'd Outline roller chain selection procedure.

5. Select the speed ratio. This is dependent on the standard sizes available as listed in Table 12.22. Ideally sprockets should have a minimum of 19 teeth. For high-speed drives subjected to transient loads, the minimum number of teeth rises to 25. Note that the maximum number of teeth should not exceed 114.
6. Establish the application and tooth factors: Table 12.23 gives values for the application factor f_1. The tooth factor is given by $f_2=19/N_1$ assuming that the selection rating charts are based on a 19-tooth sprocket.
7. Calculate the selection power. Selection power = Power $\times f_1 \times f_2$.
8. Select the chain drive pitch. Use power speed rating charts as supplied by chain manufacturers (see Fig. 12.34). The smallest pitch of a simple chain should be used as this normally gives the most economical drive. If the power requirement at a given speed is beyond the

Table 12.22 Chain reduction ratios as a function of the standard sprockets available

Number of teeth in the driven sprocket (N_2)	Number of teeth in the drive sprocket (N_1)					
	15	**17**	**19**	**21**	**23**	**25**
25	–	–	–	–	–	1.00
38	2.53	2.23	2.00	1.80	1.65	1.52
57	3.80	3.35	3.00	2.71	2.48	2.28
76	5.07	4.47	4.00	3.62	3.30	3.04
95	6.33	5.59	5.00	4.52	4.13	3.80
114	7.60	6.70	6.00	5.43	4.96	4.56

Reproduced from Renold Power Transmission, 2010. Transmision chain. Installation, maintenance and designer guide. http://www.renold.com/media/165418/Transmission-I-and-M-REN12-ENG-10-10.pdf (Accessed 10 June 2018).

Table 12.23 Application factor

Driven machine characteristics	Characteristics of driver		
	Smooth running, e.g. electric motors, IC engines with hydraulic coupling	**Slight shocks, e.g. IC engines with more than six cylinders, electric motors with frequent starts**	**Heavy shocks, e.g. IC engines with less than six cylinders**
Smooth running, e.g. fans, pumps, compressors, printing machines, uniformly loaded conveyors.	1	1.1	1.3
Moderate shocks, e.g. concrete mixing machines, nonuniformly loaded conveyors, mixers	1.4	1.5	1.7
Heavy shocks, e.g. planars, presses, drilling rigs.	1.8	1.9	2.1

Reproduced from Renold Power Transmission, 2010. Transmision chain. Installation, maintenance and designer guide. http://www.renold.com/media/165418/Transmission-I-and-M-REN12-ENG-10-10.pdf (Accessed 10 June 2018).

capacity of a single strand of chain, then the use of a multistrand chain, such as duplex (2 strands), triplex (3 strands) and up to decuplex (10 strands) for the ANSI range, permits higher power to be transmitted and can be considered.

9. Calculate the chain length. Eq. (12.13) gives the chain length as a function of the number of pitches. Note that the value for the length should be rounded up to the nearest even integer.
10. Calculate the exact centre distance. This can be calculated using Eq. (12.14).
11. Specify the lubrication method.

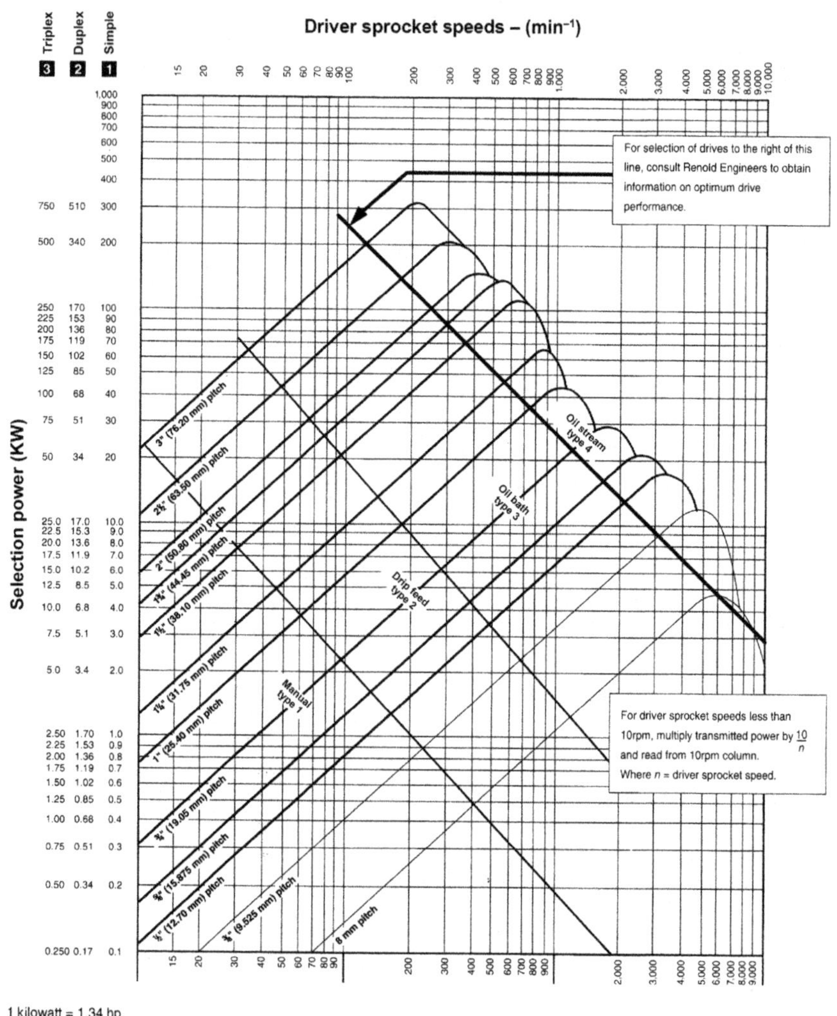

Fig. 12.34 British Standard chain drives rating chart using 19-tooth drive sprocket. Courtesy of Renold Chain.

The chain length, in pitches, is given by

$$L = \frac{N_1 + N_2}{2} + \frac{2C}{p} + \left(\frac{N_2 - N_1}{2\pi}\right)^2 \frac{p}{C} \tag{12.13}$$

where L = number of pitches, N_1 = number of teeth in the driving sprocket, N_2 = number of teeth in the driven sprocket, C = centre distance (m) and p = chain pitch (m).

The exact centre distance is given by

$$C=\frac{p}{8}\left[2L-N_2-N_1+\sqrt{(2L-N_2-N_1)^2-\frac{\pi}{3.88}(N_2-N_1)^2}\right] \tag{12.14}$$

Chain drives should be protected against dirt and moisture (tell this to a mountain biker!). Lubrication should be provided using a nondetergent mineral-based oil. For the majority of applications, multigrade SAE20/50 is suitable. There are five principle types of lubrication: manual application, drip feed, bath, stream (see Fig. 12.35) and dry lubrication. Grease lubrication is not recommended but can be used for chain

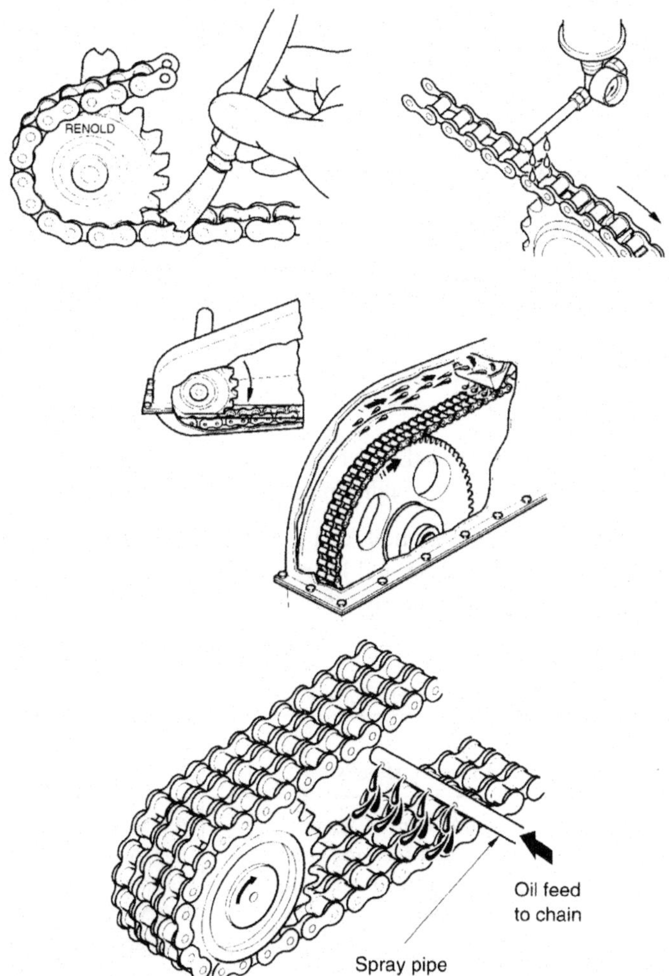

Fig. 12.35 Chain lubrication methods.
Courtesy of Renold Chain.

speeds of <4 m/s. In order to ensure the grease penetrates the working parts of the chain, it should be heated until liquid and the chain dipped into the grease until the air has been displaced. This process should be repeated at regular service intervals. For dry lubrication, solid lubricant is contained in a volatile carrier fluid. When applied to chain, the carrier transports the lubricant into the chain and then evaporates leaving the chain lubricated but dry to touch. Applications for dry lubrication include food-processing, dusty environments and fabric handling.

The pitch diameters for the driving and driven sprockets are given by

$$D_1 = \frac{N_1 p}{\pi} \quad D_2 = \frac{N_2 p}{\pi} \tag{12.15}$$

and the angle of contact (in radians) between the chain and the sprockets by

$$\theta_1 = \pi - 2\sin^{-1}\frac{p(N_2 - N_1)}{2\pi C} \tag{12.16a}$$

$$\theta_2 = \pi + 2\sin^{-1}\frac{p(N_2 - N_1)}{2\pi C} \tag{12.16b}$$

Note that the minimum angle of wrap recommended for the small sprocket is 120°.

The chain tension is given by

$$\text{Chain tension} = \frac{\text{Power}}{N_1 \omega_1 p / 2\pi} \tag{12.17}$$

Standard sprockets can be purchased. The choice of sprocket material depends on the number of teeth and the operating conditions as shown in Table 12.24.

Example 12.6. A chain drive is required for a gear pump operating at 400 rpm driven by a 5.5 kW electric motor running at 1440 rpm. The centre distance between the motor and pump shafts is approximately 470 mm.

Table 12.24 Selection of sprocket materials.

Sprocket	Smooth running	Moderate shocks	Heavy shocks
Up to 29 teeth	080M40 or 070M55	080M40 or 070M55 hardened and tempered or case hardened mild steel	080M40 or 070M55 hardened and tempered or case hardened mild steel
Over 30 teeth	Cast iron	Mild steel	080M40 or 070M55 hardened and tempered or case hardened mild steel

Reproduced from Renold Power Transmission, 2010. Transmision chain. Installation, maintenance and designer guide. http://www.renold.com/media/165418/Transmission-I-and-M-REN12-ENG-10-10.pdf (Accessed 10 June 2018).

Solution

The desired reduction ratio is $\frac{1440}{400}=3.6$.

The nearest ratio available (see Table 12.22) using standard sized sprockets is 3.62. This requires a driving sprocket of 21 teeth and a driven sprocket of 76 teeth.

The application factor from Table 12.23 is $f_1=1.0$.

The tooth factor $f_2=\frac{19}{N_1}=\frac{19}{21}=0.905$.

The selection power $=5.5\times1.0\times0.905=4.98\,\text{kW}$.

Using the BS/ISO selection chart, Fig. 12.34, a 12.7-mm pitch simple BS chain drive is suitable.

The chain length is given by

$$L=\frac{N_1+N_2}{2}+\frac{2C}{p}+\left(\frac{N_2-N_1}{2\pi}\right)^2\frac{p}{C}$$

$$=\frac{21+76}{2}+\frac{2\times470}{12.7}+\left(\frac{76-21}{2\pi}\right)^2\frac{12.7}{470}=124.6\text{ pitches}$$

Rounding up to the nearest even integer gives $L=126$ pitches.

The exact centre distance can now be calculated using

$$C=\frac{p}{8}\left[2L-N_2-N_1+\sqrt{(2L-N_2-N_1)^2-\frac{\pi}{3.88}(N_2-N_1)^2}\right]$$

$$C=\frac{12.7}{8}\left[2\times126-76-21+\sqrt{(2\times126-76-21)^2-\frac{\pi}{3.88}(76-21)^2}\right]$$
$$=479.2\text{ mm}$$

From the rating chart, Fig. 12.34, the required lubrication type is oil bath. Use of a SAE20/50 multigrade lubricant would likely suffice in the absence of more detailed knowledge concerning the operating conditions.

Example 12.7. Specify a suitable drive for a gear pump operating at 400 rpm driven by a 30 kW electric motor running at 728 rpm. The centre distance between the motor and pump shafts is approximately 1 m.

Solution

The desired reduction ratio is $\frac{728}{400}=1.82$.

From Table 12.22, the nearest ratio using standard sized sprockets is 1.8. $N_1=21$, $N_2=38$.

From Table 12.23, $f_1=1.0$.

The tooth factor $f_2=\frac{19}{N_1}=\frac{19}{21}=0.905$.

The selection power $=30\times1.0\times0.905=27.15\,\text{kW}$.

Using the BS/ISO selection chart, Fig. 12.34, a 25.4-mm pitch simple BS chain drive is suitable with oil bath lubrication

$$L = \frac{21+38}{2} + \frac{2 \times 1000}{25.4} + \left(\frac{38-21}{2\pi}\right)^2 \frac{25.4}{1000} = 108.4 \text{ pitches}$$

Rounding up to the nearest even integer gives $L = 110$ pitches.

The exact centre distance is

$$C = \frac{25.4}{8}\left[2 \times 110 - 38 - 21 + \sqrt{(2 \times 110 - 38 - 21)^2 - \frac{\pi}{3.88}(38-21)^2}\right]$$
$$= 1020 \text{ mm}$$

Example 12.8. Specify a suitable chain drive for a packaging machine operating at 75 rpm driven by a 2.2-kW electric motor running at 710 rpm. The maximum permissible centre distance between the motor and packaging machine shafts is approximately 1 m.

Solution

There is no unique solution to this question. What follows is one possible solution.

The speed ratio $710/75 = 9.47$ is too high for a single reduction and is outside the range listed in Table 12.22. A two stage reduction might be feasible.

Examination of Table 12.22 for a combination of reductions shows that choosing ratios of 3.8 and 2.48 might be suitable.

As $3.8 \times 2.48 = 9.424$ is close to the target reduction, this combination is likely to be acceptable and would give an output speed of 75.3 rpm.

Using the larger reduction on the higher-speed drive, so for a ratio of 3.8 from Table 12.22, $N_1 = 25$, $N_2 = 95$.

The application factor from Table 12.23 assuming moderate shocks is 1.4.

The tooth factor $f_2 = \frac{19}{N_1} = \frac{19}{25} = 0.76$.

The selection power $= 2.2 \times 1.4 \times 0.76 = 2.34$ kW.

Using the BS/ISO selection chart, Fig. 12.34, the following drives would be suitable:

12.7-mm pitch in oil bath (simple),

9.525 mm drip feed (duplex).

Selecting the simple chain, $p = 12.7$ mm.

Take the distance between the sprocket centres as the minimum recommended, that is $30 \times p = 30 \times 12.7 = 381$ mm to ensure the design fits in the space available. This constraint can be relaxed at a later stage if appropriate.

The chain length is given by

$$L = \frac{25+95}{2} + \frac{2 \times 381}{12.7} + \left(\frac{95-25}{2\pi}\right)^2 \frac{12.7}{381} = 124.1 \text{ pitches.}$$

Rounding up to the nearest even integer gives $L = 126$ pitches.

The exact centre distance is given by

$$C=\frac{12.7}{8}\left[2\times 126-95-25+\sqrt{(132)^2-\frac{\pi}{3.88}(70)^2}\right]=393.7\ \text{mm}$$

For the second chain with a ratio of 2.48, $N_1=23$, $N_2=57$.

The application factor from Table 12.23 is 1.4.

The tooth factor $f_2=\frac{19}{N_1}=\frac{19}{23}=0.8261$.

The selection power $=2.2\times 1.4\times 0.8261=2.544\,\text{kW}$.

Using the BS/ISO selection chart, Fig. 12.34, the following drives would be suitable:

12.7-mm pitch in oil bath (simple),

9.525-mm drip feed (duplex).

Selecting the simple chain, $p=12.7\,\text{mm}$.

Again the distance between the sprocket centres can be taken as the minimum recommended, that is $30\times p=30\times 12.7=381\,\text{mm}$.

The chain length is given by

$$L=\frac{23+57}{2}+\frac{2\times 381}{12.7}+\left(\frac{57-23}{2\pi}\right)^2\frac{12.7}{381}=101.0\ \text{pitches}$$

Rounding up to the nearest even integer gives $L=102$ pitches.

The exact centre distance is given by

$$C=\frac{12.7}{8}\left[2\times 102-57-23+\sqrt{(124)^2-\frac{\pi}{3.88}(34)^2}\right]=387.6\ \text{mm}$$

The combination of the two centre distances fits in the space available.

An alternative might be to arrange the chain sprockets vertically or at an angle to the horizontal in order to accommodate the centre distance restriction.

12.4 Conclusions

Belt and chain drives can be used for transmission of mechanical power between two rotating shafts. Belt drives are often cheaper than the equivalent gears and useful for transmitting power between shafts that are widely separated or nonparallel drives. Chain drives are usually more compact than the equivalent belt drive and can be used in oily environments where the equivalent belt would be prone to slipping. There is a wide range of belt and chain drives and this chapter has served to introduce the technology and the selection and specification of wedge and flat belt, and roller chain drives. The technology is constantly developing with new materials and surface treatments, improvements in understanding of kinematics and wear, and associated modelling procedures, and belts and chain drives represent an innovation opportunity area, particularly for new applications, extended life and improved reliability, as well as miniaturisation.

References

American Chain Association, 2005. Standard Handbook of Chains: Chains for Power Transmission and Material Handling, second ed. CRC Press, Boca Raton.

Cali, M., Sequenzia, G., Oliveri, S.M., Fatuzzo, G., 2016. Meshing angles evaluation of silent chain drive by numerical analysis and experimental test. Meccanica 51, 475–489.

Dalgarno, K.W., Day, A.J., Childs, T.H.C., 1994. Synchronous belt materials and belt life correlation. Proc. Inst. Mech. Eng. D 208, 37–48.

Firbank, T.C., 1970. Mechanics of the belt drive. Int. J. Mech. Sci. 12, 1053–1063.

Iizuka, H., Watanabe, K., Mashimo, S., 1994. Observations of fatigue failure in synchronous belts. Fatigue Fract. Eng. Mater. Struct. 17, 783–790.

Jagtap, M.D., Gaikwad, B.D., Pawar, P.M., 2014. Study of roller conveyor chain strip under tensile loading. IJMER. 4(5)2249-6645.

Meng, F., Li, C., Cheng, Y., 2007. Proper conditions of meshing for Hy-Vo silent chain and sprocket. Chin. J. Mech. Eng. 20, 57–59.

Perneder, R., Osbourne, I., 2012. Handbook Timing Belts: Principles, Calculations, Applications. Springer, Berlin.

Ulsoy, A.G., Whitesell, J.E., Hooven, M.D., 1985. Design of belt-tensioner systems for dynamic stability. J. Vib. Acoust. Stress. Reliab. Des. 107, 282–290.

Standards

BS 2969: 1980. Specification for high-tensile steel chains (round link) for chain conveyors and coal ploughs.

BS 2969:1980, ISO 610-1979. Specification for high-tensile steel chains (round link) for chain conveyors and coal ploughs.

BS 3790:2006. Specification for belt drives. Endless wedge belts, endless V-belts, banded wedge belts, banded V-belts and their corresponding pulleys.

BS 4116-4:1992. Conveyor chains, their attachments and associated chain wheels. Specification for chains and attachments (British series).

BS 5801:1979, ISO 4348-1978. Specification for flat-top chains and associated chain wheels for conveyors.

BS ISO 10190:2008. Motor cycle chains. Characteristics and test methods.

BS ISO 10823:1996. Guidance on the selection of roller chain drives.

BS ISO 1275:2006. Double-pitch precision roller chains, attachments and associated chain sprockets for transmission and conveyors.

BS ISO 155:1998. Belt drives. Pulleys. Limiting values for adjustment of centres.

BS ISO 1977:2000. Conveyor chains, attachments and sprockets.

BS ISO 2790:2004. Belt drives. V-belts for the automotive industry and corresponding pulleys. Dimensions.

BS ISO 487:1998. Steel roller chains, types S and C, attachments and sprockets.

BS ISO 5287:2003. Belt drives. Narrow V-belts for the automotive industry. Fatigue test.

BS ISO 5294:2012. Synchronous belt drives. Pulleys.

BS ISO 5296:2012. Synchronous belt drives. Belts with pitch codes MXL, XXL, XL, L, H, XH and XXH. Metric and inch dimensions.
BS ISO 606:2015. Short-pitch transmission precision roller and bush chains, attachments and associated chain sprockets.
BS ISO 9633:2001. Cycle chains. Characteristics and test methods.

Websites

At the time of going to press, the world-wide-web contained useful information relating to this chapter at the following sites.
http://www.beijingthc.com.
http://www.beltingindustries.com.
http://www.chiorino.co.uk.
http://www.chjc-chain.com.
http://www.daidocorp.com.
http://www.fennerdrives.com.
http://www.fptgroup.com/fenner.asp.
http://www.gates.com.
http://www.goodyearep.com/.
http://www.habasitiakoka.com.
http://www.hmcross.com.
http://www.hutchinsontransmission.com.
http://www.jonsonrubber.com.
http://www.kettenwulf.com/de/.
http://www.kilangrantai.com.
http://www.kkrubber.com.
http://www.martinsprocket.com.
http://www.mpta.org/.
http://www.ngfeurope.com.
http://www.omfarubbers.com.
http://www.powertransmission.com.
http://www.renold.com.
http://www.samplabelting-sa.com.
http://www.siegling.com.
http://www.sprocketco.com.
http://www.tsubakimoto.com.
http://www.visusa.com.
http://www.vmchain.com.
http://www.websterchain.com.

Further reading

Fenner Power Transmission UK. (2011) Belt Drives Design Manual.
Hamilton, P. Belt drives. In Hurst, K. (Ed.), Rotary Power Transmission Design. McGraw Hill, Maidenhead, 1994.
Renold Power Transmission, 2010. Transmision chain. Installation, maintenance and designer guide. http://www.renold.com/media/165418/Transmission-I-and-M-REN12-ENG-10-10.pdf (Accessed 10 June 2018).

Clutches and brakes

13

Chapter Outline

Nomenclature

Generally, preferred SI units have been stated.

a	distance (m)
b	distance (m)
c	distance (m)
A	area (m^2)
F	force (N)
F_f	frictional force (N)
F_n	normal force (N)
F_1	tension in the tight side of the band (N)
F_2	tension in the slack side of the band (N)
h	heat transfer coefficient (W/m^2 K)
I	mass moment of inertia (kg m^2)
L	length or thickness (m)
m	mass (kg)
M	moment (N m)
M_f	frictional moment (N m)
M_f'	moment due to the frictional forces for a self-deenergising brake (N m)
M_n	normal moment (N m)
M_n'	moment of the normal forces for a self-deenergising brake (N m)
N	number of disc faces

Mechanical Design Engineering Handbook. https://doi.org/10.1016/B978-0-08-102367-9.00013-5

p pressure (N/m^2)
p_{av} average pressure (N/m^2)
p_{max} maximum permissible pressure (N/m^2)
p'_{max} maximum pressure on the self-deenergising brake (N/m^2)
Q heat transfer rate (W)
r radius (m)
r_e effective radius (m)
r_i inner radius (m)
r_o outer radius (m)
R pad radius, reaction (N)
T torque (N m), temperature (°C)
T_f temperature of the surrounding fluid (K or °C)
T_s temperature of the surface (K or °C)
V_i initial velocity (m/s)
V_f final velocity (m/s)
w width of the brake shoe (m)
α acceleration (m/s^2)
δ ratio of effective radius to radius
δF elemental axial force (N)
δr elemental radial distance (m)
δT elemental torque (N m)
μ coefficient of friction
θ angle (rad)
ρ density (kg/m^3)
ω angular velocity (rad/s)
ΔT temperature difference (°C)

Abbreviations

ABS antilock braking system
ECU engine control unit
IC internal combustion

13.1 Introduction

A clutch is a device that permits the smooth, gradual connection of two components moving relative to each other, such as shafts rotating at different speeds. A brake enables the controlled dissipation of energy to slow down, stop or control the speed of a system. This chapter describes the basic principles of frictional clutches and brakes and outlines design and selection procedures for disc clutches, disc and drum brakes.

When a rotating machine is started it must be accelerated from rest to the desired speed. A clutch is a device used to connect or disconnect a driven component from a prime mover such as an engine or motor. A familiar application is the use of a clutch between an engine's crankshaft and the gearbox in automotive applications. The need for the clutch arises from the relatively high torque requirement to get a vehicle moving and the low torque output from an internal combustion engine at low levels of rotational speed. The disconnection of the engine from the drive enables an engine to speed up unloaded to about 1000 rpm where it is generating sufficient torque to drive the transmission. The clutch can then be engaged, allowing power to be transmitted to the gearbox, transmission shafts and wheels. A brake is a device used to reduce or control the speed of a system or bring it to rest. Typical applications of a clutch and brake are illustrated in Figs. 13.1 and 13.2. Clutches and brakes are similar devices providing frictional, magnetic or mechanical connection between two components. If one component rotates and the other is fixed to a nonrotating plane of reference the device will function as a brake and if both rotate then as a clutch. Clutches and brakes can also be used for nonrotating applications where, for example, a linearly moving plate needs to be slowed down.

Whenever the speed or direction of motion of a body is changed there is force exerted on the body. If the body is rotating, a torque must be applied to the system to speed it up or slow it down. If the speed changes, so does the energy, either by addition or absorption.

The acceleration, α, of a rotating machine is given by

$$\alpha = \frac{T}{I} \tag{13.1}$$

where T is the torque (N m) and I is the mass moment of inertia (kg m^2).

The mass moment of inertia can often be approximated by considering an assembly to be made up of a series of cylinders and discs and summing the individual values for

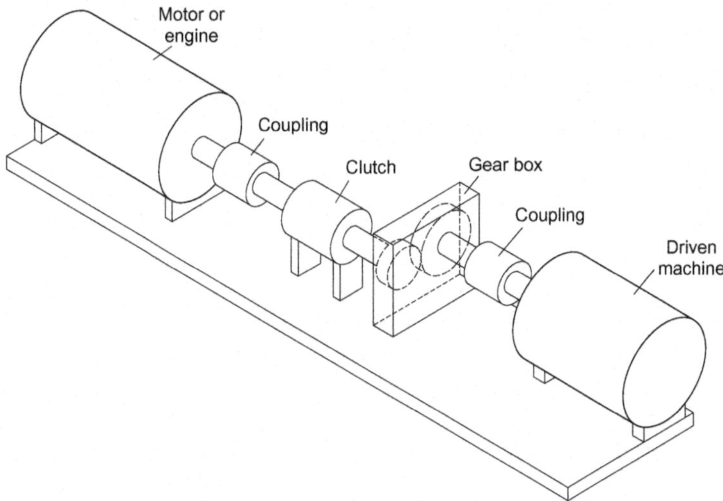

Fig. 13.1 Typical application of a clutch.

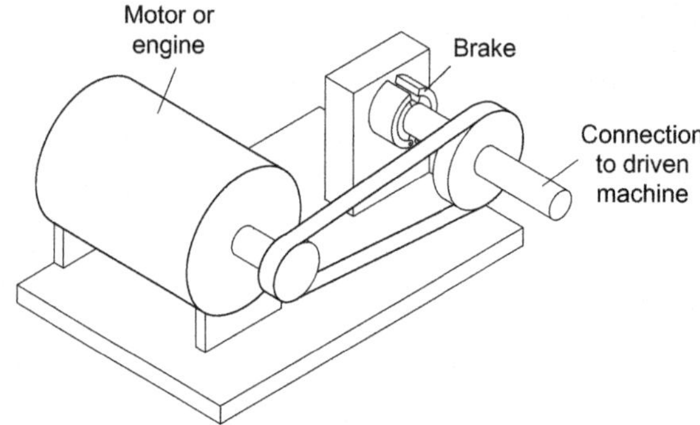

Fig. 13.2 Typical application of a brake.

the disc and cylinder mass moments of inertia. The mass moments of inertia for a cylinder and a disc are given by Eqs (13.2) and (13.3) respectively.

$$I_{\text{cylinder}} = \frac{1}{2}\rho\pi L\left(r_o^4 - r_i^4\right) \tag{13.2}$$

$$I_{\text{disc}} = \frac{1}{2}\rho\pi L r_o^4 \tag{13.3}$$

The desired level of acceleration will depend on the application and, once the acceleration has been set and I determined, the approximate value of torque required to accelerate or brake the load can be estimated. This can be used as the principal starting point for the design or selection of the clutch or brake geometry.

Torque is equal to the ratio of power and angular velocity. In other words torque is inversely proportional to angular velocity. This implies that it is usually advisable to locate the clutch or brake on the highest speed shaft in the system so that the required torque is a minimum. Size, cost and response time are lower when the torque is lower. The disadvantage is that the speed differential between the components can result in increased slipping and associated frictional heating potentially causing overheating problems.

Friction type clutches and brakes are the most common. Two or more surfaces are pushed together with a normal force to generate a friction torque (Fig. 13.3). Normally, at least one of the surfaces is metal and the other a high friction material referred to as the lining. The frictional contact can occur radially, as for a cylindrical arrangement, or axially as in a disc configuration (Fig. 13.4). The function of a frictional clutch or brake surface material is to develop a substantial friction force when a normal force is applied. Ideally a material with a high coefficient of friction, constant properties, good resistance to wear and chemical compatibility is required. Clutches and brakes transfer or dissipate significant quantities of energy and their design must enable the absorption and dissipation of this heat without damage to the component parts of the surroundings. As an example, the braking of bullet trains involves the dissipation of considerable quantities of energy and energised coils can be used to induce

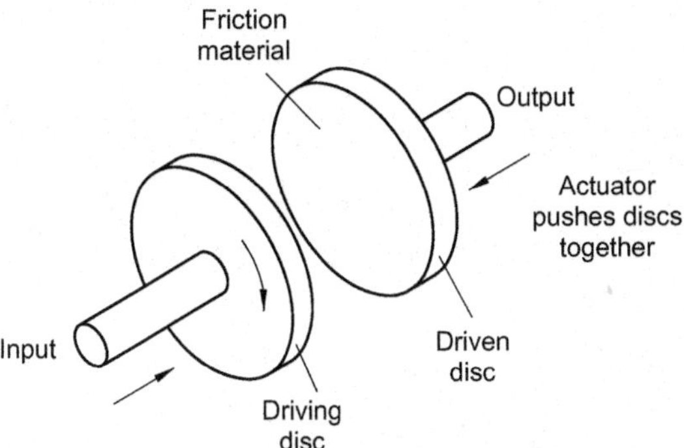

Fig. 13.3 Idealised friction disc clutch or brake.

Fig. 13.4 Heavy duty axle and associated brake system.
Figure courtesy of Astra Rail Industries, Greenbrier Europe.

eddy currents in a rotor disc to generate sufficiently strong electromagnetic fields to dissipate the train's kinetic energy into heat (Fig. 13.5).

With the exception of high volume automotive clutches and brakes, engineers rarely need to design a clutch or a brake from scratch. Clutch and brake assemblies can be purchased from specialist suppliers, and the engineer's task is to specify the torque and speed requirements, the loading characteristics and the system inertias and to select an appropriately sized clutch or brake and the lining materials.

13.2 Clutches

The function of a clutch is to permit the connection and disconnection of two shafts, either when both are stationary or when there is a difference in the relative rotational speeds of the shafts. Clutch connection can be achieved by a number of techniques from direct mechanical friction, electromagnetic coupling, hydraulic or pneumatic means or by some combination. There are various types of clutches as outlined in

Fig. 13.5 Axial flux eddy current disc brakes for a high speed electric train. Figure courtesy of Image Courtesy of Toshinori Baba (2014), CC ASA 4.0.

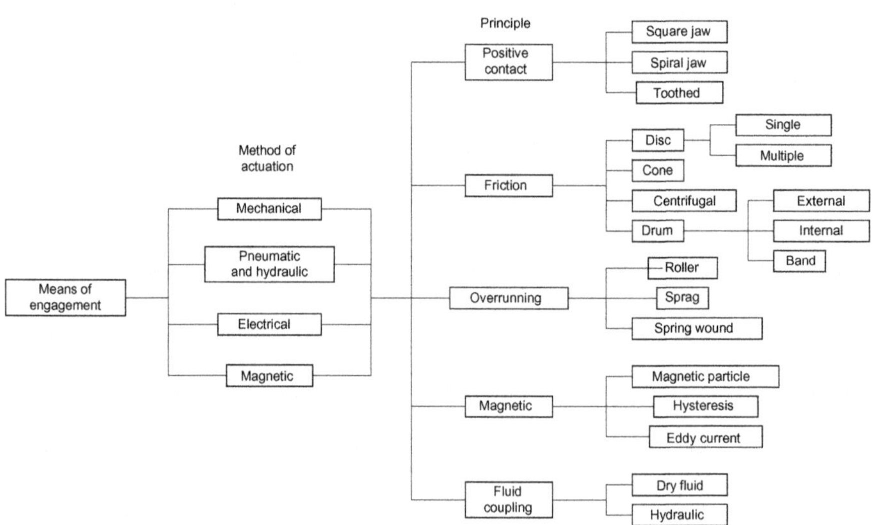

Fig. 13.6 Clutch classification.

Fig. 13.6, developed following a classification originally developed by Hindhede et al. (1983). The devices considered here are of the friction type.

Clutches must be designed to satisfy four main requirements:

1. the necessary actuation force should not be excessive,
2. the coefficient of friction should be constant,
3. the energy converted to heat must be dissipated,
4. wear must be limited to provide reasonable clutch life.

Alternatively the objective in clutch design can be stated as maximisation of a maintainable friction coefficient and minimisation of wear. Correct clutch design and selection is critical because a clutch that is too small for an application will slip and overheat and a clutch that is too large will have a high inertia and may overload the drive.

Example 13.1. Calculate the torque a clutch must transmit to accelerate a pulley with a moment of inertia of 0.25 kg m^2 to

(a) 500 rpm in 2.5 s.
(b) 1000 rpm in 2 s.

Solution

The torque is given by $T = I\alpha$.
The angular acceleration is $\alpha = \Delta n/\mathrm{t}$.
The mass moments of inertia $I = 0.25\,\mathrm{kg\,m^2}$.

$$T = 0.25\left(\frac{500(2\pi/60)}{2.5}\right) = 5.236\ \mathrm{Nm}$$

$$T = 0.25\left(\frac{1000(2\pi/60)}{2}\right) = 13.10\ \mathrm{Nm}$$

Positive contact clutches have teeth or serrations, which provide mechanical interference between mating components. Fig. 13.7 shows a square jaw and Fig. 13.8 shows a multiple serration positive contact clutch. This principle can be combined with frictional surfaces as in an automotive synchromesh clutch. As helical gears cannot be shifted in and out of mesh easily, the pragmatic approach is to keep the gears engaged in mesh and allow the gear to rotate freely on the shaft when no power is required but provide positive location of the gear on the shaft when necessary (Section 8.3.1).

Friction clutches consist of two surfaces, or two sets of surfaces, which can be forced into frictional contact. A frictional clutch allows gradual engagement between two shafts. The frictional surfaces can be forced together by springs, hydraulic pistons or magnetically. Various forms exist such as disc, cone and radial clutches. The design of the primary geometry for disc clutches is described in Section 13.2.1.

Over-running clutches operate automatically based on the relative velocity of the mating components. They allow relative motion in one direction only. If the rotation attempts to reverse the constituent components of the clutch grab the shaft and lock up. Applications include backstops, indexing and freewheeling, as on a bicycle when the wheel speed is greater than the drive sprocket. The range of overrunning clutches the simple ratchet and pawl (Fig. 13.9) roller, sprag and spring wound clutches.

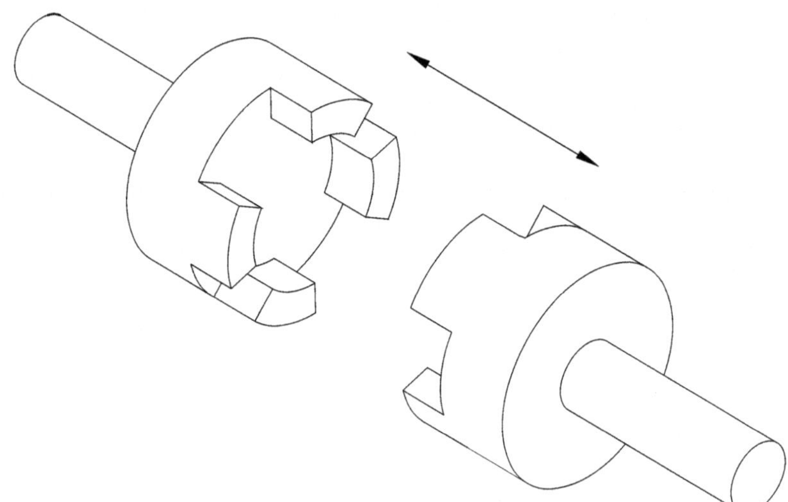

Fig. 13.7 Square jaw clutch.

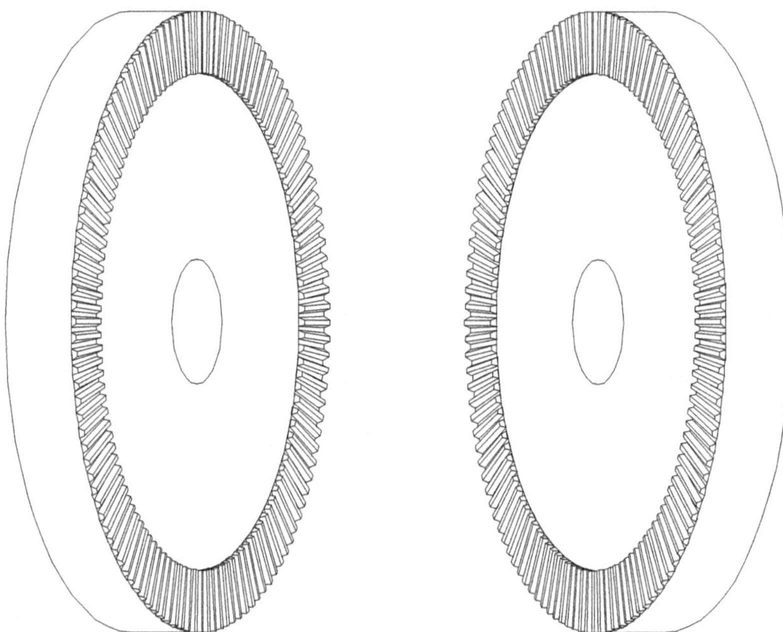

Fig. 13.8 Multiple Serration clutch.

Roller clutches consists of balls or cylindrical rollers located in wedge shaped chambers between inner and outer races (Fig. 13.10). Rotation is possible in one direction but when this is reversed the rollers wedge themselves between the races jamming the inner and outer races together. Fig. 13.11 illustrates the principal components of a

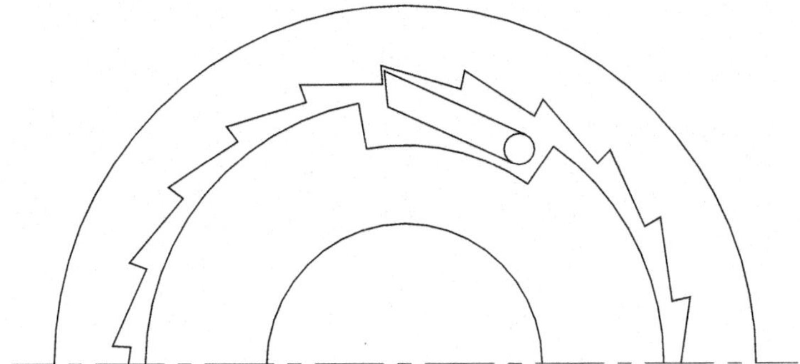

Fig. 13.9 Ratchet and pawl.
From Neale, M.J. (Ed.), 1995. The Tribology Handbook. Butterworth Heinemann.

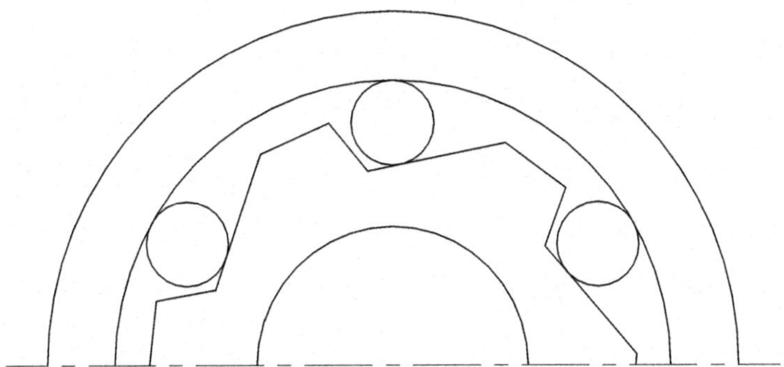

Fig. 13.10 Roller clutch.
From Neale, M.J. (Ed.), 1995. The Tribology Handbook. Butterworth Heinemann.

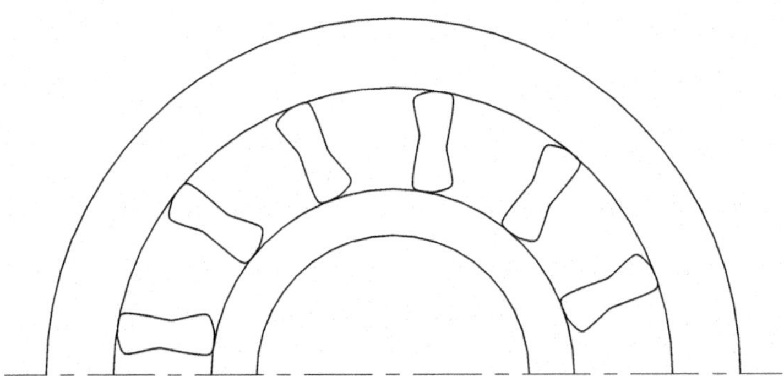

Fig. 13.11 Sprag clutch.
From Neale, M.J. (Ed.), 1995. The Tribology Handbook. Butterworth Heinemann.

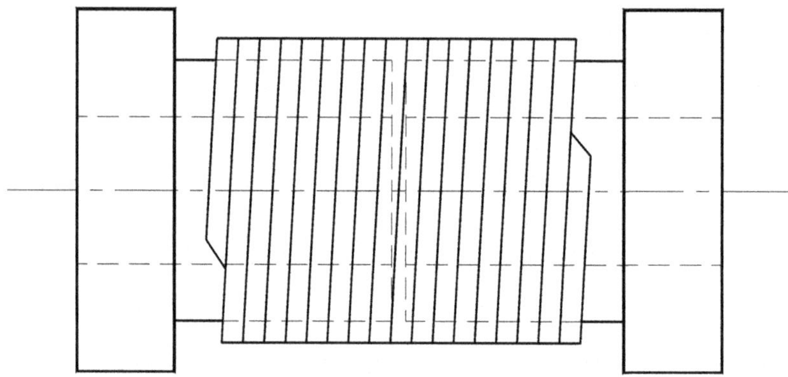

Fig. 13.12 Spring wound clutch.
From Neale, M.J. (Ed.), 1995. The Tribology Handbook. Butterworth Heinemann.

sprag clutch which consists of an inner and outer race, like a bearing, but instead of balls the space between the races contains specially shaped elements call 'sprags' which permit rotation in one direction only. If the clutch motion is reversed, the sprags jam and lock the clutch. A spring wound clutch is shown in Fig. 13.12. Friction between hubs and the coil spring causes the spring to tighten onto hubs and drive in one direction. In the opposite direction the spring unwraps and slips easily on the hubs.

Magnetic clutches use a magnetic field to couple the rotating components together. Magnetic clutches are smooth, quiet and have a long life, as there is no direct mechanical contact and hence no wear except at the bearings.

Fluid couplings transmit torque through a fluid such as oil. A fluid coupling always transmits some torque because of the swirling nature of the fluid contained within the device. For this reason it is necessary to maintain some braking force in automatic transmission automobiles that utilise fluid coupling torque converters in the automatic gearbox.

Centrifugal clutches engage automatically when the shaft speed exceeds some critical value. Friction elements are forced radially outwards and engage against the inner radius of a mating cylindrical drum (Fig. 13.13). Common applications of centrifugal clutches include chainsaws (Figs 13.14 and 13.15) overload-releases and go-karts.

The selection of clutch type and configuration depends on the application. Table 13.1 can be used as a first step in determining the type of clutch to be used. Clutches are rarely designed from scratch. Either an existing design is available and is being modified for a new application or a clutch can be bought in from a specialist manufacturer. In the latter case the type, size and the materials for the clutch lining must be specified. This requires determination of the system characteristics such as speed, torque, loading characteristic (e.g. shock loads) and operating temperatures (Abdullah and Schlattmann, 2016). Many of these factors have been lumped into a multiplier called a service factor. A lining material is typically tested under

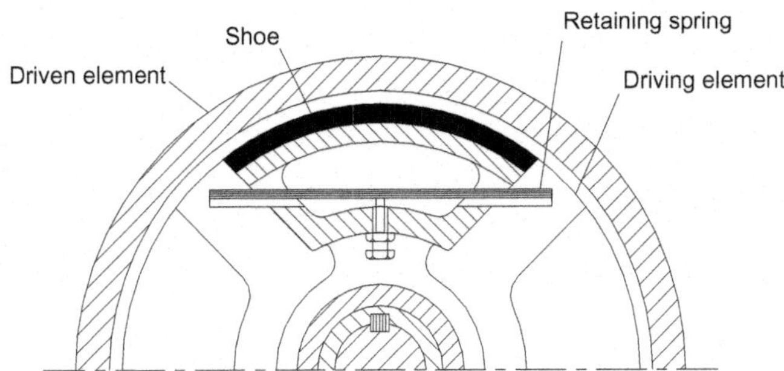

Fig. 13.13 Centrifugal clutch.
Reproduced from Neale, M.J. (Ed.), 1995. The Tribology Handbook. Butterworth Heinemann.

Fig. 13.14 Chain saw centrifugal clutch.

steady conditions using an electric motor drive. The torque capacity obtained from this test is then derated by the service factor according to the particular application to take account of vibrations and loading conditions. Table 13.2 gives an indication of the typical values for service factors.

13.2.1 Design of disc clutches

Disc clutches can consist of single or multiple discs as illustrated in Figs 13.16–13.19, respectively. Generally multiple disc clutches enable greater torque capacity but are harder to cool. Frictional clutches can be run dry or wet using oil. Typical coefficients of friction are 0.07 for a wet clutch and 0.45 for a dry clutch. While running a clutch

Fig. 13.15 Chain saw centrifugal clutch.

Table 13.1 Clutch selection criteria

Type of clutch	Characteristics	Typical applications
Sprag	One way clutch. Profiled elements jam against the outer edge to provide drive. High torque capacity	One way operation (e.g. backstop for hoists)
Roller	One way clutch. Rollers ride up ramps and drive by wedging into place	One way operation
Cone clutch	Embodies the mechanical principle of the wedge which reduces the axial force required to transmit a given torque	Contractor's plant. Feed drives for machine tools
Single disc clutch	Used when diameter is not restricted. Simple construction	Automobile drives
Multiple disc clutch	The power transmitted can be increased by using more plates allowing a reduction in diameter	Machine tool head stocks. Motorcycles
Centrifugal clutch	Automatic engagement at a critical speed	Electric motor drives. Industrial diesel drives
Magnetic	Compact. Low wear	Machine tool gearboxes. Numerical control machine tools

Table 13.2 Service factors (Vedamuttu, 1994).

		Type of driver			
Description of general system	Typical driven system	Small electric motors, turbine	IC engines (4 to 6 cylinders). Medium to large electric motors	IC engines (2 or 3 cylinders)	Single cylinder engine
Steady power source, steady load, no shock or overload	Belt drive, small generators, centrifugal pumps, fans, machine tools	1.5	1.7	1.9	2.2
Steady power source with some irregularity of load up to 1.5 times nominal power	Light machinery for wood, metal and textiles, conveyor belts	1.8	2.0	2.4	2.7
	Larger conveyor belts, larger machines, reciprocating pumps	2.0	2.2	2.4	2.7
Frequent start-stops, overloads, cycling, high inertia starts, high power, pulsating power source	Presses, Punches, piston pumps, Cranes, hoists	2.5	2.7	2.9	3.2
	Stone crushers, roll mills, heavy mixers, single cylinder compressors	3.0	3.2	3.4	3.7

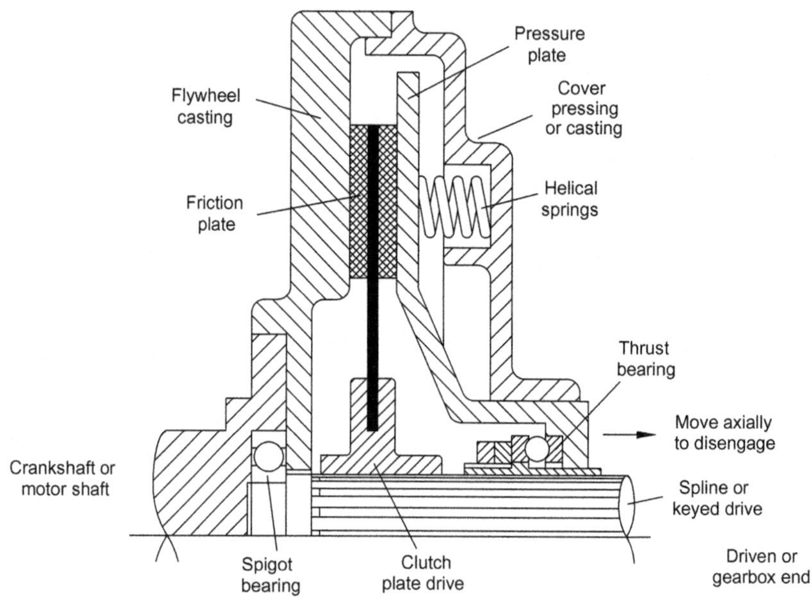

Fig. 13.16 Disc clutch.
Reproduced with adaptations from Neale, M.J., 1994. Drives and Seals a Tribology Handbook. Butterworth Heinemann.

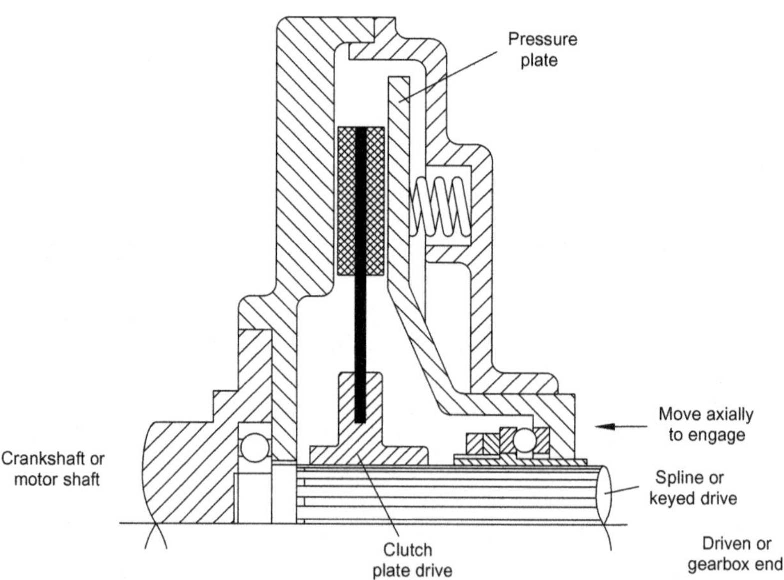

Fig. 13.17 Disc clutch shown disengaged. Little or no torque is transferred from the crankshaft to the driven load.
Reproduced with adaptations from Neale, M.J., 1994. Drives and Seals a Tribology Handbook. Butterworth Heinemann.

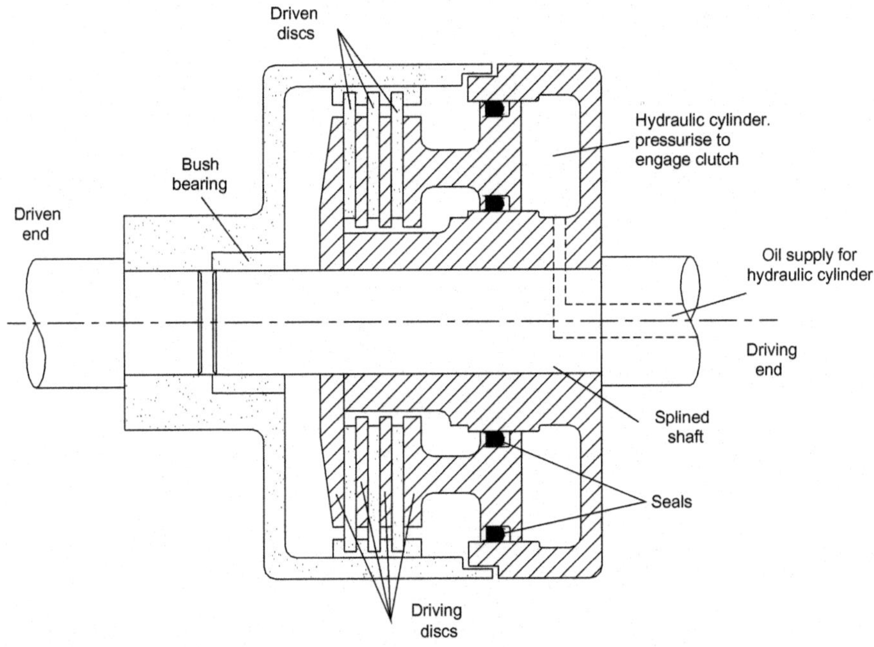

Fig. 13.18 Multiple disc clutch.
Based on Juvinall, R.C., Marshek, K.M., 1991. Fundamentals of Machine Component Design. Wiley.

wet in oil reduces the coefficient of friction it enhances heat transfer and the potential for cooling of the components. The expedient solution to the reduction of the friction coefficient is to use more discs and hence the use of multiple disc clutches.

Two basic assumptions are used in the development of procedures for disc clutch design based upon a uniform rate of wear at the mating surfaces or a uniform pressure distribution between the mating surfaces. The equations for both of these methods are outlined in this section.

The assumption of a uniform pressure distribution at the interface between mating surfaces is valid for an unworn accurately manufactured clutch with rigid outer discs.

The area of an elemental annular ring on a disc clutch (Fig. 13.20) is $\delta A = 2\pi r \delta r$. Now $F = pA$, where p is the assumed uniform interface pressure, so $\delta F = 2\pi r p \delta r$. For the disc the normal force acting on the entire face is:

$$F = \int_{r_i}^{r_o} 2\pi r p dr = 2\pi p \left[\frac{r^2}{2}\right]_{r_i}^{r_o} \tag{13.4}$$

$$F = \pi p \left(r_o^2 - r_i^2\right) \tag{13.5}$$

Note that F is also the necessary force required to clamp the clutch discs together.

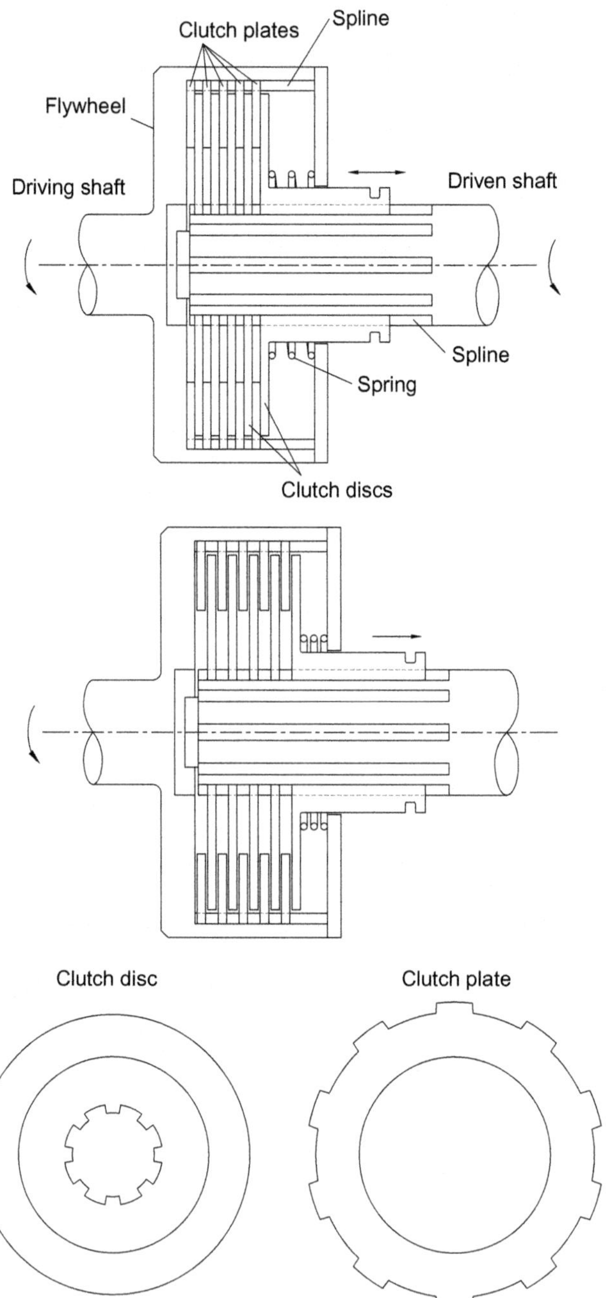

Fig. 13.19 Multiple disc clutch.

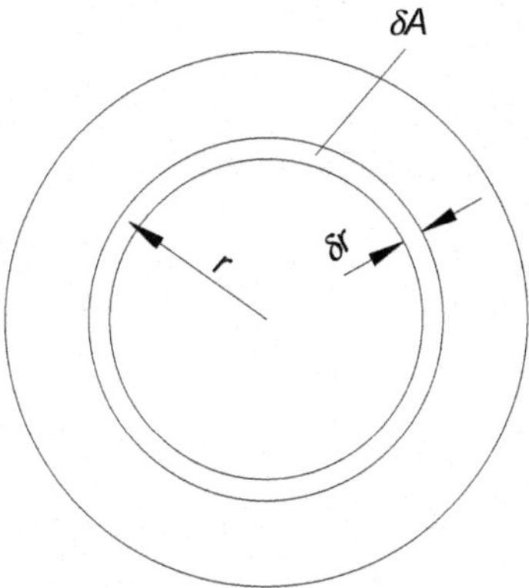

Fig. 13.20 Area of an elemental ring on a disc clutch.

The friction torque δT that can be developed on an elemental ring is the product of the elemental normal force, given by $\mu\delta F$ and the radius:

$$\delta T = r\mu\delta F = 2\mu\pi r^2 p\delta r \tag{13.6}$$

where μ is the coefficient of friction which models the less than ideal frictional contact which occurs between two surfaces.

The total torque is given by integration of Eq. (13.6) between the limits of the annular ring, r_i and r_o:

$$T = \int_{r_i}^{r_o} 2\mu\pi r^2 p dr = 2\pi p\mu \left[\frac{r^3}{3}\right]_{r_i}^{r_o} = \frac{2}{3}\pi p\mu\left(r_o^3 - r_i^3\right) \tag{13.7}$$

This equation represents the torque capacity of a clutch with a single frictional interface. In practice clutches use an even number of frictional surfaces as illustrated in Figs 13.16 and 13.18.

For a clutch with N faces the torque capacity is given by:

$$T = \frac{2}{3}\pi p\mu N\left(r_o^3 - r_i^3\right) \tag{13.8}$$

Using Eqs (13.5) and (13.8) and substituting for the pressure, p, gives an equation for the torque capacity as a function of the axial clamping force.

$$T = \frac{2}{3}\mu FN\left(\frac{r_o^3 - r_i^3}{r_o^2 - r_i^2}\right) \tag{13.9}$$

The equations assuming uniform wear are developed below. The wear rate is assumed to be proportional to the product of the pressure and velocity. So

$$pr\omega = \text{constant} \quad (13.10)$$

where ω is the angular velocity (rad/s).

For a constant angular velocity the maximum pressure will occur at the smallest radius.

$$p_{max} r_i \omega = \text{constant} \quad (13.11)$$

Eliminating the angular velocity and constant from the above two equations gives a relationship for the pressure as a function of the radius:

$$p = p_{max} \frac{r_i}{r} \quad (13.12)$$

The value for the maximum permissible pressure is dependent on the clutch lining material and typical values are listed in Table 13.3.

The elemental axial force on an elemental annular ring is given by

$$\delta F = 2\pi pr \delta r \quad (13.13)$$

Integrating to give the total axial force:

$$F = \int_{r_i}^{r_o} 2\pi prdr = \int_{r_i}^{r_o} 2\pi p_{max} \frac{r_i}{r} rdr = 2\pi p_{max} r_i (r_o - r_i) \quad (13.14)$$

The elemental torque is given by

$$\delta T = \mu r \delta F \quad (13.15)$$

So

Table 13.3 Typical values for dynamic friction coefficients, permissible contact pressures and temperature limits

Material	μ_{dry}	μ_{oil}	p_{max} (MN/m)	T (°C)
Moulded compounds	0.25–0.45	0.06–0.10	1.035–2.07	200–260
Woven materials	0.25–0.45	0.08–0.10	0.345–0.69	200–260
Sintered metal	0.15–0.45	0.05–0.08	1.035–2.07	230–680
Cork	0.30–0.50	0.15–0.25	0.055–0.1	80
Wood	0.20–0.45	0.12–0.16	0.345–0.62	90
Cast iron	0.15–0.25	0.03–0.06	0.69–1.725	260
Paper based	–	0.10–0.17		
Graphite/resin	–	0.10–0.14		

$$T = \int_{r_i}^{r_o} 2\mu\pi p_{max} r_i r dr = p_{max}\mu\pi r_i \left(r_o^2 - r_i^2\right) \tag{13.16}$$

Rearranging Eq. (13.16) gives

$$p_{max} = \frac{F}{2\pi r_i(r_o - r_i)} \tag{13.17}$$

Substituting Eq. (13.17) into Eq. (13.16) gives

$$T = \frac{\mu F}{2}\left(\frac{r_o^2 - r_i^2}{r_o - r_i}\right) = \frac{\mu F}{2}(r_o + r_i) \tag{13.18}$$

$$\text{For } N \text{ frictional surfaces } T = \int_{r_i}^{r_o} 2\mu\pi N p_{max} r_i r dr = p_{max}\mu\pi N r_i \left(r_o^2 - r_i^2\right) \tag{13.19}$$

gives

$$T = \frac{\mu N F}{2}(r_o + r_i) \tag{13.20}$$

By differentiating Eq. (13.16) with respect to r_i and equating the result to zero, the maximum torque for any outer radius r_o is found to occur when

$$r_i = \sqrt{1/3}\, r_o \tag{13.21}$$

This useful formula can be used to set the inner radius if the outer radius is constrained to a particular value.

Note that Eq. (13.20) indicates a lower torque capacity than the uniform pressure assumption. This is because the higher initial wear at the outer diameter shifts the centre of pressure towards the inner radius.

Clutches are usually designed based on uniform wear. The uniform wear assumption gives a lower torque capacity clutch than the uniform pressure assumption. The preliminary design procedure for disc clutch design requires the determination of the torque and speed, specification of space limitations, selection of materials (i.e. the coefficient of friction and the maximum permissible pressure) and the selection of principal radii, r_o and r_i. Common practice is to set the value of r_i between $0.45r_o$ and $0.8r_o$. This procedure for determining the initial geometry is itemised below, and illustrated in Fig. 13.21.

1. Determine the service factor.
2. Determine the required torque capacity, $T = \text{power}/\omega$.
3. Determine the coefficient of friction μ.
4. Determine the outer radius r_o.
5. Find the inner radius r_i.
6. Find the axial actuation force required.

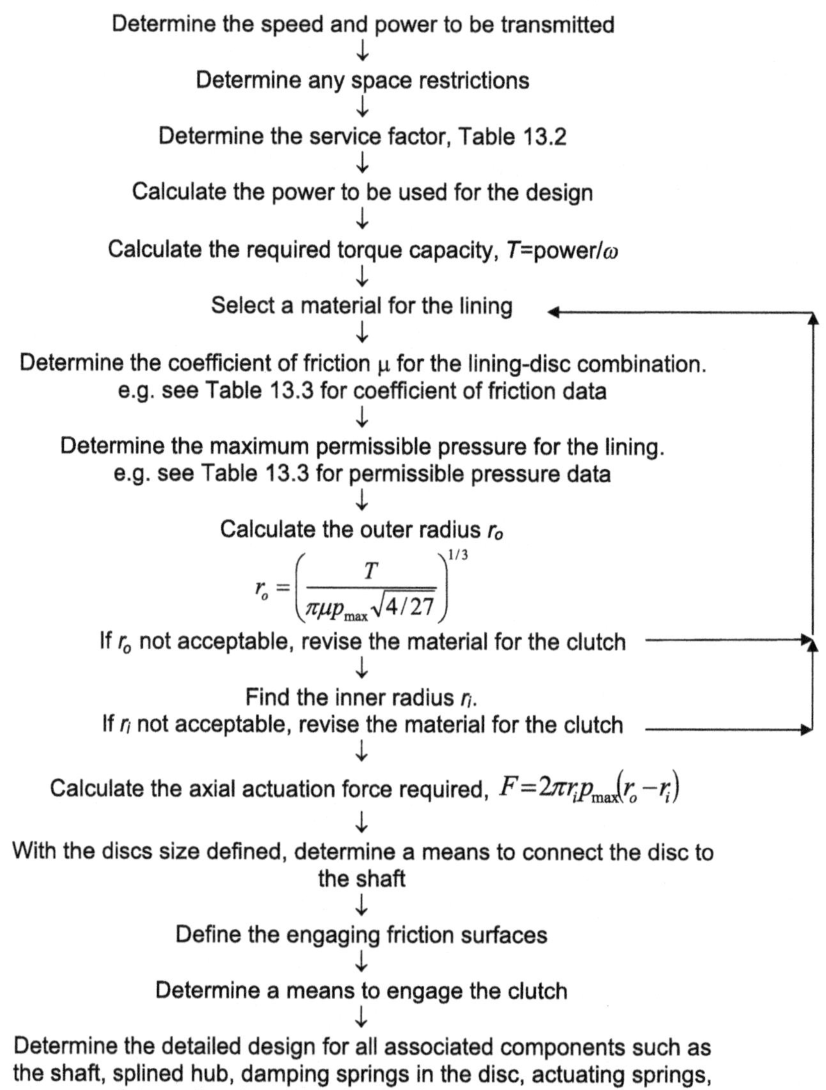

Determine the speed and power to be transmitted
Determine any space restrictions
Determine the service factor, Table 13.2
Calculate the power to be used for the design
Calculate the required torque capacity, T=power/ω
Select a material for the lining
Determine the coefficient of friction μ for the lining-disc combination. e.g. see Table 13.3 for coefficient of friction data
Determine the maximum permissible pressure for the lining. e.g. see Table 13.3 for permissible pressure data
Calculate the outer radius r_o
$r_o = \left(\frac{T}{\pi \mu p_{max} \sqrt{4/27}} \right)^{1/3}$
If r_o not acceptable, revise the material for the clutch
Find the inner radius r_i.
If r_i not acceptable, revise the material for the clutch
Calculate the axial actuation force required, $F = 2\pi r_i p_{max}(r_o - r_i)$
With the discs size defined, determine a means to connect the disc to the shaft
Define the engaging friction surfaces
Determine a means to engage the clutch
Determine the detailed design for all associated components such as the shaft, splined hub, damping springs in the disc, actuating springs, cover plate, spline, bearings, flywheel, crankshaft and gearbox connections and release mechanism

Fig. 13.21 Uniform wear outline procedure for disk clutch design.

The material used for clutch plates is typically grey cast iron or steel. The friction surface will consist of a lined material which may be moulded, woven, sintered or solid. Moulded linings consist of a polymeric resin used to bind powdered fibrous material and brass and zinc chips. Table 13.3 lists typical values for the performance of friction linings.

Example 13.2. A clutch is required for transmission of power between a four cylinder internal combustion engine and a small machine. Determine the radial dimensions for a single face dry disc clutch with a moulded lining which should transmit 5 kW at 1800 rpm (Fig. 13.22). Base the design on the uniform wear assumption.

Solution

From Table 13.2 a service factor of two should be used. The design will therefore be undertaken using a power of 2×5 kW $= 10$ kW.

The torque is given by $T = \frac{\text{Power}}{\omega} = \frac{10{,}000}{1800 \times (2\pi/60)} = 53\,\text{Nm}$.

From Table 13.3 taking mid-range values for the coefficient friction and the maximum permissible pressure for moulded linings gives $\mu = 0.35$ and $p_{max} = 1.55$ MN/m^2.

Taking $r_i = \sqrt{1/3}\,r_o$ and substituting for r_i in Eq. (13.16) gives:

$$T = \pi\mu\sqrt{1/3}\,r_o p_{max}\left(r_o^2 - \left(\sqrt{1/3}\,r_o\right)^2\right) = \pi\mu\sqrt{1/3}\,p_{max}\left(r_o^3 - \frac{1}{3}r_o^3\right)$$
$$= \pi\mu\sqrt{4/27}\,p_{max}\,r_o^3$$

$$r_o = \left(\frac{T}{\pi\mu p_{max}\sqrt{4/27}}\right)^{1/3} = \left(\frac{53.05}{\pi 0.35 \times 1.55 \times 10^6\sqrt{4/27}}\right)^{1/3} = 0.04324\,\text{m}$$

$$r_i = \sqrt{1/3}\,r_o = 0.02497\,\text{m}$$

$$F = 2\pi r_i p_{max}(r_o - r_i) = 2\pi \times 0.02497 \times 1.55 \times 10^6 (0.04324 - 0.02497)$$
$$= 4443\,\text{N}$$

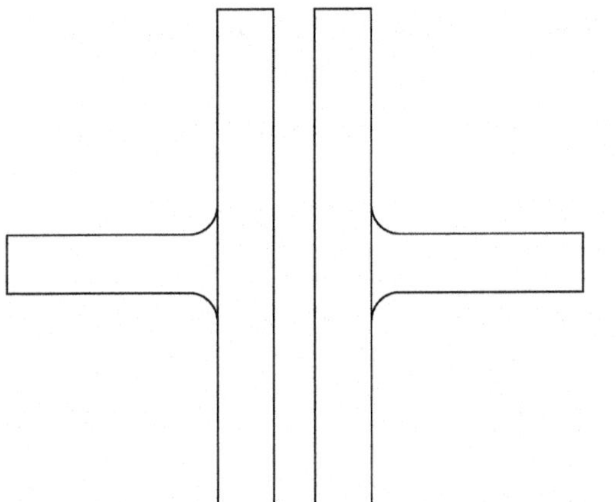

Fig. 13.22 Single face clutch.

So the clutch consists of a disc of inner and outer radius 25 and 43 mm, respectively, with a moulded lining having a coefficient of friction value of 0.35 and a maximum permissible contact pressure of 1.55 MPa and an actuating force of 4.4 kN.

Example 13.3. A disc clutch, running in oil, is required for a vehicle with a four-cylinder engine. The design power for initial estimation of the clutch specification is 90 kW at 4500 rpm. Determine the radial dimensions and actuating force required. Base the design on the uniform wear assumption.

Solution

From Table 13.2 a service factor of 2.7 should be used to account for starts and stops and the four cylinder engine. The design will therefore be undertaken using a power of $2.7 \times 90 = 243$ kW.

The torque is given by $T = \frac{\text{Power}}{\omega} = \frac{243{,}000}{4500 \times (2\pi/60)} = 515.7$ N m

From Table 13.3 taking midrange values for the coefficient friction and the maximum permissible pressure for moulded linings gives $\mu = 0.35$ and $p_{max} = 1.55$ MN/m^2.

Taking $r_i = \sqrt{1/3}r_o$ and substituting for r_i in Eq. (13.19) gives:

$$r_o = \left(\frac{T}{\pi\mu N p_{max}\sqrt{4/27}}\right)^{1/3} = \left(\frac{515.7}{\pi 0.35 \times 2 \times 1.55 \times 10^6\sqrt{4/27}}\right)^{1/3}$$
$$= 0.07325 \text{ m}$$

$$r_i = \sqrt{1/3}r_o = 0.04229 \text{ m}$$

$$F = 2\pi N r_i p_{max}(r_o - r_i) = 2\pi \times 2 \times 0.04229 \times 1.55 \times 10^6(0.07325 - 0.04229)$$
$$= 25{,}500 \text{ N}$$

So the clutch consists of a disc of inner and outer radius 42.3 and 73.3 mm respectively, with a moulded lining having a coefficient of friction value of 0.35 and a maximum permissible contact pressure of 1.55 MPa and an actuating force of 25.5 kN.

Example 13.4. A multiple disc clutch, running in oil, is required for a motorcycle with a three-cylinder engine. The power demand is 75 kW at 8500 rpm. The preliminary design layout indicates that the maximum diameter of the clutch discs should not exceed 100 mm. In addition previous designs have indicated that a moulded lining with a coefficient of friction of 0.068 in oil and a maximum permissible pressure of 1.2 MPa is reliable. Within these specifications determine the radii for the discs, the number of discs required and the clamping force.

Solution

The torque is given by

$$T = \frac{\text{Service factor} \times \text{Power}}{\omega} = \frac{3.4 \times 75000}{8500 \times (2\pi/60)} = 286.5 \text{ N m.}$$

Select the outer radius to be the largest possible, i.e. $r_o = 50\,\text{mm}$. Using $r_i = \sqrt{1/3}r_o$, $r_i = 28.87\,\text{mm}$. From Eq. (13.19), the number of frictional surfaces, N, can be determined.

$$N = \frac{T}{\pi p_{\max} r_i \mu \left(r_o^2 - r_i^2\right)} = \frac{286.5}{\pi 1.2 \times 10^6 \times 0.02887 \times 0.068\left(0.05^2 - 0.02887^2\right)}$$

$$= 23.23$$

This must be an even number, so the number of frictional surfaces is taken as $N = 24$. This requires 13 driving discs and 12 driven discs to implement.

Using Eq. (13.14) the clamping force can be calculated:

$$F = \frac{2T}{\mu N(r_o + r_i)} = \frac{2 \times 286.5}{0.068 \times 24(0.05 + 0.02887)} = 4452\,\text{N}.$$

As well as the disc or discs, detailed design of a disc clutch such as the automotive variant illustrated in Fig. 13.23, requires specification and consideration of all the associated components such as the shaft, splined hub, damping springs in the disc (Fig. 13.24) actuating springs, cover plate, spline, bearings, flywheel, crankshaft and gearbox connections and release mechanism (SAE, 1997). The choice of whether to use coil springs or a diaphragm spring (Fig. 13.25) depends on the clamping force required and cost considerations. Diaphragm springs have been developed from Belleville springs and are made from a steel disc. The inner portion of the disc has a number of radial slots to form actuating fingers. Stress relieving holes are included at the outer end of the fingers to prevent cracking. Multicoil spring units are used in heavy commercial vehicles where it is difficult to provide the required clamping force from a single diaphragm.

13.3 Brakes

The basic function of a brake is to absorb kinetic energy and dissipate it in the form of heat. An idea of the magnitude of energy that must be dissipated can be obtained from considering the familiar example of a car undergoing an emergency stop in 7 s from 60 mph (96 km/h). If the car's mass is 1400 kg and assuming that 65% of the car's weight is loaded onto the front axles during rapid braking then the load on the front axle is

$$1400 \times 9.81 \times 0.65 = 8927\,\text{N}$$

This will be shared between two brakes so the energy that must be absorbed by one brake is

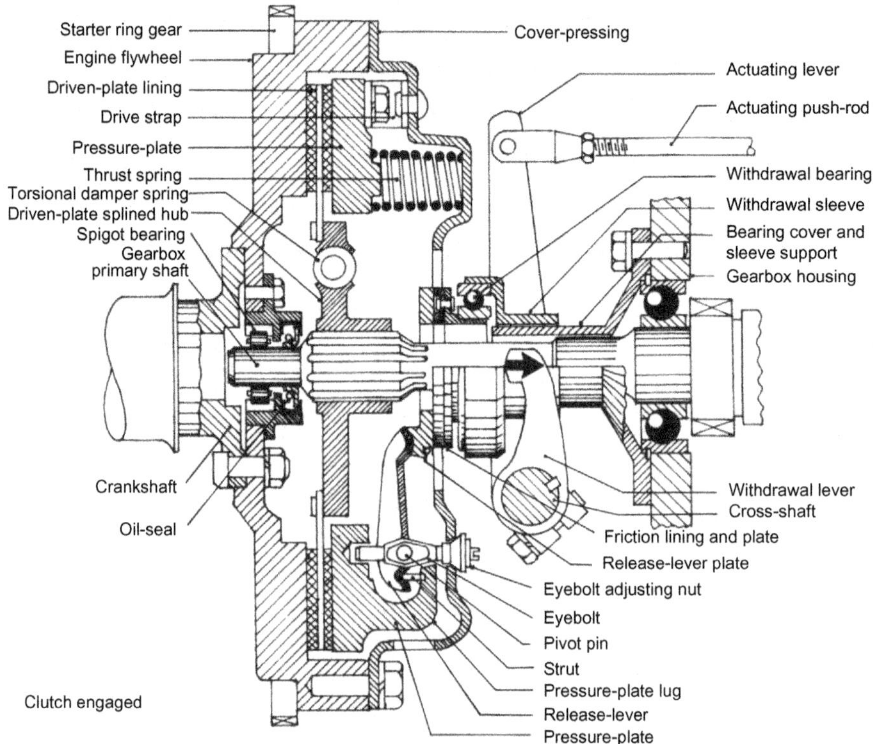

Fig. 13.23 Automotive clutch.
From Heisler, H., 1999. Vehicle and Engine Technology, second ed. Butterworth Heinemann.

$$E = \frac{1}{2}m\left(V_i^2 - V_f^2\right) \tag{13.22}$$

where m = mass (kg), V_i = initial velocity (m/s), V_f = final velocity (m/s).

$$E = \frac{1}{2}m\left(V_i^2 - V_f^2\right) = \frac{1}{2} \times \left(\frac{8927}{9.81} \times 0.5\right) \times \left(\left(\frac{96 \times 10^3}{3600}\right)^2 - 0^2\right) = 161.8\,\text{kJ}$$

If the car brakes uniformly in 7 s then the heat that must be dissipated is $161.8 \times 10^3/7 = 23.1\,\text{kW}$. From your experience of heat transfer from say 1 kW domestic heaters you will recognise that this is a significant quantity of heat to transfer away from the relatively compact components that make up brake assemblies.

For the case of an F_1 vehicle, of 800 kg, braking say for the final chicane at the Gilles Villeneuve Circuit from 340 km/h to 135 km/h in 1.64 s, the corresponding force on the front axle assuming 60% loading on the front axle is $800 \times 9.81 \times 0.6 = 4709\,\text{N}$.

$$E = \frac{1}{2}m\left(V_i^2 - V_f^2\right) = \frac{1}{2} \times (800 \times 0.6 \times 0.5) \times \left((94.44)^2 - (37.5)^2\right) = 0.9016\,\text{MJ}$$

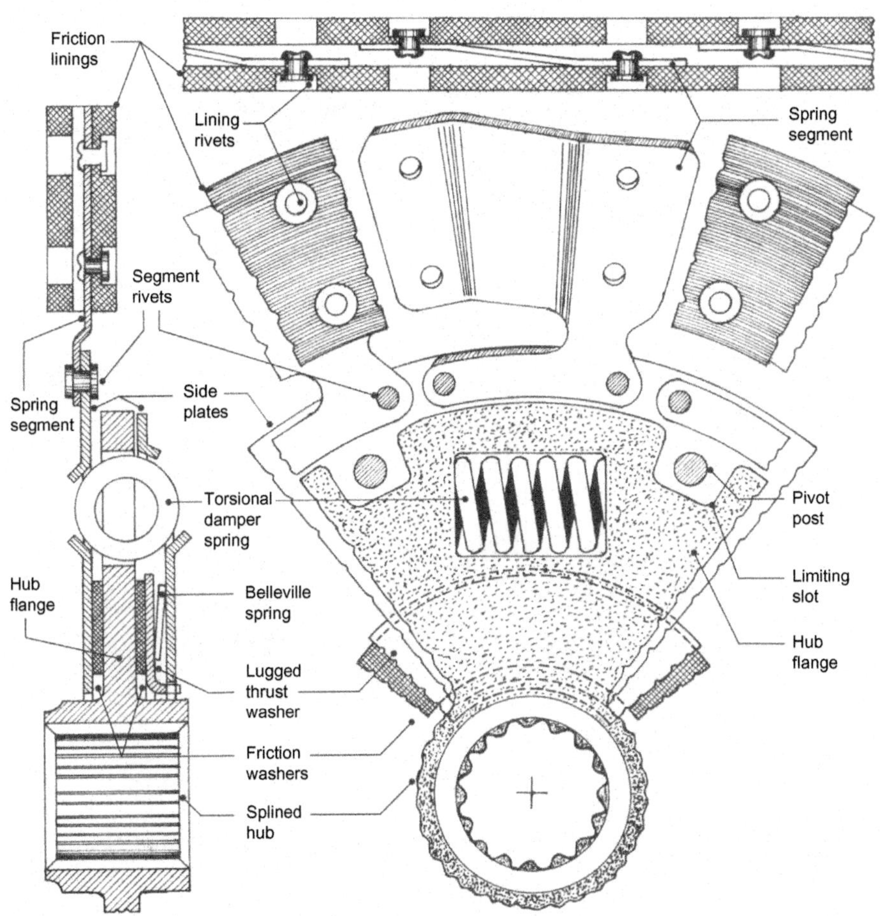

Fig. 13.24 Automotive disc illustrating the use of torsional damping springs.
From Heisler, H., 1999. Vehicle and Engine Technology, second ed. Butterworth Heinemann.

If the car brakes uniformly in 1.64 s then the heat that must be dissipated is $0.9016 \times 10^6/1.64 = 0.5498$ MW. This high value provides an indication why F_1 discs tend to glow on heavy braking with temperatures reaching ~1000°C.

Convective heat transfer can be modelled by Fourier's equation:

$$Q = hA\Delta T = hA\left(T_s - T_f\right) \tag{13.23}$$

where Q = heat transfer rate (W), h = the heat transfer coefficient (W/m^2 K), T_s = temperature of the surface (K or °C), T_f = temperature of the surrounding fluid (K or °C), A = surface area (m^2).

This equation indicates that the ability of a brake to dissipate the heat generated increases as the surface area increases or as the heat transfer coefficient rises. For air, the heat transfer coefficient is usually dependent on the local flow velocity and

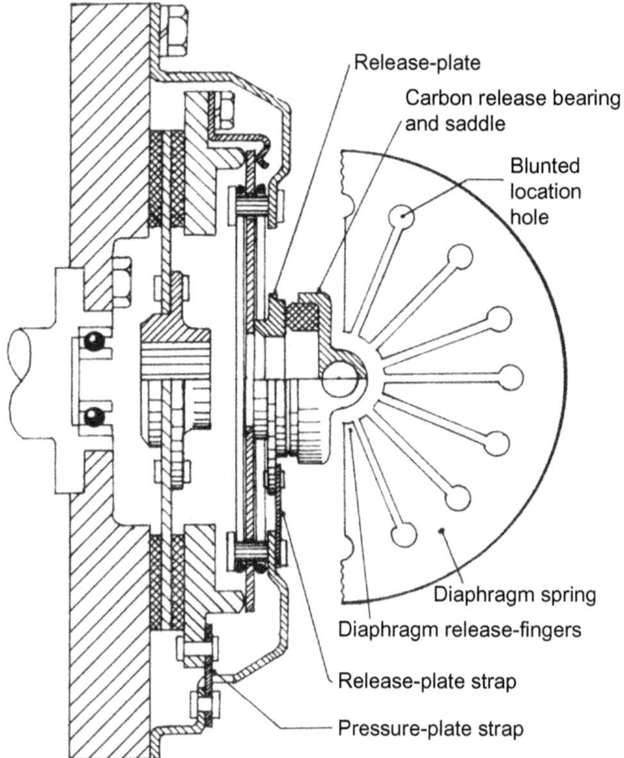

Fig. 13.25 Diaphragm spring.
From Heisler, H., 1999. Vehicle and Engine Technology, second ed. Butterworth Heinemann.

on the geometry. A method often used for disc brakes to increase both the surface area and the local flow is to machine multiple axial or radial holes in the disc (this also reduces the mass and inertia) see Limpert (1975), Prabhakar et al. (2015), Chopade and Valavade (2017), Newcomb (1960).

Example 13.5. Calculate the energy that must be absorbed in stopping a 300 tonne jumbo airliner travelling at 290 km/h in an aborted take-off, stopping in 900 m.
Solution

$$E=\frac{1}{2}m\left(V_i^2-V_f^2\right)=\frac{1}{2}\times 300\times 10^3\left(\left(\frac{290\times 10^3}{3600}\right)^2-0\right)=973.5\ \text{MJ}$$

Assuming that the aircraft brakes uniformly then the deceleration is

$$=-\frac{80.56^2}{2\times 900}=-3.606\,\text{m/s}^2$$

The time to stop is given by

$$t = \frac{V_i - V_f}{a} = \frac{80.56}{3.606} = 22.34\,\text{s}$$

The power that must be dissipated is:

$$\frac{973.5 \times 10^6}{22.34} = 43.57\,\text{MW}$$

This is a significant quantity of power, equivalent to over 43 thousand 1 kW electric heaters and even divided between multiple brakes gives an indication why aborted take-offs can result in burnt out brakes (Schmidt, 2015a, b). Note that the effects of any thrust reversal and aerodynamic drag has been ignored. Both of these effects would reduce the heat dissipated within the brakes.

There are numerous brake types as shown in Fig. 13.26. The selection and configuration of a brake depends on the requirements. Table 13.4 gives an indication of brake operation against various criteria. The brake factor listed in Table 13.4 is the ratio of frictional braking force generated to the actuating force applied.

Brakes can be designed so that, once engaged the actuating force applied is assisted by the braking torque. This kind of brake is called a self-energising brake and is useful for braking large loads. Great care must be exercised in brake design (Limpert, 2012). It is possible and sometimes desirable to design a brake, which once engaged, will grab and lock up (called self-locking action).

A critical aspect of all brakes is the material used for frictional contact. Normally one component will comprise a steel or cast iron disc or drum and this is brought into frictional contact against a geometrically similar component with a brake lining made up of one of the materials listed in Table 13.3.

Section 13.3.1 gives details about the configuration design of disc brakes and Section 13.3.2 introduces the design of drum brakes.

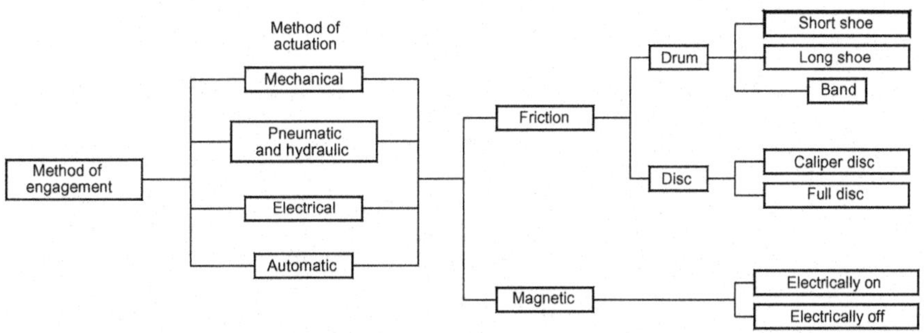

Fig. 13.26 Brake classification.

Table 13.4 Comparative table of brake performance

Type of brake	Maximum operating temperature	Brake factor	Stability	Dryness	Dust and dirt	Typical applications
Differential band brake	Low	High	Low	Unstable but still effective	Good	Winches, hoist, excavators, tractors
External drum brake (leading trailing edge)	Low	Medium	Medium	Unstable if humid, poor if wet	Good	Mills, elevators, winders
Internal drum brake (leading trailing edge)	Higher than external brake	Medium	Medium	Unstable if humid, ineffective if wet	Good if sealed	Vehicles (rear axles on passenger cars)
Internal drum brake (two leading shoes)	Higher than external brake	High	Low	Unstable if humid, ineffective if wet	Good if sealed	Vehicles (rear axles on passenger cars)
Internal drum brake (duo-servo)	Low	High	Low	Unstable if humid, ineffective if wet	Good if sealed	Vehicles (rear axles on passenger cars)
Calliper disc brake	High	Low	High	Good	Poor	Vehicles and industrial machinery
Full disc brake	High	Low	High	Good	Poor	Machine tools and other industrial machinery

Reproduced from Neale, M.J., 1994. Drives and Seals: A Tribology Handbook. Butterworth Heinemann.

13.3.1 Disc brakes

Disc brakes are familiar from automotive applications where they are used extensively for car and motorcycle wheels (Fig. 13.27), as well as bicycles (Fig. 13.28). In the case of an automotive disc brake, these typically consist of a cast iron disc, bolted to the wheel hub. This is sandwiched between two pads actuated by pistons supported in a calliper mounted on the stub shaft (Fig. 13.29). When the brake pedal is pressed, hydraulically pressurised fluid is forced into the cylinders pushing the opposing pistons and brake pads into frictional contact with the disc. The advantages of this form of braking are steady braking, easy ventilation, balancing thrust loads and design simplicity (Abhang and Bhaskar, 2014; Tirovic and Day, 1991). There is no self-energising action so the braking action is proportional to the applied force. The use of a discrete pad allows the disc to cool as it rotates, enabling heat transfer between the cooler disc and the hot brake pad. As the pads located either side of the disc are pushed on to the disc by the pistons with equal forces the net thrust load on the disc cancels. For the case of heavier loads, for example landing gear for airliners, then multiple discs can be used to accommodate the loads (Schmidt, 2015a, b) (Fig. 13.30).

With reference to Fig. 13.31, the torque capacity per pad is given by

$$T = \mu F r_e \tag{13.24}$$

where r_e is an effective radius.

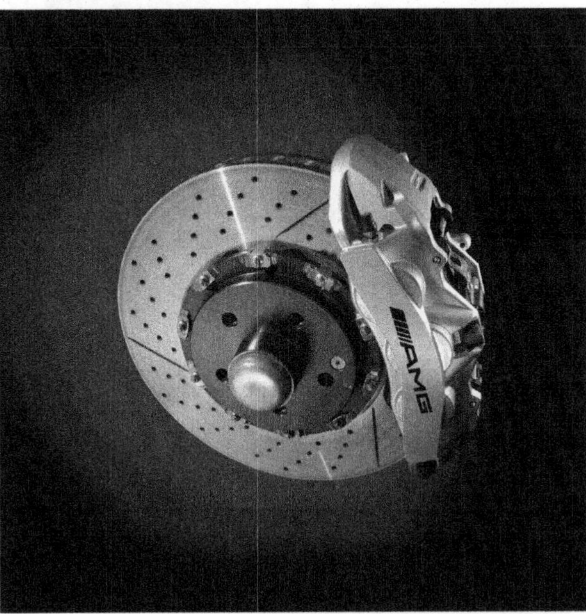

Fig. 13.27 An example of an automotive disc brake. (Image courtesy of Daimler AG)

Fig. 13.28 Bicycle disc brake.

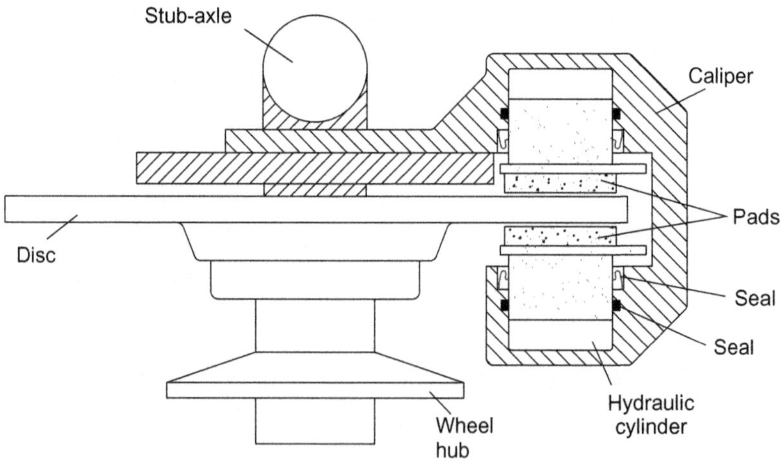

Fig. 13.29 Automotive disc brake.

Fig. 13.30 Multiple disc brake for aircraft landing gear.
(Image courtesy of Julian Herzog CC A 4.0 (2013))

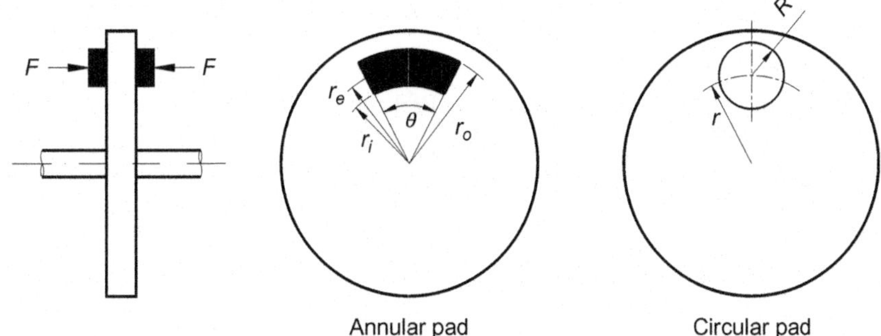

Fig. 13.31 Calliper disc brake.

The actuating force assuming constant pressure is given by

$$F = p_{av}\theta\frac{r_o^2 - r_i^2}{2} \tag{13.25}$$

or assuming uniform wear by

$$F = p_{\max}\theta r_i(r_o - r_i) \tag{13.26}$$

where θ (in radians) is the included angle of the pad, r_i is the inner radius of the pad and r_o is the outer radius of the pad.

The relationship between the average and the maximum pressure for the uniform wear assumption is given by

$$\frac{p_{av}}{p_{\max}} = \frac{2r_i/r_o}{1+r_i/r_o} \tag{13.27}$$

For an annular disc brake the effective radius is given by Eq. (13.28) assuming constant pressure and Eq. (13.29) assuming uniform wear.

$$r_e = \frac{2\left(r_o^3 - r_i^3\right)}{3\left(r_o^2 - r_i^2\right)} \tag{13.28}$$

$$r_e = \frac{r_i + r_o}{2} \tag{13.29}$$

For circular pads the effective radius is given by $r_e = r\delta$, where values for δ are given in Table 13.5 as a function of the ratio of the pad radius and the radial location, R/r. The actuating force for circular pads can be calculated using:

$$F = \pi R^2 p_{av} \tag{13.30}$$

Example 13.6. A calliper brake is required for the front wheels of a sport's car with a braking capacity of 820 N m for each brake. Preliminary design estimates have set the brake geometry as $r_i = 100\,\text{mm}$, $r_o = 160\,\text{mm}$ and $\theta = 45$ degrees. A pad with a coefficient of friction of 0.35 has been selected. Determine the required actuating force and the average and maximum contact pressures.

Solution

The torque capacity per pad = 820/2 = 410 N m.

The effective radius is $r_e = \dfrac{0.1 + 0.16}{2} = 0.13\,\text{m}$.

The actuating force is given by $F = \dfrac{T}{\mu r_e} = \dfrac{410}{0.35 \times 0.13} = 9.011\,\text{kN}$.

The maximum contact pressure is given by

Table 13.5 Circular pad disk brake design values (Fazekas, 1972)

R/r	$\delta = r_e/r$	p_{max}/p_{av}
0	1.000	1.000
0.1	0.983	1.093
0.2	0.969	1.212
0.3	0.957	1.367
0.4	0.947	1.578
0.5	0.938	1.875

$$p_{\max} = \frac{F}{\theta r_i(r_o - r_i)} = \frac{9.011 \times 10^3}{45 \times (2\pi/360) \times 0.1 \times (0.16 - 0.1)} = 1.912\,\text{MN/m}^2$$

The average pressure is given by

$$p_{av} = p_{\max} \frac{2r_i/r_o}{1 + r_i/r_o} = 1.471\,\text{MN/m}^2$$

Example 13.7. A calliper brake is required for the front wheels of a passenger car with a braking capacity of 320 N m for each brake. Preliminary design estimates have set the brake geometry as $r_i = 100\,\text{mm}$, $r_o = 140\,\text{mm}$ and $\theta = 40$ degrees. Pads with a coefficient of friction of 0.35 have been selected. Each pad is actuated by means of a hydraulic cylinder of nominal diameter 25.4 mm. Determine the required actuating force, the average and the maximum contact pressures and the required hydraulic pressure for brake actuation.

Solution

The torque capacity per pad is 320/2 = 160 N m.

The effective radius is given by $r_e = \dfrac{0.1 + 0.14}{2} = 0.12\,\text{m}$.

The actuation force required is $F = \dfrac{T}{\mu r_e} = \dfrac{160}{0.35 \times 0.12} = 3810\,\text{N}$.

The maximum pressure is

$$p_{\max} = \frac{F}{\theta r_i(r_o - r_i)} = \frac{3810}{40(2\pi/360) \times 0.1(0.14 - 0.1)} = 1.364 \times 10^6\,\text{N/m}^2$$

The average pressure is

$$p_{av} = \frac{2r_i/r_o}{1 + r_i/r_o} p_{\max} = \frac{2 \times 0.1/0.14}{1 + 0.1/0.14} 1.36 \times 10^6 = 1.137 \times 10^6\,\text{N/m}^2$$

The area of one of the hydraulic cylinders is $\pi \times 0.0127^2 = 5.067 \times 10^{-4}\,\text{m}^2$.

The hydraulic pressure required is given by

$$p_{hydraulic} = \frac{F}{A_{cylinder}} = \frac{3810}{5.067 \times 10^{-4}} = 7.519 \times 10^6\,\text{N/m}^2$$

i.e. $p_{hydraulic} \approx 75$ bar.

An outline procedure for establishing some of the principal parameters for a calliper disc brake is given in Fig. 13.32.

Full disc brakes, consisting of a complete annular ring pad, are principally used for industrial machinery. The disc clutch equations developed in Section 13.2.1 are applicable to their design. The disc configuration can be designed to function as either a clutch or a brake (a clutch-brake combination) to transmit a load or control its speed.

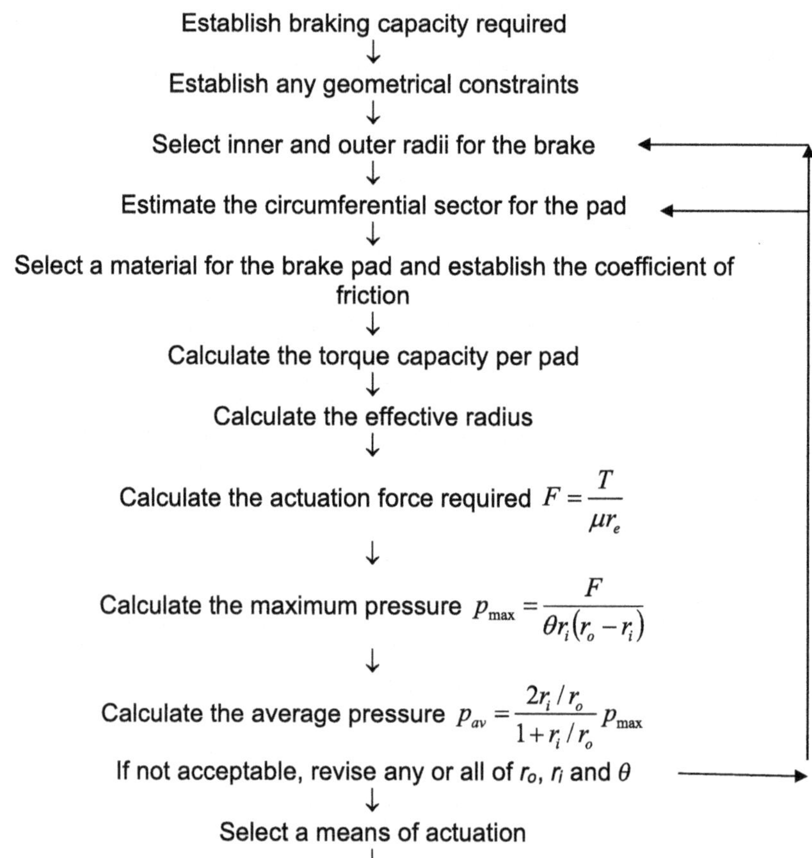

Fig. 13.32 Outline procedure for calliper disc brake design.

13.3.2 Drum brakes

Drum brakes apply friction to the external or internal circumference of a cylinder. A drum brake consists of the brake shoe (Fig. 13.33) which has the friction material bonded to it, and the brake drum. For braking, the shoe is forced against the drum developing the friction torque (Day et al., 1984). Drum brakes can be divided into two groups depending on whether the brake shoe is external or internal to the drum. A further classification can be made in terms of the length of the brake shoe: short, long or complete band.

Fig. 13.33 Railway bogie illustrating the use of external drum brakes
Figure courtesy of Astra Rail Industries, Greenbrier Europe

Short shoe internal brakes are used for centrifugal brakes that engage at a particular critical speed. Long shoe internal drum brakes are used principally in automotive applications. Drum brakes (or clutches) can be designed to be self-energising. Once engaged the friction force increases the normal force nonlinearly, increasing the friction torque as in a positive feedback loop. This can be advantageous in braking large loads but makes control much more difficult. One problem associated some drum brakes is stability. If the brake has been designed so that the braking torque is not sensitive to small changes in the coefficient of friction, which would occur if the brake is worn or wet, then the brake is said to be stable. If a small change in the coefficient of friction causes a significant change to the braking torque the brake is unstable and will tend to grab if the friction coefficient rises or the braking torque will drop noticeably if the friction coefficient reduces.

13.3.3 Short shoe external drum brakes

A schematic for a short shoe external drum brake is given in Fig. 13.34. If the included angle of contact between the brake shoe and the brake drum is <45 degrees, the force between the shoe and the drum is relatively uniform and can be modelled by a single concentrated load F_n at the centre of the contact area. If the maximum permissible pressure is $p_{\max}$ the force F_n can be estimated by

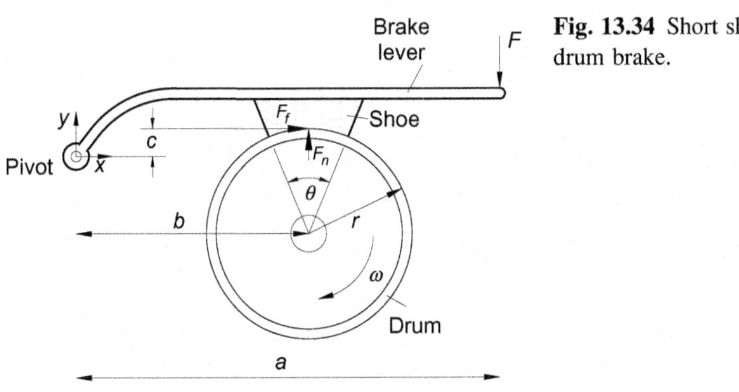

Fig. 13.34 Short shoe external drum brake.

$$F_n = p_{max} r\theta w \tag{13.31}$$

where w = width of the brake shoe (m), θ = angle of contact between the brake shoe and the lining (rad).

The frictional force, F_f, is given by

$$F_f = \mu F_n \tag{13.32}$$

where μ is the coefficient of friction.

The torque on the brake drum is

$$T = F_f r = \mu F_n r \tag{13.33}$$

Summing moments, for the shoe arm, about the pivot gives:

$$\sum M_{pivot} = aF_a - bF_n + cF_f = 0$$

$$F_a = \frac{bF_n - cF_f}{a} = F_n \frac{b - \mu c}{a} \tag{13.34}$$

Resolving forces gives the reactions at the pivot:

$$R_x = -F_f \tag{13.35}$$

$$R_y = F_a - F_n \tag{13.36}$$

Note that for the configuration and direction of rotation shown in Fig. 13.34, the friction moment $\mu F_n c$ adds or combines with the actuating moment aF_a. Once the actuating force is applied the friction generated at the shoe acts to increase the braking torque. This kind of braking action is called self-energising. If the brake direction is reversed the friction moment term $\mu F_n c$ becomes negative and the applied load F_a must be maintained to generate braking torque. This combination is called self-deenergising. From Eq. (13.34), note that if the brake is self-energising and if $\mu c > b$ then the force required to actuate the brake is zero or negative and the brake action is called self-locking. If the shoe touches the drum it will grab and lock. This is usually undesirable with exceptions being hoist stops or over-running clutch type applications.

Example 13.8. A double short-shoe external brake is illustrated in Fig. 13.35. The actuating force required to limit the drum rotation to 100 rpm is 2.4 kN. The coefficient of friction for the brake lining is 0.35. Determine the braking torque and the rate of heat generation.

Solution

For the top shoe:

$$F \times 0.56 - F_n 0.3 + \mu F_n 0.05 = 0$$

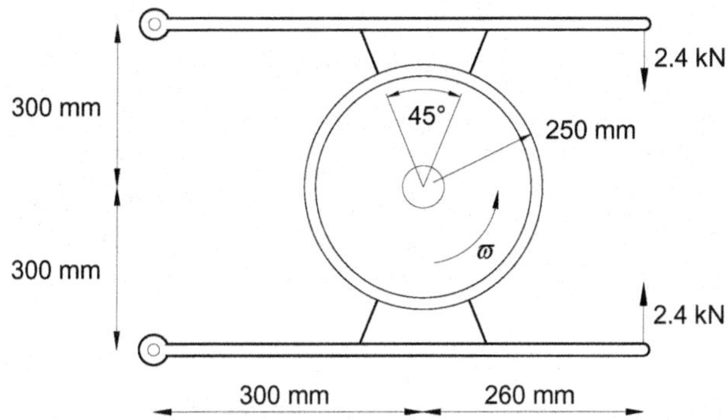

Fig. 13.35 Double short-shoe external brake.

$$F_n = \frac{2400 \times 0.56}{0.3 - 0.35 \times 0.05} = 4758\,\text{N}$$

For the bottom shoe:

$$-F \times 0.56 + F_n 0.3 + \mu F_n 0.05 = 0$$

$$F_n = \frac{2400 \times 0.56}{0.3 + 0.35 \times 0.05} = 4233\,\text{N}$$

$$\text{Torque} = \mu(F_{n\,total}) \times 0.125 = 0.35(4758 + 4233) \times 0.125 = 393.4\,\text{Nm}$$

$$\text{Heat generation} = \omega \times T = 100 \times \frac{2\pi}{60} \times 393.4 = 4119\,\text{W}$$

13.3.4 Long shoe external drum brakes

If the included angle of contact between the brake shoe and the drum is >45 degrees then the pressure between the shoe and the brake lining cannot be regarded as uniform and the approximations made for the short shoe brake analysis are inadequate. Most drum brakes use contact angles >90 degrees. Brake shoes are not rigid and the local deflection of the shoe affects the pressure distribution. Detailed shoe analysis is possible using finite element software. For initial synthesis/specification of brake geometry a simpler analysis suffices.

For a single block brake (Fig. 13.36) the force exerted on the drum by the brake shoe must be supported by the bearings. To balance this load and provide a compact braking arrangement two opposing brake shoes are usually used in a calliper arrangement as shown in Fig. 13.37.

The following equations can be used (with reference to Fig. 13.36) to determine the performance of a long shoe brake.

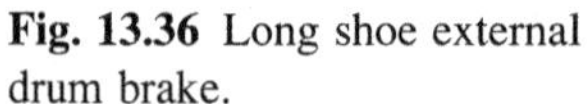

Fig. 13.36 Long shoe external drum brake.

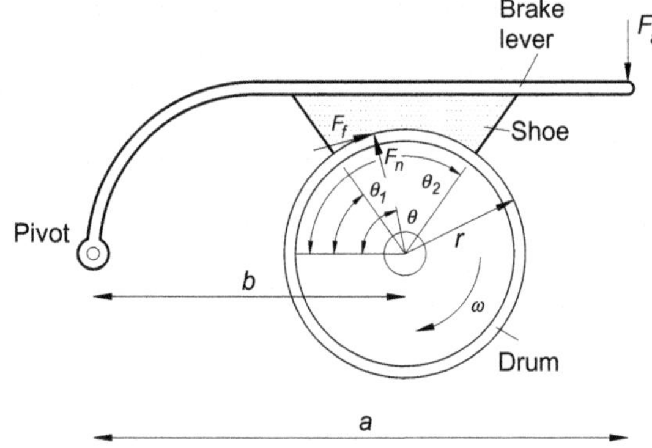

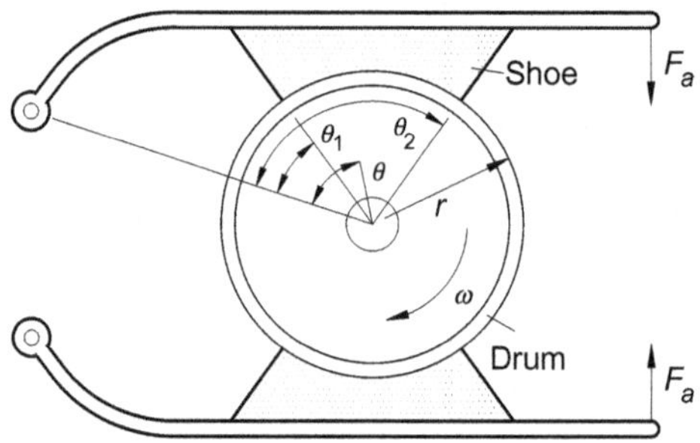

Fig. 13.37 Double long shoe external drum brake.

The braking torque T is given by

$$T = \mu w r^2 \frac{p_{max}}{(\sin\theta)_{max}} (\cos\theta_1 - \cos\theta_2) \tag{13.37}$$

where μ = coefficient of friction, w = width of the brake shoe (m), r = radius of drum (m), p_{max} = maximum allowable pressure for the lining material (N/m^2), θ = angular location (rad), $(\sin\,\theta)_{max}$ = maximum value of $\sin\,\theta$, θ_1 = centre angle from the shoe pivot to the heel of the lining (rad), θ_2 = centre angle from the shoe pivot to the toe of the lining (rad).

This is based on the assumption that the local pressure p at an angular location θ is related to the maximum pressure, p_{max}, by

$$p = \frac{p_{max}\sin\theta}{(\sin\theta)_{max}} \tag{13.38}$$

The local pressure p will be a maximum when $\theta = 90$ degrees. If $\theta_2 < 90$ degrees then the pressure will be a maximum at the toe, θ_2. The relationship given in Eq. (13.38) assumes that there is no deflection at the shoe or the drum, no wear on the drum and that the shoe wear is proportional to the frictional work and hence the local pressure. Note that if $\theta=0$ the pressure is zero. This indicates that frictional material near the pivot or heel of the brake does not contribute significantly to the braking action. For this reason common practice is to leave out the frictional material near the heal, and start it at an angle typically between $\theta_1=10$ degrees and $\theta_1=30$ degrees.

With the direction of rotation shown in Fig. 13.36 (i.e. the brake is self-energising), the magnitude of the actuation force is given by

$$F_a=\frac{M_n-M_f}{a} \tag{13.39}$$

where M_n = the moment of the normal forces (N m), M_f = the moment due to the frictional forces (N m), a = the orthogonal distance between the brake pivot and the line of action of the applied force (m).

The normal and frictional moments can be determined using Eqs (13.40) and (13.41) respectively. If the geometry and materials are selected such that $M_f=M_n$ then the actuation force becomes zero. Such a brake would be self-locking. The slightest contact between the shoe and drum would bring the two surfaces into contact and the brake would snatch giving rapid braking. Alternatively values for the brake geometry and materials can be selected to give different levels of self-energisation depending on the relative magnitudes of M_n and M_f.

$$M_n=\frac{wrbp_{\max}}{(\sin\theta)_{\max}}\left[\frac{1}{2}(\theta_2-\theta_1)-\frac{1}{4}(\sin 2\theta_2-\sin 2\theta_1)\right] \tag{13.40}$$

$$M_f=\frac{\mu wrp_{\max}}{(\sin\theta)_{\max}}\left[r(\cos\theta_1-\cos\theta_2)+\frac{b}{4}(\cos 2\theta_2-\cos 2\theta_1)\right] \tag{13.41}$$

If the direction of rotation for the drum shown in Fig. 13.36 is reversed, the brake becomes self-deenergising and the actuation force is given by

$$F_a=\frac{M_n+M_f}{a} \tag{13.42}$$

The pivot reactions can be determined by resolving the horizontal and vertical forces. For a self-energising brake they are given by

$$R_x=\frac{wrp_{\max}}{(\sin\theta)_{\max}}\left[\left(0.5\left(\sin^2\theta_2-\sin^2\theta_1\right)\right)-\mu\left(\frac{\theta_2}{2}-\frac{\theta_1}{2}-\frac{1}{4}(\sin 2\theta_2-\sin 2\theta_1)\right)\right]-F_x \tag{13.43}$$

$$R_y=\frac{wrp_{\max}}{(\sin\theta)_{\max}}\left[\left(0.5\left(\sin^2\theta_2-\sin^2\theta_1\right)\right)+\mu\left(\frac{\theta_2}{2}-\frac{\theta_1}{2}-\frac{1}{4}(\sin 2\theta_2-\sin 2\theta_1)\right)\right]+F_y \tag{13.44}$$

and for a self-deenergising brake by

$$R_x = \frac{wrp_{max}}{(\sin\theta)_{max}}\left[\left(0.5\left(\sin^2\theta_2 - \sin^2\theta_1\right)\right) + \mu\left(\frac{\theta_2}{2} - \frac{\theta_1}{2} - \frac{1}{4}(\sin 2\theta_2 - \sin 2\theta_1)\right)\right] - F_x \quad (13.45)$$

$$R_y = \frac{wrp_{max}}{(\sin\theta)_{max}}\left[\left(0.5\left(\sin^2\theta_2 - \sin^2\theta_1\right)\right) - \mu\left(\frac{\theta_2}{2} - \frac{\theta_1}{2} - \frac{1}{4}(\sin 2\theta_2 - \sin 2\theta_1)\right)\right] + F_y \quad (13.46)$$

Example 13.9. Design a long shoe drum brake to produce a friction torque of 75 N m to stop a drum rotating at 140 rpm. Initial design calculations have indicated that a shoe lining with $\mu = 0.25$ and using a value of $p_{max} = 0.5 \times 10^6$ N/m^2 in the design will give suitable life.

Solution

First propose trial values for the brake geometry, say $r = 0.1$ m, $b = 0.2$ m, $a = 0.3$ m, $\theta_1 = 30$ degrees, $\theta_2 = 150$ degrees. Using Eq. 13.37 and solving for the width of the shoe,

$$w = \frac{T(\sin\theta)_{max}}{\mu r^2 p_{max}(\cos\theta_1 - \cos\theta_2)} = \frac{75 \sin 90}{0.25 \times 0.1^2 \times 0.5 \times 10^6(\cos 30 - \cos 150)} = 0.0346\,\text{m}$$

Select the width to be 35 mm as this is a standard size. The actual maximum pressure experienced, will be

$$p_{max} = 0.5 \times 10^6 \frac{0.0346}{0.035} = 494{,}900\,\text{N/m}^2.$$

From Eq. (13.40) the moment of the normal force with respect to the shoe pivot is

$$M_n = \frac{0.035 \times 0.1 \times 0.2 \times 0.4949 \times 10^6}{\sin 90}\left[\frac{1}{2}\left(120 \times \frac{2\pi}{360}\right) - \frac{1}{4}(\sin 300 - \sin 60)\right]$$
$$= 512.8\,\text{Nm}$$

From Eq. (13.41) the moment of the frictional forces with respect to the shoe pivot is

$$M_f = \frac{0.25 \times 0.035 \times 0.1 \times 0.4949 \times 10^6}{\sin 90}\left[0.1(\cos 30 - \cos 150) + \frac{0.2}{4}(\cos 300 - \cos 60)\right]$$
$$= 75\,\text{Nm}$$

From Eq. (13.39) the actuation force is

$$F_a = \frac{M_n - M_f}{a} = \frac{512.8 - 75}{0.3} = 1459\,\text{N}$$

For the double long shoe external drum brake illustrated in Fig. 13.37, the left hand shoe is self-energising and the frictional moment reduces the actuation load. The right hand shoe, however, is self-deenergising and its frictional moment acts to reduce the maximum pressure which occurs on the right hand brake shoe. The normal and frictional moments for a self-energising and self-deenergising brake are related by

$$M_n{}' = \frac{M_n p'_{\max}}{p_{\max}} \tag{13.47}$$

$$M_f{}' = \frac{M_f p'_{\max}}{p_{\max}} \tag{13.48}$$

where $M_n{}'$ = the moment of the normal forces for the self-deenergising brake (N m), M_f' = the moment due to the frictional forces for the self-deenergising brake (N m), $p_{\max}$ = the maximum pressure on the self-deenergising brake (N/m^2).

Example 13.10. For the double long shoe external drum brake illustrated in Fig. 13.38 determine the limiting force on the lever such that the maximum pressure on the brake lining does not exceed 1.4 MPa and determine the torque capacity of the brake. The face width of the shoes is 30 mm and the coefficient of friction between the shoes and the drum can be taken as 0.28.

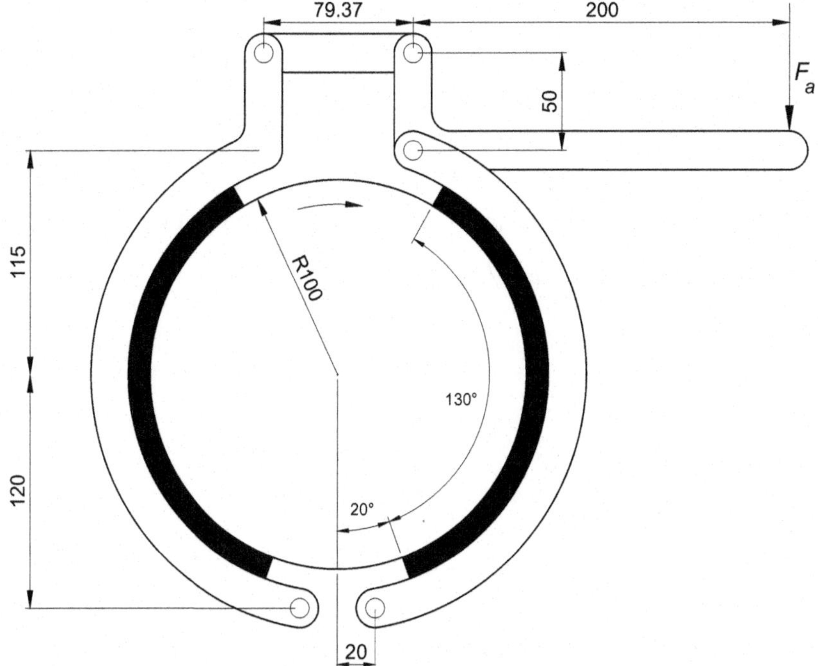

Fig. 13.38 Double long shoe external drum brake.

Solution

First it is necessary to calculate values for θ_1 and θ_2 as these are not indicated directly on the diagram.

$$\theta_1 = 20\,\text{deg} - \tan^{-1}\left(\frac{20}{120}\right) = 10.54\,\text{deg}$$

$$\theta_2 = 20\,\text{deg} + 130\,\text{deg} - \tan^{-1}\left(\frac{20}{120}\right) = 140.5\,\text{deg}$$

The maximum value of $\sin\theta$ would be $\sin 90 = 1$.

The distance between the pivot and the drum centre, $b = \sqrt{0.02^2 + 0.12^2} = 0.1217\,\text{m}$.

The normal moment is given by

$$M_n = \frac{wrbp_{\max}}{(\sin\theta)_{\max}}\left[\frac{1}{2}(\theta_2 - \theta_1) - \frac{1}{4}(\sin 2\theta_2 - \sin 2\theta_1)\right]$$

$$= \frac{0.03 \times 0.1 \times 0.1217 \times 1.4 \times 10^6}{\sin 90}\left[\frac{1}{2}\left((140.5 - 10.54) \times \frac{2\pi}{360}\right) - \frac{1}{4}(\sin 281 - \sin 21.08)\right]$$

$$= 751.1\,\text{Nm}$$

$$M_f = \frac{\mu wrp_{\max}}{(\sin\theta)_{\max}}\left[r(\cos\theta_1 - \cos\theta_2) + \frac{b}{4}(\cos 2\theta_2 - \cos 2\theta_1)\right]$$

$$= \frac{0.28 \times 0.03 \times 0.1 \times 1.4 \times 10^6}{\sin 90}\left[0.1(\cos 10.54 - \cos 140.5) + \frac{0.1217}{4}(\cos 281 - \cos 21.08)\right]$$

$$= 179.8\,\text{N m}$$

The orthogonal distance between the actuation force and the pivot, a, is given by

$$a = 0.12 + 0.115 + 0.05 = 0.285\,\text{m}$$

The actuation load on the left hand shoe is given by

$$F_{a\ left\ shoe} = \frac{M_n - M_f}{a} = \frac{751.1 - 179.8}{0.285} = 2004\,\text{N}$$

The torque contribution from the left hand shoe is given by

$$T_{left\ shoe} = \mu wr^2 \frac{p_{\max}}{(\sin\theta)_{\max}}(\cos\theta_1 - \cos\theta_2)$$

$$= 0.28 \times 0.03 \times 0.1^2 \times 1.4 \times 10^6(\cos 10.54 - \cos 140.5) = 206.4\,\text{Nm}$$

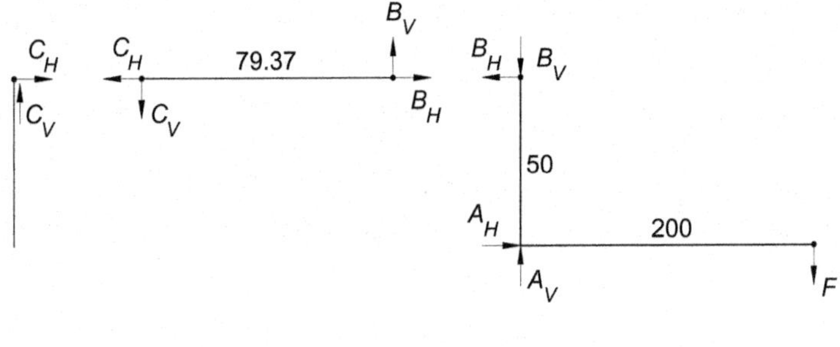

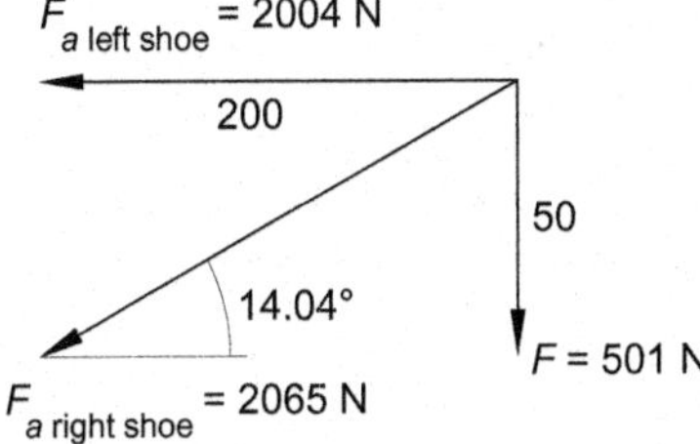

Fig. 13.39 Free body diagrams.

The actuation force on the right hand shoe can be determined by considering each member of the lever mechanism as a free body (Fig. 13.39).

$$F - A_V + B_V = 0$$

$$A_H = B_H$$

$$B_H = C_H$$

$$A_H = C_H$$

$$0.2F = 0.05B_H, F = B_H/4, B_H = 2004\text{N}, F = 2004/4 = 501\text{N}$$

So the limiting lever force is $F = 501$ N.

$$C_V = 0, B_V = 0$$

The actuating force for the right hand lever is the resultant of F and B_H. The resultant angle is given by $\tan^{-1}(0.05/0.2) = 14.04$ degrees.

$$F_{a\ right\ shoe} = \frac{2004}{\cos 14.04} = 2065\text{N}$$

The orthogonal distance between the actuation force vector and the pivot (Fig. 13.40) is given by

$$a = (0.235 - 0.01969 \tan 14.04) \times \cos 14.04 = 0.2232\text{m}$$

The normal and frictional moments for the right hand shoe can be determined using Eqs (13.47) and (13.48).

$$M'_n = \frac{M_n p'_{\max}}{p_{\max}} = \frac{751.1 p'_{\max}}{1.4 \times 10^6}$$

$$M'_f = \frac{M_f p'_{\max}}{p_{\max}} = \frac{179.8 p'_{\max}}{1.4 \times 10^6}$$

For the right hand shoe the maximum pressure can be determined from

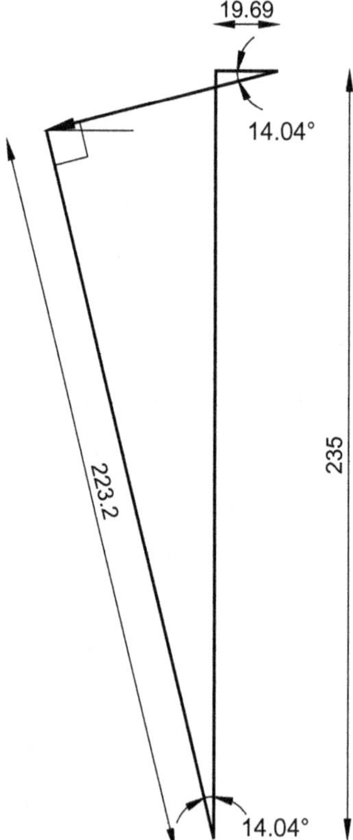

Fig. 13.40 Orthogonal distance.

$$F_{a\ right\ shoe} = \frac{M'_n + M'_f}{a} = 2065 = \frac{751.1p'_{\max} - 179.8p'_{\max}}{1.4 \times 10^6 \times 0.2232}$$

$$p'_{\max} = 1.130 \times 10^6\ \mathrm{N/m^2}.$$

The torque contribution from the right hand shoe is

$$T_{right\ shoe} = \mu w r^2 \frac{p'_{\max}}{(\sin\theta)_{\max}}(\cos\theta_1 - \cos\theta_2)$$

$$= 0.28 \times 0.03 \times 0.1^2 \times 1.13 \times 10^6(\cos 10.54 - \cos 140.5)$$

$$= 166.6\mathrm{Nm}$$

The total torque is given by.

$$T_{total} = T_{leftshoe} + T_{rightshoe} = 206.4 + 166.6 = 373\mathrm{Nm}$$

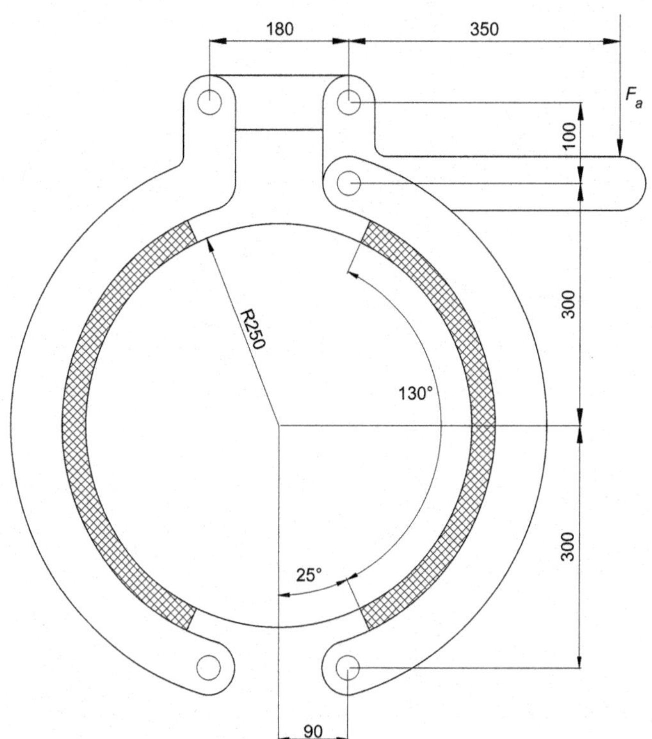

Fig. 13.41 Double external long shoe drum brake.

Example 13.11. A double external long shoe drum brake is illustrated in Fig. 13.41. The face width of the shoes is 50 mm and the maximum permissible lining pressure is 1 MPa. If the coefficient of friction is 0.32 determine the limiting actuating force and the torque capacity.

Solution

First it is necessary to calculate values for θ_1 and θ_2 as these are not indicated directly on the diagram.

$$\theta_1 = 25\,\text{deg} - \tan^{-1}\left(\frac{90}{300}\right) = 8.301\text{ deg.}$$

$$\theta_2 = 25\,\text{deg} + 130\,\text{deg} - \tan^{-1}\left(\frac{90}{300}\right) = 138.3\,\text{deg.}$$

The maximum value of $\sin\theta$ would be $\sin 90 = 1$.

The distance between the pivot and the drum centre,

$$b = \sqrt{0.09^2 + 0.3^2} = 0.3132\,\text{m}$$

The normal moment is given by

$$M_n = \frac{wrbp_{max}}{(\sin\theta)_{max}}\left[\frac{1}{2}(\theta_2 - \theta_1) - \frac{1}{4}(\sin 2\theta_2 - \sin 2\theta_1)\right]$$

$$= \frac{0.05 \times 0.25 \times 0.3132 \times 1 \times 10^6}{\sin 90}\left[\frac{1}{2}\left((138.3 - 8.301) \times \frac{2\pi}{360}\right) - \frac{1}{4}(\sin 276.6 - \sin 16.6)\right]$$

$$= 5693\,\text{Nm}$$

$$M_f = \frac{\mu wrp_{max}}{(\sin\theta)_{max}}\left[r(\cos\theta_1 - \cos\theta_2) + \frac{b}{4}(\cos 2\theta_2 - \cos 2\theta_1)\right]$$

$$= \frac{0.32 \times 0.05 \times 0.25 \times 1 \times 10^6}{\sin 90}\left[0.25(\cos 8.301 - \cos 138.3) + \frac{0.3132}{4}(\cos 276.6 - \cos 16.6)\right]$$

$$= 1472\,\text{Nm}$$

The orthogonal distance between the actuation force and the pivot, $a = 0.7$ m.

The actuation load on the left hand shoe is given by

$$F_{a\ left\ shoe} = \frac{M_n - M_f}{a} = \frac{5693 - 1472}{0.7} = 6031\,\text{N}$$

The torque contribution from the left hand shoe is given by

$$T_{left\ shoe} = \mu w r^2 \frac{p_{max}}{(\sin\theta)_{max}}(\cos\theta_1 - \cos\theta_2)$$

$$= 0.32 \times 0.05 \times 0.25^2 \times 1 \times 10^6(\cos 8.301 - \cos 138.3)$$

$$= 1736\text{Nm}$$

The actuation force on the right hand shoe can be determined by considering each member of the lever mechanism as a free body.

$$F - A_V + B_V = 0$$

$$A_H = B_H$$

$$B_H = C_H$$

$$A_H = C_H$$

$$0.35F = 0.1B_H$$

$$F = 0.1B_H/0.35$$

$$B_H = 6031$$

$$F = 6031 \times 0.1/0.35 = 1723\text{N}$$

So the limiting lever force is $F = 1723$N.

$$C_V = 0$$

$$B_V = 0$$

The actuating force for the right hand lever is the resultant of F and B_H.

The resultant angle is given by $\tan^{-1}(0.1/0.35) = 15.95$ degrees.

$$F_{a\ right\ shoe} = \frac{1723}{\cos 15.95} = 6272\text{N}$$

The perpendicular distance between the actuation force vector and the pivot is given by

$$a = 0.6 \times \cos 15.95 = 0.5769\text{m}$$

The normal and frictional moments for the right hand shoe can be determined using Eqs (13.47) and (13.48).

$$M'_n = \frac{M_n p'_{\max}}{p_{\max}} = \frac{5693 p'_{\max}}{1 \times 10^6}$$

$$M'_f = \frac{M_f p'_{\max}}{p_{\max}} = \frac{1472 p'_{\max}}{1 \times 10^6}$$

For the right-hand shoe the maximum pressure can be determined from

$$F_{a\ right\ shoe} = \frac{M'_n + M'_f}{a} = 6272 = \frac{5693 p'_{\max} - 1472 p'_{\max}}{1 \times 10^6 \times 0.5769}$$

$$p'_{\max} = 0.8571 \times 10^6 \mathrm{N/m^2}$$

The torque contribution from the right hand shoe is

$$T_{right\ shoe} = \mu w r^2 \frac{p'_{\max}}{(\sin\theta)_{\max}} (\cos\theta_1 - \cos\theta_2)$$

$$= 0.32 \times 0.05 \times 0.25^2 \times 0.8571 \times 10^6 (\cos 8.301 - \cos 138.3) = 1488 \mathrm{Nm}$$

The total torque is given by

$$T_{total} = T_{leftshoe} + T_{rightshoe} = 1736 + 1488 = 3224 \mathrm{Nm}$$

13.3.5 Long shoe internal drum brakes

Most drum brakes use internal shoes that expand against the inner radius of the drum. Long shoe internal drum brakes are principally used in automotive applications. An automotive drum brake typically comprises two brake shoes and linings supported on a back plate bolted to the axle casing. The shoes are pivoted at one end on anchor pins or abutments fixed onto the back plate (Fig. 13.42). The brake can be actuated by a double hydraulic piston expander, which forces the free ends of the brake apart so that the nonrotating shoes come into frictional contact with the rotating brake drum. A leading and trailing shoe layout consists of a pair of shoes pivoted at a common anchor point as shown in Fig. 13.42. The leading shoe is identified as the shoe whose expander piston moves in the direction of rotation of the drum. The frictional drag between the shoe and the drum will tend to assist the expander piston in forcing the shoe against the drum and this action is referred to as self-energising or the self-servo action of the shoe. The trailing shoe is the one whose expander piston moves in the direction opposed the rotation of the drum. The frictional force opposes the expander and hence a trailing brake shoe provides less braking torque than an equivalent leading shoe actuated by the same force. The equations developed for external long shoe drum brakes are also valid for internal long shoe drum brakes.

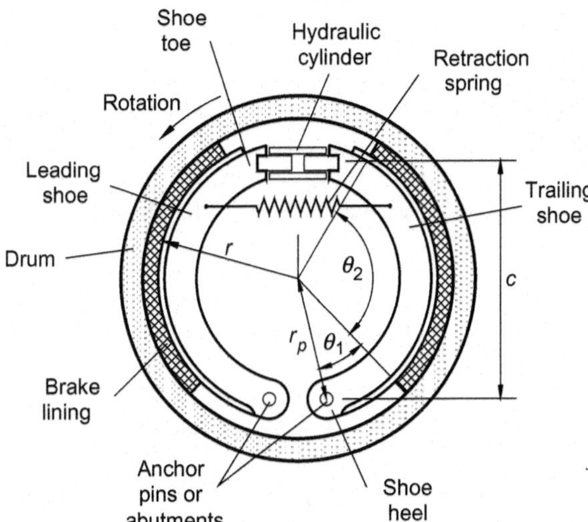

Fig. 13.42 Double long shoe internal drum brake.

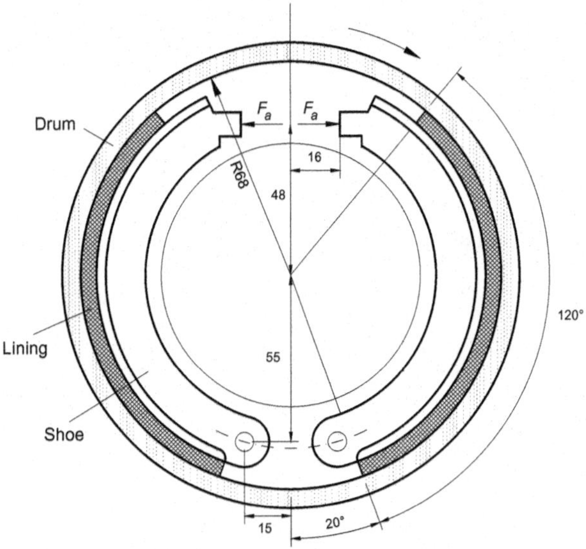

Fig. 13.43 Double long shoe internal drum brake.

Example 13.12. Determine the actuating force and the braking capacity for the double internal long shoe brake illustrated in Fig. 13.43. The lining is sintered metal with a coefficient of friction of 0.32 and the maximum lining pressure is 1.2 MPa. The drum radius is 68 mm and the shoe width is 25 mm.

Solution

$$b = \sqrt{0.015^2 + 0.055^2} = 0.05701\,\text{m}$$

As the brake lining angles relative to the pivot, brake axis line, are not explicitly shown on the diagram, they must be calculated.

$$\theta_1 = 4.745\,\text{deg}, \theta_2 = 124.7\,\text{deg}$$

As $\theta_2 > 90$ degrees, the maximum value of $\sin\theta$ is $\sin 90 = 1 = (\sin\theta)_{max}$.

For this brake with the direction of rotation as shown the right hand shoe is self-energising.

For the right hand shoe:

$$M_n = \frac{0.025 \times 0.068 \times 0.05701 \times 1.2 \times 10^6}{1}$$

$$\left[\frac{1}{2}\left((124.7 - 4.745) \times \frac{2\pi}{360}\right) - \frac{1}{4}(\sin 249.4 - \sin 9.49)\right]$$

$$= 153.8\,\text{Nm}$$

$$M_f = \frac{0.32 \times 0.025 \times 0.068 \times 1.2 \times 10^6}{1}$$

$$\left[0.068(\cos 4.745 - \cos 124.7) + \frac{0.05701}{4}(\cos 249.4 - \cos 9.49)\right]$$

$$= 57.1\,\text{N m}$$

$$a = 0.055 + 0.048 = 0.103\,\text{m}$$

$$F_a = \frac{M_n - M_f}{a} = \frac{153.8 - 57.1}{0.103} = 938.9\,\text{N}$$

The actuating force is 938.9 N.

The torque applied by the right hand shoe is given by

$$T_{right\,shoe} = \frac{\mu w r^2 p_{max}}{(\sin\theta)_{max}}(\cos\theta_1 - \cos\theta_2)$$

$$= \frac{0.32 \times 0.025 \times 0.068^2 \times 1.2 \times 10^6}{1}(\cos 4.745 - \cos 124.7) = 69.54\,\text{Nm}$$

The torque applied by the left hand shoe cannot be determined until the maximum operating pressure p'_{max} for the left hand shoe has been calculated.

As the left hand shoe is self-deenergising the normal and frictional moments can be determined using Eqs (13.47) and (13.48).

$$M'_n = \frac{M_n p'_{max}}{p_{max}} = \frac{153.8p'_{max}}{1.2 \times 10^6}$$

$$M'_f = \frac{M_f p'_{max}}{p_{max}} = \frac{57.1p'_{max}}{1.2 \times 10^6}$$

The left hand shoe is self-deenergising, so,

$$F_a = \frac{M_n + M_f}{a}$$

$F_a = 938.9\,\text{N}$ as calculated earlier.

$$938.9 = \frac{153.8p'_{max} + 57.1p'_{max}}{1.2 \times 10^6 \times 0.103}$$

$$p'_{max} = 0.5502 \times 10^6\,\text{N/m}^2.$$

The torque applied by the left hand shoe is given by

$$T_{left\,shoe} = \frac{\mu w r^2 p'_{max}}{(\sin\theta)_{max}}(\cos\theta_1 - \cos\theta_2)$$

$$= \frac{0.32 \times 0.025 \times 0.068^2 \times 0.5502 \times 10^6}{1}(\cos 4.745 - \cos 124.7) = 31.89\,\text{Nm}$$

The total torque applied by both shoes is

$$T_{total} = T_{right\,shoe} + T_{left\,shoe} = 69.54 + 31.89 = 101.4\,\text{Nm}$$

From this example the advantage in torque capacity of using self-energising brakes is apparent. Both the left hand and the right hand shoes could be made self-energising by inverting the left hand shoe, having the pivot at the top. This would be advantageous if rotation occurred in just one direction. If, however, drum rotation is possible in either direction, it may be more suitable to have one brake self-energising for forward motion and one self-energising for reverse motion.

13.3.6 Band brakes

One of the simplest types of braking device is the band brake. This consists of a flexible metal band lined with a frictional material wrapped partly around a drum. The brake is actuated by pulling the band against the drum as illustrated in Fig. 13.44.

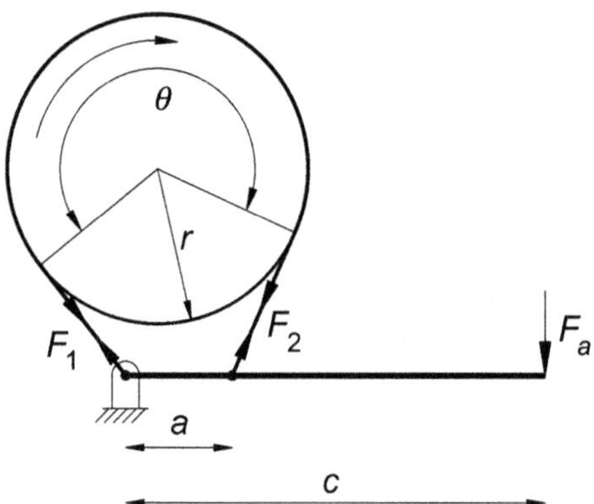

Fig. 13.44 Band brake.

For the clockwise rotation shown in Fig. 13.44, the friction forces increase F_1 relative to F_2. The relationship between the tight and slack sides of the band is given by

$$\frac{F_1}{F_2} = e^{\mu\theta} \tag{13.49}$$

where F_1 = tension in the tight side of the band (N), F_2 = tension in the slack side of the band (N), μ = coefficient of friction, θ = angle of wrap (rad).

The point of maximum contact pressure for the friction material occurs at the tight end and is given by

$$p_{max} = \frac{F_1}{rw} \tag{13.50}$$

where w is the width of the band (m).

The torque braking capacity is given by

$$T = (F_1 - F_2)r \tag{13.51}$$

The relationship, for the simple band brake shown in Fig. 13.44, between the applied lever force F_a and F_2 can be found by taking moments about the pivot point.

$$F_a c - F_2 a = 0$$

$$F_a = F_2 \frac{a}{c} \tag{13.52}$$

The brake configuration shown in Fig. 13.44 is self-energising for clockwise rotation. The level of self-energisation can be enhanced by using the differential band brake configuration shown in Fig. 13.45. Summation of the moments about the pivot gives

$$F_a c - F_2 a + F_1 b = 0 \tag{13.53}$$

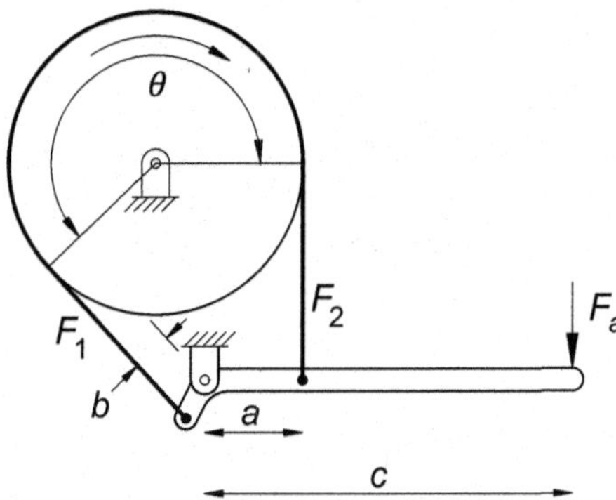

Fig. 13.45 Self-energising band brake.

So the relationship between the applied load F_a and the band brake tensions is given by

$$F_a = \frac{F_2 a - F_1 b}{c} \tag{13.54}$$

Note that the value of b must be less than a so that applying the lever tightens F_2 more than it loosens F_1. Substituting for F_1 in Eq. 13.54 gives

$$F_a = \frac{F_2\left(a - be^{\mu\theta}\right)}{c} \tag{13.55}$$

The brake can be made self-locking if $a < be^{\mu\theta}$ and the slightest touch on the lever would cause the brake to grab or lock abruptly. This principle can be used to permit rotation in one direction only as in hoist and conveyor applications.

Example 13.13. Design a band brake to exert a braking torque of 85 N m. Assume the coefficient of friction for the lining material is 0.25 and the maximum permissible pressure is 0.345 MPa.

Solution

Propose a trial geometry, say $r = 150\,\text{mm}$, $\theta = 225$ degrees and $w = 50\,\text{mm}$.

$$F_1 = p_{\max} r w = 0.345 \times 10^6 \times 0.15 \times 0.05 = 2587\,\text{N}$$

$$F_2 = \frac{F_1}{e^{\mu\theta}} = \frac{2587.5}{e^{0.25(225\times 2\pi/360)}} = 969\,\text{N}$$

$$T = (F_1 - F_2)r = (2587.5 - 969)0.15 = 242.7\,\text{Nm}$$

This torque is much greater than the 80 N m desired, so try a different combination of r, θ and w until a satisfactory design is achieved.

Try $r = 0.1\,\text{m}$, $\theta = 225$ degrees and $w = 50\,\text{mm}$.

$$F_1 = p_{max} rw = 0.345 \times 10^6 \times 0.1 \times 0.05 = 1725\,\text{N}$$

$$F_2 = \frac{F_1}{e^{\mu\theta}} = \frac{1725}{e^{0.25(225 \times 2\pi/360)}} = 646.3\,\text{N}$$

$$T = (F_1 - F_2)r = (1725 - 646.3)0.1 = 107.9\,\text{Nm}$$

Try $r = 0.09\,\text{m}$, $\theta = 225$ degrees and $w = 50\,\text{mm}$.

$$F_1 = p_{max} rw = 0.345 \times 10^6 \times 0.09 \times 0.05 = 1552.5\,\text{N}$$

$$F_2 = \frac{F_1}{e^{\mu\theta}} = \frac{1552.5}{e^{0.25(225 \times 2\pi/360)}} = 581.7\,\text{N}$$

$$T = (F_1 - F_2)r = (1552.5 - 581.7)0.09 = 87.4\,\text{Nm}$$

The actuating force is given by $F_a = F_2 a/c$. If $a = 0.08\,\text{m}$ and $c = 0.15\,\text{m}$ then,

$$F_a = 581.7 \times \frac{0.08}{0.15} = 310.2\,\text{N}$$

Practical braking systems comprise an energy supplying device, a control device, a transmission device for controlling the braking force and the brakes themselves. For example in the case of a braking system for passenger cars the braking system could control the brake pedal, a vacuum booster, master hydraulic fluid cylinder, brake fluid reservoir, a device to warn the driver of a brake circuit failure, a sensor to warn of low brake fluid level, valves, hydraulic cylinders, springs, pads and discs. For heavy vehicles, say 2–3 tonnes and over, the force required for braking can become greater than a person can exert. In such cases some assistance needs to be given to the driver. This can be achieved by using a servo mechanism that adds to the driver's effort, for example a vacuum assisted brake servo unit, or by using power operation in which case the driver's effort is simply for control purposes and is not transmitted directly to the brakes.

The performance of braking systems in, say, automotive applications has been significantly enhanced over the purely mechanical variants described so far by the use of sensors, actuators and sophisticated control systems. An example is antilock braking systems (ABS), which are closed-loop control devices within the braking system. The objective of ABS is to prevent wheel lock-up during braking and as a result retain greater steering control and vehicle stability. The principal components of an ABS system are a hydraulic modulator, wheel-based sensors and the engine control unit (ECU). The ECU processes the signals and controls and triggers the hydraulics. The control loop for an ABS is shown in Fig. 13.46. ABS systems are considered in more detail in the Bosch Automotive Handbook (2000), Garrett et al. (2001), Breuer and Dausend (2003), and Halderman and Mitchell (2016).

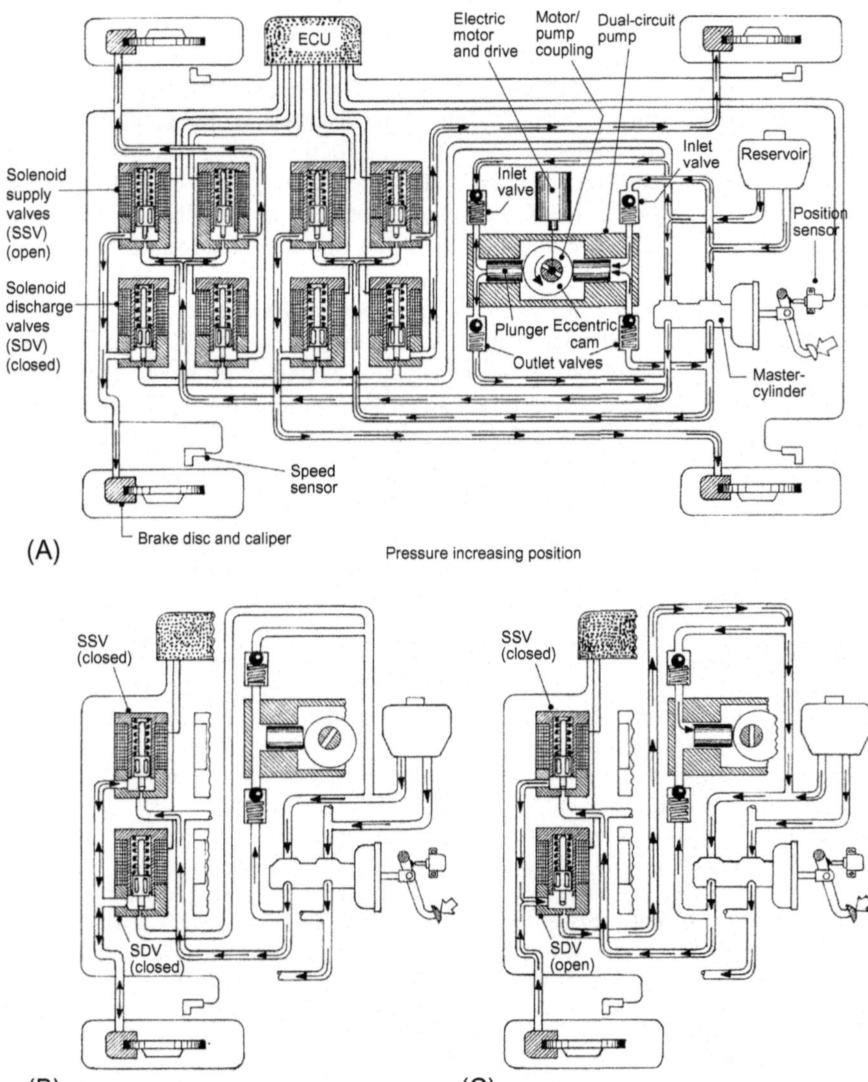

Fig. 13.46 Control loop for an antilock braking system (ABS).
Reproduced from Heisler, H., 1999. Vehicle and Engine Technology, second ed. Butterworth Heinemann.

It should be noted that the design of braking systems in the case of automotive applications is subject to extensive regulation and requirements are detailed in documents such as StVZO, EC Directive 71,320 and ECE Directive 13.

13.4 Conclusions

Clutches are designed to permit the smooth, gradual engagement or disengagement of a prime mover from a driven load. Brakes are designed to decelerate a system. Clutches and brakes are similar devices providing frictional, magnetic or direct positive connection between two components. In principle one device could function as either a clutch or a brake. If one component rotates and the other is fixed to a nonrotating plane of reference the device will function as a brake and if both rotate then as a clutch.

This chapter has concentrated on rotating clutches and brakes and specifically on the design of friction-based devices. In the case of brakes, control systems enable significant advantages over purely mechanical devices enabling the prevention of wheel lock-up during braking and as a result retain greater driveability. The detailed design of a clutch or braking system involves integration of a wide range of skills such as bearings, shafts, splines, teeth, flywheels, casings, frictional surfaces, hydraulics, sensors and control algorithms. Both brakes and clutches can be purchased from specialist suppliers or alternatively key components such as brake pads or clutch discs can be specified and bought in from specialist suppliers and integrated into a fit for purpose machine design.

References

Abdullah, O.I., Schlattmann, J., 2016. Thermal behavior of friction clutch disc based on uniform pressure and uniform wear assumptions. Friction 4, 228–237.

Abhang, S.R., Bhaskar, D.P., 2014. Design and analysis of disc brake. Int. J. Eng. Trends Technol. 8, 165–167.

Bosch, 2000. Automotive Handbook, fifth ed. Bentley Publishers.

Breuer, B.J., Dausend, U., 2003. Advanced Brake Technology. SAE.

Chopade, M., Valavade, A., 2017. Experimental investigation using CFD for thermal performance of ventilated disc brake rotor. Int. J. Automot. Technol. 18, 235–244.

Day, A.J., Harding, P.R.J., Newcomb, T.P., 1984. Combined thermal and mechanical analysis of drum brakes. Proc. Inst. Mech. Eng. D: Transp. Eng. 198, 287–294.

Fazekas, G.A., 1972. On circular spot brakes. J. Eng. Ind. Trans. ASME, 859–863.

Garrett, T.K., Newton, K., Steeds, W., 2001. The Motor Vehicle, 13th ed. Butterworth Heinemann.

Halderman, J.D., Mitchell, C.D., 2016. Automotive Brake Systems. Prentice Hall.

Limpert, R., 1975. Cooling Analysis of Disc Brake Rotors. SAE Technical Paper, No. 751014.

Limpert, R., 2012. Brake Design and Safety, third ed. SAE.

Newcomb, T.P., 1960. Temperatures reached in disc brakes. J. Mech. Eng. Sci. 2, 167–177.

Prabhakar, S., Prakash, S., Kumar, M.S., Annamalai, K., 2015. Performance analysis of ventilated brake disc for its effective cooling. J. Chem. Pharm. Sci. 974, 2115.

SAE, 1997. Manual Transmission Clutch Systems. SAE.

Schmidt, R.K., 2015a. Advances in Aircraft Brakes and Tires. SAE.

Schmidt, R.K., 2015b. Advances in Aircraft Landing Gear and Advances in Aircraft Brakes and Tires. SAE.

Tirovic, M., Day, A.J., 1991. Disc brake interface pressure distributions. Proc. Inst. Mech. Eng. D: J. Automob. Eng. 205, 137–146.

Vedamuttu, P., 1994. Clutches. In: Hurst, K. (Ed.), Rotary Power Transmission Design. McGraw Hill.

Standards

British Standards Institution. BS AU 180: Part 4: 1982. Brake linings. Method for determining effects of heat on dimensions and form of disc brake pads'.
British Standards Institution. BS 4639: 1987. Specification for brakes and braking systems for towed agricultural vehicles.
British Standards Institution. BS AU 249:1993, ISO 7649:1991. Specification for clutch housings for reciprocating internal combustions engines. Nominal dimensions and tolerances.

Further reading

Baker, A.K., 1992. Industrial Brake and Clutch Design. Pentech Press.
Heisler, H., 1999. Vehicle and Engine Technology, second ed. Butterworth Heinemann.
Hindhede, U., Zimmerman, J.R., Hopkins, R.B., Erisman, R.J., Hull, W.C., Lang, J.D., 1983. Machine Design Fundamentals. A Practical Approach. Wiley.
Juvinall, R.C., Marshek, K.M., 1991. Fundamentals of Machine Component Design. Wiley.
Neale, M.J., 1994. Drives and Seals: A Tribology Handbook. Butterworth Heinemann.
Neale, M.J. (Ed.), 1995. The Tribology Handbook. Butterworth Heinemann.
Proctor, J., 1961. Selecting clutches for mechanical drives. Prod. Eng. 43–58.

Websites

At the time of going to press the world-wide-web contained useful information relating to this chapter at the following sites.

www.boschusa.com/AutoParts/BrakePads/.
www.clutchfacing.com.
www.cross-morse.co.uk.
www.dba.com.au.
www.frendisa.com.
www.iatcoinc.com.
www.jbrakes.com.
www.kfabsc.com.
www.luk.de.
www.mondelengineering.com/mondProductsMain.htm.
www.nascoaircraft.com.
www.onassisauto.com.
www.trwauto.com.
www.warnernet.com.
www.winnard.co.uk.
www.xyd-autoparts.com.

Seals

14

Chapter Outline

Nomenclature

Generally, preferred SI units have been stated.

a	outer radius (m),
A	annular cross-sectional flow area of the seal (m^2),
b	inner radius (m),
c	seal clearance (m),
C_d	discharge coefficient (–),
$C_{d,r}$	discharge coefficient for a labyrinth fin with rounded tip edge (–),
$C_{d,se}$	discharge coefficient for a labyrinth fin with a sharp tip edge (–),
D	diameter of outer fin (m),
D_h	hydraulic diameter, defined as $D_h = 2(r_o - r_i)$ (m),
e	groove depth (m),
f	friction factor, defined as $f = \Delta p / [(L/D) 0.5 \rho u_{\text{mean}}^2]$ (–),

Mechanical Design Engineering Handbook. https://doi.org/10.1016/B978-0-08-102367-9.00014-7

g acceleration due to gravity (m/s^2),
h fin height (m),
H step height (m),
k carry-over factor (–),
k_1 Hodkinson's carry-over factor (–),
k_2 Zimmerman's carry-over factor (–),
k_i carry-over factor (–),
k_s correction factor for C_d (–),
l_{pitch} pitch (m),
L total length of seal (m),
L_{cell} honeycomb cell dimension (m),
m mass flow rate across the seal clearance area A, defined as $m = \rho\, u_{mean}\, A$ (kg/s),
m_{actual} actual mass flow rate (kg/s),
m_{ideal} ideal mass flow rate (kg/s),
Ma Mach number (–),
n number of labyrinth fins (–),
p static pressure (Pa),
p_n downstream static pressure (Pa),
p_t total pressure (Pa),
p_{t0} upstream total pressure (Pa),
Q mass flow function ($kg\,K^{0.5}/(N\,s)$), volumetric flow rate (m^3/s),
r radial coordinate (m),
r_{fillet} fillet radius (m),
r_i radius of labyrinth fin tip (m),
r_o inner radius of outer cylinder (m),
r_{round} rounding radius (m),
R characteristic gas constant (J/(kg K)),
Re axial Reynolds number (–),
Re_{D_h} Reynolds number based on hydraulic diameter (–),
Re_ϕ rotational Reynolds number (–),
S_i pressure factor (–),
t fin-tip land-width (m),
t_G groove width (m),
T fluid temperature (K),
T_t fluid total temperature (K),
T_{t0} upstream fluid total temperature (K),
Ta Taylor number (–),
u_{mean} mean velocity across the seal clearance (m/s),
$u_{\varnothing, r_i}$ fin tip angular velocity component or fin tip speed (m/s),
x axial coordinate (m),
ϕ diameter of the shaft (m),
ϕ_{fin} fin angle (degree),
γ isentropic index (–),
μ fluid dynamic viscosity (Pa s),
ρ fluid density (kg/m^3),
ρ_0 upstream fluid density (kg/m^3),
Δm mass flow change induced by honeycomb (kg/s),
Δp pressure difference (Pa),
Ω shaft angular velocity (rad/s).

Subscripts

$_0$ refers to flow conditions upstream of a labyrinth seal,
$_i$ refers to flow conditions in the ith chamber or ith fin,
$_n$ refers to flow conditions in the nth chamber or nth fin or downstream of a labyrinth seal.

14.1 Introduction to seals

Seals are devices used to prevent or limit leakage of fluids or particulates. Seals are widely used for applications where components do not move significantly relative to each other, and also for applications where there is significant movement between components as a result of rotation or reciprocation. The aims of this chapter are to introduce the variety of seal configurations, and to provide guidelines for the selection of seals and introduce calculation methods for the quantification of leakage rates for labyrinth, axial and bush seals.

The purpose of a seal is to prevent or limit flow between components. Seals are an important aspect of machine design where pressurised fluids must be contained within an area of a machine such as a hydraulic cylinder, contaminants excluded or lubricants retained. Seals fall into two general categories.

(1) Static seals, where sealing takes place between two surfaces that do no not move relative to each other.
(2) Dynamic seals, where sealing takes place between two surfaces that move relative to each other by, for example, rotary or reciprocating motion.

Any clearance between two components will permit the passage of fluid molecules in either direction, the direction depending on the pressures and momentum associated with the fluid. The basic sealing problem is illustrated in Fig. 14.1, where either boundary shown may be stationary or moving. Fluid can move between different regions of space by means of diffusion, free convection or forced convection. The size of a typical gas or vapour molecule is of the order of 10^{-9} m in diameter. They can therefore diffuse through very small gaps such as pores in a machine casing or seal component. Convection involves the mixing of one portion of fluid with another by means of gross movements of the mass of fluid. The fluid motion may be caused by external mechanical means such as by a fan or pump in which case the process is called forced convection. Alternatively, if the fluid motion is caused by density differences, due to, for instance, temperature differences, the process is called natural or free convection.

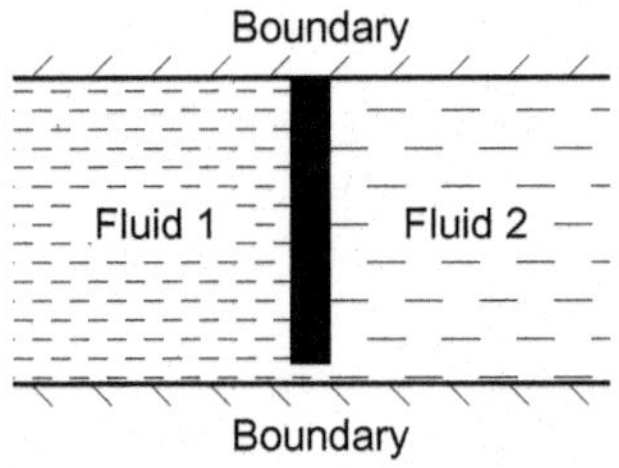

Fig. 14.1 The basic sealing geometry.

Some of the typical considerations in selecting the type of seal include the following:

(i) The nature of the fluid to be contained or excluded, whether liquid, gaseous, two phase, multiphase, multicomponent flow or particulate laden.
(ii) Static pressure levels either side of the seal.
(iii) The nature of any relative motion between the seal and mating components, e.g. rotating, reciprocating, contacting or noncontacting, high speed or low speed.
(iv) The level of sealing required. E.g. can some leakage be tolerated? Will a filter be used downstream to capture any particulates leaked?
(v) Operating temperatures.
(vi) Life expectancy.
(vii) Serviceability.
(viii) Total cost considering components and life-time service and replacement costs.

The variety of seals and sealing systems is extensive. Fig. 14.2 shows a general classification diagram for seals detailing the principal types, divided according to whether there is relative movement between the sealing surfaces. Several of these will be introduced in Section 14.2 for static seals and Section 14.3 for dynamic seals. The specific subject of labyrinth seals is introduced in Section 14.4. It should be noted, however, that once a general sealing requirement and possible solution has been identified, the best source of specific information is often one of the seal manufacturers, or for sealing bearings, bearing manufacturers. Fig. 14.3 provides a general guide to the selection of seal type, using the condition of relative motion to provide a starting point for selection.

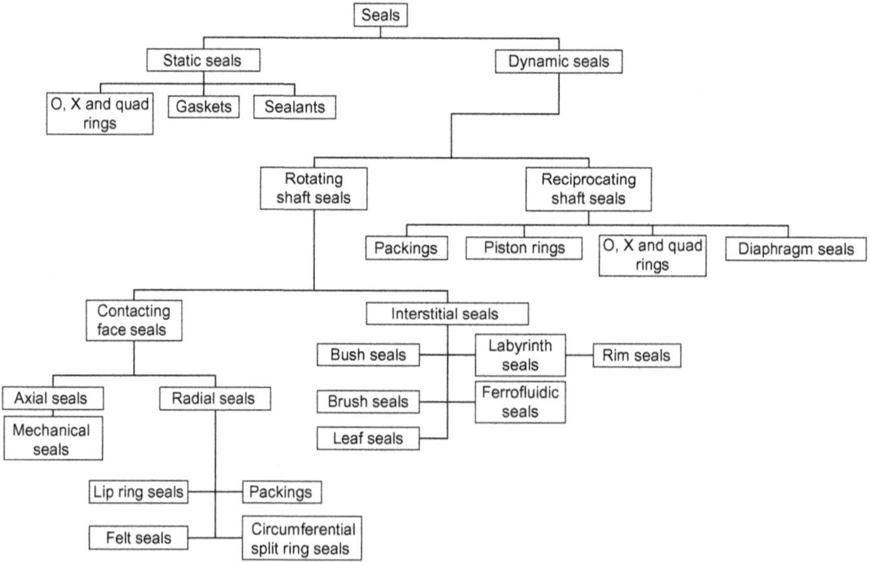

Fig. 14.2 General seal classification chart. (ESDU (2009)).

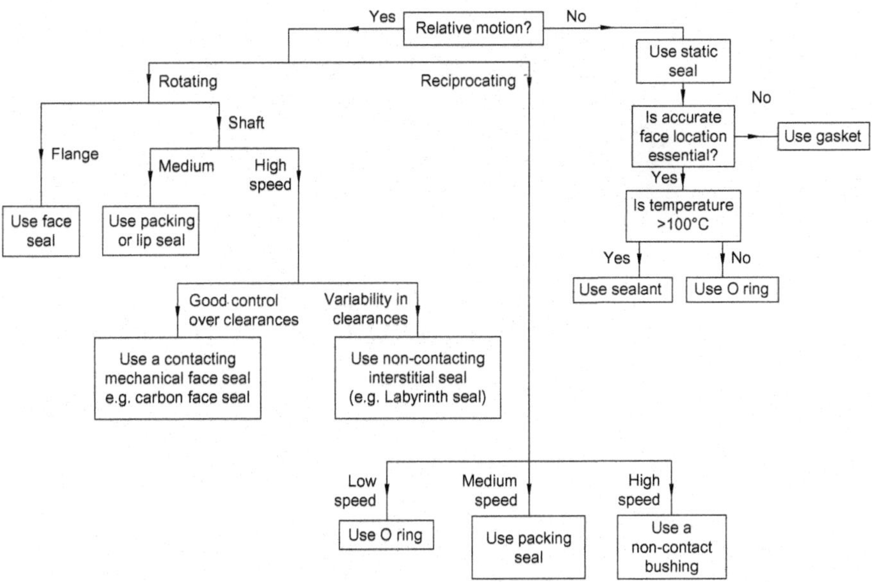

Fig. 14.3 Seal selection procedure.
Adapted from Hamilton, P., 1994. Seals. In: Hurst, K. (Ed.), Rotary Power Transmission. McGraw Hill.

14.2 Static seals

Static seals aim at providing a complete physical barrier to leakage flow. To achieve this, the seal material must be resilient enough to flow into and fill any irregularities in the surfaces being sealed and at the same time remain rigid enough to resist extrusion into clearances. Elastomeric seals and gaskets fulfil these criteria and are described in Sections 14.2.1 and 14.2.2, respectively. The particular demands of sealing of foodstuffs are introduced in Section 14.2.3.

14.2.1 Elastomeric seal rings

The 'O' ring, see Figs. 14.4 and 14.5, is a simple and versatile type of seal with a wide range of applications for both static and dynamic sealing, and even as a drive belt. An 'O' ring seal is a moulded elastomeric ring 'nipped' in a cavity in which the seal is located. The principle of operation for an 'O' ring sealing against a fluid at various pressures is illustrated in Fig. 14.6, and they can be used for pressures up to ca. 350 bar. Applications are diverse ranging from sealing casings, reciprocating, oscillating and rotating components, seats in solenoid valves and plumbing, pneumatics, vacuum and cryogenics, cushioning of impact loads, and low power transmission drive belt drives.

Elastomeric seal rings require the seal material to have an interference fit with one of the mating parts of the assembly. 'O' rings are available in a wide range of sizes with internal diameters from 3.1 mm to 249.1 mm and section diameters of 1.6, 2.4, 3.0, 4.1, 5.7 and 8.4 mm as defined in British Standard BS 4518. Table 14.1 shows

Fig. 14.4 An O ring.

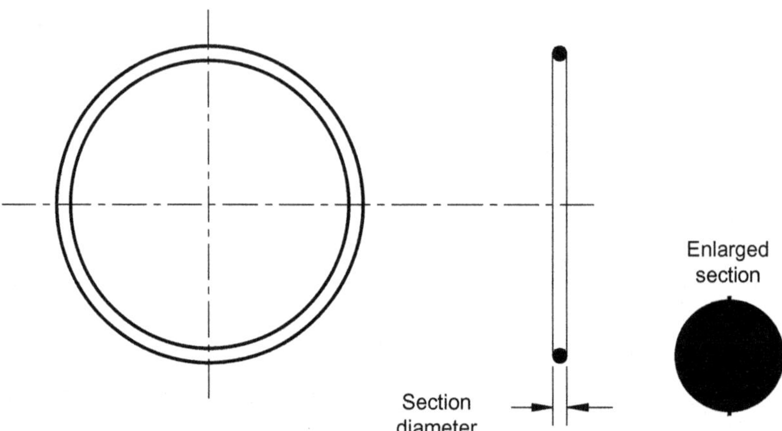

Fig. 14.5 General 'O' ring geometry.

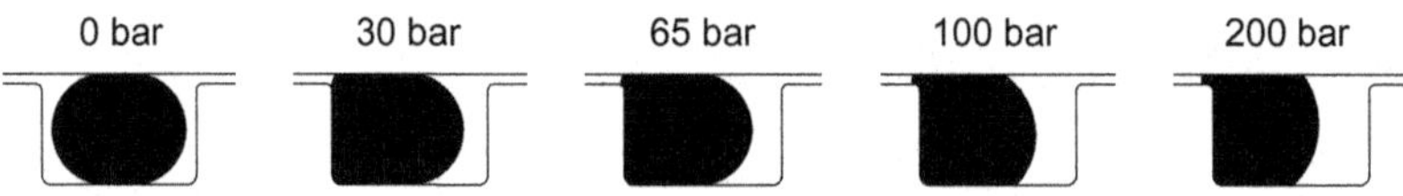

Fig. 14.6 Principle of operation of 'O' rings sealing a fluid against pressure.

the dimensions for a small number of the seal sizes available. A more extensive table is normally available in the form of sales catalogues from manufacturers or in BS 4518, ISO 3601/1 to ISO 3601/5 and DIN 3771. Fig. 14.7 shows the groove dimensions which must be specified to house the 'O' ring seal and ensure the seal is nipped or compressed sufficiently to enable effective sealing. The groove dimensions and corresponding data for static face sealing are given in Fig. 14.8 and Table 14.2, respectively.

Table 14.1 'O' ring seal dimensions for diametral sealing (mm) (limited range tabulated only for illustration)

Reference number	Internal diameter (mm)	Section diameter (mm)	$d_{nominal}$ (mm) (Fig. 14.7A)	$D_{nominal}$ (mm) (Fig. 14.7A)	$D_{nominal}$ (mm) (Fig. 14.7B)	$d_{nominal}$ (mm) (Fig. 14.7B)	B (mm)	R (mm)
0031-16	3.1	1.6	3.5	5.8	6	3.7	2.3	0.5
0041-16	4.1	1.6	4.5	6.8	7	4.7	2.3	0.5
0051-16	5.1	1.6	5.5	7.8	8	5.7	2.3	0.5
0061-16	6.1	1.6	6.5	8.8	9	6.7	2.3	0.5
0071-16	7.1	1.6	7.5	9.8	10	7.7	2.3	0.5
0081-16	8.1	1.6	8.5	10.8	11	8.7	2.3	0.5
0091-16	9.1	1.6	9.5	11.8	12	9.7	2.3	0.5
0101-16	10.1	1.6	10.5	12.8	13	10.7	2.3	0.5
0111-16	11.1	1.6	11.5	13.8	14	11.7	2.3	0.5
0036-24	3.6	2.4	4	7.7	8	4.3	3.1	0.5
0046-24	4.6	2.4	5	8.7	9	5.3	3.1	0.5
0195-30	19.5	3.0	20	24.8	25	20.2	3.7	1.0
0443-57	44.3	5.7	45	54.7	55	45.3	6.4	1.0
1441-84	144.1	8.4	145	160	160	145	9.0	1.0
2491-84	249.1	8.4	250	265	265	250	9.0	1.0

British Standards Institution, BS 4518: 1982. Specification for metric dimensions of toroidal sealing rings ('O' rings) and their housings.

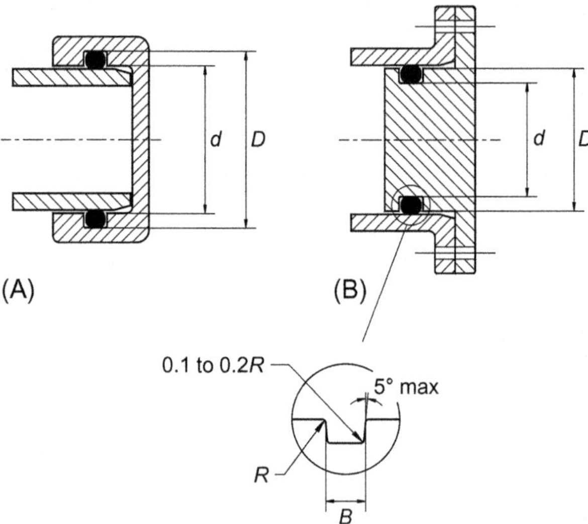

Fig. 14.7 'O' ring groove dimensions for diametral sealing: (A) External, (B) Internal.

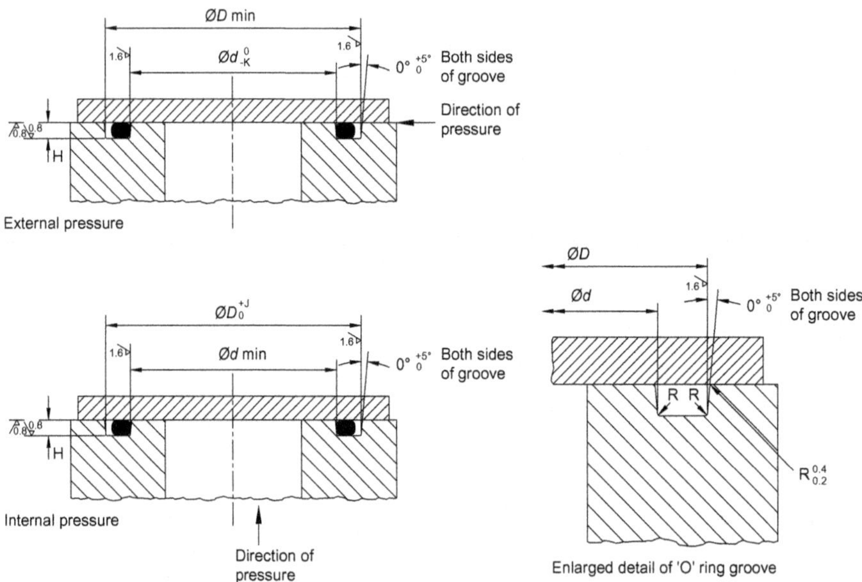

Fig. 14.8 Groove for static face sealing.
British Standards Institution, BS 4518: 1982. Specification for metric dimensions of toroidal sealing rings ('O' rings) and their housings.

Table 14.2 Groove dimensions for static face sealing (selection only)

O ring ref. no.	*d* (max) (mm)	*D* (mm)	*J* (mm)	*D* (min)	*d* (mm)	*K* (mm)	*H* (mm)	*R* (mm)
0593-30	57	65	0.19	68	60	0.19	$2.2\pm^{0.1}_{0}$	1.0
0625-30	60	68	0.19	71	63	0.19	$2.2\pm^{0.1}_{0}$	1.0
0645-30	62	70	0.19	73	65	0.19	$2.2\pm^{0.1}_{0}$	1.0
0695-30	67	75	0.19	78	70	0.19	$2.2\pm^{0.1}_{0}$	1.0
0745-30	72	80	0.19	83	75	0.19	$2.2\pm^{0.1}_{0}$	1.0
0795-30	77	85	0.22	88	80	0.19	$2.2\pm^{0.1}_{0}$	1.0
0845-30	82	90	0.22	93	85	0.22	$2.2\pm^{0.1}_{0}$	1.0
0895-30	87	95	0.22	98	90	0.22	$2.2\pm^{0.1}_{0}$	1.0
0945-30	92	100	0.22	103	95	0.22	$2.2\pm^{0.1}_{0}$	1.0
0995-30	97	105	0.22	108	100	0.22	$2.2\pm^{0.1}_{0}$	1.0
1045-30	102	110	0.22	113	105	0.22	$2.2\pm^{0.1}_{0}$	1.0
1095-30	107	115	0.22	118	110	0.22	$2.2\pm^{0.1}_{0}$	1.0
1145-30	112	120	0.22	123	115	0.22	$2.2\pm^{0.1}_{0}$	1.0
1195-30	117	125	0.25	128	120	0.22	$2.2\pm^{0.1}_{0}$	1.0
1245-30	122	130	0.25	133	125	0.25	$2.2\pm^{0.1}_{0}$	1.0
1295-30	127	135	0.25	138	130	0.25	$2.2\pm^{0.1}_{0}$	1.0
1345-30	132	140	0.25	143	135	0.25	$2.2\pm^{0.1}_{0}$	1.0
1395-30	137	145	0.25	148	140	0.25	$2.2\pm^{0.1}_{0}$	1.0
1445-30	142	150	0.25	153	145	0.25	$2.2\pm^{0.1}_{0}$	1.0
1495-30	147	155	0.25	158	150	0.25	$2.2\pm^{0.1}_{0}$	1.0
1545-30	152	160	0.25	163	155	0.25	$2.2\pm^{0.1}_{0}$	1.0
1595-30	157	165	0.25	168	160	0.25	$2.2\pm^{0.1}_{0}$	1.0
1645-30	162	170	0.25	173	165	0.25	$2.2\pm^{0.1}_{0}$	1.0
1695-30	167	175	0.25	178	170	0.25	$2.2\pm^{0.1}_{0}$	1.0
1745-30	172	180	0.25	183	175	0.25	$2.2\pm^{0.1}_{0}$	1.0
1795-30	177	185	0.29	188	180	0.25	$2.2\pm^{0.1}_{0}$	1.0
1845-30	182	190	0.29	193	185	0.29	$2.2\pm^{0.1}_{0}$	1.0

Continued

Table 14.2 Continued

O ring ref. no.	*d* (max) (mm)	*D* (mm)	*J* (mm)	*D* (min)	*d* (mm)	*K* (mm)	*H* (mm)	*R* (mm)
1895-30	187	195	0.29	198	190	0.29	$2.2\pm_{0}^{0.1}$	1.0
1945-30	192	200	0.29	203	195	0.29	$2.2\pm_{0}^{0.1}$	1.0
1995-30	197	205	0.29	208	200	0.29	$2.2\pm_{0}^{0.1}$	1.0
2095-30	207	215	0.29	218	210	0.29	$2.2\pm_{0}^{0.1}$	1.0
2195-30	217	225	0.29	228	220	0.29	$2.2\pm_{0}^{0.1}$	1.0
2295-30	227	235	0.29	238	230	0.29	$2.2\pm_{0}^{0.1}$	1.0
2395-30	237	245	0.29	248	240	0.29	$2.2\pm_{0}^{0.1}$	1.0
2445-30	242	250	0.29	253	245	0.29	$2.2\pm_{0}^{0.1}$	1.0
2495-30	247	255	0.29	258	250	0.29	$2.2\pm_{0}^{0.1}$	1.0

British Standards Institution, BS 4518: 1982. Specification for metric dimensions of toroidal sealing rings ('O' rings) and their housings

Example 14.1. Specify suitable groove dimensions for a 0195-30 'O' ring to seal against a solid cylinder.

Solution

From Table 14.1 and with reference to Fig. 14.7, $B = 3.7\,mm$, $R = 1\,mm$, groove fillet radius $= 0.2\,mm$.

In addition to their availability as a ring, O ring sections are available as a cord, sometimes called cording, which can be cut to any length. The ends can be glued or joined to form a ring if required.

A particular problem associated with O rings is ability to cope with small movements of the housings and sealing faces. A wide range of solutions have been developed to produce seals that are resistant to for instance rotation within the seal groove. An X ring, also known as a quadring, and a rectangular seal are illustrated in Fig. 14.9.

Aperture seals used, for example, for doors, windows and cabriolet bodies are typically made from elastomeric extrusions as production costs are low relative to fabricated mechanical seals and as their assembly can be automated. In the case of automobiles, the requirements are demanding with the need to seal against differential pressure, exclude dust, air, water and noise. In addition, the components need to have a life compatible with that of the vehicle, function over a wide range of operating temperatures, the cost of production and installation must be low, the function of the seal must not deteriorate significantly with time and the seal must not significantly influence the drag of the vehicle. A typical seal for an automotive window application is illustrated in Fig. 14.10.

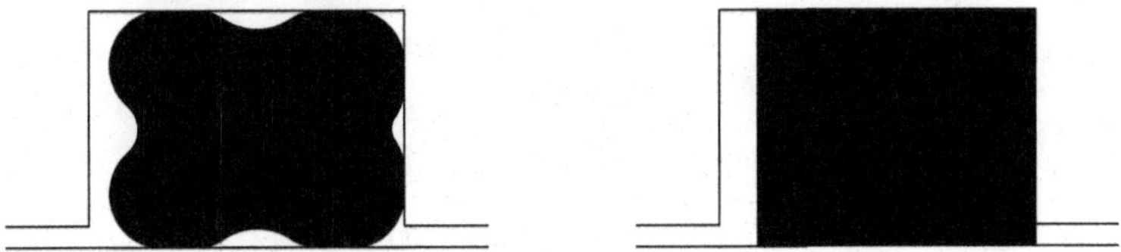

Fig. 14.9 X ring and rectangular seal.

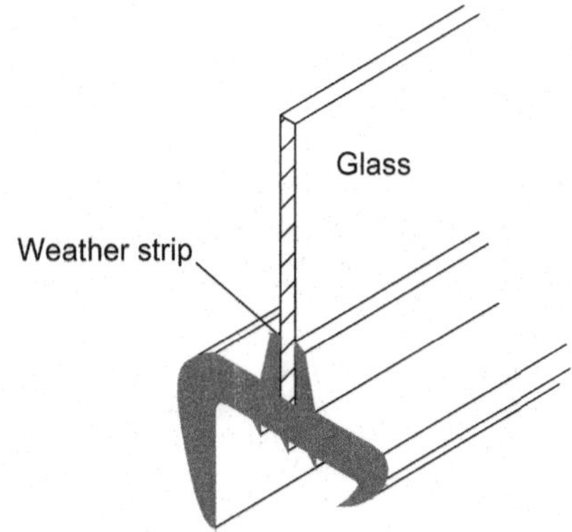

Fig. 14.10 Aperture seal for automotive applications.

14.2.2 Gaskets

A gasket is a material or composite of materials clamped between two components with the purpose of preventing fluid flow. Some examples of gaskets for sealing faces in an internal combustion engine are illustrated in Fig. 14.11. Gaskets are typically made up of spacer rings, a sealing element, internal reinforcement, a compliant surface layer and possibly some form of surface antistick treatment as shown in Fig. 14.12.

Fig. 14.13 shows a typical application for a gasket seal. When first closed a gasket seal is subject to compressive stresses produced by the assembly. Under working conditions, however, the compressive load may be relieved by the pressures generated

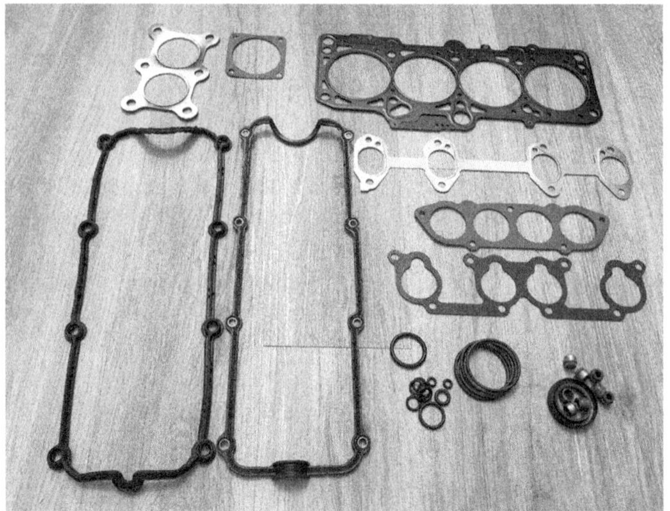

Fig. 14.11 Examples of gaskets for an internal combustion engine.

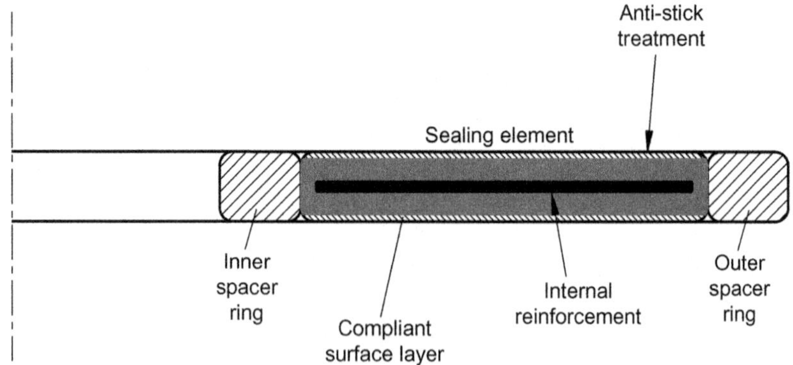

Fig. 14.12 The constituent parts of a typical gasket.

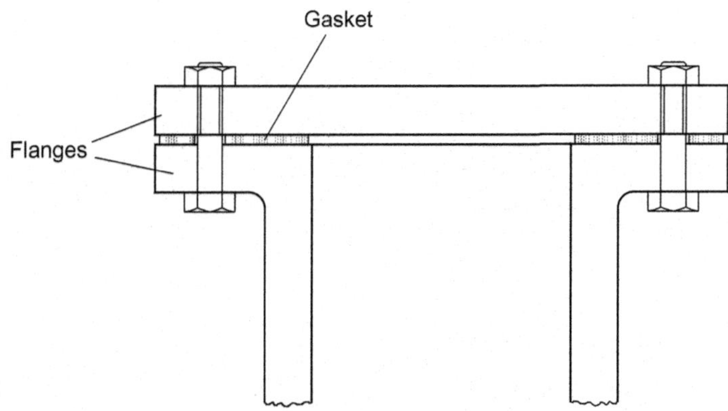

Fig. 14.13 Typical gasket application.

within the assembly or machine. This must be accounted for in the detailed design or by use of a factor to allow for the relaxation of gasket compression. Typical gasket designs are illustrated in Table 14.3. The choice of material depends on the temperature of operation, the type of fluid being contained and the leakage rate that can be tolerated. The order of magnitude of leakage for a range of materials is illustrated in Fig. 14.14. Gasket selection is considered in further detail by Winter (1998) and Czernik (1996a).

14.2.3 Foodstuffs containers

The sealing of foodstuffs amounts for a significant proportion of seal designs. Sealing of foodstuffs containers involves consideration of leakage of the contents, sealing against biological contamination by for instance bacteria and chemical odours. The typical diameter of bacteria is of the order of 1 μm and the challenge in designing foodstuffs containers is to exclude bacteria for the shelf life of the product, which may be a number of years. For many designs, the joint must be reusable as with carbonated drink and sauce bottle. A variety of typical joints for food containers are illustrated in Fig. 14.15.

14.3 Dynamic seals

The term 'dynamic seal' is used to designate a device used to limit flow of fluid between surfaces that move relative to each other. The range of dynamic seals is extensive with devices for both rotary and reciprocating motion. The requirements of dynamic seals are often conflicting and require compromise. Effective sealing may require high contact pressure between a stationary component and a rotating component but minimal wear is also desired for long seal life.

Table 14.3 Typical gasket designs

Type	Cross section	Comment
Flat	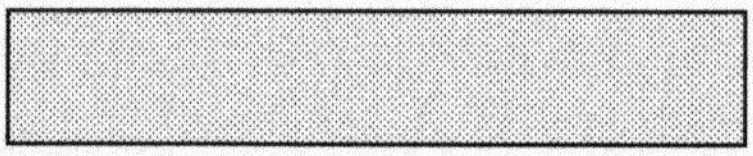	Available in a wide variety of materials. Easily formed into other shapes
Reinforced	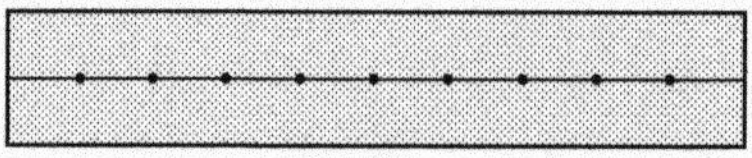	Fabric or metal reinforced. Improves torque retention and blowout resistance in comparison with flat types
Flat with rubber beads		Rubber beads located on flat or reinforced material. Gives high unit sealing pressure
Flat with metal grommet		The metal grommet gives protection to the base material
Plain metal jacket	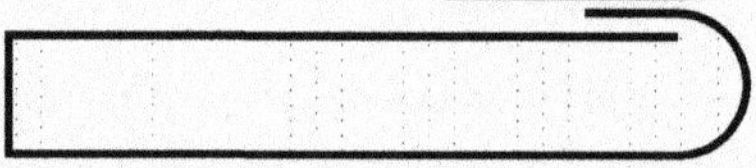	The metal jacket gives protection to the filler on one edge and across the surface
Corrugated or embossed	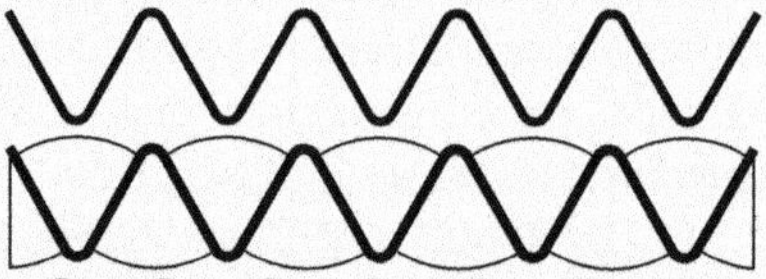	Corrugations provide increased sealing pressure capability
Profile	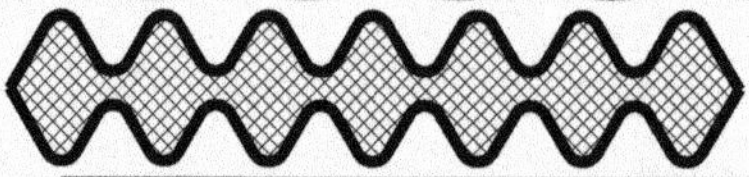	Multiple sealing surfaces
Spiral wound		Interleaving pattern of metal and filler

Adapted with alterations from Czernik, D.E., 1996. Gaskets. In: Shigley, J.E., Mischke, C.R. (Eds.), Standard Handbook of Machine Design. McGraw Hill.

14.3.1 Seals for rotating machinery

The functions of seals on rotating shafts include retaining working fluids, retaining lubricants and excluding contaminants such as dirt and dust. The selection of seal type depends on the shaft speed, working pressure and desired sealing effectiveness. Seals for rotary motion include 'O' rings, lip seals, face seals, sealing rings, compression packings and noncontacting seals such as bush and labyrinth seals.

'O' rings were considered in Section 14.2.1. Their application to rotating shafts is generally limited to use when the shaft speed is below 3.8 m/s and seal pressures below 14 bar.

The typical geometry for a radial lip seal, commonly known as an oil ring, is shown in Fig. 14.16 These seals are used to retain lubricants and exclude dirt, and are well suited to moderate-speed and low-pressure applications. The use of a radial lip seal for

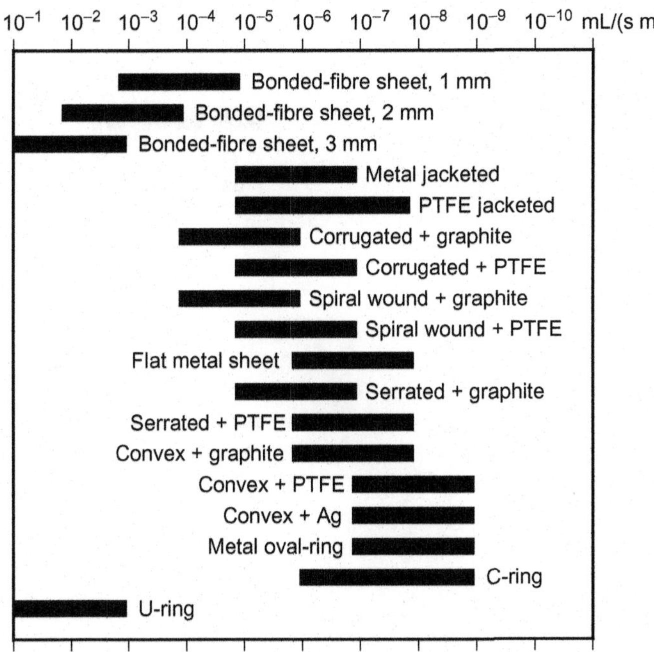

Fig. 14.14 Typical order of magnitude of gasket leakage rate per unit diameter for a range of gasket materials.
Adapted from Muller, H.K., Nau, B.S., 1998. Fluid Sealing Technology. Marcel Dekker.

a bearing is illustrated in Fig. 14.17. The outer case should be retained in the housing by an interference fit. The purpose of the garter spring is to maintain a uniform radial force on the shaft ensuring contact between the elastomeric sealing ring and the rotating shaft. The design of radial lip seals is detailed by Horve (1996) and SAE (1996) and their function in Stakenborg (1988) and Jia et al. (2014).

For applications where low leakage, high reliability and low wear are important mechanical face seals may be appropriate. A mechanical face seal consists of two sealing rings, one attached to the rotating member and one attached to the stationary component to form a sealing surface, usually perpendicular to the shaft axis as illustrated in Fig. 14.18. During rotation, the primary sealing ring attached to the shaft rubs with its seal face against the counter-seal face of the stationary ring. The two interface contact areas function like bearings and are subject to frictional wear. Any leakage flow must pass across this interface. The section of a practical mechanical face seal is shown in Fig. 14.19.

The overall assembly for a typical mechanical face seal is illustrated in Fig. 14.20. Frictional rubbing contact between the sealing faces is maintained by forces acting axially caused by hydraulics or mechanically using, for example, a spring. The rubbing action between the surfaces produces heat and wear. In order to minimise this, lubrication is used, which as well as limiting wear and heat build-up also serves to generate a

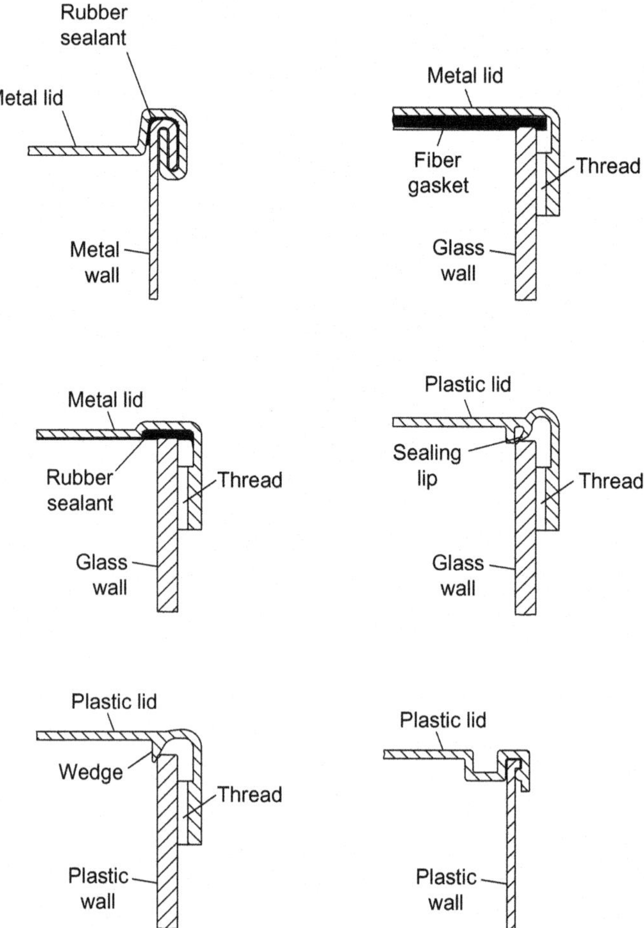

Fig. 14.15 Seals for food containers.
Adapted from Muller, H.K., Nau, B.S., 1998. Fluid Sealing Technology. Marcel Dekker.

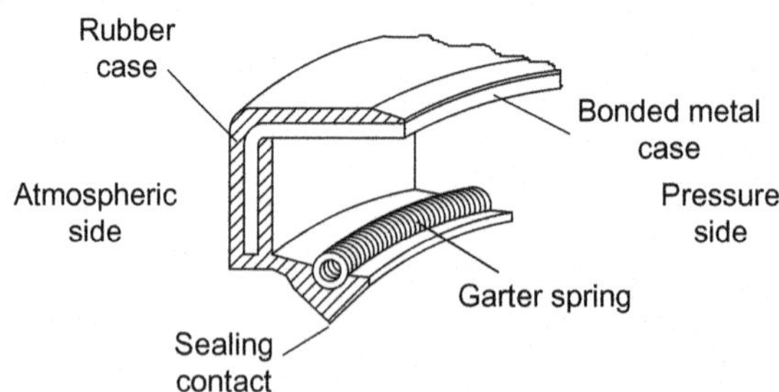

Fig. 14.16 Radial lip seal.

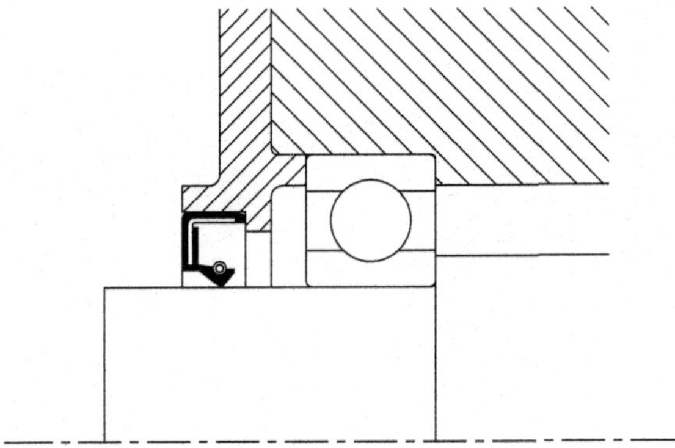

Fig. 14.17 Use of a radial lip seal for a bearing.

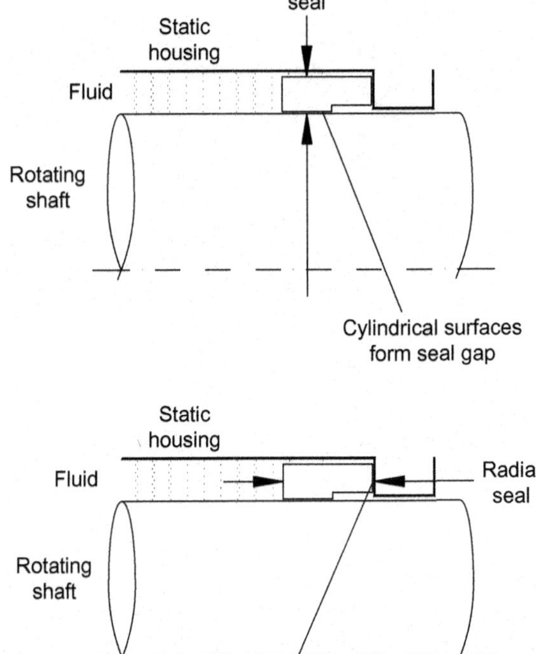

Fig. 14.18 Operating principle of mechanical face seals.

Fig. 14.19 Mechanical Seal.

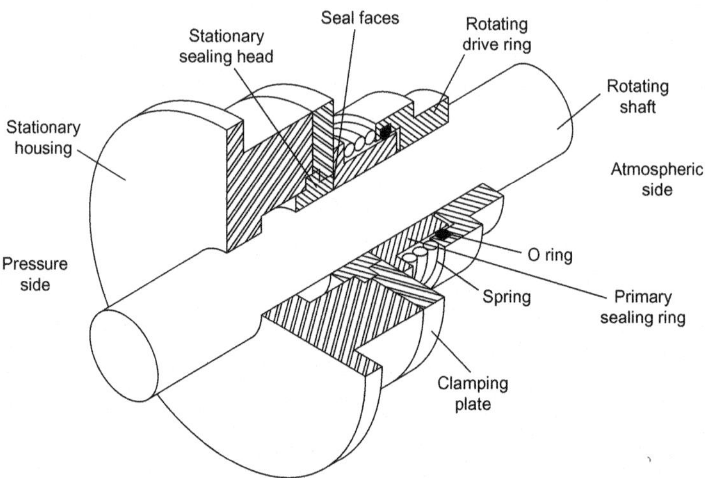

Fig. 14.20 A typical mechanical face seal.

fluid film which assists in producing a tight seal. The range of face seal designs is extensive and they are widely used in pumps, compressors, gearboxes and domestic washing machines. The design of face seals is reviewed by Dolan (1992), Summers-Smith (1992), Muller and Nau (1998), Lebeck (1991) and Chupp et al. (2006a,b). The use of textured faces for mechanical seals is described by Etsion et al. (1999) and Etsion (2004), Yu et al. (2002), Bai et al. (2010) and Adjemout et al. (2017).

The term interstitial seal is used for seals that allow unrestricted relative motion between the stationary and moving components (i.e. no seal to shaft contact). Types include labyrinth, brush and bush seals. Labyrinth seals and bush seals are described in Sections 14.4 and 14.5, respectively.

14.4 Labyrinth seals

In its simplest form, a labyrinth seal consists of a series of fins and corresponding chambers forming a restriction to the flow and a volume for expansion, respectively, as illustrated in Fig. 14.21. Labyrinth seals work by throttling the flow through successive openings in series. In each throttle, static pressure difference accelerates the flow and some of the kinetic energy associated with the flow is dissipated by turbulence induced by the intense shear stress and eddy motion in the next chamber as illustrated in Fig. 14.22. The use of a tortuous path between high and low static pressure regions incorporating a series of noncontacting restrictions and chambers in the form of a labyrinth seal was implemented by Parsons for his steam turbine concept in 1892 (see Parsons (1892) and The Engineer (1938)), Fig. 14.23. Labyrinth seals have a long

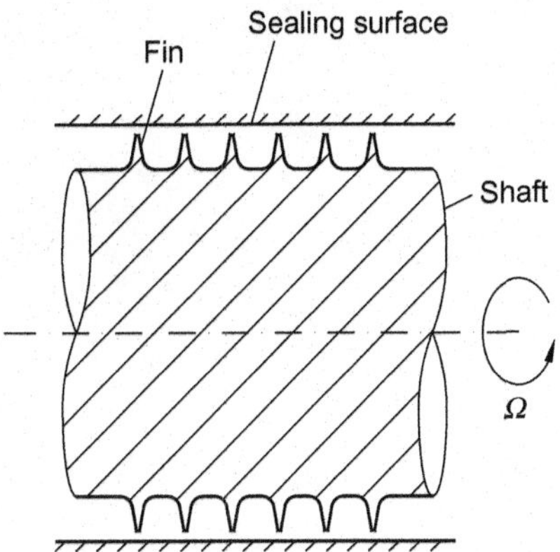

Fig. 14.21 Typical straight labyrinth seal.

Fig. 14.22 Characteristic flow through a labyrinth seal.
After Tipton, D.L., Scott, T.E., Vogel, R.E., 1986. Labyrinth seal analysis. Volume III - Analytical and experimental development of a design model for labyrinth seals. AF Wright Aeronautical Laboratories AFWAL-TR-85-2103 Volume III.

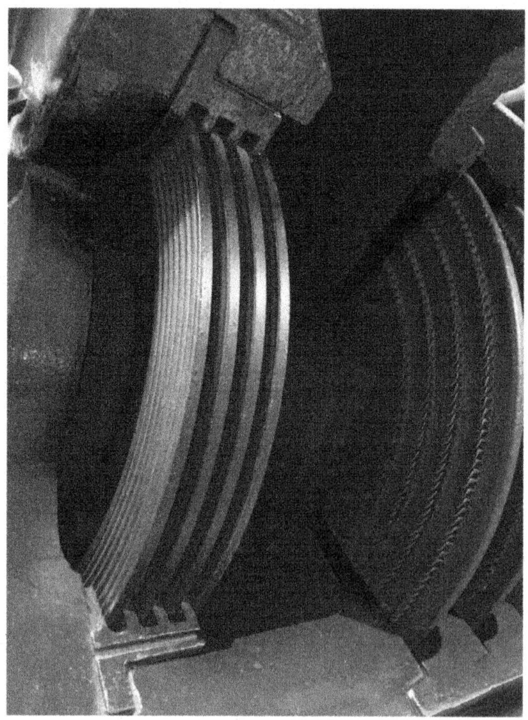

Fig. 14.23 A section of a Parson steam turbine (1891) illustrating an early example of the use of radial labyrinth seals.

history of proven reliability in a wide range of applications, including bearing chambers, see Fig. 14.24, in gas turbine engines as well as discs and stator wells (see Chupp et al., 2006a,b), with robust operation and developed technology. Labyrinth seals are considered in detail in ESDU 09008 (ESDU (2009)), upon which this section is based, Sneck (1974), Trutnovsky and Komotori (1981), Kim and Cha (2009), and Cremanns et al. (2016).

The principal geometric parameters of a typical straight labyrinth seal are shown in Figs. 14.25 and 14.26. The terms fin, knife, tooth and blade are used interchangeably in industry and the technical literature. Here, the term fin is principally used.

Common forms of labyrinth seal include the following:

- straight through;
- stepped and
- staggered.

Each of these forms of labyrinth seal can be used in axial rotating flow applications, where the axis of the fins is coincident with the axis of the shaft, or in radial sealing applications, where the fins are perpendicular to the shaft axis. Schematic diagrams for each of these are illustrated in Fig. 14.27. Fins can rotate, move in a reciprocating motion or be stationary. In general, for the case of an axial application, it is desirable to have the fins on the inner member as this improves the stability of the system and heat pickup.

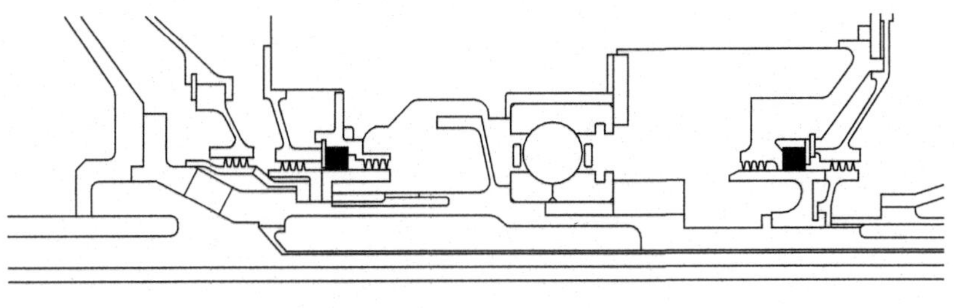

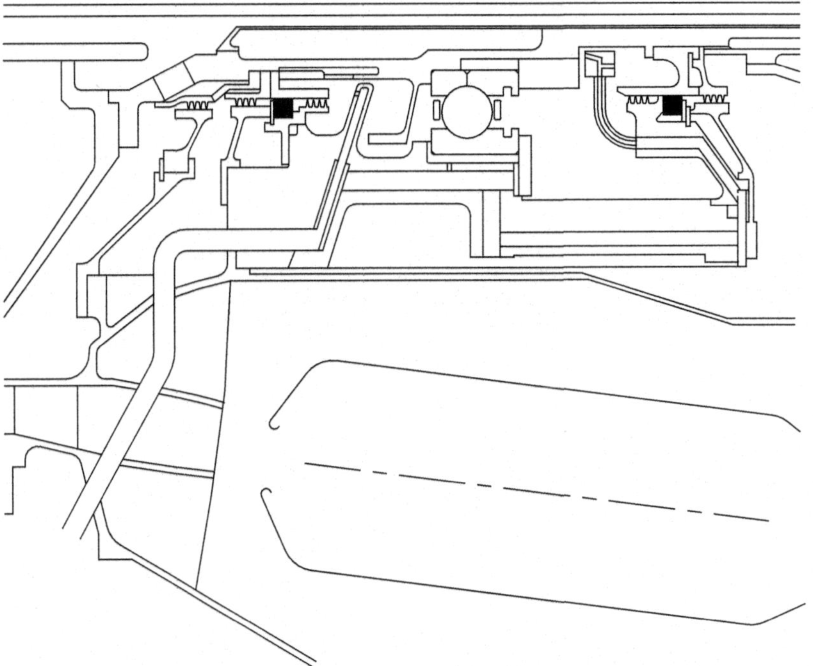

Fig. 14.24 Use of labyrinth seals for a bearing chamber in a gas turbine engine.

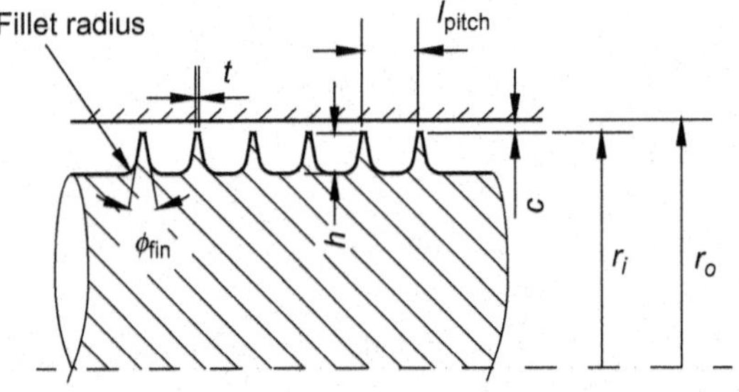

Fig. 14.25 Principal labyrinth seal geometric parameters.

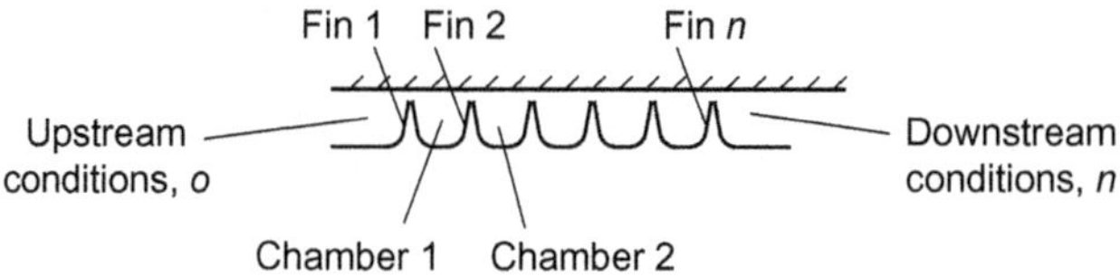

Fig. 14.26 Labyrinth fin and chamber notation.

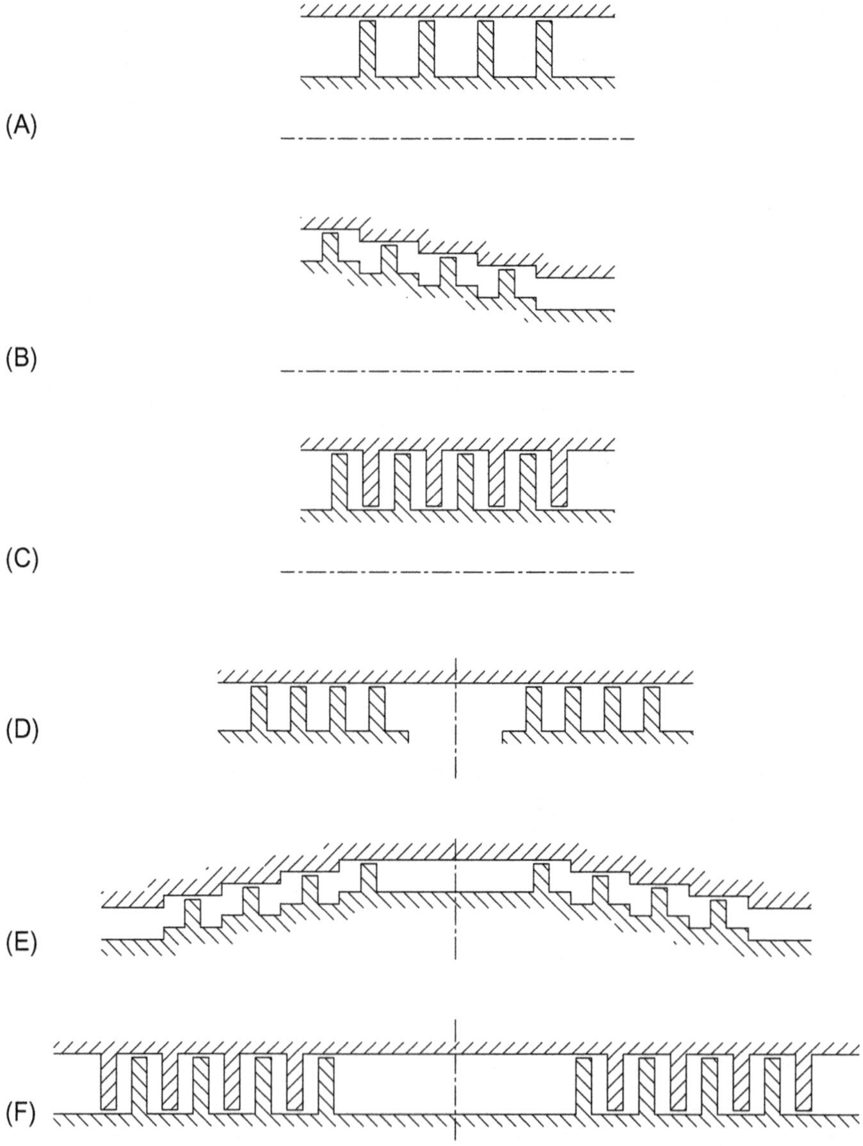

(A)

(B)

(C)

(D)

(E)

(F)

Fig. 14.27 Schematic diagrams illustrating a selection of labyrinth seals. Axial applications: (A) straight through, (B) stepped, (C) staggered. Radial applications: (D) straight through, (E) stepped, (F) staggered.

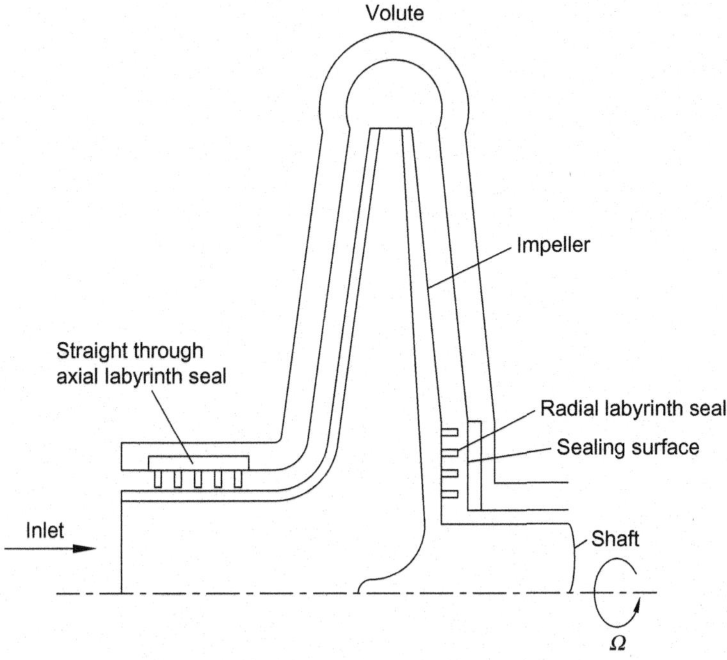

Fig. 14.28 Application of labyrinth seals in a pump with a straight through seal at the pump inlet and a radial labyrinth seal on the pump rear face.

Labyrinth seals do not provide a complete seal and if there is a static pressure difference across the seal, then there will normally be a net flow of fluid across the seal. A typical application for a labyrinth seal for limiting the leakage of flow in a centrifugal pump is illustrated in Fig. 14.28. In this application, a straight-through labyrinth seal is used for limiting the flow of leakage flow on the inlet side of the impeller shaft and a radial labyrinth is used on the rear face of the impeller.

The design of a straight-through labyrinth seal in most practical applications typically involves compromise between the number of fins and a pitch that is large enough to ensure that the kinetic energy of the flow is dissipated in the cavity (chamber). There are a significant number of performance and design parameters to consider, including (see Fig. 14.25 for a visual indication of the geometric parameters):

- clearance, c;
- pitch, l_{pitch};
- the number of fins, n;
- fin tip land width, t;
- rotational speed, Ω;
- cavity volume;
- fin angle, ϕ_{fin};
- land surface roughness and porosity;
- fin height, h;

- eccentricity;
- pressure ratio and
- axial Reynolds number, *Re*.

For a stepped seal additional parameters also include the following:

- step height, *H*;
- step configuration;
- distance from seal fin to step face and
- flow direction.

The principal fluid-dynamic parameters that determine labyrinth seal performance and their effect are listed in Table 14.4. The corresponding geometric parameters are given in Table 14.5. In terms of modelling, the parameters identified do not operate independently and therefore a significantly large matrix of parameter combinations affects labyrinth seal performance.

A particular challenge in determining labyrinth seal performance parameters arises with in-service and worn seals. In service and worn seals can involve issues with

- erosion of coatings on either the fin land or on the sealing surface;
- grooves in abradable coatings caused by wear;
- uneven fin wear;
- fin rounding;
- manufacturing errors;
- thermal distortions and
- axial movements.

Care must be taken in the design of labyrinth seals. If the cavity following a radial fin is too small, the flow will pass straight through without expanding and without the subsequent pressure drop. In any case despite careful labyrinth design and specification of a labyrinth seal parameters, flow will inevitably be carried straight over from one fin to another. In order to reduce this effect, steps can be incorporated into a labyrinth seal design and a wide range of labyrinth variants are possible.

Table 14.4 Principal fluid-dynamic parameters for labyrinth seals

Parameter	Definition	Characteristic effect
Mass flow rate function	$Q=\frac{m\sqrt{T_{t0}}}{Ap_{t0}}$	Dependent variable
Pressure ratio	p_{t0}/p_n	Strong
Axial Reynolds number	$\mathrm{Re}=\frac{(m/A)c}{\mu}$	Moderate
Fin tip speed	$u_{\varphi,\ r_i}$	Moderate
Rotational Reynolds number	$\mathrm{Re}_{\phi}=\frac{\rho\Omega rc}{\mu}$	Weak
Taylor number	$\mathrm{Ta}=\frac{\rho u_{\phi,r_i}c}{\mu}\sqrt{\frac{c}{r_i}}$	Weak

After Tipton, D.L., Scott, T.E., Vogel, R.E., 1986. Labyrinth seal analysis. Volume III—Analytical and experimental development of a design model for labyrinth seals. AF Wright Aeronautical Laboratories AFWAL-TR-85-2103 Volume III.

Table 14.5 Principal geometric parameters for labyrinth seals

Parameter	Affected parameter	Characteristic influence on			
		Straight seals	**Stepped seals**	**Staggered seals**	**Radial seals**
Number of fins, n	Number of throttles	Strong	Strong	Strong	Strong
Fin angle	Orifice geometry	Moderate	Moderate	Moderate	Moderate
Fin tip land width, t	Relative throat length	Moderate	Moderate	Moderate	Moderate
Clearance, c	Fin relative sharpness	Moderate	Moderate	Moderate	Moderate
Fin shape	Orifice geometry	Weak to moderate	Weak	Weak	Weak to moderate
Pitch, l_{pitch}	Relative throttle spacing	Moderate	Weak	Moderate	Moderate
Fin height, h	Relative chamber depth	Weak	Weak	Weak	Weak
Land surface roughness	Land relative roughness	Moderate	Weak	Moderate	Moderate
Land surface porosity	Land relative porosity	Moderate	Weak to moderate	Moderate	Moderate
Step height	Relative step height	–	Weak	–	–
Distance to contact	Rotor relative axial location	–	Weak to moderate	–	–
Flow direction	Flow-down stator step/ Flow up stator step	–	Weak	–	–

Adapted from Tipton, D.L., Scott, T.E., Vogel, R.E., 1986. Labyrinth seal analysis. Volume III—Analytical and experimental development of a design model for labyrinth seals. AF Wright Aeronautical Laboratories AFWAL-TR-85-2103 Volume III.

Methods available for quantifying the leakage through a labyrinth seal include empirical, analytical and numerical computational fluid dynamics (CFD) techniques. Quantification of fluid flow rates in various applications is fraught with difficulty and labyrinth seals are no exception.

As described previously, a straight labyrinth seal consists of a series of radial fins forming a restriction to an annular flow of fluid, see Figs. 14.21 and 14.22. In order for the fluid to pass through the annular restriction, it must accelerate and, in this process, part of the energy associated with the static pressure of the fluid entering the restriction is converted into kinetic energy. Downstream of the restriction, the fluid jet decelerates and expands with the formation of separation eddies in the cavity downstream of the fin as indicated in Fig. 14.22. These eddies dissipate some of the energy of the flow by small scale turbulence viscosity action, reducing the total pressure. This process is repeated in subsequent cavities until the flow finally exhausts through the last restriction and the pressure reaches downstream conditions.

Under design operation, labyrinth seals are essentially a controlled-clearance seal without rubbing contact. As there is no surface-to-surface contact, very high relative speeds are possible and the geometry can be arranged to limit leakage to tolerable levels. Leakage is, however, inevitable and can be quantified using the empirical methods, based on discharge coefficients and carry-over correction factors or using CFD. Typically, the leakage through a labyrinth is modelled as an adiabatic throttling process. The neglect of real gas and heat transfer effects is generally of secondary importance in practical applications and an acceptable assumption. The ideal throttling assumption has led to two principal types of flow model:

(1) Modelling the labyrinth as a series of discrete restrictions with associated local pressure losses;
(2) Modelling the seal as a rough pipe with uniformly distributed wall friction with summation of the frictional losses of the fluid as it passes through the geometric configuration of the seal. The losses due to friction are typically described in terms of a friction factor which represents the seal's resistance to leakage.

In order to avoid damage to the components of the labyrinth, typically the fins and the adjacent seal surface, as a result of contact between the rotating and stationary components, a minimum clearance needs to be maintained. Various configurations for the seal throttlings and cavities have been developed that attempt to reduce the kinetic-energy carry-over effect, with similar benefits to reducing the clearance. These include step-up and step-down configurations, configurations that use inclined fins, staggered or interlocking configurations where successive labyrinth fins project from the rotating and stationary surfaces, respectively, and configurations that use of abradable surfaces, see Fig. 14.29.

An abradable surface enables tighter clearances to be maintained and is typically used on the stationary component rather than the fin tips. Any wear (see Dogu et al. (2016a,b, 2017)) of the abradable surface can be mitigated against by replacement at service intervals in order to maintain a specific design clearance. One type of surface that is used regularly in labyrinth seals is the open-celled honeycomb, mounted with cell openings facing the labyrinth fins, Fig. 14.30. The honeycomb surface has an enhanced frictional effect that reduces the kinetic-energy carry-over effect of the seal. Some honeycomb seal surfaces are capable of running at high temperatures and have relatively stiff structural properties. Honeycombs however contribute to windage heating in labyrinth seals and this needs to be accounted for in a given application.

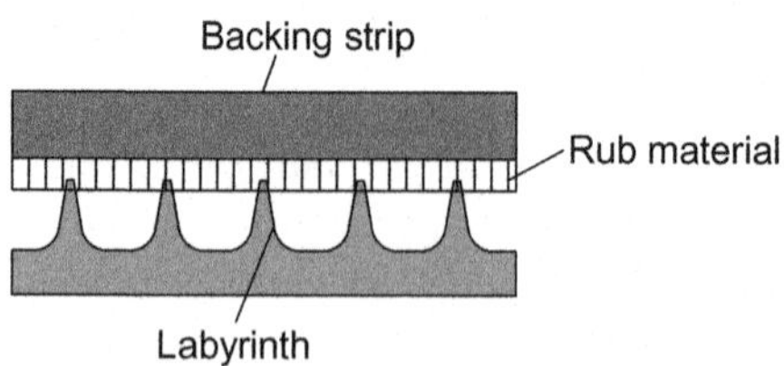

Fig. 14.29 Labyrinth seal with abradable rub-in surface.

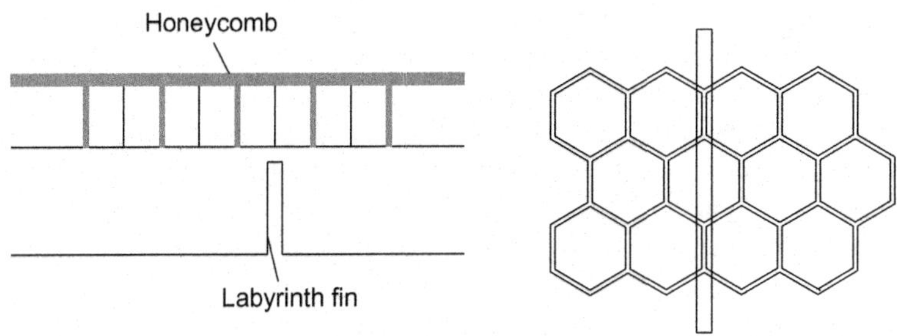

Fig. 14.30 Open-celled honeycomb mounted with cell openings facing a labyrinth fin.

In order to minimise the leakage flow in a straight through labyrinth seal, the radial clearance should be made as small as possible. However, in selecting and specifying the clearance, allowances must be made for thermal expansion, growth due to centripetal acceleration and shaft misalignment. If allowance for these factors is not made, then there is a risk of fin contact with the sealing surface with the potential for accompanying damage to the components. Abradable surfaces can be used in order to accommodate the possibility of contact between the seal fins and the seal surface. The clearance between the seal fin and abradable surface may be set such that contact between the two is deliberately caused with the fins 'cutting' a running clearance. Such an approach, provided any debris from this and the power required to overcome the frictional contact in the running-in period can be accommodated, can be used to provide a seal with the smallest practical running clearance for the conditions imposed.

The following general guidelines are noted by Morrison et al. (1983):

- Normally, a narrow fin width is preferable to wider fins.
- The fin width should always be less than one half of the value of the pitch, especially for application where the angular velocity is less than circa 50 rad/s (about 500 rpm).
- The fin height should generally be at least twice the value of the radial clearance. Measurements from Yamada (1962) showed that the flow coefficient is relatively constant for fin height to clearance ratios between 2 and 11.

Stocker's (1975) and Stocker's et al. (1977) data for stepped labyrinth seals, indicated that the optimum fin angle for minimum leakage was between 50° and 70°, and that the optimum pitch for minimum leakage tended to decrease as the fin angle increased. Stocker's data also indicated that the leakage decreased as the surface roughness was increased up to a certain amount and then began to increase as the roughness was increased further.

In addition to consideration of the geometric parameters, i.e. use of abradable materials and quantification of the seal leakage, the operating temperature of the seal needs to be assessed and the contributions to windage heating of the seal flow due to viscous dissipation of energy as a result of rotation of the seal's surfaces need to be quantified.

A further consideration to take into account in the design of straight through labyrinth seals is rotodynamic instability. Fluid flow through the cavities of a seal can create net pressure and shear forces acting on the rotor. These forces may contribute to the destabilization of marginally stable rotors rotating at high speeds. A typical analysis for labyrinth seals involves determining the rotordynamic force coefficients. These coefficients distinguish position and speed-dependent force components in two directions, thereby describing the stiffness and damping effects of the fluid. In practical applications, the tendency of labyrinth seals to produce flow instabilities can be controlled by swirl brakes or intracavity slots or blocks and drum dampers. A review on the rotodynamic instability issues is given in (Childs, 1993); specific applications are considered in (Childs and Rhode (1984)) and (Picardo and Childs, 2005).

Designing a seal to minimise the leakage of flow for specific seal requires an iterative optimisation technique. For a seal of fixed overall length, there is a trade-off between selecting the pitch and the number of throttles. As the number of throttles increases, the flow coefficient decreases with accompanying reduction in the mass flow but as the number of throttles increases, the pitch decreases, causing the carry-over factor to increase. The increase in carry-over factor causes an increase in the leakage, off-setting some or even all of the decrease in leakage achieved by decreasing the flow coefficient. A guideline for the initial specification of geometric parameters for a labyrinth seal is given in Table 14.6 including typical values given in the open literature. The design principles are applicable to the majority of types of labyrinth seal. An outline flow chart for labyrinth seal design is given in Fig. 14.31.

Table 14.6 Guideline for the initial specification of straight labyrinth seal geometric parameters

Geometric parameters	Guidelines	Typical values
Clearance, c	Minimise subject to practical constraints of manufacture and machine movements.	$0.2\,\text{mm} < c < 2\,\text{mm}$
Pitch, l_{pitch}	For a seal of fixed overall length, there is a trade-off between selecting the pitch and the number of throttles.	$2 < l_{pitch} < 10\,\text{mm}$
Number of fins, n	For a seal of fixed overall length, there is a trade-off between pitch and number of throttles. As the number of throttles increases, the flow coefficient decreases with accompanying reduction in the mass flow but as the number of throttles increases, the pitch decreases, causing the carry-over factor to increase and therefore raising the mass flow.	Trade-off with pitch, l_{pitch}
Fin width, t	Normally, a narrow fin width is preferable to wider fins. The fin width should always be less than one half of the value of the pitch: $t < 0.5\, l_{pitch}$	$0.1 < t < 0.4$
Fin height, h	The fin height should generally be at least twice the value of the radial clearance: $h > 2c$	$2 < h < 8$

Table 14.6 Continued

Geometric parameters	Guidelines	Typical values
Fin angle, ϕ_{fin}	A practical engineering angle should be defined	$7^{o} \leq \phi_{fin} \leq 40^{o}$
Fillet radius, r_{fillet}	A fillet is necessary in order to relieve stresses at the fin root	$0.3 \leq r_{\text{fillet}} \leq 1.2\,\text{mm}$
Land roughness	A practical engineering roughness should be defined to minimise windage and maximise component integrity	8.3 μm
Eccentricity	Minimise	

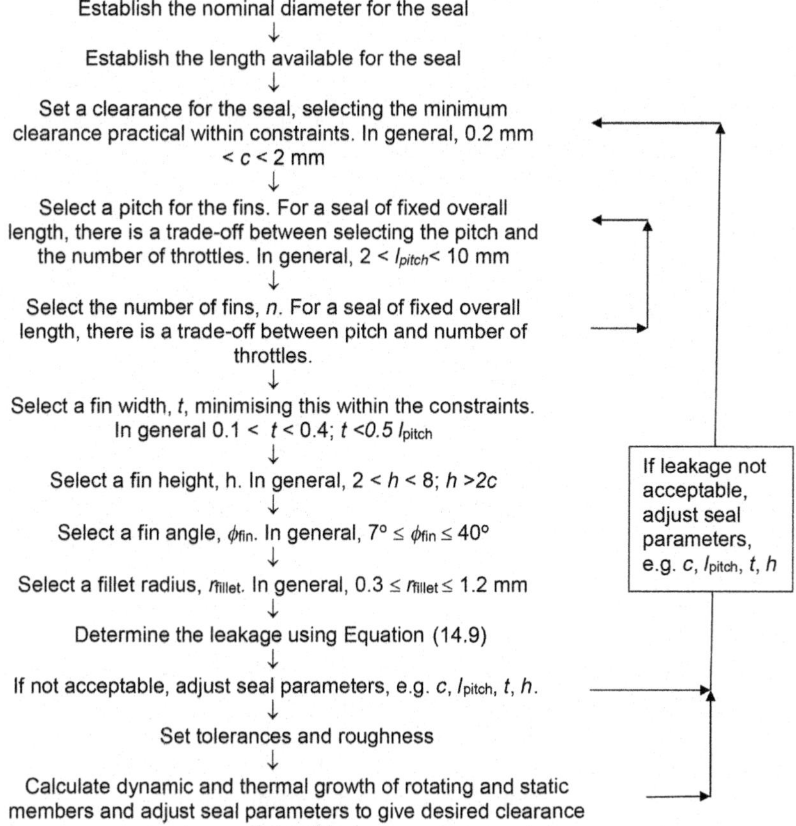

Establish the nominal diameter for the seal
Establish the length available for the seal
Set a clearance for the seal, selecting the minimum clearance practical within constraints. In general, 0.2 mm < c < 2 mm
Select a pitch for the fins. For a seal of fixed overall length, there is a trade-off between selecting the pitch and the number of throttles. In general, 2 < l_{pitch} < 10 mm
Select the number of fins, n. For a seal of fixed overall length, there is a trade-off between pitch and number of throttles.
Select a fin width, t, minimising this within the constraints. In general 0.1 < t < 0.4; t < 0.5 l_{pitch}
Select a fin height, h. In general, 2 < h < 8; h > 2c
Select a fin angle, ϕ_{fin}. In general, 7° ≤ ϕ_{fin} ≤ 40°
Select a fillet radius, r_{fillet}. In general, 0.3 ≤ r_{fillet} ≤ 1.2 mm
Determine the leakage using Equation (14.9)
If not acceptable, adjust seal parameters, e.g. c, l_{pitch}, t, h.
Set tolerances and roughness
Calculate dynamic and thermal growth of rotating and static members and adjust seal parameters to give desired clearance if possible
If leakage not acceptable, adjust seal parameters, e.g. c, l_{pitch}, t, h

Fig. 14.31 Labyrinth seal design procedure for a straight labyrinth.

Table 14.7 Selected correlations for determining the mass flow for straight labyrinth seals

Reference	Equation	Flow type	Application
Martin (1908, 1919)	$m=Ap_{t0}\sqrt{\frac{1-(p_n/p_{t0})^2}{RT_{t0}[n-\ln(p_n/p_{t0})]}}$	Compressible flow	Straight through seals
Egli (1935)	$m=\alpha\varepsilon A\sqrt{\frac{p_{t0}\rho_0\left[1-(p_n/p_{t0})^2\right]}{n+\ln(p_{t0}/p_n)}}$ where α is a flow coefficient and ε is a carry-over correction factor	Compressible flow	Straight through and staggered seals
Hodkinson (1940)	$m=\alpha k_1 A\sqrt{\frac{p_{t0}\rho_0\left[1-(p_n/p_{t0})^2\right]}{n}}$ $k_1=\left\{1-\frac{(n-1)(c/l_{pitch})}{n\left[(c/l_{pitch})+0.02\right]}\right\}^{-0.5}$ where α is a flow coefficient	Compressible flow	Straight through seals
Vermes (1961)	$m=5.76C_d kA\sqrt{\frac{p_{t0}\rho_0\left[1-(p_n/p_{t0})^2\right]}{n+\ln(p_{t0}/p_n)}}$ $k=\frac{8.52}{\left[(l_{pitch}-t)/c\right]+7.23}$	Compressible flow	Straight through and stepped seals.
Komotori and Mori (1971)	$m=\alpha\varepsilon_1 AF(n,p_{t0}/p_n,\gamma)\sqrt{\rho_0 p_0}$ where α is a flow coefficient, ε_1 is a carry-over factor and $F(n,p_{t0}/p_n,\gamma)$ is defined in the paper	Compressible flow	Straight through and radial seals
Zimmermann and Wolff (1998)	$m=k_2 C_d A p_{t0}\sqrt{\frac{1-(p_n/p_{t0})^2}{RT_{t0}[n+\ln(p_{t0}/p_n)]}}$	Compressible flow	Straight through and stepped seals

Labyrinth seals have been the subject of extensive study and a number of empirical and analytical models have been developed. A selection of these are summarised in Table 14.7.

Martin's equation, which has been widely used for determining the mass flow in a labyrinth seal is discussed in Section 14.4.1. The use of compressible flow functions for modelling a labyrinth seal by a series of orifices is introduced in Section 14.4.2. The model recommended, in this item, due to its simplicity, accuracy and versatility in accommodating practical parameters for straight labyrinth seals, the Zimmermann Wolff model is described in Section 14.4.3.

Numerical solution of the Navier–Stokes equations using CFD can provide valuable insight into detailed labyrinth seal flow characteristics and quantification of global quantities such as mass flow rate. The value of a CFD solution is that it can analyse the effect of most geometric and fluid dynamic parameters. In addition to the correlation-based approaches indicated in Table 14.7 and CFD models, various application specific software codes have been produced for modelling labyrinth seals. These include the knife-to-knife (KTK) and Labflow codes produced by the former

Allison Gas Turbine Division (now Rolls-Royce) and the Rotating Machinery and Controls group, respectively. The KTK model is based on (Tipton et al. (1986), Chupp et al. (1986), Shapiro and Chupp (2004)). The KTK provides modelling capability for both straight and stepped labyrinth seals.

14.4.1 Martin's equation

An analytical approach based on a simplified gas dynamics approach can be applied to a labyrinth seal which models the leakage in the seal as flow of an ideal gas through a series of orifices and was originally developed by Martin (1908, 1919). Martin assumed that all of the kinetic energy in each cavity or throttling chamber was completely dissipated before the flow passed into the next cavity and that the flow was isothermal. The resulting equation following this analysis is presented in Eq. (14.1) and is generally known as Martin's equation:

$$m = Ap_{t0}\sqrt{\frac{1-(p_n/p_{t0})^2}{RT_{t0}[n - \ln(p_n/p_{t0})]}}, \tag{14.1}$$

where m is the mass flow rate, A is the annular clearance area of the seal, p_{t0} is the upstream total pressure, p_n is the downstream static pressure, R is the characteristic gas constant, T_{t0} is the upstream total temperature, and n is the number of labyrinth fins.

The mass flow function, Q, provides a convenient means for comparing the relative performance of different seals. This is defined by

$$Q = \frac{m\sqrt{T_{t0}}}{Ap_{t0}}. \tag{14.2}$$

Substitution for the mass flow function in Eq. (14.1) gives

$$Q = \sqrt{\frac{1-(p_n/p_{t0})^2}{R[n - \ln(p_n/p_{t0})]}}. \tag{14.3}$$

The flow characteristics for a straight labyrinth for compressible flow, using Eq. (14.3), are illustrated in Fig. 14.32 for a range of numbers of fins, where the mass flow function, given in Eq. (14.2), is plotted against the pressure ratio, p_{t0}/p_n.

In real practical conditions, labyrinth seal flow deviates from the assumptions made in the development of Eq. (14.1). For example, typically, only a fraction of a dynamic head is lost across an individual fin. The effects of contractions and coatings also need to be taken into account. These factors can be modelled by use of a carry-over factor and a discharge coefficient. As a result of such factors the Martin equation tends to overestimate the mass flow in a labyrinth seal. Zimmermann and Wolff (1987) demonstrated that this can be by approximately 35% for a typical gas turbine engine labyrinth.

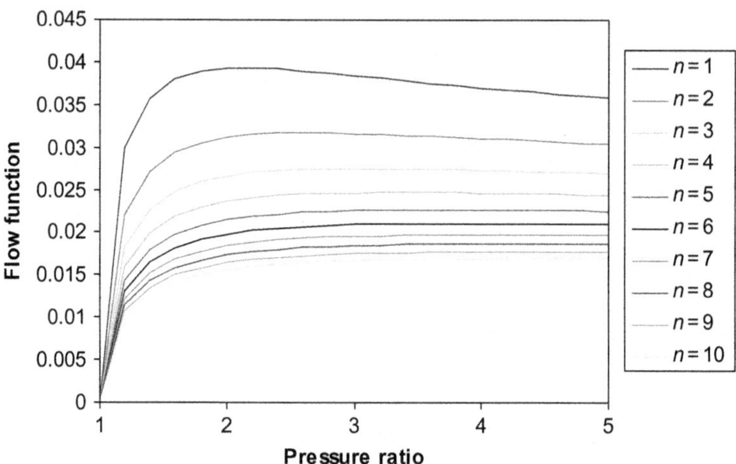

Fig. 14.32 Labyrinth mass flow function modelled using Eq. (14.3) with $R = 287\,\text{J/kg K}$.

In modelling the flow through a labyrinth seal, it is often assumed that all of the dynamic pressure is dissipated as the flow expands into a cavity downstream of a fin. A carry-over factor is used to account for the practical condition where not all of the dynamic pressure is dissipated and some kinetic energy is transferred to the following constriction.

The discharge coefficient, C_d, represents the ratio of the actual and ideal mass flow rate through the seal clearance:

$$C_d = \frac{m_{\text{actual}}}{m_{\text{ideal}}}, \tag{14.4}$$

Taking practical factors into account using a carry-over factor and a discharge coefficient, the mass flow in a labyrinth seal can be estimated by

$$m = kC_d A p_{t0} \sqrt{\frac{1 - (p_n/p_{t0})^2}{\text{RT}_{t0}[n - \ln(p_n/p_{t0})]}}, \tag{14.5}$$

where m is the mass flow rate through the seal clearance, k is the carry-over factor, and C_d is the discharge coefficient.

14.4.2 Orifice flow approach

A real labyrinth seal can be modelled as a series of successive nozzles and chambers. If the nozzles are small and the chambers large then it can be assumed that there is little or no carry-over of kinetic energy from one nozzle to the next. The flow through each nozzle can be modelled by an adiabatic polytropic expansion and a discharge coefficient can be applied. In each cavity, the flow is modelled as isentropic and the kinetic energy of the flow is recovered. Following these assumptions, use can be made of the standard compressible flow relationships for each nozzle.

The flow function in terms of the static pressure either side of a single orifice is given by

$$Q = C_{d,i}\sqrt{\frac{2\gamma_{i-1}}{R(\gamma_{i-1}-1)}\left[\left(\frac{p_i}{p_{i-1}}\right)^{2/\gamma_{i-1}} - \left(\frac{p_i}{p_{i-1}}\right)^{(\gamma_{i-1}+1)/\gamma_{i-1}}\right]}, \tag{14.6}$$

where $C_{d,i}$ is the discharge coefficient for the ith chamber.

When the pressure ratio of the last stage, p_n/p_{n-1}, is less than the critical pressure ratio then the flow will be choked and the Mach number, Ma, through the last restriction will be unity. The critical pressure ratio, $\lambda_c = \left(\frac{p_n}{p_{n-1}}\right)_{\text{critical}}$, is given by

$$\lambda_c = \left(\frac{p_n}{p_{n-1}}\right)_{\text{critical}} = \left(\frac{2}{\gamma_{n-1}+1}\right)^{\gamma_{n-1}/(\gamma_{n-1}-1)} \tag{14.7}$$

and the flow function becomes

$$Q = C_{d,n}\sqrt{\frac{2\gamma_{n-1}}{R(\gamma_{n-1}-1)}\left[(\lambda_c)^{2/\gamma_{n-1}} - (\lambda_c)^{(\gamma_{n-1}+1)/\gamma_{n-1}}\right]}. \tag{14.8}$$

Such a scheme has been applied to a labyrinth seal and the results illustrated in Fig. 14.33 where $C_{d,1} = C_{d,2} = C_{d,\text{i}} = C_{d,n} = 1$ is assumed.

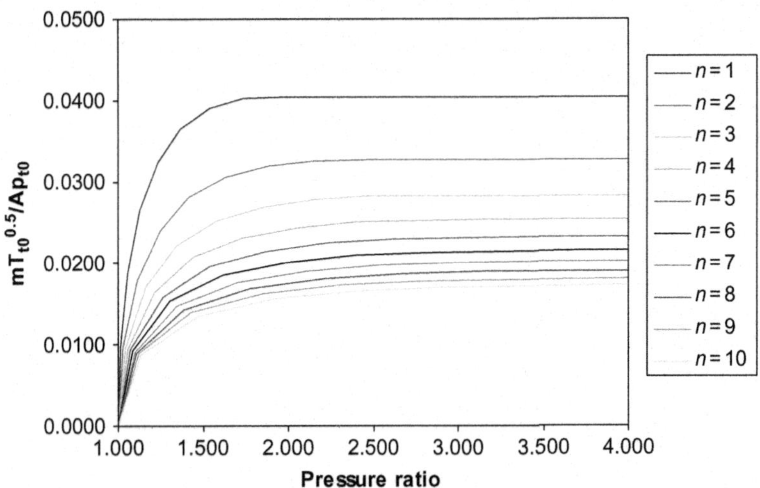

Fig. 14.33 Flow function versus pressure ratio for an isentropic throttle-based model based on Eqs. (14.6)–(14.8), with $C_d = 1$.

14.4.3 Zimmermann and Wolff's model

As indicated by the preceding discussion and the methods presented in Table 14.7 there has been a large number of modelling approaches proposed for labyrinth seals. Certain of these models have been validated for a specific application, such as gas turbine engines or steam turbines. For practical applications where an estimate for flow leakage in a labyrinth is required and accurate experimental data is not available, the model proposed by Zimmermann and Wolff (1998) for straight seals, Eq. (14.9), is suggested as the model has been validated for a wide range of turbomachinery applications and is comparatively simple to use.

$$m = k_2 C_d A p_{t0} \sqrt{\frac{1-(p_n/p_{t0})^2}{RT_{t0}[n - \ln(p_n/p_{t0})]}}. \tag{14.9}$$

In this equation, k_2 is the carry-over factor and is defined in Eq. (14.11).

The model accounts for a wide range of practical situations encountered in labyrinth seal applications including variation of

- carry-over factor with pitch and clearance;
- discharge coefficient with tip width and clearance;
- grooves;
- honeycombs;
- corner radius and
- chamber depth.

In an ideal straight labyrinth, one unit of dynamic pressure is dissipated in each of the fin seal cavities. This does not occur in practice and only a fraction of dynamic pressure is lost downstream of each fin. The carry-over factor is used to account for some of the dynamic pressure not being dissipated as the flow expands in the seal cavity. Hence, the leakage flow is greater than expected in the ideal model and the carry-over factor is greater than unity.

Measurements and modelling results have revealed that the proportion of dynamic pressure carried over is a function of the number of fins. The carry-over factor can be modelled by Hodkinson (1939):

$$k_1 = \sqrt{\frac{1}{1-\left(\frac{n-1}{n}\right)\left[\frac{c/l_{\text{pitch}}}{(c/l_{\text{pitch}})+0.02}\right]}}, \tag{14.10}$$

where k_1 is Hodkinson's carry-over factor, c is the seal clearance and l_{pitch} is the pitch of the fins.

A revised carry-over factor, k_2, recommended by Zimmermann and Wolff (1998), is given in Eq. (14.11).

$$k_2 = \sqrt{\frac{n/(n-1)}{\left(\frac{n}{n-1}\right) - \left[\frac{c/l_{\text{pitch}}}{(c/l_{\text{pitch}}) + 0.02}\right]}}, \tag{14.11}$$

where k_2 is Zimmerman's carry-over factor.

It is recommended that Zimmerman's carry-over factor, k_2, in Eq. (14.11) is used to determine the labyrinth seal flow rates in Eqs. (14.4) and (14.9). Values for the carry-over factor, given by Eq. (14.11) are plotted in Fig. 14.34.

For the case of modelling a labyrinth seal by means of a series of successive expansion chambers, Yucel and Kazakia (2001) recommended taking the carry-over factor as unity for the first fin, i.e. $i=1$, and as given in Eq. (14.12) for subsequent fins, i.e. for $i=2$ to n.

$$k_i = 1 + 0.0179(i-1) \tag{14.12}$$

The discharge coefficient is principally used to model the effects of flow contraction in a passage. Except at very low Reynolds numbers (*Re* less than about 10) and at conditions for which compressibility is significant, flow through a sharp edged orifice will separate from the sharp entrance edge. This separation restricts the available flow area and the effective reduction in flow area can be modelled by means of a discharge coefficient (see ESDU 82009 (ESDU, 1982), ESDU TN 07007 (ESDU, 2007)).

The flow contraction of a jet from a high pressure chamber through a planar orifice was studied by Chaplygin (see Gurevich, 1966) who derived the formula for the discharge coefficient in Eq. (14.13). This equation has been used by several investigators

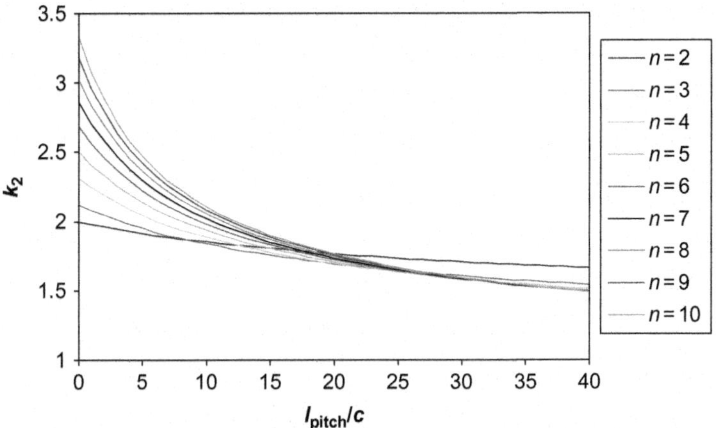

Fig. 14.34 Zimmerman's carry-over factor as a function of the ratio l_{pitch}/c for different numbers of fins.

(e.g. see Eser and Kazakia (1995) and Yucel and Kazakia (2001)) to model the flow coefficient for a particular fin constriction. The discharge coefficient C_d depends on the pressures in the two cavities adjacent to a particular fin i and is given by:

$$C_d = \frac{\pi}{\pi + 2 - 5S_i + 2S_i^2}, \tag{14.13}$$

where

$$S_i = \left(\frac{p_{i-1}}{p_i}\right)^{(\gamma-1)/\gamma} - 1. \tag{14.14}$$

In order to use Eq. (14.13) in conjunction with Eq. (14.4) or (14.9) an iterative procedure must be employed to predict the leakage flow and associated through the seals.

Plots for the discharge coefficient for a two fin labyrinth and a labyrinth seal with more than two fins are presented in Figs. 14.35 and 14.36. These are based on the data based on measurements presented by Zimmermann and Wolff (1998) attained from a proprietary database. As indicated in these figures typical values for the discharge coefficient for multiple fin labyrinths lie in the range $0.59 < C_d < 0.63$ at high Reynolds numbers, where the Reynolds number, Re_{D_h}, is based on the seal leakage based and hydraulic diameter, D_h:

$$Re_{D_h} = \frac{\rho u_{mean} D_h}{\mu}, \tag{14.15}$$

u_{mean} is the mean velocity across the sealing clearance and D_h is defined as

$$D_h = 2(r_o - r_i). \tag{14.16}$$

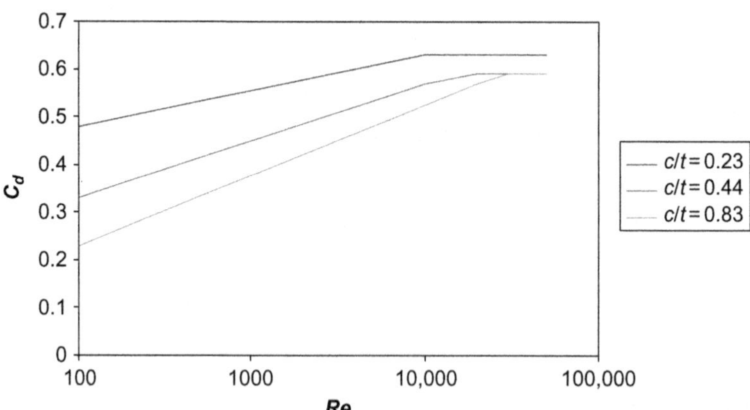

Fig. 14.35 Discharge coefficient for a straight labyrinth seal with two fins ($n = 2$).

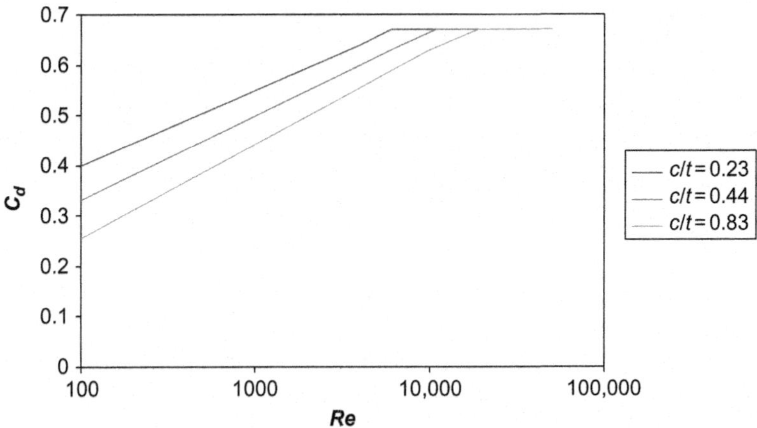

Fig. 14.36 Discharge coefficient for a straight labyrinth seal with more than two fins.

The discharge coefficient reduces with decreasing Reynolds number. For low values of Reynolds number, where the flow is dominated by friction, a friction factor would actually provide a more suitable match to the flow physics.

During transient operation of, for example, gas turbine engines, the rotor and stator assemblies expand at different rates altering the relative radial clearance and axial position of the fins in any labyrinth seal used. A change in radial clearance and axial position, in the case of for example a stepped seal, can significantly alter the leakage flow rate through a labyrinth seal. Rubbing of the fins on the stator can potentially occur in many rotating flow applications.

The inner surface of a labyrinth can be coated with an abradable coating (see Fig. 14.29), in order to accommodate potential rubs between the labyrinth fins and stator surface or in order to provide tight running clearances by allowing the fins to "machine" a running clearance in service or at a machine commissioning phase. Contact between the fin tips and stator surface results in a series of grooves along the length of the seal, as illustrated in Fig. 14.37, and a correction to the discharge coefficient needs to be made.

Rhode and Adams (2004) investigated the effects of varying the prerub radial clearance, rub-groove width and rub-groove depth in order to determine the effect of rub groove geometry on compressible flow in straight labyrinth seals. In the cases considered with each fin tip outside of its rub groove, the presence of a rub-groove increases the effective fin clearance. It was found that both the size of the rub-groove and the prerub radial clearance are important in determining changes in the local and global flow pattern as well as leakage. The presence of the rub-groove generally increases the leakage, primarily due to an increased effective fin radial clearance. For cases with a large prerub radial clearance and narrow rub-groove width, however, the increase was negligible. The negligible increase, for the case of a narrow rub-groove width, was attributed to low through-flow deflection into the rub-groove as a result of a minimized effective fin clearance. Rhode and Adams surmised that for operating conditions for which the fin tip is located inside of its rub groove that the presence of

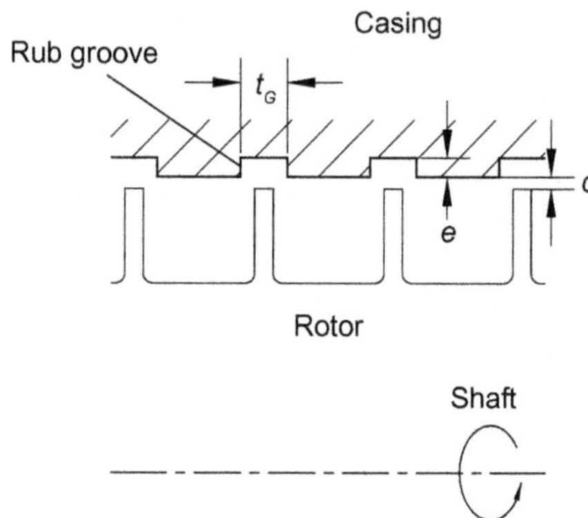

Fig. 14.37 Straight labyrinth seal with rub grooves.

the groove may act to decrease the leakage. A large width rub-groove generally gives slightly more leakage than an intermediate or narrow width, with intermediate and large width giving an almost identical leakage. The presence of the large or the intermediate width rub grooves allows the rub-groove depth to exert a fairly large effect on the leakage, especially for the smallest prerub clearance considered. The presence of a narrow rub-groove with the smallest prerub clearance considered allowed the rub-groove depth to exert a moderate effect on the leakage, but a dramatic effect on the overall flow pattern.

The ratio between the discharge coefficient with and without stator grooves as a function of fin tip to groove width ratio and groove depth and clearance is plotted in Fig. 14.38 following the data of Zimmermann and Wolff (1998).

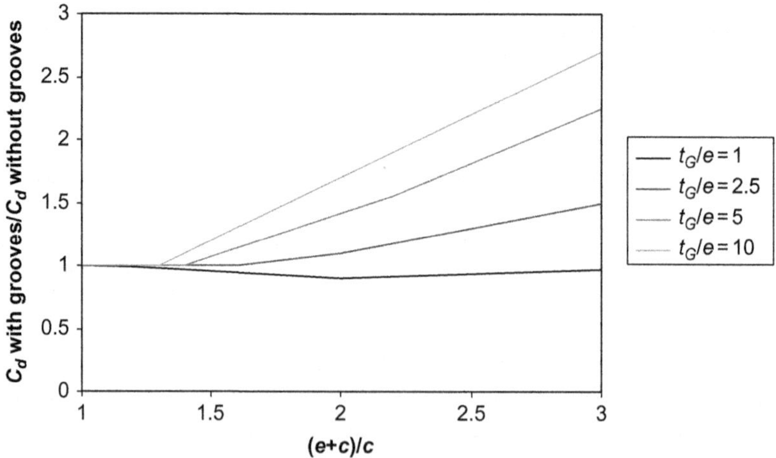

Fig. 14.38 Ratio of the discharge coefficient with and without stator grooves as a function of fin tip to groove width ratio and groove depth and clearance, $(e + c)/c$.

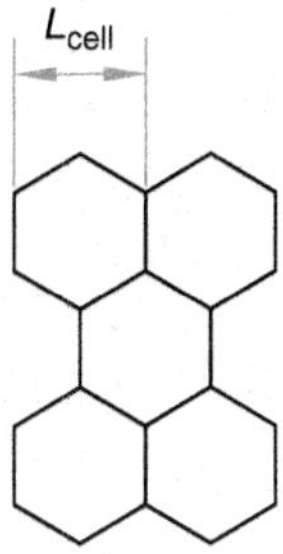

Fig. 14.39 Honeycomb cell definition.

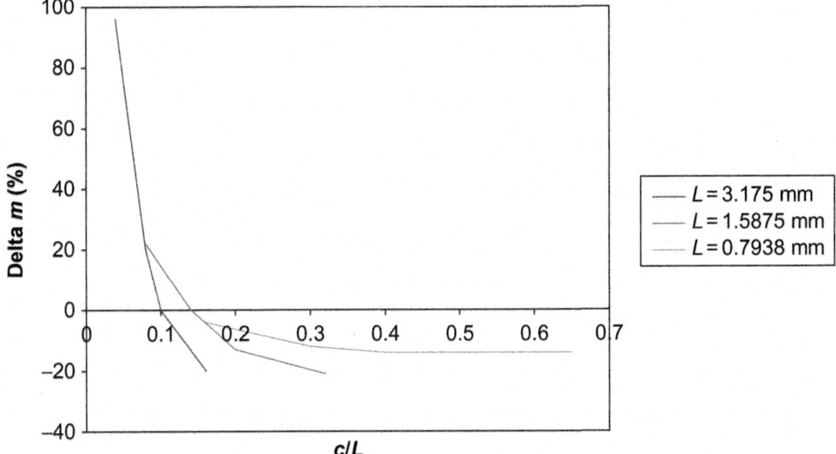

Fig. 14.40 Variation of leakage with cell size induced by a honeycomb layer on a labyrinth stator.

In order to improve the leakage characteristics, the inner surface of the labyrinth seal, i.e. the stator surface, can be treated with a honeycomb layer. The leakage is a function if the clearance and the honeycomb cell dimension, L_{cell}, see Fig. 14.39. A correction factor, Δm, see Eq. (14.17), for the change in the leakage flow in a labyrinth seal with a honeycomb on the stator, relative to that for a smooth land, is illustrated in Fig. 14.40 using experimentally derived data from Stocker (1978).

$$\Delta m = \frac{Q_{\text{with honeycomb}} - Q_{\text{smooth}}}{Q_{\text{smooth}}} \times 100\%, \tag{14.17}$$

where $Q_{\text{with honeycomb}}$ is the mass flow function with a honeycomb surface and Q_{smooth} is the mass flow function without the honeycomb.

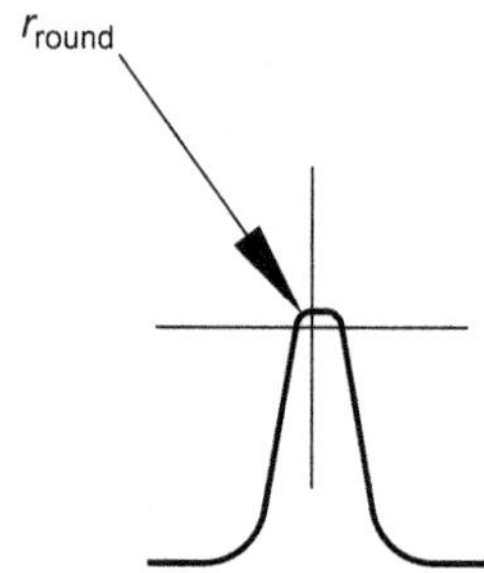

Fig. 14.41 Corner rounding definition.

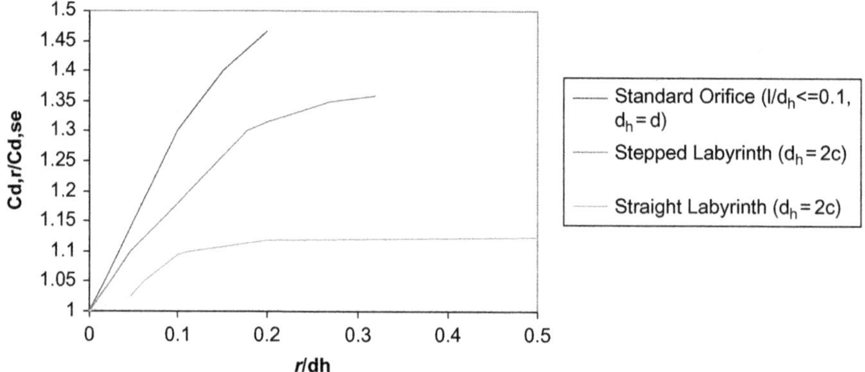

Fig. 14.42 Effect of rounding radius.

Chougule et al. (2008) report the use of CFD to optimise the configuration of a labyrinth seal with a honeycomb land and a staggered honeycomb land and straight teeth with an inclined notch gave approximately a 17% reduction in seal leakage in comparison to their baseline conditions. The performance of honeycombs to reduce labyrinth seal leakage and on wear is reported in the theses by Allcock (1999) and Collins (2006), and in Fraczek et al. (2017).

Corner rounding, see Fig. 14.41, also known as fin tip rounding, or rounded knife tips due to wear and abrasion, has a relatively minor impact (up to 12%) on the discharge coefficient for flow through a straight labyrinth seal. Corner rounding has been studied by Zimmermann et al. (1994), Rhode and Allen (2001) and Dogu et al. (2016b). Data, derived from experimental studies and proprietary information, from Zimmermann et al. (1994) for the ratio of the discharge coefficient with rounding to that with a sharp edge is plotted in Fig. 14.42 for both straight and stepped labyrinth seals. Because there is more carry-over in a straight labyrinth in comparison to a stepped labyrinth the effect of corner radius is much smaller for a straight labyrinth.

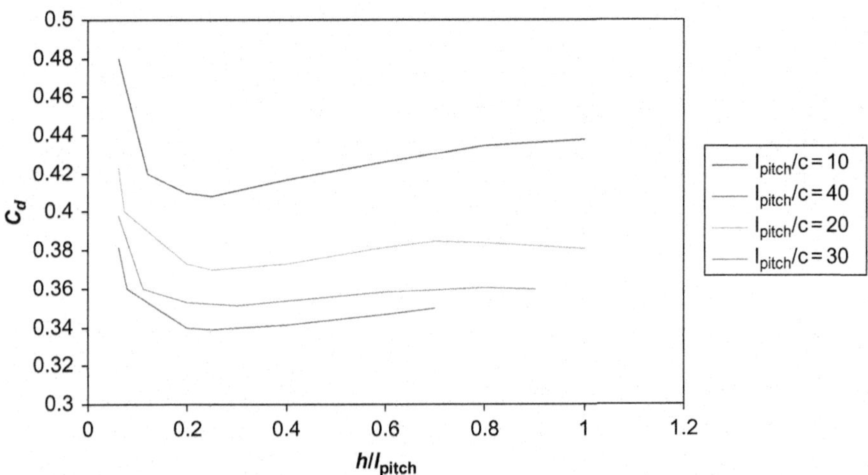

Fig. 14.43 Effect of labyrinth chamber depth on discharge coefficient for different pitch to clearance ratios.

The discharge coefficient is a function of the labyrinth chamber depth which is defined as the ratio of the fin height, h, and pitch l_{pitch}. This is plotted in Fig. 14.43 based on the experimental data from Trutnovsky and Komotori (1981). A minimum in the value of the discharge coefficient occurs approximately for $h/l_{pitch} = 0.25$ and this can therefore be used as a guideline for the specification of the optimal fin height in preliminary labyrinth design:

$$h = 0.25 l_{pitch}. \tag{14.18}$$

In general, a narrow fin-tip land-width (or fin width), t, is preferable. Yamada (1962) found that the flow coefficient generally decreased slightly with either the fin width to clearance ratio or the fin width to pitch ratio. At very low rotational speeds, about 300 rpm and less in the tests concerned, a significant dependence of the flow coefficient on the fin width to clearance ratio was identified. As a maximum, it is recommended that the fin width should be less than half of the pitch ($t < 0.5 l_{pitch}$), especially for low rotation rates (Morrison et al., 1983).

Stocker et al. (1977) performed tests for labyrinth seals with smooth, medium (8.3 μm) and rough (22.9 μm) sealing surfaces. It was identified that as the roughness increases up to 8.3 μm, leakage decreased. If the roughness is increased further, to 22.9 μm in the tests concerned, the leakage increased. Specifically, a medium surface roughness of 8.3 μm reduced the leakage in a straight labyrinth seal by approximately 23% in comparison to a seal with a smooth land for a 0.13 mm clearance and by 5.0% for a clearance of 0.51 mm.

The effect of cylindrical eccentricity of the seal shaft has been investigated by Yeh and Cochran (1970), Bell and Bergelin (1957), Nikitin and Ipatov (1973), and Gamal and Vance (2007). In general, it was found that eccentricity did not have a significant impact on the flow coefficient, with $<5\%$ change in the flow coefficient for the cases considered. Hodkinson (1939), however, found that full eccentricity of the seal increased the flow rate by a factor of 2.5 for some of the tests. The available data is therefore inconsistent. Gamal and Vance (2007) found that leakage through the seals they tested increased with increasing eccentricity and that this effects was more pronounced at lower supply pressures.

The influence of rotation on leakage flow in a labyrinth seal is negligible at axial Reynolds numbers $Re > 10^4$ (e.g. see Waschka et al. (1992)). For laminar flow ($Re \langle 200$), however, the effect is significant. Waschka et al. (1992) reported results for the variation of the discharge coefficient with Reynolds number for a range of rotational speeds and clearances. In these experiments, the discharge coefficient was found to increase from about 0.34 to 0.45 at $Re = 2000$ for a clearance of 1 mm as the rotational speed was increased to 10,000 rpm for the seal tested. Pressure losses for laminar flow result from increasing friction with decreasing Reynolds number and depend on surface area and form. Miyake and Duh (1990) investigated the relationship between mass flow rate and rotation. They found that the mass flow rate decreased in straight through labyrinth seals with increasing peripheral velocity. This decrease was found to be larger when the restriction tip thickness was large and the chamber depth was shallow. The effects of rotational speed were found to be largest for straight as opposed to stepped seals with the step-down seal performing better than the step-up seal.

Rotation of a labyrinth seal also results in frictional heating of the fluid through viscous dissipation of energy known as windage heating. This raises the temperature of the flow through the labyrinth and needs to be accounted for in any labyrinth design procedure in order to ensure that operating temperature limits are acceptable. The windage in labyrinth seals is primarily dependent upon rotational velocity and surface roughness and methods for its quantification have been studied by McGreehan and Ko (1989) and Millward and Edwards (1996).

Example 14.2. Determine the mass flow rate through a straight labyrinth seal with the following geometric characteristics:

- four fins, $n = 4$,
- outer diameter $D = 2 \times r_i = 215.9$ mm,
- fin height $h = 2.286$ mm,
- pitch $l_{pitch} = 3.175$ mm,
- fin taper angle $\phi_{fin} = 7.5°$,
- radial clearance $c = 0.127$ mm, and
- fin tip land width $t = 0.305$ mm.

The total pressure drop across the seal is 200 kPa, the downstream pressure is 1.01 bar and the average seal inlet temperature is $T_{t0} = 300$ K. Take the gas constant R as $R = 287$ J/(kg K) and the dynamic viscosity $\mu = 1.8 \times 10^{-5}$ Pa s. Example after ESDU 09004 (ESDU, 2009).

Solution

From Eq. (14.11), the carry-over factor, k_2, is given by

$$k_2 = \sqrt{\frac{n/(n-1)}{1-\left(\frac{n-1}{n}\right)\left[\frac{c/l_{\text{pitch}}}{(c/l_{\text{pitch}})+0.02}\right]}}$$

$$= \sqrt{\frac{4/(4-1)}{1-\left(\frac{4-1}{4}\right)\left[\frac{0.127/3.175}{(0.127/3.175)+0.02}\right]}} = 1.633.$$

The discharge coefficient, C_d, depends on the Reynolds number, Re, and, thus, on the leakage mass flow rate which is not known at this stage of the calculation.

From Fig. 14.36, the value of the discharge coefficient, C_d, can be taken as $C_d = 0.67$, using the value for which the discharge coefficient becomes invariant with Reynolds number; this value can be revised once an estimate of the Reynolds number in the seal has been obtained.

$r_i = D/2 = 215.9/2 = 107.95\,\text{mm}$,

$r_o = r_i + c = 107.95 + 0.127 = 108.077\,\text{mm}$.

The annular seal cross-sectional flow area is given by

$$A = \pi\left(r_o^2 - r_i^2\right) = \pi\left(0.108077^2 - 0.10795^2\right) = 8.619 \times 10^{-5}\ \text{m}^2$$

Substituting into Eq. (14.9), the mass flow rate through the clearance area can then be calculated as

$$m = k_2 C_d A p_{t0} \sqrt{\frac{1-(p_n/p_{t0})^2}{RT_{t0}[n - \ln(p_n/p_{t0})]}}$$

$$= 1.633 \times 0.67 \times 8.619 \times 10^{-5} \times 3.01$$

$$\times 10^5 \sqrt{\frac{1-\left(1.01 \times 10^5/3.01 \times 10^5\right)^2}{287 \times 300\left[4 - \ln\left(1.01 \times 10^5/3.01 \times 10^5\right)\right]}} = 0.04038\ \text{kg/s}$$

The Reynolds number now needs to be determined.

The hydraulic diameter for an annulus is given by

$$D_h = 2(r_o - r_i) = 2.54 \times 10^{-4}\ \text{m}$$

The Reynolds number based on the hydraulic diameter, D_h, is given by

$$\text{Re}_{D_h} = \frac{mD_h}{A\mu} = \frac{0.04038 \times 2.54 \times 10^{-4}}{8.619 \times 10^{-5} \times 1.8 \times 10^{-5}} = 6611.$$

The clearance to fin tip land width ratio is

$$c/t = 0.127/0.305 = 0.4164.$$

Calculating C_d. From Fig. 14.36, the revised value of the discharge coefficient, C_d, is approximately 0.64.

The calculations for the mass flow now need to be repeated with this updated value for the discharge coefficient and the results examined to see whether they have converged.

Recalculating the mass flow, m, and Re_{Dh}.

Substituting these values into Eq. (14.9), the revised mass flow rate is calculated to be

$$m = 1.633 \times 0.64 \times 8.619 \times 10^{-5} \times 3.01 \times 10^5 \sqrt{\frac{1-\left(1.01 \times 10^5/3.01 \times 10^5\right)^2}{287 \times 300\left[4 - \ln\left(1.01 \times 10^5/3.01 \times 10^5\right)\right]}} = 0.03857 \text{ kg/s}.$$

Hence, the Reynolds number is given by

$$\text{Re}_{D_h} = \frac{mD_h}{A\mu} = \frac{0.03857 \times 2.54 \times 10^{-4}}{8.619 \times 10^{-5} \times 1.8 \times 10^{-5}} = 6315.$$

These revised values for the mass flow rate and Reynolds numbers are significantly different from those obtained previously. Therefore, further calculations with the updated estimate for the Reynolds number are necessary.

From Fig. 14.36, with $\text{Re}_{Dh} = 6315$, the value of the discharge coefficient, C_d, is approximately 0.63. The calculations for the mass flow, velocity and Reynolds number give.

$$m = 0.03797 \text{ kg/s},$$

$$\text{Re}_{D_h} = 6216.$$

Examination of Fig. 14.36 with $\text{Re}_{D_h} = 6216$ gives a value of the discharge coefficient, C_d, of approximately 0.63. Since this value is close to that found previously, the solution, therefore can be assumed to have converged. The estimate for the mass flow through the seal for these conditions is approximately 0.038 kg/s.

14.4.4 Stepped labyrinth seals

A step can be incorporated in a labyrinth seal design in order to prevent carry-over of kinetic energy from occurring as illustrated in Fig. 14.44. This provides improved sealing characteristics in comparison to straight labyrinth seals. However, radial space must be available in machines where sealing is required to accommodate a stepped

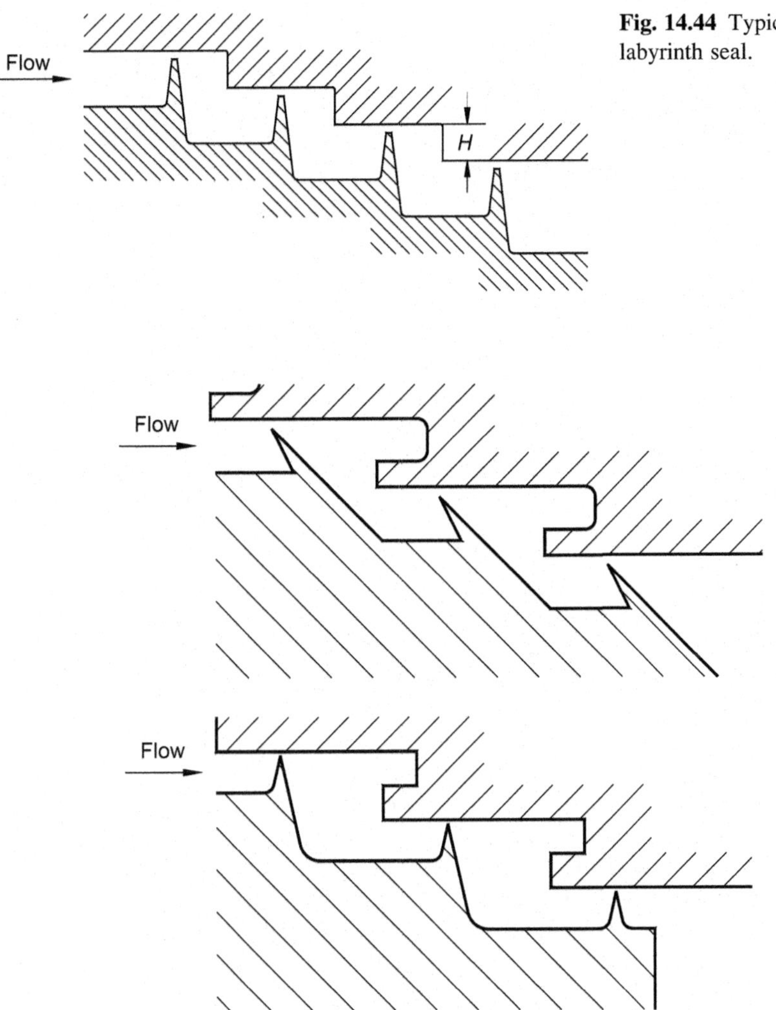

Fig. 14.44 Typical stepped labyrinth seal.

Fig. 14.45 Alternative configurations for stepped labyrinth seals.

seal. In addition, axial pressure gradients acting on step faces can cause undesirable axial loads. The geometric form of the seal can be defined to enhance the dissipation of kinetic energy and examples of this are illustrated in Fig. 14.45 (see Stocker, 1978). Such stepped seals have been applied to gas turbine engine applications.

Studies on stepped labyrinth seals have been reported by a number of researchers including Stocker (1975), Stocker et al. (1977), Vermes (1961), Benckert and Wachter (1979), Stepanoff (1931), Isaacson (1957) and Yamada (1962) and a variety of flow models have been developed for stepped labyrinth seals as indicated in Table 14.8. The detailed flow structure of flow in stepped labyrinth configurations has been studied by Rhode et al. (1997a and 1997b).

Table 14.8 Selected correlations for determining the mass flow for stepped labyrinth seals

Reference	Equation	Flow type	Application
Stocker (1975), Stocker et al. (1977)	$m=\frac{Ap_{t0}}{\sqrt{T}}\times F(\text{geometry, flow})$ $F(geometry, flow)$ from paper.	Compressible flow	Stepped seals
Vermes (1961)	$m=5.76C_d kA\sqrt{\frac{p_{t0}\rho_0\left[1-(p_n/p_{t0})^2\right]}{n+\ln(p_{t0}/p_n)}}$ $k=\frac{8.52}{\left[\left(l_{\text{pitch}}-t\right)/c\right]+7.23}$	Compressible flow	Straight through and stepped seals
Zimmermann and Wolff (1998)	$m=kC_d Ap_{t0}\sqrt{\frac{1-(p_n/p_{t0})^2}{RT_{t0}[n+\ln(p_{t0}/p_n)]}}$	Compressible flow	Straight through and stepped seals

In the absence of a validated database or code, the model proposed by Zimmermann and Wolff (1998) is suggested for stepped labyrinths, based on the following equation:

$$m=k_s C_d Ap_{t0}\sqrt{\frac{1-(p_n/p_{t0})^2}{\mathrm{RT}_{t0}[n+\ln(p_{t0}/p_n)]}}, \tag{14.19}$$

where C_d is the discharge coefficient (Fig. 14.46) and k_s is a correction factor for C_d (see Fig. 14.47).

For a stepped labyrinth seal Yucel (2004) reported using a discharge coefficient of unity for the first fin and 0.7 for subsequent fins.

Willenborg et al. (2001) provided experimental data, from a test rig representative of a labyrinth seal, for the discharge coefficient as a function of c/t, pressure ratio and Reynolds number, where the Reynolds number was defined by $Re = 2mc/A\mu$. Due to the combined influence of contraction and friction effects, a general dependence of the discharge coefficient on Reynolds number was identified. For low Reynolds numbers, the discharge coefficient increased by up to 9% for the 1:1 scale model tested as the

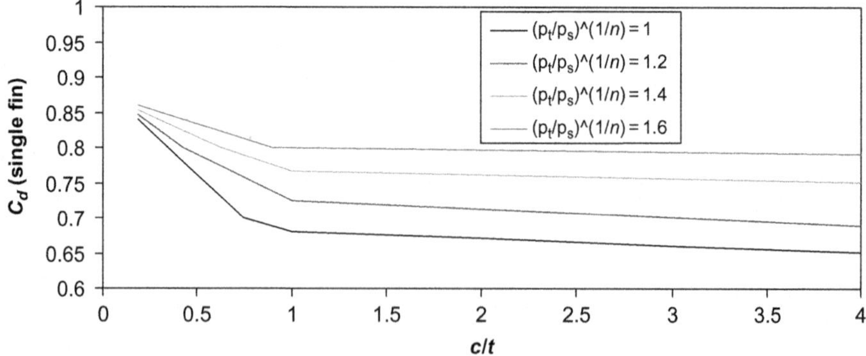

Fig. 14.46 Discharge coefficients for a stepped labyrinth seal (single fin).

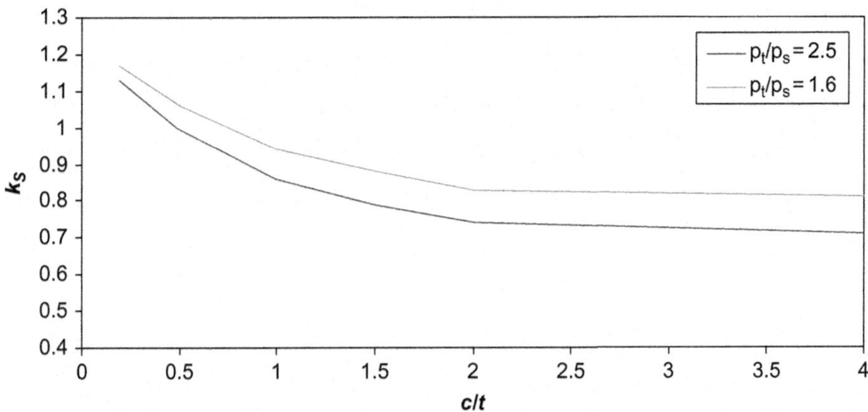

Fig. 14.47 C_d correction factor.

Reynolds number increased from 2500 to 8050. For the 1:1 model tested, the following range of values of discharge coefficient were found:

$$0.53 \le C_d \le 0.64 \text{ for } c/t = 0.909;$$

$$0.46 \le C_d \le 0.57 \text{ for } c/t = 1.515;$$

$$0.46 \le C_d \le 0.53 \text{ for } c/t = 2.424.$$

At sufficiently high Reynolds numbers, between about 30,000 and 40,000, the discharge coefficients for a particular seal configuration were found to depend only on the pressure ratio.

Zimmermann and Wolff (1998) proposed use of discharge coefficient data from Snow (1952) for a single fin (Fig. 14.46), in combination with a correction factor, k_s. Their data based on measurements and proprietary information for the discharge coefficient correction factor, k_s, are in Fig. 14.47. Zimmermann and Wolff reported that these data represent test results within ±5%.

Corner rounding (see Fig. 14.41) can have a significant impact (up to 36%) on the discharge coefficient for flow through a stepped labyrinth seal. The data based on measurements and proprietary information from Zimmermann et al. (1998) for the ratio of the discharge coefficient with rounding to that with a sharp edge, plotted in Fig. 14.42 for both straight and stepped labyrinth seals, can be used to model the effect of corner rounding.

Zimmermann and Wolff (1998) presented proprietary data for the influence of step height (Fig. 14.37). A minimum at $H/l_{\text{pitch}} = 0.075$ is recommended if practicable.

14.4.5 Staggered labyrinths

In a staggered labyrinth seal, the fins are alternatively attached to the shaft and the casing. Staggered labyrinths give lower leakage than corresponding straight labyrinth seals with the same clearance, as illustrated in Fig. 14.48. The improved sealing

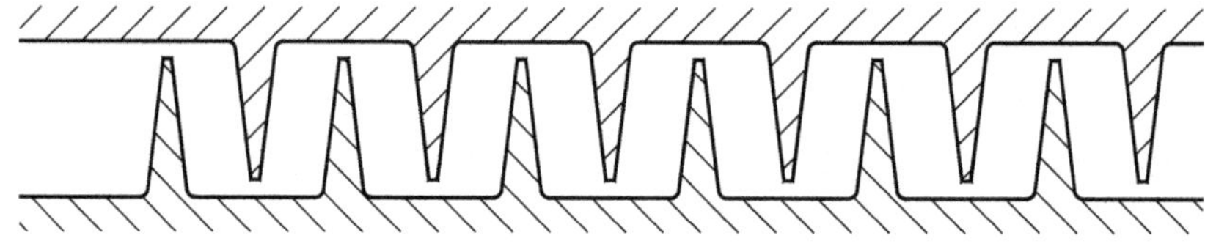

Fig. 14.48 Staggered labyrinth seal.

characteristics must, however, be assessed against the increased complexity of their assembly and limitations on axial movement. In order to assemble a staggered labyrinth seal, the case must normally be split, typically into two parts. Axial growth of the shaft or casing must be accommodated in the design to ensure no contact between alternate fins.

Studies on staggered labyrinth seals have been reported by a number of researchers including Martin (1908), Stodola (1927), Dollin and Brown (1937), Egli (1935), Kearton and Keh (1952), Komotori and Mori (1971), Deych et al. (1980a, 1980b), Samsonov et al. (1978) and Dodge (1963). A variety of flow models have been developed for staggered labyrinth seals as indicated in Table 14.9 enabling an appropriate model for a given type of flow and application to be identified.

Table 14.9 Selected correlations for determining the mass flow for staggered labyrinth seals

Reference	Equation	Flow type	Application
Martin (1908, 1919)	$m = Ap_{t0}\sqrt{\frac{1-(p_n/p_{t0})^2}{\mathrm{RT}_{t0}[n+\ln(p_{t0}/p_n)]}}$	Compressible flow	Staggered seals
Stodola (1927)	$m = Ap_{t0}\sqrt{\frac{1-(p_n/p_{t0})^2}{\mathrm{RT}_{t0}n}}$	Compressible flow	Staggered seals
Dollin and Brown (1937)	$m = A\sqrt{\frac{2\gamma\rho_0 p_{t0}\left[1-(p_n/p_{t0})^{(1+\gamma)/\gamma}\right]}{(\gamma+1)[n+(1/\gamma)\ln(p_n/p_{t0})]}}$ $m = A\sqrt{\frac{2\rho(p_{t0}-p_n)}{n}}$	Polytropic process; incompressible flow	Staggered seals
Egli (1935)	$m = \alpha\varepsilon A\sqrt{\frac{p_{t0}\rho_0\left[1-(p_n/p_{t0})^2\right]}{n+\ln(p_{t0}/p_n)}}$	Compressible flow	Straight through and staggered seals
Kearton and Keh (1952)	$m = C_d A\sqrt{\frac{F(p_0^2-p_n^2)}{n\mathrm{RT}_{t0}}}$ for $p_n/p_{t0} >$ critical F is a function of the pressure ratio and is defined in Kearton and Keh (1952) $m = \frac{3.873C_d Ap_{n-1}}{\sqrt{\mathrm{RT}_{t0}}}$ for $p_n/p_{t0} >$ critical (air)	Compressible flow	Staggered seals

Table 14.9 Continued

Reference	Equation	Flow type	Application
Komotori and Mori (1971)	$m = \alpha\varepsilon_1 A_0 F(n,\beta,\gamma)\sqrt{\rho_0 p_{t0}}$ $F(n,\beta,\gamma)$ from paper.	Compressible flow	Straight through and radial seals
Deych et al. (1980a) and (1980b)	$m = \left[\frac{\sigma}{x_0 + x_0^2}\right]^{-0.5} A\sqrt{\frac{\rho_o p_{t0}\left[1-(p_n/p_{t0})^2\right]}{n}}$ Where x_o is the steam quality and σ can be found from a graph in the original paper	Superheated and wet (two-phase) steam flow	Staggered seals
Samsonov et al. (1978)	$m = \alpha A\sqrt{\frac{\rho_0 p_{t0}\left[1-(p_n/p_{t0})^2\right]}{n}}$	Superheated and wet (two-phase) steam flow	Staggered seals
Dodge (1963)	$m = A\sqrt{\frac{2\rho(p_{t0}-p_n)}{fL/[2c+1.5+k(n-1)]}}$ $k = 1$ straight through seals $k = 2$ staggered seals. $f = 64/Re$ for $Re < 2320$ $f = 0.316/\mathrm{Re}^{0.25}$ for $Re > 2320$	Incompressible flow	Straight through and staggered seals

14.4.6 Radial labyrinths

In a radial labyrinth seal, the flow direction is orthogonal to the machine axis or shaft as illustrated in Fig. 14.27D–F. The flow can be radially inward, toward the shaft axis, or radially outward, away from the shaft axis. A radial labyrinth seal may be suitable in applications where the axial space available in a machine is limited but use of radial space can be accommodated. Radial labyrinth seals can have a straight, staggered or stepped configuration.

Studies on radial labyrinth seals have been reported by a number of researchers including Martin (1908), Stodola (1927), Kearton (1955), Solomko et al. (1979) and Ueda and Kubo (1967). A variety of flow models have been developed for radial labyrinth seals as indicated in Table 14.10 enabling an appropriate model for a given type of flow and application to be identified.

For the case of radially outward flow, the data of Kearton and Keh (1952) showed that the flow coefficient decreased slightly with pressure ratio for decreasing clearances except at small clearances where the flow coefficient maintained a fairly constant value. For the case of radial inflow, the flow coefficient decreased only slightly as the clearance was decreased across a range of pressure ratios.

14.5 Axial and bush seals

Simple axial and radial bush seals are illustrated in Figs. 14.49 and 14.50 and can be used for sealing both liquids and gases. Leakage through a concentric axial bush seal can be estimated using Eq. (14.20) for incompressible flow (flow where the density

Table 14.10 Selected correlations for determining the mass flow for radial labyrinth seals

Reference	Equation	Flow type	Application
Martin (1908, 1919)	$m=p_{t0}\sqrt{\frac{A_0A_n\left[1-(p_n/p_{t0})^2\right]}{\mathrm{RT}_{t0}\left[n+\ln\left(A_0p_{t0}/A_np_{tn}\right)\right]}}$	Compressible flow	Radial seals.
Stodola (1927)	$m=p_{t0}\sqrt{\frac{A_0A_n\left[1-(p_n/p_{t0})^2\right]}{\mathrm{RT}_{t0}n}}$	Compressible flow	Radial seals.
Kearton and Keh (1952)	$m=\alpha A\sqrt{\frac{\rho_0p_{t0}F(\beta)\left[1-(p_n/p_{t0})^2\right]}{n^{\#}}}$ $F(\beta)$ from paper. $n^{\#}=\frac{n+2l_{\text{pitch}}n(n-1)}{2r_i-l_{\text{pitch}}(2n-1)}$ inward flow $n^{\#}=\frac{n-2l_{\text{pitch}}n(n-1)}{2r_i+l_{\text{pitch}}(2n-1)}$ outward flow	Compressible flow	Radial seals.
Ueda and Kubo (1967)	$m=\alpha\sqrt{\frac{A_0A_n\rho_0p_{t0}^2g\zeta_m\left[1-(p_n/p_{t0})^2\right]}{(1-\nu^2)p_n\left[(n-1)+\frac{2\ln(p_{t0}/p_n)}{\gamma\zeta_m(1-\nu^2)}\right]}}$, where ζ_m represents the mean number of fins	Compressible flow	Radial seals.

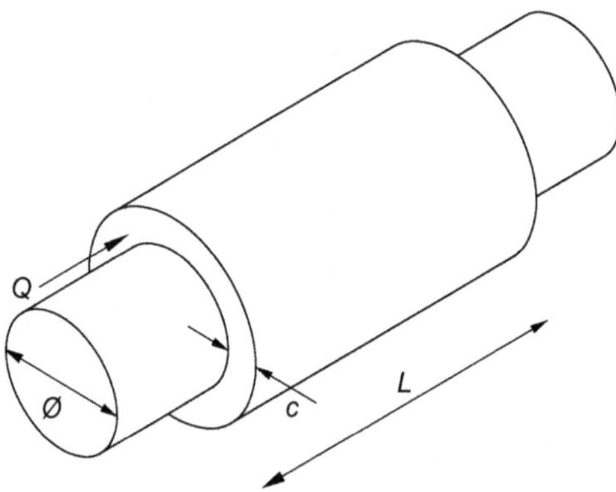

Fig. 14.49 Axial bush seal.

can be considered constant) and Eq. (14.21) for compressible flow. The leakage flow though a radial bush seal can be estimated by Eq. (14.22) for incompressible flow and Eq. (14.23) for compressible flow.

$$Q=\frac{\pi\phi c^3(p_o-p_a)}{12\mu L} \tag{14.20}$$

$$Q=\frac{\pi\phi c^3\left(p_o^2-p_a^2\right)}{24\mu Lp_a} \tag{14.21}$$

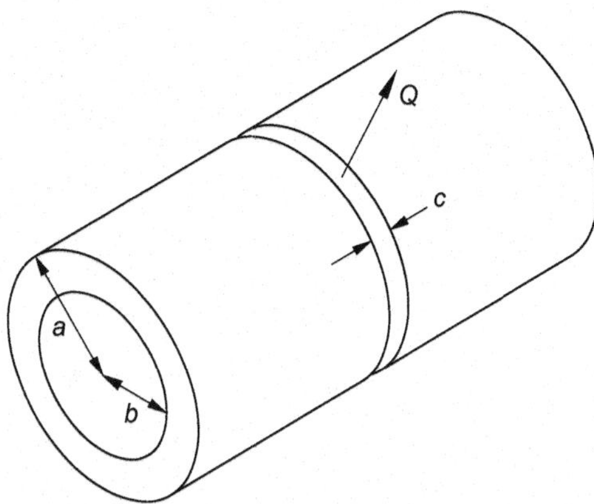

Fig. 14.50 Radial bush seal.

where Q = volumetric flow rate (m^3/s), ϕ = diameter of the shaft (m), c = radial clearance (m), p_o = upstream pressure (Pa), p_a = downstream pressure (Pa), μ = absolute viscosity (Pa s), L = length of the axial bush seal (m).

$$Q = \frac{\pi c^3 (p_o - p_a)}{6\mu \ln(a/b)}, \tag{14.22}$$

$$Q = \frac{\pi c^3 (p_o^2 - p_a^2)}{12\mu p_a}, \tag{14.23}$$

where c = axial clearance (m), a = outer radius (m) and b = inner radius (m).

Example 14.3. An axial bush seal consists of an annular gap with inner and outer radii of 50 mm and 50.5 mm, respectively. The length of the seal is 40 mm. Determine the flow rate of oil through the seal if the pressures upstream and downstream of the seal are 7 bar and 5.5 bar, respectively. The viscosity of the oil can be taken as 0.025 Pa s.

Solution

The radial clearance is $0.0505 - 0.05 = 0.0005$ m. $\phi = 0.1$ m, $L = 0.04$ m.

The volumetric flow rate is given by

$$Q = \frac{\pi \times 0.1 \times (0.5 \times 10^{-3})^3 (7 \times 10^5 - 5.5 \times 10^5)}{12 \times 0.025 \times 0.04} = 4.909 \times 10^{-4}\ \mathrm{m^3/s}.$$

14.6 Seals for reciprocating components

The seals principally used for reciprocating motion are packings and piston rings.

Packing seals are illustrated in Fig. 14.51. The seal essentially consists of a cup, V, U or X section of leather, solid rubber or fabric reinforced rubber. The sealing

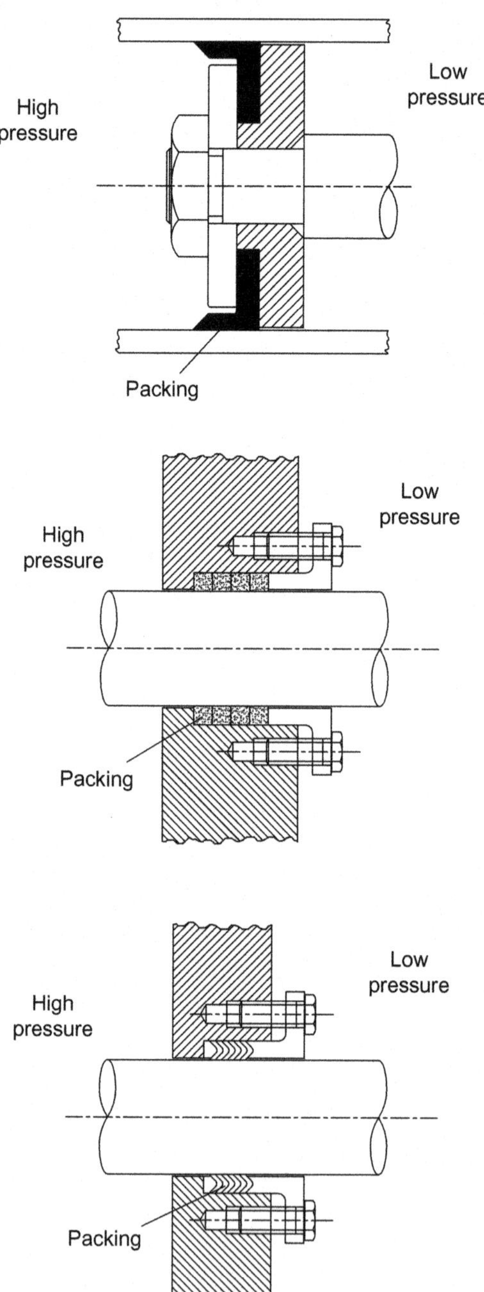

Fig. 14.51 Packing seals.

Fig. 14.52 Piston assembly.

principle is by direct contact with the reciprocating component. The contact pressure can be increased in the case of V packings by axial compression of the seals although this obviously increases the friction on the shaft and wear rate. The principal uses of cup packings are as piston seals in hydraulic and pneumatic applications, U packings for piston rods and V packings for sealing piston rods or reciprocating shafts.

Piston rings are used to seal cylinders where the operating temperature is above the limit of elastomeric, fabric or polymeric materials. Piston rings are used in automotive cylinders, e.g. Fig. 14.52, for three purposes (see Economou et al. (1982), and Andersson et al. (2002)):

(i) to seal the combustion chamber/cylinder head,
(ii) to transfer heat from the piston to the cylinder walls,
(iii) to control the flow of oil.

Piston rings are usually machined from a fine grain alloy cast iron and must be split to allow for assembly over the piston. Conventional practice is to use three piston rings with two compression rings sealing the high pressure and one to control the flow of oil. The range of piston ring available is extensive as illustrated in Figs. 14.53 and 14.54. For other applications the number of piston rings required can be determined using Table 14.11 assuming normal piston temperatures and running speeds. The design and operation of piston rings is subject to a wide number of variables and significant opportunities exist to optimise performance, reducing friction (see McGeehan (1978), Furuhama et al. (1981), Ruddy et al. (1982), Jeng (1992), Wakuri et al. (1992), Mezghani et al. (2012), and Usman and Park (2016)).

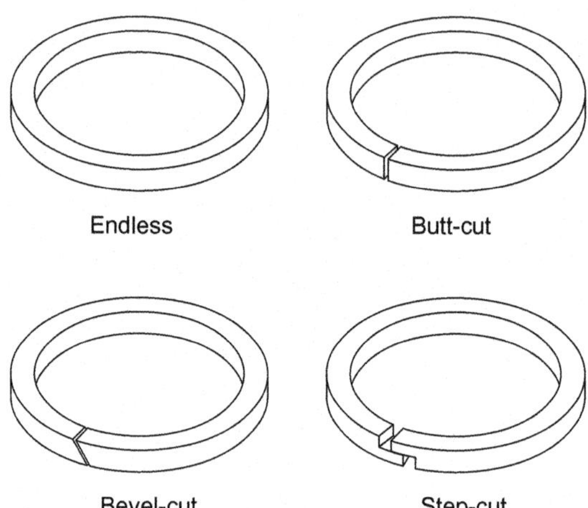

Fig. 14.53 Piston rings.

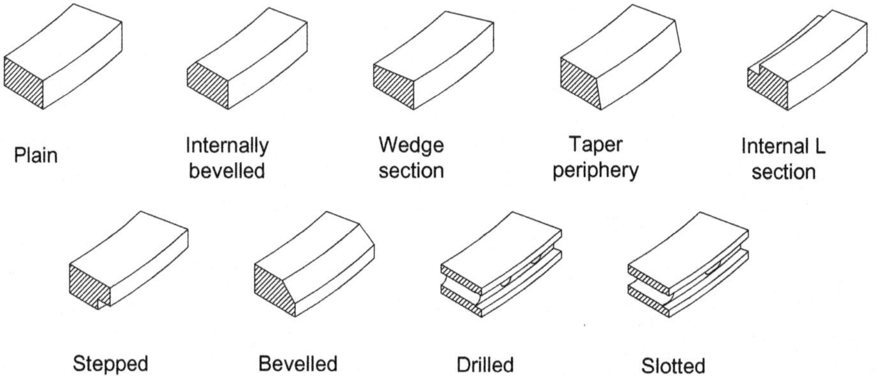

Fig. 14.54 Piston ring sections.

Table 14.11 Number of piston rings required to seal a given pressure

P_o (bar)	Number of rings
<20	2
$20 < p_o < 60$	3
$60 < p_o < 100$	4
$100 < p_o < 200$	5
>200	6+

14.7 Conclusions

It is frequently necessary in machine design to provide some means of containing or limiting the flow of fluid from one region to another. Because of the very small nature of fluid molecules this is a challenging task. This chapter has reviewed a range of static and dynamic seals. Because of the wide range of applications seals tend not to be available as stock items and instead must be designed fit for purpose. The subject of sealing for fluid applications is also introduced by Flitney (2014), Muller and Nau (1998), SAE (1998), ESDU (1980, 1983), Steinetz and Hendricks (1997), Neale (1994) and Zuk (1976) and for the specific case of labyrinth seals by ESDU (2009).

References

Books and papers

Adjemout, M., Andrieux, A., Bouyer, J., et al., 2017. Influence of the real dimple shape on the performance of a textured mechanical seal. Tribol. Int. 115, 409–416.

Allcock, D., 1999. Abradable stator gas turbine labyrinth seals. PhD Thesis, Cranfield University.

Andersson, P., Tamminen, J., Sandström, C.E., 2002. In: Piston ring tribology. A literature survey.VTT Tiedotteita-Research Notes, 2178.

Bai, S., Peng, X., Li, Y., et al., 2010. A hydrodynamic laser surface-textured gas mechanical face seal. Tribol. Lett. 38, 187–194.

Bell, K.J., Bergelin, O.P., 1957. Flow through annular orifices. Trans. ASME 79, 593–601.

Benckert, H., Wachter, J., 1979. In: Investigations on the mass flow and flow induced forces in contactless seals of turbomachines.Proceedings of the Sixth Conference on Fluid Machinery, Scientific Society of Mechanical Engineers, Budapestpp. 57–66.

Childs, D., 1993. Turbomachinery Rotordynamics: Phenomena, Modeling and Analysis. Wiley.

Childs, D., Rhode, D., 1984. Rotordynamic forces developed by labyrinth seals.Turbomachinery Laboratories Mechanical Engineering Department, AFOSR-TR-85-1070.

Chougule, H.H., Ramerth, D., Ramachandrandran, D., 2008. Low leakage designs for labyrinth seals. ASME Paper GT2008 51024.

Chupp, R.E., Holle, G.F., Scott, T.E., 1986. Labyrinth seal analysis. Volume IV—User's manual for the labyrinth seal design model. AF Wright Aeronautical Laboratories AFWAL-TR-85-2103 Volume IV.

Chupp, R.E., Hendricks, R.C., Lattime, S.B., Steinetz, B.M., 2006a. Sealing in turbomachinery. NASA/TM—2006-214341.

Chupp, R.E., Hendricks, R.C., Lattime, S.B., et al., 2006b. Sealing in turbomachinery. J. Propuls. Power 22, 313–349.

Collins, D., 2006. The effects of wear on abradable honeycomb labyrinth seals. EngD. Thesis, Department of Power, Propulsion & Aerospace Engineering, School of Engineering Cranfield University.

Cremanns, K., Roos, D., Hecker, S., et al., 2016. Efficient multi-objective optimization of labyrinth seal leakage in steam turbines based on hybrid surrogate models. ASME Turbo Expo, Seoul.

Czernik, D.E.G., 1996a. Design, Selection, and Testing. McGraw Hill.

Deych, M.Y., Supunov, O.G., Shanin, V.K., Sabri, T.I., 1980a. Effect of geometric dimensions of labyrinth seal annuli on the rate of flow of superheated and wet steam through them. Fluid Mech.-Sov. Res. 9 (4), 72–77.

Deych, M.Y., Solomko, V.I., Zezyulinskiy, G.S., 1980b. Rate of leakage of wet steam through a labyrinth seal. Fluid Mech.-Sov. Res. 9 (4), 64–71.

Dodge, L., 1963. Labyrinth shaft seals. Prod. Eng. 34, 75–79.

Dogu, Y., Sertcakan, M.C., Gezer, K., et al., 2016a. Labyrinth Seal Leakage Degradation Due to Various Types of Wear. ASME Turbo Expo, Seoul.

Dogu, Y., Sertcakan, M.C., Gezer, K., et al., 2016b. Leakage Degradation of Straight Labyrinth Seal Due to Wear of Round Tooth Tip and Acute Trapezoidal RubGroove. ASME Turbo Expo, Seoul.

Dogu, Y., Sertcakan, M.C., Gezer, K., et al., 2017. Labyrinth seal leakage degradation due to various types of wear. Trans. ASME, J. Eng. Gas Turbines Power. 139.

Dolan, P.J., 1992. Mechanical seal review. In: Nau, B.S. (Ed.), Fluid Sealing. Kluwer, pp. 413–427.

Dollin, F., Brown, W., 1937. Flow of fluids through openings in series. The Engineer 164 (4259), 223–224.

Economou, P.N., Dowson, D., Baker, A.J.S., 1982. Piston Ring Lubrication—Part 1: The Historical Development of Piston Ring Technology. J. Lubricat. Tech. 104, 118–126.

Egli, A., 1935. The leakage of steam through labyrinth seals. Trans. ASME 57, 115–122. 445–446.

ESDU. ESDU 80012 Dynamic sealing of fluids 1: guide to selection of rotary seals. Engineering Sciences Data Unit, 1980.

ESDU. ESDU 82009, Compressible flow of gases. Pressure losses and discharge coefficients of orifice plates, perforated plates and thick orifice plates in ducts, Engineering Sciences Data Unit, 1982.

ESDU. ESDU 83031 Dynamic sealing of fluids 2: guide to selection of reciprocating seals. Engineering Sciences Data Unit, 1983.

ESDU. ESDU TN 07007. Incompressible flow through orifice plates – a review of the data in the literature. Engineering Sciences Data Unit, 2007.

ESDU. ESDU 09004, Labyrinth seal flow. Engineering Sciences Data Unit, 2009.

Eser, D., Kazakia, J.Y., 1995. Air flow in cavities of labyrinth seals. Int. J. Eng. Sci. 33, 2309–2326.

Etsion, I., 2004. Improving tribological performance of mechanical components by laser surface texturing. Tribol. Lett. 17, 733–737.

Etsion, I., Kligerman, Y., Halperin, G., 1999. Analytical and experimental investigation of laser-textured mechanical seal faces. Tribol. Trans. 42, 511–516.

Flitney, R., 2014. Seals and Sealing handbook, sixth ed. Elsevier.

Fraczek, D., Wroblewski, W., Bochon, K., 2017. Influence of honeycomb rubbing on the labyrinth seal performance. Trans. ASME, J. Eng. Gas Turbines Power. 139.

Furuhama, S., Takiguchi, M., Tomizawa, K., 1981. Effect of piston and piston ring designs on the piston friction forces in diesel engines. SAE Technical Paper No. 810977.

Gamal, A.M., Vance, J.M., 2007. Labyrinth seal leakage tests: Tooth profile, tooth thickness, and eccentricity effects. ASME Paper GT2007-27223.

Gurevich, M.I., 1966. The Theory of Jets in an Ideal Fluid. Pergamon.

Hodkinson, B., 1940. Estimation of the leakage through a labyrinth gland. Proc. Inst. Mech. Eng. 141, 283–288.

Horve, L., 1996. Shaft Seals for Dynamic Applications. Marcel Dekker.

Isaacson, J. Pump impeller labyrinth seal study. DDR-712-3012, 1957.

Jeng, Y.R., 1992. Theoretical analysis of piston-ring lubrication Part I—fully flooded lubrication. Tribol. Trans. 35, 696–706.

Jia, X., Guo, F., Huang, L., Wang, L., Gao, Z., Wang, Y., 2014. Effects of the radial force on the static contact properties and sealing performance of a radial lip seal. Sci. China Technol. Sci. 57, 1175–1182.

Kearton, W.J., 1955. The flow of air through radial labyrinth glands. Proc. Inst. Mech. Eng. 169, 539–550.

Kearton, W.J., Keh, T.H., 1952. The flow of air through labyrinth glands of staggered type. Proc. Inst. Mech. Eng. 166, 180–195.

Kim, T.S., Cha, K.S., 2009. Comparative analysis of the influence of labyrinth seal configuration on leakage behaviour. J. Mech. Sci. Technol. 23, 2830–2838.

Komotori, K., Mori, H., 1971. In: Leakage characteristics of labyrinth seals.Fifth International Conference on Fluid Sealing. Paper E4, pp. 45–63.

Lebeck, A.O., 1991. Principles and Design of Mechanical Face Seals. Wiley Interscience.

Martin, H., 1908. Labyrinth packings. Engineering 85, 35–38.

Martin, H., 1919. Steam leakage in dummies of the Ljungstrom type. Engineering 107, 1–3.

McGeehan, J.A., 1978. A literature review of the effects of piston and ring friction and lubricating oil viscosity on fuel economy. SAE Technical Paper, No. 780673.

McGreehan, W., Ko, S., 1989. Power dissipation in smooth and honeycomb labyrinth seals. ASME Paper 89-GT-220.

Mezghani, S., Demirci, I., Zahouani, H., El Mansori, M., 2012. The effect of groove texture patterns on piston-ring pack friction. Precis. Eng. 36, 210–217.

Millward, J., Edwards, M., 1996. Windage heating of air passing through labyrinth seals. Trans. ASME, J. Turbomach. 118, 414–419.

Miyake, K., Duh, W.C., 1990. Leakage characteristics of labyrinth seal. J. Chin. Soc. Mech. Eng. 11, 334–348.

Morrison, G.L., Rhode, D.L., Cogan, K.C., Chi, D., Demko, J., 1983. Labyrinth seals for incompressible flow. Final Report. NASA CR 170938.

Muller, H.K., Nau, B.S., 1998. Fluid Sealing Technology. Marcel Dekker.

Neale, M.J., 1994. Drives and Seals. Butterworth Heinemann.

Nikitin, G.A., Ipatov, A.M., 1973. Design of labyrinth seals in hydraulic equipment (in Russian). Russ. Eng. J. 53, 26–30.

Parsons, C.A., 1892. Electrical lighting at Cambridge. The Engineer. November 4.

Picardo, A., Childs, D.W., 2005. Rotordynamic coefficients for a tooth-on-stator labyrinth seal at 70 bar supply pressures: measurements versus theory and comparisons to a hole-pattern stator seal. Trans. ASME, J. Eng. Gas Turbines Power 127, 843–856.

Rhode, D.L., Adams, R.G., 2004. Rub-groove width and depth effects on flow predictions for straight-through labyrinth seals. Trans. ASME, J. Tribol. 126, 781–787.

Rhode, D.L., Johnson, J.W., Broussard, D.H., 1997a. Flow visualization and leakage measurements of stepped labyrinth seals: Part 1—annular groove. Trans. ASME, J. Turbomach. 119, 839–843.

Rhode, D.L., Younger, J.S., Wernig, M.D., 1997b. Flow visualization and leakage measurements of stepped labyrinth seals: Part 2—sloping surfaces. Trans. ASME, J. Turbomach. 119, 844–848.

Ruddy, B.L., Dowson, D., Economou, P.N., 1982. In: A review of studies of piston ring lubrication. Tribology of Reciprocating Engines: Proceedings of the 9th Leeds–Lyon Symposium on Tribology, UK, September 1982 (18, p. 109).

SAE, 1996. SAE Fluid Sealing Handbook Radial Lip Seals. Society of Automotive Engineers.

SAE, 1998. Sealing for Automotive Applications. In: Society of Automotive Engineers.

Samsonov, Y.F., Solomko, V.I., Zezyulinskiy, G.S., 1978. Investigation of leakage of superheated and wet steam in labyrinth seals. Heat Trans. Sov. Res. 10, 80–87.

Shapiro, W., and Chupp, R., Numerical analytical, experimental study of fluid dynamic forces in seals. Volume 5: Description of Seal Dynamics Code Dyseal and Labyrinth Seals Code KTK. NASA/CR-2004-213199/VOL5, 2004.

Sneck, H.J., 1974. Labyrinth seal literature survey. J. Lubric. Technol. Trans. ASME 96, 579–582.

Snow, E.W. Discussion Proceedings of the Institution of Mechanical Engineers, Vol. 166, 1952.

Solomko, V.I., Sivakov, V.I., Samsonov, V.F., 1979. Investigation of the leakage of superheated steam from radial labyrinth seals. Fluid Mech. Soviet Res. 8 (2), 18–25.

Stakenborg, M.J.L., 1988. On the sealing mechanism of radial lip seals. Tribol. Int. 21, 335–340.

Steinetz, B.M., Hendricks, R.C., 1997. Engine seals. In: Zaretsky, E.V. (Ed.), Tribology for Aerospace Applications. Society of Tribologists and Lubrication Engineers.

Stepanoff, A.J., 1931. Leakage loss and axial thrust in centrifugal pumps. ASME Paper HYD-54-5.

Stocker H.L., Advanced labyrinth seal design performance for high-pressure ratio gas-turbines, ASME Paper Paper No. 75-WA/GT-22, 1975.

Stocker, H.L., 1978. Determining and improving labyrinth seal performance in current and advanced high performance gas turbines. AGARD CP-237.

Stocker, H.L., Cox, D.M., Holle, G.F., 1977. Aerodynamic performance of conventional and advanced design labyrinth seals with solid-smooth, abradable and honeycomb lands. NASA Report CR-135307.

Stodola, A., 1927. Steam and Gas Turbines, sixth ed. McGraw Hill.

Summers-Smith, J.D. (Ed.), 1992. Mechanical seal practice for improved performance. IMechE, MEP.

The Engineer, 1938. The labyrinth packing. The Engineer. 165 (4280) Jan.

Tipton, D.L., Scott, T.E., Vogel, R.E., 1986. Labyrinth seal analysis. Volume III—Analytical and experimental development of a design model for labyrinth seals. AF Wright Aeronautical Laboratories AFWAL-TR-85-2103 Volume III.

Trutnovsky, K., Komotori, K., 1981. Berührungsfreie dichtungen. VDI Verlag.

Ueda, T., Kubo, T., 1967. The leakage of air through radial labyrinth glands. Bullet JSME 1 (38), 298–307.

Usman, A., Park, C.W., 2016. Optimizing the tribological performance of textured piston ring–liner contact for reduced frictional losses in SI engine: Warm operating conditions. Tribol. Int. 99, 224–236.

Vermes, G., 1961. A fluid mechanics approach to labyrinth seal leakage problem. J. Basic Eng. 83, 161–169.

Wakuri, Y., Hamatake, T., Soejima, M., Kitahara, T., 1992. Piston ring friction in internal combustion engines. Tribol. Int. 25, 299–308.

Waschka, W., Wittig, S., Kim, S., 1992. Influence of high rotational speeds on the heat transfer and discharge coefficients in labyrinth seals. Trans. ASME, J. Turbomach. 114, 462–468.

Willenborg, K., Kim, S., Wittig, S., 2001. Effects of Reynolds number and pressure ratio on leakage loss and heat transfer in a stepped labyrinth seal. Trans. ASME, J. Turbomach. 123, 815–822.

Winter, J.R., 1998. Gasket selection—a flowchart approach. In: Bickford, J.H. (Ed.), Gaskets and Gasketed Joints. Marcel Dekker.

Yamada, Y., 1962. On the pressure loss of flow between rotating co-axial cylinders with rectangular grooves. Bullet JSME 5, 642–651.

Yeh, F.C., and Cochran, R.P. Comparison of experimental and ideal leakage flows through labyrinth seals for very small pressure differences. NASA TMX-1958, 1970.

Yu, X.Q., He, S., Cai, R.L., 2002. Frictional characteristics of mechanical seals with a laser-textured seal face. J. Mater. Process. Technol. 129, 463–466.

Yucel, U., 2004. Calculation of leakage and dynamic coefficients of stepped labyrinth gas seals. J. Appl. Math. Comput. 152, 521–533.

Yucel, U., Kazakia, J.Y., 2001. Analytical prediction techniques for axisymmetric flow in gas labyrinth seals. Trans. ASME, J. Eng. Gas Turbines Power 123, 255–257.

Zimmermann, H., Wolff, K.H., 1987. Comparison between empirical and numerical labyrinth flow calculations. ASME Paper 87-GT-86.

Zimmermann, H., Wolff, K.H., 1998. Air system correlations Part 1: Labyrinth seals. ASME Paper 98-GT-206.

Zimmermann, H., Kammerer, A., Wolff, K.H., 1994. Performance of worn labyrinth seals. ASME Paper 94-GT-131.

Zuk, J. Fundamentals of fluid sealing. NASA TN D-8151, 1976.

Standards

British Standards Institution. BS 5341: Part 5 1976. Piston rings up to 200 mm diameter for reciprocating internal combustion engines. Ring grooves.

British Standards Institution. BS 4518: 1982. Specification for metric dimensions of toroidal sealing rings ('O' rings) and their housings.

British Standards Institution. BS 2492: 1990. Specification for elastomeric seals for joints in pipework and pipelines.

British Standards Institution. BS 5341: Part 7: Section 7.4: 1992. Piston rings up to 200 mm diameter for reciprocating internal combustion engines. Designs, dimensions and designations for single piece rings. Specifications for oil control rings.

British Standards Institution. BS 7780: Part 1: 1994. Specification for rotary shaft lip type seals. Nominal dimensions and tolerances.

British Standards Institution. BS 7780: Part 2: 1994. Specification for rotary shaft lip type seals. Vocabulary.

British Standards Institution. 98/706003 DC. Fluid systems. Sealing devices. O-rings. Part 2. Housing dimensions for general applications. (ISO/CD 3601-2:1997).

British Standards Institution. 99/714538 DC. ISO/DIS 3601-1. Fluid power systems. O-rings. Part 1. Inside diameters, cross sections, tolerances and size identification codes. 1999.

British Standards Institution. BS ISO 6194-4:1999. Rotary shaft lip type seals. Performance and test procedures.

British Standards Institution. BS EN 12756:2001. Mechanical seals: Principal dimensions, designations and material codes.

British Standards Institution. BS EN 3748:2001. Aerospace series. O-ring grooves. Dimensions.

British Standards Institution. BS ISO 16589-1:2001. Rotary shaft lip-type seals incorporating thermoplastic sealing elements. Nominal dimensions and tolerances.

British Standards Institution. BS ISO 16589-2:2001. Rotary shaft lip-type seals incorporating thermoplastic sealing elements. Vocabulary.

DIN 3771 Part 1 Fluid systems, O-rings, sizes to ISO 3601-1.

ISO 3601/1 Fluid systems, O-rings—Part 1. Inner diameters, cross sections, tolerances and size coding.

ISO 3601/2 Fluid systems, O-rings—Part 2. Design criteria for O-ring grooves.
ISO 3601/3 Fluid systems. O-rings—Part 3. Quality acceptance levels.
ISO 3601/4 Fluid systems, O-rings—Part 4. O-ring grooves with back-up rings.
ISO 3601/5 Fluid systems, O-rings—Part 5. O-rings for connectors to ISO 6149.

Websites

At the time of going to press the world-wide-web contained useful information relating to this chapter at the following sites.

fluidsciences.perkinelmer.com
oringswest.com.
www.allorings.com.
www.chesterton.com.
www.china-dongya.com.
www.claron.co.uk.
www.coni-seal.com.
www.federal-mogul.com.
www.flexibox.com.
www.garlock-inc.com.
www.goodway-rubber.com.
www.jenp.com.tw.
www.mamat.com.
www.nicholsons.co.uk.
www.orings.com.
www.o-rings-ez.com.
www.pspglobal.com.
www.samsunsegman.com.
www.sealsales.com.
www.ses-seal.com.tw.
www.superseal.com.
www.totalseal.com.
www.trisunltd.com.
www.uniquerubber.com.
www.usseal.com.
www.ussealmfg.com.
www.vercoseals.com.

Further reading

Buchter, H.H., 1979. Industrial Sealing Technology. Wiley.
Czernik, D.E., 1996b. Gaskets. In: Shigley, J.E., Mischke, C.R. (Eds.), Standard Handbook of Machine Design. McGraw Hill.
Hamilton, P., 1994. Seals. In: Hurst, K. (Ed.), Rotary Power Transmission. McGraw Hill.
Han, J.T., 1979. A fluid mechanics model to estimate the leakage of incompressible fluids through labyrinth seals. ASME Paper 79-FE-4.
Hopkins, R.B., 1996. Seals. In: Shigley, J.E., Mischke, C.R. (Eds.), Standard Handbook of Machine Design. McGraw Hill.

Morse, W., 1969. Seals. Morgan-Grampian.
Parker, 2007. O Ring Handbook. Parker Hannifi n Corporation.
Rao, N.B., Sidheswar, N., 1976. Influence of straight through type of labyrinth gland parameters on the amount of air leakage. India Eng. J.-ME 56, 176–181.
Rhode, D.L., Allen, B.F., 1999. Measurement and visualization of leakage effects of rounded teeth tips and rub-grooves on stepped labyrinths. ASME Paper 99-GT-377.
Society of Automotive Engineers, 1998. Sealing for Automotive Applications. SAE.
Stone, R., 1992. Introduction to Internal Combustion Engines. Macmillan.
Warring, R.H., 1981. Seals and Sealing Handbook. Latty International.

Springs

15

Chapter Outline

Nomenclature

Generally, preferred SI units have been stated.

A	area (m^2)
b	width (m)
C	spring index
d	wire diameter (m)
D	mean coil diameter (m)
D_i	inner diameter (m)
$D_{i\ min}$	minimum inner diameter (m)
D_o	outer diameter (m)
E	Young's modulus (N/m^2)
f_n	natural frequency (Hz)
F	load (N)
F_{flat}	load at flat position (m)
F_i	installed force (N)
F_o	operating force (N)
G	modulus of rigidity (N/m^2)
h	thickness (m), deflection to flat (m)
h'	height (m)
I	second moment of area (m^4)
J	polar second moment of area (m^4)
k	spring rate (N/m)
k_{total}	total spring rate (N/m)

Mechanical Design Engineering Handbook. https://doi.org/10.1016/B978-0-08-102367-9.00015-9

K_b stress concentration factor
K_s direct shear factor
K_w Wahl factor
L length (m)
L_a assembled length (m)
L_i installed length (m)
L_f free length (m)
L_m minimum working length (m)
L_{max} maximum coil body length (m)
L_o operating length (m)
L_s shut length (m)
m mass (kg)
M applied moment (N m)
N number of coils
N_a number of active coils
N_b number of coils in the body
N_e number of equivalent coils
N_{max} maximum number of coils
p pitch (m)
PH precipitated hardened
r radius (m)
R_d diameter ratio
R_1 mean loop radius (m)
R_2 side bend radius (m)
t thickness (m)
T torque (N m)
δ deflection (m)
$\delta_{initial}$ initial deflection (m)
μ Poisson's ratio
θ angular deflection (rad)
θ' angular deflection (revolutions)
ρ density (kg/m^3)
τ shear stress (N/m^2)
τ_{inital} shear stress under initial tension (N/m^2)
τ_{max} maximum shear stress (N/m^2)
τ_{solid} shear stress at solid (N/m^2)
σ bending stress (N/m^2)
σ_{uts} ultimate tensile strength (N/m^2)

15.1 Introduction

Springs are used to store elastic energy and, when required, release it. There is a wide range of types of spring that are readily available from specialist suppliers or that can be designed and manufactured fit-for-purpose. The aim of this chapter is to introduce spring technology and to outline the principal steps in the design of helical compression, extension and torsion springs, leaf springs and Belleville washer springs.

The force produced by a spring can be compressive or tensile, and linear or radial, as in the case of a helical torsion spring clip used to hold a tube on the end of a pipe.

Alternatively springs can be configured to produce a torque with applications including door closers. The process of deflecting a spring involves the transfer of energy into stored spring energy. When the force causing the spring deflection is removed the stored spring energy will be returned.

Springs can be classified by the direction and nature of the force exerted when they are deflected. Several types of spring are illustrated in Fig. 15.1 and listed in Table 15.1 according to the nature of force or torque exerted. The principal characteristics of the various classes of springs are summarised in Table 15.2.

Virtually any material can be used to make springs. However the ideal material would have a high ultimate strength and yield point and a low modulus of elasticity to provide maximum energy storage. Ashby (1999) develops two selection charts for springs, one based on physical size and the other on weight. Both of these charts are reproduced here as Figs. 15.2 and 15.3. The highlighted regions in Figs. 15.2 and 15.3 illustrate the best choices for springs of minimum volume and weight respectively. These include high strength steels, glass-fibre reinforced plastic (GFRP), which is used for certain truck springs, titanium alloys, glass, which is used in galvanometers and nylon, which is used in some toys. The choice for most applications tends to be limited to plain carbon steels, alloy steels, stainless steels, high-nickel steels and copper-based alloys. Most spring materials are manufactured to ASTM, BS or DIN specifications. Typical properties of common spring materials are listed in Table 15.3. Round wire is the most common choice for spring material and preferred diameters are listed in Table 15.4 although any diameter can be produced if necessary.

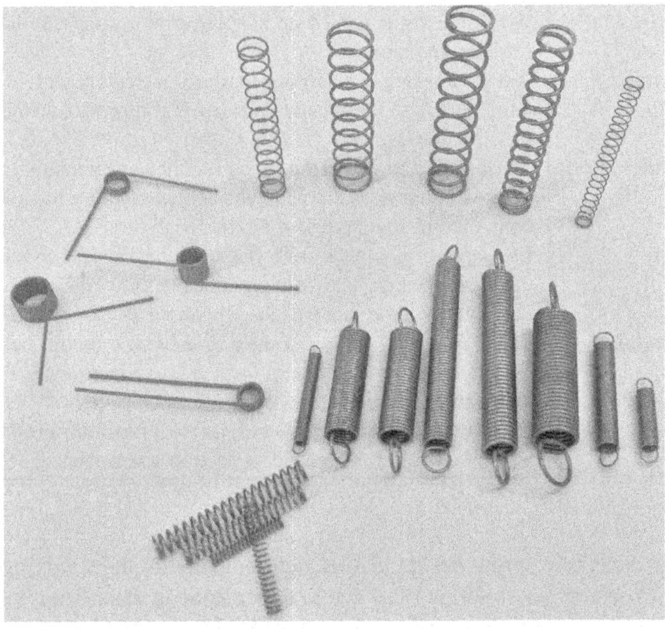

Fig. 15.1 Some examples of commercially available springs.

Table 15.1 Classification of springs according to the nature of force or torque exerted

Actuation	Type of spring
Compressive	Helical compression springs
	Belleville springs
	Flat springs (e.g. cantilever or leaf springs)
Tensile	Helical extension springs
	Flat springs (e.g. cantilever or leaf springs)
	Drawbar springs
	Constant force springs
Radial	Garter springs
	Elastomeric bands
	Spring clamps
Torque	Torsion springs
	Power springs

Table 15.2 Principal characteristics of a variety of types of spring

Type of spring	Principal characteristics
Helical compression springs	These are usually made from round wire wrapped into a straight cylindrical form with a constant pitch between adjacent coils
Helical extension springs	These are usually made from round wire wrapped into a straight cylindrical form but with the coils closely spaced in the no-load condition. As an axial load is applied the spring will extend but resisting the motion
Drawbar springs	A helical spring is incorporated into an assembly with two loops of wire. As a load is applied the spring is compressed in the assembly resisting the motion
Torsion springs	These exert a torque as the spring is deflected by rotation about their axes. A common example of the application of a torsion spring is embodied in some types of clothes peg
Leaf springs	Leaf springs are made from flat strips of material and loaded as cantilever beams. They can produce a tensile or compressive force depending on the mode of loading applied
Belleville springs	These comprise shallow conical discs with a central hole
Garter springs	These consist of coiled wire formed into a continuous ring so that they can exert a radial inward force when stretched
Volute springs	These consist of a spiral wound strip that functions in compression. They are subject to significant friction and hysteresis

Important aspects in spring design include determination of the spring material and dimensions to ensure that it will not fail due to either static or fluctuating loads for the lifetime required for the application, that it will not buckle or deform beyond allowable limits, that the natural frequencies of vibration are sufficiently in excess of the frequency of motion that they control and that cost and aesthetic aspirations are met.

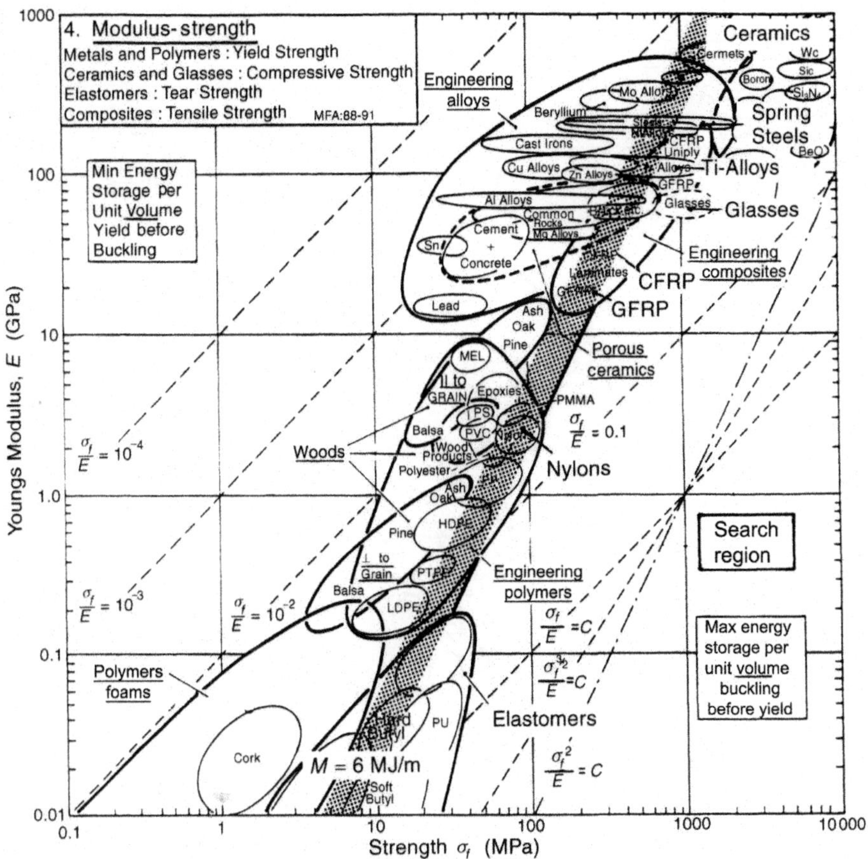

Fig. 15.2 Materials for springs of minimum volume.
Reproduced from Ashby, M.F., 1999. Materials Selection for Mechanical Design. second ed. Butterworth Heinemann.

All materials deform with load. For a spring the parameter modelling this is called the spring rate and is defined as the slope of the force deflection curve of a spring. If the slope is constant then

$$k = \frac{F}{\delta} \tag{15.1}$$

where k = spring rate (N/m), F = applied load (N) and δ = deflection (m).

The spring rate is also known as the spring constant or spring scale. Spring rates can be linear or nonlinear.

Springs can be combined in parallel (Fig. 15.4), series (Fig. 15.5) or some combination of these. For springs in parallel the spring rates add directly

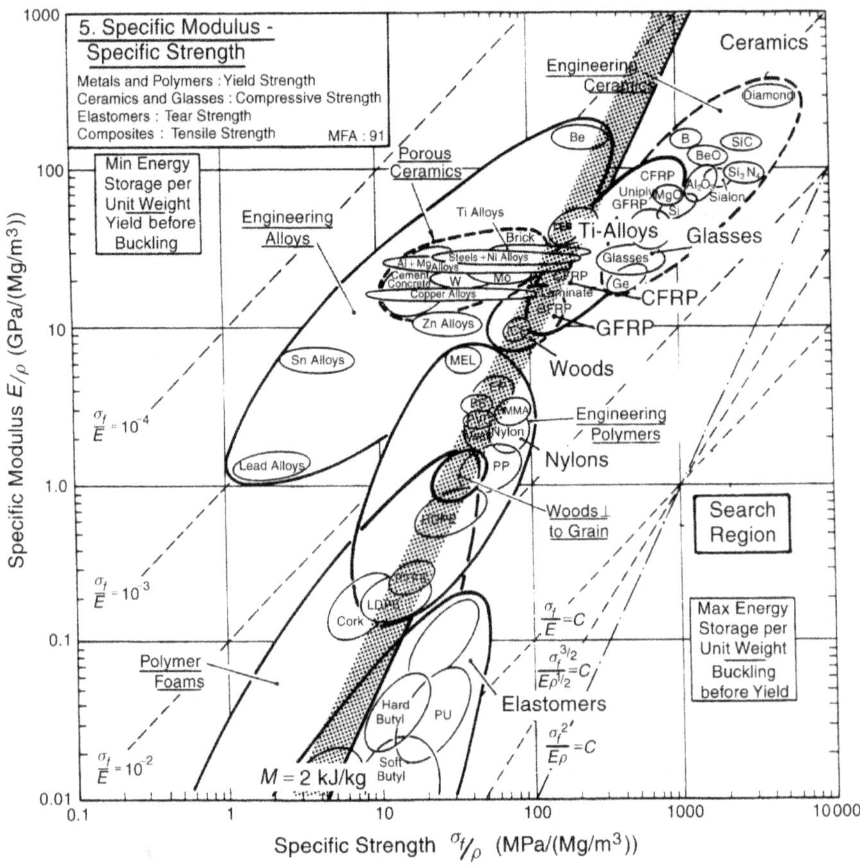

Fig. 15.3 Materials for springs of minimum weight.
Reproduced from Ashby, M.F., 1999. Materials Selection for Mechanical Design. second ed. Butterworth Heinemann.

$$k_{total} = \sum_{i=1}^{n} k_i = k_1 + k_2 + k_3 + \cdots + k_n \tag{15.2}$$

For springs in series the spring rates combine reciprocally:

$$\frac{1}{k_{total}} = \sum_{i=1}^{n} \frac{1}{k_i} = \frac{1}{k_1} + \frac{1}{k_2} + \frac{1}{k_3} + \cdots + \frac{1}{k_n} \tag{15.3}$$

15.2 Helical compression springs

The most familiar type of spring is the helical compression spring. Some common applications of compression springs are shown in Figs. 15.6–15.8 for automotive

Table 15.3 Typical properties of common spring materials

Material	Young's modulus (GPA)	Modulus of rigidity (GPA)	Density (kg/m^3)	maximum service temperature (°C)
Music wire	207	79.3	7860	120
Hard drawn wire	207	79.3	7860	150
Oil tempered	207	79.3	7860	150
Valve spring	207	79.3	7860	150
Chrome vanadium alloy steel wire	207	79.3	7860	220
Chrome silicon alloy steel wire	207	79.3	7860	245
302 Stainless steel	193	69	7920	260
17-7 pH stainless steel	203	75.8	7810	315
Phosphor bronze (A)	103	43.4	8860	95
Silicon bronze (A)	103	38.6	8530	95
Silicon bronze (B)	117	44.1	8750	95
Beryllium copper	128	48.3	8260	205
Inconel 600	214	75.8	8430	320
Inconel ×750	214	79.3	8250	595
AISI 1050	207	79.3	7860	95
AISI 1065	207	79.3	7860	95
AISI 1074	207	79.3	7860	120
AISI 1095	207	79.3	7860	120

Selected data reproduced from Joerres, R.G., 1996. Springs. In: Shigley, J.E., Mischke, C.R. (Eds.), Chapter 24 in Standard Handbook of Machine Design. McGraw Hill.

and rail suspensions. A heavy or stiff spring will perform well under a heavy load, but can result in a 'bumpy' ride under lightly loaded conditions. A compromise spring rate could be selected, but better performance can be achieved if the spring rate can be tailored to increase with the deflection. In its most common form a helical compression spring is made from constant diameter round wire with a constant pitch as shown in Fig. 15.9. Other forms are possible such as the variable pitch, barrel, hourglass and conical helical compression springs shown in Fig. 15.10. In addition to variations on pitch of the coil and diameter the formation of the end is important. A variety of common end treatments is illustrated in Fig. 15.11. Plain ends result from cutting the spring stock and leaving the spring with a constant pitch. Treatment of the end by some form of machining or pressing can facilitate alignment and this is the purpose

Table 15.4 Preferred wire diameters

Diameter (in.)	Diameter (mm)
0.004	0.10
0.005	0.12
0.006	0.16
0.008	0.20
0.010	0.25
0.012	0.30
0.014	0.35
0.016	0.40
0.018	0.45
0.020	0.50
0.022	0.55
0.024	0.60
0.026	0.65
0.028	0.70
0.030	0.80
0.035	0.90
0.038	1.00
0.042	1.10
0.045	
0.048	1.20
0.051	
0.055	1.40
0.059	
0.063	1.60
0.067	
0.072	1.80
0.076	
0.081	2.00
0.085	2.20
0.092	
0.098	2.50
0.105	
0.112	2.80
0.125	3.00
0.135	3.50
0.148	
0.162	4.00
0.177	4.50
0.192	5.00
0.207	5.50
0.225	6.00
0.250	6.50
0.281	7.00
0.312	8.00
0.343	9.00

Table 15.4 Continued

Diameter (in.)	Diameter (mm)
0.362	
0.375	
0.406	10.0
0.437	11.0
0.469	12.0
0.500	13.0
0.531	14.0
0.562	15.0
0.625	16.0

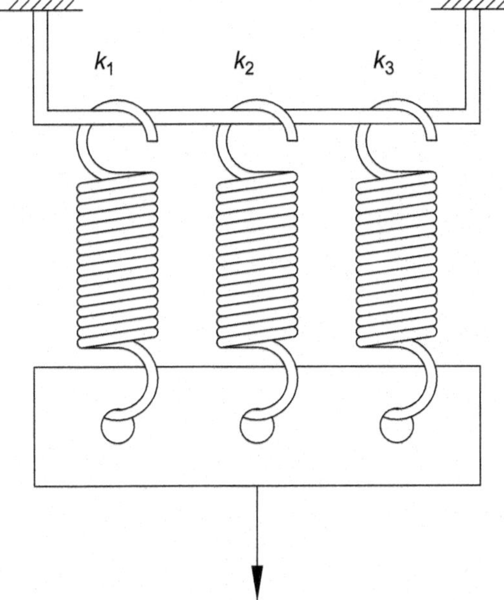

Fig. 15.4 Springs in parallel.

of options (B) to (D) illustrated in Fig. 15.11 each of which add to the cost of production of the spring and influence the performance. The end of a spring can also be formed to improve connection to mating components by the incorporation of, for example, hooks and rings.

The principal dimensions for a constant pitch helical compression spring are illustrated in Fig. 15.12. The wire diameter d, mean diameter D, free length L_f, and either the number of coils N or the pitch p are used to define a helical spring's geometry and in associated analysis. The inside and outer diameter are useful in designing the mating and locating components. The minimum diametral clearance between the outer diameter and a hole or between the inner diameter and a pin recommended, according

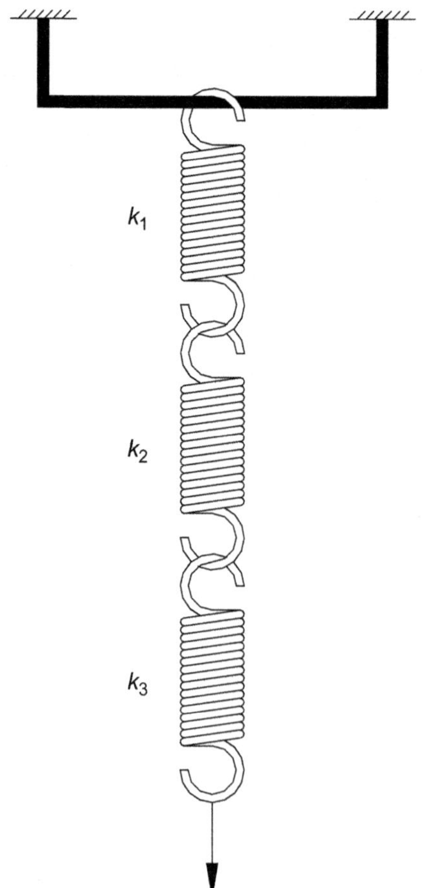

Fig. 15.5 Springs in series.

to Associated Spring (1987), is given by 0.10D for $D < 13$ mm or 0.05D for $D > 13$ mm.

As well as the geometrical parameters identified in Fig. 15.12 for an unloaded spring, there are a number of useful lengths defined for a spring in use as illustrated in Fig. 15.13.

- The installed length is the length after installation with initial deflection $\delta_{initial}$.
- The operating length is the shortest dimension to which the spring is compressed in use.
- The shut height or solid length is the length of the spring when the spring is loaded such that the coils are actually touching. This is the shortest possible length for the spring without crushing it beyond all recognition.

Springs are subject to failure by yielding due to too high a stress in the case of static loading or by fatigue in the case of dynamic loading (Carlson, 1978; Das et al., 2007; Kobelev, 2017; SAE, 1996). To determine the geometry of a spring to avoid such

Fig. 15.6 An example of the use of coil springs in vehicle suspension.
Image courtesy of Daimler AG.

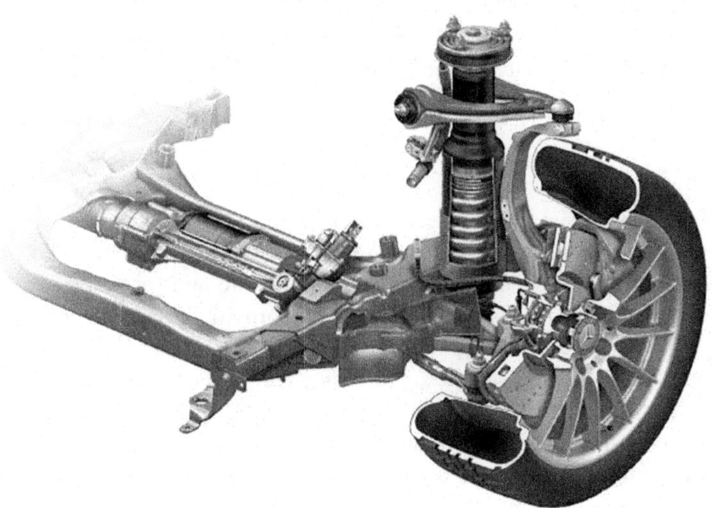

Fig. 15.7 A Mercedes-Benz SL-Class suspension assembly.
Image courtesy of Daimler AG.

failure, or to determine when failure will occur, it is necessary to consider the stresses experienced by a spring under loading.

The free-body diagram for a helical spring loaded with force F is illustrated in Fig. 15.14. There are two components of stress on any cross section of coil, a torsional shear stress due to the torque and a direct shear stress due to the force. The stresses add and the maximum shear stress will occur at the inner fibre of the wire's cross-section.

Fig. 15.8 Railway bogie.
Figure courtesy of Astra Rail Industries SA, Greenbrier Europe.

$$\tau_{\max} = \frac{Tr}{J} + \frac{F}{A} = \frac{F(D/2)(d/2)}{\pi d^4/32} + \frac{F}{\pi d^2/4} = \frac{8FD}{\pi d^3} + \frac{4F}{\pi d^2} \tag{15.4}$$

where $\tau_{\max}$ = maximum shear stress (N/m^2), T = torque (N/m), r = radius (m), J = polar second moment of area (m^4) = $\pi d^4/32$ for a solid circular cylinder, F = load (N), A = area (m^2), d = wire diameter (m) and D = mean coil diameter (m).

The spring index is defined by

$$C = \frac{D}{d} \tag{15.5}$$

Preferred values for the spring index are between 4 and 12. For values of the spring index below four, springs are difficult to manufacture and for values >12 they are prone to buckling. Buckling can be prevented by providing either internal or external support, by, for instance, placing the spring in a hole or on a rod. However, frictional contact between the spring and the support may reduce the force that can be delivered at the spring end. The tendency of a helical spring to buckle is proportional to the slenderness factor, which is the ratio of the free length to the mean coil diameter, L_f/D, the ratio of the spring's deflection to its free length, δ/L_f, and the type of end support used. Fig 15.15 provides guidance for determining whether a particular combination of these factors results in a stable spring or a spring prone to buckling.

Substituting for the spring index in Eq. (15.4) gives

$$\tau_{\max} = \frac{8FD}{\pi d^3}\left(1 + \frac{1}{2C}\right) = K_s \frac{8FD}{\pi d^3} \tag{15.6}$$

where K_s is called the direct shear factor and is given by

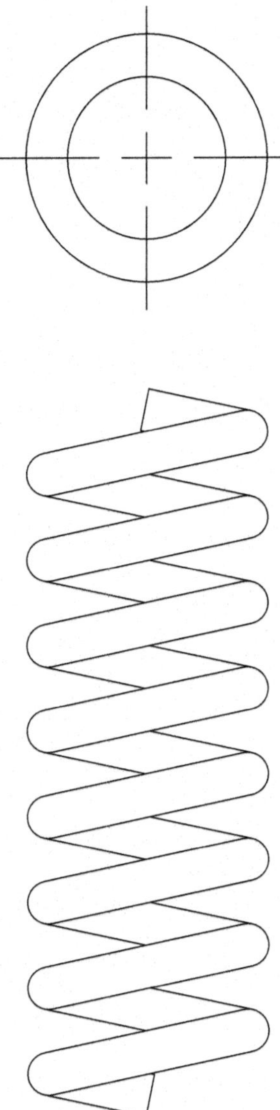

Fig. 15.9 A constant pitch helical compression spring.

$$K_s = 1 + \frac{1}{2C} \tag{15.7}$$

Curved beams have a stress concentration on the inner surface of curvature. The Wahl factor, K_w, includes both direct shear stress effects and the stress concentration factor due to curvature and is given by

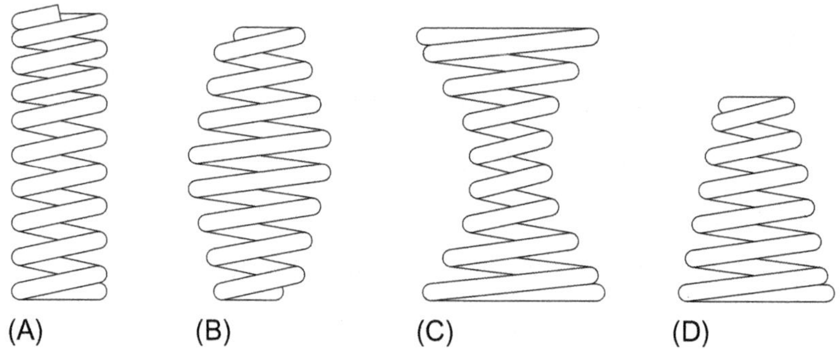

Fig. 15.10 Some additional helical spring configurations: (A) variable pitch, (B) Barrel, (C) Hourglass, (D) conical.

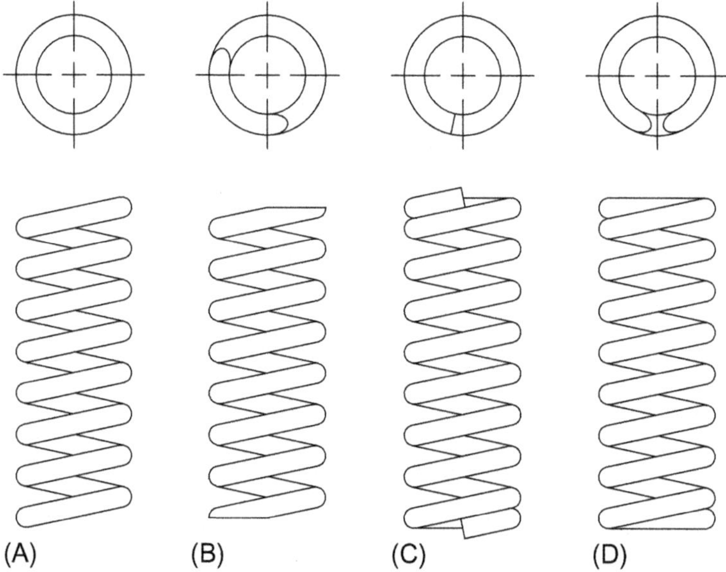

Fig. 15.11 Common styles of end treatments for helical compression springs: (A) plain ends, (B) plain ground ends, (C) squared ends, (D) squared ground ends.

$$K_w = \frac{4C-1}{4C-4} + \frac{0.615}{C} \tag{15.8}$$

$$\tau_{\max} = K_w \frac{8FD}{\pi d^3} = K_w \frac{8FC}{\pi d^2} \tag{15.9}$$

It is commonly recommended that Eq. (15.9) be used for modelling fatigue and Eq. (15.6) for modelling a spring under static loading only.

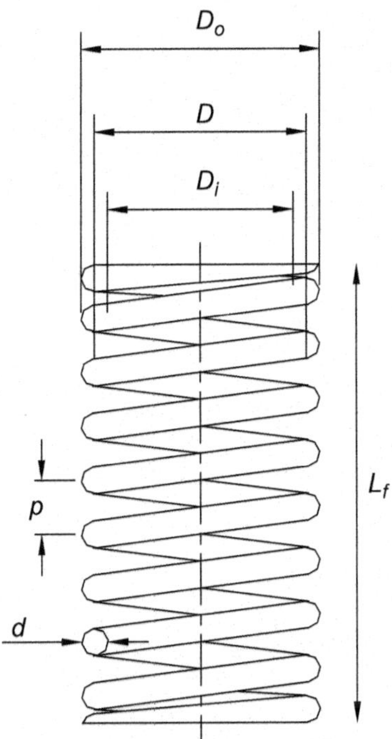

Fig. 15.12 Dimensional parameters for helical compression springs.

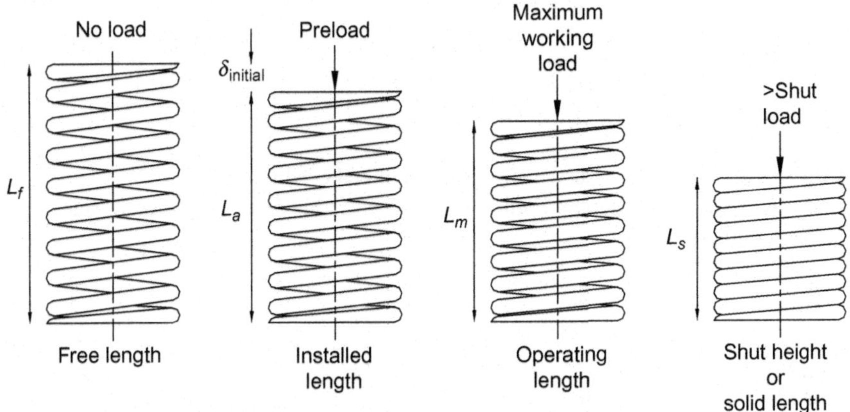

Fig. 15.13 Various lengths associated with a spring in use.

For springs with end treatments it is necessary to account for whether some of the coils are inactive and can be neglected in calculations for stress and deflection. For a spring with squared and ground ends or simply squared ends, each end coil is inactive and the number of active coils is given by

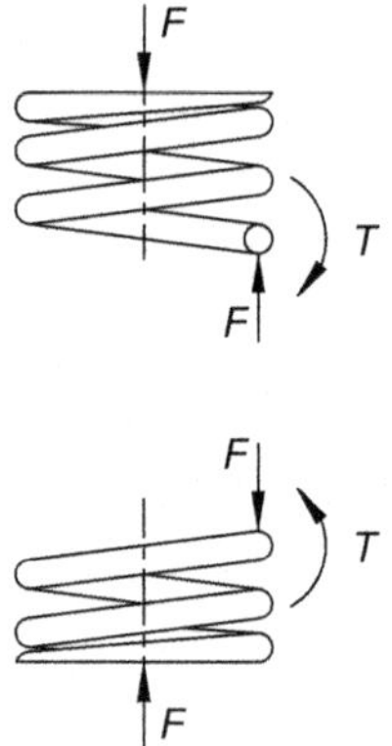

Fig. 15.14 Free body diagram for a helical compression spring loaded with force F.

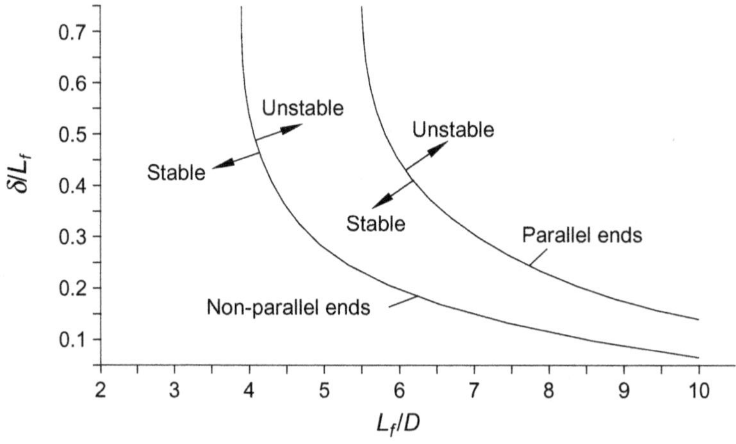

Fig. 15.15 Critical buckling curves for two types of end condition.
Adapted from Associated Spring, Barnes Group. Engineering Guide to Spring Design. Associated Spring, 1987.

$$N_a = N - 2 \tag{15.10}$$

where N is the total number of coils in the spring.

For plain coils with ground ends

$$N_a = N - 1 \tag{15.11}$$

The deflection of a helical spring under load F is given by

$$\delta = \frac{8FD^3N_a}{d^4G} = \frac{8FC^3N_a}{dG} \tag{15.12}$$

where N_a is the number of active coils in the spring.

Thus the spring rate for a helical spring is given by

$$k = \frac{F}{\delta} = \frac{d^4 G}{8D^3 N_a} = \frac{dG}{8C^3 N_a} \tag{15.13}$$

It should be noted that it is unlikely that a load on a spring will be directly on its geometric axis. Eccentric loading of a spring causes the stresses on one side of the spring to be higher than indicated in Eqs (15.6) and (15.9).

Springs can vibrate both laterally and longitudinally when excited near their natural frequencies. If a helical spring, fixed at one end, is given a sufficiently rapid compression at the other, the end coil will be pushed against its neighbour before the remaining coils have time to respond to the displacement. This compression then propagates down the spring with first coils one and two in contact, then coils two and three in contact and so on until the compression wave reaches the other end where the disturbance will be reflected back. This process repeats itself until the motion is damped out. This phenomenon is known as spring surge and causes very high stresses in the spring, which are approximately equal to those when the spring is compressed to its solid length. The natural frequency, f_n, of spring surge depends on the boundary conditions. For the fixed-fixed case:

$$f_n = \frac{1}{2}\sqrt{\frac{k}{m}} \tag{15.14}$$

where f_n = natural frequency (Hz), k = spring rate (N/m) and m = mass (kg).

The mass of a helical spring is equal to the product of density and volume, so, for N coils in a spring this is given by

$$m = \rho V = \rho N(\pi D)\left(\pi d^2/4\right) = \rho N \pi^2 d^2 D/4 \tag{15.15}$$

Substitution for the mass and spring rate in Eq. (15.14) gives

$$f_n = \frac{d}{2\pi D^2 N}\sqrt{\frac{G}{2\rho}} \tag{15.16}$$

For steel springs with a modulus of rigidity of 79.3 GPa and density 7860 kg/m^3,

$$f_n = \frac{357d}{D^2 N} \tag{15.17}$$

where d = wire diameter (m), D = coil diameter (m) and N = number of coils.

To avoid surge the spring should not be cycled at a frequency close to its natural frequency. The natural frequency of spring surge should usually be made higher than the highest significant harmonic of the motion involved, which is typically about the thirteenth. So the natural frequency should be at least 13 times the forcing frequency of the load on the spring to avoid resonance. Designing springs with high natural frequencies typically involves operating at the highest possible stress levels as this minimises the mass of the spring and thereby maximises f_n which is proportional to $m^{-0.5}$.

Springs are subject to creep under load. This is sometimes evident in old cars where the sustained weight on the suspension springs over the years has caused a permanent shortening of the spring's overall length and the car body's ground clearance is reduced. This shortening by creep is known as set. Set is directly related to yield stress. Unfortunately data for yield stress is not as readily available as that for the ultimate tensile strength. Values for the ultimate tensile strength of wire are found to vary with diameter and are illustrated in Fig. 15.16 for a range of materials. Juvinall and Marshek (1991) recommend the approximations given in Table 15.5 for shear stress to limit set in compression coil springs to 2%.

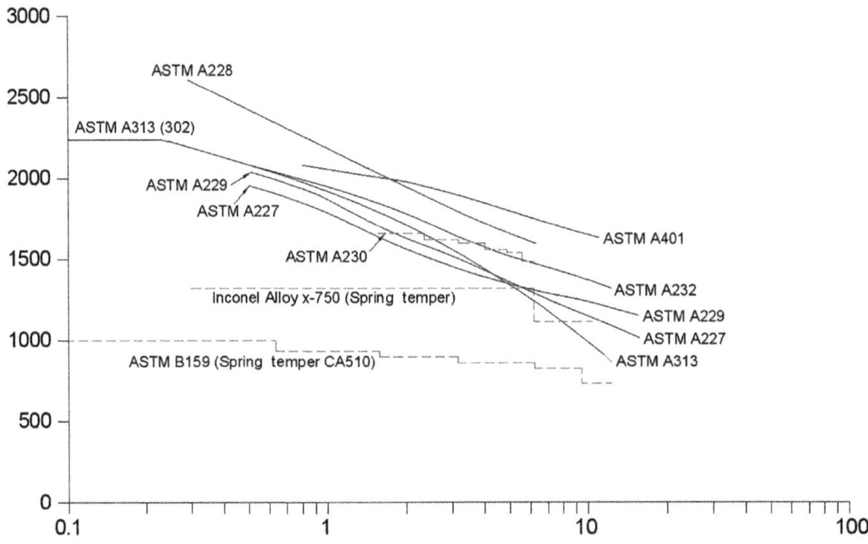

Fig. 15.16 Minimum tensile strengths of spring wire.
After Associated Spring, Barnes Group. Engineering Guide to Spring Design. Associated Spring, 1987.

Table 15.5 Solid shear stress to limit long-term set in compression coil springs to <2% (Juvinall and Marshek, 1991)

τ_{solid}	**Material condition**
$0.45\sigma_{uts}$	Ferrous without presetting
$0.35\sigma_{uts}$	Nonferrous and austenitic stainless without presetting
$0.65\sigma_{uts}$	Ferrous with presetting
$0.55\sigma_{uts}$	Nonferrous and austenitic stainless with presetting

Prestressing, also known as presetting, of a spring can be used to improve a spring's ability to withstand stress, increasing its load carrying capability and fatigue resistance (Carlson, 1982). For example compression springs manufactured from BS EN10270–1 cold drawn carbon steel without prestressing can to be loaded to 49% of the material's ultimate tensile strength. With prestressing it can be loaded to 70% of its ultimate tensile strength. Prestressing takes place after the spring has been coiled, stress relieved and ground. It involves compressing the spring to its solid length or a fixed position that is greater than its maximum working length. This process is repeated a number of times typically no less than three. During prestressing the spring's dimensions will alter. For a particular desired final length, the spring's dimensional changes must be accounted for by the manufacturer.

There are a number of strategies that can be followed in designing helical compression springs (Mott, 1999). One approach, knowing the force and length of the spring, is to specify a material, guess a trial diameter for the spring considering the space available, check the values calculated for spring rate and free length and if necessary try a new wire diameter. This approach is outlined in Fig. 15.17 and in the following example. The design procedure requires access to tables of data for material properties and wire diameters.

Example 15.1. A helical compression spring is required to exert a force of 35 N when compressed to a length of 60 mm. At a length of 48 mm the force must be 50 N. The spring is to be installed in a hole with a diameter of 24 mm. The application involves slow cycling and a total life of 250,000 cycles is required. The maximum temperature of operation is 80°C.

Solution

A standard material such as ASTM A232 chromium vanadium steel wire is suggested as an initial starting point for this application. The maximum service temperature for this material is approximately 120°C, which is greater than the operating temperature for the application. The application involves slow cyclic variation of the load so Eq. (15.9) can be used for determining the spring's stress level.

From Table 15.3 for chrome vanadium steel, $G = 79.3$ GPa.

The maximum operating force, F_o, is 50 N with an operating length, L_o, of 48 mm. The installed force, F_i, is 35 N at an installed length, L_i, of 60 mm.

The spring rate k is given by

$$k = \frac{F_o - F_i}{L_i - L_o} = \frac{50 - 35}{0.06 - 0.048} = 1250 \text{ N/m}$$

The free length L_f is given by

$$L_f = L_i + \frac{F_i}{k} = 0.06 + \frac{35}{1250} = 0.088 \text{ m}$$

A mean diameter of 18 mm is suggested, as this should locate in the 24 mm diameter hole available allowing room for the diameter of the wire itself, which is currently undetermined.

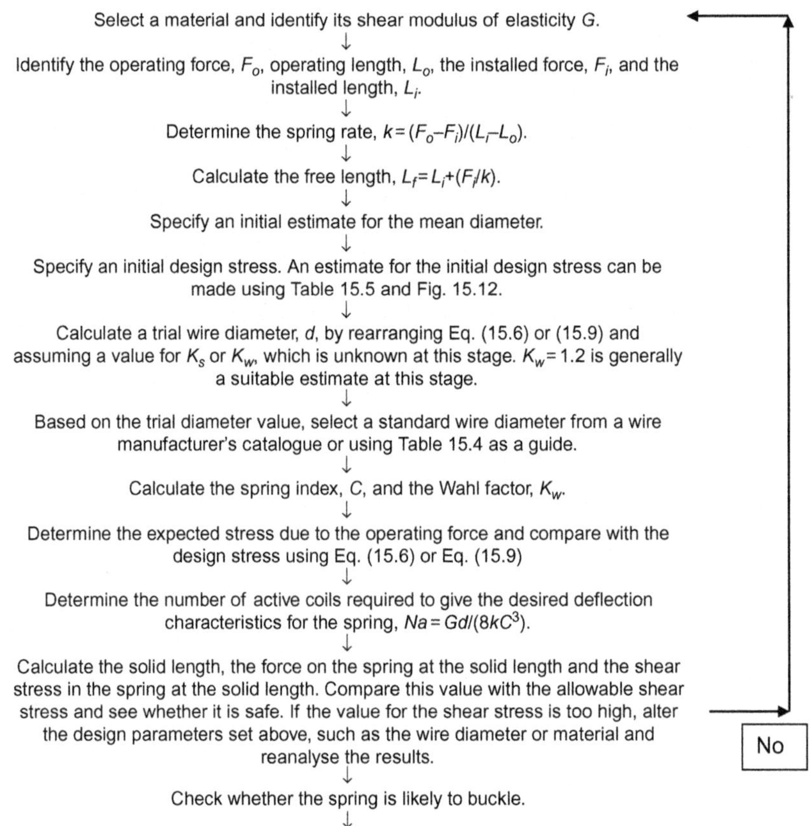

Select a material and identify its shear modulus of elasticity G.
Identify the operating force, F_o, operating length, L_o, the installed force, F_i, and the installed length, L_i.
Determine the spring rate, $k=(F_o-F_i)/(L_i-L_o)$.
Calculate the free length, $L_f=L_i+(F_i/k)$.
Specify an initial estimate for the mean diameter.
Specify an initial design stress. An estimate for the initial design stress can be made using Table 15.5 and Fig. 15.12.
Calculate a trial wire diameter, d, by rearranging Eq. (15.6) or (15.9) and assuming a value for K_s or K_w, which is unknown at this stage. $K_w=1.2$ is generally a suitable estimate at this stage.
Based on the trial diameter value, select a standard wire diameter from a wire manufacturer's catalogue or using Table 15.4 as a guide.
Calculate the spring index, C, and the Wahl factor, K_w.
Determine the expected stress due to the operating force and compare with the design stress using Eq. (15.6) or Eq. (15.9)
Determine the number of active coils required to give the desired deflection characteristics for the spring, $Na=Gd/(8kC^3)$.
Calculate the solid length, the force on the spring at the solid length and the shear stress in the spring at the solid length. Compare this value with the allowable shear stress and see whether it is safe. If the value for the shear stress is too high, alter the design parameters set above, such as the wire diameter or material and reanalyse the results.
No
Check whether the spring is likely to buckle.
Specify the spring dimensions.

Fig. 15.17 An approach to helical compression spring design if the spring diameter is constrained.

A design shear stress of $0.45\sigma_{uts}$ is suggested. σ_{uts} for ASTM A232 chrome vanadium steel with an assumed wire diameter of 2mm is approximately 1700MPa (Fig. 15.16). The design shear stress will therefore be $0.45\times 1700=765$ MPa. If the wire diameter calculated in the design procedure is significantly different then the estimate for the design shear stress will need to be modified.

The equation for shear stress requires a value for K_w. K_w is a function of $C=d/D$, but the wire diameter d is unknown at this stage. A value of $K_w=1.2$ can be used as an initial estimate here, although this will be checked and modified later in the procedure

$$d=\left(\frac{8FDK_w}{\pi\tau_{max}}\right)^{1/3}=\left(\frac{8\times 50\times 0.018\times 1.2}{\pi 765\times 10^6}\right)^{1/3}=1.532\times 10^{-3}\ \text{m}$$

The estimated wire diameter is therefore 1.53mm. Examination of Table 15.3 or a stock spring supplier's catalogue will identify a suitably close wire diameter to select. In this case from Table 15.3, the nearest larger wire diameter is 1.6mm.

The spring index can now be calculated

$$C = \frac{D}{d} = \frac{0.018}{0.0016} = 11.25$$

The Wahl factor is

$$K_w = \frac{4C-1}{4C-4} + \frac{0.615}{C} = \frac{4 \times 11.25 - 1}{4 \times 11.25 - 4} + \frac{0.615}{11.25} = 1.128$$

The maximum shear stress is given by

$$\tau_{\max} = K_w \frac{8FD}{\pi d^3} = 1.128 \frac{8 \times 50 \times 0.018}{\pi (0.0016)^3} = 631 \times 10^6 \text{ Pa}$$

This is significantly lower than the allowable maximum shear stress so the design seems acceptable at this stage.

The number of active coils can be determined from

$$N_a = \frac{Gd}{8kC^3} = \frac{79.3 \times 10^9 \times 0.0016}{8 \times 1250 \times (11.25)^3} = 8.91 \text{ coils}$$

The solid length of a spring occurs when all the coils are touching. If the spring is squared and ground then there will be two inactive coils.

$$L_s = d(N+2) = 0.0016(8.91+2) = 0.0175 \text{ m}$$

The force at the solid length would be

$$F_s = k(L_f - L_s) = 1250(0.088 - 0.0175) = 88.2 \text{ N}$$

As stress in a spring is directly proportional to force, the solid length shear stress can be determined by

$$\tau_{\text{solid}} = \tau_{\max} \frac{F_s}{F_o} = 631 \times 10^6 \frac{88.2}{50} = 1113 \times 10^6 \text{ Pa}$$

This value is greater than the maximum allowable stress of 765 MPa indicating that failure is likely. It is necessary to alter one or more of the design parameters to develop a more suitable design. One approach is to try a different wire diameter.

If $d = 0.0018$ m is tried then

$$C = \frac{D}{d} = \frac{0.018}{0.0018} = 10$$

$$K_w = \frac{4 \times 10 - 1}{4 \times 10 - 4} + \frac{0.615}{10} = 1.145$$

$$\tau_{\max} = 1.145\frac{8 \times 50 \times 0.018}{\pi(0.0018)^3} = 450 \times 10^6 \text{ Pa}$$

This is significantly lower than the allowable maximum shear stress so the design seems acceptable at this stage.

$$N_a = \frac{79.3 \times 10^9 \times 0.0018}{8 \times 1250 \times (10)^3} = 14.27 \text{ coils}$$

$$L_s = 0.0018(14.27 + 2) = 0.0293 \text{ m}$$

$$F_s = 1250(0.088 - 0.0293) = 73.4 \text{ N}$$

$$\tau_{solid} = 450 \times 10^6 \frac{73.4}{50} = 660 \times 10^6 \text{ Pa}$$

This is below the maximum allowable stress indicating an acceptable design.

The ratio of free length to the mean coil diameter, $L_f/D = 0.088/0.018 = 4.89$. The deflection of the spring under the operating force is 40 mm. Thus the ratio of deflection to free length, $\delta/L_f = 0.04/0.088 = 0.455$. Examination of Fig. 15.15 for squared and ground ends shows that this design is stable and unlikely to buckle.

The outer diameter of the spring can be determined from

$$D_o = d + D = 0.0018 + 0.018 = 0.0198 \text{ m}$$

The inner diameter of the spring will be

$$D_i = D - d = 0.018 - 0.0018 = 0.0162 \text{ m}$$

Nominal specification:

ASTM A232 chromium vanadium steel wire helical compression spring.
$k = 1250$ N/m.
$L_f = 0.088$ m.
$L_s = 0.0293$ m.
Squared and ground ends, 16.27 coils.
$d = 1.8$ mm.
$D = 18$ mm, $D_i = 16.2$ mm, $D_o = 19.8$ mm.

If there are no constraints on the spring diameter then the procedure can be modified as outlined following and shown in Fig. 15.18.

(1) Select a material and identify its shear modulus of elasticity G and an estimate of the design stress.

(2) Calculate a value for the wire diameter based on the spring material properties and assuming approximate values for C and K_w. Typical estimates for C and K_w of 7 and 1.2 respectively are generally suitable.

Select a material and identify its shear modulus of elasticity G and an estimate of the design stress.

↓

Calculate a value for the wire diameter based on the spring material properties and assuming approximate values for C and K_w. Typical estimates for C and K_w of 7 and 1.2 respectively are generally suitable.

$$d = \sqrt{\frac{8K_w F_o C}{\pi \tau_{max}}}$$

↓

Select a standard wire diameter.

↓

Determine the maximum number of active coils possible for the spring. The rational here is that the solid length must be less than the operating length. The relationship between the solid length and number of coils is a function of the end treatment.

For squared and ground ends,

$$L_s = d(N_a + 2)$$

$$N_{max} = \frac{L_o}{d} - 2$$

↓

Select any number of active coils less than the maximum calculated value. Choosing a small value will result in larger clearances between adjacent coils and will use less wire per spring but will entail higher stresses for a given load. One approach is to try progressively fewer coils until the maximum permissible design stress is approached.

↓

Calculate the spring index using

$$C = \left(\frac{Gd}{8kN_a}\right)^{1/3}$$

↓

The mean diameter can now be calculated from the definition of the spring index,

$$D = Cd$$

↓

Calculate the Wahl factor, K_w.

↓

Determine the expected stress due to the operating force and compare with the design stress.

↓

Calculate the solid length, the force on the spring at the solid length and the shear stress in the spring at the solid length. Compare this value with the allowable shear stress and see whether it is safe.

↓

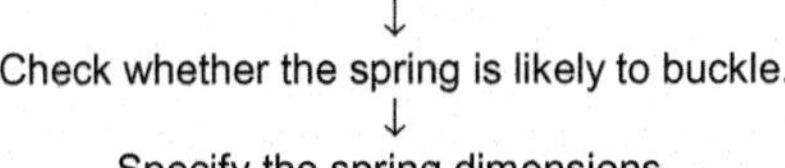

Check whether the spring is likely to buckle.

↓

Specify the spring dimensions.

Fig. 15.18 An approach to helical compression spring design if the spring diameter is not constrained.

$$d = \sqrt{\frac{8K_w F_o C}{\pi \tau_{max}}} \tag{15.18}$$

(3) Select a standard wire diameter.

(4) Determine the maximum number of active coils possible for the spring. The rationale here is that the solid length must be less than the operating length. The relationship between the solid length and number of coils is a function of the end treatment. For squared and ground ends,

$$L_s = d(N_a + 2) \tag{15.19}$$

Substituting $L_s = L_o$

$$N_{max} = \frac{L_o}{d} - 2 \tag{15.20}$$

(5) The designer can now select any number of active coils less than the maximum calculated value. Choosing a small value will result in larger clearances between adjacent coils and will use less wire per spring but will entail higher stresses for a given load. One approach is to try progressively fewer coils until the maximum permissible design stress is approached.

(6) Calculate the spring index using

$$C = \left(\frac{Gd}{8kN_a}\right)^{1/3} \tag{15.21}$$

(7) The mean diameter can now be calculated from the definition of the spring index, $D = Cd$

(8) Calculate the Wahl factor, K_w.

(9) Determine the expected stress due to the operating force and compare with the design stress.

(10) Calculate the solid length, the force on the spring at the solid length and the shear stress in the spring at the solid length. Compare this value with the allowable shear stress and see whether it is safe.

(11) Check whether the spring is likely to buckle.

(12) Specify the spring dimensions.

Example 15.2. A helical spring is required to exert a force of 160N at a length of 170mm and 200N at a length of 150mm. Specify a suitable spring for this application. The maximum operating temperature is 50°C and the load varies slowly with time and a total life of 200,000 cycles is required.

Solution

A standard material such as ASTM A232 chromium vanadium steel wire is suggested as an initial starting point for this application. The maximum service temperature for this material is approximately 120°C, which is greater than the operating temperature for the application. The application involves slow cyclic variation of the load so Eq. (15.9) should be used for determining the spring's stress level.

From Table 15.3 for chromium vanadium steel, $G = 79.3\,\text{GPa}$.

The maximum operating force, F_o, is 200 N with an operating length, L_o, of 150 mm. The installed force, F_i, is 160 N at an installed length, L_i, of 170 mm.

The spring rate k is given by

$$k = \frac{F_o - F_i}{L_i - L_o} = \frac{200 - 160}{0.17 - 0.15} = 2000\ \text{N/m}$$

The free length L_f is given by

$$L_f = L_i + \frac{F_i}{k} = 0.17 + \frac{160}{2000} = 0.25\ \text{m}$$

A design shear stress of $0.45\sigma_{uts}$ is suggested. σ_{uts} for ASTM A232 chrome vanadium steel with an assumed wire diameter of 3 mm is approximately 1600 MPa (Fig. 15.16). The design shear stress will therefore be $0.45 \times 1700 = 720\,\text{MPa}$.

$$d = \sqrt{\frac{8K_w F_o C}{\pi \tau_{max}}} = \sqrt{\frac{8 \times 1.2 \times 200 \times 7}{\pi 7.2 \times 10^8}} = 2.438 \times 10^{-3}\ \text{m}$$

Choosing $d = 3\,\text{mm}$,

$$N_{max} = \frac{L_o}{d} - 2 = \frac{0.15}{0.003} - 2 = 48$$

Trying $N = 35$ gives

$$C = \left(\frac{Gd}{8kN_a}\right)^{1/3} = \left(\frac{79.3 \times 10^9 \times 0.003}{8 \times 2000 \times 35}\right)^{1/3} = 7.517$$

$D = Cd = 0.02255\,\text{m}$.

$K_w = 1.197$

$$\tau_{max} = K_w \frac{8FD}{\pi d^3} = 1.197 \frac{8 \times 200 \times 0.02255}{\pi (0.003)^3} = 509 \times 10^6\ \text{Pa}$$

This is significantly lower than the allowable maximum shear stress so the design seems acceptable at this stage.

$$L_s = d(N_a + 2) = 0.003(35 + 2) = 0.111\ \text{m}$$

The force at the solid length would be

$$F_s = k(L_f - L_s) = 2000(0.25 - 0.111) = 278\ \text{N}$$

The solid length shear stress can be determined by

$$\tau_{solid} = \tau_{max}\frac{F_s}{F_o} = 509 \times 10^6 \frac{278}{200} = 708 \times 10^6 \text{ Pa}$$

This value is just below than the maximum allowable stress of 720 MPa indicating that failure is unlikely.

The ratio of free length to the mean coil diameter, $L_f/D = 0.25/0.02255 = 11.09$. The deflection of the spring under the operating force is 100 mm. Thus the ratio of deflection to free length, $\delta/L_f = 0.1/0.25 = 0.4$. Examination of Fig. 15.15 for squared and ground ends shows that this design is unstable and likely to buckle.

Therefore the design parameters need to be changed to ensure the spring does not buckle. Decreasing the number of coils will increase the stress, which is already approaching the limit at the solid length. A different wire diameter is proposed.

If $d = 0.004$ m

$$N_{max} = \frac{L_o}{d} - 2 = \frac{0.15}{0.004} - 2 = 35.5$$

Trying a low value for the number of active coils, say $N_a = 16$ gives

$$C = \left(\frac{Gd}{8kN_a}\right)^{1/3} = \left(\frac{79.3 \times 10^9 \times 0.004}{8 \times 2000 \times 16}\right)^{1/3} = 10.74$$

$D = Cd = 0.04296$ m.

$K_w = 1.134$.

σ_{uts} for ASTM A232 chrome vanadium steel with an assumed wire diameter of 4 mm is approximately 1550 MPa (Fig. 15.16). The design shear stress will therefore be $0.45 \times 1550 = 698$ MPa

$$\tau_{max} = K_w\frac{8FD}{\pi d^3} = 1.134\frac{8 \times 200 \times 0.04296}{\pi(0.004)^3} = 388 \times 10^6 \text{ Pa}$$

This is significantly lower than the allowable maximum shear stress so the design seems acceptable at this stage.

$$L_s = d(N_a + 2) = 0.004(16 + 2) = 0.072 \text{ m}$$

The force at the solid length would be

$$F_s = k(L_f - L_s) = 2000(0.25 - 0.072) = 356 \text{ N}$$

The solid length shear stress can be determined by

$$\tau_{solid} = \tau_{max}\frac{F_s}{F_o} = 388 \times 10^6 \frac{356}{200} = 691 \times 10^6 \text{ Pa}$$

This value is below than the maximum allowable stress of 720 MPa indicating that failure is unlikely.

The ratio of free length to the mean coil diameter, $L_f/D = 0.25/0.04296 = 5.819$. The deflection of the spring under the operating force is 100 mm. Thus the ratio of deflection to free length, $\delta/L_f = 0.1/0.25 = 0.4$. Examination of Fig. 15.15 for squared and ground ends shows that this design is stable and unlikely to buckle.

The outer diameter of the spring can be determined from

$$D_o = d + D = 0.004 + 0.04296 = 0.04696 \text{ m}$$

The inner diameter of the spring will be

$$D_i = D - d = 0.04296 - 0.004 = 0.03896 \text{ m}$$

Nominal specification:

ASTM A232 chromium vanadium steel wire helical compression spring.
$k = 2000$ N/m.
$L_f = 0.25$ m.
$L_s = 0.072$ m.
Squared and ground ends, 18 coils.
$d = 4$ mm.
$D = 42.96$ mm, $D_i = 38.96$ mm, $D_o = 46.96$ mm.

15.3 Helical extension springs

Helical extension springs are loaded in tension and exert a pulling force as well as storing energy. They are similar in appearance to helical compression springs but tend to be manufactured so that adjacent coils touch when at their free length. Hooks or loops are incorporated as part of the spring to facilitate attachment. The principal dimensions of an extension spring are illustrated in Fig. 15.19.

The coils of extension springs are tightly wound together and the wire is twisted as it is wound creating a preload in the coils that must be overcome in order to separate them. The spring rate for an extension coil is linear after the initial coil separation section and can be modelled by

$$k = \frac{F_o - F_i}{\delta} = \frac{Gd^4}{8N_a D^3} \tag{15.22}$$

The spring index is given by $C = D/d$ and should be kept in the range of 4–12.

The deflection of an extension spring can be determined by modifying Eq. (15.10) to account for the preload.

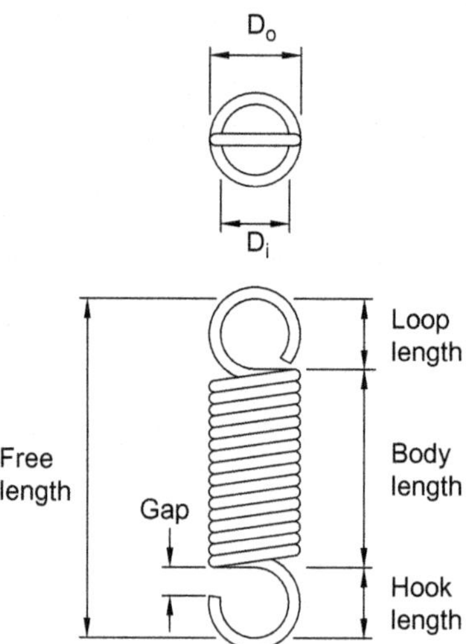

Fig. 15.19 Principal dimensions for a helical extension spring with hook and loop end configurations.

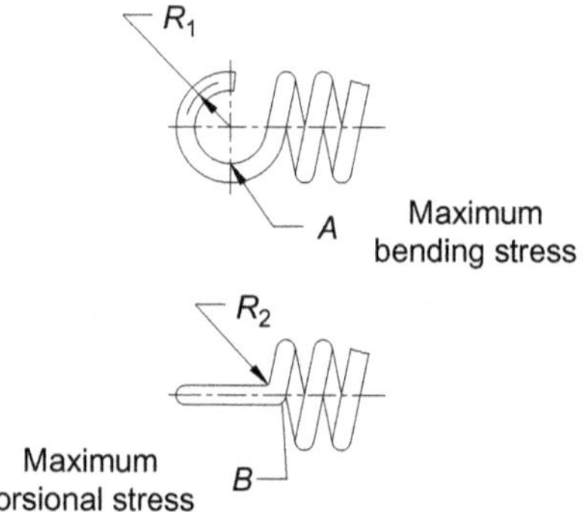

Fig. 15.20 Locations of maximum stress in the loop of an extension spring.

$$\delta = \frac{8(F_o - F_i)D^3 N_a}{d^4 G} \tag{15.23}$$

As well as accounting for stresses in the spring coils it is also necessary to consider the stresses in the hooks and loops used in an extension spring. The standard loop end illustrated in Fig. 15.20 has a high bending stress at point A and a high torsional stress at point B.

The bending stress at point A can be determined using

$$\sigma_A = K_b \frac{16DF}{\pi d^3} + \frac{4F}{\pi d^2} \tag{15.24}$$

where

$$K_b = \frac{4C_1^2 - C_1 - 1}{4C_1(C_1 - 1)} \tag{15.25}$$

$$C_1 = \frac{2R_1}{d} \tag{15.26}$$

The torsional stress at B can be found from Wahl (1963):

$$\tau_B = K_{w2} \frac{8DF}{\pi d^3} \tag{15.27}$$

where

$$K_{w2} = \frac{4C_2 - 1}{4C_2 - 4} \tag{15.28}$$

$$C_2 = \frac{2R_2}{d} \tag{15.29}$$

C_2 should be greater than four to avoid high stresses.

In a helical extension spring, all the coils are considered active and, in addition, a fictitious coil is added to the number of coils to calculate the required body length.

Design procedures for extension springs can be developed in a similar fashion to the ones outlined for compression springs. The procedure given here in the list below and in Fig. 15.21, relies upon the designer having access to data for the maximum allowable torsional shear stress. If at any step the values arrived at are deemed unacceptable, then parameters previously determined need to be reconsidered.

(1) Select a suitable material for the spring and determine the relevant material parameters, G, τ.
(2) Assume a trial mean diameter for the spring.
(3) Calculate a diameter for the spring wire, using Eq. (15.6) or (15.9), assuming a value for K of 1.2.
(4) Select the nearest larger diameter wire from Table 15.4 or a spring supplier.
(5) Determine the design stress for the wire diameter selected.
(6) Determine the outer, mean and inner diameters for the spring.
(7) Determine the spring index and the Wahl factor.
(8) Calculate the maximum shear stress in the spring under the operating load.
(9) Determine the number of coils necessary to produce the desired deflection characteristics.
(10) Calculate the body length of the spring, propose the end configurations and determine the free length.
(11) Calculate the deflection from the free length.

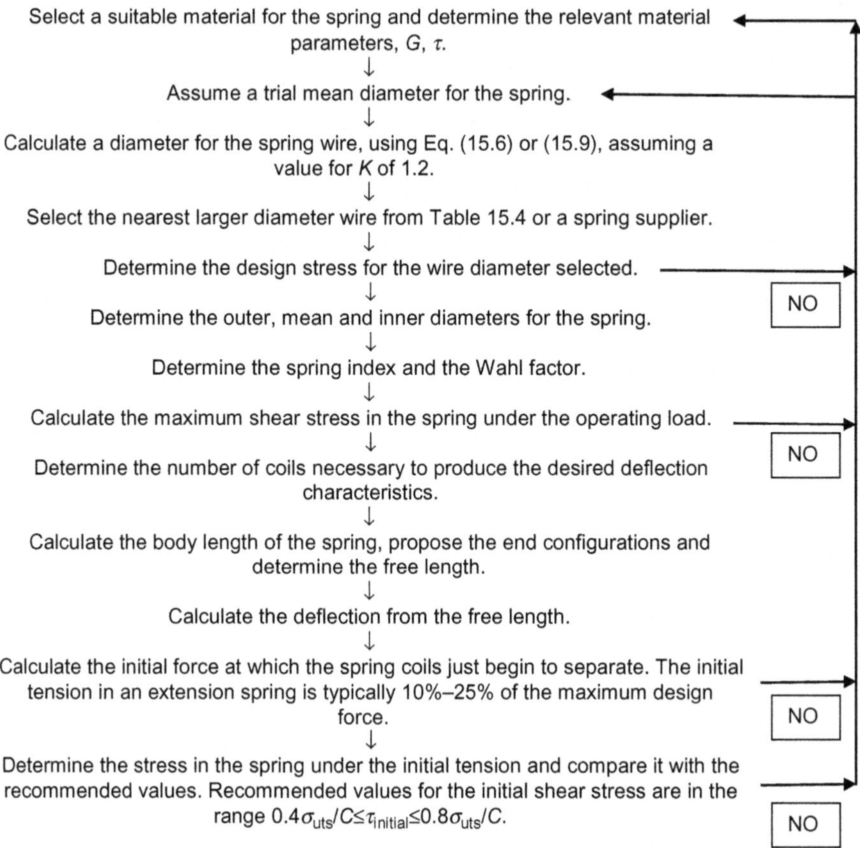

Fig. 15.21 Helical extension spring design procedure.

(12) Calculate the initial force at which the spring coils just begin to separate. The initial tension in an extension spring is typically 10%–25% of the maximum design force.

(13) Determine the stress in the spring under the initial tension and compare it with the recommended values. Recommended values for the initial shear stress are in the range $0.4\sigma_{uts}/C \leq \tau_{initial} \leq 0.8\sigma_{uts}/C$.

Example 15.3. A helical extension spring is required to exert a force of 30 N when the length between attachment locations is 70 mm and a force of 20 N at a length of 60 mm. The spring will be cycled through its load 100 times a day. ASTM A228 music wire steel has been proposed for the spring. The diameter of the spring should not exceed 20 mm. Determine suitable dimensions for the spring wire diameter, number of coils and mean diameter.

Solution

$$k = \frac{30-20}{70-60} = 1000 \text{ N/m}$$

Assuming a nominal wire diameter of 1.8 mm, the approximate ultimate tensile strength for A228 music wire is $\sigma_{uts} = 1900$ MPa.

$\tau_{max} = 0.45 \times 1900 = 855$ MPa.

A mean coil diameter of 15 mm is proposed.

$$d = \left(\frac{8 \times 30 \times 0.015 \times 1.2}{\pi 855 \times 10^6}\right)^{0.33} = 1.172 \times 10^{-3}\ \text{m}$$

Try $d = 1.4$ mm.

Design stress is $\tau_{max} = 0.45 \times 2100 = 945$ MPa. OK.

$D_i = 0.015–0.0014 = 0.0136$ m.

$D_o = 0.015 + 0.0014 = 0.0164$ m.

$C = D/d = 0.015/0.0014 = 10.71$.

$K_w = 1.135$

$$\tau = \frac{8K_w F_o D}{\pi d^3} = \frac{8 \times 1.135 \times 30 \times 0.015}{\pi 0.0014^3} = 474 \times 10^6\ \text{Pa}$$

$$N_a = \frac{Gd}{8C^3 k} = \frac{79.3 \times 10^9 \times 0.0014}{8 \times 10.71^3 \times 1000} = 11.3$$

Body length $= d(N_a + 1) = 0.0172$ m. An additional fictitious coil has been added to obtain the body length.

If we use a full loop of inside diameter 0.0136 m at each end,

$L_f = 0.0172 + 2 \times 0.0136 = 0.04442$ m.

$\delta = L_o - L_f = 0.07–0.04442 = 0.02558$ m.

$F_{initial} = F_o - k\delta = 30–1000 \times 0.02558 = 4.42$ N.

This value is between 10% and 25% of F_o so is acceptable.

$$\tau_{initial} = \tau \frac{F_{initial}}{F_o} = 474 \frac{14.42}{30} = 69.8\ \text{MPa}$$

This is $<0.4\sigma_{uts}/C$ so is acceptable.

Example 15.4. A helical compression spring is required to exert a force of 400 N when compressed to a length of 140 mm. At a length of 170 mm the force must be 160 N. A nominal spring mean diameter of 35 mm has been proposed. The application involves slow cycling and a total life of 150,000 cycles is required. The maximum temperature of operation is 60°C. Determine a suitable wire diameter for this application if ASTM A232 is to be used.

Solution

A standard material such as ASTM A232 chromium vanadium steel wire is suggested as an initial starting point for this application.

The maximum service temperature for this material is approximately 120°C, which is greater than the operating temperature for the application.

The application involves slow cyclic variation of the load so Eq. (15.9) can be used for determining the spring's stress level.

From Table 15.3 for chrome vanadium steel, $G = 79.3$ GPa.

The maximum operating force, F_o, is 400 N with an operating length, L_o, of 140 mm.

The installed force, F_i, is 160 N at an installed length, L_i, of 170 mm.

The spring rate k is given by

$$k = \frac{F_o - F_i}{L_i - L_o} = \frac{400 - 160}{0.17 - 0.16} = 8000 \text{ N/m}$$

The free length L_f is given by

$$L_f = L_i + \frac{F_i}{k} = 0.17 + \frac{160}{8000} = 0.19 \text{ m}$$

A nominal mean diameter of 35 has been proposed.

A design shear stress of $0.45\sigma_{uts}$ is suggested. σ_{uts} for ASTM A232 chrome vanadium steel with an assumed wire diameter of 5 mm is approximately 1500 MPa (Fig. 15.16).

The design shear stress will therefore be $0.45 \times 1500 = 675$ MPa.

If the wire diameter calculated in the design procedure is significantly different then the estimate for the design shear stress will need to be modified.

The equation for shear stress requires a value for K_w. K_w is a function of $C = d/D$, but the wire diameter d is unknown at this stage.

A value of $K_w = 1.2$ can be used as an initial estimate here, although this will be checked and modified later in the procedure.

$$d = \left(\frac{8FDK_w}{\pi\tau_{max}}\right)^{1/3} = \left(\frac{8 \times 400 \times 0.035 \times 1.2}{\pi 675 \times 10^6}\right)^{1/3} = 3.987 \times 10^{-3} \text{ m}$$

The estimated wire diameter is therefore 3.987 mm.

Examination of Table 15.3 or a stock spring supplier's catalogue will identify a suitably close wire diameter to select.

In this case from Table 15.3, the nearest larger wire diameter is 5 mm.

The spring index can now be calculated.

$$C = \frac{D}{d} = \frac{0.035}{0.005} = 7$$

The Wahl factor is

$$K_w = \frac{4C - 1}{4C - 4} + \frac{0.615}{C} = \frac{4 \times 7 - 1}{4 \times 7 - 4} + \frac{0.615}{7} = 1.213$$

The maximum shear stress is given by

$$\tau_{\max} = K_w \frac{8FD}{\pi d^3} = 1.213 \frac{8 \times 400 \times 0.035}{\pi (0.005)^3} = 346 \times 10^6 \text{ Pa}$$

This is significantly lower than the allowable maximum shear stress so the design seems acceptable at this stage.

The number of active coils can be determined from

$$N_a = \frac{Gd}{8kC^3} = \frac{79.3 \times 10^9 \times 0.005}{8 \times 8000 \times (7)^3} = 18.06 \text{ coils}$$

The solid length of a spring occurs when all the coils are touching.

If the spring is squared and ground then there will be two inactive coils.

$$L_s = d(N+2) = 0.005(18.06+2) = 0.1003 \text{ m}$$

The force at the solid length would be

$$F_s = k(L_f - L_s) = 8000(0.19 - 0.1003) = 717.5 \text{ N}$$

As stress in a spring is directly proportional to force, the solid length shear stress can be determined by

$$\tau_{\text{solid}} = \tau_{\max} \frac{F_s}{F_o} = 346 \times 10^6 \frac{717.5}{400} = 621 \times 10^6 \text{ Pa}$$

This is below the maximum allowable stress indicating an acceptable design.

The ratio of free length to the mean coil diameter, $L_f/D = 0.19/0.035 = 5.43$.

The deflection of the spring under the operating force is 50 mm.

Thus the ratio of deflection to free length, $\delta/L_f = 0.05/0.19 = 0.263$.

Examination of Fig. 15.15 for squared and ground ends shows that this design is stable and unlikely to buckle.

The outer diameter of the spring can be determined from

$$D_o = d + D = 0.005 + 0.035 = 0.04 \text{ m}$$

The inner diameter of the spring will be

$$D_i = D - d = 0.035 - 0.005 = 0.03 \text{ m}$$

Nominal specification:

ASTM A232 chromium vanadium steel wire helical compression spring.
$k = 8000$ N/m.

$L_f = 0.19\,\text{m}$.
$L_s = 0.1003\,\text{m}$.
Squared and ground ends, 20.06 coils.
$d = 5\,\text{mm}$.
$D = 35\,\text{mm}$, $D_i = 30\,\text{mm}$, $D_o = 40\,\text{mm}$.

Example 15.5. A helical spring is required to exert a force of 2000N at a length of 200mm and 1500N at a length of 250mm. Specify a suitable spring for this application.

Solution

As there are no constraints on the spring diameter method 2 can be used.

A standard material such as ASTM A232 chromium vanadium steel wire is suggested as an initial starting point for this application.

The maximum service temperature for this material is approximately 120°C, which is greater than the operating temperature for the application.

The application involves slow cyclic variation of the load so Eq. (15.9) should be used for determining the spring's stress level.

From Table 15.3 for chromium vanadium steel, $G = 79.3\,\text{GPa}$.

The maximum operating force, F_o, is 2000N with an operating length, L_o, of 200mm.

The installed force, F_i, is 1500N at an installed length, L_i, of 250mm.

The spring rate k is given by

$$k = \frac{F_o - F_i}{L_i - L_o} = \frac{2000 - 1500}{0.25 - 0.2} = 10000\ \text{N/m}$$

The free length L_f is given by

$$L_f = L_i + \frac{F_i}{k} = 0.25 + \frac{1500}{10000} = 0.4\ \text{m}$$

A design shear stress of $0.45\sigma_{\text{uts}}$ is suggested.

σ_{uts} for ASTM A232 chrome vanadium steel with an assumed wire diameter of 4mm is approximately 1350MPa (Fig. 15.16).

The design shear stress will therefore be $0.45 \times 1350 = 608\,\text{MPa}$

$$d = \sqrt{\frac{8K_wF_oC}{\pi\tau_{\max}}} = \sqrt{\frac{8 \times 1.2 \times 2000 \times 7}{\pi 6.08 \times 10^8}} = 8.392 \times 10^{-3}\ \text{m}$$

The next standard diameter available is $d = 9\,\text{mm}$.

However selecting such a small diameter for the spring wire could result in very high stresses at solid height.

A larger diameter is suggested for consideration initially.

Choosing $d = 11\,\text{mm}$,

$$N_{max} = \frac{L_o}{d} - 2 = \frac{0.2}{0.011} - 2 = 16.18$$

Trying $N = 12$ gives

$$C = \left(\frac{Gd}{8kN_a}\right)^{1/3} = \left(\frac{79.3 \times 10^9 \times 0.011}{8 \times 10000 \times 12}\right)^{1/3} = 9.685$$

$D = Cd = 0.1065\,\text{m}$.
$K_w = 1.15$

$$\tau_{max} = K_w \frac{8FD}{\pi d^3} = 1.15 \frac{8 \times 2000 \times 0.1065}{\pi (0.011)^3} = 469 \times 10^6 \text{ Pa}$$

This is significantly lower than the allowable maximum shear stress so the design seems acceptable at this stage.

$$L_s = d(N_a + 2) = 0.011(12 + 2) = 0.154 \text{ m}$$

The force at the solid length would be

$$F_s = k(L_f - L_s) = 10000(0.4 - 0.154) = 2460 \text{ N}$$

The solid length shear stress can be determined by

$$\tau_{solid} = \tau_{max} \frac{F_s}{F_o} = 469 \times 10^6 \frac{2460}{2000} = 577 \times 10^6 \text{ Pa}$$

This value is just below than the maximum allowable stress of 608 MPa indicating that failure is unlikely.

The ratio of free length to the mean coil diameter, $L_f/D = 0.4/0.1065 = 3.75$.

The deflection of the spring under the operating force is 200 mm.

Thus the ratio of deflection to free length, $\delta/L_f = 0.2/0.4 = 0.5$.

Examination of Fig. 15.15 for squared and ground ends shows that this design is stable and unlikely to buckle.

The outer diameter of the spring can be determined from

$$D_o = d + D = 0.011 + 0.1065 = 0.1175 \text{ m}$$

The inner diameter of the spring will be

$$D_i = D - d = 0.1065 - 0.011 = 0.09554 \text{ m}$$

Nominal specification:

ASTM A232 chromium vanadium steel wire helical compression spring.
$k = 10{,}000\,\text{N/m}$.
$L_f = 0.4\,\text{m}$.
$L_s = 0.154\,\text{m}$.
Squared and ground ends, 14 coils.
$d = 11\,\text{mm}$.
$D = 106.5\,\text{mm}$, $D_i = 95.54\,\text{mm}$, $D_o = 117.5\,\text{mm}$.

Example 15.6. A helical spring is required for a pogo stick to exert a force of 975 N at a length of 140 mm and 300 N at a length of 220 mm. Specify a suitable spring for this application if the spring material is ASTM A232.

Solution

As there are no constraints on the spring diameter method 2 can be used.

A standard material such as ASTM A232 chromium vanadium steel wire is suggested as an initial starting point for this application.

The maximum service temperature for this material is approximately 120°C, which is greater than the operating temperature for the application.

The application involves slow cyclic variation of the load so Eq. (15.9) should be used for determining the spring's stress level.

From Table 15.3 for chromium vanadium steel, $G = 79.3\,\text{GPa}$.

The maximum operating force, F_o, is 975 N with an operating length, L_o, of 140 mm.

The installed force, F_i, is 300 N at an installed length, L_i, of 220 mm.

The spring rate k is given by

$$k = \frac{F_o - F_i}{L_i - L_o} = \frac{975 - 300}{0.22 - 0.14} = 8438\ \text{N/m}$$

The free length L_f is given by

$$L_f = L_i + \frac{F_i}{k} = 0.22 + \frac{300}{8438} = 0.2556\ \text{m}$$

A design shear stress of $0.45\sigma_{\text{uts}}$ is suggested. σ_{uts} for ASTM A232 chrome vanadium steel with an assumed wire diameter of 7 mm is approximately 1450 MPa (Fig. 15.16).

The design shear stress will therefore be $0.45 \times 1450 = 653\,\text{MPa}$

$$d = \sqrt{\frac{8K_w F_o C}{\pi \tau_{\max}}} = \sqrt{\frac{8 \times 1.2 \times 975 \times 7}{\pi 6.53 \times 10^8}} = 5.654 \times 10^{-3}\ \text{m}$$

Choosing $d = 7\,\text{mm}$,

$$N_{\max} = \frac{L_o}{d} - 2 = \frac{0.14}{0.007} - 2 = 18$$

Trying $N = 14$ gives

$$C = \left(\frac{Gd}{8kN_a}\right)^{1/3} = \left(\frac{79.3 \times 10^9 \times 0.007}{8 \times 8438 \times 14}\right)^{1/3} = 8.375$$

$D = Cd = 0.05862$ m.

$K_w = 1.175$

$$\tau_{\max} = K_w \frac{8FD}{\pi d^3} = 1.175 \frac{8 \times 975 \times 0.05862}{\pi (0.007)^3} = 499 \times 10^6 \text{ Pa}$$

This is significantly lower than the allowable maximum shear stress so the design seems acceptable at this stage.

$$L_s = d(N_a + 2) = 0.007(14 + 2) = 0.112 \text{ m}$$

The force at the solid length would be

$$F_s = k\left(L_f - L_s\right) = 8438(0.2556 - 0.112) = 1211 \text{ N}$$

The solid length shear stress can be determined by

$$\tau_{\text{solid}} = \tau_{\max} \frac{F_s}{F_o} = 499 \times 10^6 \frac{1211}{975} = 630 \times 10^6 \text{ Pa}$$

This value is just below than the maximum allowable stress of 653 MPa indicating that failure is unlikely.

The ratio of free length to the mean coil diameter, $L_f/D = 0.2556/0.05862 = 4.359$.

The deflection of the spring under the operating force is 116 mm.

Thus the ratio of deflection to free length, $\delta/L_f = 0.116/0.2556 = 0.452$.

Examination of Fig. 15.15 for squared and ground ends shows that this design is stable and unlikely to buckle.

The outer diameter of the spring can be determined from

$$D_o = d + D = 0.007 + 0.05862 = 0.06562 \text{ m}$$

The inner diameter of the spring will be

$$D_i = D - d = 0.05862 - 0.007 = 0.05162 \text{ m}$$

Nominal specification:

ASTM A232 chromium vanadium steel wire helical compression spring.
$k = 8438$ N/m.
$L_f = 0.2556$ m.
$L_s = 0.112$ m.
Squared and ground ends, 16 coils.
$d = 7$ mm.
$D = 58.62$ mm, $D_i = 51.62$ mm, $D_o = 65.62$ mm.

15.4 Helical torsion springs

Many applications require a spring that exerts a torsional moment. A helical coil spring can be loaded in torsion to fulfil this requirement. Various configurations of torsional springs are shown in Fig. 15.22. The ends of the coil tend to be extended to provide levers to apply the torque.

The basic equation governing the angular deflection of a torsional spring is

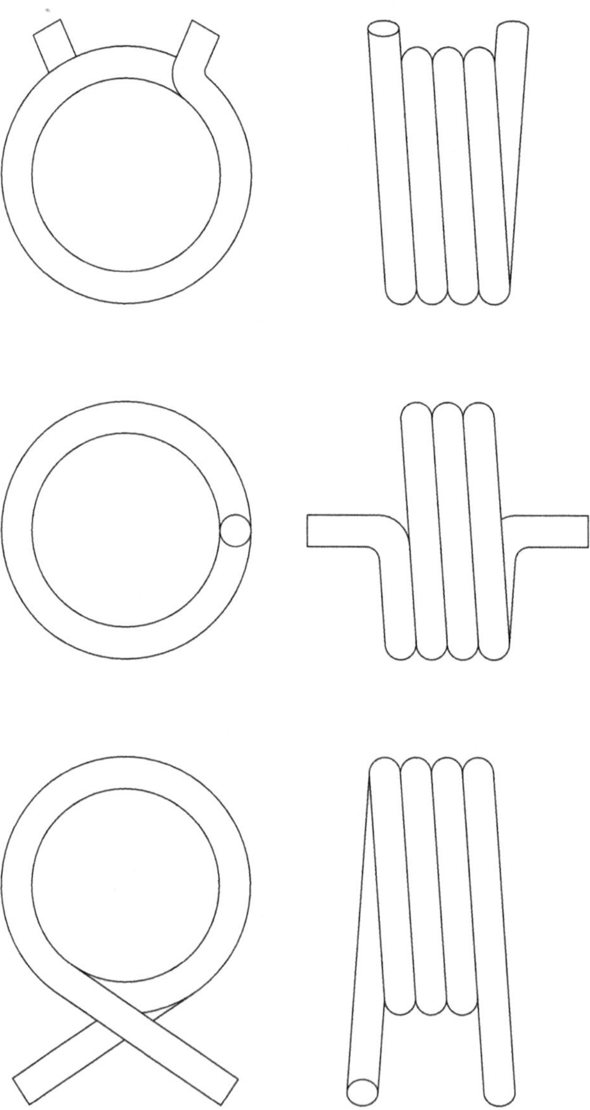

Fig. 15.22 Various configurations for torsional springs.

$$\theta = \frac{ML}{EI} \tag{15.30}$$

where M = applied moment (N m), L = length of wire in the spring (m), E = Young's modulus (N/m^2), I = second moment of area (m^4) and θ = angular deflection (rad).

Substituting for the second moment of area and the length and converting the angular deflection to revolutions gives a more convenient relationship:

$$\theta' = \frac{10.2MDN_a}{d^4E} \tag{15.31}$$

where θ' = angular deflection (revolutions).

The spring rate for a torsional spring is defined as the moment per unit of angular deflection and is given by

$$k = \frac{M}{\theta'} \cong \frac{d^4E}{10.2DN} \tag{15.32}$$

When a coil is loaded, the coil diameter decreases and its length increases as the coil is wound up. The minimum inner diameter of the coil at full deflection is given by

$$D_{i\ \min} = \frac{DN_b}{N_b + \theta'} - d \tag{15.33}$$

where $D_{i\ \min}$ = minimum inner diameter (m) and N_b = number of coils in the body.

The maximum coil body length when the spring is fully wound up is given by

$$L_{\max} = d(N_b + 1 + \theta) \tag{15.34}$$

where $L_{\max}$ = maximum coil body length (m).

The number of active coils in a torsion spring is equal to the number of body turns, N_b plus some contribution from the ends, which also bend. For straight ends this contribution can be expressed by the number of equivalent coils, N_e.

$$N_e = \frac{L_1 + L_2}{3\pi D} \tag{15.35}$$

where L_1 and L_2 are the lengths of the ends. The number of active coils is given by

$$N_{\mathrm{a}} = N_{\mathrm{b}} + N_{\mathrm{e}} \tag{15.36}$$

15.5 Leaf springs

Leaf springs consist of one or more flat strips of material loaded as cantilevers or simple beams as illustrated in Fig. 15.23. They can be designed to provide a compressive

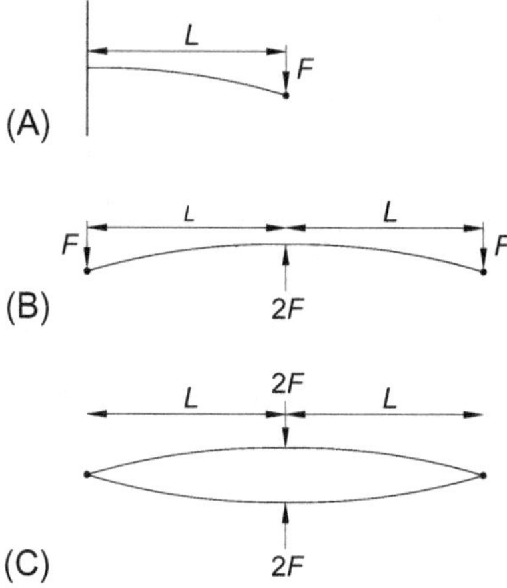

Fig. 15.23 Basic types of leaf spring: (A) simple cantilever, (B) semi-elliptic, (C) full-elliptic.

Fig. 15.24 Leaf springs for automotive applications.
Figure courtesy Owen Springs Ltd.

or tensile force as they are deflected from their free condition. Leaf springs are capable of exerting large forces within comparably small spaces. A practical leaf spring design is illustrated in Fig. 15.24 and a typical application shown in Fig. 15.25.

Particular force–deflection characteristics can be achieved for a spring by careful dimensioning of the strips and nesting of a number of components (Rajendran and Vijayarangan, 2001). Various forms for cantilever beams are illustrated in Fig. 15.26. Each type shown has special characteristics. Fig. 15.26A illustrates a generalised beam. If it is desired to maintain uniform bending-stresses over the length of the beam then the width of the cantilever needs to vary linearly with location as

Fig. 15.25 Example of the application of leaf springs to an automotive axle. Image courtesy of Tennen-Gas, GFDL and cc-by-sa-2.5,2.0,1.0

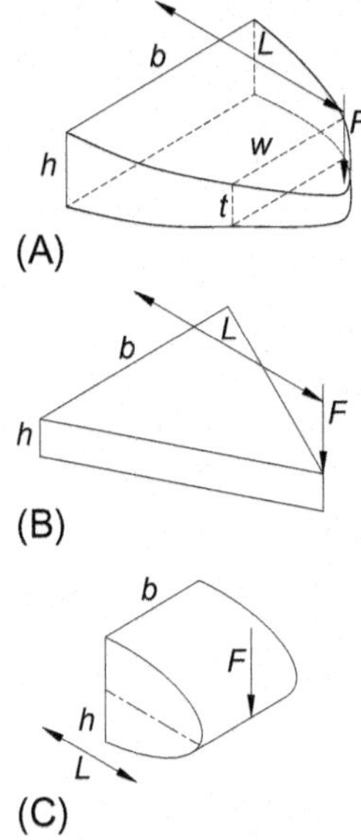

Fig. 15.26 Cantilever beams: (A) general form, (B) uniform bending stress, (C) uniform stress cantilever of constant width.

illustrated in Fig. 15.26B. This triangular form for the cantilever is the basic building block for leaf springs in common use. An alternative requirement might be to maintain uniform stress in a cantilever of constant width. In this case the profile of the beam would need to be parabolic as illustrated in Fig. 15.26C. This is the basis used for analysing the bending strength of teeth in spur gears.

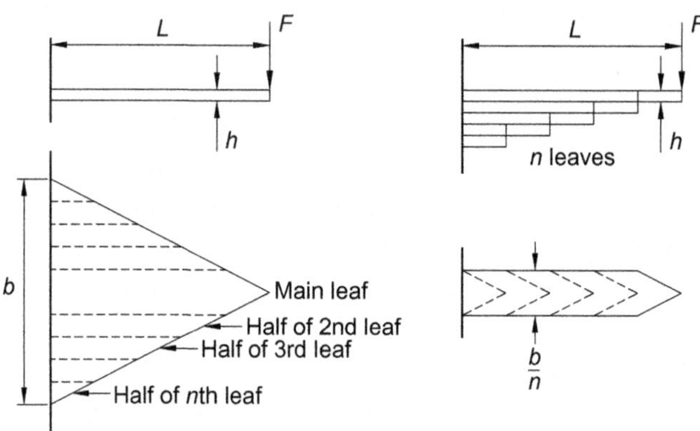

Fig. 15.27 Triangular plan view cantilever and equivalent multileaf spring.

The concept used in producing compact cantilever springs of uniform bending stress is to chop the triangular form illustrated in Fig. 15.26B into a number of strips and recombine them as illustrated in Fig. 15.27. The multileaf spring shown and the single triangular section beam both have the same stress and deflection characteristics with the exceptions that the multileaf spring is subject to additional damping due to friction between the leaves and that the multileaf spring can carry a full load in only one direction due to the tendency for the leaves to separate. Leaf separation can be partially overcome by the provision of clips around the leaves.

The deflection of a triangular leaf spring is given by

$$\delta = \frac{FL^3}{2EI} \tag{15.37}$$

where F = force (N), L = length (m), E = Young's modulus (N/m^2) and I = second moment of area (m^4).

For a rectangular cross-section,

$$I = \frac{bh^3}{12} \tag{15.38}$$

where b = width (m) and h = thickness (m).

The spring rate is given by

$$k = \frac{F}{\delta} = \frac{Ebh^3}{6L^3} \tag{15.39}$$

The corresponding bending stress is given by

$$\sigma = \frac{6FL}{bh^2} \tag{15.40}$$

For a semielliptic beam the maximum deflection at the centre is given by

$$\delta = \frac{6FL^3}{Ebh^3} \tag{15.41}$$

The corresponding bending stress for a semielliptic beam is given by Eq. (15.40).

For a full-elliptic beam the maximum deflection at the centre is given by

$$\delta = \frac{12FL^3}{Ebh^3} \tag{15.42}$$

The corresponding bending stress for a full-elliptic beam is given by Eq. (15.40).

15.6 Belleville spring washers

Belleville spring washers comprise a conical shaped disc with a hole through the centre as illustrated in Figs. 15.28 and 15.29. They have a nonlinear force deflection characteristic, which makes them useful in certain applications. Belleville springs are compact and are capable of large compressive forces but their deflections are limited. Examples of their use include gun recoil mechanisms, pipe flanges, small machine tools and applications where differential expansion could cause relaxation of a bolt (Fig. 15.30).

The force–deflection characteristics for a Belleville washer depend on material properties and the dimensions illustrated in Fig. 15.31. The force–deflection relationship is nonlinear so it is not stated in the form of a spring rate and is given (Norton, 1996) by:

$$F = \frac{4E\delta}{K_1 D_o^2 (1-\mu^2)} \left[(h-\delta)\left(h - \frac{\delta}{2}\right)t + t^3 \right] \tag{15.43}$$

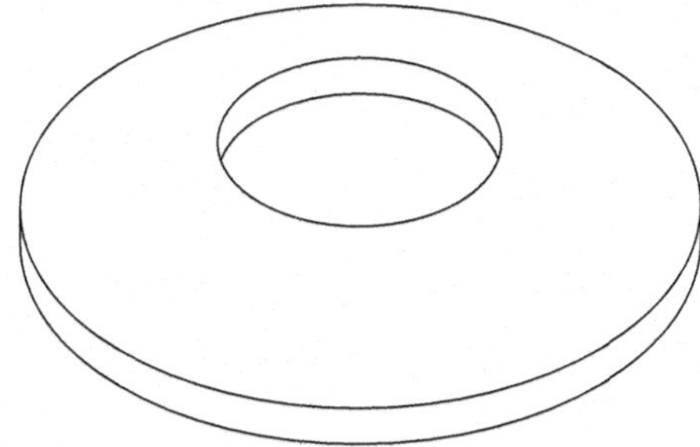

Fig. 15.28 A Belleville spring washer.

Fig. 15.29 Belleville spring stack.

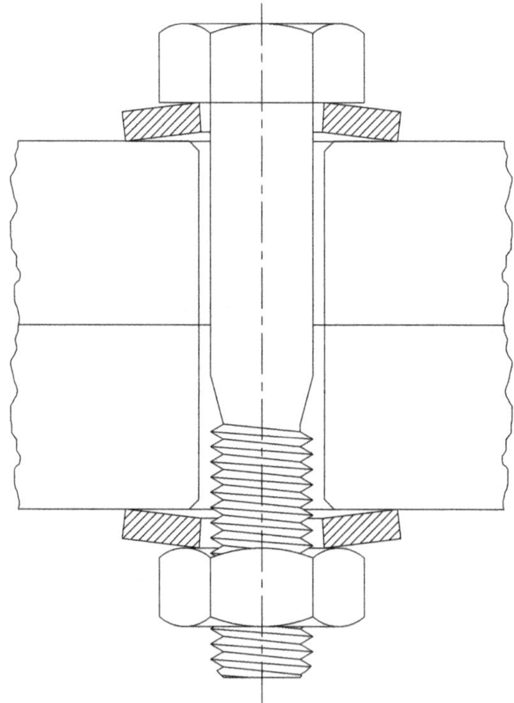

Fig. 15.30 Use of Belleville springs to maintain bolt preload.

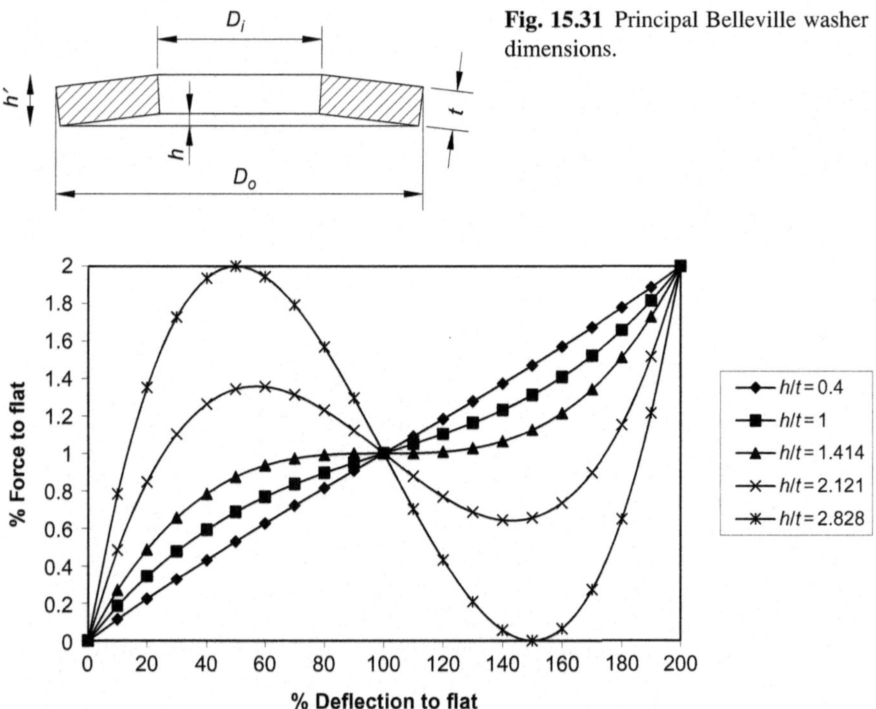

Fig. 15.31 Principal Belleville washer dimensions.

Fig. 15.32 Normalised force deflection characteristics for Belleville washer springs for various h/t ratios. Calculations performed with $E = 207\,\text{GPa}$ and $\mu = 0.3$.

where

$$K_1 = \frac{6}{\pi \ln R_d}\left[\frac{(R_d - 1)^2}{R_d^2}\right] \tag{15.44}$$

$$R_d = \frac{D_0}{D_i} \tag{15.45}$$

The load at the flat position is given by

$$F_{\text{flat}} = \frac{4Eht^3}{K_1 D_o^2 (1 - \mu^2)} \tag{15.46}$$

Eq. (15.43) has been used to illustrate the variation of force with deflection in Fig. 15.32 for a variety of h/t ratios. The curves have been normalised to the spring's condition when flat for a steel Belleville spring. For a ratio of h/t of 0.4 the ratio of force to deflection is close to linear. As the h/t ratio is increased it becomes increasingly nonlinear. For an h/t ratio of 1.414 there is a nearly constant region for the force.

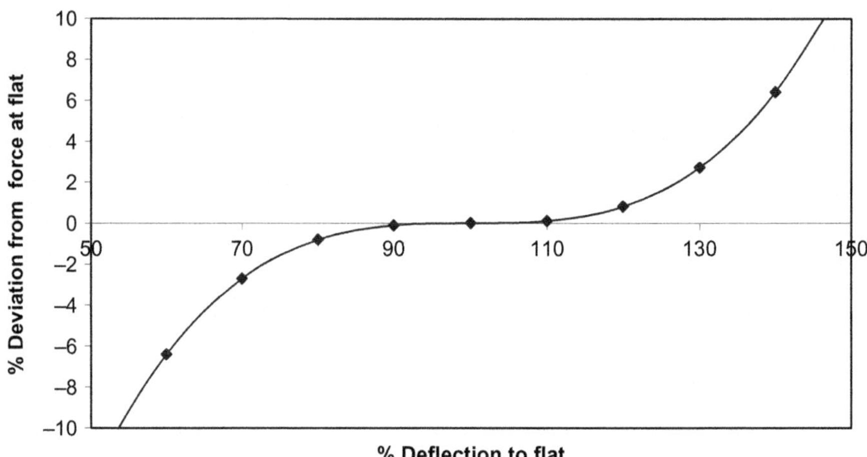

Fig. 15.33 Percentage error in a constant force Belleville spring around its flat position. $R_d = 2$, $h/t = 1.414$.

For values of $h/t > 1.414$ the curves become bimodal where a given force could cause more than one possible deflection. Once installed and in the flat position a trivial force would be required to trip it in one direction or the other. The deviation of force from the force to flat value with deflection is shown in Fig. 15.33 for a spring with $R_d = 2$ and $h/t = 1.414$. The force produced by a $h/t = 1.414$ ratio Belleville spring is within ±10% of the force to flat value for deflections of 55% to 145% of the deflection to flat. This is a useful characteristic if a nearly constant force is required over this range of deflection.

The stresses in a Belleville spring are concentrated at the edges of the inside and outside diameters (Almen and Laszlo, 1936). The largest stress is compressive and occurs at the inside radius on the convex side. The edges on the concave side have tensile stresses given by Eqs (15.48) and (15.49). These equations are quoted from Norton (1996).

$$\sigma_c = -\frac{4E\delta}{K_1 D_o^2 (1-\mu^2)}\left[K_2\left(h - \frac{\delta}{2}\right) + K_3 t\right] \tag{15.47}$$

$$\sigma_{ti} = \frac{4E\delta}{K_1 D_o^2 (1-\mu^2)}\left[-K_2\left(h - \frac{\delta}{2}\right) + K_3 t\right] \tag{15.48}$$

$$\sigma_{to} = \frac{4E\delta}{K_1 D_o^2 (1-\mu^2)}\left[K_4\left(h - \frac{\delta}{2}\right) + K_5 t\right] \tag{15.49}$$

where

$$K_2 = \frac{6}{\pi \ln R_d}\left[\frac{R_d - 1}{\ln R_d} - 1\right] \tag{15.50}$$

$$K_3 = \frac{6}{\pi \ln R_d}\left[\frac{R_d - 1}{2}\right] \tag{15.51}$$

$$K_4 = \left[\frac{R_d \ln R_d - (R_d - 1)}{\ln R_d}\right]\left[\frac{R_d}{(R_d - 1)^2}\right] \tag{15.52}$$

$$K_5 = \frac{R_d}{2(R_d - 1)} \tag{15.53}$$

The maximum deflection of a single Belleville spring tends to be small. They can, however, be stacked in combinations as illustrated in Fig. 15.34. If stacked in series the total force will be the same as for a single Belleville spring but the deflections will add. If they are stacked in parallel the total deflection will be the same as for a single spring but the forces will add. Series and parallel combinations are also possible. It should be noted that Belleville spring towers need some form of support, either by inserting them into a hole or over a rod. This however will reduce the available load due to friction.

The design or selection of a Belleville spring for a given application requires consideration of the diameter and h/t ratio as well as the type of material to be used to give the desired force–deflection characteristics (SAE, 1988). This process invariably involves some iteration. A material is selected, values for the diameter ratio are proposed, often based on given constraints, and values of the h/t ratio proposed either for a single spring or some combination.

In designing Belleville springs it can be useful to estimate the thickness required to give a particular force in the flat position which can be found, substituting $E = 207\,\text{GPa}$, $\mu = 0.3$ and $K_1 = 0.69$ in Eq. (15.46), from:

$$t = \frac{1}{1072}\left(\frac{D_o^2 F_{\text{flat}}}{h/t}\right)^{0.25} \tag{15.54}$$

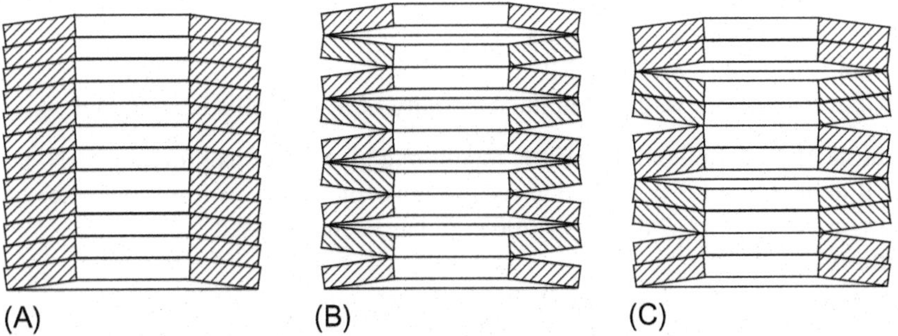

Fig. 15.34 Stacking combinations for Belleville springs: (A) parallel, (B) series, (C) series-parallel.

where t = thickness (m), h = deflection to flat (m), D_o = outer diameter (m) and F_{flat} = load at flat position (N).

Dimensions for a selection of Belleville springs manufactured to DIN 6796 are given in Table 15.6.

Example 15.7. A Belleville spring is required to give a constant force of 200 N ± 10 N over a deflection of ±0.3 mm. The spring must fit within a 62 mm diameter hole. A carbon spring steel with $\sigma_{uts} = 1700$ MPa has been proposed.

Solution

Assume a 60 mm outer diameter to allow some clearance in the hole.

To provide a constant force, an h/t ratio of 1.414 is selected.

The variation of force of ±5% can be met by choosing an appropriate deflection range to operate in from Fig. 15.33. If the deflection is limited to between 65% and 135% of the flat deflection, then the tolerance on force can be achieved. The nominal force of 200 N will occur in the flat position and the spring will provide a similar force, between 210 N and 190 N, operating on both sides of its centre.

From Eq. (15.54),

$$t = \frac{1}{1072}\left(\frac{D_o^2 F_{\text{flat}}}{h/t}\right)^{0.25} = \frac{1}{1072}\left(\frac{0.06^2 \times 200}{1.414}\right)^{0.25} = 7.88 \times 10^{-4}\ \text{m}$$

$h = 1.414\, t = 1.414 \times 0.788 = 1.114$ mm.

The minimum and maximum deflections are.

$\delta_{min} = 0.65\,\text{h} = 0.65 \times 1.114 = 0.724$ mm.

$\delta_{max} = 1.35\,\text{h} = 1.35 \times 1.114 = 1.504$ mm.

$\delta_{max} - \delta_{min}$ is greater than the required deflection range of 0.6 mm so the force tolerance can be met.

From Eqs (15.44) and (15.50)–(15.53),

$K_1 = 0.689$,
$K_2 = 1.220$,
$K_3 = 1.378$,
$K_4 = 1.115$,
$K_5 = 1$.

From Eq. (15.47), $\sigma_c = -840$ MPa.

From Eq. (15.48), $\sigma_{ti} = 355$ MPa.

From Eq. (15.49), $\sigma_{to} = 658$ MPa.

These stresses are well within the capability of a 1700 MPa uts material.

Example 15.8. A Belleville spring is required to give a constant force of 50 N ± 5 N over a deflection of ±0.2 mm. The spring must fit within a 40 mm diameter hole. A carbon spring steel with $\sigma_{uts} = 1700$ MPa has been proposed.

Solution

Assume a 35 mm outer diameter to allow some clearance in the hole.

To provide a constant force, a h/t ratio of 1.414 is selected.

Table 15.6 Dimensions for a selection of Belleville washer springs manufactured to DIN 6796 from DIN 17222 spring steel

Notation	D_i (mm)	D_o (mm)	h' max (mm)	h' min (mm)	t (mm)	Force (N)[a]	Test force (N)[b]	Mass kg/1000	Core diameter (mm)
2	2.2	5	0.6	0.5	0.4	628	700	0.05	2
2.5	2.7	6	0.72	0.61	0.5	946	1100	0.09	2.5
3	3.2	7	0.85	0.72	0.6	1320	1500	0.14	3
3.5	3.7	8	1.06	0.92	0.8	2410	2700	0.25	3.5
4	4.3	9	1.3	1.12	1	3770	4000	0.38	4
5	5.3	11	1.55	1.35	1.2	5480	6550	0.69	5
6	6.4	14	2	1.7	1.5	8590	9250	1.43	6
7	7.4	17	2.3	2	1.75	11,300	13,600	2.53	7
8	8.4	18	2.6	2.24	2	14,900	17,000	3.13	8
10	10.5	23	3.2	2.8	2.5	22,100	27,100	6.45	10
12	13	29	3.95	3.43	3	34,100	39,500	12.4	12
14	15	35	4.65	4.04	3.5	46,000	54,000	21.6	14
16	17	39	5.25	4.58	4	59,700	75,000	30.4	16
18	19	42	5.8	5.08	4.5	74,400	90,500	38.9	18
20	21	45	6.4	5.6	5	93,200	117,000	48.8	20
22	23	49	7.05	6.15	5.5	113,700	145,000	63.5	22
24	25	56	7.75	6.77	6	131,000	169,000	92.9	24
27	28	60	8.35	7.3	6.5	154,000	221,000	113	27
30	31	70	9.2	8	7	172,000	269,000	170	30

[a]Force applies to the pressed flat condition and corresponds to twice the calculated value at a deflection $h' - t$.
[b]Test force applies for loading tests to DIN 6796.

The variation of force of $\pm10\%$ can be readily met by choosing an appropriate deflection range to operate in from Fig. 15.33.

If the deflection is limited to between 65% and 135% of the flat deflection, then the tolerance on force can be achieved.

The nominal force of 50 N will occur in the flat position and the spring will provide a similar force, between 55 N and 45 N, operating on both sides of its centre.

From Eq. (15.52),

$$t=\frac{1}{1072}\left(\frac{D_o^2 F_{\text{flat}}}{h/t}\right)^{0.25}=\frac{1}{1072}\left(\frac{0.035^2\times 50}{1.414}\right)^{0.25}=4.26\times 10^{-4}\ \text{m}$$

$h=1.414\text{t}=1.414\times 0.426=0.602\,\text{mm}$

The minimum and maximum deflections are

$\delta_{\text{min}}=0.65\,\text{h}=0.65\times 0.602=0.391\,\text{mm}$

$\delta_{\text{max}}=1.35\,\text{h}=1.35\times 0.602=0.812\,\text{mm}$

$\delta_{\text{max}}-\delta_{\text{min}}$ is greater than the required deflection range of 0.4 mm so the force tolerance can be met.

From Eqs (15.44) and (15.50)–(15.53),

$K_1=0.689$
$K_2=1.220$
$K_3=1.378$
$K_4=1.115$
$K_5=1$

From Eq. (15.47), $\sigma_c=-720\,\text{MPa}$

From Eq. (15.48), $\sigma_{\text{ti}}=305\,\text{MPa}$

From Eq. (15.49), $\sigma_{\text{to}}=564\,\text{MPa}$.

These stresses are well within the capability of a 1700 MPa uts material.

Example 15.9. A Belleville spring is required to give a constant force of $10\,\text{N}\pm1\,\text{N}$ over a deflection of $\pm0.15\,\text{mm}$. The spring must fit within a 16 mm diameter hole. A carbon spring steel with $\sigma_{\text{uts}}=1700\,\text{MPa}$ has been proposed.

Solution

Assume a 14 mm outer diameter to allow some clearance in the hole.

To provide a constant force, a *h/t* ratio of 1.414 is selected.

The variation of force of $\pm10\%$ can be met by choosing an appropriate deflection range to operate in from Fig. 15.33.

If the deflection is limited to between 65% and 135% of the flat deflection, then the tolerance on force can be achieved.

The nominal force of 10 N will occur in the flat position and the spring will provide a similar force, between 11 N and 9 N, operating on both sides of its centre.

From Eq. (15.52),

$$t=\frac{1}{1072}\left(\frac{D_o^2 F_{\text{flat}}}{h/t}\right)^{0.25}=\frac{1}{1072}\left(\frac{0.014^2\times 10}{1.414}\right)^{0.25}=1.8\times 10^{-4}\,\text{m}$$

$h = 1.414\,t = 1.414 \times 0.18 = 0.255\,\text{mm}$

The minimum and maximum deflections are

$\delta_{\min} = 0.65\,h = 0.65 \times 0.255 = 0.165\,\text{mm}$

$\delta_{\max} = 1.35\,h = 1.35 \times 0.255 = 0.344\,\text{mm}$

From Eqs (15.44) and (15.50)–(15.53),

$K_1 = 0.689$
$K_2 = 1.220$
$K_3 = 1.378$
$K_4 = 1.115$
$K_5 = 1$

From Eq. (15.47), $\sigma_c = -810\,\text{MPa}$

From Eq. (15.48), $\sigma_{ti} = 341\,\text{MPa}$

From Eq. (15.49), $\sigma_{to} = 630\,\text{MPa}$

These stresses are well within the capability of a 1700 MPa uts material.

$\delta_{\max} - \delta_{\min}$ is not however greater than the required deflection range of 0.3 mm so the force tolerance is not met.

Altering the diameter of the spring however to say 15 mm or 13 mm does not give $\delta_{\max} - \delta_{\min} > 0.3\,\text{mm}$. So a different h/t ratio would need to be considered.

15.7 Conclusions

This chapter has introduced the function and characteristics of a range of spring technologies. Design methods have been outlined for helical compression, extension and torsion springs, leaf springs and Belleville spring washers. There is no single design procedure that is suitable for all types of spring and the procedures outlined here have not considered all of the important design parameters, which also include cost, appearance and environmental considerations. Spring design invariably becomes a process of optimisation where various trade-offs need to be balanced to provide the best possible total solution.

References

Books and papers

Almen, J.O., Laszlo, A., 1936. The uniform section disk spring. Trans. ASME 58, 305–314.

Ashby, M.F., 1999. Materials Selection for Mechanical Design, second ed. Butterworth Heinemann.

Associated Spring, Barnes Group, 1987. Engineering Guide to Spring Design. Associated Spring.

Carlson, H., 1978. Spring Designers Handbook. Marcel Dekker.

Carlson, H., 1982. Spring Manufacturing Handbook. Marcel Dekker.

Das, S.K., Mukhopadhyay, N.K., RaviKumara, B., Bhattacharyac, D.K., 2007. Failure analysis of a passenger car coil spring. Eng. Fail. Anal. 14, 158–163.

Juvinall, R.C., Marshek, K.M., 1991. Fundamentals of Machine Component Design. Wiley.

Kobelev, V., 2017. Durability of Springs. Springer.
Mott, R.L., 1999. Machine Elements in Mechanical Design, third ed. Prentice Hall.
Norton, R.L., 1996. Machine Design. Prentice Hall.
Rajendran, I., Vijayarangan, S., 2001. Optimal design of a composite leaf spring using genetic algorithms. Comput. Struct. 79, 1121–1129.
SAE, Society of Automotive Engineers, 1988. Manual on Design and Manufacture of Coned Disk Springs (Belleville Springs) and Spring Washers. SAE.
SAE, Society of Automotive Engineers, 1996. SAE Spring Committee. Spring Design Manual, second ed. SAE.
Wahl, A.M., 1963. Mechanical Springs, second ed. McGraw Hill.

Standards

ASTM International. ASTM A125-96(2001). Standard specification for steel springs, helical heat treated.
ASTM International. ASTM A227/A227M-99. Standard specification for steel wire, cold-drawn for mechanical springs.
ASTM International. ASTM A228/A228M-02. Standard specification for steel wire, music spring quality.
ASTM International. ASTM A229/A229M-99. Standard specification for steel wire, oil-tempered for mechanical springs.
ASTM International. ASTM A230/A230M-99. Standard specification for steel wire, oil tempered carbon valve spring quality.
ASTM International. ASTM A231/A231M-96(2002). Standard specification for chromium vanadium alloy steel spring wire.
ASTM International. ASTM A232/A232M-99. Standard specification for chromium vanadium alloy steel valve spring quality wire.
ASTM International. ASTM A313/A313M-98. Standard specification for stainless spring wire.
ASTM International. ASTM A401/A401M-98. Standard specification for steel wire, chromium silicon alloy.
ASTM International. ASTM B159M-01. Standard specification for phosphor bronze wire.
British Standards Institution. BS EN 10270-1:2001. Steel wire for mechanical springs. Patented cold-drawn unalloyed steel spring wire.
British Standards Institution. BS EN 10270-2:2001. Steel wire for mechanical springs. Oil hardened and tempered spring steel wire.
British Standards Institution. BS EN 10270-3:2001. Steel wire for mechanical springs. Stainless spring steel wire.
British Standards Institution. BS EN 12166:1988. Copper and copper alloys. Wire for general purposes.
British Standards Institution. BS 1429: 1980. Specification for annealed round steel wire for general engineering springs.
British Standards Institution. BS 1726-1:2002. Cylindrical helical springs made from round wire and bar. Guide to methods of specifying, tolerances and testing. Compression springs.
British Standards Institution. BS 1726-2:2002. Cylindrical helical springs made from round wire and bar. Guide to methods of specifying, tolerances and testing. Extension springs.
British Standards Institution. BS 1726-3:2002. Cylindrical helical springs made from round wire and bar. Guide to methods of specifying, tolerances and testing. Torsion springs.

British Standards Institution. BS 8726-1:2002. Cylindrical helical springs made from rectangular and square section wire and bar. Guide to calculation and design. Compression springs.

British Standards Institution. BS 8726-2:2002. Cylindrical helical springs made from rectangular and square section wire and bar. Guide to calculation and design. Torsion springs.

British Standards Institution. BS EN ISO 2162-1:1996. Technical product documentation. Springs. Simplified representation.

British Standards Institution. BS EN 13906-1:2002. Cylindrical helical springs made from round wire and bar. Calculation and design. Compression springs.

British Standards Institution. BS EN 13906-2:2002. Cylindrical helical springs made from round wire and bar. Calculation and design. Extension springs.

British Standards Institution. BS EN 13906-3:2002. Cylindrical helical springs made from round wire and bar. Calculation and design. Torsion springs.

DIN. DIN Handbook 140 Fasteners 4. Dimensional standards for nuts and accessories for bolt/nut assemblies. 2001.

DIN. DIN 6796. Conical spring washers for bolt/nut assemblies. 1987.

Further reading

Joerres, R.G., 1996. Springs. In: Shigley, J.E., Mischke, C.R. (Eds.), Chapter 24 in Standard Handbook of Machine Design. McGraw Hill.

Websites

At the time of going to press the world-wide-web contained useful information relating to this chapter at the following links.

www.acewirespring.com.
www.asbg.com.
www.ashfield-springs.com.
www.assocspring.co.uk.
www.astm.org.
www.automaticsprings.com.au.
www.bestspring.com.
www.bsi.org.uk.
www.harris-springs.com.
www.hzap.cz.
www.indiamart.com/paragonindustries/.
www.leespring.co.uk.
www.sae.org.
www.smihq.org.
www.springmasters.com.
www.uzelotomotiv.com.
www.wmhughes.co.uk.
www.wolverinecoilspring.com.
www2.din.de

Fastening and power screws

16

Chapter Outline

Nomenclature

Generally, preferred SI units have been stated.

A_r	cross sectional area of rivet (m^2)
A_b	projected bearing area of rivet (m^2)
A_p	cross sectional area of plate between rivet holes (m^2)
A_t	tensile stress area of the bolt (m^2)
Csk.	countersunk
d	diameter (m)
d_c	diameter of the thrust collar (m)
f	frictional force (N)
F	force (N)
F_b	final force in the bolt (N)
F_c	final force on the clamped components (N)
F_e	externally applied load (N)
F_i	preload (N)
h	maximum cantilever thickness (m)
$\underline{K}$	constant
k_b	stiffness of the bolt (N/m)
k_c	stiffness of the clamped components (N/m)
L	length (m), safe load (N)
MIG	metal inert gas
MMA	manual metal arc
n	number

Mechanical Design Engineering Handbook. https://doi.org/10.1016/B978-0-08-102367-9.00016-0

N	reaction (N)
P	pitch (m)
r, R	radius (m)
SAW	submerged arc welding
t	thickness (m)
T	torque (N m)
T_c	torque required to turn the thrust collar (N m)
T_d	torque for lowering a load (N m)
T_s	screw torque (N m)
T_u	torque for lifting a load (N m)
TIG	tungsten inert gas
UNC	Unified coarse pitch thread series
UNF	Unified fine pitch thread series
W	work done (N m)
y	deflection (m)
α	angle
ε	permissible strain
η	efficiency
λ	angle
μ	coefficient of friction
μ_c	coefficient of friction in the thrust bearing
σ	stress (N/m^2)
σ_p	proof strength of the bolt (N/m^2)
σ_t	tensile stress (N/m^2)
σ_y	yield stress (N/m^2)
τ	shear stress (N/m^2)
τ_d	allowable shear stress (N/m^2)

16.1 Introduction to permanent and nonpermanent fastening

A fastener is a device used to connect, or join, two or more components. Traditional forms of fastening include nuts, bolts, screws and rivets. In addition welding and adhesives can be used to form permanent joins between components. The aim of this chapter is to introduce a wide selection of fastening techniques. Whilst considering threaded fasteners the subject of power screws, which are used for converting rotary motion to linear motion, is also presented.

The joining of components is a frequent necessity in the design of products. For example the Boeing 747 has over 2.5 million fasteners, although this number is reducing with the introduction of composite components. Fastening techniques can also be a major feature in design, as indicated in Figs. 16.1–16.3 for various automotive wheel hubs, although current styling for automotive vehicles dictates an absence, under a cursory inspection, of the means of fastening components together (Fig. 16.4). The range of fastening techniques is extensive including adhesives, welding, brazing,

Fig. 16.1 Prominent fastener features by Bentley.

Fig. 16.2 Wheel eyebrow feature on the Range Rover Evoque.

Fig. 16.3 Defender 90 wheel eyebrow with chunky nuts.

Fig. 16.4 A vehicle facia—few if any indications of how components are held in place are evident.
Image courtesy of Mercedez-Benz.

Fig. 16.5 An example of an automotive trim fastener.

soldering, threaded and unthreaded fasteners, snap fits, special purpose fasteners (Fig. 16.5) and friction joints. Some of these techniques are permanent in nature and some allow the joint to be dismantled and are nonpermanent. A variety of basic types of joints are illustrated in Fig. 16.6.

A further means of classifying a joint is in terms of its design objective. If the objective is to locate a component then a method that does not produce a clamp load can be used, and is known as a position critical joint. If the objective is to clamp components together then a clamp load critical joint is needed, which is externally loaded and gives

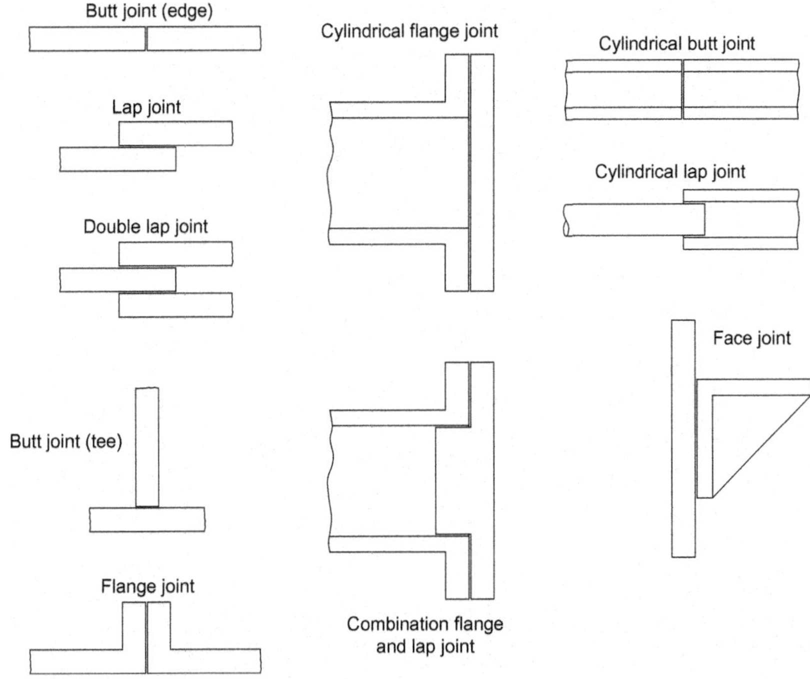

Fig. 16.6 A variety of types of joint.

enough clamping force to ensure that the components do not slide relative to each other or separate.

Design considerations for joining include:

- whether the joint should be permanent or nonpermanent
- cost
- loads in the fastener or power screw and the associated components
- life
- tooling
- assembly
- tolerances
- aesthetics
- size
- corrosion

Table 16.1 provides an indication of the merits of some types of fastener. It should be noted that the table is not exhaustive given the scope and range of specialist fastener possible.

Threaded fasteners are introduced in Section 16.2, power screws in Section 16.3, rivets in Section 16.4, adhesives in Section 16.5, welding and the related subjects of soldering and brazing in Section 16.6 and snap joints in Section 16.7.

Table 16.1 Fastener attributes

	Nut and bolt	Sheet metal screw	Solid rivet	Blind rivet	Tubular rivet	Weld	Brazing	Snap fits	Plastic fastener	Hook and loop	Adhesive
Join dissimilar materials	X	X	X	X	X		X	X	X	X	X
Vibration resistant			X	X	X	X	X			X	X
Tamper resistant			X	X	X	X	X				X
Removable	X	X		X	X			X	X	X	
No heat distortion	X	X	X	X	X			X	X	X	
Join plastic	X	X		X	X			X	X	X	X
Clamp load critical	X										
Position critical	X	X	X	X	X	X	X	X	X	X	X
Low capital investment	X	X	X	X			X	X	X	X	X
Simple assembly	X	X	X	X				X	X	X	X
Safety equipment required	X	X		X				X	X	X	
Portable tools	X	X	X	X				X	X	X	X

Single operation		X		X	X			X	X	X	
Surface preparation required						X	X			X	X
Permits alignment of work surface	X	X	X	X			X	X	X	X	X

16.2 Threaded fasteners

There is a large variety of fasteners available using a threaded form to produce connection of components, as indicated in Figs. 16.7 and 16.8. The common element of screw fasteners is the helical thread that causes a screw to advance into a component or nut when rotated.

Fig. 16.7 Examples of male threaded fastener (screws and bolts).

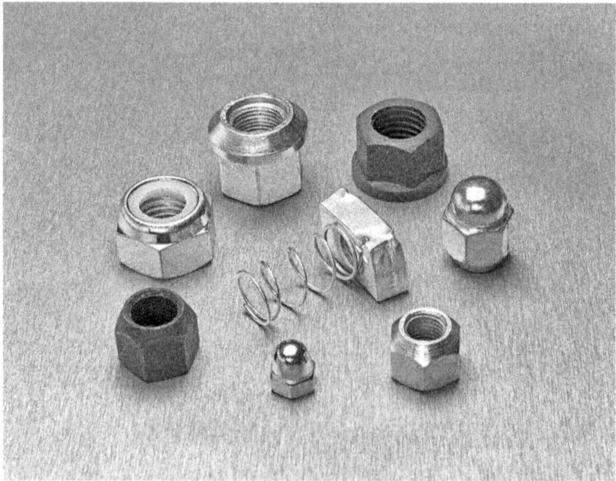

Fig. 16.8 Examples of threaded female fasteners (nuts).

In the case of a two part threaded fastener, common notation used for the two components is:

- A male component, the screw, with a helical thread or groove formed on its outer diameter.
- A female component, the nut, with a helical thread or groove formed on its inner diameter.

For a two part threaded fastener the two helical grooves are defined with the same pitch, hand and nominal diameter. A slight clearance is required to allow assembly between the diameter for the screw and nut. Rotation of, for example, the screw component causes it to advance into the nut component forming a threaded assembly. Screw threads can be either left-handed or right-handed depending on the direction of rotation desired for advancing the thread as illustrated in Fig. 16.9. Generally right hand threads are normally used, with left-handed threads being reserved for specialist applications. The shape of the groove forming the helix is called the thread form or profile. A large number of thread forms exist and are available, but the vast majority of threads and threaded fasteners use the so-called vee thread. The detailed aspects of a thread and the specialist terminology used are illustrated in Fig. 16.10 and defined in Table 16.2.

Thread forms, angle of helix and so on, vary according to specific standards. Common standards developed include UNS (unified national standard series threads) and ISO threads. Both of these use a 60° included angle but are not interchangeable. The form for an ISO metric thread for a nut is illustrated in Fig. 16.12. In practice the root of the nut and the crest of the mating bolt are rounded. Both male and female ISO threads are subject to manufacturing tolerances, which are detailed in BS 3643. A coarse series thread and fine series threads are defined in the ISO standard but fine series threads tend to be more expensive and may not be readily available from all stockists. Table 16.3 gives the standard sizes for a selection of ISO coarse series hexagon bolts, screws and nuts. ISO metric threads are designated by the letter M, followed by the nominal diameter and the pitch required, for example, M6 × 1.5.

The Unified system of screw threads was originally introduced in the United Kingdom, Canada and the United States to provide a common standard for use by the three nations. Types of unified threads in common use include the Unified coarse pitch

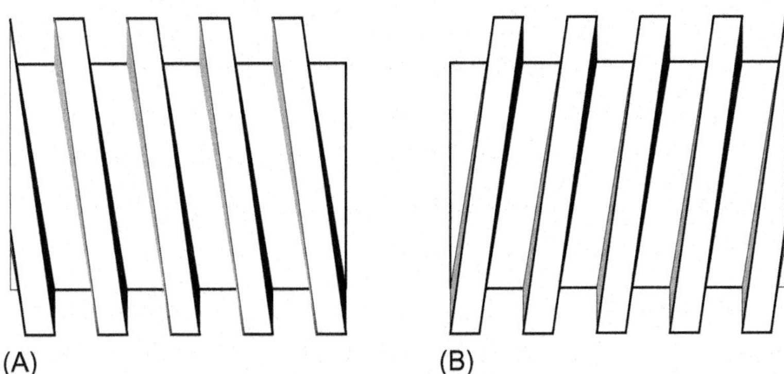

Fig. 16.9 (A) Right hand thread; (B) left hand thread.

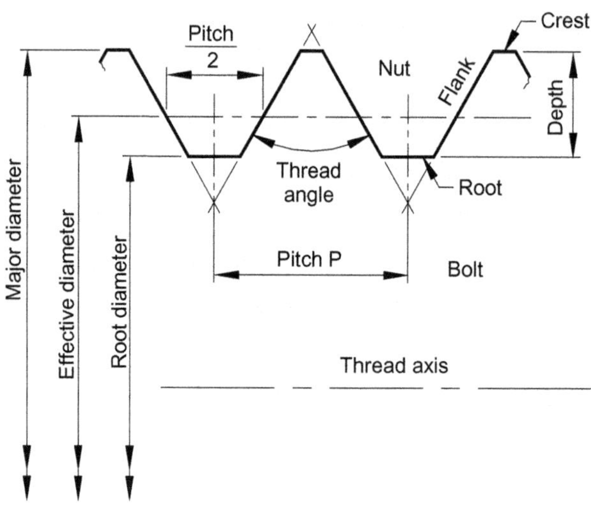

Fig. 16.10 Specialist terminology used for describing threads.

Table 16.2 Thread terminology

Term	Description
Pitch	The thread pitch is the distance between corresponding points on adjacent threads. Measurements must be made parallel to the thread axis
Outside diameter	The outside or major diameter is the diameter over the crests of the thread measured at right angles to the thread axis
Crest	The crest is the most prominent part of thread; either external or internal
Root	The root lies at the bottom of the groove between two adjacent threads
Flank	The flank of a thread is the straight side of the thread between the root and the crest
Root diameter	The root, minor or core diameter is the smallest diameter of the thread measured at right angles to the thread axis
Effective diameter	The effective diameter is the diameter on which the width of the spaces is equal to the width of the threads. It is measured at right angles to the thread axis
Lead	The lead of a thread is the axial movement of the screw in one revolution. Lead and pitch are the same for single start screws, but different for double or multiple start screws (Fig. 16.11)

thread series (UNC) and the Unified fine pitch thread series (UNF). Pertinent dimensions for selected UNC and UNF threads are given in Tables 16.4 and 16.5. Unified threads are specified by notation, in the case of a ½ bolt, in the form '½ in.-13UNC' or '½ in.20UNF' depending on whether a coarse or a fine thread is being used.

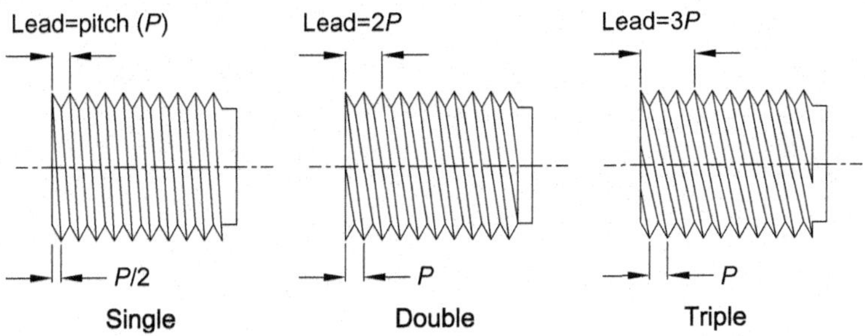

Fig. 16.11 Single, double and triple start screws.

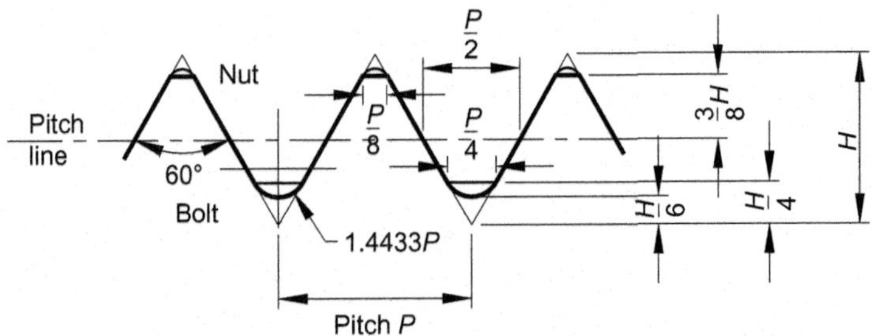

Fig. 16.12 ISO metric thread.

The range of threaded fasteners available is extensive including nuts and bolts, machine screws, set screws and sheet metal screws. A variety of machine screws is illustrated in Fig. 16.13.

With such an array of types of fasteners, the task of selection of the appropriate type for a given application can be time-consuming.

Washers can be used either under the bolt head or the nut or both to distribute the clamping load over a wide area and to provide a bearing surface for rotation of the nut. The most basic form of a washer is a simple disc with a hole through which the bolt or screw passes. There are, however, many addition types with particular attributes such as lock washers, which have projections that deform when compressed produced additional forces on the assembly decreasing the possibility that the fastener assembly will loosen in service (Izumi et al., 2005). Various forms of washer are illustrated in Fig. 16.14.

Selection of a particular fastener will depend upon many different criteria such as:

- strength at the operating temperatures concerned
- weight
- cost
- corrosion resistance
- magnetic properties
- life expectancy
- assembly considerations

Table 16.3 Selected dimensions for a selection of British Standard ISO Metric Precision Hexagon Bolts; BS 3692:1967 (all dimensions in mm)

Nominal size and thread diameter	Pitch of thread (coarse pitch series)	Width across flats		Height of head		Tapping drill	Clearance drill
		(Max)	(Min)	(Max)	(Min)		
M1.6	0.35	3.2	3.08	1.225	0.975	1.25	1.65
M2	0.4	4.0	3.88	1.525	1.275	1.60	2.05
M2.5	0.45	5.0	4.88	1.825	1.575	2.05	2.60
M3	0.5	5.5	5.38	2.125	1.875	2.50	3.10
M4	0.7	7.0	6.85	2.925	2.675	3.30	4.10
M5	0.8	8.0	7.85	3.650	3.35	4.20	5.10
M6	1	10.0	9.78	4.15	3.85	5.00	6.10
M8	1.25	13.0	12.73	5.65	5.35	6.80	8.20
M10	1.5	17.0	16.73	7.18	6.82	8.50	10.20
M12	1.75	19.0	18.67	8.18	7.82	10.20	12.20
M14	2	22.0	21.67	9.18	8.82	12.00	14.25
M16	2	24.0	23.67	10.18	9.82	14.00	16.25
M18	2.5	27.0	26.67	12.215	11.785	15.50	18.25
M20	2.5	30.0	29.67	13.215	12.785	17.50	20.25
M22	2.5	32.0	31.61	14.215	13.785	19.50	22.25
M24	3	36.0	35.38	15.215	14.785	21.00	24.25
M27	3	41.0	40.38	17.215	16.785	24.00	27.25
M30	3.5	46.0	45.38	19.26	18.74	26.50	30.50
M33	3.5	50.0	49.38	21.26	20.74	29.50	33.50
M36	4	55.0	54.26	23.26	22.74	32.00	36.50
M39	4	60.0	59.26	25.26	24.74	35.00	39.50
M42	4.5	65.0	64.26	26.26	25.74	37.50	42.50

M45	4.5		70.0	69.26	28.26	27.74	40.50	45.50
M48	5		75.0	74.26	30.26	29.74	43.00	48.75
M52	5		80.0	79.26	33.31	32.69	47.00	52.75
M56	5.5		85.0	84.13	35.31	34.69	50.50	56.75
M60	5.5		90.0	89.13	38.31	37.69	54.50	60.75
M64	6		95.0	94.13	40.31	39.69	58.00	64.75
M68	6		100.0	99.13	43.31	42.96	62.00	68.75

Table 16.4 American Standard thread dimensions for UNC screw threads

Size designation	Nominal major diameter (in.)	Threads per inch	Tensile stress area (in.^{2})
0	0.0600		
1	0.0730	64	0.00263
2	0.0860	56	0.00370
3	0.0990	48	0.0487
4	0.1120	40	0.00604
5	0.1250	40	0.00796
6	0.1380	32	0.00909
8	0.1640	32	0.0140
10	0.1900	24	0.0175
12	0.2160	24	0.0242
Fractional sizes			
¼	0.2500	20	0.0318
$^5/_{16}$	0.3125	18	0.0524
$^3/_8$	0.3750	16	0.0775
$^7/_{16}$	0.4375	14	0.1063
½	0.5000	13	0.1419
$^9/_{16}$	0.5625	12	0.182
$^5/_8$	0.6250	11	0.226
¾	0.7500	10	0.334
$^7/_8$	0.8750	9	0.462
1	1.000	8	0.606
$1^1/_8$	1.125	7	0.763
$1^1/_4$	1.250	7	0.969
$1^3/_8$	1.375	6	1.155
$1^1/_2$	1.500	6	1.405
$1^3/_4$	1.750	5	1.90
2	2.000	4.5	2.50

Table 16.5 American Standard thread dimensions for UNF screw threads

Size designation	Nominal major diameter (in.)	Threads per inch	Tensile stress area (in.^{2})
0	0.0600	80	0.00180
1	0.0730	72	0.00278
2	0.0860	64	0.00394
3	0.0990	56	0.00523
4	0.1120	48	0.00661
5	0.1250	44	0.00830
6	0.1380	40	0.01015
8	0.1640	36	0.01474
10	0.1900	32	0.0200
12	0.2160	28	0.0258

Table 16.5 Continued

Size designation	Nominal major diameter (in.)	Threads per inch	Tensile stress area (in.^{2})
Fractional sizes			
1/4	0.2500	28	0.0364
5/16	0.3125	24	0.0580
3/8	0.3750	24	0.0878
7/16	0.4375	20	0.1187
1/2	0.5000	20	0.1599
9/16	0.5625	18	0.203
5/8	0.6250	18	0.256
3/4	0.7500	16	0.373
7/8	0.8750	14	0.509
1	1.000	12	0.663
1 1/8	1.125	12	0.856
1 1/4	1.250	12	1.073
1 3/8	1.375	12	1.315
1 1/2	1.500	12	1.581

Bolted joints tend to consist of a series of components, such as flanges or plates that need to be restrained from moving relative to one another. To prevent or limit movement, a tension member, the bolt or screw, is tightened generating a tensile preload in the bolt. The preload causes an interface pressure between the components. Friction between the components as a result of the pressure then prevents, or limits, relative movement of the components. Although several or many components can be clamped in this way, generally the number is just two as shown in Fig. 16.15. Typically for this configuration the bolts are in tension and the components being clamped are in tension. The components concerned can be considered to be helical springs as indicated in Fig. 16.16.

For a bolted joint the tensile stresses formed in the minimum section of the bolt's shank need to be considered. The minimum section is normally taken as the core area of the thread, although the necking down area of the shank can be used in high stress or safety critical applications. A further issue is the inherent stress concentration caused by the thread itself. In most applications the nut is in compression, while the bolt is in tension. This has the effect of slightly shortening the nut and hence the pitch of its thread. Conversely the bolt stretches slightly and there is a consequential increase in its thread pitch. The combined effect of these small changes in pitch is that most of the bolt load is taken on the first few threads of the bolt, or perhaps just one thread, even if the nut is many bolt diameters long. This effect greatly increases the stress intensity. The maximum tensile stress in fasteners is therefore often high, and in cyclically loaded machines methods of avoiding fatigue of male threaded components (and occasionally of the female threaded component too) is often required. For any

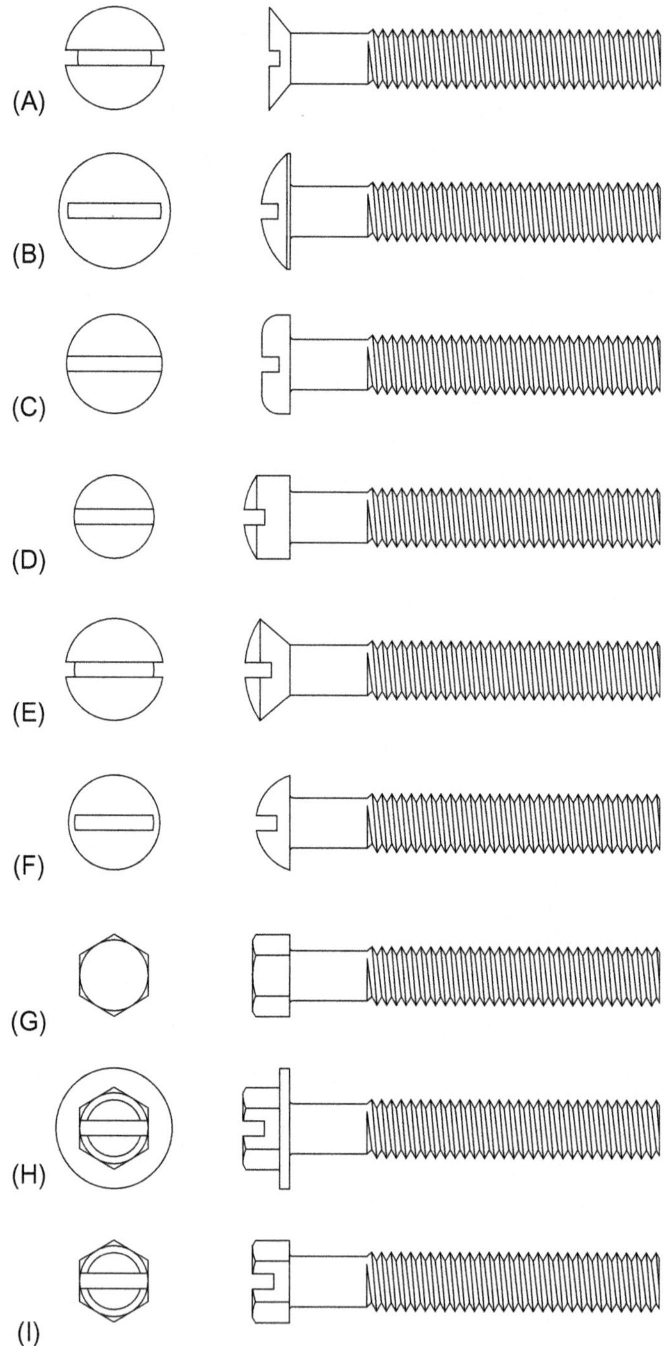

Fig. 16.13 Various machine screw styles: (A) flat countersunk head, (B) slotted truss head, (C) slotted pan head, (D) slotted fillister head, (E) slotted oval countersunk, (F) round head, (G) Hex, (H) Hex washer, (I) slotted hexagon head.

Fig. 16.14 Washers.

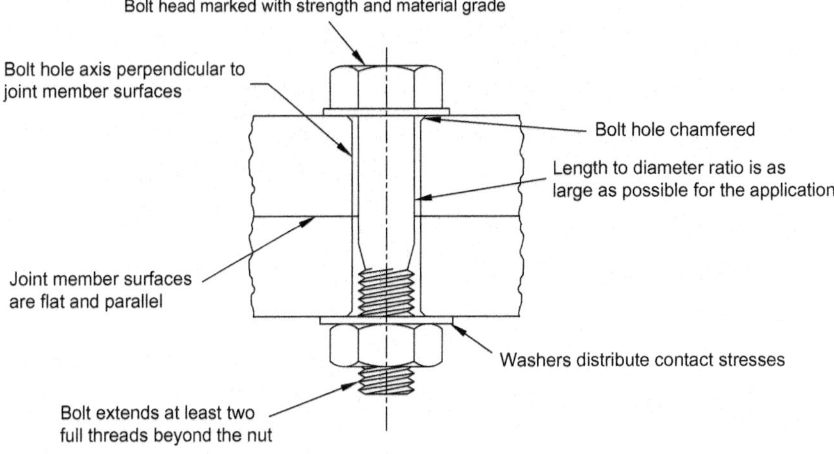

Fig. 16.15 A typical bolted joint features.

structure, once the stress concentration effect is established, the likelihood of fatigue occurring can be determined by both the stress amplitude and average stress. As a general rule, it is desirable to minimise the stress amplitude. In bolted joints this can be achieved by a number of techniques, for example, by maximising the flexibility of the

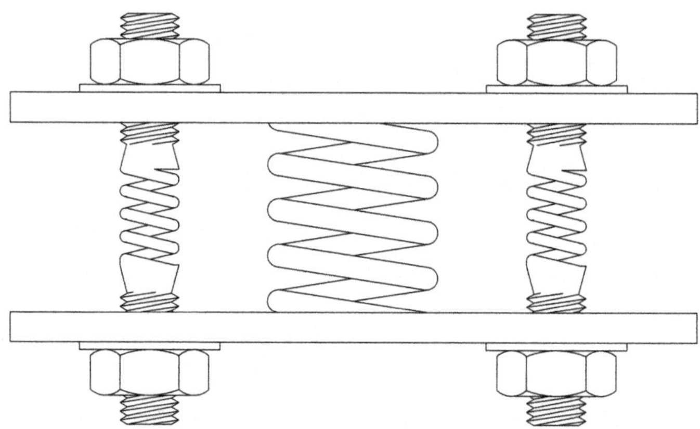

Fig. 16.16 Loading associated with a typical joint.

Fig. 16.17 Pressurised engine casing showing the use of multiple bolts.

bolt compared to the joint. In practice, this means using lots of, or several thin, long bolts, rather than a few short and fat bolts of equivalent strength. An example of this for a pipe flange is illustrated in Fig. 16.17.

The compressive stresses in a bolted joint caused by the preload of the bolt can be assumed to act over a restricted volume of material. This is usually taken as a conical form as shown in Fig. 16.18, and the surrounding material ignored in the analysis. Assuming the bolt and conical slugs of material are of similar Young's modulus, the bolt will be more flexible than the joint due to the bolt's smaller cross-sectional area.

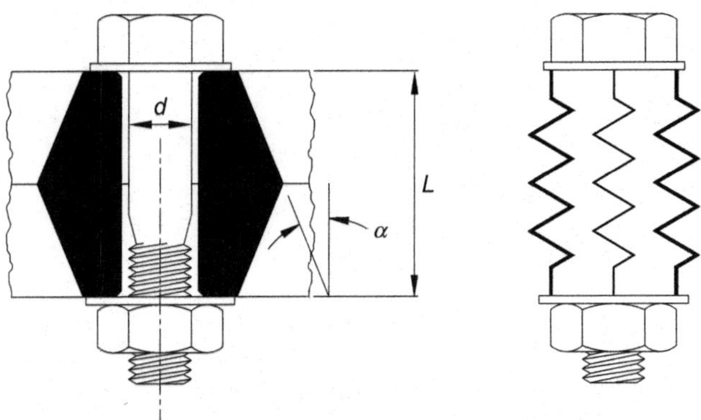

Fig. 16.18 Bolted joint modelling.

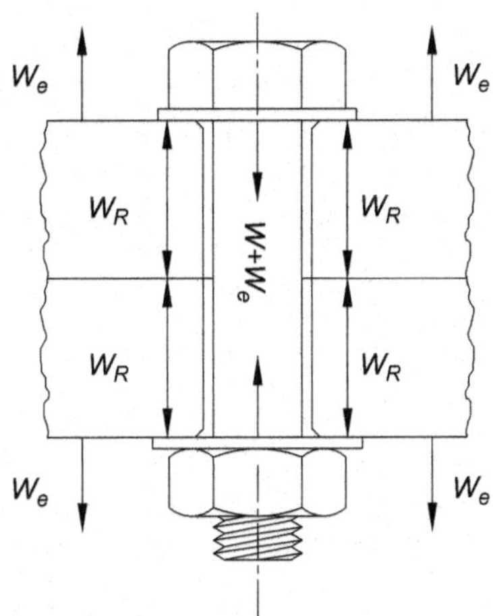

Fig. 16.19 Typical bolted joint forces.

Once tightened, any strain of the joint, due to cyclic loading, will be equal to the strain of the bolt; and due to the lower stiffness of the bolt, the stress amplitude in the bolt is lower than the applied load variation. To analyse this, the forces acting on the bolt and joint (Fig. 16.19) can be equated.

$$W_e = W_b + W_R \tag{16.1}$$

where

W_e = applied external load (N)
W_b = force on the bolt (N)
W_R = force on the clamped members (N).

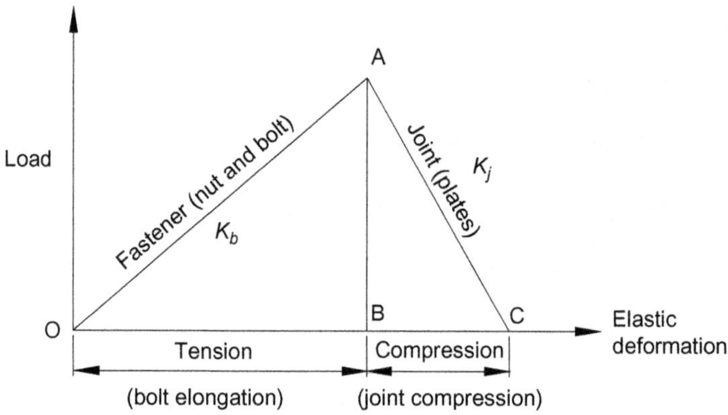

Fig. 16.20 Load and deflection in a bolt and joint respectively.

The load and deflection for the bolted joint shown in Fig. 16.19, when there is no external load, $W_e = 0$, is illustrated in Fig. 16.20. The loads are equal in magnitude and opposite in sign. The initial load is called the preload and this is an important quantity specified by the engineer or designer. The deflection of the bolt tends to be much greater than that of the joint. This is because the bolt is normally less stiff than the joint. The stiffness of the bolt and joint, given by the slope of the load/deflection lines, are stated in Eqs (16.2) and (16.3).

For a bolt, the stiffness is given by

$$k_b = \frac{\pi d^2 E}{4L} \tag{16.2}$$

where

k_b = axial spring rate for the bolt (N/m).
d = nominal bolt diameter (m).
E = Young's modulus for the bolt material (N/m^2).
L = effective load carrying length (m).

For flanges of equal thickness, using $\alpha = 30°$, the stiffness of the members is given by

$$k_j = \frac{0.5774\pi E d}{2\ \ln\left(5 \times \frac{0.5774L + 0.5d}{0.5774L + 2d}\right)} \tag{16.3}$$

where k_j is the spring rate for the members (N/m).

For the case of an external load, $W_e > 0$, applied to the joint, additional deflection will occur, altering the stresses and strains of both bolt and joint. If the applied load, W_e, is sufficiently high, separation of the two flanges will occur potentially leading to failure of the machine or degradation of machine function by, for example, in the case of a flanged pipe containing fluid, leakage. The resulting change in bolt tension for a

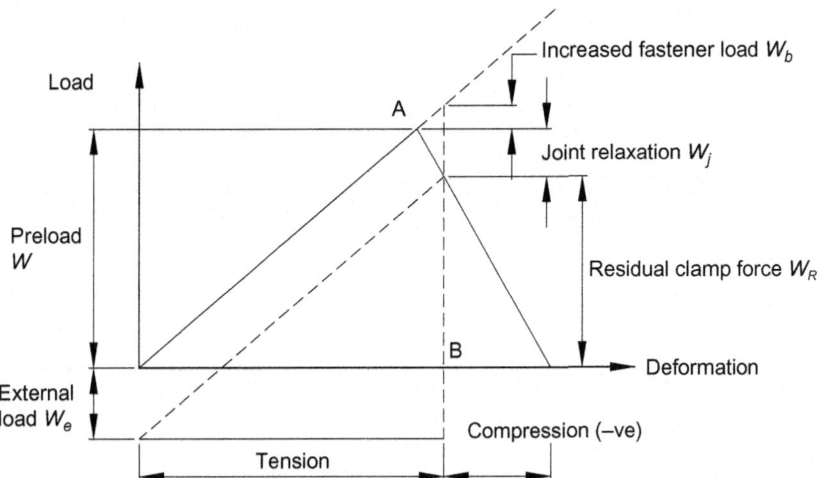

Fig. 16.21 Load and deflection in a bolt and joint respectively.

given change in W_e, is also of interest as this will enable calculation of the stress amplitude and hence characterisation of the fatigue of the bolt, if the external load varies cyclically.

The changes in loads and strains are illustrated in Fig. 16.21. The change in bolt load W_b, is much smaller than the applied load, W_e. It is also worth noting that to cause separation of the joint, a load higher than the preload must be applied. The so-called separation force, W_{sep}, is given by Eq. (16.4). Once the threshold value W_{sep} is reached, the bolt will see all of the applied load and the joint none. W_{sep} represents the highest load the bolt should be allowed to experience assuming that separation is not allowable.

$$W_{sep} = \left(1 + \frac{k_b}{k_j}\right) W_{preload} \tag{16.4}$$

where

W_{sep} = separation force (N).
$W_{preload}$ = preload force (N).

As previously indicated, threaded fasteners tend to be used such that they are predominantly loaded in tension. An example is the bolts shown in Fig. 16.22 used to fasten a flanged joint. As the fastener is tightened the tension on the bolt will increase. It might be envisaged that the strength of a threaded fastener would be limited by the area of its minor diameter. Testing, however, shows that the tensile strength is better defined using an area based on an average of the minor and pitch diameters.

$$A_t = \frac{\pi}{16}(d_p + d_r)^2 \tag{16.5}$$

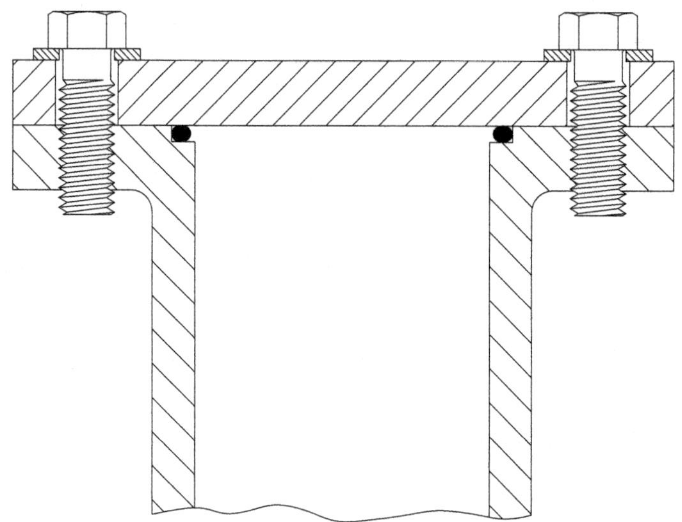

Fig. 16.22 Flanged joint.

For UNS threads,

$$d_p = d - \frac{0.649519}{N} \text{ and } d_r = d - \frac{1.299038}{N} \tag{16.6}$$

For ISO threads,

$$d_p = d - 0.649519p \tag{16.7}$$

and

$$d_r = d - 1.226869p \tag{16.8}$$

The stress in a threaded rod due to a tensile load is

$$\sigma_t = \frac{F}{A_t} \tag{16.9}$$

Theoretically one might think that when a nut engages a thread that all the threads in engagement would share the load. However, inaccuracies in thread spacing cause virtually all the load to be taken by the first pair of threads.

Bolts are normally tightened by applying torque to the head or nut, which causes the bolt to stretch. The stretching results in bolt tension, known as preload, which is the force that holds a joint together. Torque is relatively easy to measure using a torque meter during assembly so this is the most frequently used indicator of bolt tension. High preload tension helps to keep bolts tight, increases the strength of a joint, generates friction between parts to resist shear and improves the fatigue resistance of bolted connections. The recommended preload for reusable connections can be determined by

$$F_i = 0.75 A_t \sigma_p \tag{16.10}$$

and for permanent joints by

$$F_i = 0.9 A_t \sigma_p \tag{16.11}$$

where

A_t = tensile stress area of the bolt (m^2),
σ_p = proof strength of the bolt (N/m^2).

Material properties for steel bolts are given in SAE standard J1199 and by bolt manufacturers.

If detailed information concerning the proof strength is unavailable then it can be approximated by

$$\sigma_p = 0.85 \sigma_y \tag{16.12}$$

Once the preload has been determined the torque required to tighten the bolt can be estimated from

$$T = K F_i d \tag{16.13}$$

where

T = wrench torque (N m).
K = constant.
F_i = preload (N).
d = nominal bolt diameter (m).

The value of K depends on the bolt material and size. In the absence of data from manufacturers or detailed analysis, values for K are given in Table 16.6 for a variety of materials and bolt sizes.

Example 16.1. An M10 bolt has been selected for a reuseable application. The proof stress of the low carbon steel bolt material is 310 MPa. Determine the recommended preload on the bolt and the torque setting.

Solution

From Table 16.3 the pitch for a coarse series M10 bolt is 1.5 mm.
$d_p = 10 - 0.649519 \times 1.5 = 9.026$ mm.
$d_r = 10 - 1.226869 \times 1.5 = 8.160$ mm

Table 16.6 Values for the constant K for determining the torque required to tighten a bolt (Oberg et al. (1996), Machinery's Handbook 25)

Conditions	***K***
¼ in. to 1 in. mild steel bolts	0.2
Nonplated black finish steel bolts	0.3
Zinc plated steel bolts	0.2
Lubricated steel bolts	0.18
Cadmium plated steel bolts	0.16

$$A_t = \frac{\pi}{16}(9.026 + 8.16)^2 = 57.99\ \text{mm}^2$$

For a reusable connection, the recommended preload is
$F_i = 0.75A_t\sigma_p = 13.48\,\text{kN}$.
From Table 16.6, $K = 0.2$.
The torque required to tighten the bolt is given by,

$$T = KF_i d = 0.2 \times 13.48 \times 10^3 \times 0.01 = 26.96\ \text{N m}$$

The principal applications of threaded fasteners such as bolts and nuts is clamping components together. In such situations the bolt is predominantly in tension. Both the bolt and the clamped components will behave as elastic members, provided material limits are not exceeded. If a load is applied to a bolted joint that is above the clamping load, then the behaviour of the joint itself needs to be considered. As the bolt stretches, the compressive load on the joint will decrease, alleviating some of the load on the bolt. If a very stiff bolt is used to clamp a flexible member, such as a soft gasket, most of the additional force, above the clamping load, is taken by the bolt and the bolt should be designed to take the clamping force and any additional force. Such a joint can be classified as a soft joint. If, however, the bolt is relatively flexible compared to the joint, then the nearly all the externally applied load will initially go towards decreasing the clamping force until the components separate. The bolt will then carry all of the external load. This kind of joint is classified as a hard joint.

Practical joints normally fall between the two extremes of hard and soft joints. The clamped components of a typical hard joint have a stiffness of approximately three times that of the bolt. An externally applied load will be shared by the bolt and the clamped components according to the relative stiffnesses, which can be modelled by

$$F_b = F_i + \frac{k_b}{k_b + k_c}F_e \tag{16.14}$$

$$F_c = F_i - \frac{k_c}{k_b + k_c} F_e \tag{16.15}$$

where

F_b = final force in the bolt (N),
F_i = initial clamping load (N),
k_b = stiffness of the bolt (N/m),
k_c = stiffness of the clamped components (N/m),
F_e = externally applied load (N),
F_c = final force on the clamped components (N).

Example 16.2. A set of six M8 bolts are used to provide a clamping force of 20 kN between two components in a machine. If the joint is subjected to an additional load of 18 kN after the initial preload of 8.5 kN per bolt has been applied, determine the stress in the bolts. The stiffness of the clamped components can be assumed to be three times that of the bolt material. The proof stress of the low carbon steel bolt material is 310 MPa.

Solution

Taking $k_c = 3k_b$,

$$F_b = F_i + \frac{k_b}{k_b + k_c} F_e = F_i + \frac{k_b}{k_b + 3k_b} F_e = F_i + \frac{1}{4} F_e = 8500 + \frac{18000/6}{4} = 9250\,\text{N}$$

$$F_c = F_i - \frac{k_c}{k_b + k_c} F_e = F_i - \frac{3k_b}{k_b + 3k_b} F_e = F_i - \frac{3}{4} F_e = 8500 - \frac{3(18000/6)}{4}$$
$$= 6250\ \text{N}$$

As F_c is greater than zero, the joint remains tight. The tensile stress area for the M8 bolt can be determined from

$d_p = 8–0.649519 \times 1.25 = 7.188\,\text{mm}$.
$d_r = 8–1.226869 \times 1.25 = 6.466\,\text{mm}$

$$A_t = \frac{\pi}{16}(7.188 + 6.466)^2 = 36.61\,\text{mm}^2$$

The stress in each bolt is given by

$$\sigma = \frac{F_b}{A_t} = \frac{9250}{36.61 \times 10^{-6}} = 252.7\ \text{MPa}$$

This is 82% of the proof stress. The bolts are therefore safe.

A guideline for the analysis and selection of bolted joint is illustrated in Fig. 16.23.

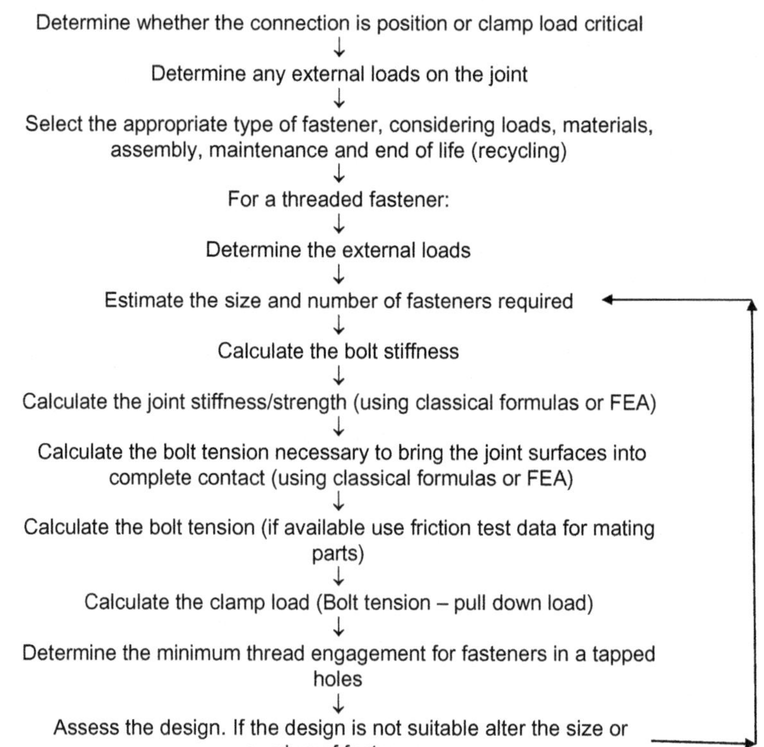

Determine whether the connection is position or clamp load critical
Determine any external loads on the joint
Select the appropriate type of fastener, considering loads, materials, assembly, maintenance and end of life (recycling)
For a threaded fastener:
Determine the external loads
Estimate the size and number of fasteners required
Calculate the bolt stiffness
Calculate the joint stiffness/strength (using classical formulas or FEA)
Calculate the bolt tension necessary to bring the joint surfaces into complete contact (using classical formulas or FEA)
Calculate the bolt tension (if available use friction test data for mating parts)
Calculate the clamp load (Bolt tension – pull down load)
Determine the minimum thread engagement for fasteners in a tapped holes
Assess the design. If the design is not suitable alter the size or number of fasteners.

Fig. 16.23 Guideline for the analysis and selection of bolted joint.

16.3 Power screws

Power screws, which are also known as lead screws, are used to convert rotary motion into linear movement. With suitably sized threads they are capable of large mechanical advantage and can lift or move large loads. Applications include screw jacks (Fig. 16.24) and traverses in production machines. Although suited to fasteners thread forms such as the ISO metric standard screw threads, UNC and UNF series described in Section 16.2 may not be strong enough for power screw applications. Instead square, Acme and buttress thread forms have been developed and standardised (Figs. 16.25–16.27) respectively. Some of the principal dimensions for standard Acme threads are given in Table 16.7.

Self-locking in power screws refers to the condition in which a screw cannot be turned by the application of axial force on the nut. This is very useful in that a power screw that is self-locking will hold its position and load unless a torque is applied. As an example most screw jacks for cars are self-locking and do not run down when the handle is let go. The opposite condition to self-locking is a screw that can be back-driven. This means that an axial force applied to the nut will cause the screw to turn.

Fig. 16.24 Screw jack.
Image courtesy Kelston Actuation Ltd.

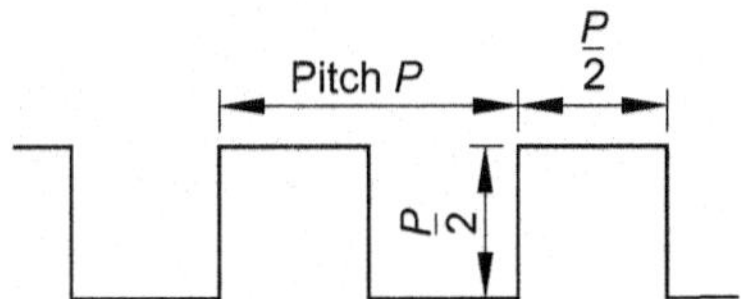

Fig. 16.25 Square thread.

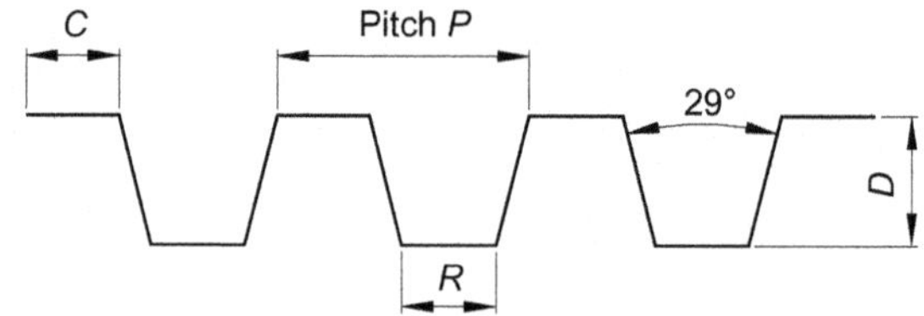

Fig. 16.26 Acme thread.

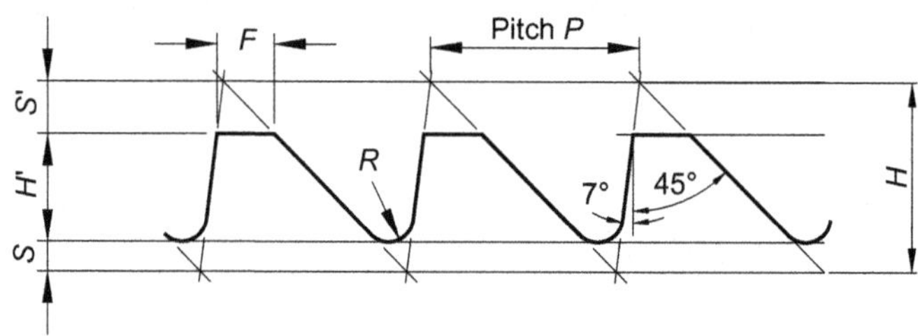

Fig. 16.27 Buttress thread.

Table 16.7 Principal dimensions for ACME threads

Major diameter (in.)	Threads per inch	Thread pitch (in.)	Pitch diameter (in.)	Minor diameter (in.)	Tensile stress area (in.^{2})
0.25	16	0.063	0.219	0.188	0.032
0.313	14	0.071	0.277	0.241	0.053
0.375	12	0.083	0.333	0.292	0.077
0.438	12	0.083	0.396	0.354	0.110
0.500	10	0.100	0.450	0.400	0.142
0.625	8	0.125	0.563	0.500	0.222
0.750	6	0.167	0.667	0.583	0.307
0.875	6	0.167	0.792	0.708	0.442
1.000	5	0.200	0.900	0.800	0.568
1.125	5	0.200	1.025	0.925	0.747
1.250	5	0.200	1.150	1.050	0.950
1.375	4	0.250	1.250	1.125	1.108
1.500	4	0.250	1.375	1.250	1.353
1.750	4	0.250	1.625	1.500	1.918
2.000	4	0.250	1.875	1.750	2.580
2.250	3	0.333	2.083	1.917	3.142
2.500	3	0.333	2.333	2.167	3.976
2.750	3	0.333	2.583	2.417	4.909
3.000	2	0.500	2.750	2.500	5.412
3.500	2	0.500	3.250	3.000	7.670
4.000	2	0.500	3.750	3.500	10.321
4.500	2	0.500	4.250	4.000	13.364
5.000	2	0.500	4.750	4.500	16.800

A product application of this is the Yankee screwdriver, which has a high-lead thread on its barrel that is attached to the screwdriver bit. The handle acts as the nut and when pushed the barrel will turn driving the bit round. Back-driveable power screws are a useful form of turning linear motion into rotary motion.

A screw thread can be considered to be an inclined plane wrapped around a cylinder to form a helix. The forces on a single thread of a power screw are illustrated for the case of lifting a load and lowering a load in Fig. 16.28.

The inclination of the plane is called the lead angle λ.

$$\tan\lambda = \frac{L}{\pi d_p} \tag{16.16}$$

In the case of lifting a load, summing the forces gives

$$\sum F_x = 0 = F - f\cos\lambda - N\sin\lambda = F - \mu N\cos\lambda - N\sin\lambda \tag{16.17}$$

So

$$F = N(\mu\cos\lambda + \sin\lambda) \tag{16.18}$$

$$\sum F_y = 0 = N\cos\lambda - f\sin\lambda - P = N\cos\lambda - \mu N\sin\lambda - P \tag{16.19}$$

So

$$N = \frac{P}{\cos\lambda - \mu\sin\lambda} \tag{16.20}$$

where μ is the coefficient of friction between the screw and the nut. The coefficient of friction can typically be taken as 0.15.

Solving Eqs (16.18) and (16.20) to give an expression for F gives

$$F = \frac{P(\mu\cos\lambda + \sin\lambda)}{\cos\lambda - \mu\sin\lambda} \tag{16.21}$$

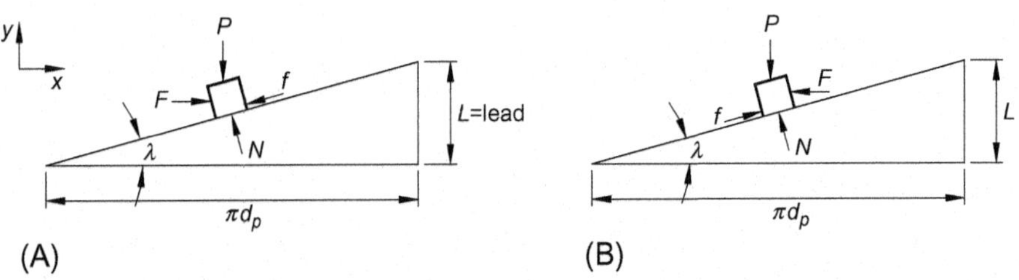

Fig. 16.28 Force analysis at the interface of a lead screw and nut: (A) lifting a load, (B) lowering a load.

The screw torque required to lift the load is given by

$$T_s = F\frac{d_p}{2} = \frac{Pd_p(\mu\cos\lambda + \sin\lambda)}{2(\cos\lambda - \mu\sin\lambda)} \tag{16.22}$$

It is convenient to rewrite this equation in terms of the lead L, so substituting with Eq. (16.12) and rearranging gives:

$$T_s = \frac{Pd_p(\mu\pi d_p + L)}{2(\pi d_p - \mu L)} \tag{16.23}$$

The thrust collar also contributes a friction torque that must be accounted for. The torque required to turn the thrust collar is given by

$$T_c = \mu_c P\frac{d_c}{2} \tag{16.24}$$

where

d_c = diameter of the thrust collar (m),
μ_c = coefficient of friction in the thrust bearing.

The total torque to lift the load for a square thread is

$$T_u = \frac{Pd_p(\mu\pi d_p + L)}{2(\pi d_p - \mu L)} + \mu_c P\frac{d_c}{2} \tag{16.25}$$

Similar analysis can be performed for lowering a load, in which case:

$$T_d = \frac{Pd_p(\mu\pi d_p - L)}{2(\pi d_p + \mu L)} + \mu_c P\frac{d_c}{2} \tag{16.26}$$

For Acme threads the equivalent torque relationship for lifting a load is:

$$T_u = \frac{Pd_p(\mu\pi d_p + L\cos\alpha)}{2(\pi d_p\cos\alpha - \mu L)} + \mu_c P\frac{d_c}{2} \tag{16.27}$$

and for lowering:

$$T_d = \frac{Pd_p(\mu\pi d_p - L\cos\alpha)}{2(\pi d_p\cos\alpha + \mu L)} + \mu_c P\frac{d_c}{2} \tag{16.28}$$

The work done on a power screw is the product of the torque and angular displacement. For one revolution of a screw,

$$W_{in} = 2\pi T \tag{16.29}$$

The work delivered for one revolution is

$$W_{out} = PL \tag{16.30}$$

The efficiency of the system defined as work out/work in is

$$\eta = \frac{PL}{2\pi T} \tag{16.31}$$

Substituting for the torque for an Acme thread using Eq. (16.23), neglecting collar friction, gives

$$\eta = \frac{L(\pi d_p \cos\alpha - \mu L)}{\pi d_p (\pi \mu d_p + L\cos\alpha)} \tag{16.32}$$

Simplifying using Eq. (16.16) gives

$$\eta = \frac{\cos\alpha - \mu\tan\lambda}{\cos\alpha + \mu\cot\lambda} \tag{16.33}$$

For a square thread, $\alpha = 0$, so $\cos\alpha = 1$, giving

$$\eta = \frac{1 - \mu\tan\lambda}{1 + \mu\cot\lambda} \tag{16.34}$$

Standard Acme screws have lead angles between about 2° and 5°. Assuming a coefficient of friction of 0.15 gives the efficiency between 18% and 36%. This shows the disadvantage of power screws. Higher efficiencies can, however, be attained by reducing the friction and one of the ways of doing this has been found to be the use of ball screws (Fig. 16.29) although these potentially add to the cost of production.

A screw will self-lock if

$$\mu \geq \frac{L}{\pi d_p}\cos\alpha \tag{16.35}$$

or

$$\mu \geq \tan\lambda\cos\alpha \tag{16.36}$$

For a square thread, where $\cos\alpha = 1$,

$$\mu \geq \frac{L}{\pi d_p} \tag{16.37}$$

or

Fig. 16.29 Ball screw.

$$\mu \geq \tan\lambda \tag{16.38}$$

It should be noted that Eqs (16.35)–(16.38) are for a statically loaded screw. Dynamic loading such as vibration can reduce the effective friction and cause a screw to back-drive.

In the design of a power screw consideration should also be given to buckling of the screw and choice of material for the screw and nut.

Example 16.3. A self-locking power screw is required for a screw jack. An initial proposal is to use a single start 1.25–5 Acme power screw. The axial load is 4000 N and collar mean diameter is 1.75 in. Determine the lifting and lowering torques, the efficiency of the power screw and whether the design proposal is self-locking.

Solution

Single start thread, so the lead, L, is equal to the pitch, P.

$N = 5$ teeth per inch

$$P = \frac{1}{N} = \frac{1}{5} = 0.2''$$

$L = 0.2'' = 5.08\,\text{mm}$.

$d_p = 1.15'' = 29.21\,\text{mm}$.

$d_c = 1.75'' = 44.45\,\text{mm}$.

Assume sliding friction, $\mu = 0.15$.

The torque to lift the load is given by Eq. (16.27).

$$T = \frac{4000 \times 0.02921}{2}\left(\frac{0.15\pi 0.02921 + 0.00508\cos 14.5}{\pi 0.02921\cos 14.5 - 0.15 \times 0.00508}\right) + 0.15 \times 4000\frac{0.04445}{2}$$

$$= 58.42\frac{0.01376 + 4.9182 \times 10^{-3}}{0.08884 - 7.62 \times 10^{-4}} + 13.34 = 12.39 + 13.34 = 25.73 \text{ Nm}$$

The torque to lower the load is given by Eq. (16.28).

$$T = 58.42\frac{0.01376 - 4.9182 \times 10^{-3}}{0.08884 + 7.62 \times 10^{-4}} + 13.34 = 5.765 + 13.34 = 19.11 \text{ N m}$$

$$\eta = \frac{PL}{2\pi T}$$

$$\eta_{screw} = \frac{4000 \times 0.00508}{2\pi 12.39} = 0.261$$

$$\eta_{both} = \frac{4000 \times 0.00508}{2\pi 25.73} = 0.126$$

The design will be self-locking if

$$\mu \geq \frac{L}{\pi d_p}\cos\alpha = \frac{0.00508}{\pi 0.02921}\cos 14.5 = 0.05359$$

$\mu = 0.15$ so design is self-locking.

16.4 Rivets

Rivets are nonthreaded fasteners that are usually manufactured from steel or aluminium. They consist of a preformed head and shank, which is inserted into the material to be joined and the second head that enables the rivet to function as a fastener is formed on the free end by a variety of means known as setting. A conventional rivet before and after setting is illustrated in Fig. 16.30.

Rivets are widely used to join components in aircraft (e.g. see Fig. 16.31) boilers, ships and boxes and other enclosures. Rivets tend to be much cheaper to install than bolts and the process can be readily automated with single riveting machines capable of installing thousands of rivets an hour.

Rivets can be made from any ductile material such as carbon steel, aluminium and brass. A variety of coatings are available to improve corrosion resistance. Care needs to be taken in the selection of material and coating to avoid the possibility of corrosion by galvanic action. In general a given size rivet will be not as strong as the equivalent threaded fastener.

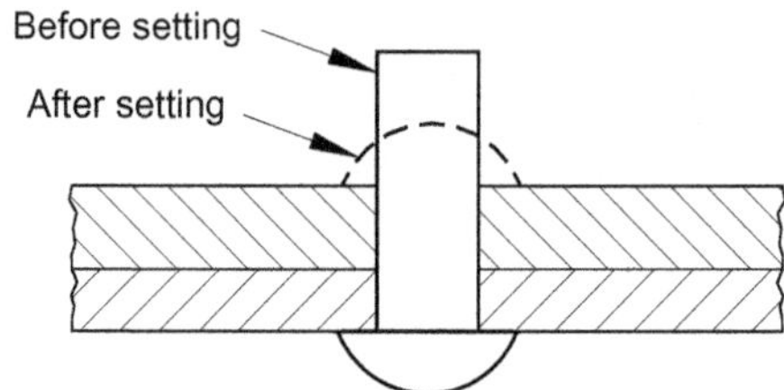

Fig. 16.30 Conventional rivet before and after setting.

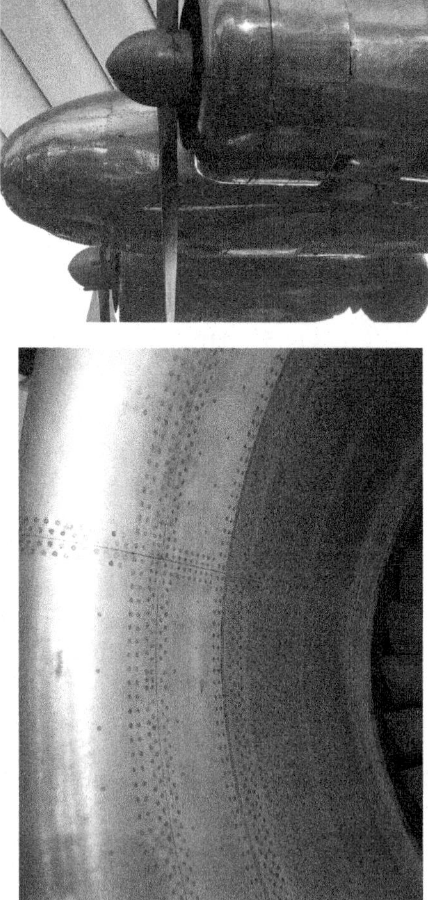

Fig. 16.31 Two historical examples of the use of rivets on the Lockheed Electra and RB211 engine nacelle.

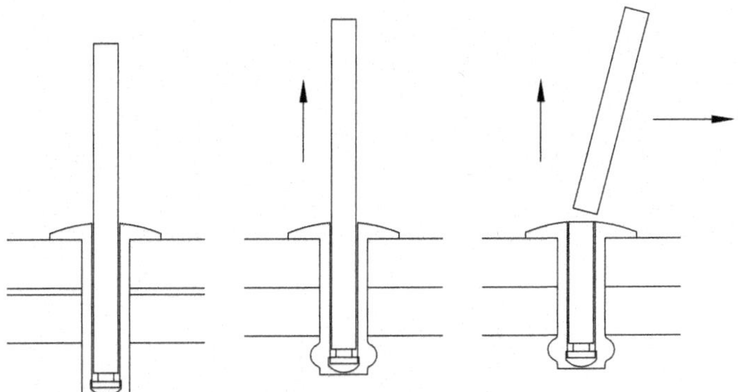

Fig. 16.32 An example of the application of a closed end blind rivet.

The two main types of rivet are tubular and blind and each type are available in a multitude of varieties. The advantage of blind rivets (Fig. 16.32) is that they require access to only one side of the joint. A further type of rivet with potentially many overall advantages, from the production perspective, is the self-piercing rivet that does not require a predrilled hole. The rivet is driven into the target materials with high force, piercing the top sheets and spreading outwards into the bottom sheet of material under the influence of an upsetting die to form the joint.

Factors in the design and specification of rivets include the size, type and material for the rivet, the type of joint, and the spacing between rivets. There are two main types of riveted joint: lap-joints and butt-joints(Fig. 16.33). In lap joints the components to be joined overlap each other, while for butt joints an additional piece of material is used to bridge the two components to be joined which are butted up against each other. Rivets can fail by shearing through one cross-section known as single shear, shearing through two cross-sections known as double shear, and crushing. Riveted plates can fail by shearing, tearing and crushing.

For many applications the correct use of rivets is safety critical and their use is governed by construction codes. For information and data concerning joints for pressure vessels reference to the appropriate standards should be made, such as the ASME Boiler Code.

Riveted joints can be designed using a simple procedure (Oberg et al., 1996; Machinery's Handbook) assuming that:

the load is carried equally by the rivets
no combined stresses act on a rivet to cause failure
the shearing stress in a rivet is uniform across the cross-section
the load that would cause failure in single shear would have to be double to cause failure in double shear
the bearing stress of the rivet and plate is distribute equally over the projected area of the rivet
the tensile stress is uniform in the section of metal between the rivets.

The allowable stress for a rivet is generally defined in the relevant standard. For example the ASME Boiler code lists an ultimate tensile stress for rivets of 379 MPa,

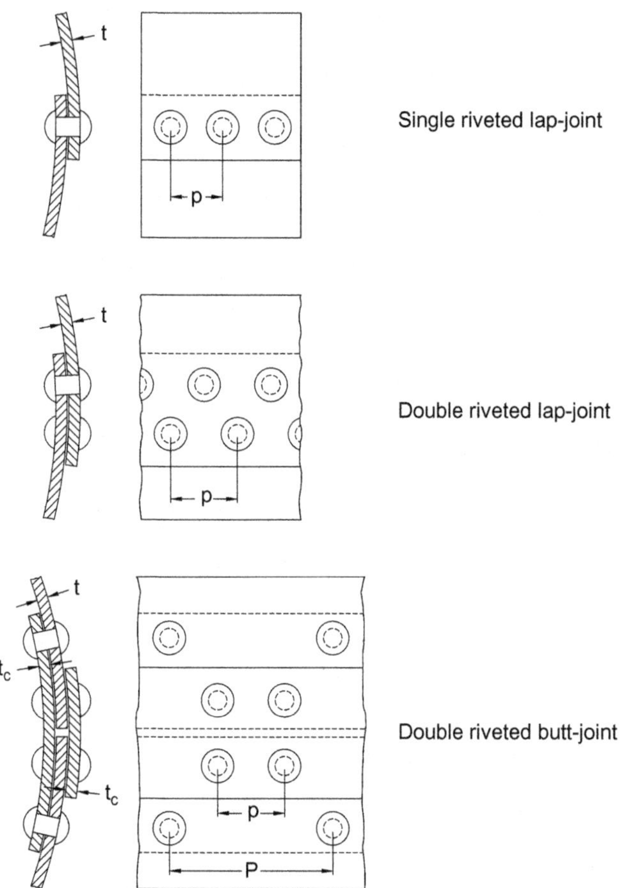

Fig. 16.33 Some types of riveted joints.

ultimate shearing stress of 303 MPa and an ultimate compressive or bearing stress of 655 MPa. Design stresses are usually 20% of these values, that is for tensile, shear and bearing stresses the design limits are 75 MPa, 60 MPa and 131 MPa respectively.

For a single lap joint, the safe tensile load based on shear is given by

$$L = nA_r\tau_d \tag{16.39}$$

For a single lap joint, the safe tensile load based on compressive or bearing stress is given by

$$L = nA_b\sigma_c \tag{16.40}$$

$$A_b = \text{td} \tag{16.41}$$

For a single lap joint, the safe tensile load based on tensile stress is given by

$$L = A_p \sigma_t \tag{16.42}$$

where

L = load (N),
A_r = cross sectional area of rivet (m^2),
A_b = projected bearing area of rivet (m^2),
A_p = cross sectional area of plate between rivet holes (m^2),
t = thickness of plate (m),
d = diameter of rivet (m),
τ_d = allowable shear stress (N/m^2),
σ_c = allowable bearing or compressive stress (N/m^2),
σ_t = allowable tensile stress (N/m^2).

The efficiency of a riveted joint is given by

$$\eta = \frac{\text{Least safe load}}{\text{Ultimate tensile strength of unperforated section}} \tag{16.43}$$

A selection of rivets specified in BS 4620:1970 and ANSI B18.1.2-1972 are included in Tables 16.8–16.11 as examples. Rivets however are available as stock items from specialist manufacturers and suppliers in a much wider variety than the small selection presented in Tables 16.8–16.11. For any given application the relevant standard should be referenced and a range of manufacturer's products considered.

Example 16.4. Determine the safe tensile, shear and bearing loads and the efficiency for a 300 mm section of single-riveted lap joint made from ¼″ plates using six 16 mm diameter rivets. Assume that the drilled holes are 1.5 mm larger in diameter than the rivets. The values for the design limits for tensile, shear and bearing stress can be taken as 75 MPa, 60 MPa and 131 MPa respectively.

Solution

The safe tensile load, L, based of shear of the rivets is given by

$$L = nA_r\tau_d = 6\pi\frac{0.016^2}{4} \times 75 \times 10^6 = 90.48 \text{ kN}$$

The safe tensile load based on bearing or compressive stress is given by

$$L = 6 \times 0.016 \times 6.35 \times 10^{-3} \times 131 \times 10^6 = 79.86 \text{ kN}$$

The safe load based on tensile load is given by $L = A_p\sigma_t$.

The area of the plate between the rivet holes, A_p is given by

$$A_p = 0.00635\left(0.3 - 6\left(16 \times 10^{-3} + 1.5 \times 10^{-3}\right)\right) = 1.238 \times 10^{-3} \text{ m}^2$$

Table 16.8 British Standard hot forged rivets for general engineering purposes (extracted from BS 4620:1970; note see the standard for full ranges)

Hot forged rivets

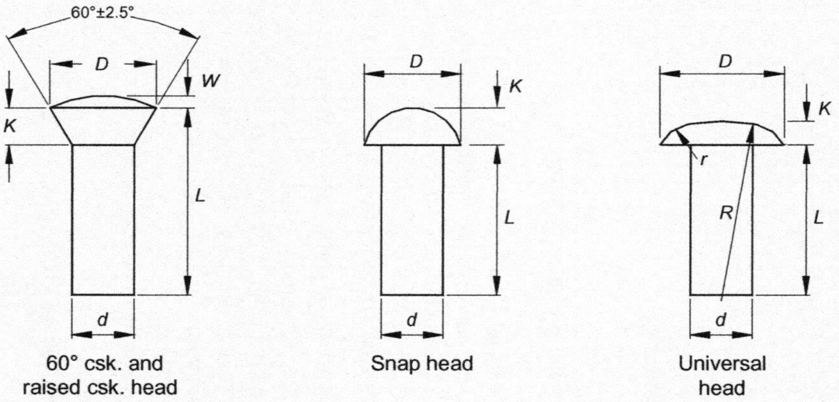

		60° Csk. and raised Csk. head		Snap head		Universal head			
		Head dimensions							
Nom. shank diam. *d* (mm)	Tol. on diam. *d* (mm)	Nom. diam. *D* (mm)	Height of raise *W* (mm)	Nom. diam. *D* (mm)	Nom. depth *K* (mm)	Nom. diam. *D* (mm)	Nom. depth *K* (mm)	Rad. *R* (mm)	Rad. *r* (mm)
(14)	±0.43	21	2.8	22	9	28	5.6	42	8.4
16		24	3.2	25	10	32	6.4	48	9.6
(18)		27	3.6	28	11.5	36	7.2	54	11

20	±0.52	30	4.0	32	13	40	8.0	60	12
(22)		33	4.4	36	14	44	8.8	66	13
24		36	4.8	40	16	48	9.6	72	14
(27)	±0.62	40	5.4	43	17	54	10.8	81	16
30		45	6.0	48	19	60	12.0	90	18
(33)		50	6.6	53	21	66	13.2	99	20
36		55	7.2	58	23	72	14.4	108	22
(39)		59	7.8	62	25	78	15.6	117	23

All dimensions are in millimetres. Sizes shown in parentheses are nonpreferred.

Table 16.9 British Standard cold forged rivets for general engineering purposes (extracted from BS 4620:1970, note see the standard for full ranges)

Nom. shank diam. *d* (mm)	Tol. on diam. *d* (mm)	90° Csk. Head	Snap head		Universal head				Flat head	
		Head dimensions								
		Nom. diam. *D* (mm)	Nom. diam. *D* (mm)	Nom. depth *K* (mm)	Nom. diam. *D* (mm)	Nom. depth *K* (mm)	Rad. *R* (mm)	Rad. *r* (mm)	Nom. diam. *D* (mm)	Nom. depth *K* (mm)
1	±0.07	2	1.8	0.6	2	0.4	3.0	0.6	2	0.25
1.2		2.4	2.1	0.7	2.4	0.5	3.6	0.7	2.4	0.3
1.6		3.2	2.8	1.0	3.2	0.6	4.8	1.0	3.2	0.4
2		4	3.5	1.2	4	0.8	6.0	1.2	4	0.5
2.5		5	4.4	1.5	5	1.0	7.5	1.5	5	0.6
3		6	5.3	1.8	6	1.2	9.0	1.8	6	0.8
(3.5)	±0.09	7	6.1	2.1	7	1.4	10.5	2.1	7	0.9
4		8	7	2.4	8	1.6	12	2.4	8	1.0
5		10	8.8	3.0	10	2.0	15	3.0	10	1.3
6		12	10.5	3.6	12	2.4	18	3.6	12	1.5

(7)	±0.11	14	12.3	4.2	14	2.8	21	4.2	14	1.8
8		16	14	4.8	16	3.2	24	4.8	16	2
10		20	18	6.0	20	4.0	30	6	20	2.5
12	±0.14	24	21	7.2	24	4.8	36	7.2	–	–
(14)		–	25	8.4	28	5.6	42	8.4	–	–
16		–	28	9.6	32	6.4	48	9.6	–	–

All dimensions are in millimetres. Sizes shown in parentheses are nonpreferred.

Table 16.10 Selected American National Standard large button, high button, cone and pan head rivets (extracted from ANSI B18.1.2-1972, R1989; note see the standard for full ranges)

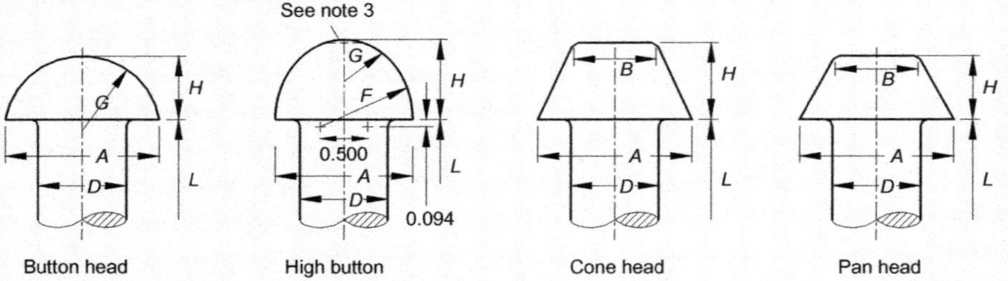

Nom. body diam. D^a	Head diam. A (in.)		Height H (in.)		Head diam. A (in.)		Height H (in.)	
	Mfd. note 1	Driven note 2	Mfd. note 1	Driven note 2	Mfd. note 1	Driven note 2	Mfd. note 1	Driven note 2
	Button head				**High button head (Acorn)**			
$^1/_2$	0.875	0.922	0.375	0.344	0.781	0.875	0.500	0.375
$^5/_8$	1.094	1.141	0.469	0.438	0.969	1.062	0.594	0.453
$^3/_4$	1.312	1.375	0.562	0.516	1.156	1.250	0.688	0.531
$^7/_8$	1.531	1.594	0.656	0.609	1.344	1.438	0.781	0.609
1	1.750	1.828	0.750	0.688	1.531	1.625	0.875	0.688
$1^1/_8$	1.969	2.062	0.844	0.781	1.719	1.812	0.969	0.766
$1^1/_4$	2.188	2.281	0.938	0.859	1.906	2.000	1.062	0.844
$1^3/_8$	2.406	2.516	1.031	0.953	2.094	2.188	1.156	0.938
$1^1/_2$	2.625	2.734	1.125	1.031	2.281	2.375	1.250	1.000
$1^5/_8$	2.844	2.969	1.219	1.125	2.469	2.562	1.344	1.094
$1^3/_4$	3.062	3.203	1.312	1.203	2.656	2.750	1.438	1.172

	Cone head				Pan head			
$^{1}/_{2}$	0.875	0.922	0.438	0.406	0.800	0.844	0.350	0.328
$^{5}/_{8}$	1.094	1.141	0.547	0.516	1.000	1.047	0.438	0.406
$^{3}/_{4}$	1.312	1.375	0.656	0.625	1.200	1.266	0.525	0.484
$^{7}/_{8}$	1.531	1.594	0.766	0.719	1.400	1.469	0.612	0.578
1	1.750	1.828	0.875	0.828	1.600	1.687	0.700	0.656
$1^{1}/_{8}$	1.969	2.063	0.984	0.938	1.800	1.891	0.788	0.734
$1^{1}/_{4}$	2.188	2.281	1.094	1.031	2.000	2.094	0.875	0.812
$1^{3}/_{8}$	2.406	2.516	1.203	1.141	2.200	2.312	0.962	0.906
$1^{1}/_{2}$	2.625	2.734	1.312	1.250	2.400	2.516	1.050	0.984
$1^{5}/_{8}$	2.844	2.969	1.422	1.344	2.600	2.734	1.138	1.062
$1^{3}/_{4}$	3.062	3.203	1.531	1.453	2.800	2.938	1.225	1.141

All dimensions are given in inches.
Note 1: Basic dimensions of head as manufactured.
Note 2: Dimensions of manufactured head after driving and also of driven head.
Note 3: Slight flat permissible within the specified head-height tolerance.
The following formulae give the basic dimensions for manufactured shapes: *Button Head*, $A = 1.750D$; $H = 0.750D$; $G = 0.885D$. *High Button Head*, $A = 1.500D + 0.031$; $H = 0.750D + 0.125$; $F = 0.750D + 0.281$; $G = 0.750D - 0.281$.
Cone Head, $A = 1.750D$; $B = 0.938D$; $H = 0.875D$. *Pan Head*, $A = 1.600D$; $B = 1.000D$; $H = 0.700D$. Length L is measured parallel to the rivet axis, from the extreme end to the bearing surface plane for flat bearing surface head type rivets, or to the intersection of the head top surface with the head diameter for countersunk head type rivets.
[a]Tolerance for diameter of body is plus and minus from nominal and for $^{1}/_{2}$-in. size equals +0.020, −0.022; for sizes $^{5}/_{8}$ to 1-in., incl., equals +0.030, −0.025; for sizes $1^{1}/_{8}$ and $1^{1}/_{4}$-in. equals +0.035, −0.027; for sizes $1^{3}/_{8}$ and $1^{1}/_{2}$-in. equals +0.040, −0.030; for sizes $1^{5}/_{8}$ and $1^{3}/_{4}$-in. equals +0.040, −0.037

Table 16.11 Selected American National Standard large flat and oval countersunk rivets (extracted from ANSI B18.1.2-1972, R1989; note see the standard for full ranges)

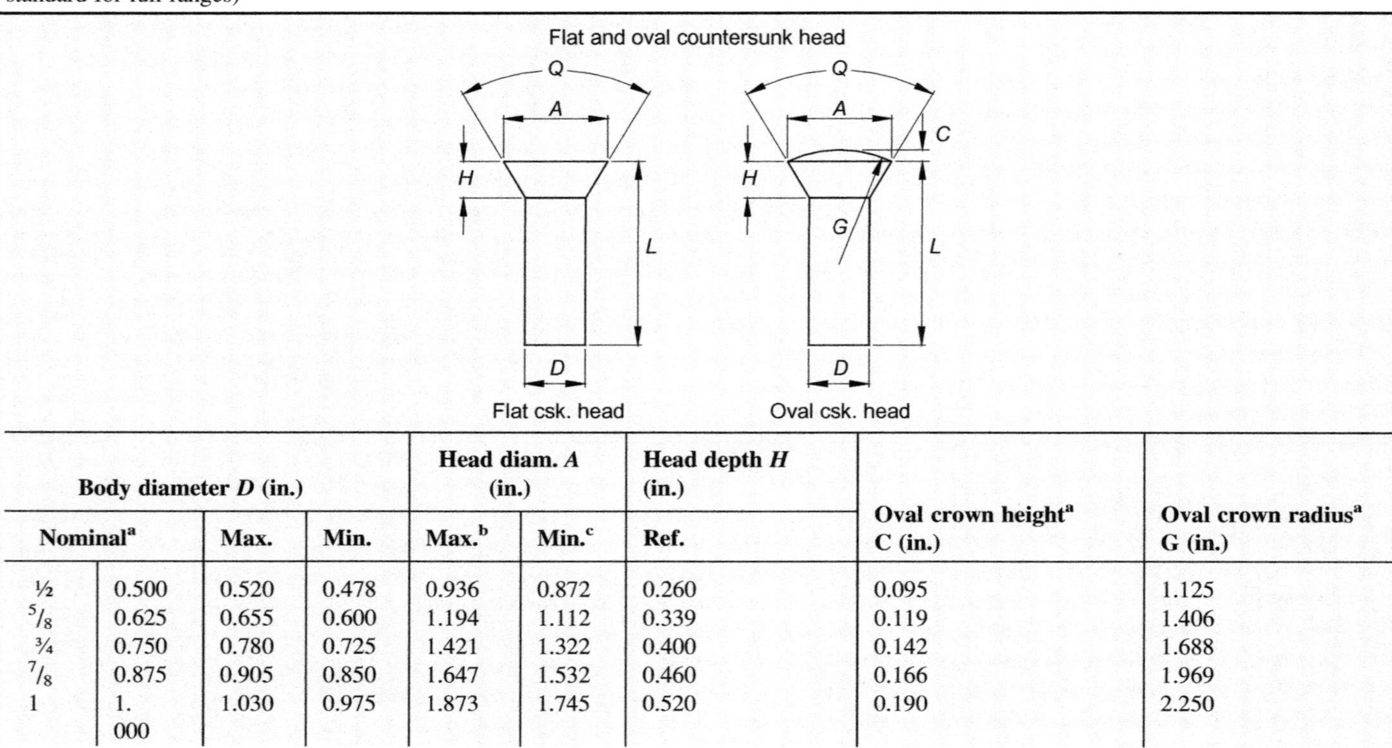

Body diameter D (in.)				Head diam. A (in.)		Head depth H (in.)	Oval crown height[a] C (in.)	Oval crown radius[a] G (in.)
Nominal[a]		Max.	Min.	Max.[b]	Min.[c]	Ref.		
½	0.500	0.520	0.478	0.936	0.872	0.260	0.095	1.125
5/8	0.625	0.655	0.600	1.194	1.112	0.339	0.119	1.406
¾	0.750	0.780	0.725	1.421	1.322	0.400	0.142	1.688
7/8	0.875	0.905	0.850	1.647	1.532	0.460	0.166	1.969
1	1. 000	1.030	0.975	1.873	1.745	0.520	0.190	2.250

$1^1/_8$	1.125	1.160	1.098	2.114	1.973	0.589	0.214	2.531
$1^1/_4$	1.250	1.285	1.223	2.340	2.199	0.650	0.238	2.812
$1^3/_8$	1.375	1.415	1.345	2.567	2.426	0.710	0.261	3.094
$1^1/_2$	1.500	1.540	1.470	2.793	2.652	0.771	0.285	3.375
$1^5/_8$	1.625	1.665	1.588	3.019	2.878	0.831	0.309	3.656
$1^3/_4$	1.750	1.790	1.713	3.262	3.121	0.901	0.332	3.938

All dimensions are given in inches.
[a]Basic dimension as manufactured. For tolerances see table footnote on Table 16.10.
[b]Sharp edged head.
[c]Rounded or flat edged irregularly shaped head (heads are not machined or trimmed).

$$L = A_p \sigma_t = 1.238 \times 10^{-3} \times 131 \times 10^6 = 162.2 \text{ kN}$$

The safe tensile load would be the least of the three values determined, that is $L = 79.86\,\text{kN}$.

$$\eta = \frac{79.86 \times 10^3}{0.00635 \times 0.3 \times 75 \times 10^6} = 0.56$$

The efficiency is 56%.

Example 16.5. Determine the maximum safe tensile load that can be supported by a 1 m section of double riveted butt joint with 15 mm thick main plates and two 8 mm thick cover plates. There are six rivets in each of the outer rows and seven rivets in each of the inner rows. The rivets are all 20 mm in diameter. Assume that the drilled holes are 1.5 mm larger in diameter than the rivets. The values for the design limits for tensile, shear and bearing stress can be taken as 75 MPa, 60 MPa and 131 MPa, respectively.

Solution

In analysing a double riveted joint, it is only necessary to analyse one side due to symmetry.

The safe tensile load based on double shearing of the rivets is equal to the number of rivets times the number of shearing planes per rivet times the cross-sectional area of the rivet times the allowable shearing stress.

$$L = n \times 2 \times A_r \tau_d$$

$$n = 6 + 7 = 13$$

$$L = 13 \times 2 \times \pi \frac{0.02^2}{4} \times 60 \times 10^6 = 490.1 \text{ kN}$$

The safe tensile load based on bearing stress is given by $L = nA_b\sigma_c$

$$L = 13 \times 0.02 \times 0.015 \times 131 \times 10^6 = 510.9 \text{ kN}$$

The cover plates have a combined thickness greater than that of the main plates so detailed analysis need not be considered here.

The safe tensile load based on tensile stress, $L = A_p\sigma_t$

$$A_p = 0.015(1 - 6(0.02 + 0.0015)) = 0.01307 \text{ m}^2$$

$$L = 0.01307 \times 75 \times 10^6 = 980.3 \text{ kN}$$

To complete the analysis it is necessary to consider the sum of the load that would cause tearing between rivets in the inner section plus the load carried by the rivets in the outer section. The sum of the loads is used because if the joint is to fail it must fail at both sections simultaneously.

$$L = nA_b\sigma_c = 6 \times 0.02 \times 0.015 \times 131 \times 10^6 = 235.8 \text{ kN}$$

The safe tensile load based on the tensile strength of the main plate between the holes in the inner section is

$$L = A_p\sigma_t = 0.015(1 - 7(0.02 + 0.0015)) \times 75 \times 10^6 = 955.7 \text{ kN}$$

The total safe tensile load based on the sum of the above two values is 1.191 MN.

The safe tensile load for the riveted joints would be the least of the values calculated, that is 490.1 kN.

The efficiency of the riveted joint is given by

$$\eta = \frac{490.1 \times 10^3}{0.015 \times 1 \times 75 \times 10^6} = 0.436 = 43.6\%$$

16.5 Adhesives

Adhesive bonding involves the joining of similar or dissimilar materials using natural or synthetic substances, which form a rigid or semirigid interface without the use of mechanical fastening. Adhesives can be used to bond metals to nonmetals such as plastics, rubber, glass and other ceramics and in applications where welding would cause adverse effects on material properties. The advantages of adhesives in comparison to discrete fastening mechanisms include more uniform distribution of stresses over the bonded area, stiffer structures, smooth surface finishes, fatigue resistance, low temperature fabrication, connection of dissimilar materials and sealing. The limitations of adhesives include strength, decreasing strength of the joint with increasing temperature, oxidation or other forms of chemical reaction reducing the strength of the adhesive joint over time, necessity to cure and/or clamp components during assembly and bonded structures are not easy to dismantle.

The range of adhesives is extensive including acrylics, cyanoacrylates, epoxies, anaerobics, silicones, phenolics, plastisols, polyester hot melts, polyvinyl acetates and polyurethane. The merits of a variety of adhesives are listed in Table 16.12. Adhesives can be purchased from specialist manufacturers and suppliers, however, it should be noted that no one manufacturer is likely to produce all the different types available. The choice of which adhesive to use for a given application depends on the type of materials to be joined, the type of joint to be used, the environmental factors for the application such as temperature and chemicals, the method of production and costs.

Table 16.12 Types of adhesives and principal characteristics

Adhesive	Principal characteristics	Typical applications
Anaerobics	Cures in the absence of oxygen	Locking, and sealing of close fitting components such as threaded fasteners
Cyanoacrylates	Cures through reaction with moisture held on surfaces to be bonded. Flows easily between close fitting components. Usually solidify within seconds	Small plastic components, metals, ceramics, rubber and wood
Acrylics	Fast curing. High strength. Tough. Supplied in two parts, resin and catalyst	Metals and plastics
Ceramics	High temperature capability	Metals and ceramics
Epoxies	Consist of an epoxy resin and a hardener. Form strong and durable bonds	Metals, glass, plastics and ceramics
Hot melts	Based on thermoplastics and epoxy and polyurethane thermosetting adhesives	Fast assembly of lightly loaded components
Phenolics	Requires heat and pressure to cure bond	Metal to metal, metal to wood and metal to brake-liners
Plastisols	Modified PVC dispersion that requires heat to harden. Resilient and tough	Mouldings, coatings
Polyurethanes	Two part and fast curing. Form strong, resilient joints that are resistant to impact	Glass-fibre reinforced plastic. Certain thermoplastics
Polyvinyl acetates (PVA)	Flexible, permanent, neutral PH, water soluble	Porous materials such as paper or wood and general packaging
Rubber adhesives	Based on solutions or latexes, they solidify through loss of the solvent or water medium	Foamed plastics, plastic laminates, wood, plywood, veneer, plasterboard, canvas, leather
UV curable adhesives	Can be cured rapidly by exposure to ultraviolet radiation	Plastics, metal, glass-ware, acrylic
Silicones	Flexible adhesive with good high temperature performance (up to approximately 200°C)	Metals, glass, plastics, rubber, wood and ceramics

Bonded joints can fail by means of tensile, compressive, shear or peel stresses or some combination of these (Fig. 16.34). Bonded joints tend to perform poorly under peel or cleavage loading and generally a bonded joint needs to be designed so that the loading is along the lines of the adhesive's strongest bond. Examples of joint designs and their relative performance are illustrated in Fig. 16.35 and practical assemblies in

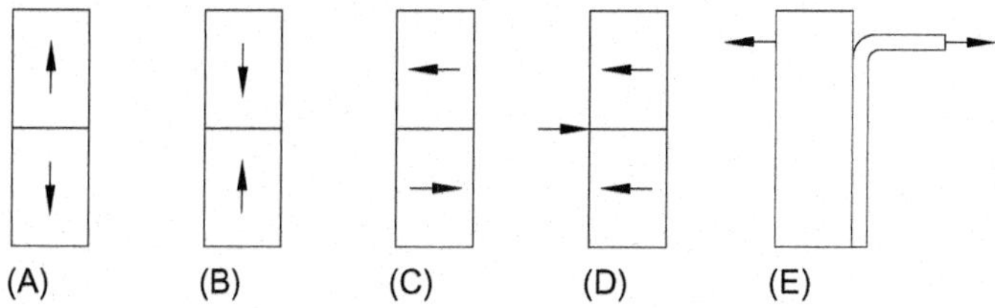

Fig. 16.34 Loading of adhesive joints: (A) tension, (B) compression, (C) shear, (D) cleavage, (E) peel.

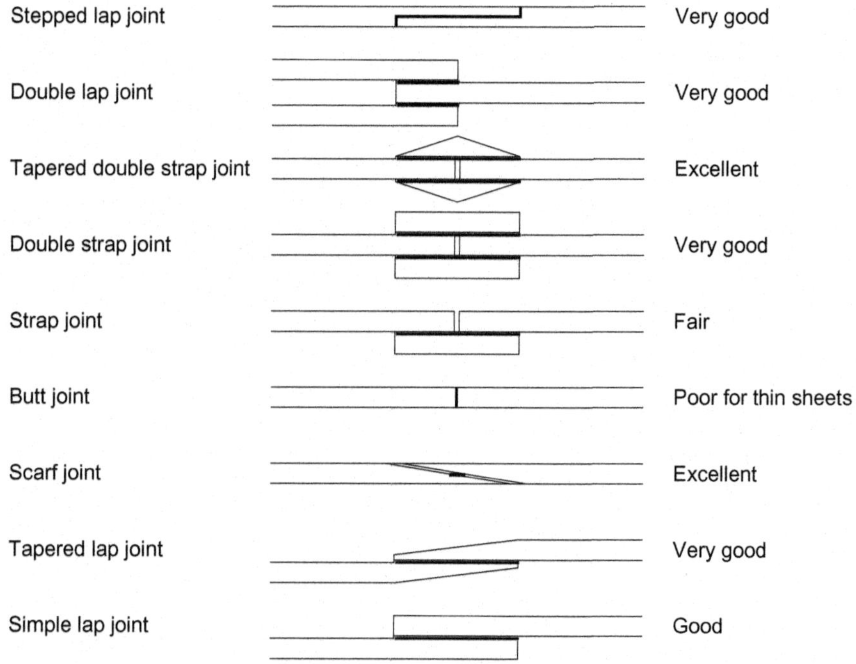

Fig. 16.35 Relative performance of a variety of joint designs.

Fig. 16.36. Adhesives can be useful for adding stiffeners to a sheet of material or adding strength to an aperture as illustrated in Figs. 16.37 and 16.38.

The strength of a lap joint depends on the nature of the materials, the type of adhesive, the thickness of the sheets and the area of the overlap. Test data is available for a variety of combinations of these variables as illustrated in Fig. 16.39. The tensile stress for the sheet material is given by

$$\sigma = \frac{P}{tw} \tag{16.44}$$

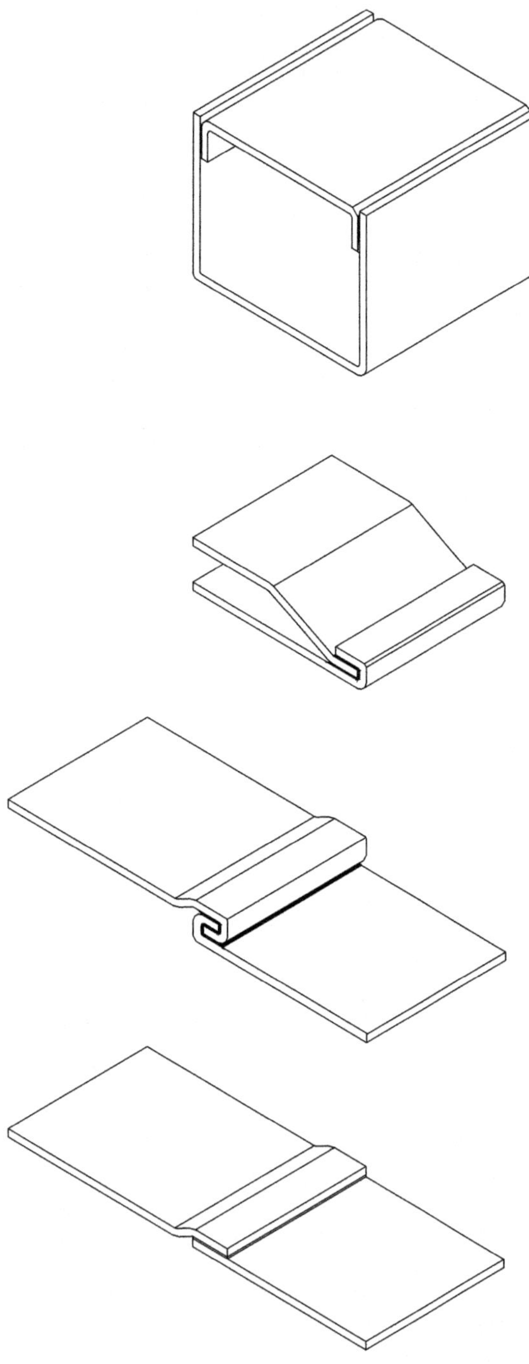

Fig. 16.36 Practical joins for sheet materials.

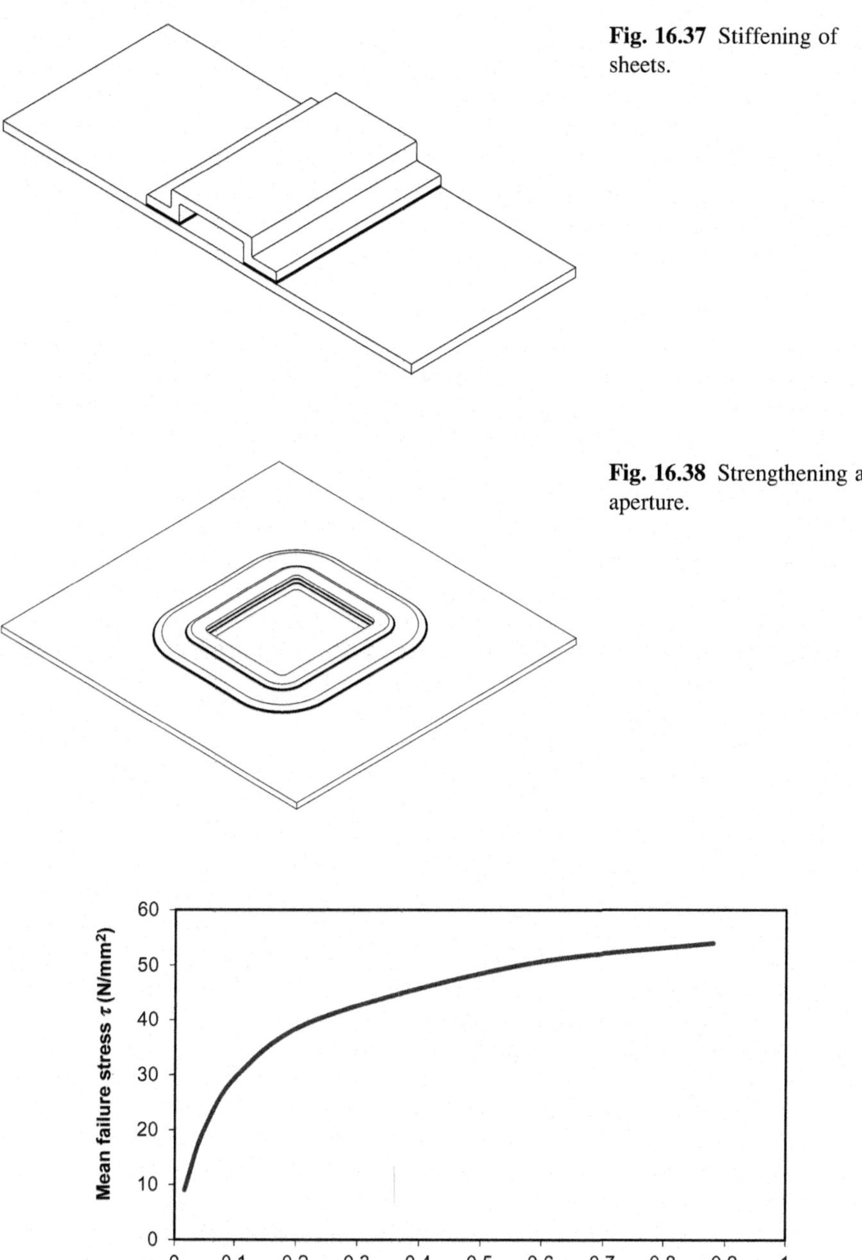

Fig. 16.37 Stiffening of sheets.

Fig. 16.38 Strengthening an aperture.

Fig. 16.39 Shear strength versus *t/L* for simple lap joints. Data relates to Araldite bonded BS 1470-HS30 lap joints.
Data from Ciba-Geigy. User's Guide to Adhesives, 1995.

The shear stress in the joint is given by

$$\tau = \frac{P}{Lw} \tag{16.45}$$

Substituting for the load gives

$$\tau = \sigma \frac{t}{L} \tag{16.46}$$

This equation can be used to determine the overlap required for a given sheet thickness, load and joint strength.

A substantial source of data and design hints for the use of adhesives is given in the Loctite World Design Handbook (1998), and Ebnesajjad and Landrock (2014).

Example 16.6. Determine the optimum overlap for a simple lap joint for 1/16″ sheet metal if the maximum load is 250 N/mm width. Use the araldite for which data is given in Fig. 16.39.

Solution

Selecting a safety factor of two to accommodate uncertainty in the loading conditions, the design load becomes 500 N/mm. The tensile stress in the metal is given by

$$\sigma = \frac{P}{t} = \frac{500 \times 10^3}{(1/16) \times 25.4 \times 10^{-3}} = 315.0 \text{ MPa}$$

$$\sigma = \frac{\tau}{t/L} = 315 \text{ MPa}$$

$$\tau = 315t/L$$

Plotting this line on Fig. 16.39 gives the intercept shear stress of approximately 28 MPa and $t/L = 0.09$.

With $t = 1.59$ mm, $L = 17.6$ mm.

The optimum overlap is therefore 17.6 mm.

Example 16.7. The araldite for which data is given in Fig. 16.39 is to be used for joining two sheets of metal. The joint should support loads of up to 180 N/mm width. If an overlap of 10 mm can be accommodated, determine the optimum sheet thickness.

Solution

Selecting a safety factor of two to accommodate uncertainty in the loading conditions, the design load becomes 360 N/mm.

The failure stress in the adhesive is given by

$$\tau = \frac{F/w}{L} = \frac{360 \times 10^3}{0.01} = 36 \text{ MPa}$$

Examination of Fig. 16.39 gives $t/L = 0.16$ for $\tau = 36$ MPa.

If $L = 0.01$ m then $t = 1.6 \times 10^{-3}$ m $= 1.6$ mm.

16.6 Welding

Welding can be described as the process of joining material together by raising the temperature of the surfaces to be joined so that they become plastic or molten.

For metals, welding involves the metallurgical bonding of components usually by application of heat by an electric arc, gas flame or resistance heating under heavy pressure. There are numerous types of welding including tungsten inert gas welding (TIG), metal inert gas welding (MIG), manual metal arc welding (MMA), submerged arc welding (SAW), resistance welding and gas welding (Swift and Booker, 2013). A wide number of types of joint are possible with welding a few of which are illustrated in Fig. 16.40. The design of component joints is considered by Sampath (1997).

Thermoplastics can also be welded. Heat can be applied by means of a hot gas, which is usually inert. Other types of welding for plastics include inertia, ultrasonic and vibration welding (Oberg et al., 1996).

Soldering and brazing involve the joining of heated metals while the components are in their solid state by means of molten filler metals (solders) (Beitz and Kuttner, 1994).

16.7 Snap fasteners

Snap fasteners comprise a feature that deforms during assembly and then positively locates against a feature. Various examples are illustrated in Fig. 16.41. Common structural features of snap fit joints include hooks, knobs, protrusions or bulges on one of the parts and corresponding depressions, undercuts, detents and openings on the other part to be joined. Snap fasteners are extremely popular because they can be moulded directly into a component reducing part count and the need to purchase separate fasteners and the ease of assembly and associated savings in costs.

A cantilevered beam with a hook and corresponding undercut is commonly used in snap fit joints, as illustrated in Fig. 16.42. The joining angle α aids the assembly of components and as the components engage a bending moment will be experienced by the cantilever, until the hook snaps into its assembled position. Other common forms of snap fit include full cylindrical undercut and mating lip (Fig. 16.43) and spherical undercut (Fig. 16.44; BASF, 2007; Bayer, 2013). Cylindrical snap fits tend to be stronger and more robust, but require greater assembly force than their cantilevered counterparts.

The retaining force for cantilevered snap fits is a function of the bending stiffness. In an application under working conditions, the lugs may be unloaded or partially loaded to form a tight assembly. The lugs need to be designed to ensure that allowable stresses are not exceeded during assembly. Recommended practice is to dimension lugs to give a constant stress distribution over the whole length. This can be achieved using a tapered cross section, or adding a rib.

The design of snap fits can be undertaken by considering the deflection such that the strain does not exceed permissible limits for the material and application concerned, and determination of the deflection force and mating force and checking that

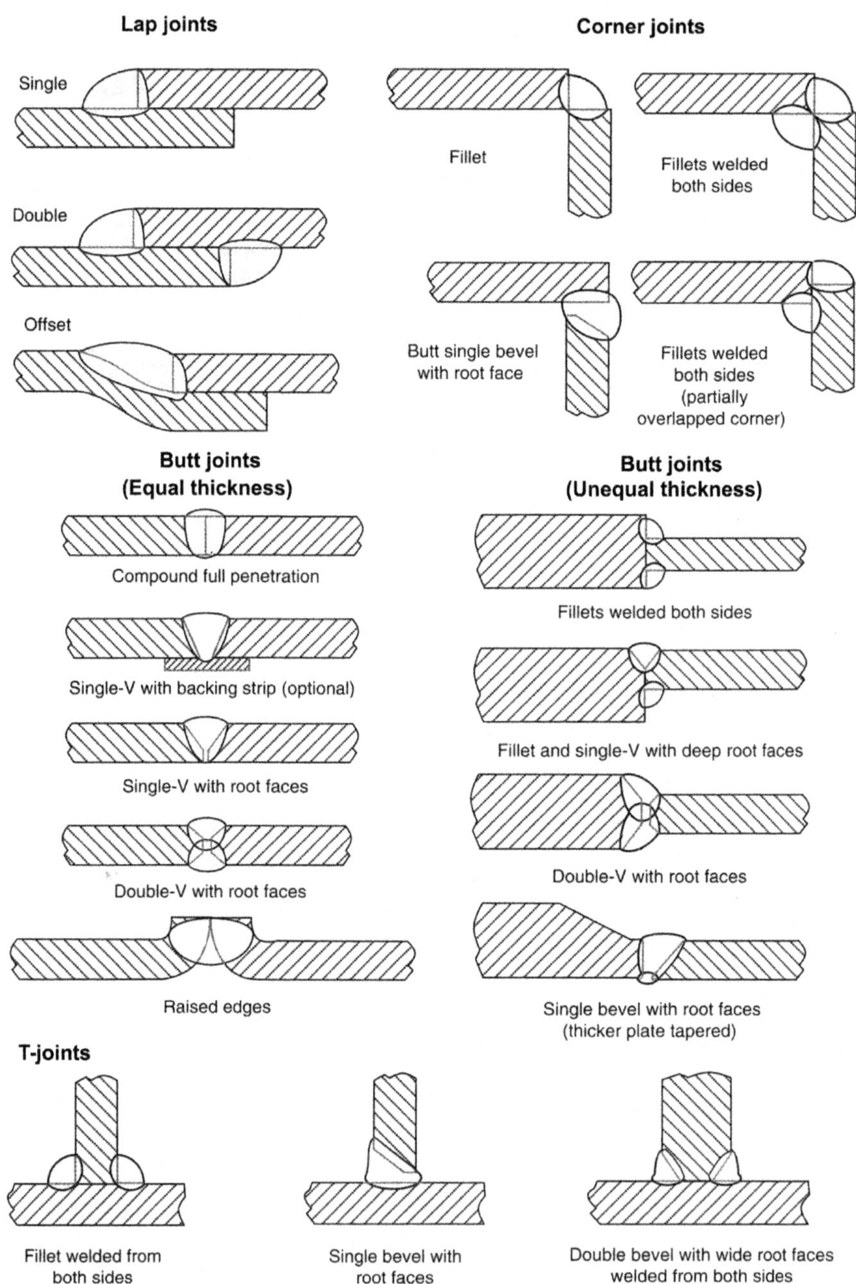

Fig. 16.40 Weld joint design.
Reproduced with permission from Swift, K.G., Booker, J.D., 2013. Manufacturing Process Selection Handbook. Butterworth Heinemann.

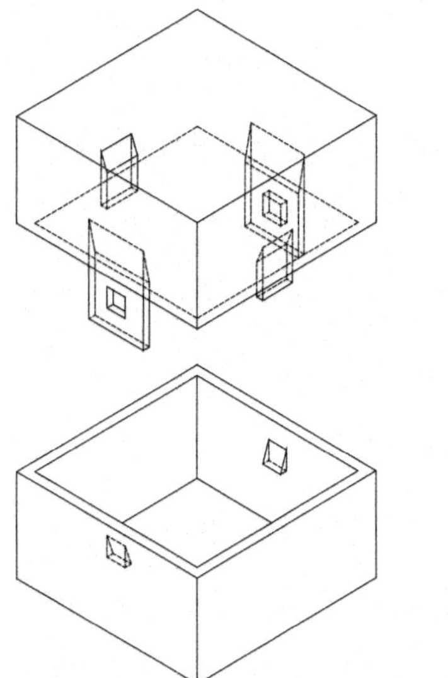

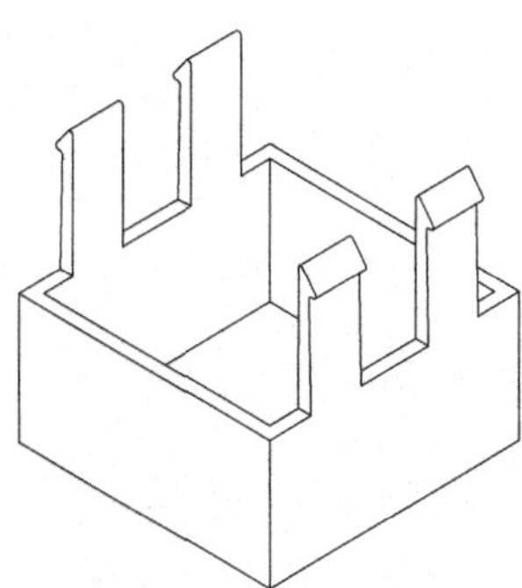

Fig. 16.41 Examples of snap fasteners integrated onto mouldings.

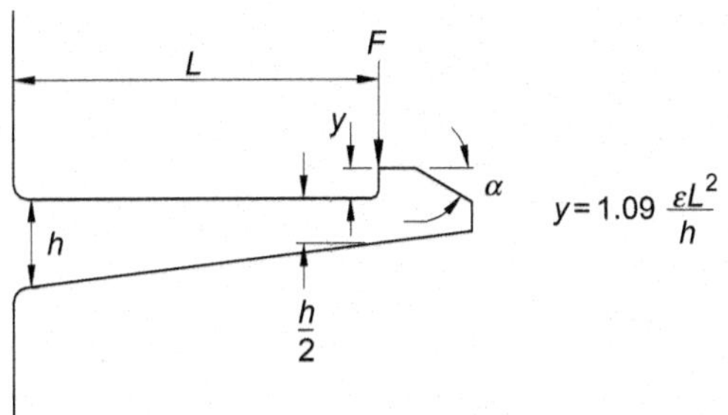

Fig. 16.42 Permissible deflection for a rectangular-section cantilever snap joint.

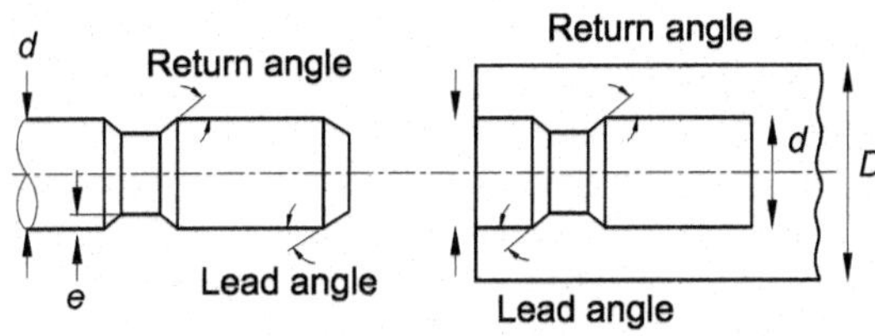

Fig. 16.43 Full cylindrical undercut and mating lip.

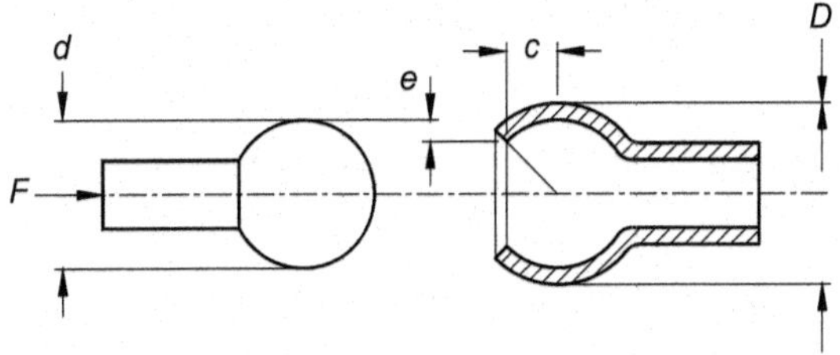

Fig. 16.44 Spherical undercut.

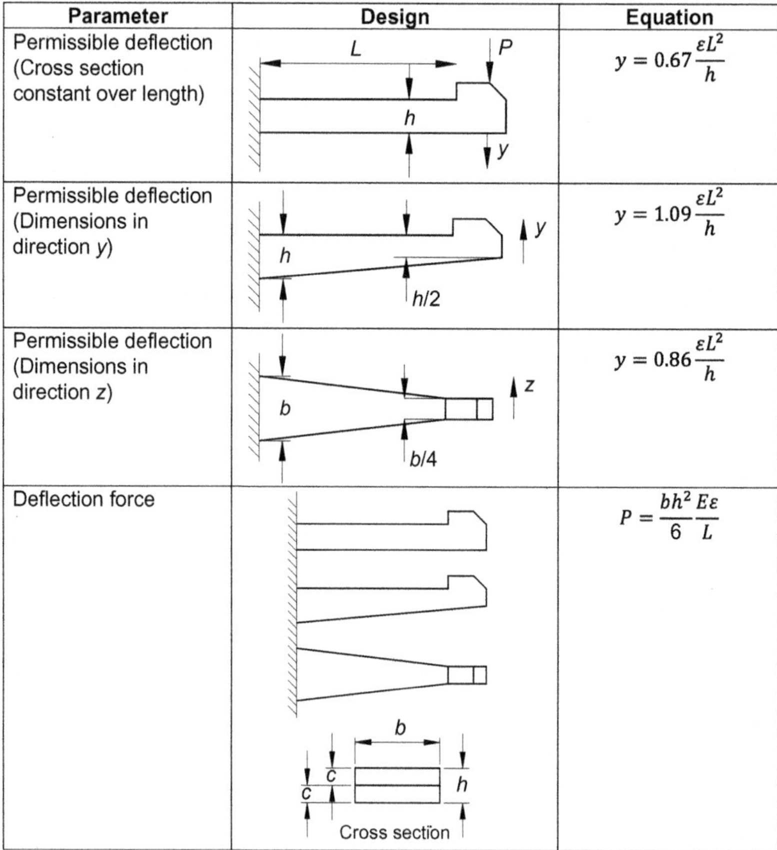

Parameter	Design	Equation
Permissible deflection (Cross section constant over length)	L, P, h, y	$y = 0.67\frac{\varepsilon L^2}{h}$
Permissible deflection (Dimensions in direction y)	h, h/2, y	$y = 1.09\frac{\varepsilon L^2}{h}$
Permissible deflection (Dimensions in direction z)	b, b/4, z	$y = 0.86\frac{\varepsilon L^2}{h}$
Deflection force	b, c, c, h, Cross section	$P = \frac{bh^2}{6}\frac{E\varepsilon}{L}$

Fig. 16.45 Deflection formulas for a variety of cantilever snap fit joints.

these are appropriate. The deflection for a variety of types of design is given in Fig. 16.45. Values for permissible strain can be obtained from experience and material manufacturers (e.g. BASF, Bayer). A guideline for some materials is given in Table 16.13. Friction coefficients for a variety of materials is given in Table 16.14. It should be noted that care needs to be taken in establishing Young's modulus and Secant modulus for hygroscopic materials such as nylon. Material data sheets may give the dry moulded state (DAM), but under, for example, normal 50% relative humidity conditions the physical properties alter, and in the case of materials such as nylon, the stiffness reduces and deflections increase. Calculations need to be performed for both conditions.

The mating force, required to overcome friction and deflection for the simple cantilever illustrated in Fig. 16.42 is given by

$$W_{mating} = F\left(\frac{\mu + \tan\alpha}{1 - \tan\alpha}\right) \tag{16.47}$$

Table 16.13 Allowable short-term strain for snap fit joints guide data (single joining operation)

Material	%
High heat PC	4
PC/ABS	2.5
Polycarbonate blends	3.5
PC	4
Glass-fibre-reinforced (%glass) **PC (10%)**	2.2
Glass-fibre-reinforced (%glass) PC **(20%)**	2

Data after Bayer, 2013. Bayer Snap-Fit Joints for Plastics A Design Guide. Bayer Polymers Division.

Table 16.14 Friction coefficient, μ(Guide data from literature for the coefficients of friction of plastics on steel)

Material	Coefficient of friction
PTFE	0.12–0.22
PE rigid	0.20–0.25 (×2.0)
PP	0.25–0.30 (×1.5)
POM	0.20–0.35 (×1.5)
PA	0.30–0.40 (×1.5)
PBT	0.35–0.40
PS	0.40–0.50 (×1.2)
SAN	0.45–0.55
PC	0.45–0.55 (×1.2)
PMMA	0.50–0.60 (×1.2)
ABS	0.50–0.65 (×1.2)
PE flexible	0.55–0.60 (×1.2)
PVC	0.55–0.60 (×1.0)

Data after Bayer, 2013. Bayer Snap-Fit Joints for Plastics A Design Guide. Bayer Polymers Division.

Example 16.8. Determine the thickness for snap fit necessary to ensure that that the strain does not exceed 50% of the permissible value, the deflection force and the mating force, for a rectangular cross section snap fit hook which has a constant decrease in thickness from h at the root to $h/2$ at the hook. The material is PC, length 15 mm, width 8 mm, undercut 2 mm and angle of inclination 30°. The Secant modulus can be taken as 1.815 GPa. The coefficient of friction is 0.4.

Solution

From Fig. 16.45, $y = 1.09\frac{\varepsilon L^2}{h}$

Hence,

$$h = 1.09\frac{\varepsilon L^2}{y} = \frac{1.09 \times 0.02 \times 0.015^2}{0.002} = 0.002453 \text{ m}$$

The deflection force, from Fig. 16.45, is given by

$$P = \frac{bh^2}{6}\frac{E_s\varepsilon}{L} = \frac{0.01 \times (0.002453)^2}{6} \times \frac{1.815 \times 10^6 \times 0.02}{0015} = 19.41 \text{ N}$$

The mating force is given by Eq. (16.47)

$$W_{\text{mating}} = F\left(\frac{\mu + \tan\alpha}{1 - \mu\tan\alpha}\right) = 19.41\left(\frac{0.6 + \tan 30}{1 - 0.6\tan 30}\right) = 34.96 \text{ N}$$

16.8 Conclusions

The use of fasteners is critical to the majority of machine design. Their selection and design is frequently left to too late a time in the design process. This can result in the use of too many fasteners or inconvenient locations with resultant component and machining costs. Instead the means of fastening components should be considered within the total design activity and the interrelationships with other components identified and appropriate fastening methods selected at an early stage in the design process. The range of fastening techniques available to the design is extensive. Frequently they can be obtained as a stock item with accompanying economies. This chapter has provided an overview of some of the fastening techniques available. It is however merely an introduction to an important subject.

References

Books and papers

BASF, 2007. Snap fit design manual. BASF. http://www8.basf.us//PLASTICSWEB/displayanyfile?id=0901a5e1801499d5. Accessed 30 August 2018.

Bayer, 2013. Bayer Snap-Fit Joints for Plastics A Design Guide. Bayer Polymers Division.

Beitz, W., Kuttner, K.-H., 1994. Dubbel's Handbook of Mechanical Engineering. Springer Verlag, London.

Ebnesajjad, S., Landrock, A.H., 2014. Adhesives Technology Handbook, third ed. Elsevier.

Izumi, S., Yokoyama, T., Iwasaki, A., Sakai, S., 2005. Three-dimensional finite element analysis of tightening and loosening mechanism of threaded fastener. Eng. Fail. Anal. 12, 604–615.

Loctite, (Ed.), 1998. World Design Handbook, second ed. Loctite European Group.

Oberg, E., Jones, F., Horton, H.L., Ryffel, H.H., 1996. Machinery's Handbook, twenty-fifth ed. Industrial Press Inc.

Sampath, K., 1997. Design for joining. In: ASM Handbook, Vol. 20 Material Selection and Design. ASM International, pp. 762–773.

Swift, K.G., Booker, J.D., 2013. Manufacturing Process Selection Handbook. Butterworth Heinemann.

Further reading

Bickford, J.H., 1995. An Introduction to the Design and Behaviour of Bolted Joints, third ed. Marcel Dekker.

Bonenberger, P.R., 2016. The First Snap-Fit Handbook: Creating and Managing Attachments for Plastics Parts, third ed. Hanser Gardner Publications.

Erhard, G., 2006. Designing with Plastics. Hanser Gardner Publications.

Industrial Fastener Institute, (Ed.), 1999. Metric Fastener Standards, third ed. Industrial Fastener Institute, Cleveland, Ohio, USA.

McCauley, C.J., 2016. Machinery's Handbook, thirtieth ed. Industrial Press Inc.

Parmley, R.O., 1997. Standard Handbook of Fastening and Joining. McGraw Hill.

Speck, J.A., 2017. Mechanical Fastening Joining, and Assembly, second ed. CRC Press.

Standards

British Standards Institution. BS 4620:1970. Specification for rivets for general engineering purposes. BSI, 1970.

British Standards Institution. BS 3643. ISO metric screw threads. BSI.

British Standards Institution. BS 3692. Metric precision hexagon bolts, screws and nuts. BSI.

British Standards Institution. BS 4168-3/4/5. Cap and socket headed setscrews. BSI.

British Standards Institution. BS 4183. ISO metric machine screws and machine screw nuts. BSI.

British Standards Institution. BS 4320. Metric washers. BSI.

British Standards Institution. BS 4464. Metric washers. BSI.

EN 24032/3/5. ISO metric hexagon bolts, screws and nuts.

EN ISO 2009. Slotted head bolts.

EN ISO 7046-1. Recessed head set screws.

Websites

At the time of going to press the world-wide-web contained useful information relating to this chapter at the following sites.

fastenertech.com.

www.apexfasteners.com.

www.associatedfastening.com.

www.atlanticfasteners.com.

www.avdel.textron.com.

www.boltscience.com.

www.earnestmachine.com.

www.fabory.com.

www.fastenal.com.

www.fastener-manufacturers.co.uk.

www.fastenertechnology.com.

www.libertyfastener.com.

www.machinedesign.com/bde/fastening/fastening.html.
www.nutsandbolts.net.
www.plastic-fastener.com.
www.rivet.com.
www.rotaloc.com.
www.seac.uk.com.
www.sinofastener.com.
www.skybolt.com.
www.ssfast.co.uk.
www.toleeto.com.
www.vaughanjones.co.uk.
www.watai.com.
www.yerd.com.hk.
Ciba-Geigy. User's Guide to Adhesives, Ciba-Geigy, 1995.
International Organization for Standardization, 2001. ISO Standards Handbook: Fasteners and screw threads. In: Vol. 1: Terminology and nomenclature, General reference standards. Ed. 5.
Renaud, J.E., Karr, P., 1996. Development of an integrated plastic snap fastener design advisor. Eng Des Autom. 2, 291–304.

Wire rope

17

Chapter Outline

Nomenclature

Generally, preferred SI units have been used.

A_{rope} approximate metallic cross sectional area of the rope (m^2)
d_{rope} rope diameter (m)
d_{sheave} sheave or drum diameter (m)
d_{wire} wire diameter (m)
E_{rope} modulus of elasticity of the wire rope (N/m^2)
M applied moment (N m)
N_d number of cycles
p unit radial pressure (N/m^2)
R_N fatigue strength parameter
T resultant tensile force (N)
σ_b bending stress (N/m^2)
σ_d design stress (N/m^2)
σ_{uts} ultimate tensile strength (N/m^2)

17.1 Introduction

Flexible wire rope is formed by using many small diameter wires twisted around a central core. Wire rope is widely used for hoists, haulage and conveyor applications where the wire rope supports tensile loading along its length. This chapter introduces the technology and outlines a selection procedure for typical applications.

A heavy lift winch developed and used by NASA is shown in Fig. 17.1. By using lots of small diameter wires, twisted around a central core, some flexibility in the wire rope can be achieved and it is possible for the wire rope to articulate drums and other radial segments.

Mechanical Design Engineering Handbook. https://doi.org/10.1016/B978-0-08-102367-9.00017-2

Wire rope is manufactured by helically twisting several small diameter wires together to form a strand, as illustrated in Fig. 17.2. The typical number of small diameter wires used to form a strand is 7, 19 or 37. Subsequently, a number, often 6 or 8, multiwire strands are twisted about a core to form the flexible wire rope. Some common multiwire strand wire rope configurations are indicated in Table 17.1. The central core is usually saturated with a lubricant to facilitate sliding movement of the strands relative to each other.

Fig. 17.1 Use of wire rope in a winch application for the NASA 325 ton hoist.

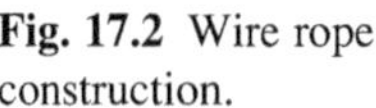

Fig. 17.2 Wire rope construction.

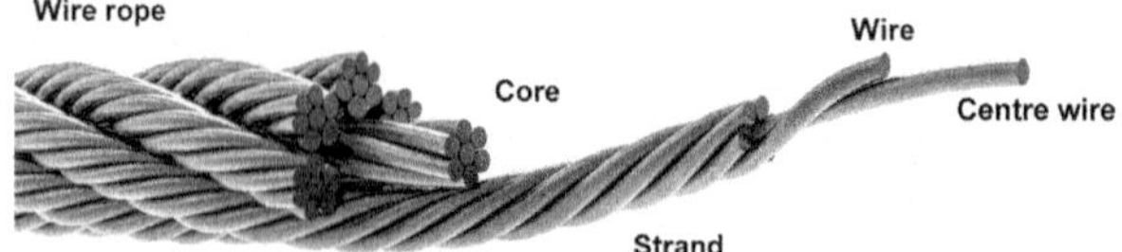

Table 17.1 Common multiwire strand wire rope configurations

Wire	Typical application	
6 × 7	Haulage	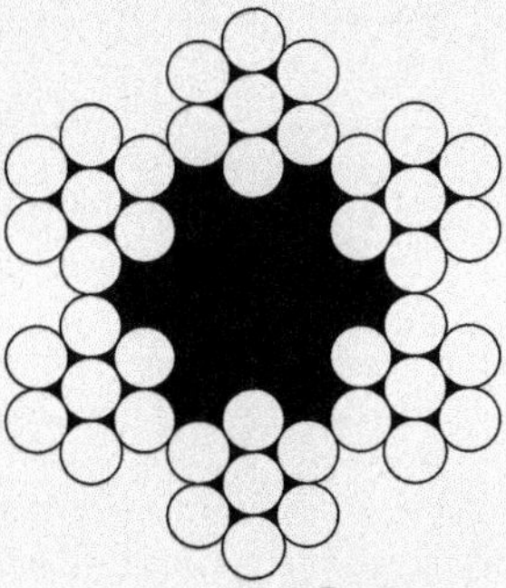
6 × 19	Standard Hoisting	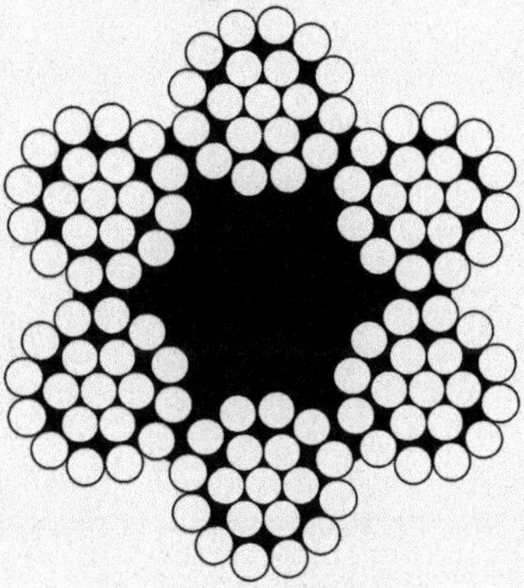
8 × 19	Extra flexible	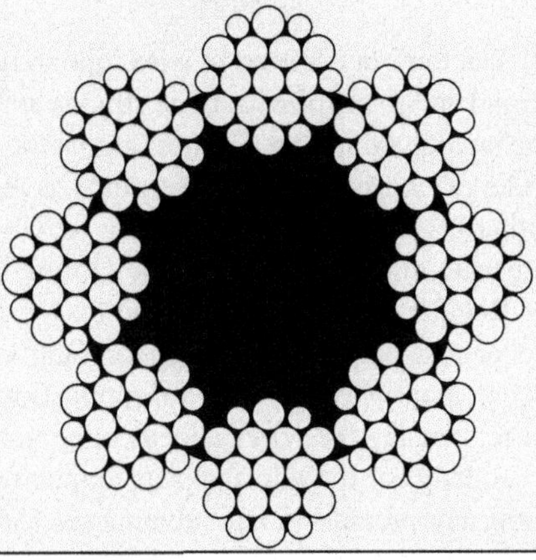

Continued

Table 17.1 Continued

Wire	Typical application	
6 × 37	Specially flexible	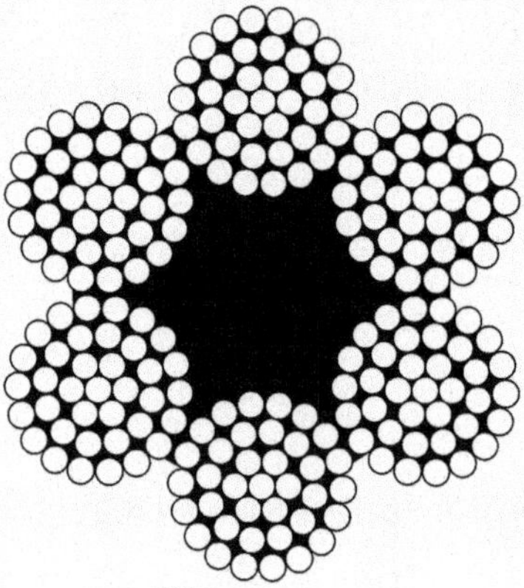
7 wire	Haulage	

The term lay is used in wire rope to refer to the direction of twist of the wires in a strand and to the direction the strands are laid in the rope, with variations according to the intended function. Some examples are illustrated in Fig. 17.3. The direction is right or left according to how the strands have been laid around the core. The type of lay is either regular or lang, depending on whether the wires in the strands are laid in the opposite direction of the strands or the same direction as the strands.

Regular lay wire rope is used for the widest range of applications. It generally has better resistance to crushing than lang lay wire rope and does not rotate as severely under load when used in an application where either end of the rope is not fixed. In regular lay the wires appear to be nominally aligned with the axis of the rope.

In lang lay the wires appear to make an angle with respect to the axis of the rope. Lang lay provides some advantages with possibly 15%–20% advantage for fatigue and wear resistance. This is because:

1. smaller strains occur in the outer layers as the rope passes over a drum or radius;
2. the contact areas between wires are larger and therefore the contact stresses are smaller.

However lang lay rope has some disadvantages. It tends to rotate when axial loads are applied, unless the rope is secured at both ends. Lang lay rope has less resistance against crushing against a drum or sheave.

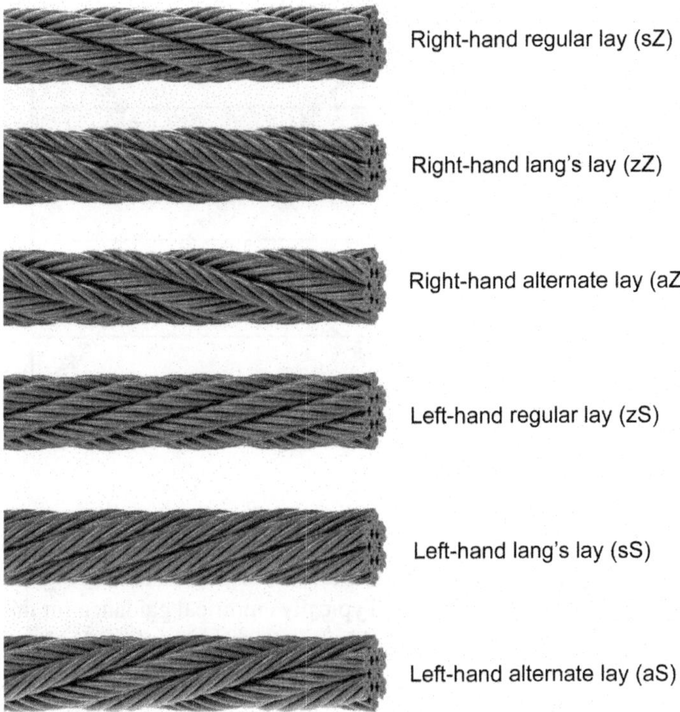

Fig. 17.3 Wire rope lay.

When a tensile load is applied to helically twisted wire rope, the wires tend to stretch and the helixes 'tighten'. This causes Hertzian contact stresses and sliding motion between the wires. With load cycling and as wire rope is bent around formers and drums, the wire rope can experience failure due to:

- tensile fatigue;
- bending fatigue;
- fretting fatigue;
- surface fatigue wear;
- abrasive wear;
- yielding or
- rupture.

17.2 Wire rope selection

Aspects that require consideration in rope wire system design include:

- materials;
- wire rope dimensions;
- construction type;
- rope attachment;

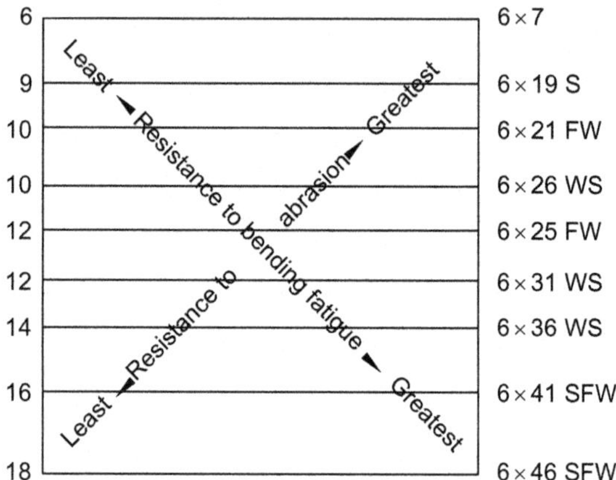

Fig. 17.4 X plot comparing bending fatigue resistance versus abrasion resistance for selected wire rope constructions. Adapted from Wire Rope Technical Board, 2005. *Wire Rope Users Manual*, fourth ed. Wire Rope Technical Board.

- drum or mandrel dimensions;
- inspection and maintenance.

Striking a balance between abrasion resistance and resistance to bending fatigue is an important aspect of wire rope selection. Typically empirical guidance for this is available in the form of use of an industry selection chart known as an X plot or X chart (Fig. 17.4).

Usually wire rope selection is an iterative process requiring selection of an initial set of parameters, followed by analysis of the proposed design and then revision of the proposal and reanalysis as necessary.

The following procedure provides a basis for wire rope selection.

1. Establish the principal design specifications for the system such as loads, failure modes, life, safety, cost, maintenance.
2. Select an initial wire rope construction using the X plot (Fig. 17.4).
3. Select a wire rope material.
4. Select a safety factor.
5. Determine the rope size using $\sigma_t = T/A$ (Eq. 17.1).
6. Using the rope diameter, determine the minimum recommended sheave diameter (see manufacturer data or Table 17.3 for indicative data).
7. Estimate the approximate bending stress in the outer wires (using Eq. 17.2).
8. Determine the fatigue strength parameter, R_N (Fig. 17.5).
9. Determine the required rope diameter based on fatigue.
10. Determine the wear-based limiting pressure for the rope class and sheave or drum material (see manufacturer data or Table 17.2 for indicative data).
11. Calculate the necessary rope diameter based on wear.
12. Identify the larger of the wire rope diameters determined from consideration of $d_{r\ \mathrm{static}}$, $d_{r\ \mathrm{fatigue}}$, $d_{r\ \mathrm{wear}}$.
13. Check all of the calculations with this new diameter and reconsider or modify the selections if necessary.

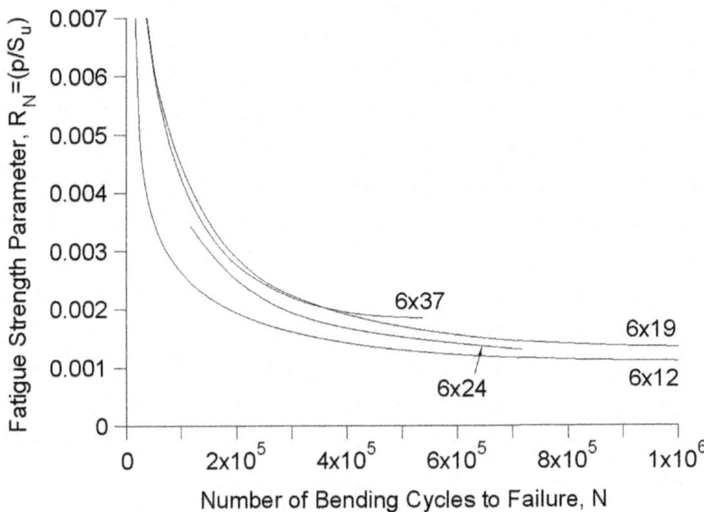

Fig. 17.5 Fatigue lives for several wire rope constructions as a function of fatigue strength parameter (after Collins et al., 2010).

Table 17.2 Maximum permissible pressures (Bridon)

Number of outer wires in strand	Cast iron (kgf/cm^2)	Low carbon steel (kgf/cm^2)	11%–13% Mn steel or equivalent (kgf/cm^2)
5–8 Ordinary lay	20	40	105
5–8 Lang's lay	25	45	120
9–13 Ordinary lay	35	60	175
9–13 Lang's lay	40	70	200
14–18 Ordinary lay	42	75	210
14–18 Lang's lay	47	85	240
Triangular strand	55	100	280

The direct tensile stress can be estimated using

$$\sigma_t = \frac{T}{A_{\text{rope}}} \tag{17.1}$$

where T is the resultant tensile force (N); A_{rope} is the approximate metallic cross sectional area of the rope (m^2) as a function of the rope diameter, d_r.

Force components that may need to be considered include:

- load to be lifted;
- weight of the wire rope;
- inertial effects from accelerating the load to the operational velocity (these forces can be a few times the static load);
- impact loading;
- frictional resistance.

The bending stress (N/m^2) can be determined from

$$\sigma_b = \frac{M\left(\frac{d_w}{2}\right)}{\left(\frac{Md_s}{2E_r}\right)} = \frac{d_w}{d_s}E_r \tag{17.2}$$

where d_{wire} is the wire diameter (m); d_{sheave} is the sheave or drum diameter (m); E_{rope} is the modulus of elasticity of the wire rope (N/m^2); M is the applied moment (N m).

The compressive stress between the wire rope and the sheave can be estimated using

$$p = \frac{2T}{d_r d_s} \tag{17.3}$$

where p is the unit radial pressure (N/m^2).

The fatigue strength parameter, R_N, can be calculated from

$$R_N = \frac{p}{\sigma_{\text{uts}}} \tag{17.4}$$

Example 17.1. Select a wire rope for a hoist required to lift 800 kg. The hoist design uses two lines to support the load attached to a vertically moving sheave, on which is mounted the swivel hook. The desired design life is 2 years, and the anticipated maximum use is 20 lifts per hour, 7.5 h a day for 220 days per year. A safety factor of 5 is required based on static ultimate strength and 1.5 based on fatigue.

Solution

Load $= 400 \times 9.81 = 3924$ N

From Fig. 17.4, going for a balance of resistance to abrasion and fatigue, a 6 × 31 WS or a 6 × 25 FW wire rope construction might be appropriate.

From Table 17.3, the nearest classifications compatible with 6 31 WS and 6 × 25 FW are 6 × 37 and 6 × 19 respectively. As a first iteration for the design a 6 × 37 WS wire rope construction will be explored.

A safety factor of 5 based on ultimate tensile strength has been defined.

For a suddenly applied load,

$$\sigma_{\text{max suddenly applied}} = 2\sigma_{\text{max static}}$$

From Eq. 17.1,

$$\sigma_{\text{max suddenly applied}} = 2\sigma_{\text{max static}} = 2\frac{T}{A_{\text{rope}}}$$

Table 17.3 Material and construction data for selected wire rope classes (Collins et al., 2010)

Classification	**6 × 7**	**6 × 19**	**6 × 37**	**8 × 19**
Number of outer strands	6	6	6	8
Number of wires per strand	3–14	15–26	27–49	15–26
Maximum number of outer wires	9	12	18	12
Approx diameter of outer wires	$d_r/9$	$d_r/13$ to $d_r/16$	$d_r/22$	$d_r/15$ to $d_r/19$
Material available (typically)—Core: FC	IPS (200)	I (80) T (130) IPS (200)	IPS (200)	I (80) T (130) IPS (200)
Material available (typically)—Core: IWRC	IPS (190)	I (190) T (220) IPS (255)	EIPS (220) EEIPS (255)	IPS (190) EIPS (200)
Approx metallic cross section of rope—Core: FC	$0.384d_r^2$	$0.404d_r^2S^4$	$0.427d_r^2(FW)^4$	$0.366d_r^2(W)^4$
Approx metallic cross section of rope—Core: IWRC	$0.451d_r^2$	$0.470d_r^2S^4$	$0.493d_r^2(FW)^4$	$0.497d_r^2(W)^4$
Standard nominal rope diameters	¼ to 5/8 by 1/16s ¾ to 1½ by 1/8s	¼ to 5/8 by 1/16s ¾ to 2¾ by 1/8s	¼ to 5/8 by 1/16s ¾ to 3¼ by 1/8s	¼ to 5/8 by 1/16s ¾ to 1½ by 1/8s
Unit weight of rope (lb/ft)	$1.5d_r^2$	$1.6d_r^2$	$1.55d_r^2$	$1.45d_r^2$
Approx modulus of elasticity of the rope (psi) 0%–20% of S_u	11.7×10^6 (FC)	10.8×10^6 (FC) 13.5×10^6 (IWRC)	9.9×10^6 (FC) 12.6×10^6 (IWRC)	8.1×10^6
Approx modulus of elasticity of the rope (psi) 21%–65% of S_u	13×10^6 (FC)	12×10^6 (FC) 15×10^6 (IWRC)	11.6×10^6 (FC) 14×10^6 (IWRC)	9×10^6
Recommended minimum sheave or drum diameter	$42d_r$	$34d_r$	$18d_r$	$26d_r$

Each wire rope carries 3924/2 = 1962 N. From Table 17.3, $A_{rope} = 0.427d_r^2$

$$\sigma_{max\ suddenly\ applied} = 2\frac{T}{A_{rope}} = 2\frac{1962}{0.427d_{rope}^2}$$

From Table 17.3, the static ultimate tensile strength of improved plow steel (IPS) is 200,000 psi ≈ 1379 MPa

With a static safety factor of 5, the design stress is given by

$$\sigma_d = 1379/5 = 275.8 \text{ MPa}$$

Equating the suddenly applied stress and the design stress, $\sigma_{maxsuddenlyapplied} = \sigma_{design}$, and solving for the rope diameter gives:

$$d_{rope} = \sqrt{\frac{3924}{275.8 \times 10^6 \times 0.427}} = 8.164 \text{ mm}$$

From Table 17.3, the nearest larger standard rope diameter is 11/32 in. or 12/32 in. Going for the larger of these, 12/32 in., $d_{r\ static} = 9.525$ mm.

From Table 17.3 the minimum recommended sheave diameter is given by

$$d_{sheave} = 18 d_{rope} = 18 \times 9.525 = 171.5 \text{ mm}$$

The bending stress in the outer wires can be estimated from

$$\sigma_b = \frac{d_{wire}}{d_{sheave}} E_{rope}$$

The approximate diameter of the outer wires, in this case, from Table 17.3 is $d_{wire} = d_{rope}/22$. So

$$\sigma_b = \frac{d_{wire}}{d_{sheave}} E_{rope} = \frac{8.164/22}{171.5} 75.8 \times 10^9 = 191.5 \text{ MPa}$$

The number of cycles is

$$N_d = 20 \times 7.5 \times 220 \times 2 = 66000$$

From Fig. 17.5, the fatigue strength parameter, $R_N = 0.0055$

The pressure corresponding to failure in 66,000 cycles is

$$p_{N_f} = R_N \sigma_{uts} = 0.0055 \times 1379 = 7.584 \text{ MPa}$$

The fatigue safety facture is 1.5, so

$$p_{d\ fatigue} = \frac{p_{N_f}}{1.5} = \frac{7.584}{1.5} = 5.056 \text{ MPa}$$

The wire rope diameter based on fatigue is given by

Table 17.4 Maximum allowable bearing pressure between rope and sheave

Rope	Sheave material			
	Wood	Cast iron	Cast steel	Manganese steel
Regular lay				
6 × 7	*1.0*	*2.1*	*3.8*	*10.1*
6 × 19	1.7	3.3	6.2	16.6
6 × 37	2.1	4.0	7.4	20.7
8 × 19	2.4	4.7	8.7	24.1
Lang lay				
6 × 7	*1.1*	*2.4*	*4.1*	*11.4*
6 × 19	1.9	3.8	6.9	19.0
6 × 37	2.3	4.6	8.1	22.8
8 × 19	2.7	5.5	10.0	27.6

Adapted from Wire Rope Users Manual.

$$d_{\text{rope fatigue}} = \frac{2T}{p_{d\,\text{fatigue}} d_{\text{sheave}}} = \frac{2 \times 1962}{5.056 \times 10^6 \times 0.1715} = 9.054 \text{ mm}$$

From Table 17.4, for a 6 × 37 wire rope on a cast carbon steel sheave (BHN-160), the allowable bearing pressure load based on wear is 8.1 MPa

The wire rope diameter based on wear is given by

$$d_{\text{rope wear}} = \frac{2T}{p_{d\,\text{wear}} d_{\text{sheave}}} = \frac{2 \times 1962}{8.1 \times 10^6 \times 0.1715} = 5.627 \text{ mm}$$

In this case, examining the wire diameters based on static loading, fatigue and wear, the largest is that for static loading.

Summarising the wire rope required is 3/8 in. 6 × 37 WS improved plow steel (IPS) fibre core (FC). The sheave is carbon steel (BHN-160) with a diameter of 171.45 mm.

17.3 Wire rope terminations

A variety of end terminations are available for wire rope. A stud can be formed on a loose end, or the wire can be looped around on itself or a former and clamped to form an eye (Figs 17.6 and 17.7). A special component known as a thimble is used for the eye when safe working loads need to be managed (Fig. 17.8). Wire rope pulleys are also readily available as stock items (Fig. 17.9).

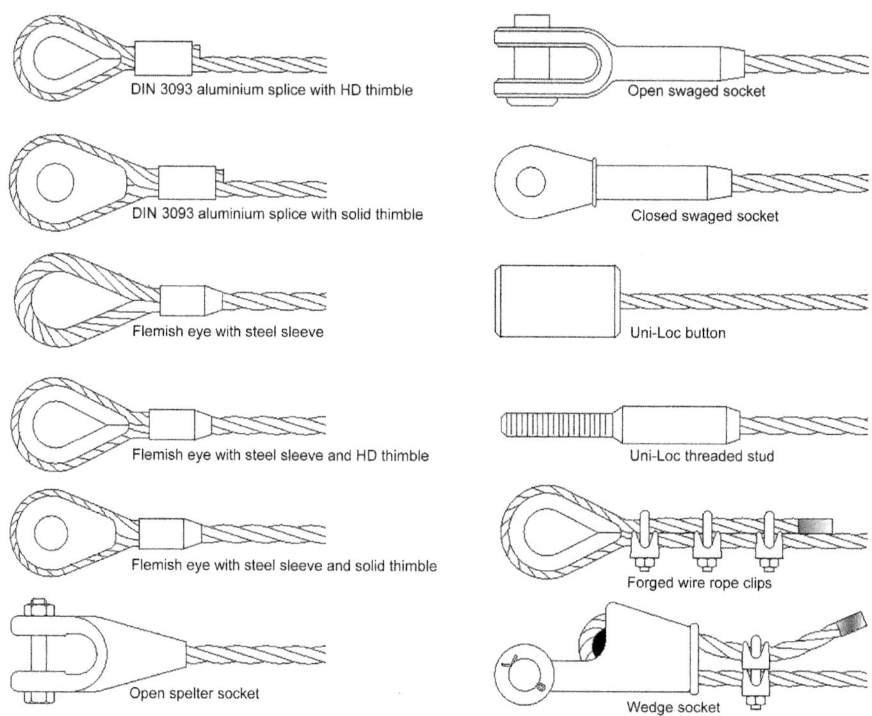

Fig. 17.6 Some examples of end termination for wire rope.

Fig. 17.7 Application of wire rope fittings.
Image courtesy Frank Vincentz CC-BY-SA-3.0.

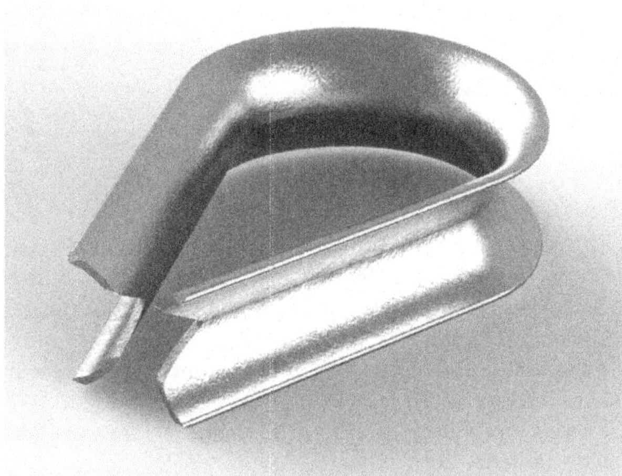

Fig. 17.8 Wire rope thimble.

Fig. 17.9 Examples of wire rope pulleys.

17.4 Conclusions

Wire rope can be useful for hoist, haulage and conveyer applications where a force needs to be applied over a large distance and some flexibility is an advantage. The flexibility of wire rope enables it to be wound on a drum or sheave. Wire rope is available in a wide range of classes as a standard stock item. When a tensile load is applied to helically twisted wire rope Hertzian contact stresses arise which may induce failure by fatigue, bending fatigue, fretting, surface wear, abrasion, yielding or rupture. Failure due to corrosion may also be an issue. Step-by-step procedures for the selection of wire rope are available.

References

Collins, J.A., Busby, H., Staab, G., 2010. Mechanical Design of Machine Elements and Machines, second ed. Wiley.

Standards

ASTM A 1023/A 1023M. Standard specification for stranded carbon steel wire ropes for general purposes.

BS 302-1: 1987. Stranded steel wire ropes—Part 1: Specification for general Requirements. British Standards Institute, 1987.

BS 302-2:1987. Stranded steel wire ropes—Part 2: Specification for ropes for general purposes. British Standards Institute, 1987.

BS 302-5:1987. Stranded steel wire ropes—Part 5: Specification for ropes for hauling purposes. British Standards Institute, 1987.

BS EN 12385-4:2002 +A1:2008. Steel wire ropes—Safety—Part 4: Stranded ropes for general lifting applications. British Standards Institute, 2008.

BS EN12385-2:2002 +A1:2008. Steel wire ropes—Safety—Part 2: Definitions, designation and classification. British Standards Institute, 2008.

BS EN12385-3:2004 +A1:2008. Steel wire ropes—Safety—Part 3: Information for use and maintenance. British Standards Institute, 2008.

Websites

At the time of going to press, the world-wide-web contained useful information relating to this chapter at the following sites.

http://www.bridon.com.
http://www.certex.co.uk.
http://www.dlhonline.co.uk.
http://www.fuchslubricants.com.
http://www.jakob.co.uk.
http://www.juststainless.co.uk.
http://www.midlandwirerope.com.
http://www.ormiston-wire.co.uk.
http://www.s3i.co.uk.
http://www.southwestwirerope.com.
http://www.stahlcranes.co.uk.
http://www.steelwirerope.com.
http://www.tecni-cable.co.uk.
http://www.wire-rope.co.uk.
http://www.wirerope.com.
http://www.wire-rope-direct.com.
http://www.wrigging.com.
http://www.wwirerope.com.
http://www.wwwrope.com.
http://nubco.sympaconline.com.

Further reading

AISI, Wire Rope Users Manual. AISI, 1979.

Costello, G.A., 1997. Theory of Wire Rope, second ed. Springer.

Drucker, D.L., Tachau, H., 1945. A new design criterion for wire rope. Trans. ASME 67, A33–A38.

Starkey, W.L., Cress, H.A., 1959. An analysis of critical stresses and mode of failure of a wire rope. J. Eng. Ind. Trans ASME 81, 307–316.

Wire Rope Technical Board, Wire Rope Users Manual, fourth ed. Wire Rope Technical Board, 2005.

Pneumatics and hydraulics

18

Chapter Outline

Nomenclature

Generally, preferred SI units have been stated.

A	area (m^2)
A_{rod}	cross sectional area of the rod (m^2)
AC	alternating current
C	a proportionality constant, coefficient of discharge
d	diameter of orifice (d)
D	upstream internal pipe diameter (d)
DC	direct current
F	force (N)
l	length (m)
LVDT	linear variable differential transformers
$\dot{m}$	mass flow (kg/s)
n_s	synchronous speed (rpm)
N	number of teeth in the rotating toothed wheel
NPT	national pipe thread
p_o	stagnation pressure (N/m^2)
p	pressure (N/m^2)
q_m	mass flow (kg/s)
$\dot{Q}$	volumetric flow (m^3/s)

Mechanical Design Engineering Handbook. https://doi.org/10.1016/B978-0-08-102367-9.00018-4

R	characteristic gas constant (J/kg K)
TPS	throttle position sensor
T	time (s)
u	speed (m/s)
$\bar{u}$	average flow speed (m/s)
V	voltage (V), velocity (m/s)
β	diameter ratio
ε_1	expansibility factor
ρ	density (kg/m^3)
ρ_1	density of the fluid (kg/m^3)
ω	angular velocity (rad/s)
Δp	differential pressure (Pa)

18.1 Introduction

Hydraulics and pneumatics are useful for the actuation of diverse elements from brakes and valves to robotics and machine tools. Hydraulic and pneumatics represent well developed technologies where components are available from a wide range of suppliers in modular form enabling the engineering designer to specify a bespoke system for a particular application. This chapter introduces both technologies and the associated principal components.

A wide range of industrial applications require substances, objects or components to be moved from one location to another. A further typical requirement is the application of a force to locate, hold, shape or compress a component or material. These tasks can be achieved using a prime mover, with rotary motion being provided for example by an electric motor and linear motion by screw jacks, rack and pinions and solenoids. Liquids and gases can also be used to convey energy from one location to another and as a result produce rotary and linear motions and apply forces. Fluid-based systems using a liquid as the transmission media are known as hydraulic, and those using a gas are known as pneumatic. Examples of a hydraulic and a pneumatic cylinder are given in Figs. 18.1 and 18.2 respectively. For both examples a high

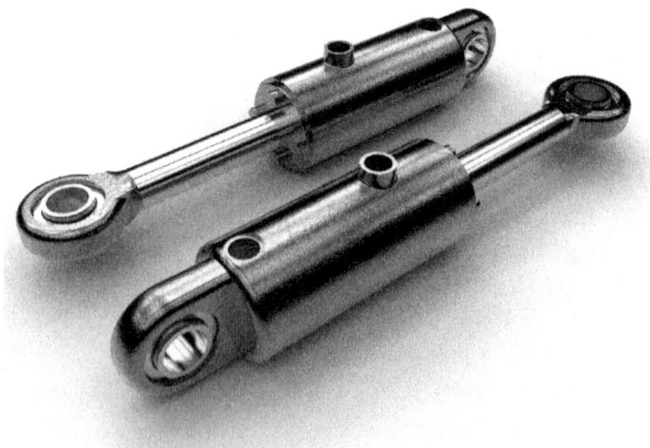

Fig. 18.1 Single acting hydraulic cylinder.

Fig. 18.2 Single acting pneumatic cylinder.

pressure fluid admitted to the cylinder acts on the surface of the piston to cause linear motion of the associated piston.

The gases used in pneumatic systems tend to be low density and compressible, in comparison with the relatively high density and incompressible liquids used in hydraulic systems. As a result pneumatic systems can be characterised as having a 'softer' action in comparison with hydraulic systems, while hydraulic systems can generally be operated at much higher pressures, producing higher forces. Equivalent pneumatic and hydraulic systems for lifting a load are illustrated in Fig. 18.3 (see Parr, 2011).

This chapter provides a brief overview of pressure in Section 18.2, along with the relevant subjects for pneumatic and hydraulic systems of pressure, flow and temperature measurement. Hydraulic pumps and air compressors are introduced in Sections 18.3 and 18.4 respectively. Filters and control valves which represent fundamental building blocks for hydraulic and pneumatic systems are considered in Sections 18.5 and 18.6. Pneumatics and hydraulic actuators are introduced in Section 18.7.

18.2 Pressure

A fluid particle will respond to a force in the same way that a solid particle will. If a force is applied to a particle an acceleration will result, as governed by Newton's second law of motion, which states that the rate of change of momentum of a body is proportional to the unbalanced force acting on it and takes place in the direction of the force. It is useful to consider the forces that a fluid particle can experience. These include:

- body forces such as gravity and electromagnetism;
- forces due to pressure;
- forces due to viscous action;
- forces due to rotation.

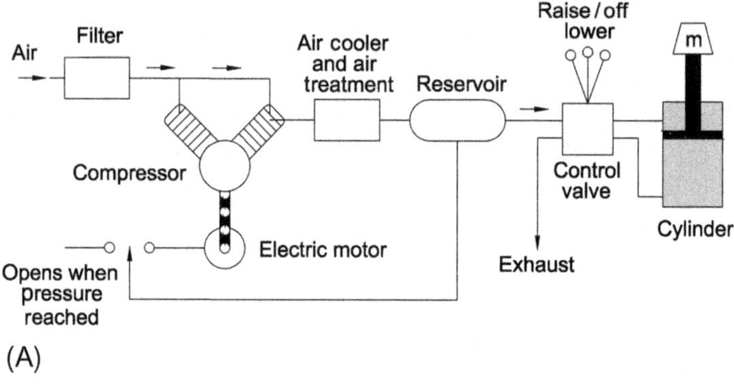

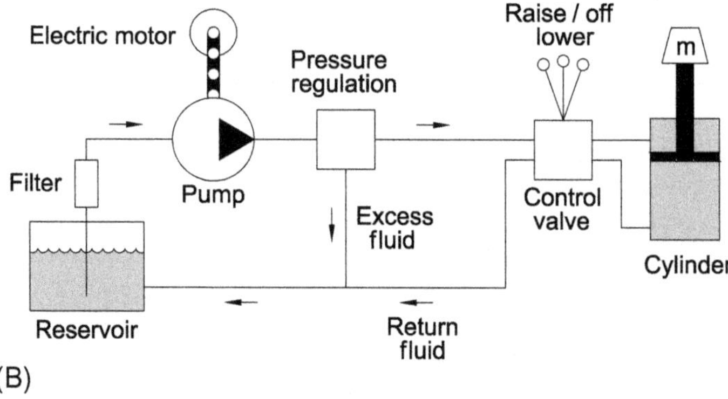

Fig. 18.3 Comparison of the physical components for equivalent pneumatic and hydraulic systems for lifting a load using a ram. (A) pneumatic; (B) hydraulic.

The usual laws of mechanics including conservation of mass and Newton's laws of motion can be applied to a fluid. The resulting formulations provide the mathematical basis for modelling fluid flow and developing understanding and insight into fluid flow phenomena.

In the case of a practical fluid the number of particles concerned is significantly high. For example, in each cubic meter of air at room temperature and pressure there are approximately 2.549×10^{25} molecules. The term pressure is used to describe the force acting per unit area (Eq. 18.1) and pressure will occur in a fluid when it is subjected to a force, whether this is as a result of gravity, compression, viscous action, rotation or a combination of these.

$$p = \frac{F}{A} \tag{18.1}$$

For a vessel containing a fluid with its surface open to the atmosphere, pressure arises in the fluid as a result of the weight of the fluid, and depends on the height of the fluid. This pressure is referred to as the hydrostatic pressure and can be stated as absolute or relative to atmospheric pressure in which case it is referred to as gauge pressure. Hydrostatic pressure is given by Eq. (18.2).

$$p = \rho g h \tag{18.2}$$

Pascal's law states that pressure at a point in a fluid is uniform. There may be slight variations in the hydrostatic pressure in a practical hydraulic or pneumatic system but these are likely to be negligible in comparison with the system operating pressure.

18.2.1 Pressure measurement

Pressure can be measured by a wide range of devices such as McLeod gauges, barometers, manometers which are suitable for laboratory use, Bourdon gauges and bellows which are suitable for some industrial applications and piezoelectric transducers that are useful for both scientific measurements and industrial control and monitoring.

The static pressure of a fluid flowing parallel to a surface can be measured by forming a small hole perpendicular to a solid wall (Fig. 18.4). The hole must be perpendicular to the flow and free of burrs and must not be chamfered or have a fillet radius.

Piezoelectric pressure transducers convert an applied pressure into an electrical signal. This signal is proportional to the applied pressure and in many cases nearly linear. The output electrical signal can be used in control, displayed on a panel meter or read by a data logger or voltmeter. Standard output signals available from transducer suppliers are a 4–20 mA two wire current loop or a voltage signals such as 0–10 V direct current. All transducers require an excitation or supply voltage to power the internal circuitry typically in the range of ±30 V DC, although any voltage within the published range for a particular transducer is acceptable.

Most pressure transducers are designed for a fixed range of pressure. The range should be equal to or greater than the highest working pressure expected for that particular application. There are four basic pressure references.

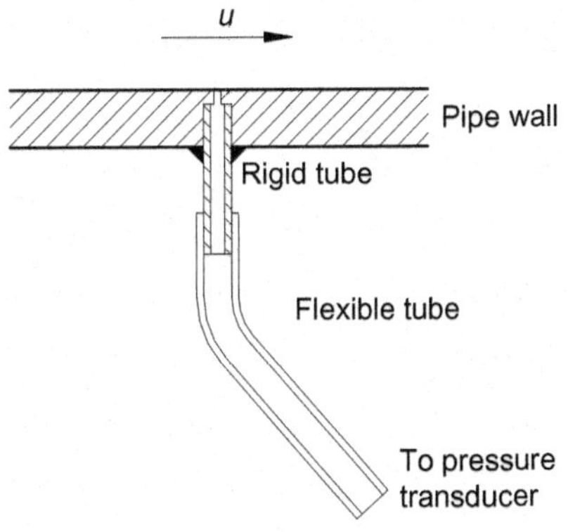

Fig. 18.4 Measuring static pressure.

- Vacuum pressure transducers are vented to atmosphere and typically produce a 20 mA output when no vacuum is applied. The output decreases as the vacuum increases.
- Gauge pressure transducers are referenced to atmospheric pressure. With no pressure applied, the output is typically 4 mA or 0 V DC.
- Compound ranges are vented to atmosphere and measure vacuum or pressure relative to atmospheric pressure.
- Absolute pressure transducers are referenced to a sealed vacuum chamber located within the sensor. These transducers sense changes in barometric pressure as the applied pressure is compared to a known vacuum. Absolute units are used where barometric changes affect the operation of the system being measured.

The maximum pressure defines the safety margin of pressure the transducer can tolerate for short periods of time without significant damage. Burst Pressure defines the maximum pressure causing permanent damage or destruction of the transducer.

General purpose and industrial grade transducers are available as packaged units comprising amplification and conditioning circuitry and a plug connector, and are usually fitted with threads, depending on the country and application concerned such as metric threads, British standard pipe fittings, 316 stainless steel NPT (national pipe thread) process connections. These transducers are for use with nonclogging media only. When measuring the pressure of a medium that is highly viscous, crystallising or contains particulates, thread connection is not suitable because the connecting orifices may clog. In such applications a flush-mounted transducer should be used. Alternatively a chemical diaphragm seal can be used to protect the transducer.

18.2.2 Fluid flow measurement

Fluid flow measurements are necessary in a very wide range of applications from the control of fuel flow in engine management systems to the regulation of drug delivery in ventilators. Fluid flow measurements involve determination of the flow velocity, the mass flow rate or volumetric flow rate.

The one dimensional continuity equation for mass flow of a fluid is given by

$$\dot{m} = \rho \bar{u} A \tag{18.3}$$

where

$\dot{m}$= mass flow rate (kg/s).
ρ = density (kg/m^3)
$\bar{u}$ = average flow speed (m/s).
A = cross sectional area of flow channel (m^2).

Dividing through by density gives the volumetric flow rate

$$\dot{Q} = \bar{u} A \tag{18.4}$$

where $\dot{Q}$ is the volumetric flow rate (m^3/s).

From Eqs (18.3) and (18.4), measurements of the average flow speed can be used to determine the mass flow, if the density and cross-sectional area are known. Alternatively if the mass flow is known the average flow speed can be determined.

Fluid velocity can be measured by a variety of techniques including Pitot tubes, thin film gauges and laser Doppler anemometry. Pitot tubes consist of an L shaped tube inserted into the flow so that the end stagnates the flow (Fig. 18.5). The flow speed is given by

$$u = \sqrt{\frac{2(p_o - p)}{\rho}} \tag{18.5}$$

where

u = speed (m/s).
p_o = stagnation pressure (N/m^2).
p = static pressure (N/m^2).
ρ = density (kg/m^3)

A number of devices measure mass flow based on the pressure loss across a constriction or flow feature. These include orifice plates and Venturi meters. The installation and use of orifice plates is documented in ISO 5167-1:1991(E) (also BS1042: Section 1.1:1992). The information given here has been sourced from this document and refers to devices with static pressure tappings at a distance of one diameter upstream of the plate and half a diameter downstream only (known as D and D/2 tappings). The mass flow rate through an orifice plate installation can be determined from:

$$q_m = \frac{C}{\sqrt{1-\beta^4}} \varepsilon_1 \frac{\pi}{4} d^2 \sqrt{2 \Delta p \rho_1} \tag{18.6}$$

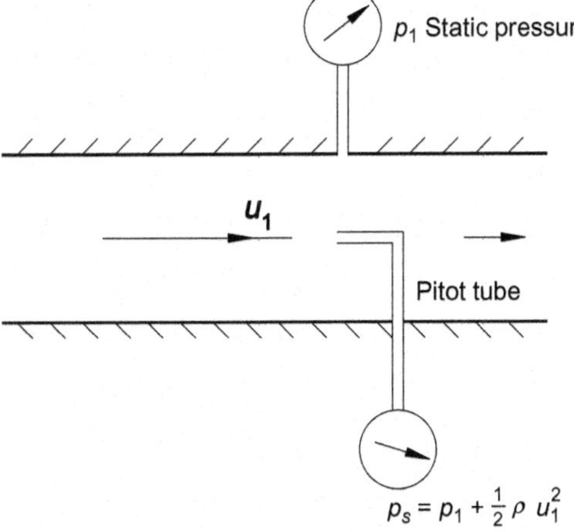

Fig. 18.5 Velocity measurement with a Pitot tube.

where

q_m = mass flow (kg/s).
C = coefficient of discharge.
β = diameter ratio = d/D.
d = diameter of orifice (d).
D = upstream internal pipe diameter (d).
ε_1 = expansibility factor.
Δp = differential pressure (Pa)
ρ_1 = density of the fluid (kg/m^3)
q_m = mass flow rate (kg/s).
$l_1 = D \pm 0.1D$.
$l_2 = 0.5D \pm 0.02D$ for $d/D \leq 0.6$.
$l_2 = 0.5D \pm 0.01D$ for $d/D > 0.6$.

The density can be evaluated from conditions at the upstream pressure tapping ($\rho_1 = P_1/RT$). The temperature should be measured downstream of the orifice plate at a distance of between 5D and 15D. The temperature of fluid upstream and downstream of the orifice plate is assumed constant. The minimum straight lengths of pipe upstream and downstream of the orifice plate are listed in ISO 5167-1:1991(E). For a wide range of conditions for air ε_1 and C can be approximated by $\varepsilon_1 = 0.99$, $C = 0.61$.

A common device for flow measurement is the convergent-divergent nozzle or Venturi meter (Fig. 18.6). This device involves the contraction of the fluid and subsequent expansion. In comparison to other flow measurement devices such as an orifice plate, the smooth contraction and expansion does not cause a significant pressure loss as fluid streamlines are able to conform to the geometry.

Assuming the flow is incompressible flow (ρ is constant) and that pressure variations due to area change dominate the effects of friction,

$$Q = u_1 A_1 = A_1 \sqrt{\frac{2(p_1 - p_t)}{\rho\left[(A_1/A_t)^2 - 1\right]}} \tag{18.7}$$

Hot film anemometry is used as the basis of mass flow measurement in automotive engine applications. A thin film of nickel is maintained at a constant temperature on a cylinder or alternative mounting located in the fluid stream (Fig. 18.7).

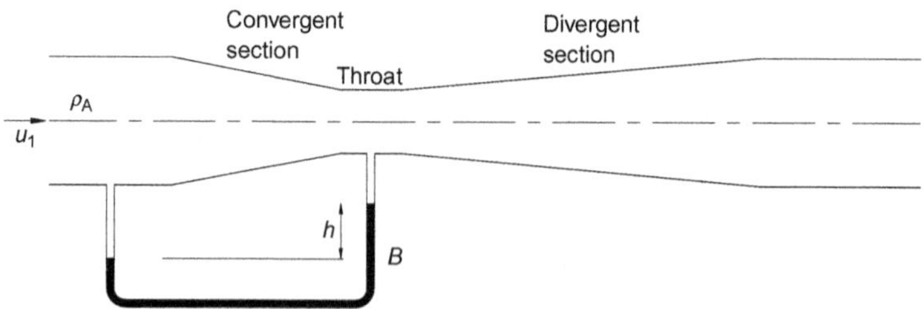

Fig. 18.6 Convergent-divergent nozzle (Venturi meter).

Fig. 18.7 Hot film air mass flow sensor.

A constant temperature is achieved by an electronic circuit, which rapidly varies the current supplied, depending on the flow conditions. The convection heat transfer characteristics for a cylinder in cross flow are well understood or alternatively a calibration can be undertaken for the specific geometry concerned to relate the current flow to the local mass flow. Such sensors can be manufactured fit-for-purpose and with care the total cost minimised, as there is only a single measurement that needs to be made.

18.2.3 Temperature measurement

Temperature cannot be measured directly. Instead its measurement requires the use of a transducer to convert a measurable quantity to temperature. Examples of temperature sensors and the associated transducer property used and the measured quantity are listed in Table 18.1. As can be seen from this table a wide variety of sensors are available and within each type listed a further selection is usually available.

The selection of a specific sensor for a temperature measurement can become a complex activity requiring consideration of a number of aspects including:

- uncertainty
- temperature range
- thermal disturbance
- level of contact
- size of the sensor
- transient response
- sensor protection
- availability
- cost

The range of application, defined as the span from the minimum temperature of operation to the maximum temperature for a particular measurement technique or for a

Table 18.1 Examples transducer properties and the physical property measured for some temperature sensors

Sensor	Transducer property	Physical quantity measured
Liquid-in-glass thermometer	Thermal expansion	Length
Constant volume gas thermometer	Thermal expansion	Pressure
Constant pressure gas thermometer	Thermal expansion	Volume
Bimetallic strip	Thermal expansion	Length
Thermocouple	Seebeck effect	Voltage
Platinum resistance thermometer	Electrical resistance	Resistance
Thermistor	Electrical resistance	Resistance
Transistor	Electrical resistance	Voltage
Capacitance thermometers	Electric permittivity	Capacitance
Noise thermometers	Johnson noise	Power
Thermochromic liquid crystals	Reflection	Colour
Thermographic phosphors	Fluorescence	Intensity
Heat sensitive crayons and paints	Chemical	Colour
Pyrometric cones	Chemical	Shape
Infrared thermometer	Thermal radiation	Radiation intensity
Thermal imager	Thermal radiation	Radiation intensity
Absorption spectroscopy	Absorption	Wavelength
Emission spectroscopy	Emission	Wavelength
Rayleigh scattering	Elastic scattering	Wavelength
Raman scattering	Inelastic scattering	Wavelength
CARS (Coherent Anti-Stokes Raman Scattering)	Scattering	Wavelength
Degenerative four wave mixing	Scattering	Wavelength
Laser-induced fluorescence	Emission	Time
Acoustic thermography	Sound velocity	Time

Based on Childs, P.R.N., 2002. Advances in temperature measurement. In: Hartnett, J.P. (Ed.), Advances in Heat Transfer, vol. 36, pp. 111–181.

specific instrument is a critical parameter in selection. Some types of measurement techniques such as thermocouples are suitable for use over a wide temperature range, in this case from approximately −273°C to 3000°C, but this can only be achieved using several different types of thermocouples as no single thermocouple is suitable for the entire range. Fig. 18.8 illustrates the approximate temperature range capability for a variety of methods.

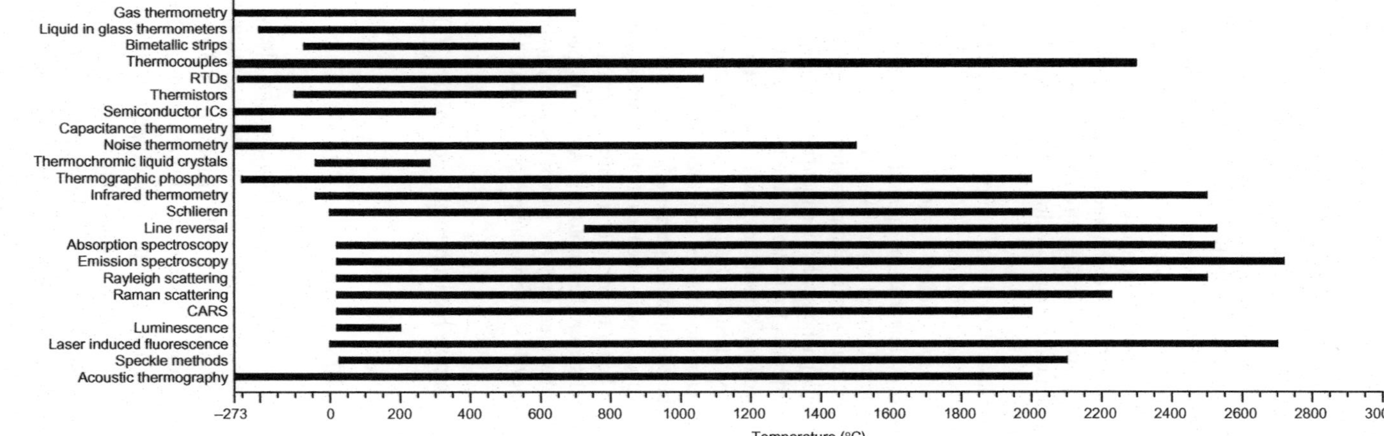

Fig. 18.8 Approximate temperature ranges for various temperature measurement techniques.

Before selecting a temperature measuring system a number of aspects have to be considered. Some of the issues that might have to be addressed are given below (see ESDU Item 02006).

(1) What size and shape of sensor will be required? If there is no special requirement most types of invasive sensor are available as cylindrical probes, which usually incorporate a protective cover. These may range from about 0.25 mm to 15.0 mm in diameter.
(2) Will an invasive, or noninvasive method be the most suitable? In most circumstances, an invasive sensor can be used and this may provide the most economical solution. However, with some more difficult measuring situations, noninvasive methods are essential and any disadvantages such as complexity, cost and increased uncertainty have to be accepted.
(3) What is the temperature range? Some techniques provide a wide temperature range with a single sensor, some a wide overall temperature range but individual sensors are limited, and with others the temperature range possible with the technique itself may be quite limited.
(4) What uncertainty can be accepted? Is the temperature needed with low uncertainty in absolute terms or are temperature differences or changes in temperature really required? With many types of sensor, the gradient of the output versus temperature characteristic may remain more stable with time than the apparent origin. If low uncertainty is required the sensors will need to be calibrated against known standards before use. If this uncertainty has to be maintained, regular substitution by new calibrated sensors may be needed.
(5) Are steady state or transient phenomena to be measured? Some types of invasive sensors are better suited to fast response, for example fine thermocouple or resistance wires.
(6) Will the sensor need protection? Sensors may be fragile and also susceptible to contamination. Many sensors can be fitted in tubes, sleeves or pockets but by so doing their uncertainty and response characteristics may be impaired.
(7) Will the sensor take up the required temperature?
(8) Will the presence of the sensor modify the local temperature significantly? If this is likely to be a problem it may be better to consider a noninvasive method.
(9) Can a standard, catalogued system be employed? Although special sensors and systems can be designed and manufactured they are inevitably likely to be more expensive.
(10) Is it a multichannel scheme? If so thermocouple or resistance temperature detectors are generally more suitable.

For further details on temperature measurement see Childs (2001 and 2013).

18.3 Hydraulic pumps

Hydraulic systems rely on a pressurised liquid to energise the various components. Pressurisation of the hydraulic fluid, normally an oil, is typically achieved using a hydraulic pump. Hydraulic pumps are usually driven by a 3 phase electrical motor running at constant speed rotating at 1500 or 1000 rpm on a 50 Hz supply or 1800 or 1200 rpm on a 60 Hz supply. The three principal types of pumps used for hydraulic systems are: gear pumps (Fig. 18.9); vane pumps; piston pumps. These can be obtained from a wide range of suppliers and most provide the motor and pump as a single unit. A typical hydraulic pump and associated motor circuit components are illustrated in Fig. 18.10.

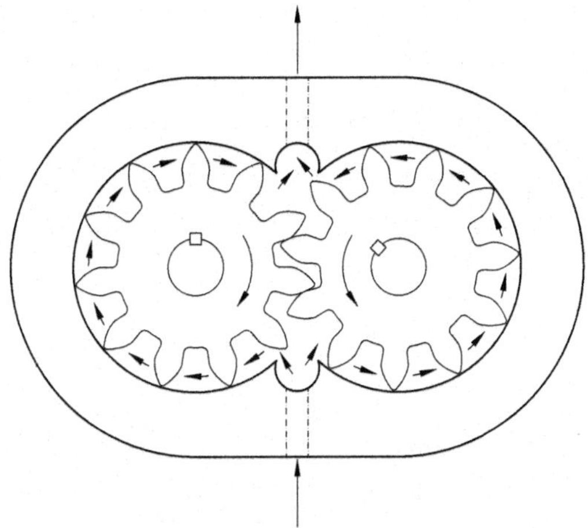

Fig. 18.9 Gear pump.

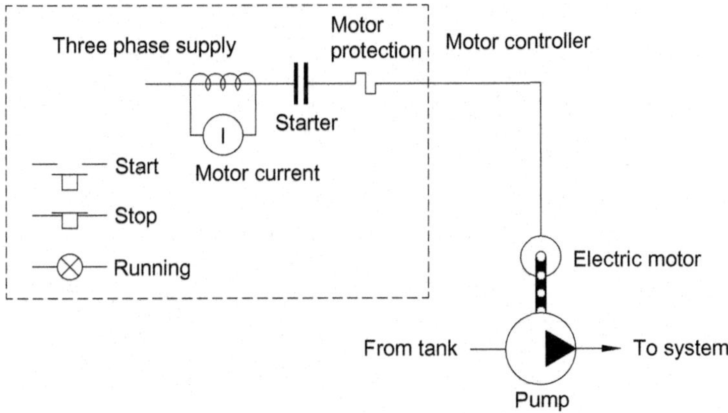

Fig. 18.10 Hydraulic pump and associated components.

A means of maintaining pressure at a safe operating level is necessary in hydraulic circuits and this can be achieved by using a pressure regulating valve or a number of valves. One scheme for maintaining pressure for a hydraulic ram circuit is illustrated in Fig. 18.11. The pressure regulating valve shown, V_3, is normally closed when the pressure is below a preset value known as the cracking pressure. When the pressure reaches the cracking pressure, the regulating valve begins to open. The valve becomes fully open when the pressure reaches a value known as the full flow pressure. When valve V_1 is closed all the fluid from the pump will return to the tank via the pressure regulating valve V_3. The operation of a simple pressure relief valve is illustrated in Fig. 18.12. Examples of pressure regulating valves are illustrated in Fig. 18.13.

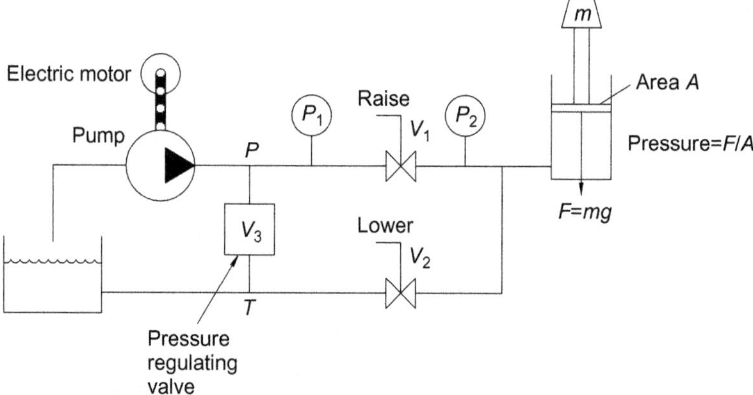

Fig. 18.11 Hydraulic circuit with a pressure regulating valve, V3.

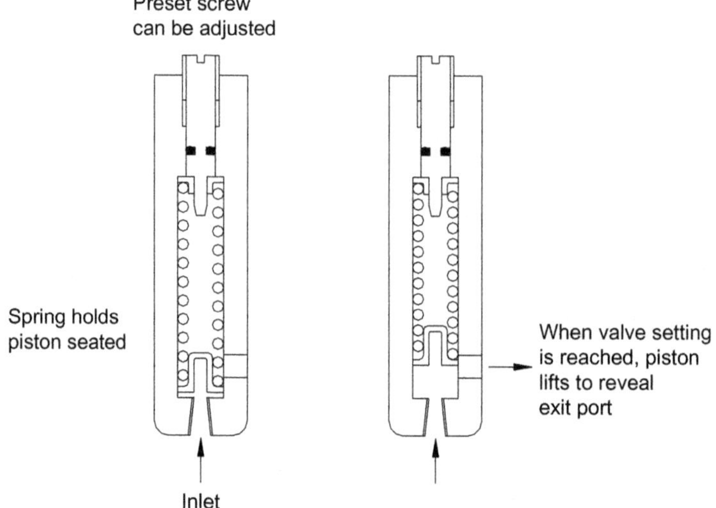

Fig. 18.12 Pressure relief valve.

18.4 Air compressors and receivers

Most pneumatic systems use air as the working fluid. The vast majority of pneumatic systems are open with the air sourced from ambient atmosphere and vented back to the atmosphere.

The ideal gas law is given by

$$pV = nRT \tag{18.8}$$

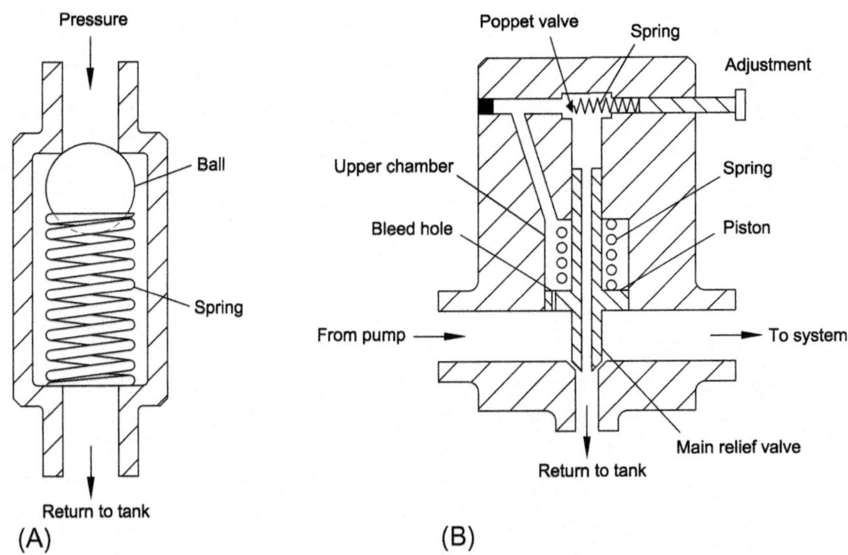

Fig. 18.13 (A) Simple pressure regulator; (B) Balance piston relief valve.

where

p = pressure (N/m^2),
V = volume (m^3),
n = number of moles of the gas ($n = m/M$ (m = mass, M = molar mass)),
R = the universal gas constant (=8.314510 J/mol K (Cohen and Taylor (1999))),
T = temperature (K).

From Eq. (18.8), the general gas equation for a gas subjected to changes in pressure, volume and temperature between two states 1 and 2 can be obtained.

$$\frac{p_1 V_1}{T_1} = \frac{p_2 V_2}{T_2} \tag{18.9}$$

where

p = absolute pressure (N/m^2),
T = temperature (K),
V = volume (m^3).

The two principal classes of air compressors are:

- Positive displacement compressors where a fixed volume of air is delivered with each rotation of the compressor shaft;
- Centrifugal and axial compressors.

The compressor must be selected such that it provides the pressure required as well as the volume of gas at the working pressure. A typical system will be designed to operate with the pressure in the receiver at a slightly higher pressure than that in the remainder

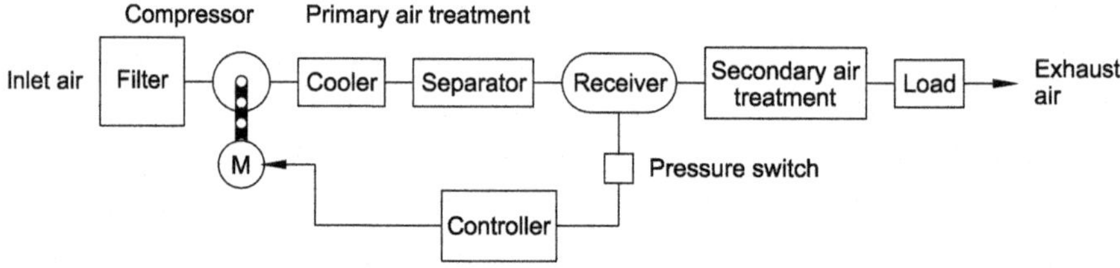

Fig. 18.14 Typical pneumatic system.

of the circuit with a pressure regulator being used. A typical pneumatic system is illustrated in Fig. 18.14.

The use of an air receiver tends to enable a smaller air compressor to be specified. The air receiver needs to have a capacity that is sufficiently large to enable to provide the peak flow requirements for sufficient periods compatible with the application. The volume of the receiver tends to reduce fluctuations in pressure in comparison to a system directly connected to a compressor. Air exiting from a compressor will be hotter than the inlet air and the receiver surface volume, sometimes finned, is used to dissipate the thermal energy by natural convection. Water will tend to condense within the receiver and regular drainage of this is necessary. Some applications require drying of air in which case a simple water trap (Fig. 18.15), refrigerator dryer, deliquescent dryer or adsorption dryer may be necessary (see Parr, 2011). Some typical pressure regulating valves are illustrated in Fig. 18.16 (see Parr, 2011).

18.5 Filters

Dirt in hydraulic and pneumatic systems can cause sticking of valves, failure of seals and wear. Filters can be installed to limit or prevent dirt from contaminating

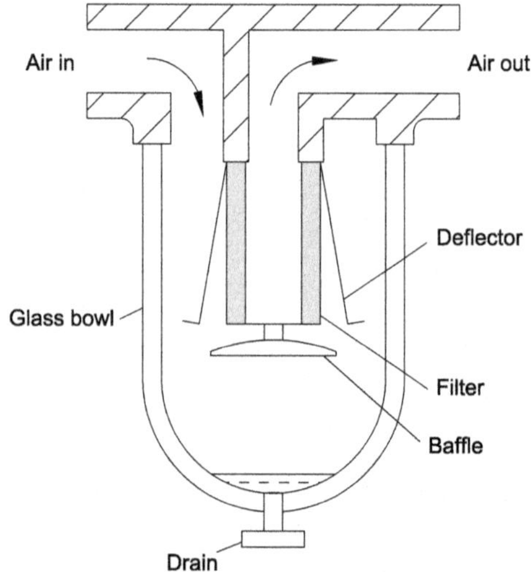

Fig. 18.15 Air filter and water trap. After Parr, A., 2011. Hydraulics and Pneumatics. third ed. Butterworth Heinemann.

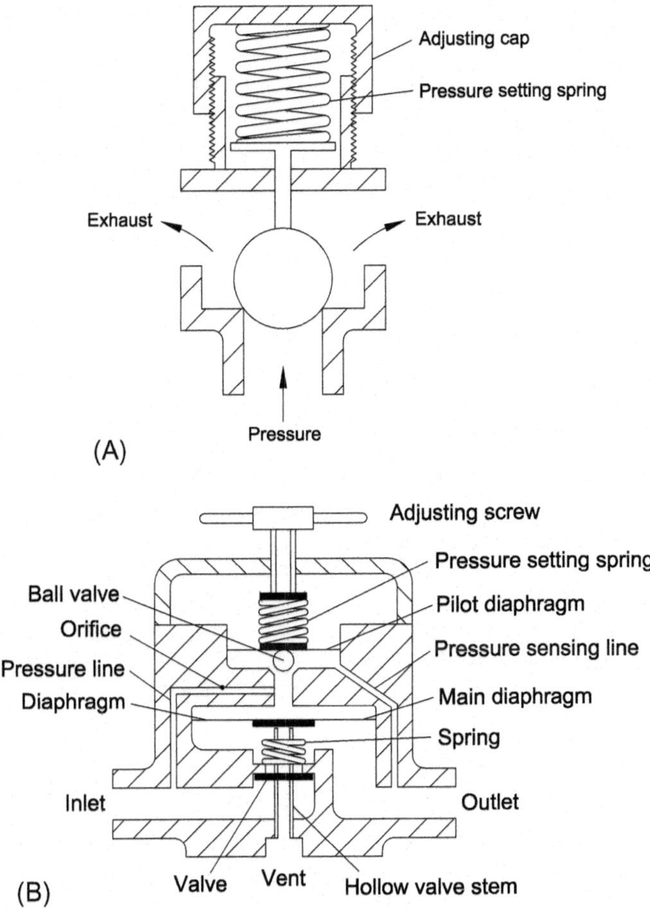

Fig. 18.16 (A) Relief valve; (B) pilot operated regulator.
Figures after Parr, A., 2011. Hydraulics and Pneumatics. third ed. Butterworth Heinemann.

vulnerable components. Simple course mesh strainers are usually fitted inside tanks to trap large particles. In addition separate filters are necessary to remove finer particles with typical filter positions as shown in Fig. 18.17.

18.6 Control valves

The building blocks for a hydraulic and pneumatic circuit are the pump or compressor, valves, the hydraulic actuators, sensors and the control circuit. Valves are used to control the flow direction, pressure and flow rate. There are two principal types of valve:

- Infinite position valves which can take up any position between open and closed
- Finite position valves which are used to block or allow flow

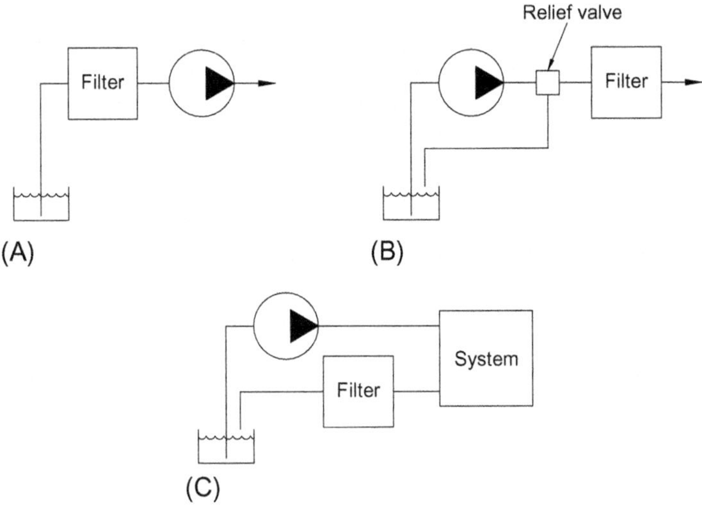

Fig. 18.17 Filter positions. (A) Inlet line filter, (B) pressure line filter, and (C) return line filter.

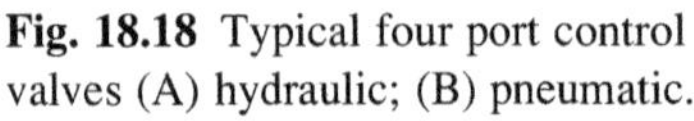

Fig. 18.18 Typical four port control valves (A) hydraulic; (B) pneumatic.

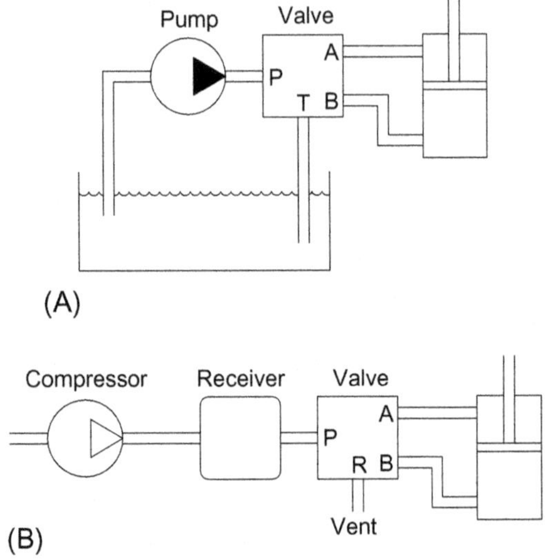

Connections to a valve are called ports. A simple on or off valve will have two ports. Most valves however have four ports as illustrated in Fig. 18.18. In the hydraulic valve application illustrated, hydraulic fluid is returned to the tank from port T. In the pneumatic valve application, air is vented to atmosphere from port R. Finite position valves are commonly described using a 'number of ports'/'number of positions' notation. As an example a 2/2 valve would be a two port, two position valve and a 4/3 valve would be a four port, three position valve. Fig. 18.19 shows the possible valve action for a 4/3 valve.

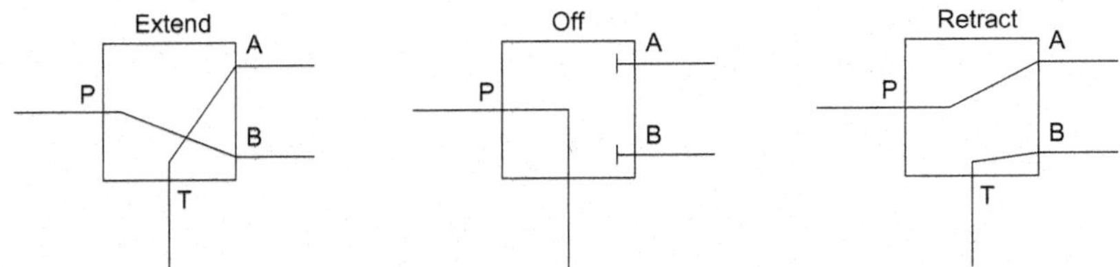

Fig. 18.19 Possible valve action for a 4/3 valve.

The scope of valves available is extensive. A set of graphic symbols has evolved to indicate the function of the valve with the need to show the physical construction of the valve and have been standardised in for example DIN24300, BS2917, ISO1219, ISO5599, CETOP (P.3). A valve is represented with a square for each of its switching positions. So if a valve has three different possible switching positions, it will be represented by three squares (Fig. 18.20B). The valve positions can be indicated with letters, with 0 being used for a central neutral position.

The general designations used are:

- Ports are shown on the outside of boxes;
- Arrows are used to represent the direction of flow;
- A T is used to show shut off positions;
- A series of specific symbols are used to designate whether the valve is actuated by a button, spring, pressure line, level, solenoid, detent and so on. Some examples are given in Fig. 18.21.

There are three principal types of control valve:

- Poppet valves—discs, cones or balls are used in combination with simple valve seats to control the flow;
- spool valves—a cylindrical spool with raised lands moves horizontally within the valve housing to block or open ports;
- rotary valves—a rotating spool aligns with holes in the casing.

Fig. 18.22 illustrates a two-way valve along with the corresponding symbolic representation. In two-way valves the flow path can be controlled by the position of a spool. Two-way valves are useful for directing flow to either of two different parts of a flow circuit depending on the spool position. The diagrams in Fig. 18.22 illustrate the flow

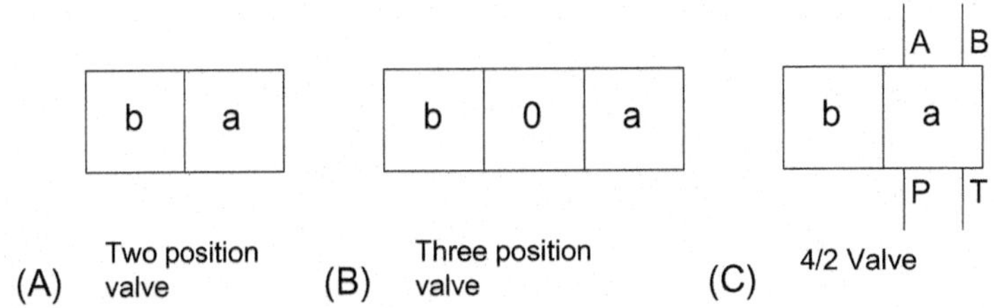

Fig. 18.20 Example valve symbols.

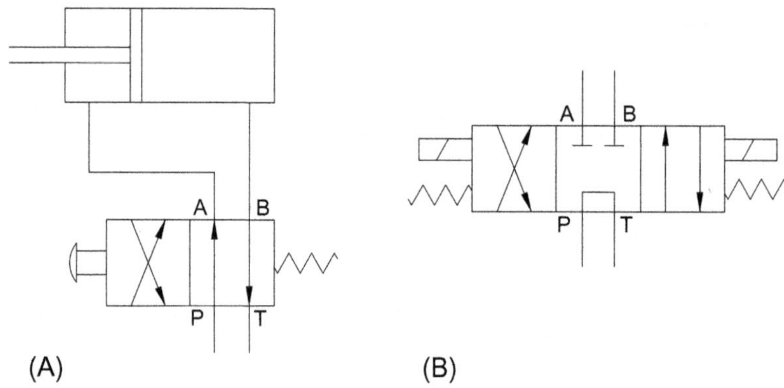

Fig. 18.21 Example symbols (A) Push button extend, spring retract when pushbutton released; (B) 4/3 valve, solenoid operated, spring return to centre. Pressure line uploads to tank and load locked in centre position.

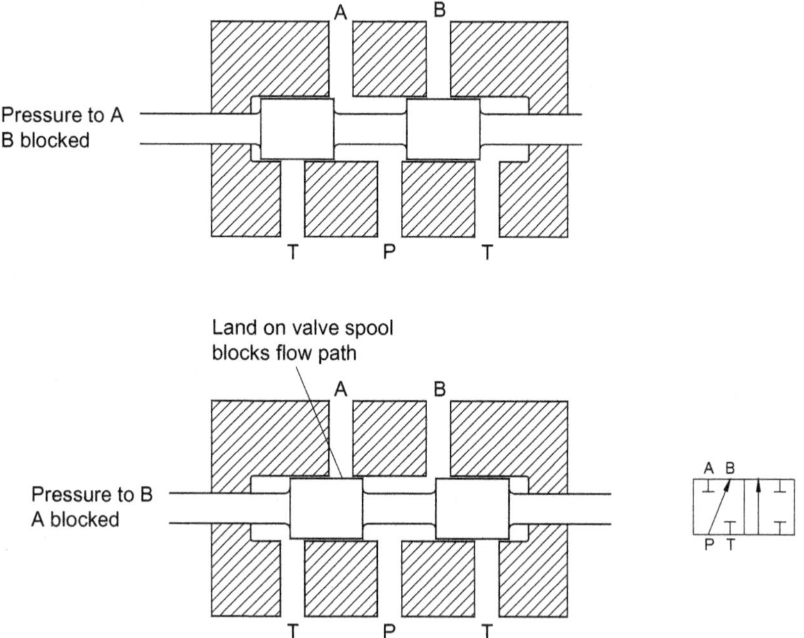

Fig. 18.22 Spool positions inside a two-way valve.

path options. In the first position, flow can pass from P to A. In the second position, flow can pass from P to B.

A very common method of actuating a spool valve is by using a solenoid. A solenoid is a specially designed type of electromagnet comprising a coil and plunger connected to an application of interest (Fig. 18.23). They are used very widely in

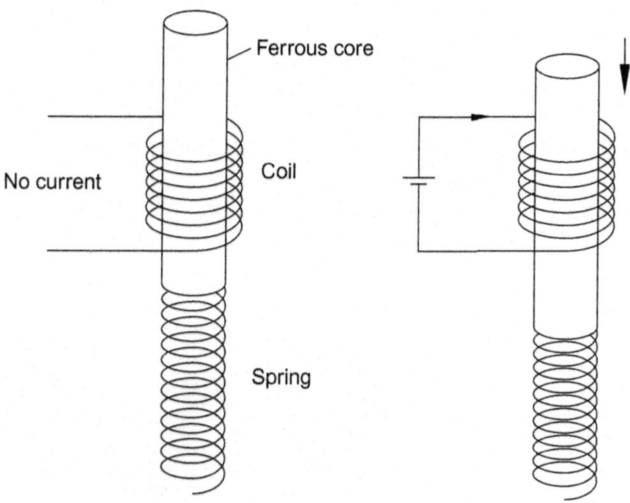

Fig. 18.23 Principle of operation for a solenoid.

industry with applications including the actuation of valves and locks. When a current flows through a wire an electromagnetic field is set up. If a wire is formed into a coil the magnetic field becomes many times stronger. Although a magnetic field can be formed in air it flows much more easily through iron or steel. This is the principle in adding an iron path or C stack around the coil, which concentrates the field to give high power densities and therefore more powerful and compact devices. If an iron path, known as a T plunger, is also added at the centre of the coil, this will concentrate the magnetic field even further. Because iron provides a more attractive path for the magnetic field than air a moveable T plunger will be drawn in further to the coil so that the maximum possible magnetic flux can pass through the plunger. To limit stray eddy currents in an electromagnet, the C stack and plunger can be made of laminated sheets of alternate conducting and insulating material. The magnetic field can still flow through its path but the eddy currents are limited to a two-dimensional plane. Unfortunately for electromagnets, although increasing the current will increase the power of the magnetic field, the heat generated in a solenoid increases approximately with the square of the current. Double the current will increase the heat generated by a factor of four. Each time an electromagnet is cycled its temperature rises and its force decreases. If cycled too fast it will overheat and burn out. Solenoid operated pneumatic and hydraulic control valves (Fig. 18.24) are available as stock items from a wide range of suppliers and represent a well-developed technology.

18.7 Pneumatic and hydraulic actuators

An actuator is a device that puts a parameter into action or physical motion. In the case of machine elements actuators include:

Fig. 18.24 Solenoid actuated valve.

- pneumatics
- hydraulics
- magnets
- electromagnets and solenoids
- electric motors
- piezoelectric devices

Fluid power actuators extract energy from a fluid and convert it to a mechanical output to perform a useful task. Fluid actuators can be classified as hydraulic or pneumatic depending on whether a liquid or gas is used as the working fluid. Fluid actuators include devices for producing either linear motion such as hydraulic cylinders, which extend and retract, or rotary motion such as hydraulic motors.

The simplest type of linear actuator is the single acting hydraulic or pneumatic cylinder as illustrated in Fig. 18.25. This consists of a piston inside a cylindrical housing called a barrel. Attached to one end of the piston is a rod that extends through a hole and seal in the cylinder. At the other end is a port for the admission and exit of hydraulic fluid such as oil. Single acting cylinders can only provide actuation in the extending direction. Retraction is achieved by means of gravity or by including a compression spring in the rod end.

The double acting hydraulic cylinder illustrated in Fig. 18.26 can be extended and retracted under hydraulic power. Pneumatic and hydraulic cylinders are available in a wide variety of sizes and end configurations and the rod end is usually threaded so that it can be connected directly to a load, a clevis, yoke or some other mating device.

The force in the extension stroke from a double-acting cylinder is given by

$$F = pA \tag{18.10}$$

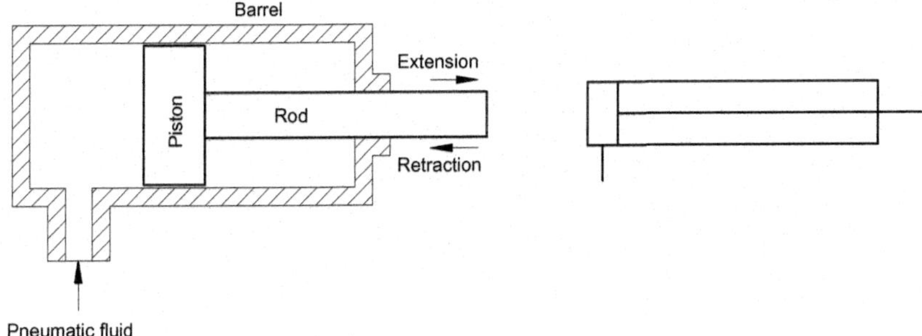

Fig. 18.25 Single-acting hydraulic cylinder and corresponding symbolic representation.

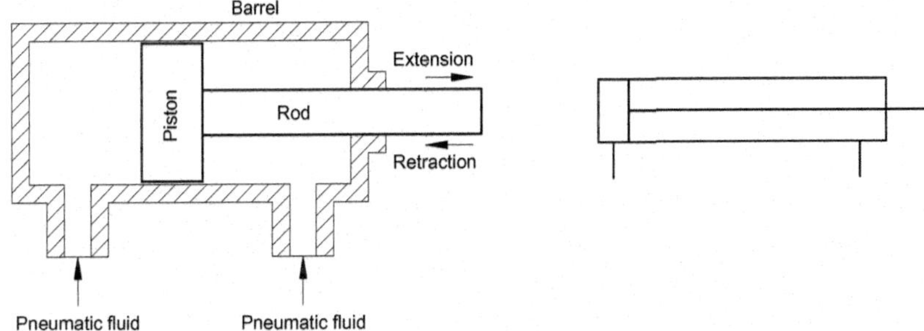

Fig. 18.26 Double-acting hydraulic cylinder and corresponding symbolic representation.

where

F = force (N),
p = pressure (N/m^2),
A = piston area (m^2).

The velocity is given by

$$V = \frac{Q}{A} \tag{18.11}$$

where Q = volumetric input flow (m^3/s).

In the retraction stroke, the force and speed are moderated by the effects of the rod area:

$$F = p(A - A_{rod}) \tag{18.12}$$

where A_{rod} = cross sectional area of the rod (m^2).

$$V = \frac{Q}{A - A_{rod}} \tag{18.13}$$

The power delivered by a hydraulic cylinder can be determined from

$$Power = FV = pQ \tag{18.14}$$

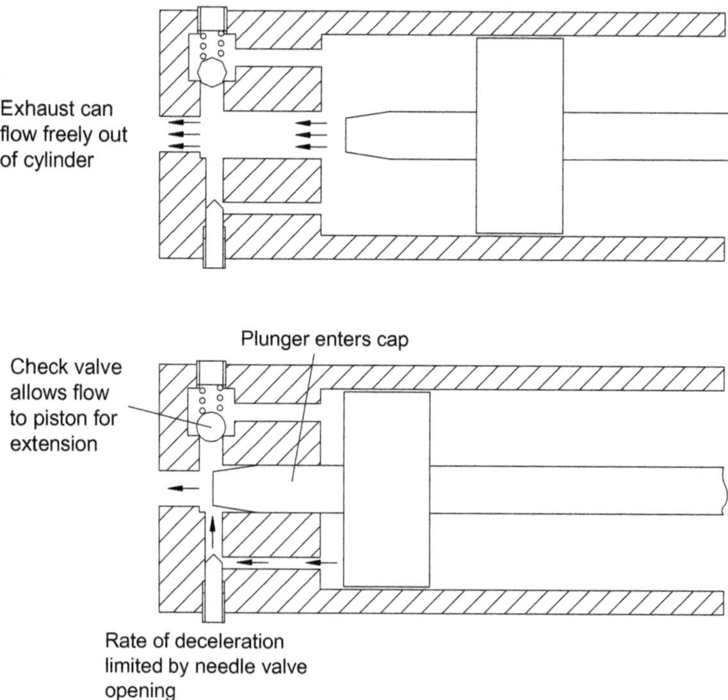
Exhaust can
flow freely out
of cylinder
Plunger enters cap
Check valve
allows flow
to piston for
extension
Rate of deceleration
limited by needle valve
opening

Fig. 18.27 Cylinder cushioning.

Some double-acting cylinders include cylinder cushions to decelerate the piston near the end of the stroke. These typically comprise channels and associated pressure relief valves as illustrated in Fig. 18.27.

18.8 Conclusions

Pneumatic and hydraulic systems can be used in a wide range of applications to apply a force to a component or material. Due to the compressible nature of air, pneumatic systems tend to enable softer action in comparison with hydraulic systems which can generally be operated at much higher pressures, producing higher forces. For both pneumatic and hydraulic systems, the pressurisation of the fluid can be remote from the actuators, thereby providing a compact solution in comparison, say, to the use of an electric motor.

References

Childs, P.R.N., 2001. Practical Temperature Measurement. Butterworth Heinemann.

Childs, P.R.N., 2013. Temperature measurement. In: Kutz, M. (Ed.), Handbook of Measurement in Science and Engineering. In: Vol. 1. Wiley.

Cohen, E.R., Taylor, B.N., 1999. The Fundamental Physical Constants. Physics Today, pp. BG5–BG9.

Parr, A., 2011. Hydraulics and Pneumatics, third ed. Butterworth Heinemann.

Standards

BS 2917:1977, ISO 1219-1976. Specification for graphical symbols used on diagrams for fluid power systems and components.

BS ISO 5599-1:2001. Pneumatic fluid power. Five-port directional control valves. Mounting interface surfaces without electrical connector.

BS ISO 5599-2:2001. Pneumatic fluid power. Five-port directional control valves. Mounting interface surfaces with optional electrical connector.

CETOP RP3.

DIN24300, Schaltsymbole für Ölhydraulik und Pneumatik.

ISO 5167-1:2003 Measurement of fluid flow by means of pressure differential devices inserted in circular cross-section conduits running full – Part 1: General principles and requirements.

Websites

At the time of going to press the world-wide-web contained useful information relating to this chapter at the following sites.

www.airlines-pneumatics.co.uk.
www.airtec-usa.com.
www.atlascopco.co.uk.
www.boschrexroth.co.uk.
www.cpi-pneumatics.co.uk/.
www.crane-bel.com/.
www.drakepneumatics.co.uk/.
www.eaton.com.
www.festo.com.
www.globalhydraulics.co.uk/.
www.harrisonpneumatics.co.uk/.
www.hydrastore.co.uk/.
www.hydrauliccomponents.co.uk/.
www.hydraulicpartsdirect.co.uk/.
www.hydrovaneproducts.com/.
www.johnguest.com.
www.metalwork.co.uk/.
www.mhpneumatics.co.uk/.
www.mhs.co.uk/.
www.norgren.com.
www.parker.com.
www.phoenixhydraulics.co.uk.
www.pneumaticsnorthern.co.uk/.
www.pneumatic-source.com/.
www.sauer-danfoss.com/.
www.smc.com.
www.spiraxsarco.com.
www.thoritedirect.co.uk.

Further reading

Barber, A. (Ed.), 1997. Pneumatic Handbook, eighth ed. Elsevier Science Ltd.

Childs, P.R.N., 2002. Advances in temperature measurement. In: Hartnett, J.P. (Ed.), Advances in Heat Transfer. Vol. 36. Academic Press, pp. 111–181.

ESDU, 2002. Temperature Measurement: Techniques. ESDU 02006. Engineering Sciences Data Unit.

ESDU, 2012. Pressure and Flow Measurement. ESDU 12003. Engineering Sciences Data Unit.

Forster, I., 2002. Electronics and engine control, Visteon Advanced Powertrain Systems. Institution of Mechanical Engineers Automotive Division Southern Centre presentation. Private communication.

Tolerancing and precision engineering

19

Chapter Outline

Nomenclature

Generally, preferred SI units have been stated.

a	internal radius of inner cylinder (m), inner radius (m)
b	outer radius of inner cylinder (m), interface radius (m), outer radius (m)
c	outer radius of outer cylinder (m)
C	length of the clearance volume (m)
d	wire diameter (m)
D	coil diameter (m), piston diameter (m)
E	Young's modulus (N/m^2)
f	coefficient of friction
$f(x)$	probability density function of the continuous random variable x
G	modulus of rigidity (N/m^2)
H	distance from gudgeon pin to piston crown (m)
k	spring stiffness (N/m)

Mechanical Design Engineering Handbook. https://doi.org/10.1016/B978-0-08-102367-9.00019-6

L	length of interference fit (m), connecting rod length (m)
M	distance from crank centre to underside of cylinder head (m)
n	number of coils, expansion stroke polytropic exponent
p	pressure (N/m^2)
P	cylinder pressure at start of stroke (N/m^2)
Q	ratio of clearance to clearance-plus-swept volumes
r	radius (m)
R	crank radius (m)
T	torque (N m)
u	radial growth (m)
$u_{thermal}$	radial growth due to temperature (m)
W	work done during the expansion stroke (N m)
α	coefficient of linear expansion ($°C^{-1}$)
δ	total diametral interference (m)
μ	Poisson's ratio, mean
ρ	density (kg/m^3)
σ	standard deviation
ω	angular velocity (rad/s)
ΔT	temperature difference (°C)

19.1 Introduction

The aims of this chapter are to introduce the concepts of component and process variability and the allocation of component tolerances. The selection of suitable tolerances for the assembly of components is a key requirement in the development of a machine design for given functionality, and this requires an understanding of process capability as well as careful attention to the detailing of the allowable variability component dimensions. The principles of engineering tolerancing can also be applied to model the variability of a process or function.

A solid is defined by its surface boundaries. Designers typically specify a component's nominal dimensions such that it fulfils its requirements. In reality, components cannot be made repeatedly to nominal dimensions due to surface irregularities and the intrinsic surface roughness. Some variability in dimensions must be allowed to ensure manufacture is possible. However, the variability permitted must not be so great that the performance of the assembled parts is impaired. The allowed variability on the individual component dimensions is called the tolerance.

The term tolerance applies not only to the acceptable range of component dimensions produced by manufacturing techniques but also to the output of machines or processes. For example, the power produced by a given type of internal combustion (IC) engine varies from one engine to another depending on the clearances and running fits for the specific engine concerned. In practice, the variability is usually found to be modelled by a frequency distribution curve, for example, the normal distribution (also called the Gaussian distribution). The modelling and analysis of this kind of variable output is outlined in Section 19.3. One of the tasks of the designer is to specify a dimension on a component or process and the allowable variability on this value that will give acceptable performance. A general introduction to the specification of component tolerances is given in Section 19.2.

19.2 Component tolerances

Control of dimensions is necessary in order to ensure assembly and interchangeability of components. Tolerances are specified on critical dimensions that affect clearances and interference fits. One method of specifying tolerances is to state the nominal dimension followed by the permissible variation. So a dimension could be stated as 40.00 ± 0.03 mm. This means that the dimension should be machined so that it is between 39.97 and 40.03 mm. Where the variation can vary either side of the nominal dimension, the tolerance is called a bilateral tolerance. For a unilateral tolerance, one tolerance is zero, e.g. $40.00\,^{+0.06}_{\;\;0.00}$ mm.

Most organisations have general tolerances that apply to dimensions when an explicit dimension is not specified on a drawing. For machined dimensions, a general tolerance may be ± 0.5 mm. So a dimension specified as 15.0 mm may range between 14.5 mm and 15.5 mm. Other general tolerances can be applied to features such as angles, drilled and punched holes, castings, forgings, weld beads and fillets.

When specifying a tolerance for a component, reference can be made to previous drawings or general engineering practice. Tolerances are typically specified in bands as defined in British or ISO (International Organisation for Standardisation) standards. Table 19.1 gives a guide for the general applications of tolerances. For a given tolerance, e.g. H7/s6 a set of numerical values is available from a corresponding chart for the size of component under consideration. The following section gives specific examples of this for a shaft or cylindrical spigot fitting into a hole.

19.2.1 Standard fits for holes and shafts

A common engineering task is to define tolerances for an assembly of cylindrical components, such as a shaft, fitting or rotating inside a corresponding cylindrical component or hole. The tightness of fit will depend on the application. For example, a gear located onto a shaft would require a 'tight' interference fit where the diameter of the shaft is actually slightly greater than the inside diameter of the gear hub in order to be able to transmit the desired torque. Alternatively, the diameter of a journal bearing must be greater than the diameter of the shaft to allow rotation. Given that it is not economically possible to manufacture components to exact dimensions, some variability in sizes of both the shaft and hole dimension must be specified. However, the range of variability should not be so large that the operation of the assembly is impaired. Rather than having an infinite variety of tolerance dimensions that could be specified, national and international standards have been produced defining bands of tolerances, examples of which are listed in Table 19.1, e.g. H11/c11. To turn this information into actual dimensions, corresponding tables exist defining the tolerance

Table 19.1 Example tolerance bands and typical applications

Class	Description	Characteristic	ISO code	Assembly	Application
Clearance	Loose running fit	For wide commercial tolerances	H11/c11	Noticeable clearance	Internal combustion (IC) engine exhaust valve in guide.
	Free running fit	Good for large temperature variations, high running speeds or heavy journal pressures	H9/d9	Noticeable clearance	Multiple bearing shafts, hydraulic piston in cylinder, removable levers, bearings for rollers.
	Close running fit	For running on accurate machines and accurate location at moderate speeds and journal pressures	H8/f7	Clearance	Machine tool main bearings, crankshaft and connecting rod bearings, shaft sleeves, clutch sleeves, guide blocks.
	Sliding fit	When parts are not intended to run freely, but must move and turn and locate accurately	H7/g6	Push fit without noticeable clearance	Push on gear wheels and clutches, connecting rod bearings, indicator pistons.
	Location clearance fit	Provides snug fit for location of stationary parts, but can be freely assembled	H7/h6	Hand pressure with lubrication	Gears, tailstock sleeves, adjusting rings, loose bushes for piston bolts and pipelines.

Table 19.1 Continued

Class	Description	Characteristic	ISO code	Assembly	Application
Transition	Location transition fit	For accurate location (compromise between clearance and interference fit)	H7/k6	Easily tapped with hammer	Pulleys, clutches, gears, flywheels, fixed handwheels and permanent levers.
	Location transition fit	For more accurate location	H7/n6	Needs pressure	Motor shaft armatures, toothed collars on wheels.
Interference	Locational interference fit	For parts requiring rigidity and alignment with accuracy of location	H7/p6	Needs pressure	Split journal bearings.
	Medium drive fit	For ordinary steel parts or shrink fits on light sections	H7/s6	Needs pressure or temperature difference	Clutch hubs, bearing bushes in blocks, wheels, connecting rods. Bronze collars on grey cast iron hubs.

levels for the size of dimension under consideration (see Tables 19.2 and 19.3). In order to use this information, the following list and Fig. 19.1 give definitions used in conventional tolerancing. Usually, the hole-based system is used (data sheet BS 4500A, Table 19.2) as this results in a reduction in the variety of drill, reamer, broach and gauge tooling required within a company. A series of definitions are commonly used in component tolerances as follows:

- Size: A number expressing in a particular unit the numerical value of a dimension.
- Actual size: The size of a part as obtained by measurement.
- Limits of size: The maximum and minimum sizes permitted for a feature.
- Maximum limit of size: The greater of the two limits of size.

Table 19.2 Selected ISO fits (holes basis; after BS4500 data sheet 4500A)

Nominal sizes		Clearance fits										Transition fits				Interference fits				Nominal sizes	
		Tolerance		Tolerance		Tolerance		Tolerance		Tolerance		Tolerance		Tolerance		Tolerance		Tolerance		Tolerance	
Over	To	H11	c11	H9	d10	H9	e9	H8	f7	H7	g6	H7	h6	H7	k6	H7	n6	H7	p6	H7	s6	Over	To
mm	mm	0.001 mm	0.001 mm	0.001 mm	0.001 mm	0.001 mm	0.001 mm	0.001 mm	0.001 mm	0.001 mm	0.001 mm	0.001 mm	0.001 mm	0.001 mm	0.001 mm	0.001 mm	0.001 mm	0.001 mm	0.001 mm	0.001 mm	0.001 mm	mm	mm
—	3	+60 0	−60 −120	+25 0	−20 −60	+25 0	−14 −39	+14 0	−6 −16	+10 0	−2 −8	+10 0	−6 0	+10 0	+6 0	+10 0	+10 +4	+10 0	+12 +6	+10 0	+20 +14	—	3
3	6	+75 0	−70 −145	+30 0	−30 −78	+30 0	−20 −50	+18 0	−10 −22	+12 0	−4 −12	+12 0	−8 0	+12 0	+9 +1	+12 0	+16 +8	+12 0	+20 +12	+12 0	+27 +19	3	6
6	10	+90 0	−80 −170	+36 0	−40 −98	+36 0	−25 −61	+22 0	−13 −28	+15 0	−5 −14	+15 0	−9 0	+15 0	+10 +1	+15 0	+19 +10	+15 0	+24 +15	+15 0	+32 +23	6	10
10	18	+110 0	−95 −205	+43 0	−50 −120	+43 0	−32 −75	+27 0	−16 −34	+18 0	−6 −17	+18 0	−11 0	+18 0	+12 +1	+18 0	+23 +12	+18 0	+29 +18	+18 0	+39 +28	10	18
18	30	+130 0	−110 −240	+52 0	−65 −149	+52 0	−40 −92	+33 0	−20 −41	+21 0	−7 −20	+21 0	−13 0	+21 0	+15 +2	+21 0	+28 +15	+21 0	+35 +22	+21 0	+48 +35	18	30
30	40	+160 0	−120 −280	+62	−80	+62	−50	+39	−25	+25	−9	+25	−16	+25	+18	+25	+33	+25	+42	+25	+59	30	40
40	50	+160 0	−130 −290	0	−180	0	−112	0	−50	0	−25	0	0	0	+2	0	+17	0	+26	0	+43	40	50
50	65	+190 0	−140 −330	+74	−100	+74	−60	+46	−30	+30	−10	+30	−19	+30	+21	+30	+39	+30	+51	+30 0	+72 +53	50	65

65	80	+190	−150	0	−220	0	−134	0	−60	0	−29	0	0	0	+2	0	+20	0	+32	+30	+78	65	80
		0	−340																	0	+59		
80	100	+220	−170																	+35	+93	80	100
		0	−390	+87	−120	+87	−72	+54	−36	+35	−12	+35	−22	+35	+25	+35	+45	+35	+59	0	+71		
100	120	+220	−180	0	−260	0	−159	0	−71	0	−34	0	0	0	+3	0	+23	0	+37	+35	+101	100	120
		0	−400																	0	+79		
120	140	+250	−200																	+40	+117	120	140
		0	−450																	0	+92		
140	160	+250	−210	+100	−145	+100	−84	+63	−43	+40	−14	+40	−25	+40	+28	+40	+52	+40	+68	+40	+125	140	160
		0	−460	0	−305	0	−185	0	−83	0	−39	0	0	0	+3	0	+27	0	+43	0	+100		
160	180	+250	−230																	+40	+133	160	180
		0	−480																	0	+108		
180	200	+290	−240																	+46	+151	180	200
		0	−530																	0	+122		
200	225	+290	−260	+115	−170	+115	−100	+72	−50	+46	−15	+46	−29	+46	+33	+46	+60	+46	+79	+46	+159	200	225
		0	−550	0	−355	0	−215	0	−96	0	−44	0	0	0	+4	0	+31	0	+50	0	+130		
225	250	+290	−280																	+46	+169	225	250
		0	−570																	0	+140		
250	280	+320	−300																	+52	+190	250	280
		0	−620	+130	−190	+130	−110	+81	−56	+52	−17	+52	−32	+52	+36	+52	+66	+52	+88	0	+158		
280	315	+320	−330	0	−400	0	−240	0	−108	0	−49	0	0	0	+4	0	+34	0	+56	+52	+202	280	315
		0	−650																	0	+170		
315	355	+360	−360																	+57	+226	315	355
		0	−720	+140	−210	+140	−125	+89	−62	+57	−18	+57	−36	+57	+40	+57	+73	+57	+98	0	+190		
355	400	+360	−400	0	−440	0	−265	0	−119	0	−54	0	0	0	+4	0	+37	0	+62	+57	+244	355	400
		0	−760																	0	+208		
400	450	+400	−440																	+63	+272	400	450
		0	−840	+155	−230	+155	−135	+97	−68	+63	−20	+63	−40	+63	+45	+63	+80	+63	+108	0	+232		
450	500	+400	−480	0	−480	0	−290	0	−131	0	−60	0	0	0	+5	0	+40	0	+68	+63	+292	450	500
		0	−880																	0	+252		

Table 19.3 Selected ISO fits (shafts basis; after BS4500 data sheet 4500B)

Nominal sizes		Clearance fits												Transition fits				Interference fits				Nominal sizes	
		C11 / h11		D10 / h9		E9 / h9		F8 / h7		G7 / h6		H7 / h6		h6 / K7		h6 / N7		h6 / P7		h6 / S7		Holes / Shafts	
		Tolerance		Tolerance		Tolerance		Tolerance		Tolerance		Tolerance		Tolerance		Tolerance		Tolerance		Tolerance			
Over	To	h11	C11	h9	D10	h9	E9	h7	F8	h6	G7	h6	H7	h6	K7	h6	N7	h6	P7	h6	S7	Over	To
mm	mm	0.001 mm	0.001 mm	0.001 mm	0.001 mm	0.001 mm	0.001 mm	0.001 mm	0.001 mm	0.001 mm	0.001 mm	0.001 mm	0.001 mm	0.001 mm	0.001 mm	0.001 mm	0.001 mm	0.001 mm	0.001 mm	0.001 mm	0.001 mm	mm	mm
—	3	0 −60	+120 +60	0 −25	+60 +20	0 −25	+39 +14	0 −10	+20 +6	0 −6	+12 +2	0 −6	+10 0	0 −6	0 −10	0 −6	−4 −14	0 −6	−6 −16	0 −6	−14 −24	—	3
3	6	0 −75	+145 +70	0 −30	+78 +30	0 −30	+50 +20	0 −12	+28 +10	0 −8	+16 +4	0 −8	+12 0	0 −8	+3 −9	0 −8	−4 −16	0 −8	−8 −20	0 −8	−15 −27	3	6
6	10	0 −90	+170 +80	0 −36	+98 +40	0 −36	+61 +25	0 −15	+35 +13	0 −9	+20 +5	0 −9	+15 0	0 −9	+5 −10	0 −9	−4 −19	0 −9	−9 −24	0 −9	−17 −32	6	10
10	18	0 −110	+205 +95	0 −43	+120 +50	0 −43	+75 +32	0 −18	+43 +16	0 −11	+24 +6	0 −11	+18 0	0 −11	+6 −12	0 −11	−5 −23	0 −11	−11 −29	0 −11	−21 −39	10	18
18	30	0 −130	+240 +110	0 −52	+149 +65	0 −52	+92 +40	0 −21	+53 +20	0 −13	+28 +7	0 −13	+21 0	0 −13	+6 −15	0 −13	−7 −28	0 −13	−14 −35	0 −13	−27 −48	18	30
30	40	0 −160	+280 +120	0	+180	0	+112	0	+64	0	+34	0	+25	0	+7	0	−8	0	−17	0	−34	30	40
40	50	0 −160	+290 +130	−62	+80	−62	+50	−25	+25	−16	+9	−16	0	−16	−18	−16	−33	−16	−42	−16	−59	40	50
50	65	0 −190	+330 +140	0	+220	0	+134	0	+76	0	+40	0	+30	0	+9	0	−9	0	−21	0 −19	−42 −72	50	65

65	80	0 −190	+340 +150	−74	+100	−74	+60	−30	+30	−19	+10	−19	0	−19	−21	−19	−39	−19	−51	0 −19	−48 −78	65	80
80	100	0 −220	+390 +170	0	+260	0	+159	0	+90	0	+47	0	+35	0	+10	0	−10	0	−24	0 −22	−58 −93	80	100
100	120	0 −220	+400 +180	−87	+120	−87	+72	−35	+36	−22	+12	−22	0	−22	−25	−22	−45	−22	−59	0 −22	−66 −101	100	120
120	140	0 −250	+450 +200																	0 −25	−77 −117	120	140
140	160	0 −250	+460 +210	0 −100	+305 +145	0 −100	+185 +85	0 −40	+106 +43	0 −25	+54 +14	0 −25	+40 0	0 −25	+12 −28	0 −25	−12 −52	0 −25	−28 −68	0 −25	−85 −125	140	160
160	180	0 −250	+480 +230																	0 −25	−93 −133	160	180
180	200	0 −290	+530 +240																	0 −29	−105 −151	180	200
200	225	0 −290	+550 +260	0 −115	+355 +170	0 −115	+215 +100	0 −46	+122 +50	0 −29	+61 +15	0 −29	+46 0	0 −29	+13 −33	0 −29	−14 −60	0 −29	−33 −79	0 −29	−113 −159	200	225
225	250	0 −290	+570 +280																	0 −29	−123 −169	225	250
250	280	0 −320	+620 +300	0	+400	0	+240	0	+137	0	+62	0	+52	0	+16	0	−14	0	−36	0 −32	−138 −190	250	280
280	315	0 −320	+650 +330	−130	+190	−130	+110	−52	+56	−32	+17	−32	0	−32	−36	−32	−66	−32	−88	0 −32	−150 −202	280	315
315	355	0 −360	+720 +360	0	−440	0	+265	0	+151	0	+75	0	+57	0	+17	0	−16	0	−41	0 −36	−169 −226	315	355
355	400	0 −360	+760 +400	−140	+210	−140	+125	−57	+62	−36	+18	−36	0	−36	−40	−36	−73	−36	−98	0 −36	−187 −244	355	400
400	450	0 −400	+840 +440	0	−480	0	+290	0	+165	0	+83	0	+63	0	+18	0	−17	0	−45	0 −40	−209 −272	400	450
450	500	0 −400	+880 +480	−155	+230	−155	+135	−63	+68	−40	+20	−40	0	−40	−45	−40	−80	−40	−108	0 −40	−229 −292	450	500

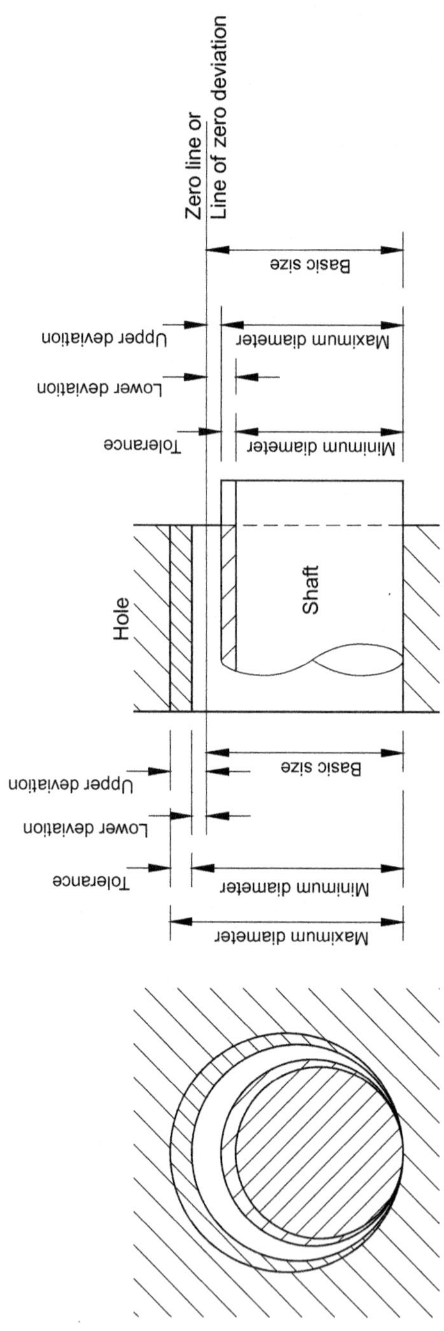

Fig. 19.1 Definitions of terms used in conventional tolerancing. After BS4500.

- Minimum limit of size: The smaller of the two limits of size.
- Basic size: The size by reference to which the limits of size are fixed.
- Deviation: The algebraic difference between a size and the corresponding basic size.
- Actual deviation: The algebraic difference between the actual size and the corresponding basic size.
- Upper deviation: The algebraic difference between the maximum limit of size and the corresponding basic size.
- Lower deviation: The algebraic difference between the minimum limit of size and the corresponding basic size.
- Tolerance: The difference between the maximum limit of size and the minimum limit of size.
- Shaft: The term used by convention to designate all external features of a part (including parts that are not cylindrical).
- Hole: The term used by convention to designate all internal features of a part.

BS4500 data sheets 4500A and 4500B list selected ISO clearance, transitional and interference fits for internal and external components. When using the Standard upper case letters refer to the hole and lower case letters to the shaft.

Example 19.1. Find the shaft and hole dimensions for a loose running fit with a 35 mm diameter basic size.

Solution

From Table 19.1, the ISO symbol is 35H11/c11.

(the 35 designating a 35 mm nominal tolerance).

From data sheet BS4500A (Table 19.2) the hole tolerance, for a 35 mm hole, is $^{+0.16}_{0.00}$ mm and the shaft tolerance, for a 35 mm shaft, is $^{-0.12}_{-0.28}$ mm.

So the hole should be given dimensions 35.00 $^{+0.16}_{0.00}$ mm and the shaft 35.00 $^{-0.12}_{-0.28}$ mm.

Example 19.2. Find the shaft and hole dimensions for a medium drive fit using a basic hole size of 60 mm.

Solution

From Table 19.1, the ISO symbol is 60H7/s6.

From data sheet BS4500A (Table 19.2), the hole tolerance is $^{+0.030}_{0.000}$ mm and the shaft tolerance is $^{+0.072}_{+0.053}$ mm.

So the hole should be given dimensions 60.000 $^{+0.030}_{0.000}$ mm and the shaft 60.000 $^{+0.072}_{+0.053}$ mm.

19.2.2 Interference fits

Interference fits are those for which, prior to assembly, the inside component is larger than the outside component. There is some deformation of the components after assembly and a pressure exists at the mating surfaces. This pressure is given by

Eq. (19.1) if both components are of the same material or by Eq. (19.2) if the two components are of differing materials (see Fig. 19.2 for definition of radii).

$$p = \frac{E\delta}{2b}\frac{(c^2 - b^2)(b^2 - a^2)}{2b^2(c^2 - a^2)}, \tag{19.1}$$

$$p = \frac{\delta}{2b\left[\frac{1}{E_o}\left(\frac{c^2 + b^2}{c^2 - b^2} + \mu_o\right) + \frac{1}{E_i}\left(\frac{b^2 + a^2}{b^2 - a^2} - \mu_i\right)\right]}, \tag{19.2}$$

where p = pressure at the mating surface (N/m^2), δ = total diametral interference (m), a = internal radius of inner cylinder (m), b = outer radius of inner cylinder (m), c = outer radius of outer cylinder (m), E = Young's modulus (N/m^2), μ = Poisson's ratio.

The subscripts i and o refer to the inner and outer components, respectively.

The tensile stress in the outer component, σ_o, can be calculated from Eq. (19.3) and the compressive stress in the inner component, σ_i, from Eq. (19.4).

$$\sigma_o = p\left(\frac{c^2 + b^2}{c^2 - b^2}\right) \tag{19.3}$$

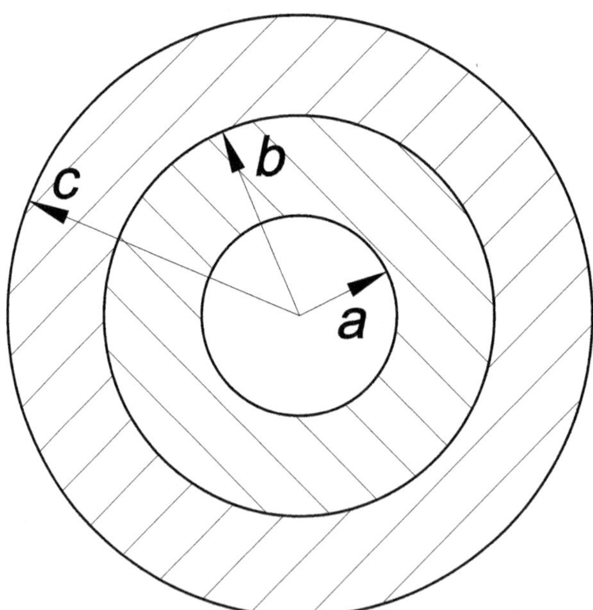

Fig. 19.2 Terminology for cylindrical interference fit.

$$\sigma_i = -p\left(\frac{b^2 + a^2}{b^2 - a^2}\right) \tag{19.4}$$

The increase in diameter of the outer component, δ_o, due to the tensile stress can be calculated from Eq. (19.5) and the decrease in diameter of the inner component due to the compressive stress, δ_i, from Eq. (19.6).

$$\delta_o = \frac{2bp}{E_o}\left(\frac{c^2 + b^2}{c^2 - b^2} + \mu_o\right) \tag{19.5}$$

$$\delta_i = -\frac{2bp}{E_i}\left(\frac{b^2 + a^2}{b^2 - a^2} - \mu_i\right) \tag{19.6}$$

The torque which can be transmitted between cylindrical components by an interference fit can be estimated by

$$T = 2fp\pi b^2 L \tag{19.7}$$

where T = torque (N m), f = coefficient of friction, p = interference pressure (N/m^2), b = interface radius (m) and L = length of interference fit (m).

Example 19.3. What interference fit is required in order to transmit 18 kN m torque between a 150 mm diameter steel shaft and a 300 mm outside diameter cast iron hub which is 250 mm long. (Take Young's modulus for the steel and cast iron to be 200 GPa and 100 GPa, respectively, assume the coefficient of friction is 0.12 and that Poisson's ratio is 0.3 for both materials).

Solution

The torque is given by $T = 2fp\pi b^2 L$ and this equation can be rearranged to evaluate the pressure required to transmit the torque:

$p = \frac{T}{2f\pi b^2 L} = \frac{18\times10^3}{2\times0.12\pi(0.075)^2\times0.25} = 16.97\times10^6\,\text{N/m}^2.$

The interference required to generate this pressure is given by

$$\begin{aligned}\delta &= 2bp\left[\frac{1}{E_o}\left(\frac{c^2 + b^2}{c^2 - b^2} + \mu_o\right) + \frac{1}{E_i}(1 - \mu_i)\right]\\ &= 2\times0.075\times16.97\times10^6\left[\frac{1}{100\times10^9}\left(\frac{0.15^2 + 0.075^2}{0.15^2 - 0.075^2} + 0.3\right) + \frac{1}{200\times10^9}(1 - 0.3)\right]\\ &= 5.89\times10^{-5}\ \text{m}\end{aligned}$$

So the required interference fit is 0.059 mm. This could be specified using the H7/s6 tolerance band with the hole diameter dimension as $150.000\,^{+0.04}_{+0}$ mm and the shaft diameter as $150.000\,^{+0.125}_{+0.100}$ mm.

19.2.3 Machine capability

A given manufacturing technique can only produce a component within a range of accuracy around the specified nominal dimension. For instance, centre lathes are capable of machining 10mm diameter components to tolerances of 0.01mm (BSI, 1991). An indication of the capability of a variety of production processes is given in Table 19.4. In addition, different manufacturing techniques produce differing levels of surface roughness, as shown in Table 19.5. Drilling, for example, cannot be normally used to give a surface roughness of better than 0.8 μm. Tolerances for drilled components should therefore not be specified smaller than 0.0008mm. When defining a tolerance, the designer should have in mind the 'total design' of the product. This encompasses efficient operation of the product, the ability of the company to manufacture the component and total costs (design, manufacture, servicing and performance, etc.). The cost of production is a key factor in the vast majority of manufactured items, and a detailed assessment of the cost of manufacture needs to be considered at the design stage. In general, for lower accuracy processes, ±0.25 to ±0.025mm, the cheapest machining operation is turning, followed by milling, boring, planning and shaping (BSI 1991). For higher accuracy processes ±0.025 to ±0.0025mm, the cheapest machining operation is generally grinding followed by

Table 19.4 Approximate tolerance capability for various production processes

Process	Material	Typical tolerance (mm)	Typical tolerance (in)
Sand casting			
	Cast iron	±1.3	(±0.050)
	Steel	±1.5	(±0.060)
	Aluminium	±0.5	(±0.020)
Die casting		±0.12	±0.005
Plastic moulding:			
	Polyethylene	±0.3	(±0.010)
	Polystyrene	±0.15	(±0.006)
Machining:			
Drilling, 6mm (0.25 in hole)		±0.08/±0.03	(+0.003/−0.001)
Drilling, 25.4mm (0.75 in to 1 in hole)		±0.25/±0.03	(+0.010/−0.001)
Milling		±0.08	(±0.003)
Turning		±0.05	(±0.002)
Abrasive			
Grinding		±0.008	(±0.0003)
Lapping		±0.005	(±0.0002)
Honing		±0.005	(±0.0002)
Nontraditional and thermal			

Table 19.4 Continued

Process	Material	Typical tolerance (mm)	Typical tolerance (in)
Chemical machining		±0.08	(±0.003)
Electric discharge		±0.025	(±0.001)
Electrochemical grinding		±0.025	(±0.001)
Electrochemical machining		±0.05	(±0.002)
Electron beam cutting		±0.08	(±0.003)
Laser beam cutting		±0.08	(±0.003)
Plasma arc cutting		±1.3	(±0.050)

Various sources including Bakerjian and Mitchell (1992), Brown and Sharpe (1992), Airforce and Navy Aeronautical Design Standard AND 10387, BSI (1991).

Table 19.5 Surface roughness values produced by common production processes

PROCESS	ROUGHNESS VALUE, R_a (μm)												
	50	25	12.5	6.3	3.2	1.6	0.8	0.4	0.2	0.1	0.05	0.025	0.0125
Flame cutting													
Sawing													
Planing, Shaping													
Drilling													
Electric discharge machining													
Milling													
Broaching													
Reaming													
Boring, Turning													
Roller burnishing													
Grinding													
Honing													
Polishing													
Lapping													
Superfinishing													
Sand casting													
Hot rolling													
Forging													
Permanent mould casting													
Investment casting													
Extruding													
Cold rolling, Drawing													
Die casting													

Average application
Less frequent application

After BSI (1991)

honing and lapping; however, precise details will depend on the component features to be machined. A methodology for assessing cost of manufacture is given in Swift and Booker (2013).

19.2.4 Geometric tolerancing

The location for a drilled hole has been specified in Fig. 19.3 by means of a dimension with bilateral tolerances. The dimensions given imply a square rectangular tolerance zone as identified in Fig. 19.4A, anywhere within which it would be permissible to set the hole centre. Examination of the tolerance zone shows that the maximum deviation in location of the hole centre actually occurs along the diagonals as shown in the right-hand part of Fig. 19.4. Thus, although the drawing has superficially specified the maximum tolerance deviation as ±0.1 mm or 0.2 mm, the maximum actual deviation within the specification of the drawing is 0.28 mm. If the deviation of 0.28 mm provides correct functioning of the device in the diagonal directions, it is likely to be acceptable in all directions. If this is the case, then a circular tolerance could be specified with a diameter of 0.28 mm as shown in Fig. 19.5. The area of this circular zone is 57% larger than the square tolerance zone and could have substantial implications for cost reduction in the manufacturing process.

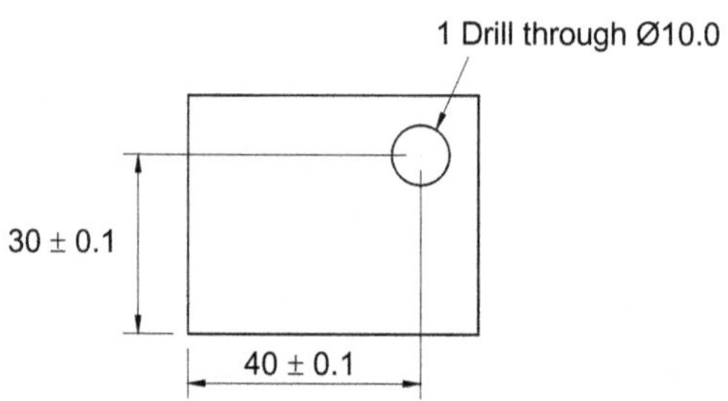

Fig. 19.3 Drilled hole location.

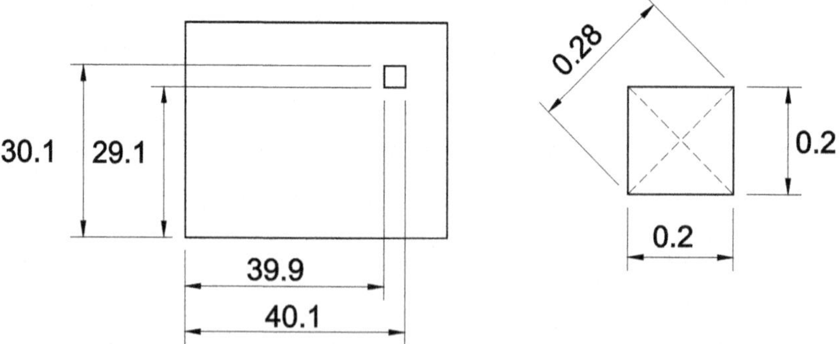

Fig. 19.4 Implied tolerance zone.

The specification of a circular tolerance zone can be achieved by specifying the exact desired location of the hole and a permissible circular tolerance zone centred at the exact location as illustrated in Fig. 19.6. Here, the exact location of the hole is indicated by the boxed dimensions and the tolerance zone is indicated by the ⌖ symbol within the rectangular box. The diameter of the tolerance zone is specified to the right of the ⌖ symbol within the rectangular box.

The symbol used for the positional tolerance is one of a series used in engineering drawing as illustrated in Table 19.6 and their use is referred to as geometric tolerancing. Geometric tolerancing allows the engineer to specify control over feature locations and deviation of form such as flatness, parallelism and concentricity. The various symbols used and a description are given in Table 19.6. Fig. 19.7 illustrates their use for defining the locating radius and shoulder for a centrifugal compressor

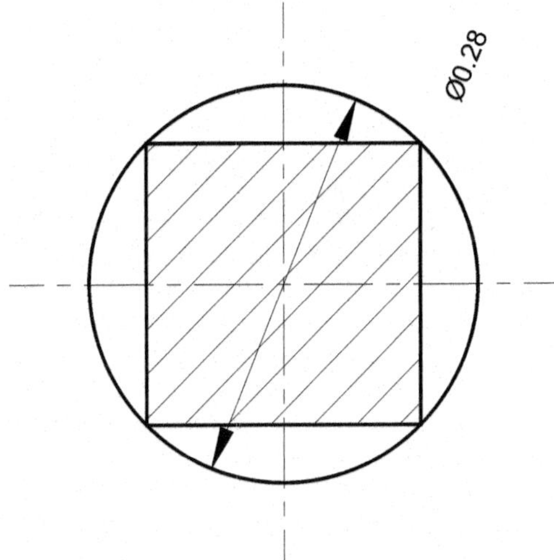

Fig. 19.5 Circular tolerance zone.

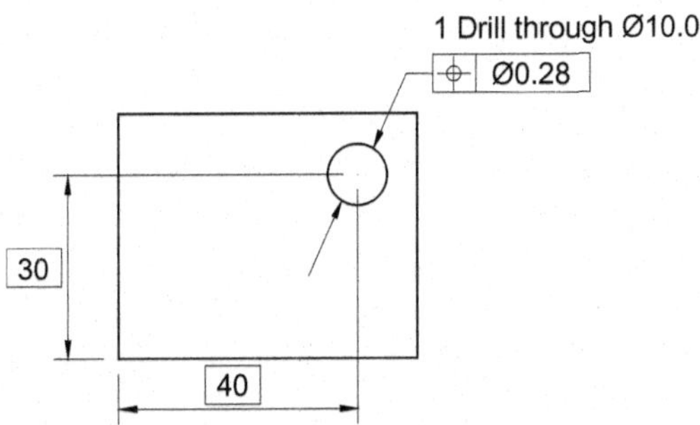

Fig. 19.6 Circular tolerance zone specification.

Table 19.6 Geometric tolerancing symbols

Symbol	Definition	Description
⏤	Straightness	This defines the straightness tolerance for an edge, axis of revolution or line on a surface. The tolerance zone is either the area between two parallel straight lines in the plane containing the toleranced edge or surface line or a cylindrical region for an axis of revolution.
⏥	Flatness	This is the condition of a surface having all elements in one plane.
○	Roundness	This is the condition on a surface of revolution (cylinder or cone or sphere) where all the points of the surface intersected by any plane (i) perpendicular to a common axis (cylinder/cone) (ii) passing through a common centre (sphere) are equidistant from the centre.
⌭	Cylindricity	A cylindricity tolerance zone is the annular space between two coaxial cylinders. The tolerance zone is the radial separation of the two cylinders.
⌒	Profile of a line	The profile of a line is the condition permitting a uniform amount of profile variation, either unilaterally or bilaterally, along a line element of a feature.
⌓	Profile of a surface	This defines the tolerance zone of variability for a surface relative to the perfect or exact form.
//	Parallelism	This is the condition of a surface, line or axis which is equidistant at all points from a datum plane or axis.
⊥	Squareness	The squareness tolerance defines the angular tolerance for perpendicularity (90°).
∠	Angularity	This defines the allowable variability for angular location. The angular tolerance zones are defined by either two parallel lines or surfaces.
⌖	Position	This defines a zone within which the axis or centre plane of a feature is permitted to vary from the true, theoretically correct, position.
◎	Concentricity	A concentricity tolerance limits the deviation of the centre or axis of a feature from its true position.
⌯	Symmetry	This typically defines the area between two parallel lines or the space between two parallel surfaces, which are symmetrically located about a datum feature.
↗	Runout	This is the deviation in position of a surface of revolution as a part is revolved about a datum axis.

impeller. In order to avoid significant out of balance forces and correct clearance between the compressor blade tips and the stationary shroud, it is critical that the impeller is located concentrically on the shaft and the back face is perpendicular to the shaft. Control over the shaft radial dimensional variation has been specified by means of the runout symbol which defines the limits of variability for the cylindrical surface and the squareness symbol for the shoulder.

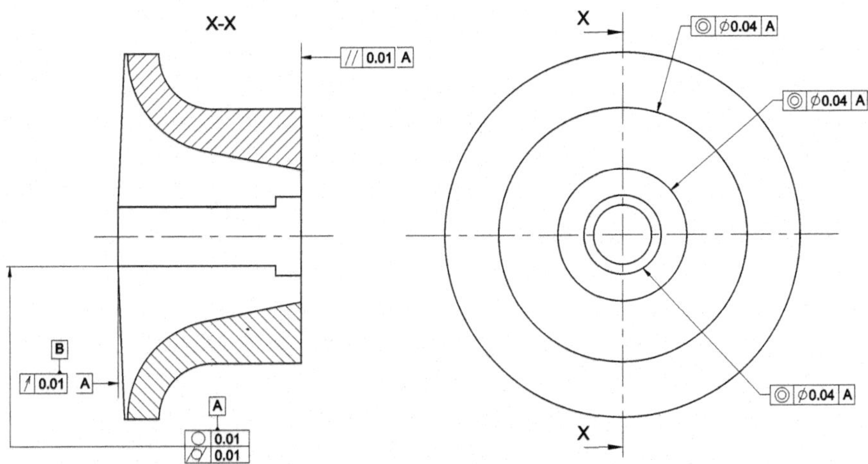

Fig. 19.7 Example of the use of geometric tolerancing for specifying control over a centrifugal compressor shaft dimensions.

BS 8888:2011 provides information on the British Standard approach to specifying and graphically representing products using engineering drawing, computer aided design (CAD) and product specification methods and symbols, which are also reviewed by Simmons et al. (2012). The texts by Spotts (1983), Drake (1999) and Krulikowski (2012) provide useful information on tolerancing of components.

19.3 Statistical tolerancing

Many processes produce outputs, such as component dimensions or power, which follow a frequency distribution such as the 'normal' or Gaussian distribution illustrated in Fig. 19.8. The average value of the output, say a component dimension, is called the mean. The spread of the output values around the mean represents the variability of the process and is represented by a quantity called the standard deviation. The normal distribution is defined such that 68.26% of all the output is within one standard deviation of the mean, 95.44% of all the output is within two standard deviations of the mean and 99.73% of all the output is within three standard deviations of the mean.

The normal distribution is defined by the equation:

$$f(x)=\frac{1}{\sigma\sqrt{2\pi}}\exp\left[\frac{1}{2}\left(\frac{x-\mu}{\sigma}\right)^2\right],\quad -\infty<x<\infty, \tag{19.8}$$

where $f(x)$ is the probability density function of the continuous random variable x, μ is the mean, σ is the standard deviation.

Eq. (19.8) can be solved numerically or by use of a table of solutions (as given in Table 19.7). As an illustration of use of the normal table with $z=3$, gives

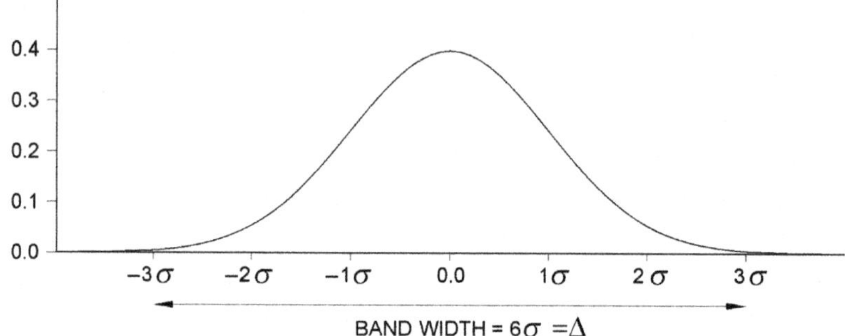

Fig. 19.8 The normal distribution.

$F(z)$=0.998650. So, $2\times(1-0.998650)=0.0027=0.27\%$ of all items can be expected to fall outside the tolerance limits. In other words, 99.73% of all items produced will have dimensions within the tolerance limits.

In practise, 80% of all processes, e.g. many machined dimensions and engine power outputs, are well modelled by the normal distribution, BSI (1983). Given a set of data, a statistical test can be undertaken to determine whether the data is normally distributed. If measurements of a dimension for a particular component are found to be normally distributed with mean μ and standard deviation σ, then the tolerance limits are often taken as $\mu \pm 3\sigma$. Tolerance limits defined as $\pm$ three standard deviations from the mean are called 'natural tolerance limits'. So if the mean and standard deviation of a component are 8.00 and 0.01, respectively, the dimension and bilateral tolerances would be defined as $8.00^{+0.03}_{-0.03}$ (i.e. the natural tolerance $=6\sigma=6\times 0.01=0.06$, so the bilateral tolerance is $0.06/2=0.03$.).

19.3.1 Sure-fit or extreme variability

The sure-fit or extreme variability of a function of several uncorrelated random variables could be found by substituting values for the variables to find the maximum and the minimum value of the function, e.g. for the function:

$$y=\frac{x_1+2x_2}{x_3^3}$$

The maximum value of y_{max} is given by

$$y_{max}=\frac{x_{1,max}+2x_{2,max}}{x_{3,min}^3}$$

Similarly for the minimum value:

$$y_{min}=\frac{x_{1,min}+2x_{2,min}}{x_{3,max}^3}$$

Table 19.7 The normal distribution

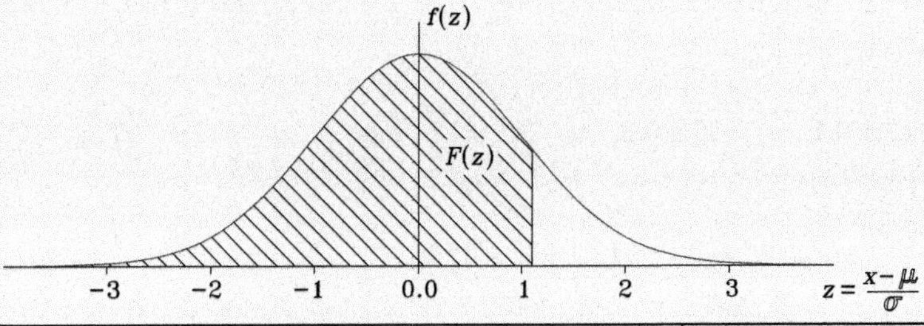

z	$f(z)$	$F(z)$
0.0	0.398942	0.5
0.1	0.396952	0.539827
0.2	0.391043	0.559260
0.3	0.381388	0.617911
0.4	0.368270	0.655422
0.5	0.352065	0.691465
0.6	0.333225	0.725747
0.7	0.312254	0.758036
0.8	0.289692	0.788145
0.9	0.266085	0.815940
1.0	0.241971	0.841345
1.1	0.217852	0.864334
1.2	0.194186	0.884930
1.3	0.171369	0.903195
1.4	0.149727	0.919243
1.5	0.129518	0.933193
1.6	0.110921	0.945201
1.7	0.094049	0.955435
1.8	0.078950	0.964069
1.9	0.065616	0.971284
2.0	0.053991	0.977250
2.1	0.043984	0.982136
2.2	0.035475	0.986097
2.3	0.028327	0.989276
2.4	0.022395	0.991803
2.5	0.017528	0.993791
2.6	0.013583	0.995339
2.7	0.010421	0.996533
2.8	0.007915	0.997495
2.9	0.005952	0.998134
3.0	0.004432	0.998650
3.5	0.000873	0.999768
4.0	0.000134	0.999968
4.5	0.000016	0.999996
5.0	0.0000015	0.9999997

The sure-fit or extreme variability can then be determined as follows:

$$\Delta y = y_{max} - y_{min}$$

If $x_1 = 12 \pm 0.1$, $x_2 = 15 \pm 0.2$, $x_3 = 2 \pm 0.15$, then $x_{1,max} = 12.1$, $x_{1,min} = 11.9$, $x_{2,max} = 15.2$, $x_{2,min} = 14.8$, $x_{3,max} = 2.15$, $x_{3,min} = 1.85$. $\bar{y} = 5.25$.

$$y_{max} = \frac{x_{1,max} + 2x_{2,max}}{x_{3,min}^3} = \frac{12.1 + (2 \times 15.2)}{1.85^3} = 6.712$$

$$y_{min} = \frac{x_{1,min} + 2x_{2,min}}{x_{3,max}^3} = \frac{11.9 + (2 \times 14.8)}{2.15^3} = 4.176$$

$\Delta y = y_{max} - y_{min} = 6.712–4.176 = 2.536$, so the variable $y = \bar{y} \pm (\Delta y/2) = 5.25 \pm 1.268$.

Alternatively the expression given in Eq. (19.13) can be used to approximate this extreme variability.

$$\Delta y \cong \left|\frac{\partial y}{\partial x_1}\right| \Delta x_1 + \left|\frac{\partial y}{\partial x_2}\right| \Delta x_2 + \left|\frac{\partial y}{\partial x_3}\right| \Delta x_3$$

where $\Delta x_1 = x_{1,max} - x_{1,min}$, $\Delta x_2 = x_{2,max} - x_{2,min}$, $\Delta x_3 = x_{3,max} - x_{3,min}$.

For the numerical example,

$$\partial y/\partial x_1 = 1/x_3^3 = 1/2^3 = 0.125,$$

$$\partial y/\partial x_2 = 2/x_3^3 = 2/2^3 = 0.25,$$

$$\partial y/\partial x_3 = -3(x_1 + 2 \times x_2)/x_3^4 = 7.875,$$

$$\Delta x_1 = 0.2,\ \Delta x_2 = 0.4,\ \Delta x_3 = 0.3,$$

$$\Delta y_{approx} = 0.125 \times 0.2 + 0.25 \times 0.4 + 7.875 \times 0.3 = 2.4875.$$

So, $y = 5.25 \pm 1.244$ (a value within a few per cent of the value calculated in the previous exact method).

19.3.2 Linear functions or tolerance chains

Often we require to know the overall tolerance in a dimension chain or knowledge of which dimensions can be slackened without a deleterious effect on performance or which tolerances need to be tightened to improve the overall tolerance.

Taking a simple linear chain of components of overall dimension z:

$$z = x_1 + x_2 + x_3 + \cdots + x_n \tag{19.9}$$

The expected value of z is given by

$$E(z) = \mu_1 + \mu_2 + \cdots + \mu_n = \mu_z \tag{19.10}$$

and the variance by

$$\text{Variance}(z) = \sigma_1^2 + \sigma_2^2 + \sigma_3^2 + \cdots + \sigma_n^2 = \sigma_z^2 \tag{19.11}$$

If the response variable is a linear function of several measured variables, each of which is normally distributed then the response variable will also be normally distributed.

Example 19.4. A product is made by aligning four components (see Fig. 19.9) whose lengths are x_1, x_2, x_3 and x_4. The overall length is denoted by z. The tolerance limits of the lengths are known to be 8.000 ± 0.030 mm, 10.000 ± 0.040 mm, 15.000 ± 0.050 mm and 7.000 ± 0.030 mm, respectively. If the lengths of the four components are independently normally distributed, estimate the tolerance limits for the length of the overall assembly.

Solution

Assuming $\pm 3\sigma$ natural tolerance limits:
$\sigma_1 = 0.03/3 = 0.01$ mm,
$\sigma_2 = 0.04/3 = 0.01333$ mm,
$\sigma_3 = 0.05/3 = 0.01667$ mm,
$\sigma_4 = 0.03/3 = 0.01$ mm,

$$\sigma_z^2 = \sigma_1^2 + \sigma_2^2 + \sigma_3^2 + \sigma_4^2 = 0.01^2 + 0.01333^2 + 0.01667^2 + 0.01^2 = 6.555 \times 10^{-4}$$
$$\sigma_z = 0.0256$$

Overall variability $= 6\sigma_z$.
Tolerance limits $= \pm(6\sigma_z/2) = \pm 3\sigma_z = \pm 0.077$.
The mean length of the overall assembly is given by

$$\mu_z = \mu_{x_1} + \mu_{x_2} + \mu_{x_3} + \mu_{x_4} = 8 + 10 + 15 + 7 = 40 \text{ mm}$$

$$z = 40.000 \pm 0.077 \text{ mm}.$$

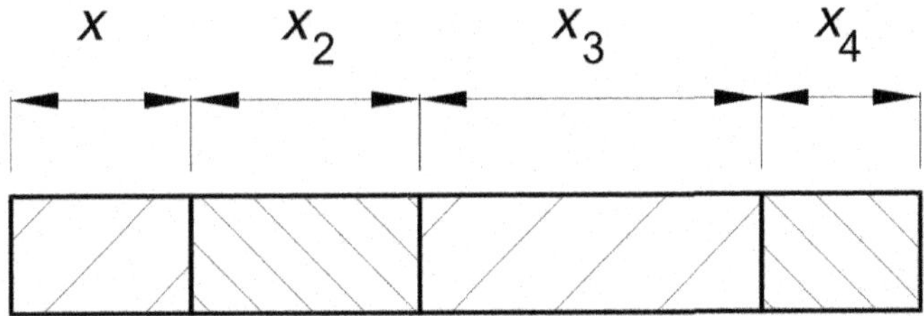

Fig. 19.9 Dimension chain.

So, 99.73% of all items would have an overall dimension within these tolerance limits.

(By comparison, the sure fit or extreme variability tolerance limits would be $z = 40.000 \pm 0.150$ mm, i.e. 100% of all items would be within this tolerance limit.)

The basic normal model indicated a better, or more attractive (certainly for marketing purposes), tolerance band. If the sure-fit model was used and an assembly tolerance of ±0.077 mm was desired, the tolerances on the individual components would need to be 8.000±0.019, 10.000±0.019, 15.000±0.019 and 7.000±0.019 mm. This is a very fine or tight tolerance and would be costly to achieve. As the statistical model takes account of 99.73% of the assemblies, use of the sure-fit approach only gives 0.27% of the assemblies with better quality and would require considerable extra effort and expense.

Example 19.5. Suppose the statistical tolerance bandwidth for the overall length of the previous example (±0.077 mm) is too large for the particular application. An overall statistical tolerance of ±0.065 mm is required. How can this be achieved?

Solution

This can be achieved in several different ways. One way would be to set the tolerance (and hence the standard deviation) equal, on all the components:

$\sigma_1 = \sigma_2 = \sigma_3 = \sigma_4 = \sigma$.

Set $\sigma_z^2 = (0.065/3)^2 = 4\sigma^2$,

Hence $\sigma = 0.0108$ mm.

Hence, the tolerance limits for each individual component would be $\pm 3\sigma = \pm 0.033$ mm.

Alternatively, the tolerance limits on say components 1 and 4 could be kept the same and the tolerance limits on components 2 and 3 could be tightened.

Suppose we decide that the tolerance limits on items 2 and 3 should be equal ($\sigma_2 = \sigma_3$).

$$\sigma_z^2 = \sigma_1^2 + \sigma_4^2 + 2\sigma_2^2$$

$$2\sigma_2^2 = 2.69444 \times 10^{-4}$$

$$\sigma_2 = 0.0116\,\text{mm}$$

So the tolerance limits on components 2 and 3 would need to be ±0.035 mm in order for the overall tolerance of the assembly to be ±0.065 mm.

Example 19.6. Which of the two following methods of manufacturing a product 150 mm in length would give the best overall tolerance?

(i) An assembly of ten components, each of length 15.000 mm with tolerance limits of ±0.050 mm.

(ii) A single component of 150 mm also with tolerance limits of ±0.050 mm.

Solution.

(i) $\sigma_1 = 0.05/3$,

$\sigma_z^2 = 10\sigma_1^2$

$\sigma_z = 0.0527$.

Tolerance limits for overall length $\pm 3\sigma = \pm 0.158\,\text{mm}$.
$\mu_z = 10\mu_1 = 150.000\,\text{mm}$,
$z = 150.000 \pm 0.158\,\text{mm}$,
(ii) $z = 150.000 \pm 0.050\,\text{mm}$.

As this method gives smaller tolerance limits, this is the most suitable way of achieving the tighter tolerance.

19.3.3 Several independent, uncorrelated random variables

Statistical analysis of tolerances can be applied to calculate the variation in physical phenomena such as stress, fluid flow rates, compression ratio and material properties, as well as component dimension chains. BS5760, Part 2, states that the normal distribution model can be considered to be an appropriate distribution for modelling 80% or more cases of distributional analysis in non-life circumstances. As such this analysis has wide ranging capability.

The theoretical preliminaries that enable analysis of functions of several independent, uncorrelated random variables are given below (after Furman, 1981).

Let y be a function of several independent, uncorrelated random variables.

$$y = f(x_1, x_2, \ldots, x_j, \ldots x_n) \tag{19.12}$$

The sure-fit or extreme variability of the variable y is given by

$$\delta y \cong \sum_{j=1}^{n} \frac{\partial f}{\partial x_j} \delta x_j \tag{19.13}$$

The statistical mean and standard deviation of a function of several independent, uncorrelated random variables are given by

$$\mu_y \cong f(\mu_1, \mu_2, \ldots, \mu_n) + \frac{1}{2} \sum_{j=1}^{n} \frac{\partial^2 f}{\partial x_j^2} \sigma_{x_j}^2 \tag{19.14}$$

$$\sigma_y^2 \cong \sum_{j=1}^{n} \left(\frac{\partial f}{\partial x_j} \right)^2 \sigma_{x_j}^2 + \frac{1}{2} \sum_{j=1}^{n} \left(\frac{\partial^2 f}{\partial x_j^2} \right)^2 \sigma_{x_j}^4 \approx \sum_{j=1}^{n} \left(\frac{\partial f}{\partial x_j} \right)^2 \sigma_{x_j}^2 \tag{19.15}$$

These equations may appear complex; however, their application is procedural and various examples are given to show their use.

Example 19.7. The stiffness of a helical compression spring (see Fig. 19.10) can be calculated from:

$$k = \frac{d^4 G}{8 D^3 n}$$

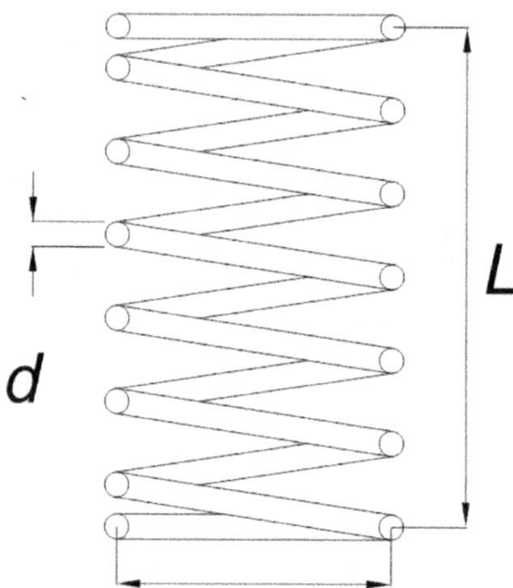

Fig. 19.10 Helical compression spring.

where d = wire diameter, G = modulus of rigidity, D = coil diameter, n = number of coils, k = spring stiffness.

The mean and standard deviation for each variable are known:

$$\mu_d = 2.34\,\text{mm},$$
$$\mu_D = 16.71\,\text{mm},$$
$$\mu_G = 79.29 \times 10^3\,\text{N/mm}^2,$$
$$\mu_n = 14\,\text{coils},$$

$$\sigma_d = 0.010\text{mm},$$
$$\sigma_D = 0.097\,\text{mm},$$
$$\sigma_G = 1.585 \times 10^3\,\text{N/mm}^2, \sigma_n = 0.0833\,\text{coils}.$$

Calculate the sure-fit extreme tolerance limits and the statistical basic normal tolerance limits for the spring stiffness. Assume $\pm 3\sigma$ natural tolerance limits: i.e. 99.73% of measurements lie within plus or minus 3 standard deviations.

Solution

Substituting values into the equation for k to give the average value of the spring constant:

$$\overline{k} = \frac{2.34^4 \times 79.29 \times 10^3}{8 \times 16.71^3 \times 14} = 4.549\text{N/mm},$$
$$\sigma_d = 0.01\,\text{mm},$$
$$\Delta d = 6 \times \sigma_d = 0.06\,\text{mm},$$
$$\sigma_G = 1.585 \times 10^3,$$
$$\Delta G = 6 \times \sigma_G = 9510\ \text{N/mm}^2,$$
$$\sigma_D = 0.097,$$
$$\Delta D = 6 \times \sigma_D = 0.582\,\text{mm},$$
$$\sigma_n = 0.0833\ \text{coils},$$
$$\Delta n = 6 \times \sigma_n = 0.5,$$

$$\frac{\partial k}{\partial d}=\frac{4d^3G}{8D^3n}=\frac{2.34^3\times 79.29\times 10^3}{2\times 16.71^3\times 14}=7.776,$$

$$\frac{\partial k}{\partial G}=\frac{d^4}{8D^3n}=5.7374\times 10^{-5},$$

$$\frac{\partial k}{\partial D}=-\frac{3d^4G}{8D^4n}=-8.1673\times 10^{-1},$$

$$\frac{\partial k}{\partial n}=-\frac{d^4G}{8D^3n^2}=-3.2494\times 10^{-1},$$

$$\Delta k_{\text{sure-fit}}=\left|\frac{\partial k}{\partial d}\right|\Delta d+\left|\frac{\partial k}{\partial G}\right|\Delta G+\left|\frac{\partial k}{\partial D}\right|\Delta D+\left|\frac{\partial k}{\partial n}\right|\Delta n$$
$$=7.776\times 0.06+5.737\times 10^{-5}\times 9510+0.8167\times 0.582+0.3249\times 0.5$$
$$=0.4666+0.5456+0.4753+0.1625=1.65\,\text{N/mm}.$$

This figure ($\Delta k_{\text{sure-fit}}$) gives the overall worst case variability of the spring stiffness (i.e. the total tolerance). The bilateral tolerance can be obtained by dividing Δk by two (i.e. 1.65/2=0.825).

So the spring stiffness can be stated as $\underline{k}=4.549\pm 0.825$ N/mm.

Calculating the basic normal variability:

$$\Delta k^2_{\text{basic normal}}=\left|\frac{\partial k}{\partial d}\right|^2\Delta d^2+\left|\frac{\partial k}{\partial G}\right|^2\Delta G^2+\left|\frac{\partial k}{\partial D}\right|^2\Delta D^2+\left|\frac{\partial k}{\partial n}\right|^2\Delta n^2$$
$$=0.2177+0.2977+0.2259+0.02640=0.7677,$$

$$\Delta k_{\text{basic normal}}=0.8762\text{N/mm}.$$

The overall variability in the spring stiffness is calculated as 0.8762 N/mm using the basic normal method. The bilateral tolerance can be calculated by dividing this value in two. i.e. the spring stiffness $k=4.549\pm 0.4381$ N/mm.

The basic normal variation model implies a narrower band width of variation $k=4.549\pm 0.438$ N/mm versus $k=4.549\pm 0.825$ N/mm for sure-fit. This is more attractive from a sales perspective and accounts for 99.73% (i.e. nearly all) of all items.

If we wanted to decrease the variation in k we would need to tighten the individual tolerances on d, G, D and n with particular attention to d and D as these are controllable and influential on the overall variability.

(The standard deviation of the spring stiffness can be calculated, if required, by $\sigma_k=\Delta k/6=0.146$ N/mm).

Example 19.8. In the design work for a model single cylinder reciprocating engine, the work done during the expansion stroke is to be represented as:

$$W=P\pi D^2C\left(1-Q^{n-1}\right)/(4(n-1))$$

where $Q=C/(C+2R)$, and $C=M-R-L-H$.

The quantities are defined as
W = work done during the expansion stroke,
P = cylinder pressure at start of stroke,
D = piston diameter,
C = length of the clearance volume,
n = expansion stroke polytropic exponent,
R = crank radius,
M = distance from crank centre to underside of cylinder head,
L = connecting rod length,
H = distance from gudgeon pin to piston crown,
Q = ratio of clearance to clearance-plus-swept volumes.

It is recognised that the independent quantities are subject to variability and it is desired to determine the consequent extreme and probable variability in the work done.

If:

$$P = 20\times 10^5 \pm 0.5\times 10^5\ \mathrm{N/m^2},$$

$$n = 1.3 \pm 0.05,$$

$$D = 4\times 10^{-2} \pm 25\times 10^{-6}\ \mathrm{m},$$

$$R = 2\times 10^{-2} \pm 50\times 10^{-6}\ \mathrm{m},$$

$$M = 10.5\times 10^{-2} \pm 50\times 10^{-6}\ \mathrm{m},$$

$$H = 2\times 10^{-2} \pm 25\times 10^{-6}\ \mathrm{m},$$

$$L = 6\times 10^{-2} \pm 50\times 10^{-6}\ \mathrm{m},$$

determine the extreme and probable (normal model) limits of the work (Ellis (1990)).

To help with your solution, certain partial derivatives have already been calculated:

$$\frac{\partial \mathrm{W}}{\partial \mathrm{R}} = -2599,$$

$$\frac{\partial \mathrm{W}}{\partial \mathrm{M}} = 2888,$$

$$\frac{\partial \mathrm{W}}{\partial \mathrm{H}} = -2888,$$

$$\frac{\partial \mathrm{W}}{\partial \mathrm{L}} = -2888.$$

Note: It may be useful to recall that if $y = a^x$ then $\frac{dy}{dx} = x\log_e(a)$.

Solution

The sure-fit 'extreme' limit of the work done is given by

$$\Delta W_{\text{sure fit}} = \left|\frac{\partial W}{\partial P}\right|\Delta P + \left|\frac{\partial W}{\partial D}\right|\Delta D + \left|\frac{\partial W}{\partial M}\right|\Delta M + \left|\frac{\partial W}{\partial R}\right|\Delta R + \left|\frac{\partial W}{\partial L}\right|\Delta L + \left|\frac{\partial W}{\partial H}\right|\Delta H + \left|\frac{\partial W}{\partial n}\right|\Delta n$$

The average values of C, Q and W are as follow: $\overline{C}=0.005$, $\overline{Q}=0.1111$, $\overline{W}=20.22$.

$$\frac{\partial W}{\partial P} = \frac{\pi D^2 C(1-Q^{n-1})}{4(n-1)} = \pi 0.04^2 0.005\frac{1-0.1111^{0.3}}{4\times 0.3} = 1.011035\times 10^{-5},$$

$$\Delta P = 2\times 0.5\times 10^5 = 1\times 10^5,$$

$$\left|\frac{\partial W}{\partial P}\right|\Delta P = 1.011035,$$

$$\frac{\partial W}{\partial D} = 2P\pi DC\frac{1-Q^{n-1}}{4(n-1)} = 2\pi 20\times 10^5\times 0.04\times 0.005\frac{1-0.1111^{0.3}}{4\times 0.3} = 1011.03,$$

$$\Delta D = 50\times 10^{-6},$$

$$\left|\frac{\partial W}{\partial D}\right|\Delta D = 0.0505517,$$

$$\frac{\partial W}{\partial M} = 2888.39,\quad \Delta M = 100\times 10^{-6},$$

$$\left|\frac{\partial W}{\partial M}\right|\Delta M = 0.288839,$$

$$\frac{\partial W}{\partial R} = -2599.46,\quad \Delta R = 100\times 10^{-6},$$

$$\left|\frac{\partial W}{\partial R}\right|\Delta R = 0.259946,$$

$$\frac{\partial W}{\partial L} = -2888.39,\quad \Delta L = 100\times 10^{-6},$$

$$\left|\frac{\partial W}{\partial L}\right|\Delta L = 0.288839,$$

$$\frac{\partial W}{\partial H} = -2888.39\quad \Delta H = 50\times 10^{-6},$$

$$\left|\frac{\partial W}{\partial H}\right|\Delta H = 0.1444195.$$

$$\text{Let } W = K_3\left(\frac{1-Q^{n-1}}{n-1}\right) \quad K_3 = \frac{P\pi D^2 C}{4},$$

$$\left|\frac{\partial W}{\partial n}\right| = K_3\left(\frac{-(n-1)\frac{\partial}{\partial n}Q^{n-1} - (1-Q^{n-1})1}{(n-1)^2}\right)$$

$$= K_3\left(\frac{-(n-1)Q^{n-1}\ \ln Q - (1-Q^{n-1})}{(n-1)^2}\right),$$

which on substituting for K_3 gives

$$= \frac{W(n-1)}{1-Q^{n-1}}\left(\frac{-(n-1)Q^{n-1}\ \ln Q - (1-Q^{n-1})}{(n-1)^2}\right),$$

$$\frac{\partial W}{\partial n} = -\frac{W}{n-1}\left(\frac{(n-1)Q^{n-1}\ \ln Q}{1-Q^{n-1}} + 1\right)$$

$$= \frac{20.22}{0.3}\left(\frac{0.3 \times 0.1111^{0.3}\ \ln 0.1111}{1-0.1111^{0.3}} + 1\right) = 19.7909.$$

$$\Delta n = 2 \times 0.05 = 0.1$$

$$\left|\frac{\partial W}{\partial \text{n}}\right|\Delta \text{n} = 1.979.$$

$$\Delta W_{\text{sure fit}} = 1.011 + 0.05055 + 0.2888 + 0.2599 + 0.2888 + 0.1444 + 1.979$$
$$= 4.022.$$

$$W = \overline{W} \pm (\Delta W/2) = 20.22 \pm 2.011 \text{ J}$$

Statistical tolerance limits:

$$\Delta W^2_{\text{normal}} = \left(\frac{\partial W}{\partial P}\right)^2 \Delta P^2 + \left(\frac{\partial W}{\partial D}\right)^2 \Delta D^2 + \left(\frac{\partial W}{\partial M}\right)^2 \Delta M^2 + \left(\frac{\partial W}{\partial R}\right)^2 \Delta R^2$$
$$+ \left(\frac{\partial W}{\partial L}\right)^2 \Delta L^2 + \left(\frac{\partial W}{\partial H}\right)^2 \Delta H^2 + \left(\frac{\partial W}{\partial n}\right)^2 \Delta n^2 = 5.196,$$

$$\Delta W = 2.2796.$$

So, $W = \overline{W} \pm (\Delta W/2) = 20.22 \pm 1.14$ J.

If required, $\sigma_W = 1.1398/3 = 0.38$.

Example 19.9. The torque capacity, T, for a multiple disc clutch can be calculated using the relationship

$$T = \frac{2}{3}\mu F_a N\left(\frac{r_o^3 - r_i^3}{r_o^2 - r_i^2}\right),$$

where μ = coefficient of friction, F_a = axial clamping force, N = number of disc faces, r_o = outer radius of clutch ring, r_i = inner radius of clutch ring.

Determine the sure-fit and basic normal probable limits for the torque capacity if.

$\mu = 0.3 \pm 0.03$,
$F_a = 4000 \pm 200$ N,
$N = 12$,
$r_o = 60 \pm 0.5$ mm,
$r_i = 30 \pm 0.5$ mm.

Which design parameter should be reduced to create a maximum reduction in the variability of the torque capacity?

Solution

$$T_{av} = \frac{2}{3} \times 0.3 \times 4000 \times 12 \times \left(\frac{0.06^3 - 0.03^3}{0.06^2 - 0.03^2}\right) = 672 \text{ N m},$$

$$\text{For sure-fit variability}: \delta y \approx \sum_{j=1}^{n} \frac{\partial f}{\partial x_j}\delta x_j,$$

$$\Delta T_{\text{sure fit}} \approx \left|\frac{\partial T}{\partial F_a}\right|\Delta F_a + \left|\frac{\partial T}{\partial \mu}\right|\Delta \mu + \left|\frac{\partial T}{\partial r_o}\right|\Delta r_o + \left|\frac{\partial T}{\partial r_i}\right|\Delta r_i,$$

$$\frac{\partial T}{\partial F_a} = \frac{2}{3}\mu N \frac{r_o^3 - r_i^3}{r_o^2 - r_i^2} = \frac{2}{3} \times 0.3 \times 12 \times \frac{0.06^3 - 0.03^3}{0.06^2 - 0.03^2} = 0.168,$$

$$\frac{\partial T}{\partial \mu} = \frac{2}{3}F_a N \frac{r_o^3 - r_i^3}{r_o^2 - r_i^2} = \frac{2}{3} \times 4000 \times 12 \times \frac{0.06^3 - 0.03^3}{0.06^2 - 0.03^2} = 2240.$$

$$\text{Quotient rule}: \left(\frac{u}{v}\right)' = \frac{vu' - uv'}{v^2},$$

$$\frac{\partial T}{\partial r_o} = \frac{2}{3}\mu F_a N \frac{(r_o^2 - r_i^2)3r_o^2 - (r_o^3 - r_i^3)2r_o}{(r_o^2 - r_i^2)^2}$$

$$= \frac{2}{3} \times 0.3 \times 4000 \times 12 \times \frac{2.7 \times 10^{-3} \times 3 \times 0.06^2 - 1.89 \times 10^{-4} \times 2 \times 0.06}{(0.06^2 - 0.03^2)^2}$$

$$= 0.8889 \times 9600 = 8533$$

$$\frac{\partial T}{\partial r_i}=\frac{2}{3}\mu F_a N\frac{-\left(r_o^2-r_i^2\right)3r_i^2+\left(r_o^3-r_i^3\right)2r_i}{\left(r_o^2-r_i^2\right)^2}$$
$$=\frac{2}{3}\times 0.3\times 4000\times 12\times\frac{-2.7\times 10^{-3}\times 3\times 0.03^2-1.89\times 10^{-4}\times 2\times 0.03}{\left(0.06^2-0.03^2\right)^2}$$
$$=0.5556\times 9600=5333$$

$$\Delta F_a=2\times 200=400\text{ N},$$

$$\Delta\mu=2\times 0.3=0.6,$$

$$\Delta r_o=2\times 0.0005=0.001\text{ m},$$

$$\Delta r_i=2\times 0.0005=0.001\text{ m},$$

$$\Delta T_{\text{sure fit}}=(2240\times 0.06)+(0.168\times 400)+(0.8889\times 0.001\times 9600)$$
$$+(0.5556\times 0.001\times 9600)=134.4+67.2+8.533+5.333=215.5\text{ N m},$$

$$T_{\text{sure fit}}=T_{\text{av}}\pm\left(\frac{\Delta T_{\text{sure fit}}}{2}\right)=672\pm 108\text{ N m},$$

$$\text{Basic normal probable limits}:\sigma_y^2\approx\sum_{j=1}^{n}\left(\frac{\partial f}{\partial x_j}\right)^2\sigma_{x_j}^2.$$

Assuming natural tolerance limits:

$$\Delta T^2_{\text{basic normal}}\approx\left|\frac{\partial T}{\partial F_a}\right|^2\Delta F_a^2+\left|\frac{\partial T}{\partial\mu}\right|^2\Delta\mu^2+\left|\frac{\partial T}{\partial r_o}\right|^2\Delta r_o^2+\left|\frac{\partial T}{\partial r_i}\right|^2\Delta r_i^2$$
$$=134.4^2+67.2^2+8.533^2+5.333^2=22680,$$

$$\Delta T_{\text{basic normal}}=150.6\text{ N m},$$

$$T_{\text{basic normal}}=T_{av}\pm\left(\frac{\Delta T_{\text{basic normal}}}{2}\right)=672\pm 75\text{ N m}.$$

The coefficient of friction contributes the most to the variability. This would require more stringent manufacture control.

Example 19.10. For a standard series 'A' Belleville spring, the spring rate or stiffness can be determined by the equation

$$k=\frac{4E}{(1-\mu^2)}\times\frac{t^3}{K_1 D_e^2},$$

where E is the Young's modulus of the washer material, t is the washer thickness, μ is the Poisson's ration, K_1 is a dimensionless constant and D_e is the external diameter.

Determine the sure-fit and probable limits for the spring stiffness stating any assumptions made if:

$t = 2.22 \pm 0.03\,\text{mm}$,
$D_e = 40.00 \pm 0.08\,\text{mm}$,
$\mu = 0.30 \pm 0.003$,
$E = 207 \times 10^9 \pm 2 \times 10^9\ \text{N/m}^2$,
$K_1 = 0.69$.

Comment on which tolerance could be altered to best reduce the overall variability.

Solution

$$k = \frac{4E}{(1-\mu^2)} \times \frac{t^3}{K_1 D_e^2}.$$

$K_1 = \text{constant}$.

Assuming quantities are subject to variability, are uncorrelated and random.

$$\Delta k_{\text{sure fit}} = \left|\frac{\partial k}{\partial E}\right| \Delta E + \left|\frac{\partial k}{\partial \mu}\right| \Delta \mu + \left|\frac{\partial k}{\partial t}\right| \Delta t + \left|\frac{\partial k}{\partial D_e}\right| \Delta D_e,$$

$$\Delta k^2_{\text{basic normal}} = \left|\frac{\partial k}{\partial E}\right|^2 \Delta E^2 + \left|\frac{\partial k}{\partial \mu}\right|^2 \Delta \mu^2 + \left|\frac{\partial k}{\partial t}\right|^2 \Delta t^2 + \left|\frac{\partial k}{\partial D_e}\right|^2 \Delta D_e^2,$$

$$\frac{\partial k}{\partial E} = \frac{4t^3}{(1-\mu^2)K_1 D_e^2} = \frac{4 \times \left(2.22 \times 10^{-3}\right)^3}{\left(1-0.3^2\right) \times 0.69 \times 0.04^2} = 4.356 \times 10^{-5},$$

$$\frac{\partial k}{\partial \mu} = -\frac{2\mu}{\left(1-\mu^2\right)^2} \times \frac{4Et^3}{K_1 D_e^2} = -\frac{2 \times 0.3}{(1-0.09)^2} \times \frac{4 \times 207 \times 10^9 \times \left(2.22 \times 10^{-3}\right)^3}{0.69 \times 0.04^2}$$
$$= -5.946 \times 10^6.$$

$$\text{Quotient rule}: \left(\frac{u}{v}\right)' = \frac{uv' - vu'}{v^2},$$

$$\frac{\partial k}{\partial t} = \frac{12Et^2}{(1-\mu^2)K_1 D_e^2} = \frac{12 \times 207 \times 10^9 \times \left(2.22 \times 10^{-3}\right)^2}{(1-0.09) \times 0.69 \times 0.04^2} = 1.219 \times 10^{10},$$

$$\frac{\partial k}{\partial D_e} = -\frac{8Et^3}{(1-\mu^2)K_1 D_e^3} = -\frac{8 \times 207 \times 10^9 \times (0.00222)^3}{(1-0.09) \times 0.69 \times (0.04)^3} = -4.509 \times 10^8.$$

Assume natural tolerance limits $\pm 3\sigma$,

$$\Delta E = 2 \times 2 \times 10^9 = 4 \times 10^9,$$

$$\Delta\mu = 2 \times 0.003 = 0.006,$$

$$\Delta t = 2 \times 0.03 \times 10^{-3} = 6 \times 10^{-5},$$

$$\Delta D_e = 2 \times 0.08 \times 10^{-3} = 1.6 \times 10^{-4}.$$

$$\Delta k_{\text{sure fit}} = \left(4.356 \times 10^{-5} \times 4 \times 10^{9}\right) + \left(5.946 \times 10^{6} \times 0.006\right) + \left(1.219 \times 10^{10} \times 6 \times 10^{-5}\right) + \left(4.509 \times 10^{8} \times 1.6 \times 10^{-4}\right) = 174240 + 35676 + 731400 + 72144 = 1013460.$$

$$\overline{k} = \frac{4 \times 207 \times 10^{9}}{1 - 0.3^{2}} \times \frac{\left(2.22 \times 10^{-3}\right)^{3}}{0.69 \times 0.04^{2}} = 9017347.3\ \text{N/m}.$$

$$k_{\text{sure fit}} = 9{,}017{,}000 \pm 506{,}700\ \text{N/m}.$$

$$\Delta k^{2}{}_{\text{basic normal}} = 174240^{2} + 35676^{2} + 731400^{2} + 72144^{2} = 5.718 \times 10^{11}.$$

$$\Delta k_{\text{basic normal}} = 756{,}200.$$

$$k_{\text{basic normal}} = 9{,}017{,}000 \pm 378{,}100\ \text{N/m}.$$

To reduce the variability, tighten tolerances on t as this has the highest sensitivity coefficient.

19.3.4 Statistical design techniques and quality assurance

The traditional approach to handling quality problems is based upon defining quality as conformance to specifications. For example if the dimension and tolerance for a component was defined as 40.0 ± 0.1 mm, it would not matter whether the component dimension was 40.1, 40.5, 40.0 or 39.9 mm, the specification would have been satisfied. Genichi Taguchi (1993) defined a new approach for solving quality problems based on product uniformity around the target value. In the previous example, the target value is 40.0 and the approach proposed by Taguchi is to endeavour to ensure as many components as possible are near to this value. Essentially, Taguchi methods are based on the premise that quality is the avoidance of financial loss to society after the article is shipped. Within this definition, quality is related to quantifiable monetary loss not emotional gut reaction. Taguchi techniques of ensuring good quality require a three step approach to product or process design consisting of system, parameter and tolerance design. These are defined as follows.

- *System design*: First, all the possible systems that can perform the required functions must be considered including new ones that have not yet been developed.
- *Parameter design*: During this stage the appropriate system parameters should be specified to improve quality and reduce costs.

- *Tolerance design*: Decisions concerning the tolerance specifications for all the components and appropriate grades of materials must be made. The objective of tolerance design is to decide trade-offs between quality levels and cost in designing new systems.

As a qualitative illustration of this technique a case study is outlined of the production of the same design of an automotive transmission by a multinational corporation but at two different sites, one in Japan and one in the USA. Both factories manufactured the transmission within a tolerance band as specified by the designer but in market surveys it was found that American customers preferred the Japanese manufactured transmissions. Various factors were investigated including the dimensional tolerance chain which is shown for the two factories in Fig. 19.11. Too tight a tolerance chain results in high wear or binding of the transmission and too loose a tolerance chain results in high noise and component impact levels. The Japanese manufactured transmission was found to be approximately normally distributed about the target value, with 0.27% of items falling outside the tolerance bandwidth. The output from the US factory was rectangularly distributed with all the items within the tolerance band. So the product from the US factory was always within the tolerance specifications unlike the output from the Japanese factory which was occasionally outside the tolerance specification. However, the market surveys showed that customers had preferred the product that had originated from the Japanese factory. This initially surprising result, considering the on-target nature of the US product, can be identified as being due to proportionally more customers getting a higher quality, or nearer the target, product if supplied from the Japanese factory. The customers' perception of quality responded to the Japanese factory paying more attention to meeting the target, as opposed to USA factory's approach of meeting the tolerances.

Application of Taguchi methods requires an understanding of statistical information along with budgetary information for each stage of the design, production and marketing process. Further information on these methods can be found in the original texts by Taguchi et al. (1989), Taguchi (1993), Phadke (1989), Krishnamoorthi and Ram Krishnamoorthi (2011) and Chao-Ton Su (2013).

Customers require products that have characteristics that satisfy their needs and expectations. Because of the competitive nature of business as well as the desire to improve service, organisations are continually driven to improve their products and processes. The ISO 9000 and 9001 series of standards describe the fundamentals of

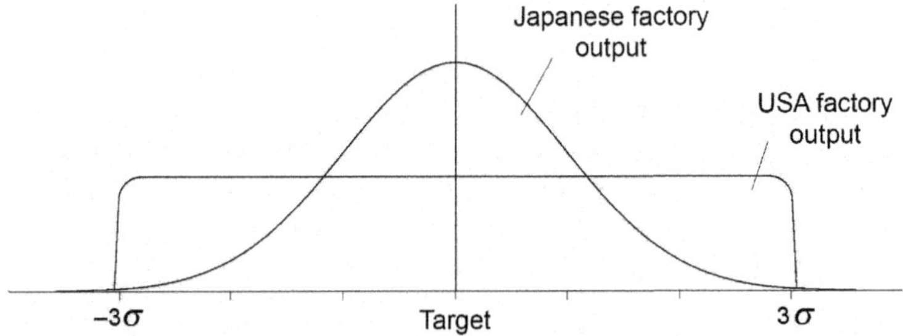

Fig. 19.11 Distribution of automotive transmission tolerance.

quality management systems and the requirements where an organisation needs to demonstrate its ability to provide products that fulfil customer and regulatory requirements. An increasing proportion of business has embraced such approaches and the reader is referred to the standards themselves for full details or Hoyle (1998).

19.4 Precision engineering and case studies

The concept of precision may be familiar from statistics. Accuracy is used to refer to how close a quantity is to a true or target value, while the term precision refers to how close the results of a sample are to each other. Precision engineering is normally concerned with the design of machinery and other physical applications that have very high tolerances or have functional attributes within tight tolerances, and are repeatable and stable over time. A common association with precision engineering is the requirement to be able to measure the quantities being produced, normally to within one part in a million.

As an example a bearing housing for the outer raceway of bearing may be specified by the manufacturer as requiring tolerances of ± 0.03 mm, with a surface roughness of 6.3 μm. In order for the performance of the bearing in service to match that of the manufacturers, it is important that the casing dimensions for all the components concerned are within these tolerance limits.

19.4.1 Axial compressor clearance case study

Axial compressors are widely used in industrial gas turbine and jet engines to provide a high pressure supply to the combustor. An axial compressor functions through the addition of rotational energy imparted by rotating blades to the flow and conversion of this energy to pressure in corresponding stationary blade rows. Axial compressor performance is a function of various parameters including casing diameter, hub-casing ratio, blade number, chord lengths, blade angles and form, radial tip clearance, rotational speed and mass flow. These parameters serve to define the operating characteristic of the compressor which can be described in terms of the pressure ratio and mass flow that can be obtained at a given rotational speed. Over tip leakage flows from the pressure to the suction side of a blade tend to act to disrupt the mainstream annulus flows and result in increased secondary flows and associated entropy rise and efficiency loss. Such flows can reduce the pressure ratio, flow range as well as efficiency. For example, Smith and Cumpsty (1984) measured a 23% reduction in maximum pressure rise and a 15% increase in flow coefficient at stall in a low speed compressor rig as the radial tip clearance was increased from 1% to 6% of chord. Wisler (1985) measured a 1.5% drop in efficiency for a low speed compressor associated with a doubling of the radial tip clearance. An example of axial compressor blading is shown in Fig. 19.12 and a schematic of the cross-section of a high by-pass ratio engine is shown in Fig. 19.13.

Various methods for assessing the efficiency loss caused by compressor blade tip radial clearance have been developed including correlations by Lokai et al. (1985)

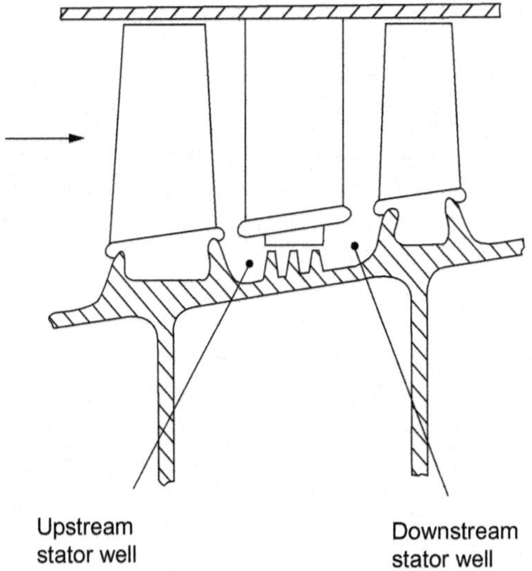

Fig. 19.12 Axial compressor blading (Bayley and Childs, 1994).

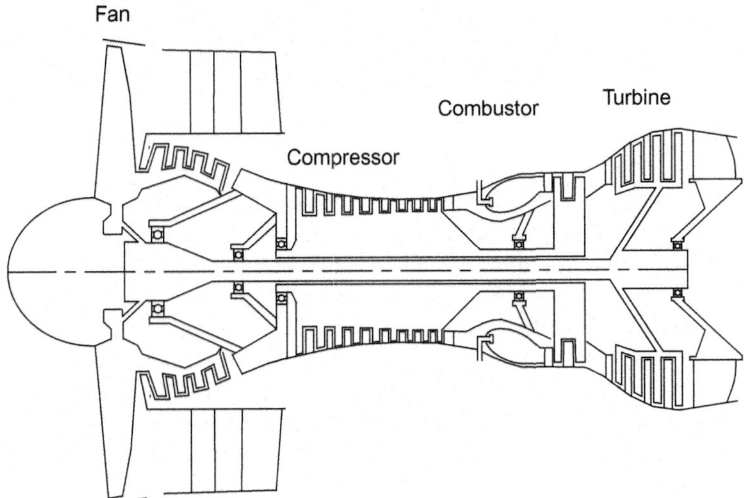

Fig. 19.13 By-pass engine cross-section schematic.

and plots showing the compressor characteristics in terms of flow coefficient versus to stage loading coefficient and contours of efficiency, sometimes known as Smith plots. As indicated one of the key parameters in compressor performance is tip clearance. For example Danish et al. (2016) indicate a drop off in efficiency for a mass flow of 12.5 kg/s from 88.5% at a minimal radial clearance of 0.003 mm to 88% at 0.3 mm, 87% at 1 mm and 86% at 2 mm. A related factor is eccentricity (see Young et al., 2017).

As efficiency is a key determining factor in fuel requirements and hence running costs, the optimisation of running clearance within practical engineering and cost constraints is worthwhile. This requires attention to the concentricity of the rotor drum within a nacelle or casing. Attention to the relative thermal growth of the compressor components is necessary. This will be a function of the running conditions, for example, whether the engine is starting from cool, under steady state design conditions, or shutting down. Under steady state running conditions the rotor drum will be warm relative to the casing. Approximate and detailed assessments of the running clearance can be estimated by simple assumptions and detailed Finite Element Analysis(FEA) models, respectively. The clearance will also be a function of the radial growth due to rotation. Again simple and detailed assessments can be made for this using equations for a rotating disc, e.g. Eq. (19.16) (den Hartog, 1952), combined with an assumption for blade growth, or more detailed FEA modelling. The eccentricity can be accounted by consideration of deflection of the bearings as well as dynamic modelling of the rotating assembly.

$$u=\rho\omega^2\frac{3+\mu}{8}\frac{1-\mu}{E}r\left[a^2+b^2-\frac{1+\mu}{3+\mu}r^2+\frac{1+\mu}{1-\mu}\frac{a^2b^2}{r^2}\right], \tag{19.16}$$

where u = radial growth (m), ρ = density (kg/m^3), ω = angular velocity (rad/s), μ = Poisson's ratio, a = inner radius (m), b = outer radius (m), r = radius (m).

A reasonable approximation for the thermal growth is to assume linear expansion.

$$u_{\text{thermal}}=r\alpha\Delta T, \tag{19.17}$$

where u_{thermal} = radial growth due to temperature (m), α = coefficient of linear expansion (°C^{-1}), ΔT = temperature difference (°C).

Here, we will examine outline calculations for one compressor disc within an axial compressor. The disc has a bore with diameter 120 mm, outer diameter of 450 mm, with blades of radial height 20 mm, and rotates at 12000 rpm. The disc is made from titanium alloy with density, Young's modulus, Poisson's ratio and linear coefficient of thermal expansion of 4500 kg/m^3, 116 GPa, 0.3 and $8\times10^{-6}\,K^{-1}$, respectively. The cold dimension of the steel casing inner diameter is 490.5 mm with a linear coefficient of thermal expansion of $11\times10^{-6}\,\text{K}^{-1}$. The bulk temperature difference between the rotating components and the casing is 350°C. There are many approaches to tackling this scenario with increasing levels of validity. At the design stage, useful insight can be obtained by considering some of the principal parameters using crude assumptions in order to establish component dimensions and form. Here considering the growth of the disc due to temperature and rotation and ignoring any radial growth of the blade gives radial growths of 0.63 mm and 0.163 mm, respectively. For the casing, the radial growth due to the temperature difference is 0.944 mm. The resulting radial clearance between the casing and the tip of the blades is 0.401 mm.

If we know the variability in properties such as Poisson's ratio, density, Young's modulus and thermal expansion coefficient, as well as the bilateral tolerances for the radii, we can then apply the methods described in Section 19.3 (and Cooke et al., 2009) to determine the bilateral tolerance for the radial clearance. More detailed analysis is possible by means of computational engineering analysis methods with for example an FEA model of the solid components, enabling consideration of both radial and axial dimensional changes. This is important for most compressors, where the radius of the casing and rotor drum varies with location along the axis of the machine. Small changes, due to thermal growth or thrust loads, in the relative position of the rotor drum relative to the casing can result in significant variation in radial clearance. It is important to consider such scenarios and this requires application of increasingly sophisticated modelling. Once a model has been developed the sensitivity of key parameters such as radial clearance as a function of various component and system variables can be explored.

19.4.2 Cordless and corded hand tools case study

Bearing alignment is a significant factor in determining the life of a bearing. In engines and many conventional machines, engineering practice has resulted in validated and standard approaches that can be followed in order to enable adequate alignment. These applications have typically been associated with casings machined or fabricated in metal. There can be significant cost and functional advantages in selecting a non-metallic material such as a polymer or composite for a casing. Common casing materials include various types of thermoplastic such as polycarbonates, acrylonitrile-butadiene-styrene (ABS) and polypropylene.

A cordless power hand tool will typically comprise an energy source, normally an electrical battery, an electrical motor, transmission and functional head housed within a casing. The location of these components within the housing requires careful consideration. The precision with which components such as batteries need to be located will be different to that necessary for a transmission element such as a bearing. Careful assessment is necessary in order to ensure that expensive manufacturing effort is not being applied unnecessarily. In contrast transmission elements may require precise location in order to ensure adequate machine life and avoid costly returns.

It is possible by careful placement of location features within a moulded component, to deliver high levels of precision. This can achieved by attention to stiffness of the casing, ensuring that it is appropriately rigid in locations where this is actually needed, while minimising material use in other locations where stiffness is not critical. Stiffness in a casing can be achieved by a combination of material choice, material thickness and form, employing for example ribbed structures. Examples of corded and cordless hand tools casings are illustrated in Figs. 19.14 and 19.15 showing particular attention to ensuring the torque associated with the transmission does not result in excessive distraction of the casing and good alignment of the transmission elements. An initial layout for such casings can be modelled using FEA and the arising distortion under loaded conditions examined in order to explore whether the design is adequate or optimum.

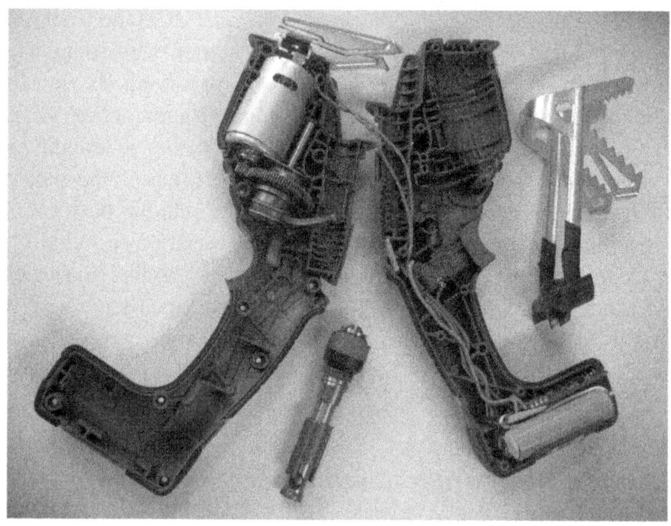

Fig. 19.14 A power hand tool casing for branch saw.

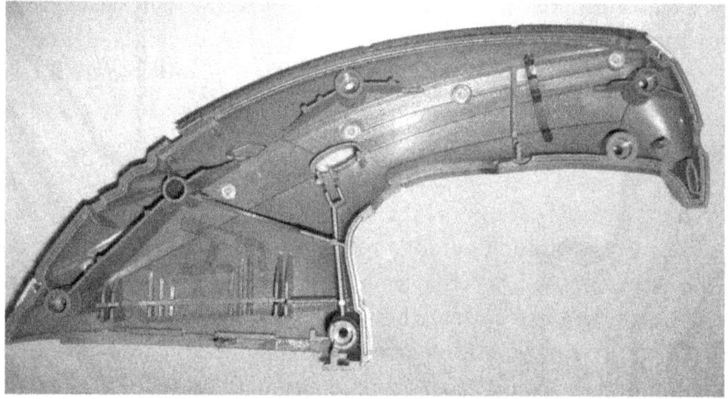

Fig. 19.15 A cordless hand tool casing for a spot sander.

19.4.3 Robot transmission case study

There are many examples of machinery where one aspect may appear to be a low precision output, but this relies upon other subsystems which need to be very high precision. QBot-Ltd. has developed a range of robots designed to spray thermal insulation on the underside of flooring in order to improve the energy efficiency of buildings (Holloway et al. (2016)). The robots, Fig. 19.16, can be fed into a small opening and then survey the void and subsequently drive to a location in the underfloor void to commence the spraying operation, Fig. 19.17.

One of the key subsystems for the robot are the wheels. A wheel hub electrical motor is used comprising an electrical motor mounted off the drive arm, driving

Fig. 19.16 The Mark 6.3 QBot developed to apply thermal insulation to the underside of flooring.

Fig. 19.17 The Mark 6.3 QBot shown spraying foam on timber flooring.

the wheel hub via an epicyclic gear train. Careful attention to the tolerances for each of the components within this assembly is necessary in order to ensure adequate function of the drive train while also avoiding over specification of non-essential features. A generic design for a wheel hub assembly is illustrated in Fig. 19.18.

19.5 Conclusions

Tolerances have to be applied to components in order to define acceptable limits for manufacture purposes. Generally the larger the tolerance, the easier it is to ensure that components will conform. The subject of tolerances is, however, not limited to components and assemblies of components, but also applies to processes. This chapter has

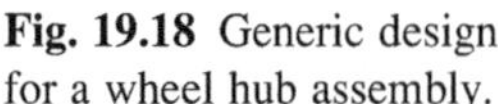

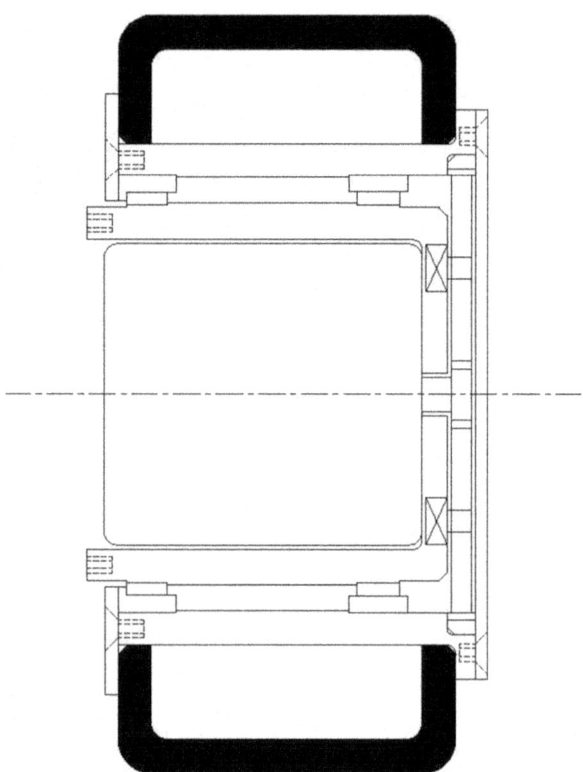

Fig. 19.18 Generic design for a wheel hub assembly.

introduced the subject of tolerances and introduced methods to model the overall tolerance for an assembly or process based on statistical methods. In addition, the subject of quality assurance has been introduced.

References

Books and papers

Bakerjian, R., Mitchell, P., 1992. Tool and Manufacturing Engineers Handbook, fourth ed. Vol. VI. Design for Manufacturability. Society of Manufacturing Engineers, Dearborn, Michigan.

Bayley, F.J., Childs, P.R.N., 1994. Air temperature rises in compressor and turbine stator wells. American Society of Mechanical Engineers, ASME Paper 94-GT-185.

Brown & Sharpe, 1992. Handbook of Metrology. North Kingston, Rhode Island.

Chao-Ton, S., 2013. Quality Engineering: Off-line Methods and Applications. CRC Press.

Cooke, A.O., Childs, P.R.N., Sayma, A., Long, C.A., 2009. A disc to air heat flux error and uncertainty analysis as part of the design of a new experimental rig. Proc. ImechE C: J. Mech. Eng. Sci. 223, 659–674.

Danish, S.N., Qureshi, S.R., Imran, M.M., Khan, S.U.D., Sarfraz, M.M., El-Leathy, A., Al-Ansary, H., Wei, M., 2016. Effect of tip clearance and rotor-stator axial gap on the efficiency of a multistage compressor. Appl. Therm. Eng. 99, 988–995.
Den Hartog, J.P., 1952. Advanced Strength of Materials. McGraw Hill.
Drake, P.J., 1999. Dimensioning and Tolerancing Handbook. McGraw Hill.
Ellis, J., 1990. Mechanical Design, BEng Final Examination. University of Sussex.
Furman, T.T., 1981. Approximate Methods in Engineering Design. Academic Press.
Holloway, M., Childs, P.R.N., Julia, M., 2016. In: A robot for spray applied insulation in underfloor voids.ISR 2016, Munich, 47st International Symposium on Robotics, Germany.
Hoyle, D., 1998. ISO 9000 Pocket Guide. Butterworth Heinemann.
Krishnamoorthi, K.S., Ram Krishnamoorthi, V.R., 2011. A First Course in Quality Engineering: Integrating Statistical and Management Methods of Quality, second ed CRC Press.
Krulikowski, A., 2012. Fundamentals of Geometric Dimensioning and Tolerancing, third ed Delmar Cengage Learning.
Lokai, V.I., Karimova, A.G., Prokop'ev, V.I., Maksimov, O.L., 1985. Generalisation of experimental data on heat transfer from working fluid to GTE compressor case elements. Izvestiya VUZ Aviatsionnaya Tekhnika 28 (1), 55–85.
Phadke, M.S., 1989. Quality Engineering Using Robust Design. Prentice Hall.
Simmons, C.H., Maguire, D.E., Phelps, N., 2012. Manual of Engineering Drawing: Technical Product Specification and Documentation to British and International Standards, fourth ed Butterworth-Heinemann.
Smith, G.D.J., Cumpsty, N.A., 1984. Flow phenomena in compressor casing treatment. ASME J Eng Gas Turbines Power 106, 532–541.
Spotts, M.F., 1983. Dimensioning and Tolerancing for Quantity Production. Prentice Hall.
Swift, K.G., Booker, J.D., 2013. Manufacturing Process Selection Handbook. Butterworth Heinemann.
Taguchi, G., 1993. Taguchi on Robust Technology Development. ASME Press.
Taguchi, G., Elsayed, E.A., Hsiang, T., 1989. Quality Engineering in Production Systems. McGraw Hill.
Wisler, D.C., 1985. Loss reduction in axial flow compressors through low-speed model testing. ASME J Turbomachinery 107, 354–363.
Young, A.M., Cao, T., Day, I.J., Longley, J.P., 2017. Accounting for eccentricity in compressor performance predictions. J Turbomachinery. 139.

Standards

BIP 3092 ISO 9000 Collection. The ISO 9000 standards collection. Quality management systems.
British Standards Institution. BS 1916: Part 1: 1953. Specification for limits and fits for engineering. Limits and tolerances, 1953.
British Standards Institution. BS 4500A: 1970. Specification for ISO limits and fits. Data sheet: selected ISO fits - holes basis.
British Standards Institution. BS 4500B: 1970. Specification for ISO limits and fits. Data sheet: selected ISO fits - shafts basis.
British Standards Institution. BS 8888:2011. Technical product documentation and specification.
British Standards Institution. BS5760: Part 2: Guide to the assessment of reliability.
British Standards Institution. BSI Handbook 22, Quality assurance, 1983.
British Standards Institution, 1991. Manual of British Standards in Engineering Drawing and Design, 2nd ed. BSI.

BS 5760–18:2010. Reliability of systems, equipment and components. Guide to the demonstration of dependability requirements. The dependability case.
BS EN ISO 9000:2005. Quality management systems. Fundamentals and vocabulary.
BS EN ISO 9001:2008. Quality management systems. Requirements.
BS EN ISO 9004:2009. Managing for the sustained success of an organisation. A quality management approach.
BS ISO 10006:2003. Quality management systems. Guidelines for quality management in projects.
BS ISO 10007:2003. Quality management systems. Guidelines for configuration management.
International Organisation for Standardisation. ISO 9000–1:1994. Quality management and quality assurance standards – Part 1: Guidelines for selection and use.
ISO 1101:2017(en). Geometrical product specifications (GPS)—Geometrical tolerancing—Tolerances of form, orientation, location and run-out. 2017.
ISO 14638:2015(en). Geometrical product specifications (GPS)—Matrix model. 2015.
MIL AND10387–6. Air force-navy aeronautical design standard drill sizes and drilled hole tolerance-twist (superseding AND10387–5). 1974.

Further reading

Chatfield, C., 1983. Statistics for Technology. Chapman and Hall.
Haugen, E.B., 1968. Probabilistic Approaches to Design. Wiley.

Mechanisms

Chapter Outline

Nomenclature

Generally, preferred SI units have been stated.

a	length (m)
c	length (m)
d	pitch diameter (mm)
F	force (N)
$F_{applied}$	applied force (N)
F_e	effort (N)
F_{load}	load (N)
L	length (m)
m	module (mm)
M	moment (N m)
n	rotational speed (rpm)
n_{arm}	rotational speed of arm (rpm)
n_{sun}	rotational speed of sun gear (rpm)
N	number of teeth, number of discs
N_R	number of teeth in ring gear
N_S	number of teeth in sun gear
p	pressure (N/m^2)
r	radius (m)
R	radius (m)
T	torque (N m)

Mechanical Design Engineering Handbook. https://doi.org/10.1016/B978-0-08-102367-9.00020-2

μ	coefficient of friction
θ	angle (rad)
ω	angular velocity (rad/s)

20.1 Introduction

The human race has been inventing and developing mechanisms for as long as our history indicates. The knowledge arising from the operation of these can be leveraged in designing variants, or developments, for modern applications. This chapter serves to introduce a range of mechanisms, some of which will have already been introduced or be familiar, in order to provide an indication of available designs and inspiration for the development of new ones.

The term machine element is commonly used for the component parts of a machine which cannot readily be disassembled into simpler components. A mechanism or machine is a device that combines two or more machine elements in order to transform energy from one form to another. A wide range of mechanisms have been developed throughout history. This knowledge can be used in the application and development of these mechanisms to modern challenges. Opportunities may exist to use new materials, replace subsystems with sensors and electrical actuators and a control algorithm, or to explore some completely new mechanism. The advent of micro- and nano-manufacturing capability, as well as biotechnology innovations, has resulted in exploration of a wide range of mechanism at the micro- and nanoscale (e.g. see, for example, Culpepper and Anderson (2004), Liu and Xu (2015), and Akbari and Pirbodaghi (2017)).

Various books describing mechanisms have been produced, with seminal examples including Chironis (1965) and Brown (1901). Both of these contain a wide range of mechanisms with illustrations and descriptions of their function. Use of a search engine with phrases such as ‘mechanical mechanism’, ‘mechanical gizmo’ or ‘machine design’ will yield many videos, images and references on a wide range of gadgets and mechanisms which can be scrutinised to learn how they work and whether they are of some use for a given application. In addition, a patent database can be a great source of information for prior art in mechanisms. As many of these patents have expired, it may be possible to use such designs directly in a modern application. For current patents, permission will need to be sought from the patent owner, if the invention is to be exploited in a commercial application.

Mechanisms can be classified amongst other attributes according to their function, application and market sector. Looking at functional attributes, it is possible to classify mechanisms according to whether they involve force amplification, power transfer, energy storage, locking and time metering. A series of machine elements and mechanisms are reviewed in this chapter. Levers, pulleys and linkages are presented in Sections 20.2, 20.3 and 20.4. A series of power transmission devices including gears, belts and chains, mechanisms for converting between rotating and linear motion, brakes and clutches are presented in Sections 20.5, 20.6 and 20.7. A selection of time-metering devices is presented in Section 20.8 and some diverse mechanisms in Section 20.9. A brief description is presented along a table given some principal attributes for the machine element or mechanism concerned with the intention of providing a prompt for the consideration of the use of the item concerned.

20.2 Levers

A lever can be used with a pivot, also known as a fulcrum, to enable a small force moving over a large distance to create a large force moving over a short distance. The fundamental equation for a lever can be determined by taking moments about the pivot point, giving for the case of a simple lever,

$$F_1 L_1 = F_2 L_2. \tag{20.1}$$

This equation can be rearranged, depending on the class of lever being considered to enable analysis of a given application or calculation of design parameters. A first-order lever is illustrated in Fig. 20.1. If the pivot is moved closer to the load, F_1, as illustrated in Fig. 20.2, the effort, F_2, required to lift the load, F_1, is reduced.

There are various classes of levers known as first-, second- and third-order levers, depending on where the pivot is relative to the load and applied force. Common examples of a first-order lever include the sea-saw, a rowing oar and scissors. Examples of a second-order lever include some nutcrackers, wheelbarrows and bicycle brake levers. Examples of third-order levers include staple removers, tweezers and your arm and elbow with the force being provided by the upper arm muscles. Some characteristics and useful equations for the three principal classes of lever are presented in Table 20.1.

20.3 Pulleys

A pulley comprises one or more wheels running on shafts over which a rope has been looped in order to make lifting a load easier. A pulley is a class of device called a simple machine that enables force amplification in a manner similar to levers.

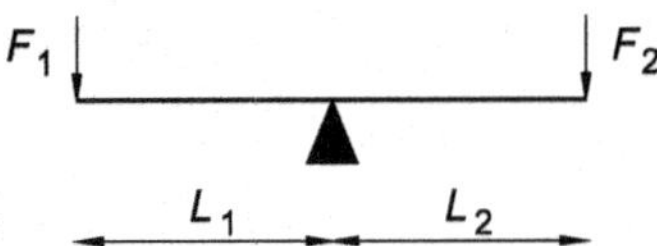

Fig. 20.1 First-order lever.

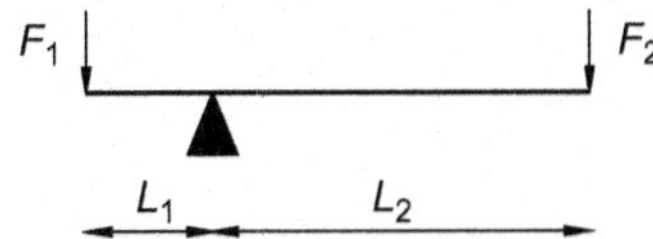

Fig. 20.2 By moving the pivot closer to the load, F_1, the effort F_2 required to move the load is reduced.

Table 20.1 Characteristics and useful equations for the three principal class of lever

Type of lever	Example	Characteristics	Useful equations and references
First-order lever	F_l, F_e, L_l, L_e	The load and the force are separated by the fulcrum. As one side moves, the other moves in the opposite direction. The magnitude of the force and the movement is proportional to the distance from the fulcrum.	$F_e = F_l L_l / L_e$ den Hartog (1961) Serway and Jewett (2018)
Second-order lever	Load F_l, Effort F_e, L_l, L_e	The load is between the force and the fulcrum. This type of lever uses mechanical advantage to ease the lifting of a large weight or the application of a large load.	$F_e = F_l L_l / L_e$ Serway and Jewett (2018)
Third-order lever	Load F_l, Effort F_e, L_e, L_l	In a third-order lever, the force is between the fulcrum and the load. Mechanical advantage is reduced but the movement at the load point is increased.	$F_e = F_l L_l / L_e$ Serway and Jewett (2018)

In the case of a two-wheel pulley, Fig. 20.3, the mass is supported by two strands of the rope. As you pull on the loose end of the rope, you only need to impart half the force of the load in order to lift it, but you will need to pull the loose end of the rope twice as far as the height through which the load is lifted. The term mechanical advantage is used to refer to how much a simple machine multiplies a force. For a two-wheel pulley, the mechanical advantage is two. Similarly with a four-wheel pulley, Fig. 20.4, the mass is supported on four strands of rope, and you need to impart a force of one-quarter of the weight of the load in order to move it, although you will need to pull the loose end through a distance four times that of the height that the load is lifted. A four-wheel pulley provides a mechanical advantage of four.

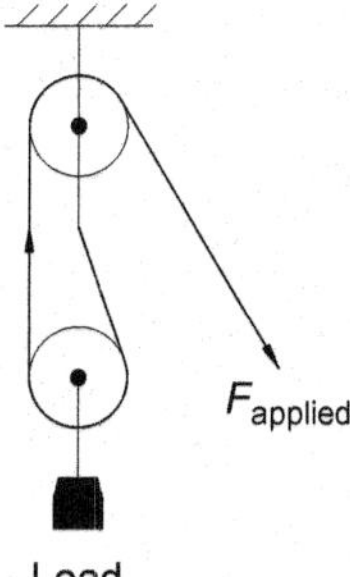

Fig. 20.3 Two-wheel pulley.

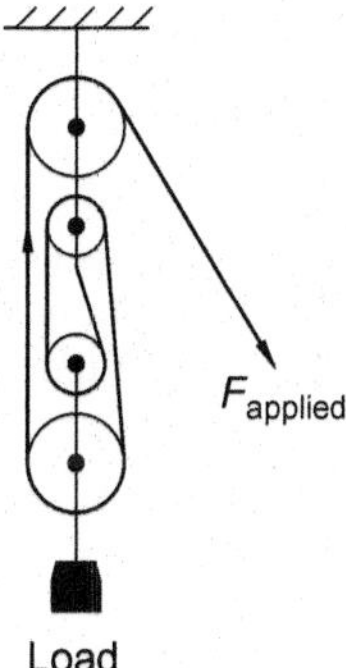

Fig. 20.4 Four-wheel pulley.

Although pulleys provide mechanical advantage, the actual energy required to move the load is slightly more than the potential energy associated with the change in elevation because of friction between the rope and pulley wheels. A wide variety of pulley configurations are possible, with a selection of these presented in Table 20.2.

Table 20.2 A selection of some pulley configurations

Type of Pulley	Example	Characteristics	Useful equations and references
Single wheel	$F_{applied}$ Load	This configuration reverses the direction of the lifting force. As you pull down, the mass will rise.	$F_{applied} = F_{load}$ Serway and Jewett (2018)

Continued

Table 20.2 Continued

Type of Pulley	Example	Characteristics	Useful equations and references
Two wheels	$F_{applied}$ Load	With two wheels, the mass is supported by two strands of rope. This configuration gives a mechanical advantage of two, and it is possible to lift the load with half the force of the load.	$F_{applied}=0.5F_{load}$ Serway and Jewett (2018)
Four wheels	$F_{applied}$ Load	With four wheels, the mass is supported by four strands of rope. This configuration gives a mechanical advantage of four, and it is possible to lift the load with one-quarter of the force of the load.	$F_{applied}=0.25F_{load}$ Serway and Jewett (2018)

20.4 Linkages

A linkage is an assembly of links and joints that translate one type of motion into another. Types of linkage include the simple lever, considered in Section 20.2, and four- and five-bar linkages. A four-bar linkage is made up of four links and four joints, with an example shown in Fig. 20.5. In the example illustrated, as the input link is rotated anticlockwise around its pivot A, the other links are forced to move as well, with the motion of the output pivot C, producing a reciprocating motion on an arc. A series of different positions for the linkage shown in Fig. 20.5 is illustrated in Fig. 20.6.

Four-bar linkages can be designed to produce a very diverse range of output motions, depending on the detailed configuration of the linkage. The lengths of the bars can be altered, and pivots or sliders can be used giving an infinity of variations. A four-bar linkage with different length bars to that of Fig. 20.5 is illustrated in Fig. 20.7. This features a planar structure on the link between B and C. The arising locus of points traced by vertex E as linkage rotates is shown in Fig. 20.8.

Four-bar linkages are very widely used in engineering, with applications ranging from earth movers, tailgates and airline luggage compartments, to packaging

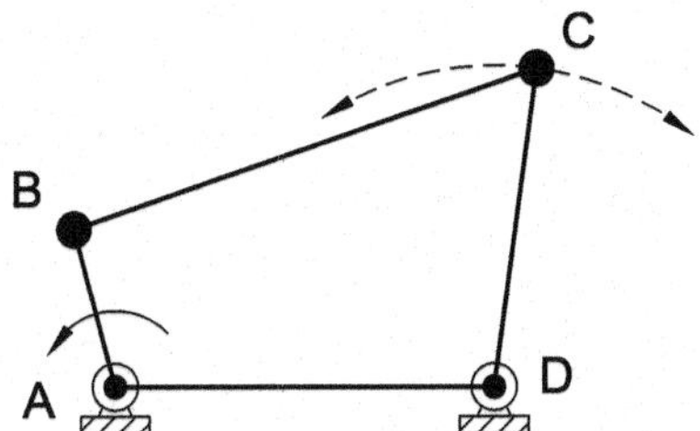

Fig. 20.5 An example of a four-bar linkage, giving in this case a reciprocating output on a circular arc.

machines and grips. A five-bar linkage, see Fig. 20.9, is used to provide mechanical advantage in bolt croppers.

There is a well-defined set of terminology commonly used for linkages as follows:

- Link—a nominally rigid body that has at least two nodes. Links can be straight, curved and have features incorporated in them.
- Node—a means of attachment to other links by joints.
- Joint—this is the connection between two or more links at their nodes, which enables motion between the links to occur.
- Revolute joint (R) or Pivot—a joint that enables rotational motion, formed by, for example, a pin passing through both links.
- Prismatic joint (P) or Slider—a joint that enables linear or sliding motion to occur between two links.

There are various types of planar four-bar linkage which can be classified by the type of joint used:

- four revolute joints (RRRR);
- three revolute and a prismatic joint, such as a slider-crank linkage (RRRP, RRPR);
- two revolute and two prismatic joints (e.g. PRRP, RRPP, RPPR, RPRP);
- Three prismatic joints and a revolute (RPPP, PRPP).

Planar RRRR four-bar linkages have rotating joints, as illustrated in Fig. 20.5. One of the links is usually fixed, and known as the ground or fixed link, or frame. The ground link serves as the reference for the motion of all of the other links. The two links connected to the frame are sometimes called the grounded links and tend to be the input and output links for the mechanism. The input drive to a link can be an oscillating or a continuous rotation. A link providing oscillating input drive to a linkage is known as a rocker. A link providing continuous rotating input to a linkage is known as a crank. The input drive can be provided by a prime mover such as an engine or electric motor. A follower is a link connected to the ground link though a joint at one end. The remaining link is called the floating link, or the coupler or connecting rod, as it connects the input and output.

The links in a four-bar linkage are, as a result of the nominally rigid geometry of the links and the rotation enabled by the joints, geometrically constrained. A linkage can be analysed, assuming rigid links to determine its configuration, and the velocity and angular velocities of the links (see, for example, Hrones and Nelson (1951), Wilson

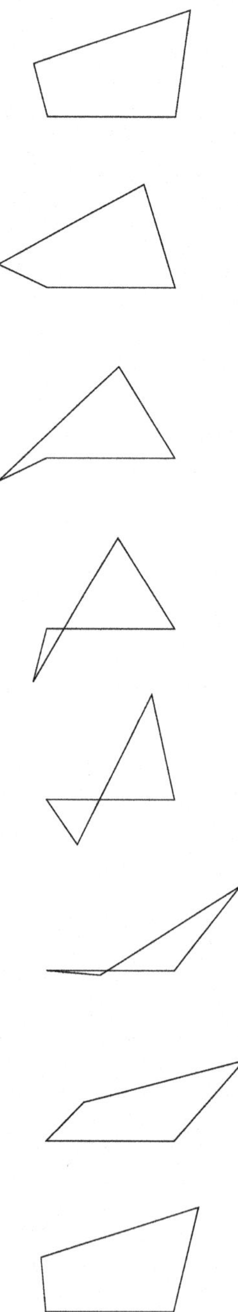

Fig. 20.6 Examples of different positions for the four-bar linkage of Fig. 20.5.

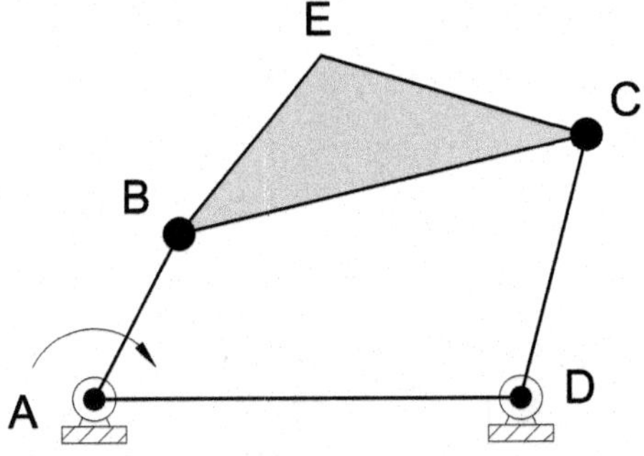

Fig. 20.7 Four-bar linkage featuring a planar structure on the link between B and C.

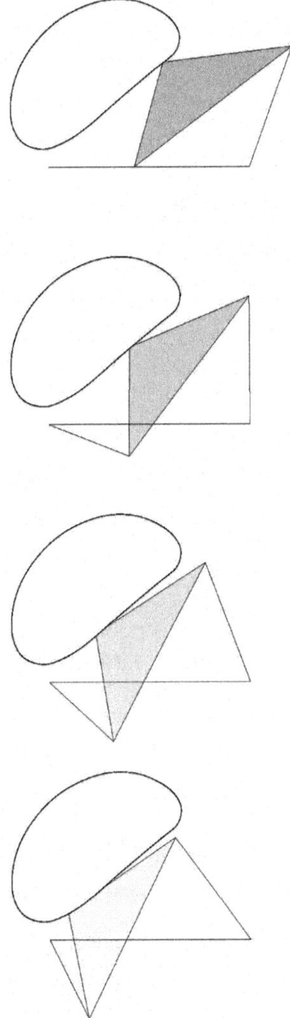

Fig. 20.8 The arising locus of points traced by a vertex on the coupler for the linkage shown in Fig. 20.7.

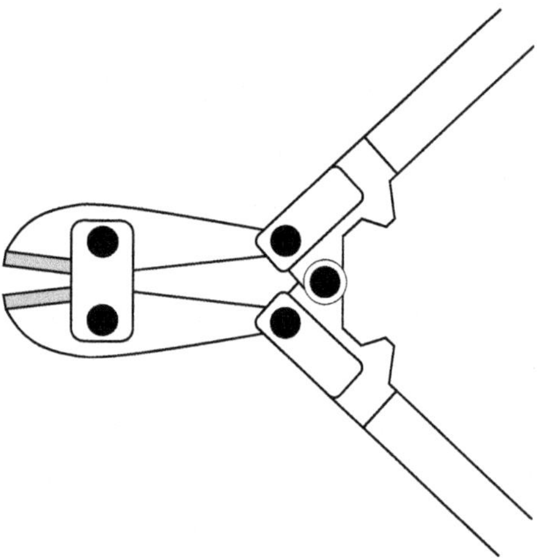

Fig. 20.9 Use of a five-bar linkage to provide mechanical advantage in bolt croppers.

and Sadler (2003), and McCarthy and Soh (2010)). Design can be defined as the transformation of an existing into a desired state. In the design of a linkage, a common requirement involves producing a defined or desired output motion. Use of prior experience and knowledge of linkages can be made in order to develop a form for a linkage that provides that motion. Alternatively software has been developed enabling the majority of four-bar linkage configurations to be modelled. A range of four-bar linkages is outlined in Table 20.3.

Table 20.3 A selection of some four-bar linkage configurations

Type of four-bar linkage	Example	Characteristic	Useful equations and references
Four-bar quadrilateral linkage (RRRR)	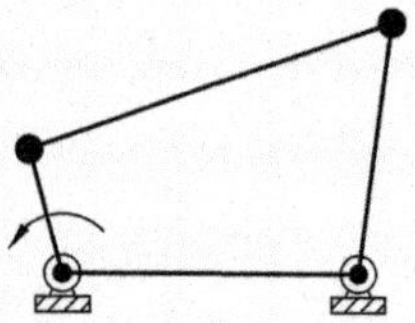	Planar RRRR four-bar linkages have rotating joints. With the configuration shown, the top right-hand joint traces out a reciprocating arc as the input link rotates.	Todd et al. (2014)
Four-bar quadrilateral linkage (RRRR)		In a parallelogram, four-bar linkage configuration, the orientation of the coupler does not change with rotation of the input link.	Todd et al. (2014)

Table 20.3 Continued

Type of four-bar linkage	Example	Characteristic	Useful equations and references
Four-bar quadrilateral linkage (RRRR)	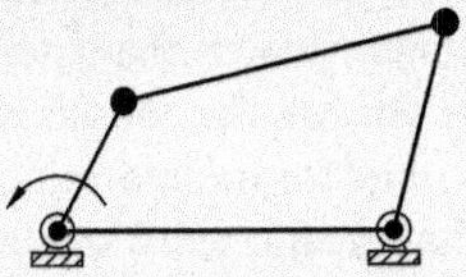	Widely differing output paths can be achieved by altering the lengths of the links in a four-bar linkage. With this particular configuration, the points on the coupler follow an approximately elliptical path.	Todd et al. (2014)
Slider crank linkage (RRRP)	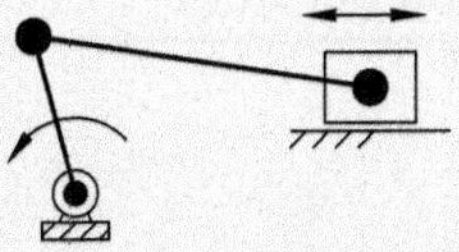	This is the classic crank slider mechanism for turning rotational motion into reciprocating linear motion.	Todd et al. (2014)
Double slider linkage (PRRP)	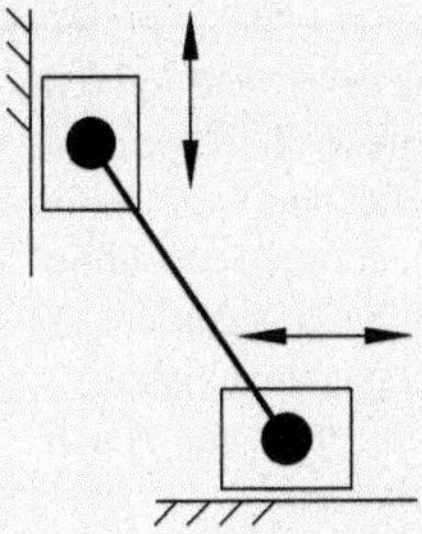	This is constructed by connected two sliders with a coupler. If the sliders are perpendicular to each other, then the trajectories of points on the coupler will trace out ellipses. This form of double slider is known as an elliptical trammel or the trammel of Archimedes.	Todd et al. (2014)
Inverted slider-crank (RRPR)	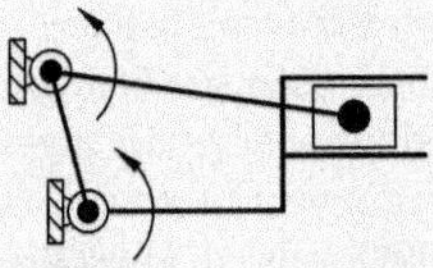	Widely used because of the relative simplicity of design and manufacture.	McCarthy and Soh (2010)
Turning block (RPRR)	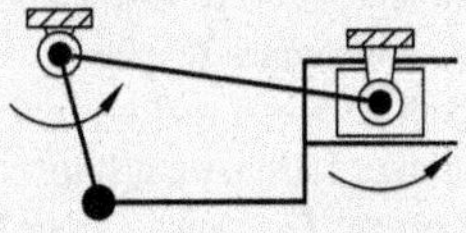	Can be used to convert the displacement of a linear motor into the swing motion of a rocker.	Simionescu (2017) McCarthy and Soh (2010)
(PRRR)	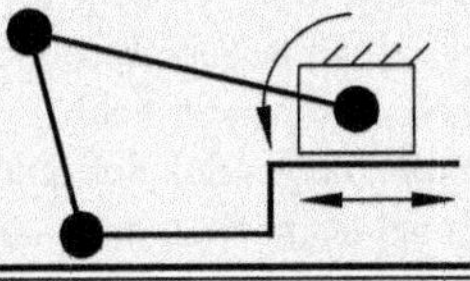	This is the same as a slider crank mechanism.	McCarthy and Soh (2010)

20.5 Gears, belts and chains

Gears, belts and chains can be used to transmit power from one rotating shaft to another. Many power-producing machines, or prime movers, produce power in the form of rotary motion. The operating torque versus speed characteristics of prime movers vary according to their type and size as do the characteristics of the loads to be driven. It is common for the torque versus speed characteristics to be mismatched, requiring the need for a transformation of the torque and speed, which can be achieved with gearing. Various options exist for transforming the torque speed characteristic of a rotating application, including gears, belts and chains. A range of types of gear and gearboxes, belts and chains are presented in Tables 20.4, 20.5 and 20.6, respectively.

Table 20.4 A selection of various types of gear and gearboxes

Type of gear and gear train	Example	Characteristics	Useful equations and references
Spur gears		Spur gears tend to be the cheapest of all types for parallel shaft applications. The straight teeth allow running engagement or disengagement using sliding shaft and clutch mechanisms.	$m=\frac{d}{N}$, $n_2=-\frac{N_1}{N_2}n_1$. See Sections 8.1 and 8.5, and Chapter 9
Helical gears		A helical gear is a cylindrical gear whose tooth traces are helixes. Helical gears are typically used for heavy-duty high-speed power transmission. Noise levels are lower than for spur gears because teeth in mesh make point contact rather than line contact. The forces arising from meshing of helical gears can be resolved into three component loads, radial, tangential and axial.	See Section 8.1 and Chapter 9
Bevel gears		Bevel gears have teeth cut on conical blanks and a gear pair can connect non-parallel intersecting shafts	See Section 8.1 and Chapter 10

Table 20.4 Continued

Type of gear and gear train	Example	Characteristics	Useful equations and references
Crossed axis helical gears	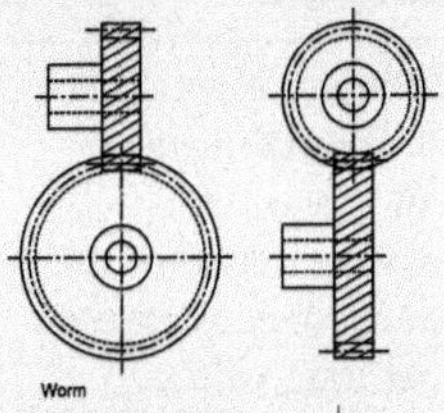	Crossed axis helical gears can offer a cost-effective solution for right-angle transmission applications requiring only limited load carrying capacity.	
Worm gears	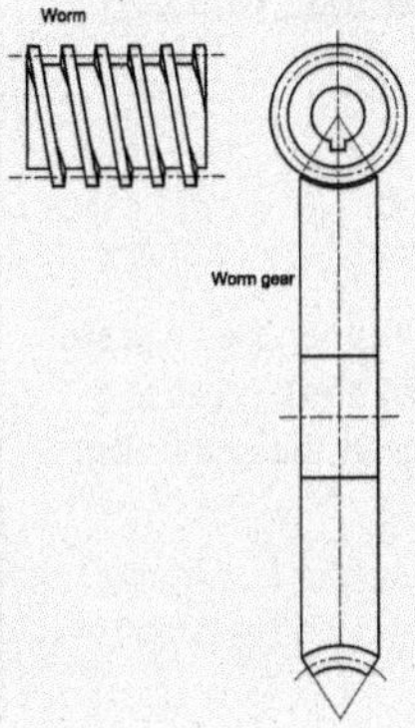 Worm Worm gear	A worm gear is a cylindrical helical gear with one or more threads. A worm wheel is a cylindrical gear with flanks cut in such a way as to ensure contact with the flanks of the worm gear. Worm gear sets are capable of high-speed reduction and high-load applications where non-parallel, non-interacting shafts are used.	See Section 8.1 and Chapter 11
Sector gears		A sector gear has teeth missing around a proportion of its circumference. They can be useful for applications where a complete rotation of a drive is not necessary or if positive drive is required for part of a revolution.	
Simple gear train	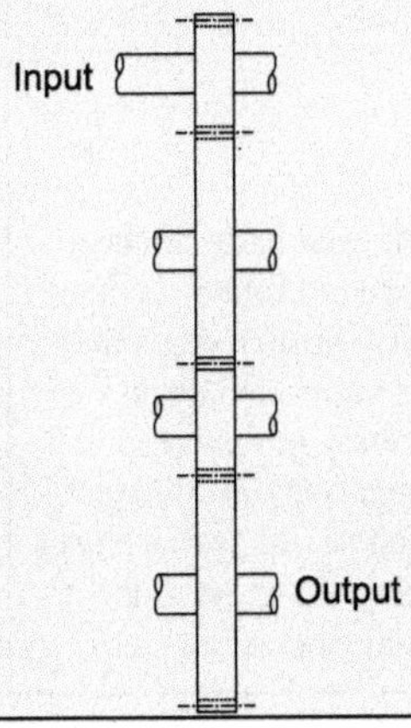 Input Output	A gear train is one or more pairs of gears operating together to transmit power.	See Section 8.3

Continued

Table 20.4 Continued

Type of gear and gear train	**Example**	**Characteristics**	**Useful equations and references**
Idler gear	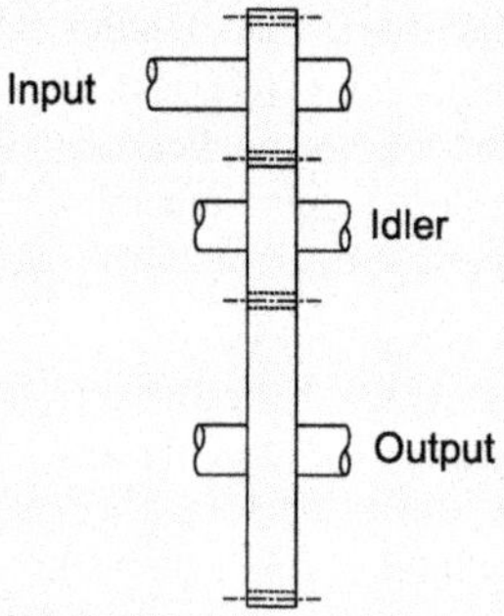 Input Idler Output	The addition of an extra gear on an intermediate shaft in a gear train reverses the output of the gear train but does not alter the gear train value.	See Section 8.3
Compound gear train	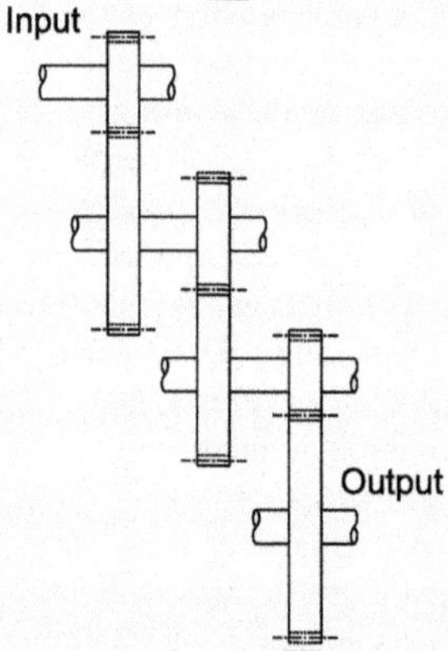 Input Output	In a compound gear train, more than one gear is located on the same shaft.	See Section 8.3
Reverted compound gear train	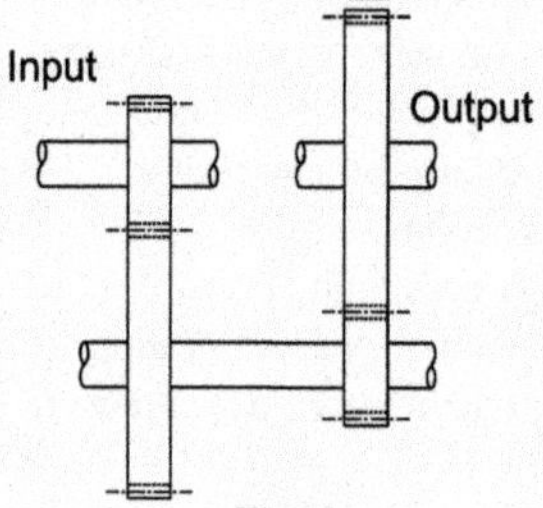 Input Output	A reverted gear train is a type of compound gear train where the input and output shafts are coaxial.	See Section 8.3
Epicyclic	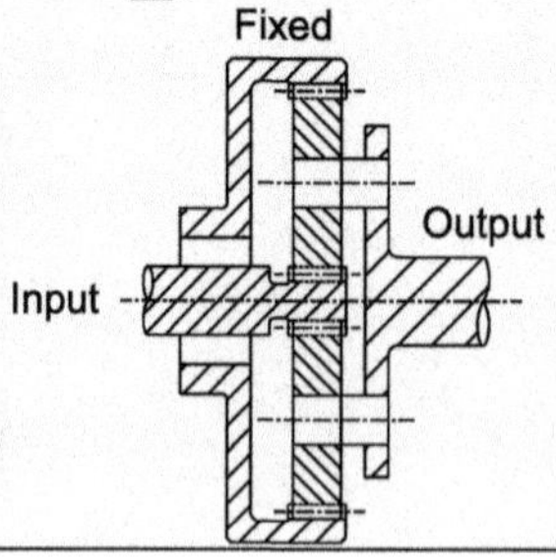 Fixed Output Input	Some gear axes can be allowed to rotate about others. In such cases the gear trains are called planetary or epicyclic. Planetary trains always include a sun gear, a planet carrier or arm, and one or more planet gears.	$\frac{n_{sun}}{n_{arm}} = \frac{N_R}{N_S} + 1$. See Sections 8.3.2 and 8.6

Table 20.4 Continued

Type of gear and gear train	Example	Characteristics	Useful equations and references
Epicyclic star arrangement	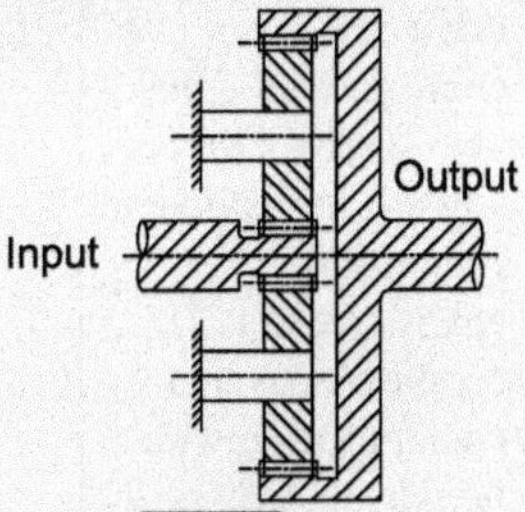 Input, Output	For a star arrangement, the sun is the driver, the ring is the driven element and the planetary carrier is fixed.	$\frac{n_A}{n_S} = -\frac{N_S}{N_R}$. See Section 8.3.2
Epicyclic solar arrangement	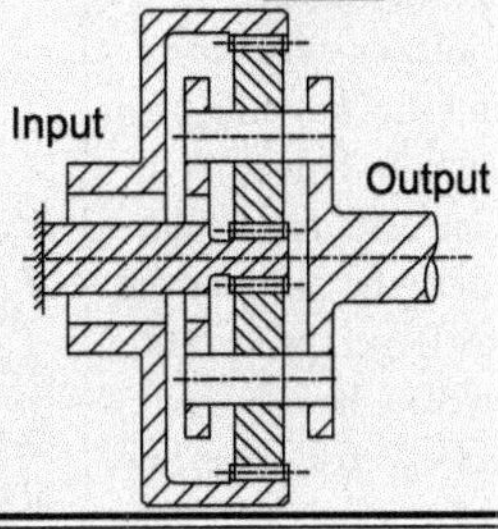 Input, Output	In the solar arrangement, the sun gear is fixed and the annulus gear, and planet carrier rotate.	$\frac{n_R}{n_{arm}} = 1 + \frac{N_S}{N_R}$. See Section 8.3.2

Table 20.5 A selection of various types of belt drive

Type of belt or belt drive.	Example	Characteristics	Useful equations and references
Belt drives	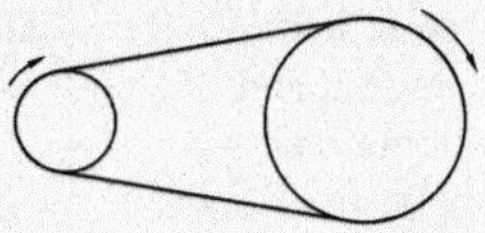 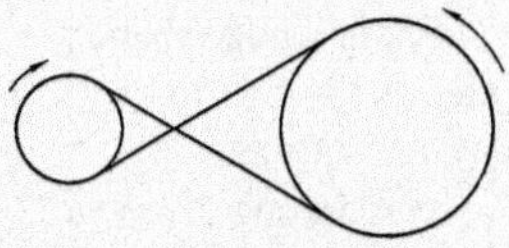	Belt drives can have numerous advantages over gear and chain drives including easy installation, low maintenance, high reliability, adaptability to non-parallel drive and high transmission speeds. The principal disadvantages of belt drives are their limited power transmission capacity	See Sections 12.1 and 12.2

Continued

Table 20.5 Continued

Type of belt or belt drive.	Example	Characteristics	Useful equations and references
		and limited speed ratio capability. Belt drives are less compact than either gear or chain drives and are susceptible to changes in environmental conditions such as contamination with lubricants. In addition, vibration and shock loading can damage belts.	
Flat		Flat belts have high strength, can be used for large speed ratios (>8:1), have a low pulley cost, give low noise levels and are good at absorbing torsional vibration. The driving force is limited by the friction between the belt and the pulley.	$\frac{F_1-F_c}{F_2-F_c}=e^{\mu\theta}$, $F_{1,\max}=\sigma_{\max}A$, $A=\frac{F_1-F_2}{\sigma_1-\sigma_2}$. See Sections 12.2 and 12.2.4
V and Wedge Belts		The most widely used type of belt in industrial and automotive applications is the V belt or wedge belt. The V or wedge shape causes the belt to wedge into the corresponding groove in the pulley increasing friction and torque capacity. Multiple belts are commonly used so that a cheaper small cross-sectional area belt can be used to transmit more power.	See Sections 12.2 and 12.2.2

Table 20.5 Continued

Type of belt or belt drive.	Example	Characteristics	Useful equations and references
Polyvee		A polyvee belts comprises several ribs across the belt width. A polyvee belt drive system tends to provide high-speed capability.	
Synchronous		Synchronous belts, also known as timing belts, have teeth which mesh with corresponding teeth on the pulleys. The positive contact between the pulleys and the belt provides angular synchronisation between the driving and the driven shafts and ensures a constant speed ratio.	See Sections 12.2 and 12.2.3

Table 20.6 A selection of various types of chain drive

Type of chain or chain drive	Example	Characteristics	Useful equations and references
Chain drive		Chains are usually more compact than belt drives for a given speed ratio and power capacity. Chain drives are generally more economical than the equivalent gear drive and are usually competitive with belt drives. Chains are inherently stronger than	

Continued

Table 20.6 Continued

Type of chain or chain drive	**Example**	**Characteristics**	**Useful equations and references**
		belt drives due to the use of steels in their manufacture and can therefore support higher tension and transmit greater power. The disadvantages of chain drives are limited speed ratios and power trans mission capability and also safety issues.	
Roller chain		Roller chain is made up of a series of links. Each chain link consists of side plates, pins, bushes and rollers. The chain runs on toothed sprockets and as the teeth of the sprockets engage with the rollers, rolling motion occurs and the chain articulates onto the sprocket.	Selection power = Power $\times f_1 \times f_2$ See Sections 12.1, 12.3 and 12.3.1 Renold Power Transmission (2010)
		Leaf chain consists of a series of pin connected side plates. It is generally used for load-balancing applications.	
Conveyer		Conveyor chain is specially designed for use in materials handling and conveyor equipment and is characterised by its long pitch, large roller diameters and high tensile strength. The sides of conveyor chains often incorporate special features to aid connection to conveyor components.	Jagtap et al. (2014) Renold Power Transmission (2010)

Table 20.6 Continued

Type of chain or chain drive	Example	Characteristics	Useful equations and references
Silent		Silent chain (also known as inverted tooth chain) has teeth formed with a notch to accommodate articulation on a specially formed sprocket. The chain consists of alternately mounted links so that the chain can articulate onto the mating sprocket teeth. Silent chains can operate at higher speeds than comparatively sized roller chain and, as the name suggests, more quietly.	Meng et al. (2007) and Cali et al. (2016)

20.6 Rotating and linear motion conversion

A wide of mechanisms are capable of converting rotational into linear motion including the crank slider, rack and pinion mechanisms, cams and Cardan gears. A selection of mechanisms capable of producing linear or reciprocating motion from a rotational input, or vice versa are presentated in Table 20.7.

20.7 Brakes and clutches

It is often necessary to control or slow down the speed of a machine or device. As described in Chapter 13, a brake enables the controlled dissipation of energy to slow down, stop or control the speed of a system, A wide range of methods of braking have been developed including the use of friction between surfaces and the interaction of magnetic fields. A series of standard brake configurations are presented in Table 20.8.

A clutch enables the smooth, gradual connection of two components moving relative to each other, such as shafts rotating at different speeds. The connection between two mechanical subsystems can be achieved by a number of means ranging from direct mechanical friction, electromagnetic coupling, use of hydraulics or pneumatics

Table 20.7 A selection of various types of devices for conversion between rotating and linear motion

Type of Cam	Example	Characteristics	Useful equations and references
Crank slider	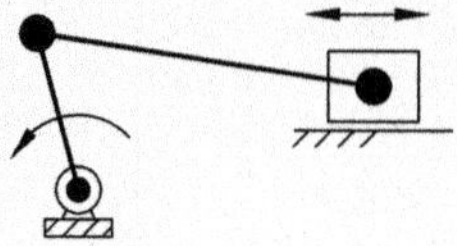	Rotation of the crank drives reciprocating motion. Alternatively reciprocation of the slider can produce rotation	Todd et al. (2014)
Cam	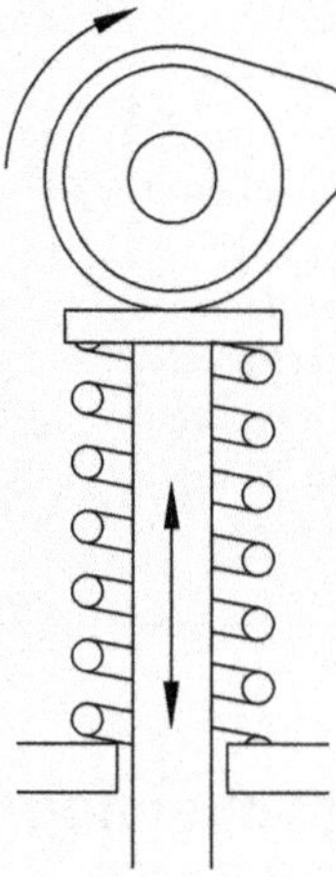	Cams are used to convert rotary motion into reciprocating motion. A cam is a rotating shape whose angular motion is converted into output motion by a cam follower which rides on the cam surface.	Norton (2009)
Rack and pinion	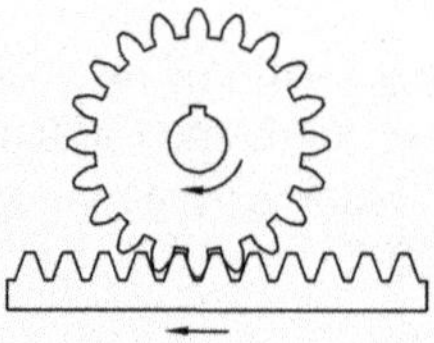	A rack comprises straight gear teeth cut on one surface of a bar. A small cylindrical gear, normally a spur or helical gear, known as the pinion with corresponding module meshes with the rack. If the pinion axis is fixed, then as the pinion rotates the rack with move linearly.	
Reciprocating rack assembly and sector gear	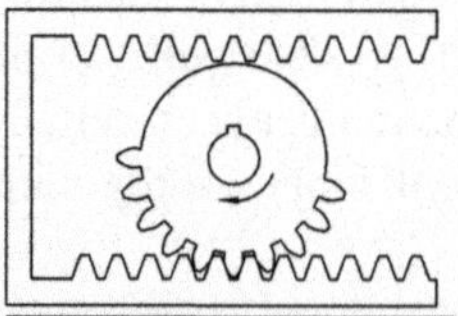	As the sector gear rotates the rack assembly produces a reciprocating motion.	
Reciprocating rack assembly and pinion gear		As the pinion rotates, the rack assembly produces a reciprocating motion.	Brown (1901)

Table 20.7 Continued

Type of Cam	Example	Characteristics	Useful equations and references
Cardan gear		Invented by Girolamo Cardano in the 16th century, these devices convert the rotation of gear trains into reciprocating linear motion without the use of linkages or slideways. In the illustration shown, the ring gear has double the number of teeth as the pinion. The dot shown will trace out a straight line as the pinion rotates around the ring gear.	

or by some combination. Clutches can also be used for non-rotating applications where, for example, a system that is moving linearly needs to be joined to another linearly moving system. Clutch principles and their design are described in Chapter 13 and a summary of some principal types of clutch is given in Table 20.9.

20.8 Time metering

A wide range of applications require a portion of a machine to be stationary for a period of time or to control the speed of a component. A rich variety of mechanisms have been developed for such purposes ranging from escapements to the Geneva stop. A selection of these devices is presented in Table 20.10.

20.9 Miscellaneous devices

The range and scope of mechanisms is boundless. A rich source of mechanisms is presented in the classic text by Brown (1901) and a contemporary collection is presented by Nguyen Duc Thang (2018). As an example, an ellipsograph, see also Table 20.3, is illustrated in Fig. 20.10. The bar has two studs which run in the grooves of the cross slide. By turning the bar, a scribe attached to the bar will trace out an ellipse. A gripper is illustrated in Fig. 20.11 showing the combination of a rack and pinion and lever to provide a means to grasp an item.

Table 20.8 A selection of brake configurations

Type of Brake	Example	Characteristics	Useful equations and references
friction disc brake		Two or more surfaces are pushed axially together with a normal force to generate a friction torque. The function of a frictional brake surface material is to develop a substantial friction force when a normal force is applied.	
Calliper disc brake	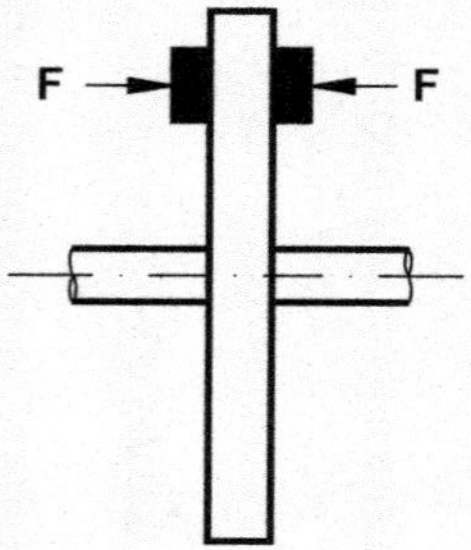 F F	The use of a discrete pad allows the disc to cool as it rotates, enabling heat transfer between the cooler disc and the hot brake pad. As the pads on either side of the disc are pushed on to the disc with equal forces, the net thrust load on the disc cancels.	$F=\frac{T}{\mu r_e}$ $p_{\max}=\frac{F}{\theta r_i(r_o-r_i)}$ $p_{av}=\frac{2r_i/r_o}{1+r_i/r_o}p_{\max}$ $p_{\text{hydraulic}}=\frac{F}{A_{\text{cylinder}}}$ See Section 13.3.1 Halderman, and Mitchell (2016)
Drum brakes	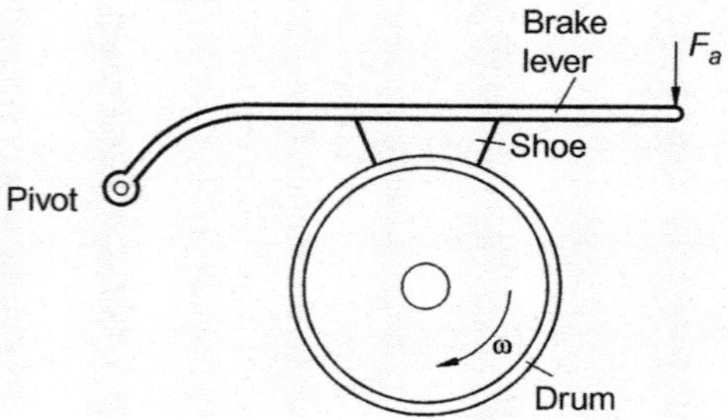 Brake lever, F_a, Shoe, Pivot, ω, Drum	Drum brakes apply friction to the external or internal circumference of a cylinder. A drum brake consists of the brake shoe, which has the friction material bonded to it, and the brake drum. Drum brakes can be designed to be self-energising. Once engaged, the friction force increases the normal force non-linearly, increasing the friction torque as in a positive feedback loop.	See Section 13.3.2

Short-shoe external drum brake	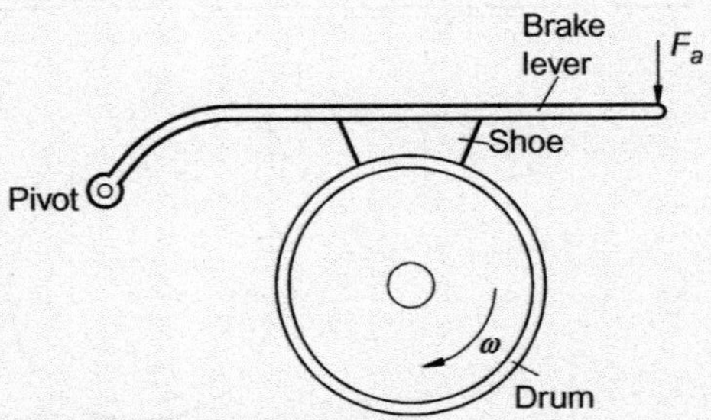	Short-shoe internal brakes are used for centrifugal brakes that engage at a particular critical speed.	See Section 13.3.3
Double short-shoe external brake	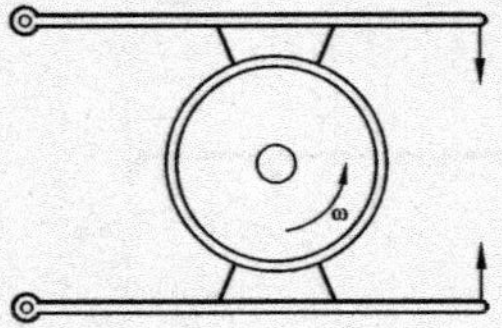	More than one brake shoe can be used to apply a braking force to a rotating drum.	
Long-shoe external drum brake	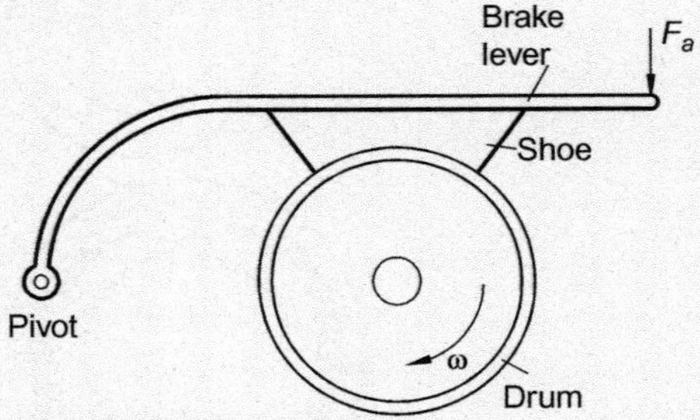	If the included angle of contact between the brake shoe and the drum is >45° then the pressure between the shoe and the brake lining cannot be regarded as uniform and the approximations made for the short-shoe brake analysis are inadequate. Most drum brakes use contact angles >90°.	Self-energising: $F_a = \frac{M_n - M_f}{a}$. Self-de-energising: $F_a = \frac{M_n + M_f}{a}$. See Section 13.3.4
Double long-shoe external drum brake	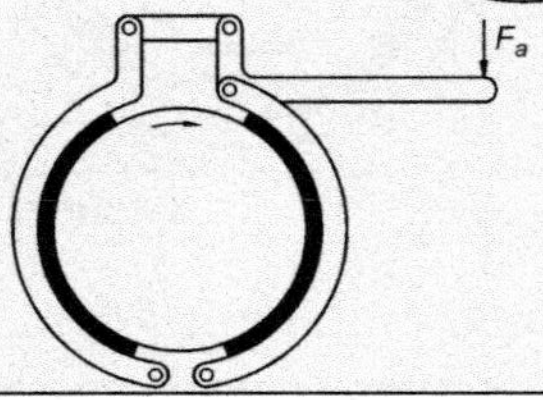	For the double long-shoe external drum brake illustrated, the left-hand shoe is self-energising and the frictional moment reduces the actuation load. The right-hand shoe is self-de-energising and its frictional moment acts to reduce the maximum pressure which occurs on the right-hand brake shoe.	$M'_n = \frac{M_n p'_{max}}{p_{max}}$ $M'_f = \frac{M_f p'_{max}}{p_{max}}$ See Section 13.3.4

Continued

Table 20.8 Continued

Type of Brake	Example	Characteristics	Useful equations and references
Double long-shoe internal drum brake		This of type of brake uses internal brake shoes that expand against the inner radius of a drum. The shoes are pivoted at one end on anchor pins. The brake can be actuated by a hydraulic piston which forces the free ends of the brake apart so that the non-rotating shoes come into frictional contact with the rotating brake drum.	$F_a=\frac{M_n-M_f}{a}$, $F_a=\frac{M_n+M_f}{a}$, $T_{\text{total}}=T_{\text{right shoe}}+T_{\text{left shoe}}$. See Section 13.3.5 Halderman, and Mitchell (2016)
Band brake		A flexible metal band lined with a frictional material wrapped partly around a drum. The brake is actuated by pulling the band against the drum. The brake configuration shown is self-energising for clockwise rotation.	$F_a=F_2\frac{a}{c}$ See Section 13.3.6
Self-energising differential band brake		The level of self-energisation can be enhanced by using a differential band brake configuration. The brake can be made self-locking if $a<be^{\mu\theta}$ and the slightest touch on the lever would cause the brake to grab or lock abruptly.	$F_a=\frac{F_2\left(a-be^{\mu\theta}\right)}{c}$ See Section 13.3.6

Table 20.9 A selection of clutch configurations

Type of Clutch	Example	Characteristics	Useful equations and references
Friction disc clutch		Friction clutches consist of two, or two sets of, surfaces, which can be forced into frictional contact. They allow gradual engagement between two shafts.	$r_o = \left(\frac{T}{\pi\mu p_{\max}\sqrt{4/27}}\right)^{1/3}$, $F = 2\pi r_i p_{\max}(r_o - r_i)$. See Section 13.2.1
Multiple disc clutch		In a multiple disc clutch, the frictional load is distributed across the faces of each disc, enabling a high surface area to be packaged within a given diameter.	$T = \frac{2}{3}\mu FN\left(\frac{r_o^3 - r_i^3}{r_o^2 - r_i^2}\right)$. See Section 13.2.1
Square jaw clutch		A square jaw clutch provides a positive drive between two shafts in either direction. They can be suitable for when engagement and disengagement does not need to take place during rotation	
Multiple Serration clutch		A series of radial serrations provide positive contact between annular rings on each disc.	

Continued

Table 20.9 Continued

Type of Clutch	Example	Characteristics	Useful equations and references
Ratchet and pawl		A ratchet and pawl allows relative motion in one direction only. If the rotation attempts to reverse the pawl catches on the serrations and relative rotational motion is not possible. A ratchet can be used to move a toothed wheel one tooth at a time.	
Roller clutch		One-way clutch. Rollers ride up ramps and drive by wedging into place.	
Sprag clutch		One-way clutch. Profiled elements jam against the outer edge to provide drive. High torque capacity.	
Spring wound clutch		Friction between hubs and the coil spring causes the spring to tighten onto hubs and drive in one direction. In the opposite direction, the spring unwraps and slips easily on the hubs.	

Magnetic clutch	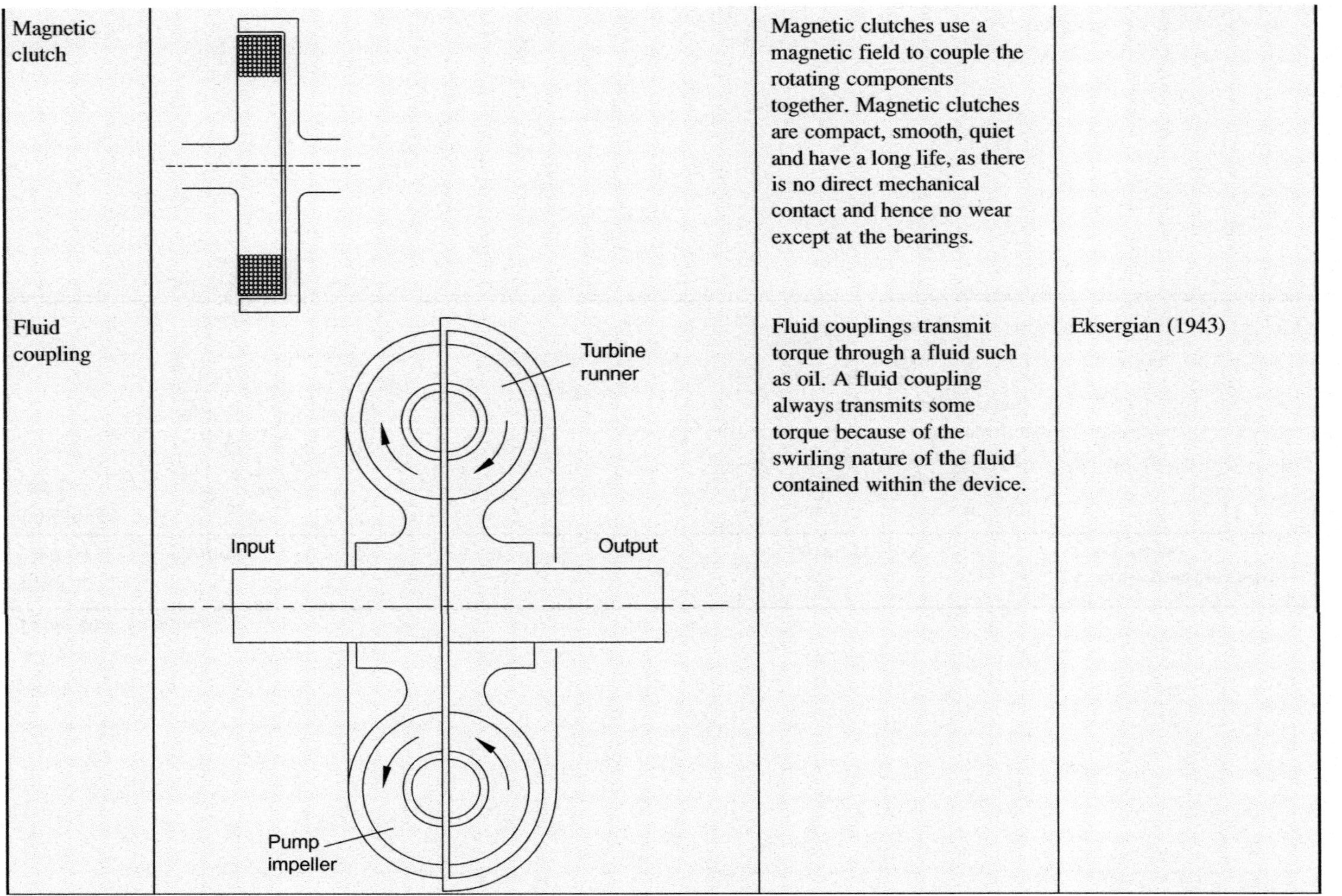	Magnetic clutches use a magnetic field to couple the rotating components together. Magnetic clutches are compact, smooth, quiet and have a long life, as there is no direct mechanical contact and hence no wear except at the bearings.	
Fluid coupling		Fluid couplings transmit torque through a fluid such as oil. A fluid coupling always transmits some torque because of the swirling nature of the fluid contained within the device.	Eksergian (1943)

Continued

Table 20.9 Continued

Type of Clutch	Example	Characteristics	Useful equations and references
Centrifugal clutch		Centrifugal clutches engage automatically when the shaft speed exceeds some critical value. Friction elements are forced radially outwards and engage against the inner radius of a mating cylindrical drum.	

Table 20.10 A selection of time metering mechanisms

Type of device	Example	Characteristics	Useful equations and references
Recoil escapement	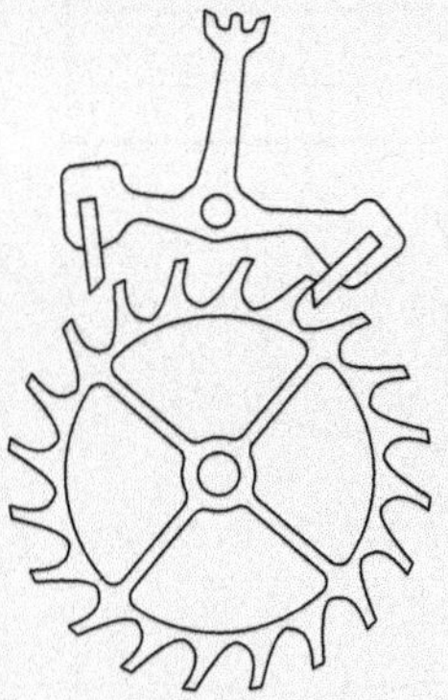	The lever escapement mechanism provides a means to control the release of energy from a mechanical energy source. The overall mechanism as used in watches and clocks comprises a balance wheel, hairspring, pallet fork and escape wheel. Using power provided from the mainspring, the escape wheel rotates with regulated movement, locking and unlocking the pallet fork as it rotates.	Brown (1901), Du and Xie (2012)
Geneva stop	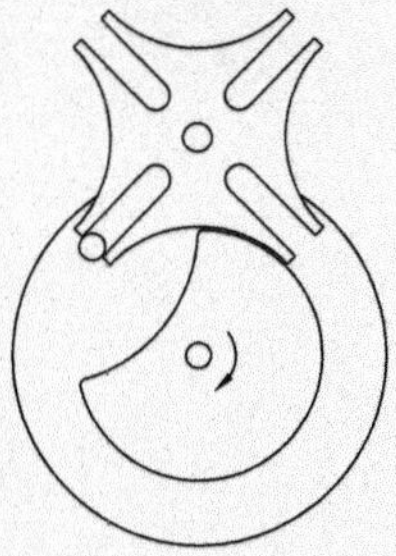	A Geneva stop mechanism provides intermitted motion. Driven by a wheel which turns continuously, the runner will turn in proportion to the number of vertices, one-quarter in the case of four vertices. The crescent cutout on the wheel lets a runner vertex past and then locks the runner in place, for the remainder of the rotation.	Sclater (2011)
Ratchet wheel		Intermittent circular motion for the larger wheel can be produced by continuous motion of the smaller wheel.	Brown (1901)

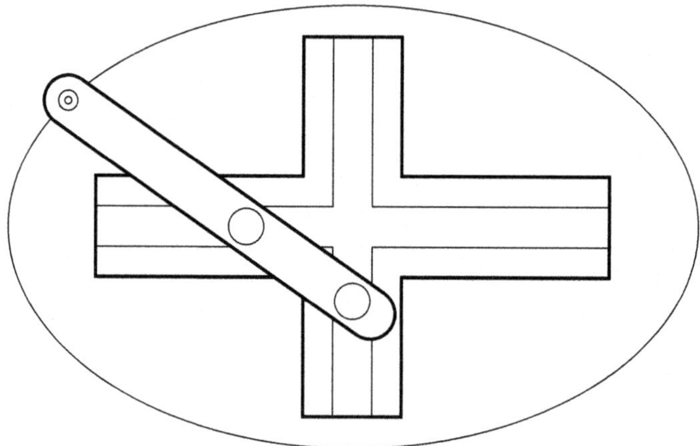

Fig. 20.10 Ellipsograph.

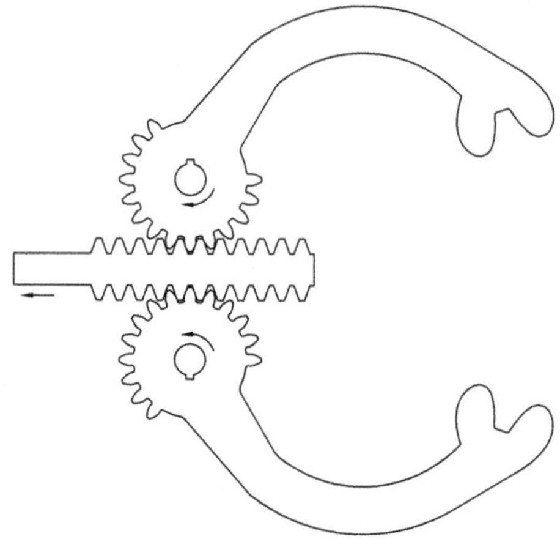

Fig. 20.11 Gripper.

20.10 Conclusions

The goal in mechanism or machine design is to determine the size, shape and selection of materials and manufacturing processes for each of the parts of the machine so that it will perform its intended function without failure. Although design work may involve concentrating on one component at a given time, it is essential to consider its interrelationship with the whole product, taking into account the overall functional requirements as well consideration of other relevant aspects such as economic and aesthetic functions as relevant. This may entail detailed consideration of market requirements, specification, concept development, detailed design and manufacture, and end of life and circular economy considerations, if following a total design methodology.

In the development of a mechanism, it can be helpful to recognise that engineering design is often a tactile, visual, verbal, cerebral and physical activity. Whether learning or developing ideas, it can be useful to play and fiddle with parts, sketch ideas, create physical and analytical models to identify opportunities and test possible strategies, detail the machine using all the skills and tools at your disposal, and to build and test your machines. Such an approach, using conceptualisation, modelling and prototyping can help ensure the viability of the mechanism in practice.

References

Books and papers

Akbari, S., Pirbodaghi, T., 2017. Precision positioning using a novel six axes compliant nanomanipulator. Microsyst. Technol. 23, 2499–2507.

Brown, H.T., 1901. 507 Mechanical Movements, 19th edition Brown and Seward. Dover edition, 2005.

Cali, M., Sequenzia, G., Oliveri, S.M., Fatuzzo, G., 2016. Meshing angles evaluation of silent chain drive by numerical analysis and experimental test. Meccanica 51, 475–489.

Chironis, N.P., 1965. Mechanisms, Linkages and Mechanical Controls. McGraw Hill.

Culpepper, M.L., Anderson, G., 2004. Design of a low-cost nano-manipulator which utilizes a monolithic, spatial compliant mechanism. Precis. Eng. 28, 469–482.

den Hartog, J.P., 1961. Mechanics. Courier Corporation. Dover edition 2003.

Du, R., Xie, L., 2012. The Mechanics of Mechanical Watches and Clocks. Springer.

Eksergian, R., 1943. The fluid torque converter and coupling. J. Franklin Inst. 235, 441–478.

Halderman, J.D., Mitchell, C.D., 2016. Automotive Brake Systems. Prentice Hall.

Hrones, J.A., Nelson, G.L., 1951. Analysis of the Four-Bar Linkage: Its Application to the Synthesis of Mechanisms. Wiley.

Jagtap, M.D., Gaikwad, B.D., Pawar, P.M., 2014. Study of roller conveyor chain strip under tensile loading. IJMER 4, 2249–6645.

Liu, Y., Xu, Q., 2015. In: Design and analysis of a large-range micro-gripper.IEEE International Conference on Manipulation, Manufacturing and Measurement on the Nanoscale (3M-NANO), pp. 55–58.

McCarthy, J.M., Soh, G.S., 2010. Geometric Design of Linkages, second ed. Springer.

Meng, F., Li, C., Cheng, Y., 2007. Proper conditions of meshing for Hy-Vo silent chain and sprocket. Chin. J. Mech. Eng. 20, 57–59.

Nguyen Duc Thang, https://www.youtube.com/user/thang010146/videos. (Accessed 10 June 2018).

Norton, R.L., 2009. Cam Design and Manufacturing Handbook, second ed. Industrial Press.

Renold Power Transmission, 2010. Transmision chain. Installation, maintenance and designer guide. http://www.renold.com/media/165418/Transmission-I-and-M-REN12-ENG-10-10.pdf. (Accessed 10 June 2018).

Sclater, N., 2011. Mechanisms and Mechanical Devices Sourcebook, fifth ed. McGraw Hill.

Serway, R.A., Jewett, J.W., 2018. Physics for Scientists and Engineers, tenth ed. Brooks Cole.

Simionescu, P., 2017. Optimum synthesis of oscillating slide actuators for mechatronic applications. J. Comput. Des. Eng. 5.

Todd, P., Mueller, D., Fichter, E., 2014. Atlas of the Four-Bar Linkage, second ed. Saltire Software.

Wilson, C.E., Sadler, J.P., 2003. Kinematics and Dynamics of Machinery, third ed. Pearson.

Websites

At the time of going to press the World Wide Web contained useful information relating to this chapter at the following sites:

507movements.com/.

mechanicaldesign101.com/.

www.geogebra.org/m/uaguEEM8.

www.hackaday.com/.

www.mekanizmalar.com/.

www.youtube.com/user/thang010146/videos.

Index

Note: Page numbers followed by *f* indicate figures and *t* indicate tables.

B

C

I

J

K

L

M

N

O

P

Q

R

T

U

V

Printed in the United States
By Bookmasters